WIND ENERGY SYSTEMS

CONTROL ENGINEERING DESIGN

WIND ENERGY SYSTEMS

CONTROL ENGINEERING DESIGN

Mario Garcia-Sanz • Constantine H. Houpis

CRC Press is an imprint of the
Taylor & Francis Group, an **informa** business

CRC Press
Taylor & Francis Group
6000 Broken Sound Parkway NW, Suite 300
Boca Raton, FL 33487-2742

Version Date: 20111208

International Standard Book Number: 978-1-4398-2179-4 (Hardback)

Library of Congress Cataloging-in-Publication Data

García-Sanz, Mario.
Wind energy systems : control engineering design / Mario García-Sanz, Constantine H. Houpis.
p. cm.
Includes bibliographical references and index.
ISBN 978-1-4398-2179-4 (hbk. : alk. paper)
1. Wind energy conversion systems--Automatic control. 2. Wind turbines--Automatic control. I. Houpis, Constantine H. II. Title.

TK1541.G37 2012
621.31'2136--dc23 2011046928

Visit the Taylor & Francis Web site at
http://www.taylorandfrancis.com

and the CRC Press Web site at
http://www.crcpress.com

To Marta, María, Pablo and Sofía

To Harry and Lin Lin and To Angella and Brian

Contents

Part II Wind Turbine Control

Preface

Harvesting energy on a global, sustainable, and economic scale is one of the major challenges of this century. With emerging markets, newly industrializing nations, and shortage of existing resources, this problem will continue to grow. *Wind energy,* in the current scenario, is playing a central role, being the fastest-growing source of energy worldwide in the last few decades.

However, long-term economic sustainability of wind energy is still to be achieved. This would imperatively require improving critical engineering and economic practices to reduce the cost of wind energy as compared to conventional energy to help increase its proliferation.

Wind turbines are complex systems, with large flexible structures working under very turbulent and unpredictable environmental conditions. Moreover, they are subject to a variable and demanding electrical grid. Their efficiency, cost, availability, and reliability strongly depend on the *applied control* strategy. As wind energy penetration in the grid increases, additional challenges such as the response to grid voltage dips, active power control and frequency regulation, reactive power control and voltage regulation, grid damping, restoration of grid services after power outages, necessity of wind prediction, etc., crop up.

Large nonlinear characteristics and high model uncertainty due to the interaction of the aerodynamic, mechanical and electrical subsystems, stability problems, energy-conversion efficiency, load reduction strategies, mechanical fatigue minimization problems, reliability issues, availability aspects, and cost reduction topics all impose the need to design advanced control systems in a *concurrent engineering approach.* This approach coordinates many variables such as pitch, torque, power, rotor speed, yaw orientation, temperatures, currents, voltages, power factors, etc. It is a multidisciplinary task and must be developed under the leadership of an experienced engineer. In this book, we claim this role for the *control engineer,* with the ultimate goal of achieving an optimum design, taking into account all the aspects of the big picture of the new energy system.

Thus, the first objective of this book is to present the latest developments in the field of applied control system analysis and design and to stimulate further research, including new advanced nonlinear multi-input multi-output robust control system design techniques: quantitative feedback theory (QFT), and nonlinear switching strategies. The second objective of this book is to *bridge the gap* between the advanced control theory and the engineering application to design, optimize, and control wind energy systems.

The book is especially useful as it combines hard-to-find industrial knowledge of wind turbines and the required control theory in a concise document from the perspective of both a practicing and real-world engineer. The wind turbine design standards and the experimental results included in this book will be vital to anyone entering the industry. The control portion of the book guides the reader through robust theories and reliable ideas successfully applied to real multimegawatt wind turbines.

This book is divided into two parts. Part I (Advanced Robust Control Techniques: QFT and Nonlinear Switching) consists of seven chapters and presents concepts of nonlinear multi-input multi-output robust control in such a way that students and practicing engineers can readily grasp the fundamentals and appreciate its *transparency* in *bridging the gap* between theory and real–world applications.

Part II (Wind Turbine Control) consists of 11 chapters and introduces the main topics as follows: (1) modern wind turbine design and control, (2) classical and advanced turbines, (3) dynamic modeling, (4) control objectives and strategies, (5) standards and certification, (6) controller design, (7) economics, (8) grid integration, and (9) a large number of applications in different areas like onshore and offshore wind turbines, floating wind turbines, airborne wind energy systems, smart blades, advanced blades manufacturing, and real experimentation with multimegawatt machines.

In addition, eight appendices and a compilation of problems enhance the technical content of this book. Appendix A describes how to calculate the QFT templates. Appendix B introduces the inequality expressions for QFT bounds. Appendix C discusses a complementary technique to calculate the QFT bounds analytically. Appendix D introduces elements for loop shaping. Appendix E discusses the analysis of the controller fragility with QFT. Appendix F describes the new QFT control toolbox (*QFTCT*), an interactive computer-aid-design (CAD) package for MATLAB®. Appendix G presents five illustrative controller design examples with the new *QFTCT*. Finally, Appendix H presents a table for conversion of units. A compilation of problems, answers to selected problems, and a comprehensive reference section about QFT robust control and wind energy systems are also included.

The new interactive and user-friendly *QFTCT* for MATLAB®, described in Appendix F and available at http://cesc.case.edu and http://www.crcpress.com/product/isbn/9781439821794, is offered to assist the student and the control engineer in applying the QFT controller design methods and is stressed throughout the book.

Many universities worldwide are currently developing a new curriculum in renewable energy. This book presents practical control design methodologies to face the main wind turbine design and control problems.

Numerous methodologies presented in this book have been developed by us during the last two decades. The methodologies are based on our experiences and our association with Professor Isaac Horowitz. The control methodologies have also been validated by us in real-world multimegawatt wind turbine and spacecraft applications, especially through our association with the M.Torres Group, the European Space Agency ESA-ESTEC and NASA-JPL.

The book is primarily intended as a textbook for undergraduate or graduate courses. It can also be used for self-study by practicing engineers. The book has been class tested at the Public University of Navarra (Spain) and at Case Western Reserve University (Ohio), thus enhancing its value for classroom use and self-study.

The book begins where a typical undergraduate control course ends. This means that it is assumed that the reader is familiar with differential equations, basic linear algebra, time-domain and frequency-domain control analysis and design, single-input single-output systems, and elementary State Space theory. This corresponds to the material in, for example, D'Azzo, Houpis, and Sheldon's (2003) *Linear Control System Analysis and Design with Matlab* (Marcel Dekker) or Dorf and Bishop's (2011) *Modern Control Systems* (Prentice Hall), or any other book that deals with similar topics.

We wish to express our appreciation for the support and encouragement of Manuel Torres, founder and president of the M.Torres Group; Professor N. Davis, head of the Department of Electrical and Computer Engineering, Air Force Institute of Technology; Dr. Christian Philippe, European Space Agency; the Milton and Tamar Maltz Family Foundation; the Cleveland Foundation; and the Spanish Ministry of Science and Technology (MCyT) for their encouragement and support in writing this book and the *QFTCT* computer-aid-design package.

We would also like to thank the following former graduate PhD and master's students of Professor Mario García-Sanz in advancing the state of the art of the QFT control field in the last 15 years: Dr. Xabier Ostolaza, Dr. Juan Carlos Guillén, Dr. Montserrat Gil, Dr. Arturo Esnoz, Dr. Igor Egaña, Dr. Marta Barreras, Dr. Juan Jose Martín, Dr. Irene Eguinoa, Dr. Alejandro Asenjo, Dr. Jorge Elso, Carlos Molins, Augusto Mauch, Pablo Vital, Javier Villanueva, Manu Motilva, Juan Antonio Osés, Javier Castillejo, Mikel Iribas, Ana Huarte, María Brugarolas, Xabier Montón, Asier Oiz, and Daniel Casajus at the Public University of Navarra (Spain), as well as Timothy Franke, Tipakorn Greigarn, Nicholas White, Nicholas Tierno, Katherine Faley, Gerasimos Houpis, Trupti Ranka, and Dr. Ion Irizar at Case Western Reserve University (United States).

Finally, we would like to acknowledge many colleagues and friends at Case Western Reserve University (Ohio), at the Public University of Navarra (Spain), at M.Torres HQ (Spain), at ESA-ESTEC (the Netherlands), and at the Air Force Institute of Technology (Ohio) for their help and cooperation in the completion of this book.

Mario García-Sanz
Cleveland, Ohio

Constantine H. Houpis
Wright-Patterson AFB, Ohio

For MATLAB® and Simulink® product information, please contact:

The MathWorks, Inc.
3 Apple Hill Drive
Natick, MA, 01760-2098 USA
Tel: 508-647-7000
Fax: 508-647-7001
E-mail: info@mathworks.com
Web: www.mathworks.com

Authors

Dr. Mario García-Sanz is an endowed chair professor at Case Western Reserve University (CWRU), Ohio; the Milton and Tamar Maltz professor in energy innovation; and director of the Control and Energy Systems Center at CWRU (http://cesc.case.edu). As senior advisor to the president of the M.Torres Group and as full professor at the Public University of Navarra, he has played a central role in the design and field experimentation of advanced multimegawatt wind turbines for industry leaders over the last two decades. Dr. García-Sanz held visiting professorships at the Control Systems Centre, UMIST (United Kingdom, 1995); at Oxford University (United Kingdom, 1996); at the Jet Propulsion Laboratory NASA-JPL (California, 2004); and at the European Space Agency ESA-ESTEC (the Netherlands, 2008). He holds 20 industrial patents, has done more than 40 large research projects for industry and space agencies, and is the author or coauthor of more than 150 research papers, including the books *Quantitative Feedback Theory: Theory and Applications* (Taylor & Francis, 2006) and *Wind Energy Systems: Control Engineering Design* (Taylor & Francis, 2012). Dr. García-Sanz is subject editor of the *International Journal of Robust and Nonlinear Control*, is a member of IFAC and IEEE technical committees (robust control, aerospace control), and has served as NATO/RTO lecture series director and as guest editor of international journals (special issues on: *Robust Control, QFT Control, Wind Turbine Control, Spacecraft Control*). He was awarded the IEE Heaviside Prize (United Kingdom) in 1995 and the BBVA Research Award (Spain) in 2001. Professor García-Sanz's main research interest focuses on bridging the gap between advanced control theory and applications, with special emphasis on energy innovation, wind energy, space, water, environmental, and industrial applications.

Dr. Constantine H. Houpis is an emeritus professor at the Air Force Institute of Technology (AFIT). Prior to this, he was a senior research associate emeritus at the Air Force Research Laboratory, Wright-Patterson Air Force Base, Ohio. Dr. Houpis is an IEEE life fellow and has served as a NATO/RTO lecture series director several times. For almost two decades, he worked very closely with Professor Isaac Horowitz at AFIT and at the Air Force Research Laboratory on the fundamentals of quantitative feedback theory and its applications to real-world projects, many of them in the field of aerospace. His textbook, *Feedback Control System Analysis and Synthesis* (1960), coauthored with his colleague, John J. D'Azzo, is recognized as a classic in its field. This textbook and its sequel, *Linear Control System Analysis and Design—Conventional and Modern*, have been translated into several languages and have had seven editions. Other well-known books written by Dr. Houpis are *Digital Control Systems—Theory, Hardware, Software* (McGraw-Hill, 1991, two editions) and *Quantitative Feedback Theory: Theory and Applications* (Taylor & Francis, 2006, two editions). Dr. Houpis has received numerous awards, the latest being the NAECON 2009 Research Visionary Award, for outstanding research and visionary contributions to the education of undergraduate and graduate students in both control theory and robust multivariable control systems.

1

Introduction

1.1 Broad Context and Motivation

With a capacity that has tripled in the last 5 years, wind energy is again the fastest growing energy source in the world. Wind turbines (WTs) are used to collect kinetic energy and to convert it into electricity. The average power output of a WT unit has increased significantly in the last few years. Most major manufacturers have developed large turbines that produce 1.5–5.0 MW of power for onshore applications and are thinking of bigger units for offshore projects. Grouped together, they generate energy equaling 2% of the global electricity consumption, with about 200 GW of wind-powered generators worldwide by the end of 2010.

Although wind energy is a clean and renewable source of electric power, many challenges still remain unaddressed. WTs are complex machines with large flexible structures working under turbulent, unpredictable, and sometimes extreme environmental weather conditions, and are connected to a constantly varying electrical grid with changing voltages, frequency, power flow, and the like. WTs have to adapt to these variations, so their *reliability, availability, and efficiency depend heavily on the control strategy* applied. As wind energy penetration in the grid increases, additional challenges are being revealed: response to grid disturbances, active power control and frequency regulation, reactive power control and voltage regulation, grid damping, restoration of grid services after power outages, and wind prediction, to name a few.

However, occasionally, you can still find very serious accidents, involving unstable situations and uncontrollable conditions, even in the largest machines and with the best international companies (see Figures 1.1 through 1.4).

With this very critical motivation, the authors claim for the *need of a truly reliable control design methodology*! In their experience in designing commercial multimegawatt WTs, the quantitative feedback theory (QFT) has been proved to be a reliable control engineering technique successfully applied to many critical systems,[151–187] including wind energy applications as seen in this book.

1.2 Concurrent Engineering: A Road Map for Energy

Control engineering plays a primary and central role in the design and development of new challenging engineering systems. In addition, *bridging the gap between advanced control theory and engineering real-world applications* is the key factor in achieving breakthroughs

FIGURE 1.1
Catastrophic accident and tower collapse of a large WT. Cause: rotor over-speed.

FIGURE 1.2
Structural failure and tower collapse of a large WT. Cause: rotor over-speed.

FIGURE 1.3
WT burning accident. Cause: overheat and fire.

FIGURE 1.4
Blade failure and tower collapse of a WT. Cause: lighting.

and transformational ideas in many engineering fields, particularly in energy innovation and wind energy.

For many years, the classical way of working has been sequential, having every engineering team (mechanical, aerodynamic, electrical, electronics, and control engineers) working independently to merge all the components of the design at a very late stage (see Figure 1.5).

On the contrary, the design philosophy presented in this book implies a new way of thinking, where the engineering teams work together in a simultaneous and concurrent manner from the very beginning. This integrated view of the design (*concurrent engineering*) requires that *control engineers* play a central role from the very beginning of the project, applying control concepts to understand the interactions of the subsystems in the entire system and coordinating the different disciplines to achieve a better system's dynamics, controllability, and optimum design (see Figure 1.5).

Any new energy system, including any new multimegawatt WT, involves many aspects that deeply affect each other in a continuous feedback loop, such as engineering concepts, economics, reliability issues, efficiency, certification, regulations, implementation, marketing, interaction with grid and environment, maintenance, etc., as illustrated in the *project road map* (see Figure 1.5). In other words, any single aspect in this big picture can affect the rest of the areas and eventually be the critical bottleneck of the entire project.

As an example, a specific airfoil for the blades could affect the rotor speed, which could require modifying the electrical design of the generator, which in turn could change the economics, reliability, efficiency, and the maintenance of the project. Analogously, a marketing issue or a regulation matter could require a noise level reduction, which eventually could result in modifying the rotor speed or the blade design, which again could require modifying the electrical design of the generator, which could change the economics, and/or the reliability, and/or the efficiency, and/or the maintenance of the project. Also a transport limitation or an economic specification could require shorter blades, which could affect the rotor speed, which could require modifying the electrical design of the generator, which in turn could change the mechanical loads, the structure of the tower, the foundation, the economics, reliability, efficiency, and maintenance of the entire project, etc. (see also Chapters 9 and 15).

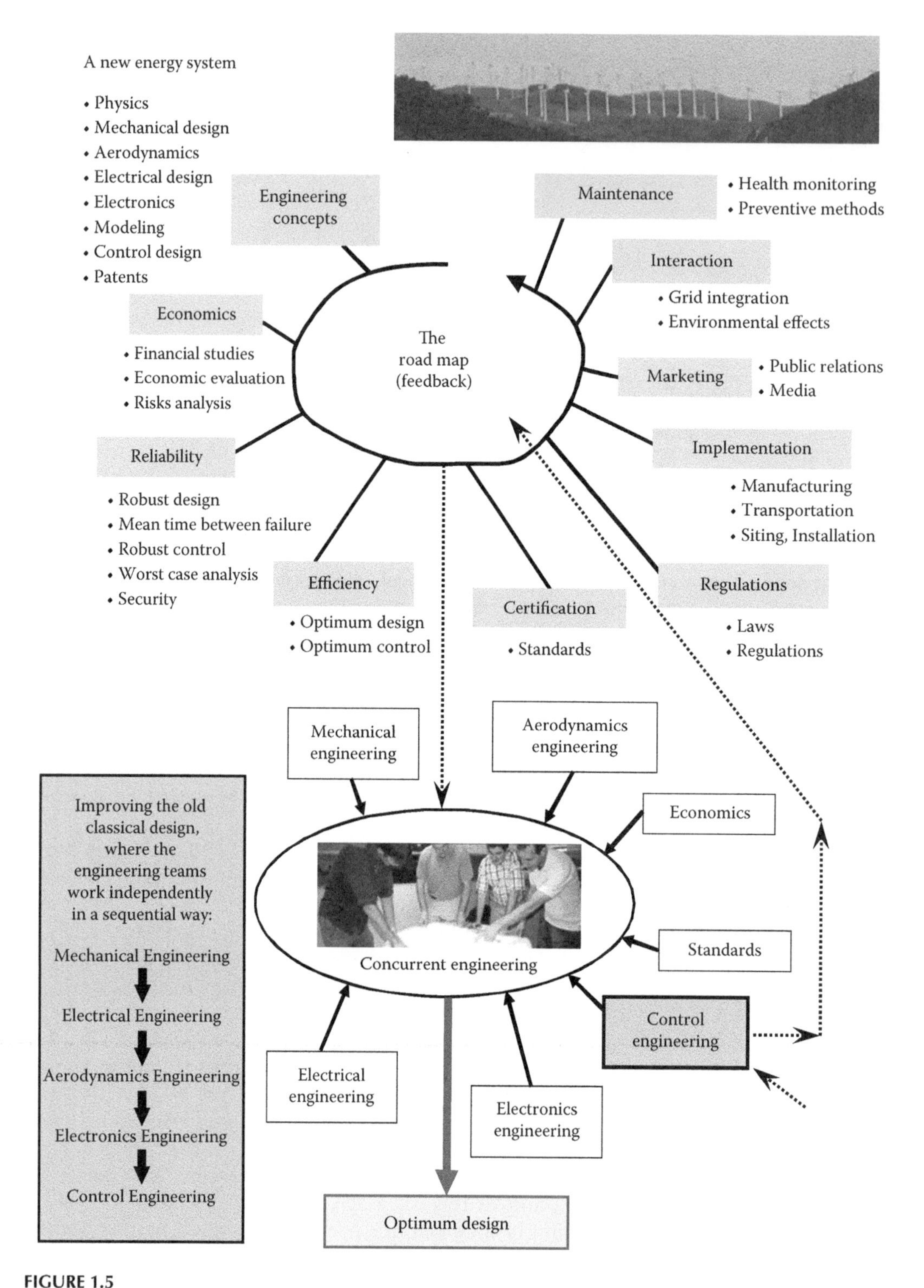

FIGURE 1.5
Concurrent engineering approach: the big picture to design a new energy system, and the primary and central role of the control engineer.

This big picture definitely requires the concurrent engineering way of thinking and working with the leadership of a *multidisciplinary engineer*, which the authors claim for the *control engineer*, to be able to achieve an optimum design, taking into account all the aspects of the road map.

1.3 Quantitative Robust Control

Many of the frequency domain fundamentals were established by Hendrik W. Bode in his original book *Network Analysis and Feedback Amplifier Design*,[1] published in 1945. It strongly influenced the understanding of automatic control theory for many years, especially where system sensitivity and feedback constraints were concerned.

But it was not until 1963 that a new book, *Synthesis of Feedback Systems*,[2] written by Isaac Horowitz, showed a formal combination of the frequency methodology with plant ignorance considerations under quantitative analysis. It addresses an extensive set of sensitivity problems in feedback control and was the first work in which a control problem was treated quantitatively in a systematic way. Any serious student of feedback control theory must eventually first study Isaac's book carefully.

Since then, and during the last decades of the twentieth century, there has been a tremendous advancement in robust frequency domain methods. One of the main techniques, also introduced by Horowitz, which characterizes closed loop performance specifications against parametric and nonparametric plant uncertainty, mapped into open loop design constraints, became known as QFT in the early 1970s.[3–187]

QFT is an engineering control design methodology that explicitly emphasizes the use of feedback to simultaneously and quantitatively reduce the effects of plant uncertainty and satisfy performance specifications (see also the book by Houpis, Rasmussen, and Garcia-Sanz).[6]

QFT is deeply rooted in classical frequency response analysis involving Bode diagrams, template manipulations, and Nichols charts. It relies on the observation that the feedback is needed principally when the plant presents model uncertainty or when there are uncertain disturbances acting on the plant. Figure 1.6 describes the big picture of QFT control system design, bridging the gap between control theory and real-world applications.[151–187]

The essential aspects of the control system design process are also illustrated in Figure 1.6. The intent of the figure is to give the reader an overview of what is involved in achieving a successful and practical control system design, presenting the factors that help in *bridging the gap* between theory and the real world. While accomplishing a practical control system design, the designer must keep in mind that the goal of the design process, besides achieving a satisfactory theoretical robust design, is to implement a control system that meets the functional requirements. In other words, during the design process one must keep the real world in mind. For instance, in performing the simulations, one must be able to interpret the results obtained, based upon a knowledge of what can be reasonably expected of the plant that is being controlled. Position saturation and, even more dramatically, rate saturation of the actuators will significantly affect the achievement of the functional requirements. Linear and nonlinear simulations are very helpful in early evaluation, but if the system is to operate in the real world, hardware-in-the-loop

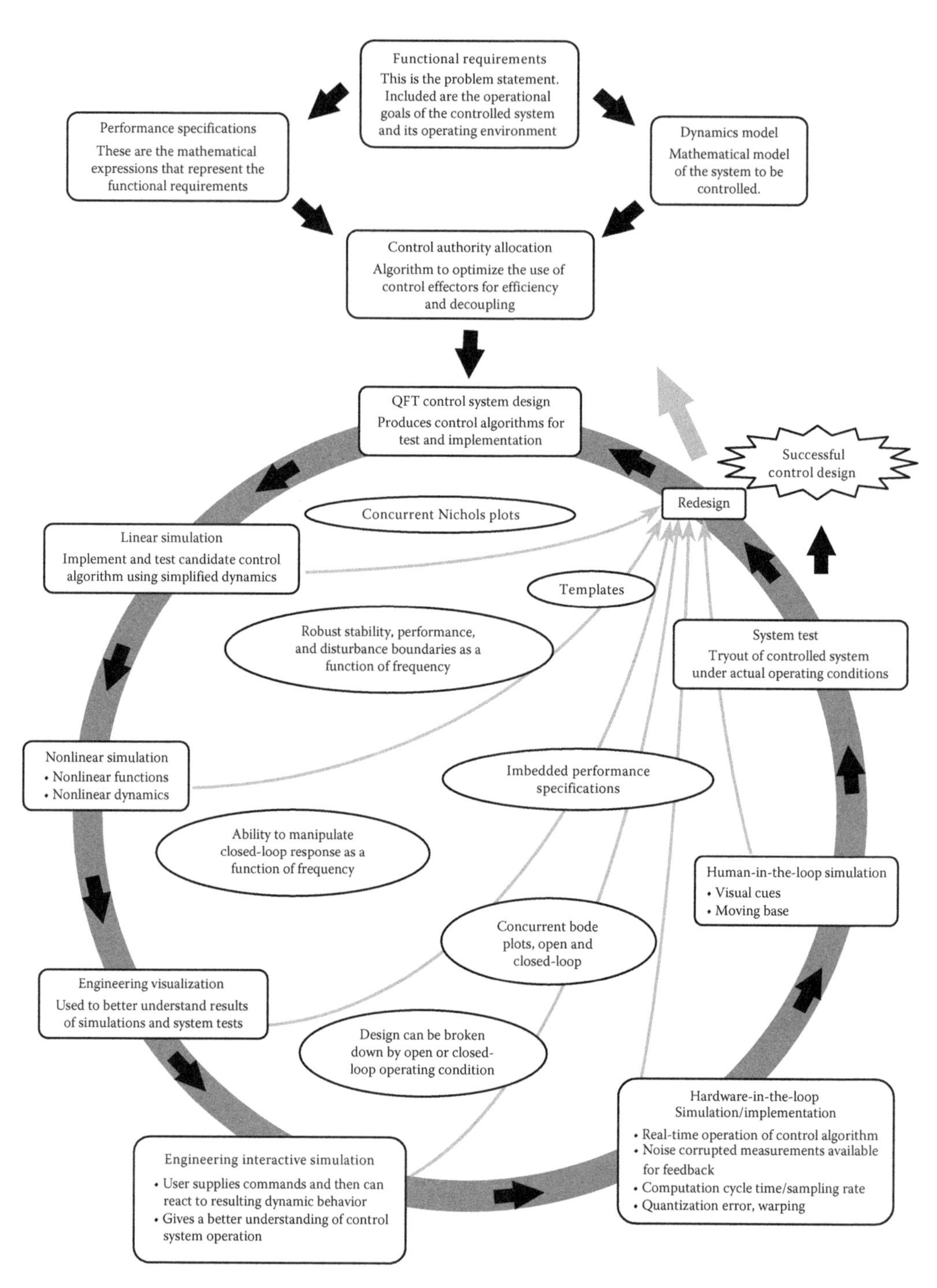

FIGURE 1.6
QFT control system design process: *bridging the gap.*

and system tests must be performed to check for unmodeled effects not taken into account during the design and implementation phases.

1.4 Novel CAD Toolbox for QFT Controller Design

The use of CAD tools has made control engineering design much simpler. This book introduces the new interactive object-oriented CAD tool for QFT controller design (*QFT Control Toolbox*), developed by Garcia-Sanz, Mauch, and Philippe (see Figure 1.7).[149,150] The toolbox can be downloaded from http://cesc.case.edu and http://www.crcpress.com/product/isbn/9781439821794. It includes the latest technical quantitative robust control achievements within a user-friendly and interactive environment. The tool runs under MATLAB® and shows a special architecture based on four windows (W1: *Plant Modeling*; W2: *Control Specifications*; W3: *Controller Design*; W4: *Analysis*) and a common memory. It also includes a library of basic and advanced functions to be selected in the corresponding Windows, allows a multitasking/threading operating system, offers a user-friendly and interactive environment by using object-oriented programming, permits to rescale the problem from SISO (single-input-single-output) to MIMO (multiple-input-multiple-output) problems easily, and uses reusable code. The toolbox has been developed at the Public University of Navarra and Case Western Reserve University under a research project for the European Space Agency, ESA-ESTEC.

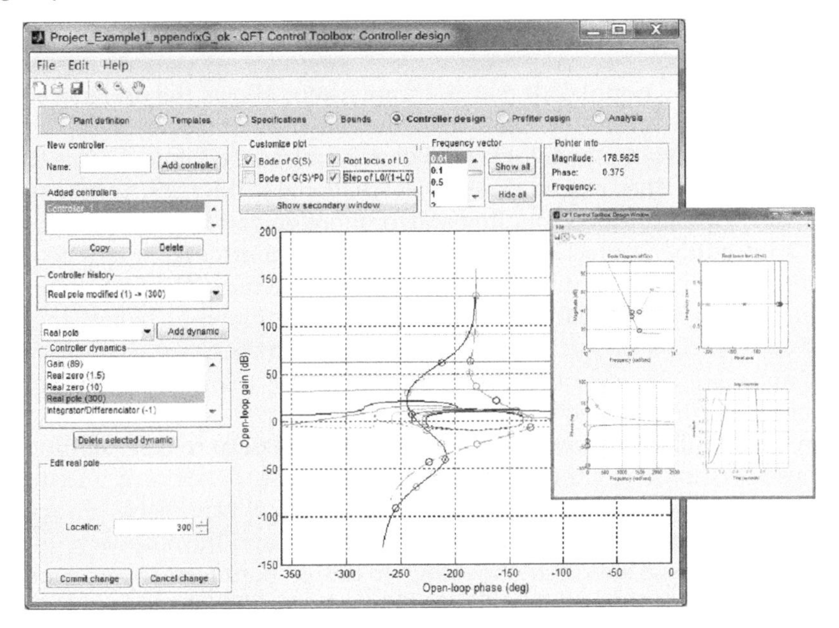

FIGURE 1.7
(See color insert.) The QFT control toolbox for MATLAB®. An interactive object-oriented tool for quantitative robust controller design. Developed by Garcia-Sanz, Mauch, Philippe et al., at the Public University of Navarra, Case Western Reserve University and the European Space Agency. (From Garcia-Sanz, M. et al., QFT control toolbox: An interactive object-oriented Matlab CAD tool for quantitative feedback theory, in *Sixth IFAC Symposium on Robust Control Design, ROCOND'09*, Haifa, Israel; Garcia-Sanz, M. et al., QFT control toolbox: An interactive object-oriented Matlab CAD tool for controller design, Research project, European Space Agency ESA-ESTEC, Public University of Navarra, Pamplona, Spain.)

1.5 Outline

The book is divided into two parts. The purpose of Part I is to present the concepts of the QFT control engineering technique in such a manner that students and practicing engineers can readily grasp the fundamentals and appreciate its *transparency* in *bridging the gap* between theory and the real world. Seven chapters and seven appendices present the technique.[3–187]

Chapters 2 and 3 present the main topics of the QFT robust control methodology for multiple-input single-output analog systems.[3–76]

Chapter 4 focuses on the application of the QFT technique to sampled-data (digital) control systems. A digital control system utilizes a digital computer or microprocessor to compute the control law or algorithm, an analog-to-digital converter (A/D C) to sample the analog sensor signals and convert them into digits or numbers for the algorithm, and a digital-to-analog converter (D/A C) to convert the result of the control algorithm (numbers) into an analog control signal for the actuators.[104–107]

Chapter 5 extends the QFT control design technique to MIMO systems, presenting two diagonal MIMO techniques: the nonsequential method (*Method 1*) and the sequential method (*Method 2*).[77–102] Control of MIMO systems with model uncertainty is one of the most difficult problems the control engineer has to face in real-world applications. Problems like the compensator/controller matrix structure, sensor–actuator pairings, input–output directions, loops coupling, transmission zeros, etc., become much more difficult when the plant presents model uncertainty. Two of the main characteristics that define a MIMO system are the input and output directionality and the coupling among control loops.[223–273]

A nondiagonal (fully populated) matrix compensator allows the designer more design flexibility to regulate MIMO systems than the classical diagonal controller structure. Chapter 6 presents two methodologies (*Method 3* and *Method 4*) to design nondiagonal matrix compensators. They extend the classical diagonal MIMO QFT compensator design to new systems with more loop coupling and/or more demanding specifications.[77–102]

Chapter 7 introduces a practical quantitative robust control technique to design one-point feedback controllers for distributed parameter systems (DPS) with uncertainty. The method considers (1) the spatial distribution of the relevant points where the inputs and outputs of the control system are applied (actuators, sensors, disturbances, and control objectives) and (2) a new set of transfer functions (TFs) that describe the relationships between those distributed inputs and outputs.

Based on the definition of such distributed TFs, the classical robust stability and robust performance specifications are extended to the DPS case, developing a new set of quadratic inequalities to define the QFT bounds. The method can deal with uncertainty in both the model and the spatial distribution of the inputs and the outputs.[108–115]

Chapter 8 introduces a hybrid methodology to design nonlinear robust control systems that are able to go beyond the classical linear limitations. Combining robust QFT controller designs and stable switching, the new system optimizes the time response of the plant by fast adaptation of the controller parameters during the transient response according to certain rules based on the amplitude of the error. The methodology is based on both a graphical frequency domain stability criterion for switching linear systems and the use of the robust QFT control system design technique.[140,285–295]

To complement the previous chapters, useful information is included in the appendices: QFT templates (Appendix A); quadratic inequalities for QFT bounds (Appendix B);

FIGURE 1.8
Multimegawatt direct-drive WT. (Courtesy of M. Torres.)

analytical formulation to compute QFT bounds (Appendix C); elements for loop-shaping and controller design (Appendix D); controller fragility analysis (Appendix E); the User's Guide for the QFT Control Toolbox (Appendix F); a collection of illustrative controller design examples to help the student understand the main concepts and procedures of the QFT Control Toolbox (Appendix G); and a list of the main units used in this book, with the conversion between the International system and the American system of units (Appendix H). Additionally, a compilation of design problems is also included at the end of this book.[33–76,149,150]

After the presentation of the QFT controller design engineering technique in the first part of this book, Part II introduces the main topics of modern WT design and control, including (1) the description of classical and advanced turbines, (2) dynamic modeling, (3) control objectives and strategies, (4) standards and certification, (5) controller design, and (6) a large number of applications like onshore and offshore WTs, floating WTs, airborne wind energy systems, and advanced manufacturing and real experimentation (see Figure 1.8). Eleven chapters present the wind energy field.[302–521]

Chapter 9 analyzes how the results achieved in the control engineering field play a central role in the success of the wind energy development. This chapter introduces a piece of the history of the birth of modern WTs, describes current and future market sizes and investments in the wind energy field, analyzes future challenges and opportunities, and finally proposes new control engineering solutions to open virgin global markets.[302–318,337–350,518–521]

Chapter 10 gives an introduction of technical standards and certification processes for WTs, including the technical standard formal concept and its strategic value, the structure and development procedures for new standards, and the fundamentals of the certification of WTs, as well as the impact of the controller design in the process.[337–350]

Chapter 11 introduces the main characteristics, objectives, and strategies of the control system of a WT, including the classical control loops and the components to regulate the machine.[302–318]

Chapter 12 describes the aerodynamic and mechanical models for the dominant dynamics of a WT as a first step to understand its dynamical behavior and to design the controllers. It includes the aerodynamic models for the ideal and real conditions and the general mechanical State Space WT models for both the direct-drive synchronous generator (DD) and the doubly fed induction generator (DFIG).[302–318,351–392]

Chapter 13 describes the electrical models for WTs connected to the grid, including squirrel cage induction generators (SCIG), DFIG, DD, and a summary of the most common power electronic converters. The chapter also analyzes the typical WT power quality characteristics, including flicker, harmonics and inter-harmonics, power peaks, reactive power and power factors, voltage dips, as well as wind farm grid integration issues such as capacity factors, limited transmission capacity, and grid control.[393–409]

Chapter 14 deals with the design of advanced pitch control systems for variable-speed pitch-controlled DFIG WTs. The application of the robust QFT control techniques and nonlinear switching methods previously introduced is developed in detail in three realistic examples, which include mechanical fatigue minimization, performance optimization, and large nonlinear parameter variation conditions.[140,149,150]

Chapter 15 presents some experimental control system results of the multimegawatt variable-speed direct-drive multipole synchronous Torres wind turbines (TWT)—TWT-1.65/70 (Class Ia), TWT-1.65/77 (Class IIa), and TWT-1.65/82 (Class IIIa) designed by M. Torres with the collaboration of Professor Garcia-Sanz. The first project was started in 1998, and the first prototype was installed in Cabanillas (Navarra, Spain) and began its operation in April 2001. Since then many WTs of this family have been installed in different countries, and a large amount of experimental data have been collected. The design of the controllers was made by using advanced QFT-robust and nonlinear-switching control strategies based on both mathematical modeling and analysis of the experimental data. This chapter introduces the main advantages of the direct-drive multipole system and shows some of the most representative experimental results with the direct-drive TWT with the QFT and switching controllers under medium and extreme wind conditions and connected to different grid situations.[174,349,366,379]

Chapter 16 applies the sequential MIMO QFT methodology to control the temperatures inside a multivariable industrial furnace for WT blades manufacturing. The designed controller is found to be effective in achieving the demanded performance specifications. It not only copes with plant uncertainties but also enhances the rejection of the typical exothermic reactions of the polymerization process and attenuates successfully the coupling between the control loops due to the heterogeneous thermal loads in the furnace.[178,375,386]

Chapter 17 presents an introduction to the state of the art of smart blades for WTs as new effective ways to reduce the mechanical loads on the WT by means of more sophisticated active load control techniques. They consist of locally distributed aerodynamic control systems with built-in intelligence on the blades.[433–498]

Chapter 18 analyzes the main projects and characteristics of offshore wind energy systems, including the history of offshore oil and natural gas platforms; the main current offshore wind energy projects with fixed platforms in the world; and the specific characteristics, challenges, and tools for offshore floating WT platforms.[410–432]

Chapter 19 describes the main projects and characteristics of airborne wind energy systems, including an overview of the field and the design of a novel airborne WT: *the EAGLE system*.[499–517]

The reference section compiles the main references used in each part of the book in both Parts I and II. This section is arranged according to subject and chronologically within each subject.

The new interactive object-oriented CAD tool for QFT controller design (*QFT Control Toolbox for MATLAB*) introduced with this book can be downloaded from http://cesc.case.edu and http://www.crcpress.com/product/isbn/9781439821794.

Part I

Advanced Robust Control Techniques: QFT and Nonlinear Switching

2

Introduction to QFT

2.1 Quantitative Feedback Theory

This chapter presents an overview of the quantitative feedback theory (QFT) to enhance the understanding of the technique, which is introduced in the following chapters. A discussion of why feedback is needed to achieve performance goals is presented first. This is followed by a QFT overview of design objectives, an explanation of structured parametric uncertainty, a design overview, and QFT basics and design. The last few sections are devoted to the insight provided by the QFT technique and to the benefits of QFT.

2.2 Why Feedback?

To answer the question "Why do you need feedback?" first consider the following system. The plant $P(s)$ responds to the input $r(t)$ $[R(s)]$ with the output $y(t)$ $[Y(s)]$ in the presence of disturbances $d_1(t)$ $[D_1(s)]$ and $d_2(t)$ $[D_2(s)]$ (see Figure 2.1).[6,22] If it is desired to achieve a specified system transfer function $T(s)$ $[=Y(s)/R(s)]$, then it is necessary to insert a prefilter whose transfer function is $T(s)/P(s)$, as shown in Figure 2.2. This compensated system produces the desired output as long as the plant does not change, that is, there is no plant uncertainty, and there are no disturbances. This type of system is sensitive to changes or uncertainty in the plant, and the disturbances are reflected directly in the output. Thus, it is necessary to feed back the output in order to reduce the output sensitivity to parameter variation and attenuate the effect of disturbances on the plant output (see Figure 2.3).

In designing a feedback control system, it is desired to utilize a technique that

1. Addresses at the onset all known plant variations
2. Incorporates information on the desired output tolerances
3. Maintains reasonably low loop gain and bandwidth (i.e., reduces the "cost of feedback")

This last item is important in order to avoid problems associated with high loop gains such as sensor noise amplification, saturation, and high frequency uncertainties.

Consider the control system of Figure 2.4a, containing a plant uncertainty set, $\mathcal{P}(s) = \{P(s)\}$, that represents a plant with variable parameters. This system has two inputs: $r(t)$ the desired input signal to be tracked and $d(t)$ an external disturbance input signal, which is to be

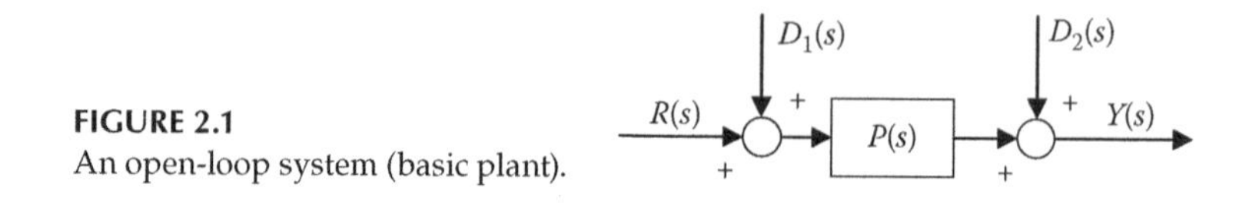

FIGURE 2.1
An open-loop system (basic plant).

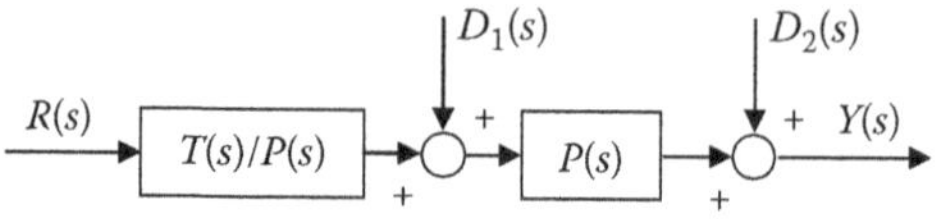

FIGURE 2.2
A compensated open-loop system.

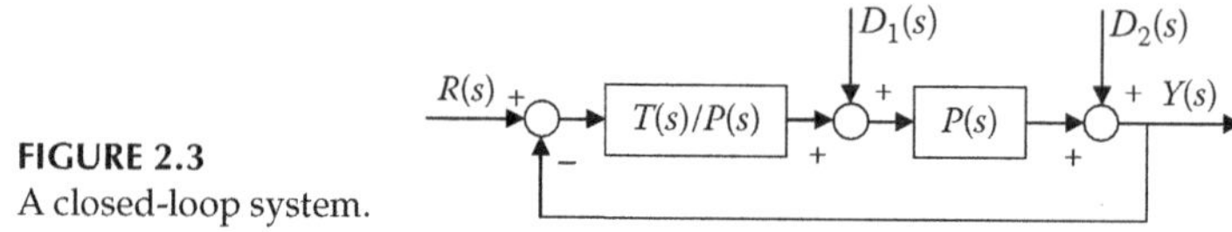

FIGURE 2.3
A closed-loop system.

attenuated to have minimal effect on $y(t)$. The tracking and disturbance control ratios of Figure 2.4a, based upon the nominal plant $P_o(s) \in \mathcal{P}(s)$, are, respectively,

$$T_R(s) = \frac{Y(s)}{R(s)} = P_o(s) \tag{2.1}$$

$$T_D(s) = \frac{Y(s)}{D_1(s)} = P_o(s) \tag{2.2}$$

The sensitivity functions[218] of the open-loop uncompensated system of Figure 2.4a for the two cases $Y_R(s)|_{d(t)=0}$ and $Y_D(s)|_{r(t)=0}$ are identical; that is,

$$S_{P_o(s)}^{Y_R(s)}(s) = S_{P_o(s)}^{Y_D(s)} = 1 \tag{2.3}$$

For the compensated system of Figure 2.4b, the tracking and disturbance control ratios are, respectively,

$$T_R = \frac{P_o G}{1 + P_o G} = \frac{L_o}{1 + L_o} \tag{2.4}$$

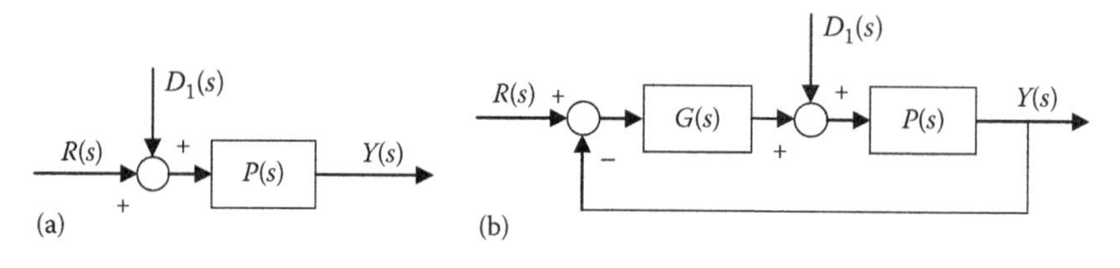

FIGURE 2.4
Control systems: (a) open-loop uncompensated system; (b) closed-loop compensated system.

$$T_D = \frac{P_o}{1+P_oG} = \frac{P_o}{1+L_o} \tag{2.5}$$

where $L_o \equiv P_oG$ is defined as the *nominal loop transmission function*. For the compensated system of Figure 2.4b, the sensitivity functions for these two cases are also identical; that is,

$$S_{P_o(s)}^{Y_R(s)}(s) = S_{P_o(s)}^{Y_D(s)}(s) = \frac{1}{1+P_oG} = \frac{1}{1+L_o} \tag{2.6}$$

Comparing Equation 2.6 with Equation 2.3 readily reveals that the effect of changes of the uncertainty set $\mathcal{P}(s) = \{P(s)\}$ upon the output of the closed-loop control system is reduced by the factor $1/[1 + P_oG]$ compared to the open-loop control system. *This reduction is an important reason why feedback systems are used.*

Horowitz has shown[22] that a robust control system design is best achieved by working with L_o and not with the sensitivity function S. The reasons for the choice of L_o are as follows:

1. The sensitivity function is very sensitive to the cost of feedback.
2. A practical optimum design requires working to the limits of the system's performance specifications.
3. The order of the compensator (controller) $G(s)$ can be minimized by incorporating[218] the nominal plant $P_o(s)$ into $L_o(s)$.

The reader is referred to the QFT literature for a more detailed analysis of the sensitivity function with respect to sensor noise and plant parameter uncertainty on system performance.[1,31,126]

2.3 QFT Overview

2.3.1 QFT Design Objective

Design and implement robust control for a system with structured parametric uncertainty that satisfies the desired performance specifications.[31]

2.3.2 Parametric Uncertainty: A Basic Explanation

2.3.2.1 Simple Example

To illustrate "What is parametric uncertainty?" consider the undergraduate laboratory experiment that involves hooking up the DC shunt motor of Figure 2.5. Further, consider that the students entered the laboratory on a cold January Monday morning to perform this experiment. The weekend room temperature was set at 50°F (10°C) but was reset to 70°F (21°C) when the students entered the room. The students hurriedly hooked up the

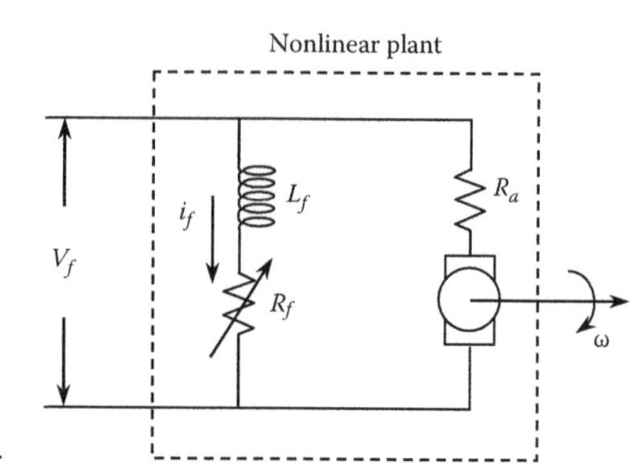

FIGURE 2.5
DC shunt motor.

motor and set the field rheostat to yield a speed of ω = 1200 r.p.m. Upon accomplishing this phase of the experiment, they took a 1 h break in order to allow the room to reach the desired temperature of 70°F (21°C). Upon their return, they found that the speed of the motor was now 1250 r.p.m. with no adjustments to the applied voltage or to the field rheostat R_f. *Why the change in speed?*

- Due to the heating of the DC shunt field by the field current i_f and the environmental conditions, the value of R_f increased.
- This, in turn, decreased the value of i_f and hence resulted in the increase in speed since speed is inversely proportional to i_f, assuming V_f is constant.
- Therefore, during the operation of the motor, the parameter R_f can vary anywhere within the range $R_{fmin} \leq R_f \leq R_{fmax}$ due to the variable environmental temperature and field current.
- As a consequence, there is *uncertainty* as to what the actual value of the parameter R_f will be at the instant a command is given to the system.
- Thus, the *parametric uncertainty is structured* because the range of the variation of R_f is known and its effects on the relationship between V_f and ω can be modeled.

2.3.2.2 Simple Mathematical Description

The transfer function of a DC servo motor, utilized as a position control device, is

$$P_\iota(s) = \frac{\Theta_m(s)}{V_f(s)} = \frac{Ka}{s(s+a)} \tag{2.7}$$

where $\Theta_m(s)$ is the angular position of the rotor, $V_f(s)$ the applied voltage, and where the parameters K and a vary, due to the operating scenario, over the following range: $K \in (K_{min}, K_{max})$ and $a \in (a_{min}, a_{max})$. Over the region of operation, in a position control system, the plant parameter variations are described by Figure 2.6. The shaded region in the figure represents the region of *structured parametric uncertainty* (region of plant uncertainty). The motor can be represented by six linear time-invariant (LTI) transfer functions P_ι ($\iota = 1, 2, \ldots, J$) at the points indicated in the figure. The Bode plots for these 6 LTI plants are shown in Figure 2.7.

2.3.3 Control System Performance Specifications

In many control systems the output $y(t)$ must lie between specified upper and lower bounds, $y(t)_U$ and $y(t)_L$, respectively, as shown in Figure 2.8a. The conventional time-domain figures

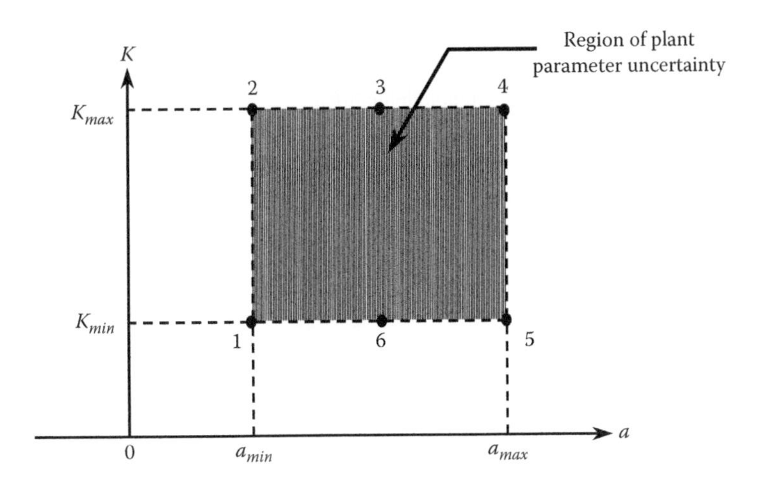

FIGURE 2.6
Region of plant parameter uncertainty.

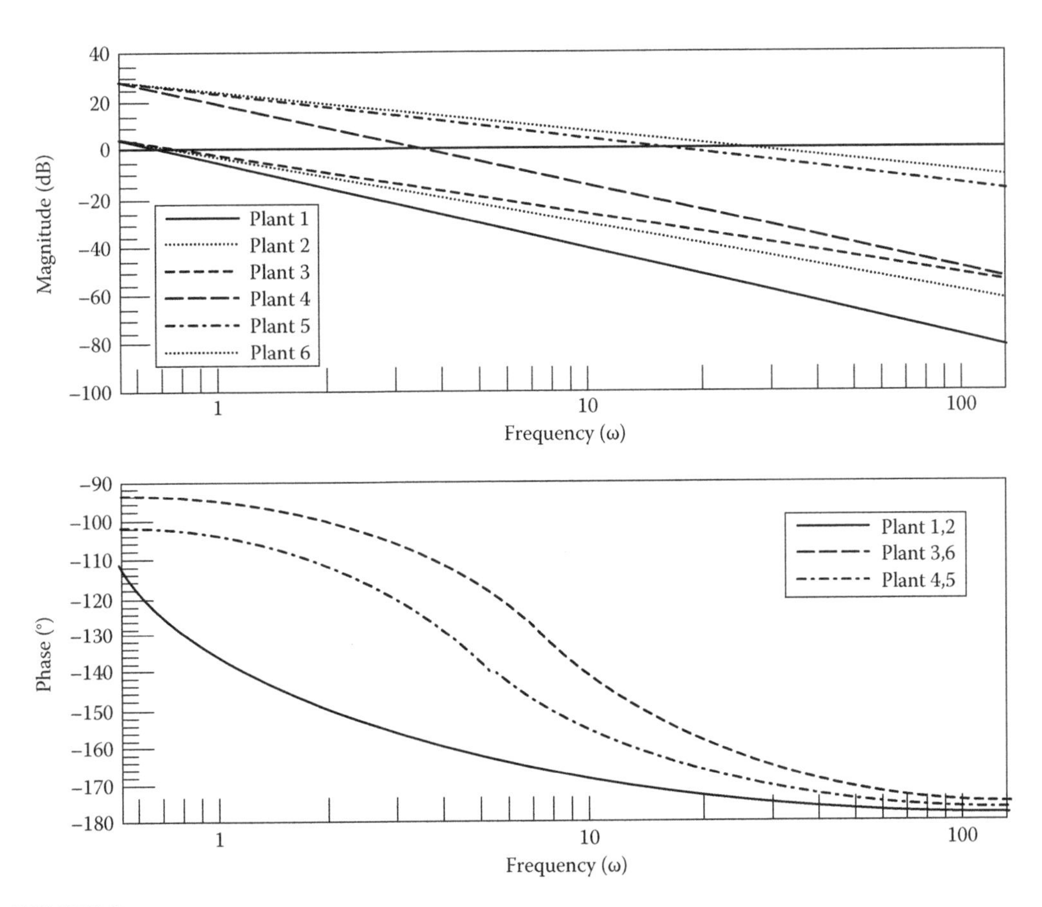

FIGURE 2.7
Bode plots of 6 LTI plants: the range of parameter uncertainty.

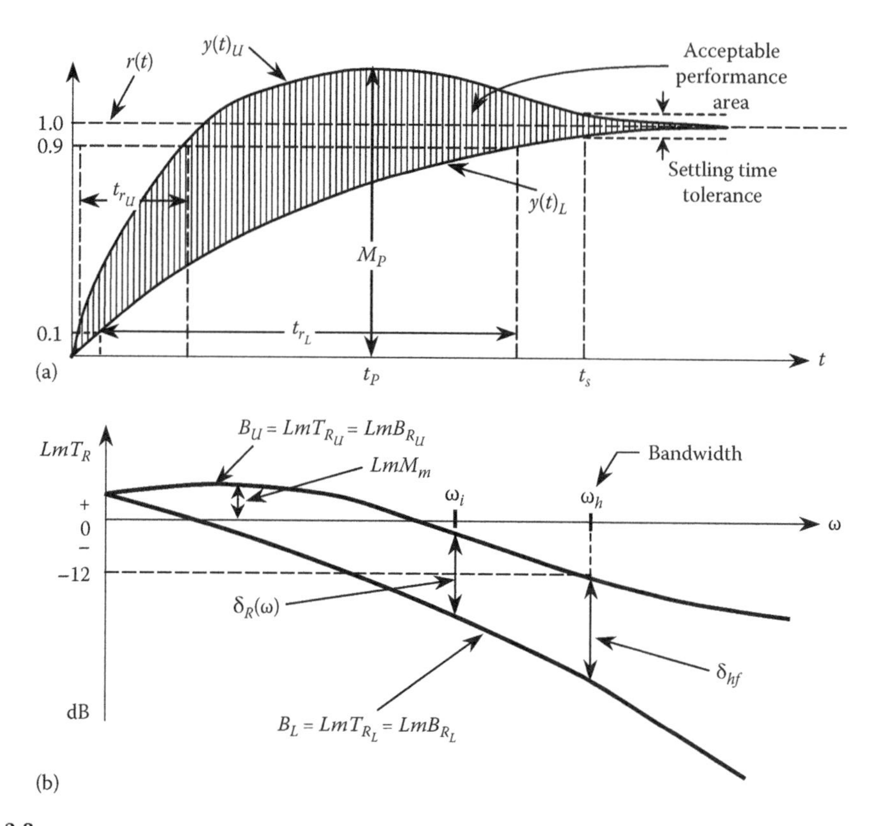

FIGURE 2.8
Desired system performance specifications: (a) time-domain response specifications; (b) frequency-domain response specifications.

of merit, based upon a *step input signal* $r(t) = R_o u_{-1}(t)$, are shown in Figure 2.8a. They are as follows: M_p, peak overshoot; t_r, rise time; t_p, peak time; and t_s, settling time. Corresponding system performance specifications in the frequency domain are B_U and B_L, the upper and lower bounds, respectively, peak overshoot LmM_m (where $Lm(\cdot) = 20 \log_{10} |\cdot|$), and the frequency bandwidth ω_h, which are shown in Figure 2.8b. Assuming that the control system has negligible sensor noise and sufficient control effort authority, for a stable *LTI minimum-phase* (*m.p.*) plant, an *LTI* compensator may be designed to achieve the desired control system performance specifications. The case of *non-minimum-phase* (*n.m.p.*) plant is discussed in Chapter 4 (see also Refs. 116 through 121).

2.3.4 QFT Design Overview

The QFT design objective is achieved by

- Representing the characteristics of the plant and the desired system performance specifications in the frequency domain
- Using these representations to design a compensator (controller)
- Representing the nonlinear plant characteristics by a set of *LTI* transfer functions that cover the range of structured parametric uncertainty
- Representing the system performance specifications (see Figure 2.8) by *LTI* transfer functions that form the upper (B_U) and lower (B_L) boundaries for the design

- Reducing the effect of parameter uncertainty by shaping the open-loop frequency responses so that the Bode plots of the J closed-loop systems fall between the boundaries B_U and B_L, while simultaneously satisfying all performance specifications
- Obtaining the stability, tracking, disturbance, and cross-coupling (for multiple-input-multiple-output [MIMO] systems) boundaries on the Nichols chart (NC) in order to satisfy the performance specifications

2.3.5 QFT Basics

Consider the control system of Figure 2.9, where $G(s)$ is a compensator, $F(s)$ is a prefilter, and $\mathcal{P}$ is the nonlinear plant. To accomplish a QFT design

- The nonlinear plant is described by a set of J *m.p. LTI* plants, that is, $\mathcal{P} = \{P_\iota(s)\}$ ($\iota = 1, 2, \ldots, J$), which define the structured plant parameter uncertainty. *Note*: QFT can work with $n \times n$ MIMO systems (Chapters 5 and 6), *n.m.p.* plants (Chapter 4), etc. This chapter is restricted to the discussion of *m.p.* plants.
- The magnitude variation due to plant parameter uncertainty, $\delta_P(\omega_i)$, is depicted by the Bode plots of the *LTI* plants shown in Figure 2.10 for the examples shown in Figures 2.6 and 2.7.
- J data points (log magnitude and phase angle), for each value of frequency $\omega = \omega_i$, are plotted on the NC. A contour is drawn through the data points, for ω_i, which describes the boundary of the region that contains all the J points. This contour is referred to as a *template*. It represents the region of structured plant parametric uncertainty on the NC and is obtained for specified values of frequency, $\omega = \omega_i$, within the bandwidth (BW) of concern. Six data points (log magnitude and phase angle) for each value of ω_i are obtained, as shown in Figure 2.11a, for the example shown in Figure 2.6, to plot the templates, for each value of ω_i, as shown in Figure 2.11b.
- The system performance specifications are represented by *LTI* transfer functions, and their corresponding Bode plots are shown in Figure 2.10 by the upper and lower bounds B_U and B_L, respectively.

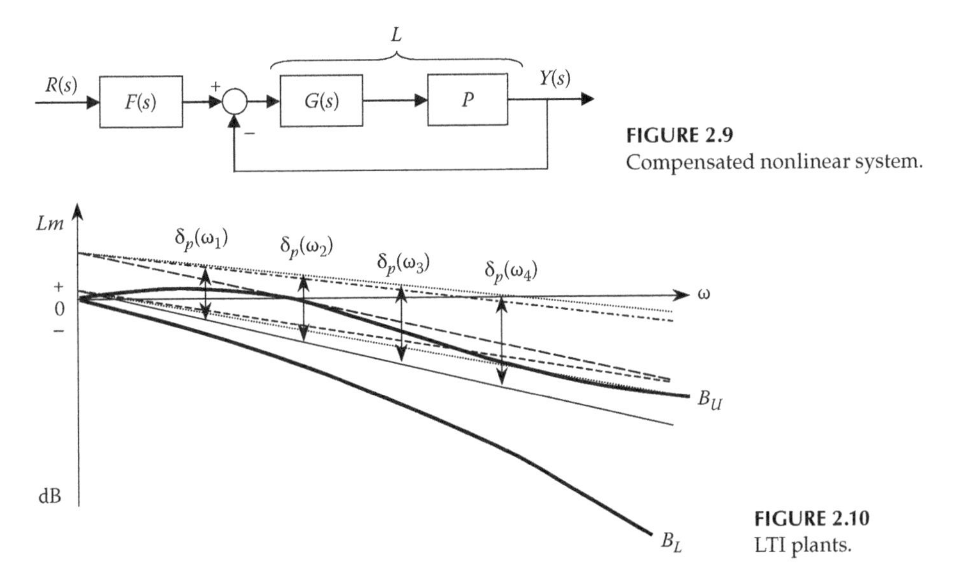

FIGURE 2.9
Compensated nonlinear system.

FIGURE 2.10
LTI plants.

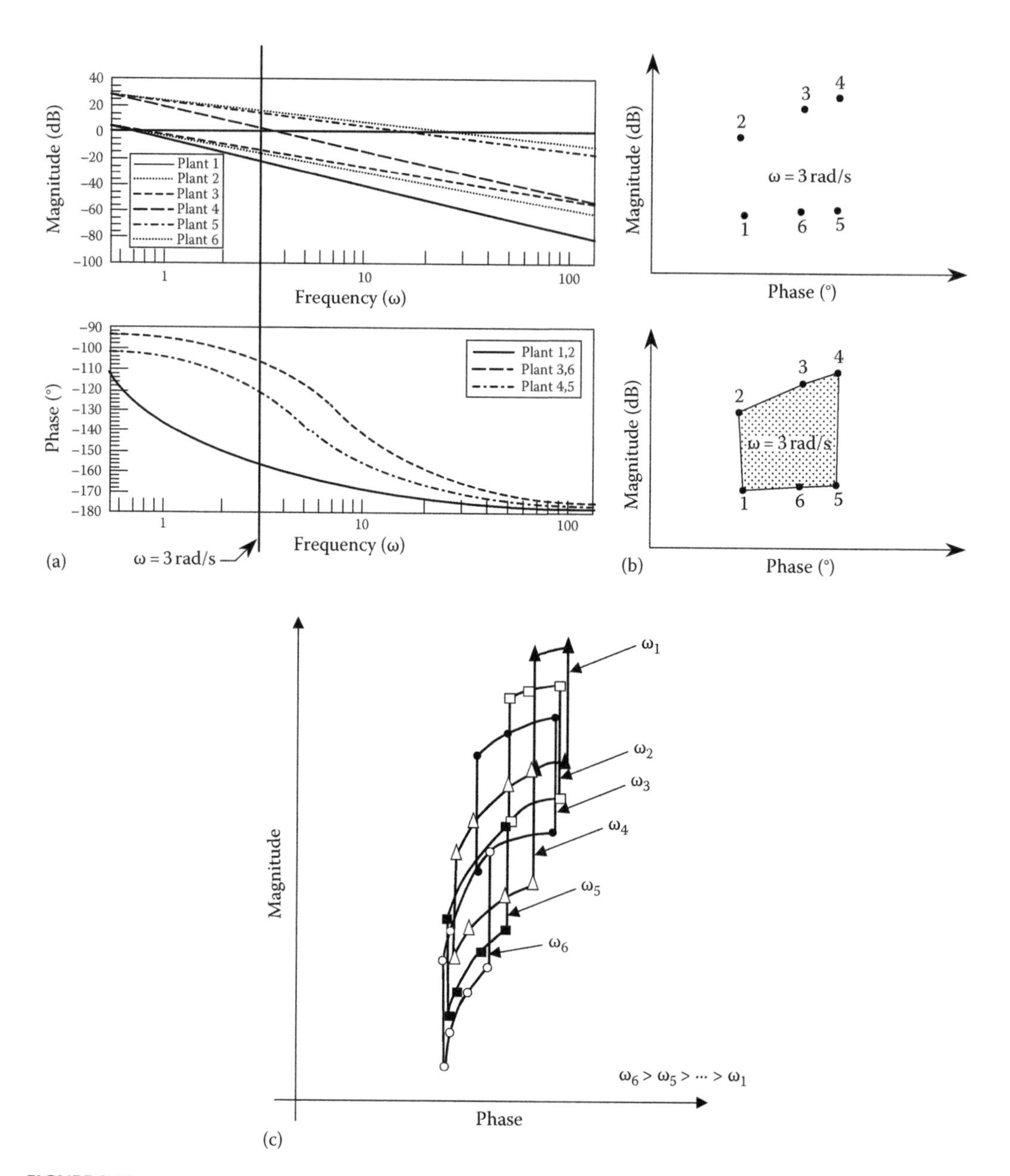

FIGURE 2.11
(a) Bode plots of 6 LTI plants; (b) template construction for $\omega_i = 3$ rad/s; (c) construction of the NC plant templates.

2.3.6 QFT Design

The tracking design objective is to

1. Synthesize a compensator $G(s)$ of Figure 2.9 that
   a. Results in satisfying the desired performance specifications of Figure 2.8
   b. Results in the closed-loop frequency responses T_{L_ι} shown in Figure 2.12

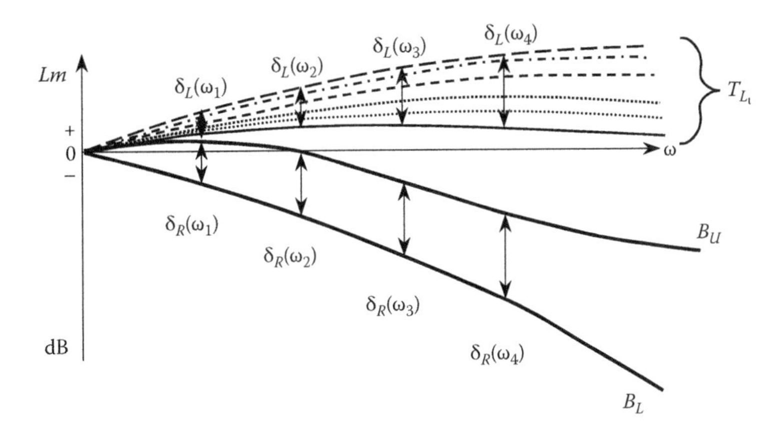

FIGURE 2.12
Closed-loop responses: LTI plants with $G(s)$.

c. Results in the $\delta_L(\omega_i)$ of Figure 2.12, for the compensated system, being equal to or less than $\delta_P(\omega_i)$ of Figure 2.10 for the uncompensated system, which is equal to or less than $\delta_R(\omega_i)$ for each value of ω_i of interest; that is,

$$\delta_L(\omega_i) \le \delta_R(\omega_i) \le \delta_P(\omega_i)$$

2. Synthesize a prefilter $F(s)$ of Figure 2.9 that
   a. Results in shifting and reshaping the T_{L_t} responses in order that they lie within the B_U and B_L boundaries in Figure 2.12 as shown in Figure 2.13

Therefore, the QFT robust design technique assures that the desired performance specifications are satisfied over the prescribed region of structured plant parametric uncertainty.

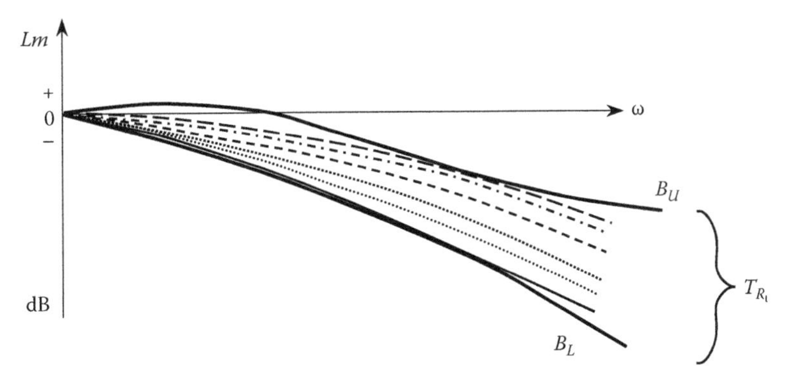

FIGURE 2.13
Closed-loop responses: LTI plants with $G(s)$ and $F(s)$.

2.4 Insight into the QFT Technique

2.4.1 Open-Loop Plant

Consider the position control system of Figure 2.9 whose plant transfer function is given by

$$P_\iota(s) = \frac{Ka}{s(s+a)} = \frac{K'}{s(s+a)} \tag{2.8}$$

where $K' = Ka$ and $\iota = 1, 2, \ldots, J$. The log magnitude variation due to the plant parameter uncertainty, for $J = 6$, is depicted by the Bode plots in Figure 2.7. The loop transmission $L(s)$ is defined as

$$L_\iota(s) = P_\iota(s)G(s) \tag{2.9}$$

2.4.2 Closed-Loop Formulation

The control ratio T_L of the unity-feedback system of Figure 2.9 is

$$T_{L_\iota} = \frac{Y}{R_L} = \frac{L_\iota}{1+L_\iota} \tag{2.10}$$

The overall system control ratio T_R is given by

$$T_{R_\iota}(s) = \frac{L_\iota(s)}{1+L_\iota(s)}F(s) \tag{2.11}$$

2.4.3 Results of Applying the QFT Design Technique

The proper application of the robust QFT design technique requires the utilization of the prescribed performance specifications from the onset of the design process and the selection of a nominal plant P_o from the J LTI plants. Once the proper loop shaping of $L_o(s) = P_o(s)$ $G(s)$ is accomplished, a synthesized $G(s)$ is achieved that satisfies the desired performance specifications. The last step of this design process is the synthesis of the prefilter $F(s)$ that ensures that the Bode plots of T_{R_ι} all lie between the upper and lower bounds, B_U and B_L.

2.4.4 Insight into the Use of the NC in the QFT Technique[31]

This section is intended to provide the reader a review of the use of the NC and an insight as to how it applies to the QFT technique.

1. *Open-loop characteristics*—For the nominal plant $P_o(j\omega)$, the nominal loop transmission function is

$$LmL_o = LmP_oG = LmP_o + LmG \tag{2.12}$$

   whereas for all other plants, $P(j\omega)$, the loop transmission function is

$$LmL = LmPG = LmP + LmG \tag{2.13}$$

Thus, for $\omega = \omega_i$, the variation $\delta_P(\omega_i)$ in $LmL(j\omega_i)$ is given by

$$\delta_P(\omega_i) = LmL(j\omega_i) - LmL_o(j\omega_i) = LmP(j\omega_i) - LmP_o(j\omega_i) \quad (2.14)$$

and its phase angle variation is given by

$$\angle\Delta P(j\omega_i) = \angle L - \angle L_o = (\angle G + \angle P) - (\angle G + \angle P_o) = \angle P - \angle P_o \quad (2.15)$$

The expression $LmP(j\omega_i) = LmP_o(j\omega_i) + \delta_P(\omega_i)$, obtained from Equation 2.14, is substituted into Equation 2.13 to yield

$$LmL(j\omega_i) = LmP_o(j\omega_i) + LmG(j\omega_i) + \delta_P(\omega_i) \quad (2.16)$$

2. *Closed-loop characteristics*—The closed-loop system characteristics can be obtained, for a given $G(j\omega)$ and $P_o(j\omega)$, from the plot of $LmL_o(j\omega)$ versus $\angle L_o$ shown on the NC in Figure 2.14.[218] Also shown is a plot of a template, $\Im P(j\omega_i)$, whose contour is

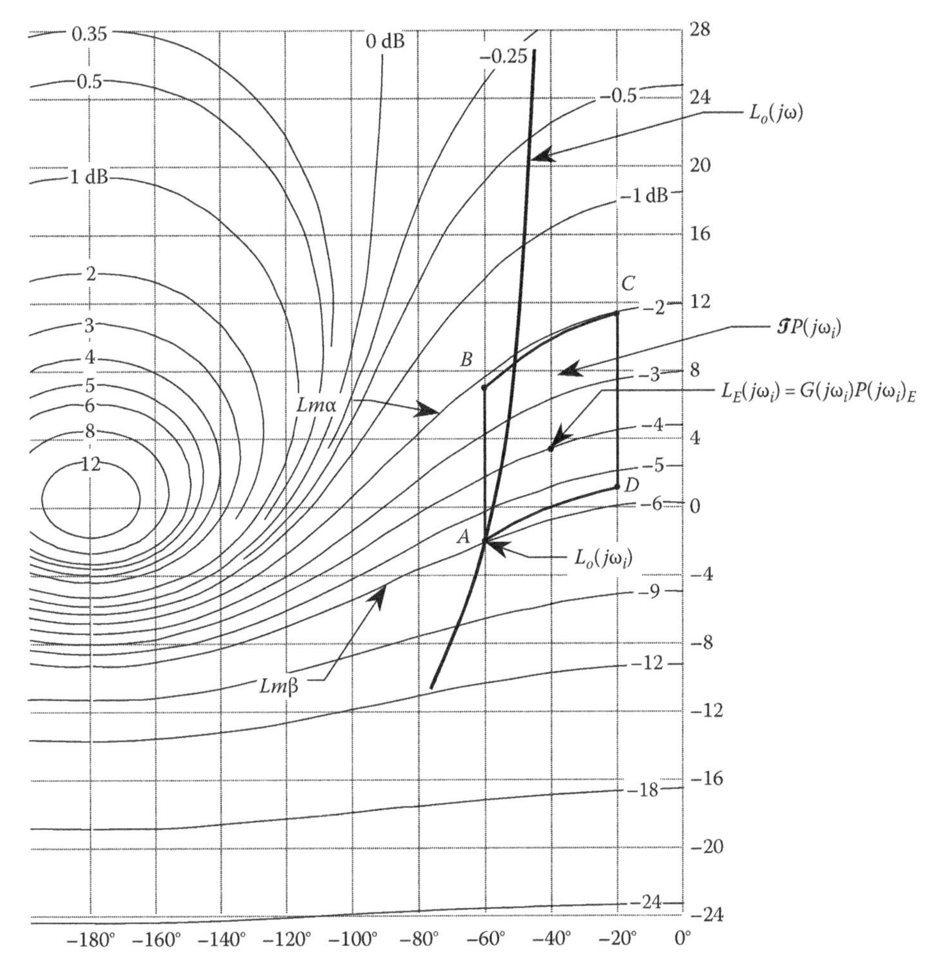

FIGURE 2.14
Nominal loop transmission plot with plant parameter area of uncertainty.

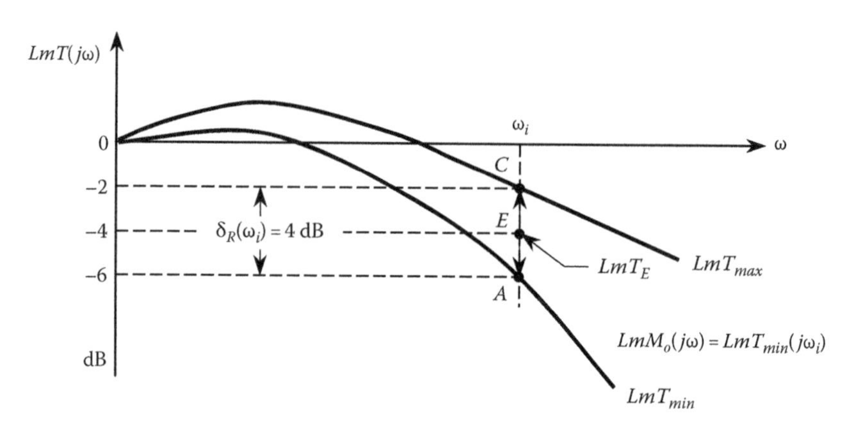

FIGURE 2.15
Closed-loop responses obtained from Figure 2.14.

based upon the data obtained for $\omega = \omega_i$ from Figure 2.7. This template represents a region of plant parameter uncertainty for ω_i as expressed mathematically by Equations 2.15 and 2.16. From the loop transmission plot and its intersections with the M- and α-*contours*, the closed-loop frequency response data may be obtained for plotting M_o and α versus ω. In Figure 2.15 a plot of M_o versus ω is shown,[218] where

$$M_o(j\omega)\angle\alpha(j\omega) = \frac{Y(j\omega)}{R(j\omega)} = \frac{L_o(j\omega)}{1 + L_o(j\omega)} \tag{2.17}$$

3. *Parametric variation NC characteristics*—As an example, consider that for Equation 2.9, $G = 1\angle 0°$. If point A on the template in Figure 2.14 represents LmP_o versus $\angle P_o$, then a variation in P results in the following:
   a. A horizontal translation in the angle of P, given by Equation 2.15
   b. A vertical translation in the log magnitude value of P, given by Equation 2.14

Translations are shown in Figure 2.14 at points B, C, and D. Variation $\delta_P(\omega_i)$ of the plant, and in turn $L(j\omega_i)$, from the nominal value for $\omega = \omega_i$, over the range of plant parameter variation is described by the template $\Im P(j\omega_i)$ shown in Figure 2.14. Consider the point A on the template, which represents the nominal plant $P = P_o(j\omega_i)$. Then, its corresponding closed-loop frequency response is given by

$$LmM_A = Lm\beta = -6\,\text{dB}$$

For $P(j\omega) = P_i(j\omega)$, point C in Figure 2.14, its corresponding closed-loop frequency response is

$$LmM_C = Lm\alpha = -2\,\text{dB}$$

These values are plotted in Figure 2.15. Note that point A represents the minimum value of $LmM(j\omega_i)$ at $\omega = \omega_i$, that is,

$$LmT_{min} = [LmM(j\omega_i)]_{min} = Lm\beta$$

Also, point C represents the largest value of $LmM(j\omega_i)$ at $\omega = \omega_i$, that is,

$$LmT_{max} = [LmM(j\omega_i)]_{max} = Lm\alpha$$

with a range of plant parameter variation described by the template. Thus, the maximum variation in LmM, denoted by $\delta_L(\omega_i)$, for this example is

$$\delta_L(\omega_i) = Lm\alpha - Lm\beta = -2-(-6) = 4 \text{ dB} \tag{2.18}$$

When $L_o(s)$ is properly synthesized, according to the QFT design technique, then $\delta_L(\omega_i) \leq \delta_R(\omega_i)$. Shown in Figure 2.15 is point E, midway between points A and C, that corresponds to a point E in Figure 2.14, which lies within the variation template $\Im P(j\omega_i)$. This procedure is repeated to obtain the maximum variation $\delta_L(\omega_i)$ for each value of frequency $\omega = \omega_i$ within the desired BW (see Figure 2.15). From Figure 2.15 it is possible to determine the variation in the control system's figures of merit due to the plant's parameter uncertainty.

This graphical description of the effect of plant parameter uncertainty on the system's performance is the basis of the QFT technique.

2.5 Benefits of QFT

The benefits of the QFT technique may be summarized as follows:

- It results in a robust design, which is insensitive to structured plant parameter variation.
- There can be one robust design for the full, operating envelope.
- Design limitations are apparent up front and during the design process.
- The achievable performance specifications can be determined early in the design process.
- If necessary, one can quickly redesign for changes in the specifications with the aid of the *QFT Control Toolbox* (*QFTCT*) CAD package (see Appendix F).
- The structure of the compensator (controller) is determined up front.
- As a consequence of the aforementioned benefits, the development time is less for a full envelope design.

TABLE 2.1

Index: Main Topics Part I

Area	References	Chapters	Appendices
Books: QFT and frequency domain	1–6		
Special issues about QFT	7–11		
International QFT symposia	12–20		
Tutorials about QFT	21–24		
About the QFT history	25–29		
First QFT papers	30–32		
QFT templates	33–41	2, 3	A
QFT bounds	42–63	2, 3	B, C
QFT loop shaping	64–73	2, 3	D
Fragility with QFT	70		E
Existence conditions for QFT controllers	74–76		
MIMO QFT	77–102	5, 6	
Time-delay systems and QFT	103		
Digital QFT	104–107	4	
Distributed parameter systems and QFT	108–115	7	
Non-minimum-phase systems and QFT	116–121	4	
Multi-loop systems and QFT	122–125		
Nonlinear systems: QFT controller design	126–136		
Linear-time-variant (LTV) systems: QFT	137–140	8	
QFT CAD tools	141–150		F and visit http://cesc.case.edu or http://www.crcpress.com/product/isbn/9781439821794
Real-world applications with QFT	151–187	14, 15, 17–19	
QFT controller design examples		14, 15, 17–19	G and visit http://cesc.case.edu or http://www.crcpress.com/product/isbn/9781439821794
Books related to control engineering	188–222		
General MIMO systems	223–273		
General fragility	70, 274–284		E
Hybrid and switching control systems	285–295	14	
Miscellaneous: control	296–301		

2.6 Summary

The purpose of this chapter was to present the concepts of the QFT technique in such a way that students and practicing engineers could readily grasp the fundamentals and appreciate its *transparency* in *bridging the gap* between theory and the real world. For a more theoretical discussion of QFT, such as existence theorems and *n.m.p.* plants, the reader is referred to the literature listed in the references section. It is highly recommended that the reader, at the conclusion of reading this text, read the excellent paper by Professor I.M. Horowitz entitled *Survey of Quantitative Feedback Theory (QFT)*.[22] Table 2.1 shows an index with the main topics and references of Part I of this book.

3

MISO Analog QFT Control System

3.1 Introduction

This chapter builds upon the introduction to quantitative feedback theory (QFT) that is presented in Chapter 2. It presents an in-depth understanding and appreciation of the power of the QFT technique. This is accomplished by first presenting an introduction of the QFT technique using the frequency-response method as applied to a single-loop MISO system and an overview of the design procedure. Next a discussion of minimum-phase (*m.p.*) performance specifications is presented. The remaining portion of the chapter is devoted to an in-depth development of the QFT technique as applied to the design of robust single-loop control systems having two inputs, a tracking and an external disturbance input, and a single output (a MISO system).[218]

The in-depth development of the QFT technique as applied to the design of robust MIMO control systems is presented in Chapters 5 and 6.

The multiple-input-multiple-output (MIMO) synthesis problem is converted into a number of single-loop feedback problems in which parameter uncertainty, cross-coupling effects, and system performance tolerances are derived from the original MIMO problem.[22] The solutions to these single-loop problems represent a solution to the MIMO plant. It is not necessary to consider the complete system characteristic equation. The design is tuned to the extent of the uncertainty and the performance tolerances. In Chapters 5 and 6 the development of a suitable mapping is presented that permits the analysis and synthesis of a MIMO control system having n inputs and n outputs by a set of n^2 equivalent single-loop multiple-input-single-output (MISO) control systems.

3.2 QFT Method (Single-Loop MISO System)

The feedback control system represented in Figure 3.1 is composed of $\mathcal{P}(s) = \{P(s)\}$, which represents the set of transfer functions that describe the region of plant parameter uncertainty, $G(s)$ which is the cascade compensator, and $F(s)$, which is the input prefilter transfer function. The output $y(t)$ is required to track the command input $r(t)$ and to reject the external disturbances $d_1(t)$, $d_2(t)$, or $d(t)$ in general and the noise $n(t)$. The compensator $G(s)$ in Figure 3.1 is to be designed so that the variation of $y(t)$ due to the uncertainty in the plant $P(s)$ is within allowable tolerances and the effects of the disturbances $d_1(t)$, $d_2(t)$, $d(t)$ and the noise $n(t)$ on $y(t)$ are acceptably small. Also, the prefilter properties of $F(s)$ must be designed to the desired tracking by the output $y(t)$ of the input $r(t)$. Since the control system

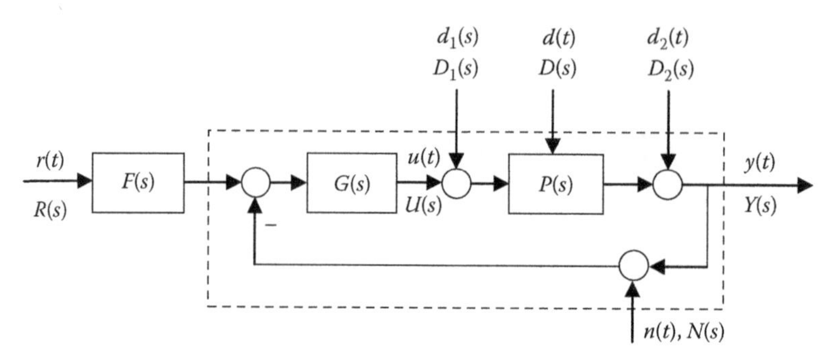

FIGURE 3.1
A 2DOF feedback control structure.

in Figure 3.1 has two measurable quantities, $r(t)$ and $y(t)$, and two controller blocks, $G(s)$ and $F(s)$, it is referred to as a two-degree-of-freedom (DOF) feedback structure. If the disturbance inputs are measurable, then it represents a bigger DOF structure. The actual design is closely related to the extent of the uncertainty and to the narrowness of the performance tolerances. The uncertainty of the plant transfer function is denoted by the set

$$\mathcal{P}(s) = \{P_\iota(s)\}, \quad \text{where } \iota = 1, 2, \ldots, J \tag{3.1}$$

and is illustrated by Example 3.1.

Example 3.1

The plant transfer function is

$$P(s) = \frac{K}{s(s+a)} \tag{3.2}$$

where
the value of K is in the range [1, 10]
a is in the range [–2, 2]

The region of parameter uncertainty of the plant is illustrated in Figure 3.2. The uncertainty of the disturbance is denoted by a set of plant disturbances:

$$\mathcal{D}(s) \equiv \{D(s)\} \tag{3.3}$$

and the acceptable closed-loop transmittances (control ratios) are denoted by

$$\mathcal{T}(s) \equiv \{T(s)\} \tag{3.4}$$

The design objective is to guarantee that $T_R(s) = Y(s)/R(s)$ and $T_D(s) = Y(s)/D(s)$ are members of the sets of acceptable $\Im_R$ and $\Im_D$, respectively, for all $P \in \mathcal{P}$ and $D \in \mathcal{D}$. In a feedback system, the principal challenge in the control system design is to relate the system performance specifications to the requirements on the *loop transmission function*

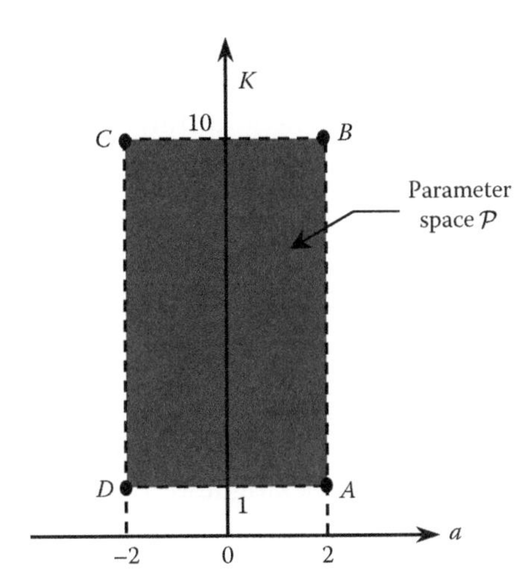

FIGURE 3.2
Plant parameter uncertainty.

$L(s) = P(s)G(s)$ in order to achieve the desired benefits of feedback, that is, the desired reduction in sensitivity to plant uncertainty and desired disturbance attenuation. The advantage of the frequency domain is that $L(s) = G(s)P(s)$ is simply the multiplication of complex numbers. In the frequency domain, it is possible to evaluate $L(j\omega)$ at every ω_i separately, and thus, at each ω_i, the optimal bounds on $L(j\omega_i)$ can be determined.

3.3 Design Procedure Outline

The basic design procedure that is to be followed in applying the QFT robust design technique is outlined in this section. This outline enables the reader to obtain an overall perspective of the QFT technique at the onset. The following sections present a detailed discussion that is intended to establish a firm understanding of the fundamentals of this technique.

The objective is to design the prefilter $F(s)$ and the compensator $G(s)$ of Figure 3.1 so that the specified robust design is achieved for the given region of plant parameter uncertainty. The design procedure to accomplish this objective is as follows:

Step 1: Synthesize the desired tracking model.

Step 2: Synthesize the desired disturbance model.

Step 3: Specify the J linear time-invariant (LTI) plant models defined by the region of plant parametric/nonparametric uncertainty.

Step 4: Obtain plant templates, at specified frequencies, that pictorially describe the region of plant uncertainty on the Nichols chart (NC).

Step 5: Select the nominal plant transfer function $P_o(s)$.

Step 6: Determine the stability contour (*U*-contour) on the NC.

Steps 7–9: Determine the disturbance, tracking, and optimal bounds on the NC.

Step 10: Synthesize the nominal loop transmission function $L_o(s) = P_o(s)G(s)$ that satisfies all the bounds and the stability contour.

Step 11: Based upon Steps 1–10, synthesize the prefilter $F(s)$.

Step 12: Simulate the system in order to obtain the time-response data for each of the J plants.

The following sections illustrate this design procedure. Note that our CAD Package, QFT Control Toolbox (*QFTCT*), can be used in all the steps of the methodology. See Appendices F and G and Refs. 149 and 150, and visit the websites http://cesc.case.edu or http://www.crcpress.com/product/isbn/9781439821794.

3.4 Minimum-Phase System Performance Specifications

In order to apply the QFT technique, it is necessary to synthesize the desired or model control ratio, based upon the system's desired performance specifications in the time domain. For the *m.p.* LTI MISO system of Figure 3.1, the control ratios for tracking and for disturbance rejection are, respectively,

$$T_R(s) = \frac{P(s)G(s)F(s)}{1+P(s)G(s)} = \frac{LF}{1+L} = TF, \quad \text{with } d_1(t) = d_2(t) = d(t) = 0 \tag{3.5}$$

$$T_{D_1} = \frac{P(s)}{1+P(s)G(s)} = \frac{P}{1+L}, \quad \text{with } r(t) = d_2(t) = d(t) = 0 \tag{3.6}$$

$$T_{D_2} = \frac{1}{1+P(s)G(s)} = \frac{1}{1+L}, \quad \text{with } r(t) = d_1(t) = d(t) = 0 \tag{3.7}$$

Note that for T_{D_1}, the specified maximum value $|y(t_p)| = \alpha_p$, due to $d_1(t) = u_{-1}(t)$, is often used as the disturbance model specification, that is, maximum $Lm\, T_{D_1} = Lm\alpha_p$, where $Lm(\cdot)$ is $20 \log_{10} |\cdot|$.

3.4.1 Tracking Models

The QFT technique requires that the desired tracking control ratios be modeled in the frequency domain to satisfy the required gain K_m and the desired time-domain performance specifications for a step input. Thus, the system's tracking performance specifications for a simple second-order system are based upon satisfying some or all of the step forcing function figures of merit (FOM) for underdamped (M_p, t_p, t_s, t_r, K_m) and overdamped (t_s, t_r, K_m) responses. These are graphically depicted in Figure 3.3. The time responses $y(t)_U$ and $y(t)_L$ in this figure represent the upper and lower bounds, respectively, of the tracking performance specifications; that is, an acceptable response $y(t)$ must lie between these bounds. The Bode plots of the upper bound B_U and the lower bound B_L for $LmT_R(j\omega)$ versus ω are shown in Figure 3.4. Note that for *m.p.* plants, only the tolerance on $|T_R(j\omega_i)|$ needs to be satisfied for a satisfactory design. For non-minimum-phase (*n.m.p.*) plants, tolerances on $\angle T_R(j\omega_i)$ must also be specified and satisfied in the design process. This chapter deals only with *m.p.* plants. The case of *n.m.p.* plants is discussed in Chapter 4. It should be noted that for *m.p.* plants any desired frequency bandwidth (BW) is achievable whereas for *n.m.p.* plants the BW that is achievable is limited.

The modeling of a desired transmittance $T(s)$ is discussed in Ref. 218. It is desirable to synthesize the control ratios corresponding to the upper and lower bounds T_{R_U} and T_{R_L}, respectively, so that $\delta_R(\omega_i)$ increases as ω_i increases above the 0-dB crossing frequency ω_{cf} (see Figure 3.4b) of T_{R_U}. This characteristic of $\delta_R(\omega_i)$ simplifies the process of synthesizing the loop transmission or open-loop transfer function $L_o(s) = P_o(s)G(s)$ as discussed in

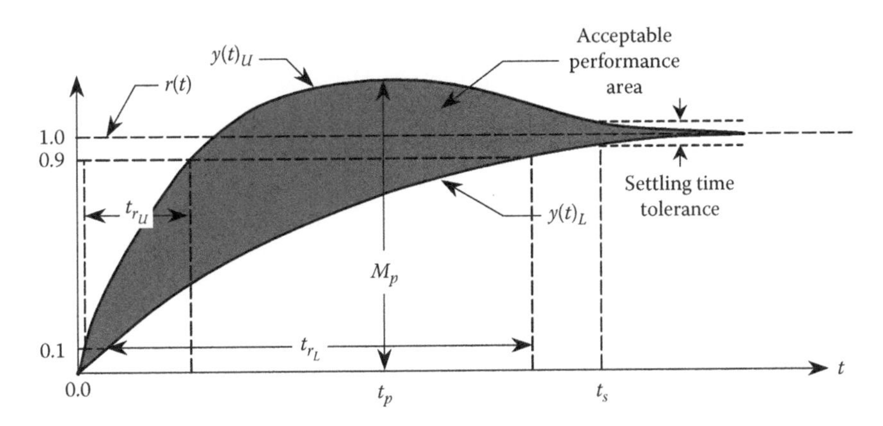

FIGURE 3.3
System time-domain tracking performance specifications.

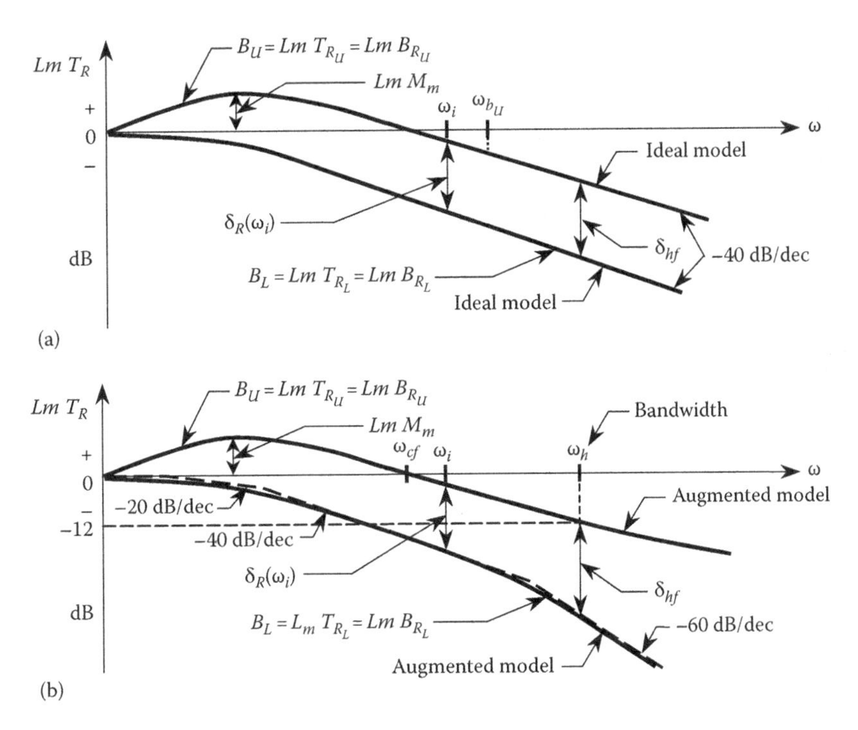

FIGURE 3.4
Bode plots of T_R: (a) ideal simple second-order models; (b) augmented models.

Section 3.12. To synthesize $L_o(s)$, it is necessary to determine the tracking bounds $B_R(j\omega_i)$ (see Section 3.9), which are obtained based upon $\delta_R(\omega_i)$. This characteristic of $\delta_R(\omega_i)$ ensures that the tracking bounds $B_R(j\omega_i)$ decrease in magnitude as ω_i increases (see Section 3.9).

An approach to the modeling process is to start with a simple second-order model of the desired control ratio T_{R_U} having the form[218]

$$T_{R_U}(s) = \frac{\omega_n^2}{s^2 + 2\zeta\omega_n s + \omega_n^2} = \frac{\omega_n^2}{(s - p_1)(s - p_2)} \tag{3.8}$$

where $\sigma_D = -\zeta\omega_n$, $p_1 = \sigma_D + j\omega_d$, $p_2 = \sigma_D - j\omega_d$, $\omega_n^2 = p_1 p_2$, and $t_s \approx T_s = 4/\zeta\omega_n = 4/|\sigma_D|$ (the desired settling time). The control ratio $T_{R_U}(s)$ of Equation 3.8 can be represented by an equivalent unity-feedback system so that

$$T_{R_U}(s) = \frac{Y(s)}{R(s)} = \frac{G_{eq}(s)}{1+G_{eq}(s)} \tag{3.9}$$

where

$$G_{eq}(s) = \frac{\omega_n^2}{s(s+2\zeta\omega_n)} \tag{3.10}$$

The gain constant of this equivalent Type 1 transfer function $G_{eq}(s)$ is $K_1 = \lim_{s\to 0}[sG_{eq}(s)] = \omega_n/2\zeta$. Equation 3.8 satisfies the requirement that $y(t)_{ss} = R_0 u_{-1}(t)$ for $r(t) = R_0 u_{-1}(t)$. The frequency ω_b for which $|T_{R_U}(j\omega_b)| = 0.7071$ is defined as the *system BW frequency* ω_{bU}.

The simplest overdamped model for $T_{R_L}(s)$ is of the form

$$T_{R_L}(s) = \frac{Y(s)}{R(s)} = \frac{K}{(s-\sigma_1)(s-\sigma_2)} = \frac{G_{eq}(s)}{1+G_{eq}(s)} \tag{3.11}$$

where

$$G_{eq}(s) = \frac{\sigma_1\sigma_2}{s[(s-(\sigma_1+\sigma_2))]}$$

and $K_1 = -\sigma_1\sigma_2/(\sigma_1 + \sigma_2)$. For this system $y(t)_{ss} = R_0$ for $r(t) = R_0 u_{-1}(t)$. Selection of the parameters σ_1 and σ_2 is used to meet the specifications for t_s and K_1. Achievement of the desired performance specification is based upon the BW, $0 < \omega < \omega_{hR}$, which is determined by the intersection of the −12 dB line and the B_U curve in Figure 3.4b.

Once the ideal models $T_{R_U}(j\omega)$ and $T_{R_L}(j\omega)$ are determined, the time- and frequency-response plots of Figures 3.3 and 3.4a, respectively, can then be drawn. Because the models for T_{R_U} and T_{R_L} are both second order, the high-frequency asymptotes in Figure 3.4a have the same slope. The high-frequency range (*hf*) in Figure 3.4a is defined as $\omega \geq \omega_b$ where ω_b is the model BW frequency of B_U. In addition to achieving the desired characteristic of an increasing magnitude of δ_R of B_U for $\omega_i > \omega_{cf}$, an increasing spread between B_U and B_L is required in the *hf* range (see Figure 3.4b); that is,

$$\delta_{hf} = B_U - B_L \tag{3.12}$$

must increase with increasing frequency. This desired increase in δ_R is achieved by changing B_U and B_L, without violating the desired time-response characteristics, by augmenting T_{R_U} with a zero (see Equation 3.13) as close to the origin as possible without significantly affecting the time response. This additional zero raises the curve B_U for the frequency range above ω_{cf}. The spread can be further increased by augmenting T_{R_L} with a negative real pole (see Equation 3.14), which is as close to the origin as possible but far enough not to significantly affect the time response. Note that the straight-line Bode plot is shown only for T_{R_L}. This additional pole lowers B_L for this frequency range:

$$T_{R_U}(s) = \frac{(\omega_n^2/a)(s+a)}{s^2 + 2\varsigma\omega_n s + \omega_n^2} = \frac{(\omega_n^2/a)(s-z_1)}{(s-\sigma_1)(s-\sigma_2)} \tag{3.13}$$

$$T_{R_L}(s) = \frac{K}{(s+a_1)(s+a_2)(s+a_3)} = \frac{K}{(s-\sigma_1)(s-\sigma_2)(s-\sigma_3)} \tag{3.14}$$

Thus, for these augmented models, the magnitude of $\delta_R(\omega_i)$ increases as ω_i increases above ω_{cf}.

The manner of achieving a $\delta_R(\omega)$ that increases with frequency is described in the following and is illustrated in Design Example 1. In order to minimize the iteration process in achieving acceptable models for $T_{R_U}(s)$ and $T_{R_L}(s)$, which have an increasing $\delta_R(\omega)$, the following procedure may expedite the design process: (a) First synthesize the second-order model of Equation 3.13 containing the zero at $|z_1| = a \geq \omega_n$ that meets the desired FOM. (b) Then, as a first trial, select all three real poles of Equation 3.14 to have the value of $|\sigma_3| = a_3 = \omega_n > a_2 = a_1 > |\sigma_D|$. For succeeding trials, if necessary, one or more of these poles are moved right and/or left until the desired specifications are satisfied. As illustrated by the slopes of the straight-line Bode plots in Figure 3.4b, selecting the value of all three poles in the range specified earlier insures an increasing δ_R. Other possibilities are as follows: (c) The specified values of t_p and t_s for T_{R_L} may be such that a pair of complex poles and a real pole need to be chosen for the model response. For this situation, the real pole *must* be more dominant than the complex poles. (d) Depending on the performance specifications (see the paragraph following Equation 3.15), $T_{R_U}(s)$ may require two real poles and a zero "close" to the origin, that is, select $|z_1|$ very much less than $|p_1|$ and $|p_2|$ in order to effectively have an underdamped response.

At high frequencies, δ_{hf} (see Figure 3.4b) must be larger than the actual variation in the plant, δ_P. This characteristic is the result of the Bode theorem, which, for stable and *m.p.* systems, states that

$$\int_0^\infty 20\log_{10}(S_P^T)\,d\omega = 0$$

Thus, the reduction in sensitivity S_P^T at the lower frequencies must be balanced by an increase in sensitivity at the higher frequencies. At some high frequency $\omega_i \geq \omega_h$ (see Figure 3.4b), since $|T_{R_L}| \approx 0$, then

$$\lim_{\omega \to \infty} \delta_R(\omega_i) \approx B_U - (-\infty) = \infty \text{ dB} \tag{3.15}$$

For the case where $y(t)$, corresponding to T_{R_U}, is to have an allowable "large" overshoot followed by a small tolerable undershoot, a dominant complex-pole pair is not suitable for T_{R_U}.

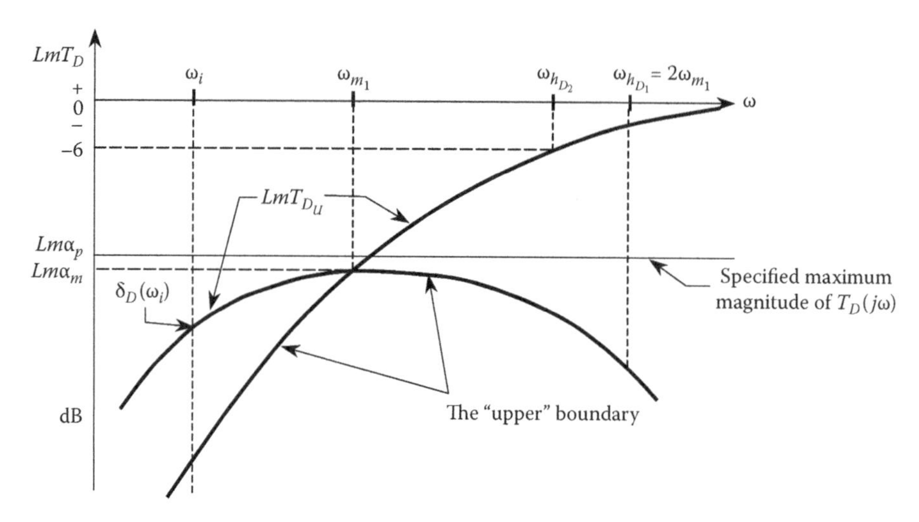

FIGURE 3.5
Bode plots of disturbance models for $T_D(j\omega)$.

An acceptable overshoot with no undershoot for T_{R_U} can be achieved by T_{R_U} having two real dominant poles $p_1 > p_2$, a dominant real zero ($z_1 > p_1$) "close" to p_1, and a far-off pole $p_3 \ll p_2$. The closeness of the zero dictates the value of M_p. Thus, a designer selects a pole-zero combination to yield the form of the desired time-domain response.

3.4.2 Disturbance-Rejection Models

The simplest disturbance control ratio model specification is $|T_D(j\omega)| = |Y(j\omega)/D(j\omega)| < \alpha_p$, a constant (the desired maximum magnitude of the output based upon a unit-step disturbance input [D of Figure 3.1]); that is, for $d_1(t)$: $|y(t_p)| \le \alpha_p$ and for $d_2(t)$: $|y(t)| \le \alpha_p$ for $t \ge t_x$. Thus, the frequency-domain disturbance specification is $Lm\,T_D(j\omega) < Lm\,\alpha_p$ over the desired specified BW (see Figure 3.5). Thus, the disturbance specification is represented by only an upper bound on the NC over the specified BW. A detailed discussion on synthesizing disturbance-rejection models for Equations 3.6 and 3.7, based upon desired performance specifications, to yield the upper bounds shown in Figure 3.5 for D_1 and D_2 is given in Ref. 218.

3.5 *J* LTI Plant Models

The simple plant of Equation 2.7, where $K \in \{1,10\}$ and $a \in \{1,10\}$, is used to illustrate the MISO QFT design procedure. The region of plant parameter uncertainty is illustrated in Figure 3.6. This region may be described by J LTI plants, where $\iota = 1, 2, \ldots, J$, which lie on its boundary. That is, the boundary points 1, 2, 3, 4, 5, and 6 are utilized, as discussed in Chapter 2, to obtain 6 LTI plant models that adequately define the region of plant parameter uncertainty. Note that the numbered points around the contour in Figure 2.6 are relabeled by letters as shown in Figure 3.6. This is done in order to simplify the labeling in the figures associated with the discussion in this chapter.

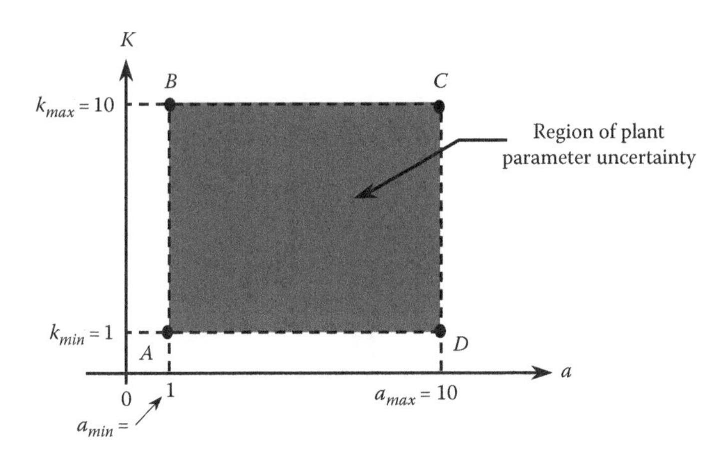

FIGURE 3.6
Region of plant parameter uncertainty.

3.6 Plant Templates of $P_\iota(s)$, $\Im P(j\omega_i)$

With $L = PG$, Equation 3.5 yields

$$LmT_R = LmF + Lm\left[\frac{L}{1+L}\right] = LmF + LmT \tag{3.16}$$

The change in T_R due to the uncertainty in P, since F is LTI, is

$$\Delta(LmT_R) = Lm\,T_R - LmF = Lm\left[\frac{L}{1+L}\right] \tag{3.17}$$

The proper design of $L = L_o$ and F must restrict this change in T_R so that the actual value of LmT_R always lies between B_U and B_L of Figure 3.4b. The first step in synthesizing an L_o is to make NC templates that characterize the variation of the plant uncertainty for various values of ω_i over a frequency range $\omega_x \leq \omega_i \leq \omega_{hR}$, where $\omega_x < \omega_{cf}$. A guideline for selecting the frequency range for the templates is to select three frequency values below and above the 0dB crossing frequency ω_{cf}, no less than an octave apart, up to approximately the −12dB value of the B_U plot in Figure 3.4b. In addition, for a Type 0 plant select $\omega_x = 0$ and for Type 1 or higher-order plants select $\omega_x \neq 0$.

To provide more details about obtaining templates, the simple plant of Equation 2.7 is used, whose region of plant uncertainty is depicted in Figure 3.6. The plant template on the NC can be obtained by mapping the plant parameter uncertainty region. A number of points on the perimeter of Figure 3.6 are selected, and the values of $LmP(j\omega_i)$ and $\angle P(j\omega_i)$ are obtained at each point. These data, for each value of frequency $\omega = \omega_i$, are plotted on an NC as illustrated in Chapter 2. A curve is drawn through these points, which becomes the template $\Im P(jw_i)$ at the frequency ω_i. A sufficient number of points must be selected so that the contour of $\Im P(jw_i)$ accurately reflects the region of plant uncertainty. In addition to the points A, B, C, D marked in Figure 3.6, it may also be necessary to include additional points on the perimeter.

For the points A, B, C, and D, at the frequency $\omega = 1$, the data obtained from Equation 2.7 are $P_A(j1) = 0.7079\ \angle{-135}°$, $P_B(j1) = 7.0795\ \angle{-135}°$, $P_C(j1) = 10\ \angle{-95.7}°$, and $P_D(j1) = 0.995\ \angle{-95.7}°$,

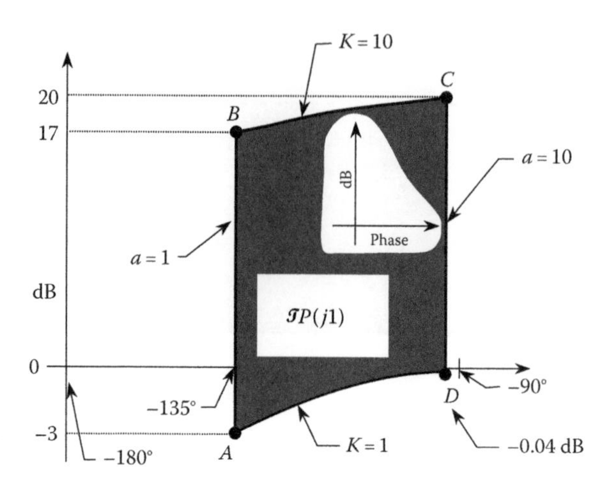

FIGURE 3.7
NC characterizing Equation 2.7 over the region of uncertainty.

respectively. These data are plotted on the NC shown in Figure 3.7. A curve is drawn through the points A, B, C, D, A and the shaded area is labeled $\Im P(j1)$. The contour $ABCD$ in Figure 3.7 may be drawn on a plastic sheet (preferably colored) so that a plastic template for $\Im P(j1)$ can be cut and labeled. The templates for other values of ω_i are obtained in a similar manner. A characteristic of these templates is that, starting from a "low value" of ω_i, the templates widen (angular width becomes larger) for increasing values of ω_i. Then, as ω_i takes on larger values and approaches infinity, the templates become narrower (see Figure 2.11) and eventually approach a straight line of height V dB (see Equation 3.20).

For an aircraft, each point A, B, C, D, etc. in Figure 3.7 represents a given flight condition (FC) at $\omega = \omega_i$, that is, $[P_{A/C}(j\omega_i)]_{FC_i}$. One of the FCs may be identified as the nominal plant P_o.

For the plant of Equation 2.7, the values $K = a = 1$ represent the lowest point of each of the templates $\Im P(jw_i)$ and may be selected as the nominal plant P_o for all frequencies. However, any plant in P can be chosen as the nominal plant.[79] With $L = PG$ and $L_o = P_oG$, as given in Equations 2.15 and 2.16,

$$\delta_P(\omega_i) = LmL - LmL_o = (LmG + LmP) - (LmG + LmP_o)$$
$$= (LmP - LmP_o) \le \delta_R(\omega_i)\ \text{dB}$$

and

$$\angle\Delta P = \angle P - \angle P_o$$

Thus, if point A in Figure 3.7 represents LmP_o, a variation in P results in a horizontal translation in the angle of P and a vertical translation in the log magnitude value of P. When $G(j\omega)$ represents a specific transfer function, the template of Figure 3.7 can be converted into a template of $L(j\omega_i)$ by translating it vertically by $LmG(j\omega_i)$ and horizontally by $\angle G(j\omega_i)$. For the template of $L(j\omega_i)$, the values of the M-contours at the intersections with the template are the values of the control ratio $LmT(j\omega_i) = Lm[L(j\omega_i)/(1 + L(j\omega_i)]$. The range of values of $T(j\omega_i)$ for the entire range of parameter variation (K and a) can therefore be determined. For complement information about the templates, see Appendix A and Refs. 33 through 41.

3.7 Nominal Plant

While any plant case can be chosen, it is a common practice to select, whenever possible, a nominal plant whose NC point is at the lower left corner of the template at a selected frequency. This nominal plant, defined by a set of parameters with specific values, must be the same for the rest of the frequencies for which templates are to be obtained (see Chapter 9 of Ref. 6 for further details).

3.8 *U*-Contour (Stability Bound)

It is well known that $|T(j\omega)| \leq M_L$ establishes a circle in the NC (see Figure 3.8). It also defines the phase margin (*PM*) and the gain margin (*GM*). Let $L(j\omega) = P(j\omega)G(j\omega) = |L|e^{j\phi}$. In that case the gain margin is $GM = 1/|L|$ at the angle $\phi = -180°$. Then, in terms of the M_L circle, $GM = 1 + 1/M_L$. Analogously, the phase margin is $PM = 180° + \phi$ at the angle where ϕ is the phase of $L(j\omega)$ at $|L(j\omega)| = 1$. Hence, in terms of the M_L circle, $PM \geq 180° - 2\cos^{-1}(0.5/M_L)$.

The specifications on system performance in the time domain (see Figure 3.3) and in the frequency domain (see Figure 3.4) identify a minimum damping ratio ζ for the dominant roots of the closed-loop system, which corresponds to a bound on the value of $M_p \approx M_m$. On the NC, this bound on $M_p = M_L$ (see Figure 3.8) establishes a region, which must not be penetrated by the templates and the loop transmission functions $L_\iota(j\omega)$ for all ω.

The boundary *abcdefa* of this region is referred to as the *universal high-frequency boundary* (*UHFB*) or *stability bound*, the *U-contour*, because this becomes the dominating constraint on $L(j\omega)$. Therefore, the top portion, *efa*, of the M_L-contour becomes part of the *U*-contour. The formation of the *U*-contour is discussed in this section.

For the two cases of disturbance rejection depicted in Figure 3.1 the control ratios are, respectively, as given in Equations 3.6 and 3.7,

$$T_{D_1} = \frac{P}{1+L} \quad \text{and} \quad T_{D_2} = \frac{1}{1+L}$$

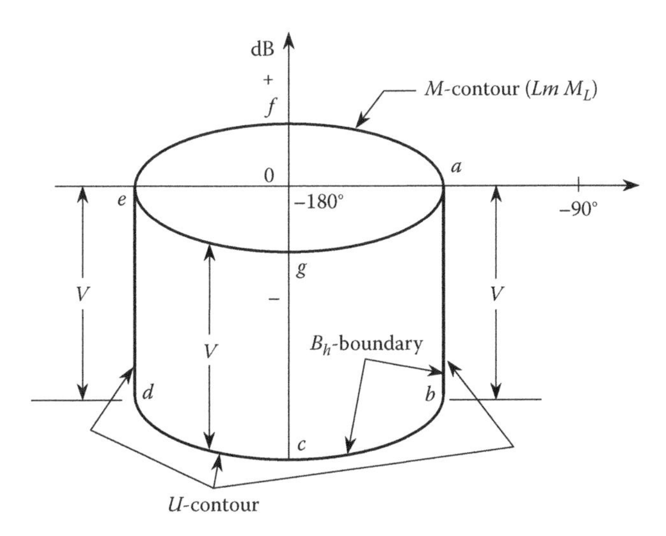

FIGURE 3.8
U-contour construction (stability contour).

Thus, it is necessary to synthesize an $L_o(s)$ so that the disturbances are properly attenuated. For the present case, only one aspect of this disturbance-response problem is considered, namely a constraint is placed on the damping ratio ζ of the dominant complex-pole pair of T_D nearest to the $j\omega$-axis.[206] This damping ratio is related to the peak value of

$$|T(j\omega)| = \left|\frac{L(j\omega)}{1+L(j\omega)}\right| \tag{3.18}$$

For example, consider the case for which the dominant complex-pole pair of Equation 3.18 results in a peak of $LmT = 8\,\text{dB}$ for $\zeta = 0.2$, and $2.7\,\text{dB}$ for $\zeta = 0.4$, etc. Although this large peak does not appear in T_R due to the design of the filter $F(s)$, it does affect the response for T_D. If $d(t)$ is very small, a peak of T_D, due to $\zeta = 0.2$, can be "very large," and it may be difficult to achieve the restriction on the peak overshoot α_p of the time response:

$$|c(t)| \le \alpha_p$$

Therefore, it is reasonable to add the requirement

$$|T| = \left|\frac{L}{1+L}\right| \le M_L \tag{3.19}$$

where M_L is a constant for all ω and over the whole range of $\mathcal{P}$ parameter values. This results in a constraint on ζ of the dominant complex-pole pair of T_D. This constraint can therefore be translated into a constraint on the maximum value T_{max} of Equation 3.18. For example, for $LmM_m = 2\,\text{dB}$, the oval, *agefa*, in Figure 3.8 corresponds to the 2-dB *M*-contour on the NC. This results in limiting the peak of the disturbance response. A value of M_L can be selected to correspond to the maximum value of T_R. Therefore, the top portion, *efa*, of the *M*-contour on the NC, which corresponds to the value of the selected value of M_L, becomes part of the *U*-contour.

For a large class of problems, as $\omega \to \infty$, the limiting value of the plant transfer function approaches

$$\lim_{\omega\to\infty}[P(j\omega)] = \frac{K'}{\omega^{\lambda}}$$

where λ represents the excess of poles over zeros of $P(s)$. The plant template, for this problem class, approaches a vertical line of length equal to

$$\begin{aligned}\Delta &\equiv \operatorname*{Lim}_{\omega\to\infty}[LmP_{max} - LmP_{min}] \\ &= LmK'_{max} - LmK'_{min} = V\,\text{dB}\end{aligned} \tag{3.20}$$

If the nominal plant is chosen at $K' = K'_{min}$, then the constraint M_L gives a boundary that approaches the *U*-contour *abcdefa* of Figure 3.8. (*Note*: For a MIMO plant $P = \{p_{ij}\}$, as $\omega \to \infty$, the templates may not approach a vertical line if the λ_{ij} are not the same for all p_{ij} elements of the plant matrix. When the λ_{ij} are different, then the widths of the templates are a multiple of 90°.)

For the simple plant of Equation 2.7, where $K \in \{1,10\}$ and $a \in \{1,10\}$ and where $K' = Ka$, applying the limiting condition, $\omega \to \infty$, to Equation 3.20 yields

$$V = \Delta LmP = \lim_{\omega \to \infty}[\{Lm(Ka)_{max} - Lm(j\omega)^2\} - \{Lm(Ka)_{min} - Lm(j\omega)^2\}]$$

$$= Lm(Ka)_{max} - Lm(Ka)_{min} = Lm100 - Lm1 = 40\,\text{dB} \qquad (3.21)$$

For the *m.p.* plant of Equation 2.7, where the poles are real, the plant templates have the typical shape of Figure 3.7.

The high-frequency boundary B_h, the *bcd* portion of the *U*-contour in Figure 3.8, is obtained by measuring down V dB from the *ega* portion of the *M*-contour as illustrated in this figure. V is determined by Equation 3.21, which, for this example, is 40 dB. The remaining portions of the *U*-contour, portions *ab* and *de*, not necessarily straight lines, are determined by satisfying the requirement of Equation 3.19 and $\delta_R(j\omega_1)$. The $\Im P(jw_i)$ is used to determine the corresponding tracking bounds $B_R(j\omega_i)$ on the NC in the manner described in Section 3.9.

3.9 Tracking Bounds $B_R(j\omega)$ on the NC

As an introduction to this section, the procedure for adjusting the gain of a unity-feedback system to achieve a desired value of M_m by using NC is reviewed. Consider the plot of $Lm\,P(j\omega)$ versus $\angle P(j\omega)$ for a plant shown in Figure 3.9 (the solid curve). With $G(s) = k_G = 1$ and $F(s) = 1$ in Figure 3.1, $L = P$. The plot of $LmL(j\omega)$ versus $\angle L(j\omega)$ is tangent to the $M = 1$ dB curve with a resonant frequency $\omega_m = 1.1$. If $LmM_m = 2$ dB is specified for LmT_R, the gain k_G is increased, raising $LmL(j\omega)$, until it is tangent to the 2-dB *M*-curve. For this example the curve is raised by $Lmk_G = 4.5$ dB ($G = k_G = 1.679$) and the resonant frequency is $\omega_m = 2.09$.

Now consider that the plant uncertainty involves only the variation in gain between the values of 1 and 1.679. It is desired to find a cascade compensator $G(s)$, in Figure 3.1, such that the specification $1 \le LmM_m \le 2$ dB is always maintained for this plant gain variation while the resonant frequency ω_m remains constant. This requires that the loop transmission $L(j\omega) = G(j\omega)P(j\omega)$ be synthesized so that it is tangent to an *M*-contour in the range of $1 \le LmM \le 2$ dB for the entire range of $1 \le \text{gain} \le 1.679$ and the resultant resonant frequency satisfies the requirement $\omega_m = 2.09 + \Delta\omega_m$. The manner of achieving this and other time-response specifications is detailed in the following.

It is assumed for Equation 3.17 that the compensators *F* and *G* are fixed (LTI), that is, they have negligible uncertainty. Thus, only the uncertainty in *P* contributes to the change in T_R given by Equation 3.17. The solution requires that the actual $\Delta LmT_R(j\omega_i) \le \delta_R(\omega_i)$ dB in Figure 3.4b. Thus, it is necessary to determine the resulting constraint, or bound $B_R(j\omega_i)$, on $L(j\omega_i)$. The procedure is to pick a nominal plant $P_o(s)$ and to derive the bounds on the resulting nominal loop transfer function $L_o(s) = G(s)P_o(s)$.

As an illustration, consider the plot of $LmP(j2)$ versus $\angle P(j2)$ for the plant of Equation 2.7 (see Figure 3.6). As shown in Figure 3.10, the plant's region of uncertainty $\Im P(j2)$ is given by the contour *ABCD*, that is, $LmP(j2)$ lies on or within the boundary of this contour. The nominal plant transfer function, with $K_o = 1$ and $a_o = 1$, is

$$P_o = \frac{1}{s(s+1)} \qquad (3.22)$$

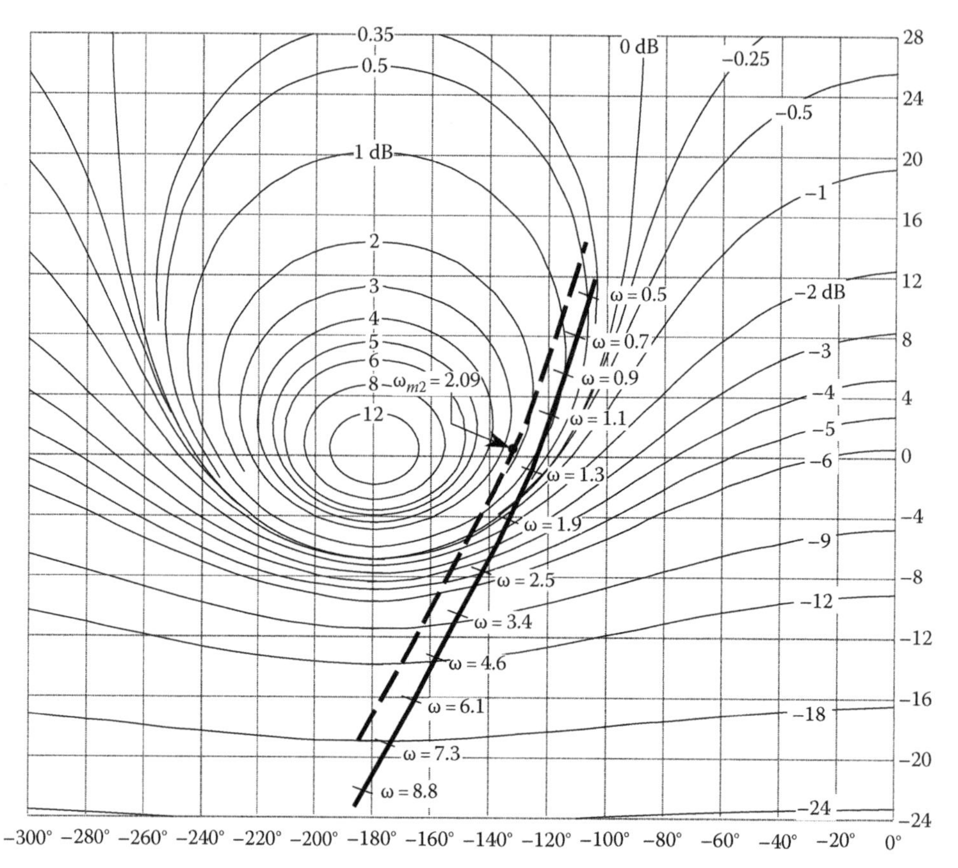

FIGURE 3.9
Log magnitude angle diagram.

and is represented in Figure 3.10 by point A for $\omega = 2$ [−13.0 dB, −153.4°]. Note, once a nominal plant is chosen, it must be used for determining all the bounds $B_R(j\omega_i)$. Since $LmL(j2) = LmG(j2) + LmP(j2)$, $\Im P(j2)$ is translated on the NC vertically by the value of $LmG(j2)$ and horizontally by the angle $\angle G(j2)$. The templates $\Im P(jw_i)$ are relocated to find the position of $L_o(j\omega)$, which satisfies the specifications in Figure 3.4b of $\delta_R(\omega_i)$ for each value of ω_i. For example, if a trial design of $L(j2)$ requires sliding $\Im P(j2)$ to the position $A'B'C'D'$ in Figure 3.10, then

$$\left|LmG(j2)\right| = \left\|Lm\,L(j2)\right|_{A'} - \left|Lm\,P(j2)\right|_{A}\right| = \left\|-2\right| - \left|-13\right\| = 11\,\text{dB} \tag{3.23}$$

$$\angle G(j2) = \angle L(j2)_{A'} - \angle P(j2)_A = -60^\circ - (-153.4^\circ) = 93.4^\circ \tag{3.24}$$

Using the contours of constant $LmM = Lm[L/(1 + L)]$ on the NC in Figure 3.10, the maximum occurs at point C' ($M = -0.49$ dB) and the minimum at point A' ($M = -6$ dB) so that the maximum change in LmT is, in this case, (−0.49) − (−6) = 5.51 dB. If the specifications tolerate a change of 6.5 dB at $\omega = 2$, the aforementioned trial position of $LmL_o(j2)$

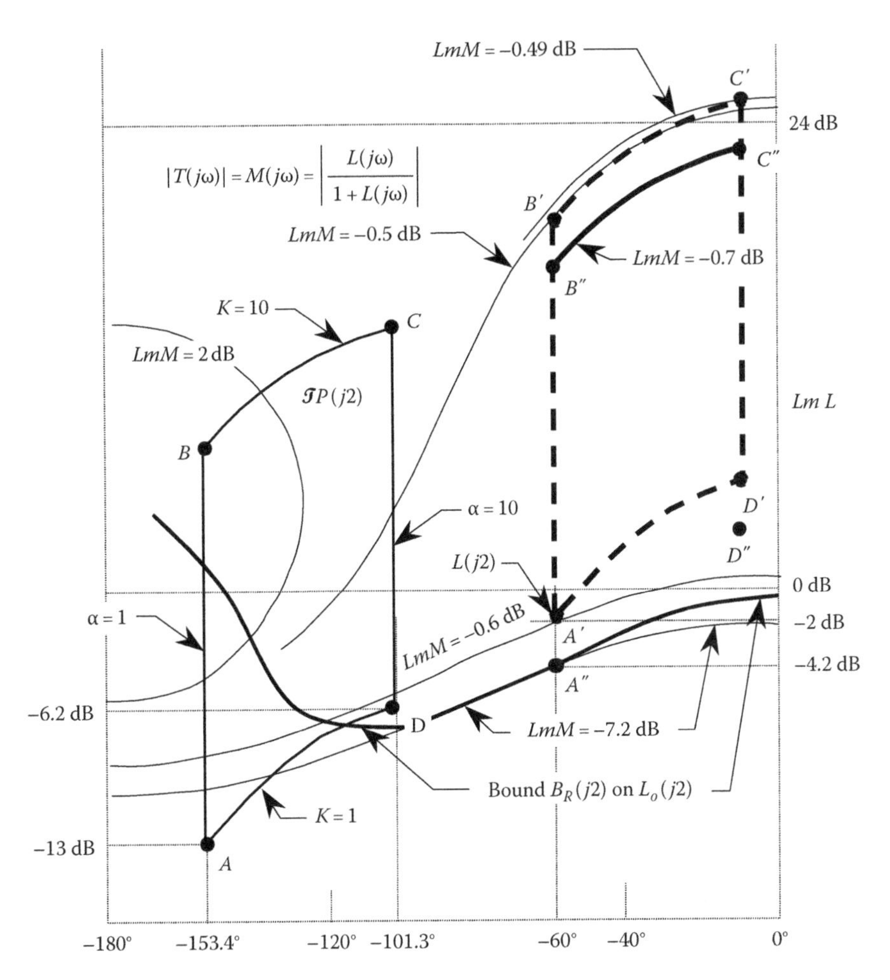

FIGURE 3.10
Derivation of bounds $B_R(j\omega_i)$ on $L_o(j\omega)$ for $\omega = 2$.

is well within the permissible tolerance. Lowering the template on the NC to $A''B''C''D''$, where the extreme values of $Lm[L/(1 + L)]$ are at $C''(-0.7\,\text{dB})$ and $A''(-7.2\,\text{dB})$, yields $LmL(j2)_{C''} - LmL(j2)_{A''} = -0.7 - (-7.2) = 6.5\,\text{dB} = \delta_R(2)$. Thus, if $\angle L_o(j2) = -60°$, then $-4.2\,\text{dB}$ is the smallest or the minimum value of $LmL_o(j2)$ that satisfies the 6.5 dB specification for $\delta_R(\omega_i)$. Any smaller magnitude is satisfactory but represents overdesign at that frequency. The manipulation of the $\omega = 2$ template, for ease of the design process, is repeated along a new angle (vertical) line, and a corresponding new minimum of $L_o(j2)$ is found. Sufficient points are obtained in this manner to permit drawing a continuous curve of the bound $B_R(j2)$ on $L_o(j2)$, as shown in Figure 3.10. The aforementioned procedure is repeated at other frequencies, resulting in a family of boundaries $B_R(j\omega_i)$ of permissible $L_o(j\omega)$. The procedure for determining the boundaries $B_R(j\omega_i)$ is summarized as follows:

1. From Figure 3.4b obtain values of $\delta_R(\omega_i)$ for a range of values of ω_i $(\omega_1, \omega_2, \ldots, \omega_h)$, preferably an octave apart, over the specified BW. The selection of ω_h results in the bound $B_R(j\omega_h)$ passing under the U-contour. (In the low-frequency range [desired tracking BW], generally, an octave apart will provide a reasonable separation of the bounds on the NC.)

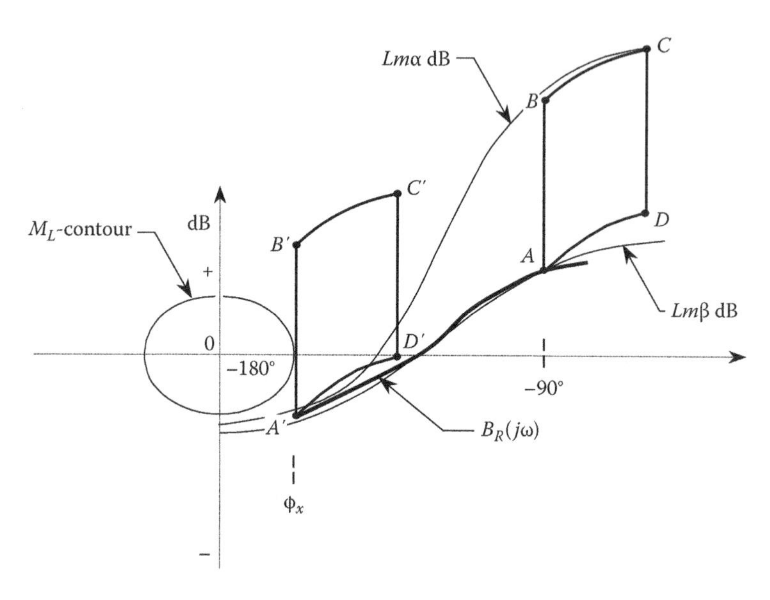

FIGURE 3.11
Graphical determination of $B_R(j\omega)$.

2. Place the template $\Im P(j\omega_i)$ on the NC containing the *U*-contour to determine the bound $B_R(j\omega_i)$ as follows:
   a. Use major angle divisions of the NC for lining up the $\Im P(j\omega_i)$.
   b. Select P_o to represent, in general, the lowest point of the templates. For example, for the plant being used in Sections 3.6 through 3.9, the templates $\Im P(j\omega_i)$ are shown in Figure 3.25, where the lowest points, at each frequency, are shown by the black dots. Thus, this design example
      i. Select point *A* in Figure 3.11 as P_o.
      ii. Line up side *A–B* of $\Im P(j\omega_i)$ on the −90° line, as shown in Figure 3.11. Move the template up or down until the difference $\Delta Lm T_R(j\omega_i)$ between the values of two adjacent *M*-contours (*Lm*α and *Lm*β) is equal to the value of $\delta_R(\omega_i)$ obtained from Figure 3.4b. Thus, in Figure 3.11, determine the locations of the templates on the −90° line where

$$\Delta Lm\, T_R(j\omega_i) = Lm\alpha - Lm\beta = \delta_R(\omega_i) \tag{3.25}$$

      When Equation 3.25 is satisfied, the point *A* on the *M*-contour (=*Lm*β) lies on the bound $B_R(j\omega_i)$. Mark this point on the NC. For other plants, the shape of the template may be such that if point *A* represents P_o, another point of the template may be the lowest point that satisfies Equation 3.25. When this equation is satisfied, point *A* still yields points of $B_R(j\omega_i)$.
3. Repeat Step 2 on the lines −100°, −110°, etc., up to −180° or until a point of the template becomes tangent to the M_L-contour. No intersection of the M_L-contour by a template is permissible. For example, in moving the template from the −90° line to the left, the template may eventually become tangent to the M_L-contour at some angle ϕ_x, as illustrated in Figure 3.11. If the template is moved further to the left, it will intersect the M_L-contour and permit a peak of $T(j\omega)$ greater than M_L. In order

to satisfy the requirement of Equation 3.19, point A' on the $\phi = \phi_x$ line becomes the left boundary or terminating point for the $B_R(j\omega_i)$ contour and is a point on the U-contour (a point on ab of Figure 3.8).[31] Draw a curve through all the points to obtain the contour for $B_R(j\omega_i)$. For the plant of this example, the U-contour is symmetrical about the −180° axis. Note that obtaining the bounds $B_R(j\omega_i)$ as described in Step (ii) only guarantees that the difference $\delta_L(\omega_i)$ between the upper bound LmT_U and the lower bound LmT_L for $LmT = Lm[L/(1 + L)]$ will satisfy

$$\delta_L(\omega_i) \le \delta_R(\omega_i) = LmT_{RU}(j\omega_i) - LmT_{RL}(j\omega_i)$$
$$= LmT_U(j\omega_i) - LmT_L(j\omega_i) \tag{3.26}$$

4. Repeat Steps 2 and 3 over the range of $\omega_x \le \omega_i \le \omega_h$, generally octaves apart, until the highest bound $B_R(j\omega_i)$ and lowest bound $B_R(j\omega_h)$ on the NC clear the U-contour. For reasonably damped plants ($\zeta > 0.6$), over the entire region of plant uncertainty, the magnitudes of the bounds $B_R(j\omega_i)$ usually decrease as ω increases. Thus, for this type of plant, it is desirable to have $\delta_R(\omega_i)$ increasing with the increase in ω_i, as discussed in Section 3.4. When this characteristic of $\delta_R(\omega_i)$ is not observed, it is possible to have $|B_R(j\omega_j)| > |B_R(j\omega_i)|$ for $\omega_j > \omega_i$. For a plant that is highly underdamped ($\zeta \le 0.6$), over some portion of the region of plant uncertainty, avoid selecting an underdamped nominal plant $P_o(s)$. For this latter situation it is desirable, but not necessary, to synthesize T_{R_U} and T_{R_L} based upon an overdamped $P_o(s)$ model. Note, if γ is specified instead of M_L it dictates the side a–b of the U-contour.

3.10 Disturbance Bounds $B_D(j\omega_i)$

Two disturbance inputs are shown in Figure 3.1. It is assumed that only one disturbance input exists at a time. Both cases are analyzed.

3.10.1 Case 1: ($d_2(t) = D_0 u_{-1}(t)$, $d_1(t) = 0$)

3.10.1.1 Control Ratio

From Figure 3.1 the disturbance control ratio for input $d_2(t)$ is

$$T_D(s) = \frac{1}{1+L} \tag{3.27}$$

Substituting $L = 1/\ell$ into Equation 3.27 yields

$$T_D(s) = \frac{\ell}{1+\ell} \tag{3.28}$$

This equation has the mathematical format required to use the NC. Over the specified BW it is desired that $|T_D(j\omega)| \ll 1$, which results in the requirement, from Equation 3.28, that $|L(j\omega)| \gg 1$ (or $|\ell(j\omega)| \ll 1$), that is,

$$|T_D(j\omega)| \approx \frac{1}{|L(j\omega)|} = |\ell(j\omega)| \tag{3.29}$$

3.10.1.2 Disturbance-Response Characteristic

A time-domain tracking response characteristic based upon $r(t) = u_{-1}(t)$ often specifies a maximum allowable peak overshoot M_p. In the frequency domain, this specification may be approximated by

$$|M_R(j\omega)| = |T_R(j\omega)| = \left|\frac{Y(j\omega)}{R(j\omega)}\right| \leq M_m \approx M_p \tag{3.30}$$

The corresponding time- and frequency-domain response characteristics, based upon the step disturbance forcing function $d_2(t) = u_{-1}(t)$, are, respectively,

$$|m_D(t)| = \left|\frac{y(t)}{d(t)}\right| \leq \alpha_p \quad \text{for } t \geq t_x \tag{3.31}$$

and

$$|M_D(j\omega)| = |T_D(j\omega)| = \left|\frac{Y(j\omega)}{D(j\omega)}\right| \leq \alpha_m \approx \alpha_p \tag{3.32}$$

3.10.1.3 Application

Let $L = KL'$ in the *tracking ratio* $T_L = L/(1 + L)$, where K is an unspecified gain, and the specification on the system performance is $M_m = 1.26$ (2 dB). By means of an NC determine the value of K required to achieve this value of M_m and obtain the data to plot $|M(j\omega)|$ versus ω. The plot of $LmL'(j\omega)$ versus $\angle L'(j\omega)$, for $K = 1$, on the NC is tangent to $LmM = 1$ dB contour, resulting in the plot of $LmL(j\omega)$ versus $\angle L(j\omega)$ in Figure 3.12 for $K = 1$.[218] Intersections of $LmL(j\omega)$ with the M-contours provide the data to plot the tracking control ratio $|M(\omega)|$ versus ω.

Now consider the corresponding disturbance control ratio for the same control system. The disturbance transfer function: $T_D = 1/(1 + L) = \ell/(1 + \ell)$ has the desired BW $0 \leq \omega \leq \omega_2$ for which $|L(j\omega)| \gg 1$ and $|\ell(j\omega)| \ll 1$. Thus Equation 3.30 applies within the BW region. Table 3.1 contains data for two points on the Nichols plot of Figure 3.12.[218] The plot of $Lm\ell(j\omega)$ versus $\angle\ell$ for these two points is also shown in this figure. The NC in Figure 3.12 is rotated by 180° and is shown in Figure 3.13. Since $Lm\ell(j\omega) = Lm[1/L(j\omega)] = -LmL(j\omega)$, a negative value of $Lm\ell$ yields a positive value for LmL as shown in Figure 3.13. Since $L = KL' = 1/\ell$,

$$\ell(j\omega) = K^{-1}\ell'(j\omega) \tag{3.33}$$

If $\ell'(j\omega)$ is given and it is required to determine K^{-1} to satisfy Equation 3.32, then the plot $Lm\ell'(j\omega)$ versus $\angle\ell'(j\omega)$ must be raised or lowered until it is tangent to the $Lm\alpha_p$-contour ($|T_D|_{max} = \alpha_m$). The amount Δ by which this plot is raised or lowered yields the value of K, that is, $Lm\, K^{-1} = \Delta$. Note that this is the same procedure used for the tracking example of Figure 3.12, except that the adjustment in $Lm\ell(j\omega)$ is K^{-1}.

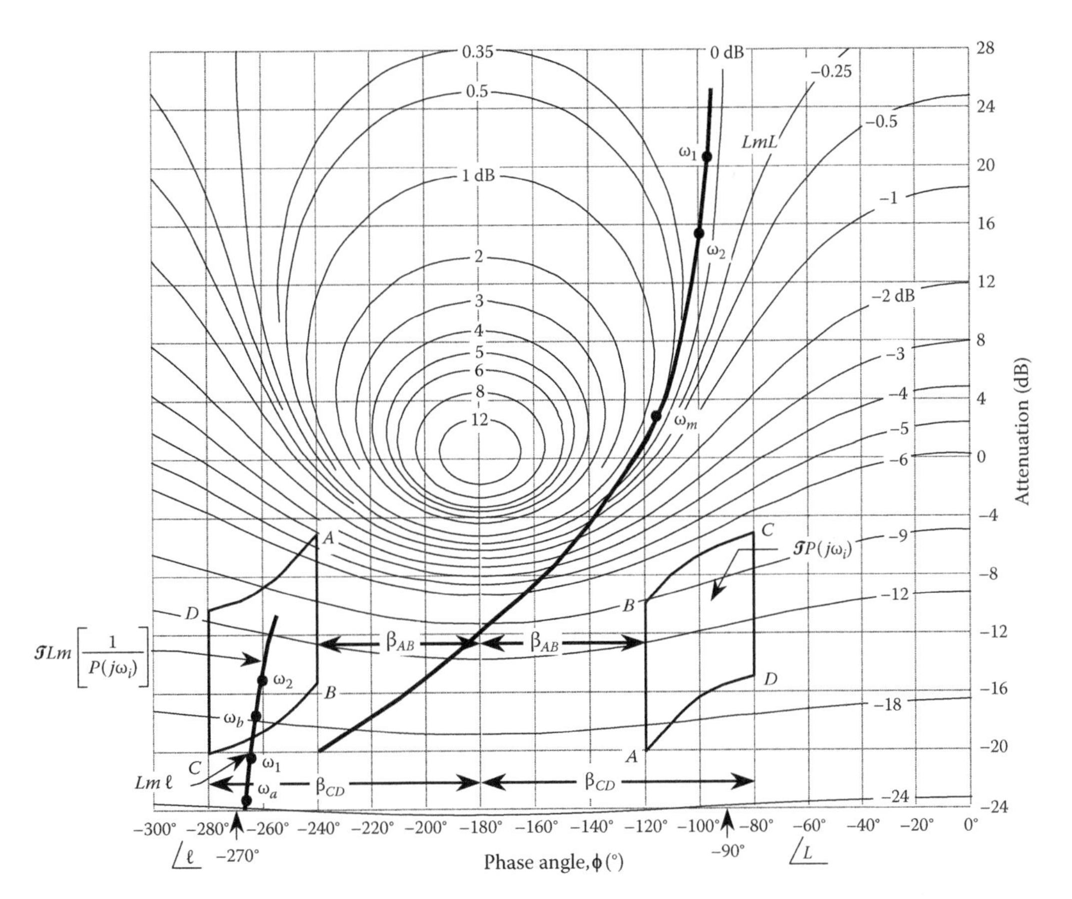

FIGURE 3.12
Regular NC.

TABLE 3.1
Data Points for a Nichols Plot

ω	LmL	$\angle L$	$Lm\ell$	$\angle \ell$
ω_1	21	−96°	−21	96° (or −264°)
ω_2	15	−98°	−15	98° (or −262°)

3.10.1.4 Templates

For a given plant P having uncertain parameters, consider that its template $\Im P(j\omega_i)$ for a given ω_i has equal dB differences along its A–B and C–D boundaries; that is, for a given $\angle P(j\omega_i)$,

$$\Delta(LmP_B - LmP_A) = \Delta(LmP_C - LmP_D) = 10 \text{ dB}$$

This template is arbitrarily set on the NC as shown in Figure 3.12. The data corresponding to the template location shown in Figure 3.12 are given in Table 3.2.

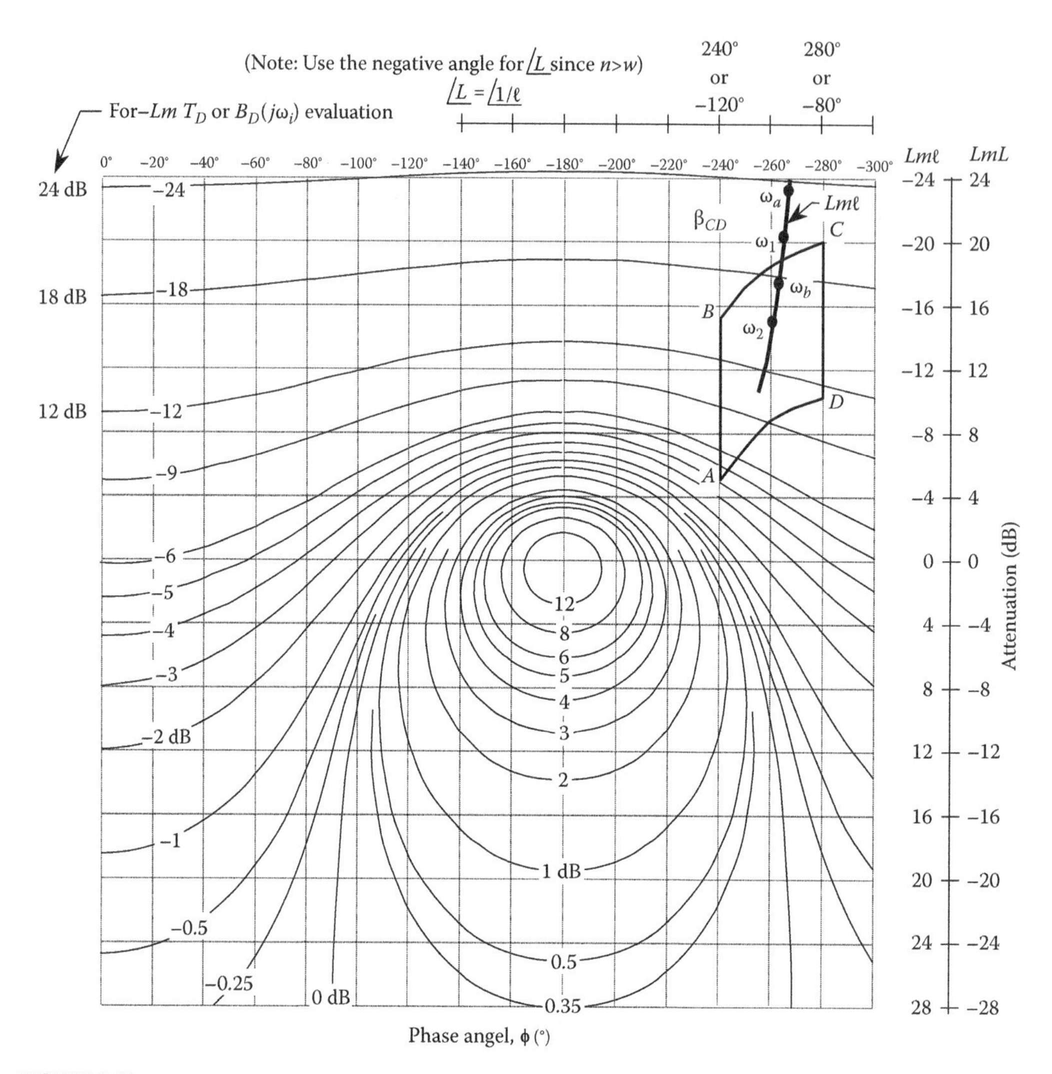

FIGURE 3.13
Rotated NC.

TABLE 3.2

Data Points for the Templates of Figure 3.12

Points	$\angle P$	$\angle 1/P$
A,B	–120°	120° (or –240°)
C,D	–80°	80° (or –280°)

The template of the reciprocal, $Lm[1/P(j\omega_i)]$, is arbitrarily set on NC in Figure 3.12 for the same frequency as for the template of $LmP(j\omega_i)$ and for the angles of Table 3.2.[6] Note that the template of $Lm[1/P(j\omega_i)]$ is the same as the template of $LmP(j\omega_i)$ but is rotated by 180°. Thus, it is located by first reflecting the template of $LmP(j\omega_i)$ about the –180° axis, "flipping it over" vertically, and then moving it up or down so that it lies between –5 and –20 dB. For the *arbitrary* location of the template of $Lm[1/P(j\omega_i)]$, note that

1. $$\beta_{AB} = 180^\circ + \angle P = 180^\circ - 120^\circ = 60^\circ$$

$$\beta_{CD} = 180^\circ + \angle P = 180^\circ - 80^\circ = 100^\circ$$

For $1/P$, the corresponding angles are

$$\angle(1/P_{AB}) = -180^\circ - \beta_{AB} = -180^\circ - 60^\circ = -240^\circ$$

$$\angle(1/P_{CD}) = -180^\circ - \beta_{CD} = -180^\circ - 100^\circ = -280^\circ$$

2. Templates of $Lm\, P(j\omega_i)$ are used for the tracker case $T = L/(1 + L)$, and the templates of $Lm\, [1/P(j\omega_i)]$ are used for the disturbance-rejection case of $T_D = 1/(1 + L) = \ell/(1 + \ell)$.

3.10.1.5 Rotated NC

Since LmL is desired for the disturbance-rejection case, Equation 3.27, the disturbance boundary $B_D(j\omega_i)$ for $L(j\omega_i) = 1/\ell(j\omega_i)$ is best determined on the rotated NC of Figure 3.13. Thus, the NC of Figure 3.12 is rotated clockwise (cw) by 180°, where the rotation of the $Lm\, [1/P(j\omega)]$ template $ABCD$ is reflected in Figure 3.13. The rotated NC is used to determine directly the boundaries $B_D(j\omega_i)$ for $L_D(j\omega_i)$. Point A for the simple plant of this design example corresponds to the nominal plant parameters and is the lowest point of the template $\Im P(jw_i)$. This point is *again used* to determine the disturbance bounds $B_D(j\omega_i)$. The lowest point of the template must be used to determine the bounds and, in general, may or may not be the point corresponding to the nominal plant parameters. Based upon Equations 3.27 and 3.28 in Figure 3.5,

$$-LmT_D = Lm[1+L] \geq -Lm\alpha(j\omega_i) > 0\,\text{dB} \tag{3.34}$$

where $\alpha(j\omega_i) < 0$. Since $|L| \gg 1$ in the BW,

$$-LmT_D \cong LmL \geq -Lm\alpha(j\omega_i) = -\delta_D(\omega_i) \tag{3.35}$$

In terms of $L(j\omega_i)$, the constant M-contours of the NC can be used to obtain the disturbance performance T_D. This requires the change of sign of the vertical axis in dB and the M-contours, as shown in Figure 3.13.

3.10.1.6 Bounds $B_D(j\omega_i)$

The procedure for determining the boundaries $B_D(j\omega_i)$ is as follows:

1. From Figure 3.5 obtain the values of $\delta_D(\omega_i) = Lm\alpha_m$ (see Equation 3.32) for the same values of frequency as for the tracking boundary $B_R(j\omega_i)$.
2. Select the lowest point of $\Im P(jw_i)$ to represent the nominal plant P_o in Figure 3.13. For the design example used in this chapter, select point A in Figure 3.13 as P_o. The same nominal point must be used in obtaining the tracking and disturbance bounds.
3. Use major angle divisions of the NC for lining up the $\Im P(jw_i)$. Line up the side A–B of $\Im P(jw_i)$, for example, on the −280° line for ℓ (or the −80° line for L). Move the template up or down until point A lies on the M-contour that represents $\delta_D(\omega_i)$. Mark this point on the NC.

4. Repeat Step 3 on the vertical lines for –100°, –120°, etc., up to the –180° line or the *U*-contour. Draw a curve through all the points to obtain the contour for $B_D(j\omega_i)$.
5. Repeat Steps 3 and 4 over the desired frequency range $\omega_x \leq \omega_i \leq \omega_h$.
6. Transcribe these $B_D(j\omega_i)$ onto the NC that contains the bounds $B_R(j\omega_i)$.

Note that when $|L| \gg 1$, then, from Equation 3.34, $|L(j\omega_i)| \gg \alpha(j\omega_i)$. Thus the *M*-contour corresponding to $T_D(j\omega_i)$ becomes the boundary $B_D(j\omega_i) = -LmM$ *for* $L(j\omega_i)$. For example, if $\alpha(j\omega_i) = 0.12$, then $L(j\omega_i) = 8.33$ and thus $Lm\alpha(j\omega_i) = -18.4$ dB and $LmL(j\omega_i) = 18.4$ dB.

3.10.2 Case 2: $(d_1(t) = D_o u_{-1}(t),\ d_2(t) = 0)$

3.10.2.1 Control Ratio

From Figure 3.1 the disturbance control ratio for the input $d_1(t)$ is

$$T_D(j\omega) = \frac{P(j\omega)}{1+G(j\omega)P(j\omega)} \tag{3.36}$$

Assuming point *A* of the template represents the nominal plant P_o, Equation 3.36 is multiplied by P_o/P_o and rearranged as follows:

$$T_D = \frac{P_o}{P_o}\left[\frac{1}{(1/P)+G}\right] = \frac{P_o}{(P_o/P)+GP_o} = \frac{P_o}{(P_o/P)+L_o} = \frac{P_o}{W} \tag{3.37}$$

where

$$W = \left(\frac{P_o}{P}\right) + L_o \tag{3.38}$$

Thus, Equation 3.37 with $LmT_D = \delta_D$ yields

$$LmW = LmP_o - \delta_D \tag{3.39}$$

3.10.2.2 Disturbance Response Characteristics

Based on Equation 3.30, the time- and frequency-domain response characteristics, for a unit-step disturbance forcing function, are given, respectively, by

$$|M_D(t)| = \left|\frac{y(t_p)}{d(t)}\right| = |y(t_p)| \leq \alpha_p \tag{3.40}$$

and

$$|M_D(j\omega)| = |T_D(j\omega)| = \left|\frac{Y(j\omega)}{D(j\omega)}\right| \leq \alpha_m \equiv \alpha_p \tag{3.41}$$

where t_p is the peak time.

3.10.2.3 Bounds $B_D(j\omega_i)$

The procedure for determining the boundaries $B_D(j\omega_i)$ is as follows:

1. From Figure 3.5 obtain the value of $\delta_D(\omega_i)$ representing the desired model specification $T_D = T_{DU} = \alpha_p$ for the same values of frequency as for the tracker boundaries $B_R(j\omega_i)$.
2. *Evaluate* in tabular form for each value of ω_i the following items in the order given:

$$LmP_o(j\omega_i),\quad \delta_D(\omega_i),\quad Lm\,W(j\omega_i),\quad |W(j\omega_i)|,\quad \frac{P_o(j\omega_i)}{P_\iota(j\omega_i)}$$

The ratio P_o/P_ι is evaluated at each of the four points in Figure 3.6 for each value of ω_i. It may be necessary to evaluate this ratio at additional points around the perimeter of the *ABCD* contour as shown in Figure 3.6.

3. Before presenting the procedure for the graphical determination of $B_d(j\omega_i)$, where $B_D = LmB_d$, it is necessary to first review, graphically, the phasor relationship between B_d, P_o/P, and W for $\omega = \omega_i$. Equation 3.38, with L_o replaced by its bound B_d, is rearranged to the form

$$W = \frac{P_o}{P} + B_d \rightarrow -B_d = \frac{P_o}{P} + (-W) \tag{3.42}$$

For arbitrary values of $P_o(j\omega_i)/P(j\omega_i)$ and $W(j\omega_i)$, Figure 3.14 presents the phasor relationship of Equation 3.42. Since the values of $P_o(j\omega_i)/P(j\omega_i)$ for $P \in \mathcal{P}$ and $|W(j\omega_i)|$ are known, the following procedure can be used to evaluate $B_d(j\omega_i)$:

a. On polar or rectangular graph paper draw $\Im[P_o/P]$ for each ω_i as shown in Figure 3.15, where point A is the nominal point.

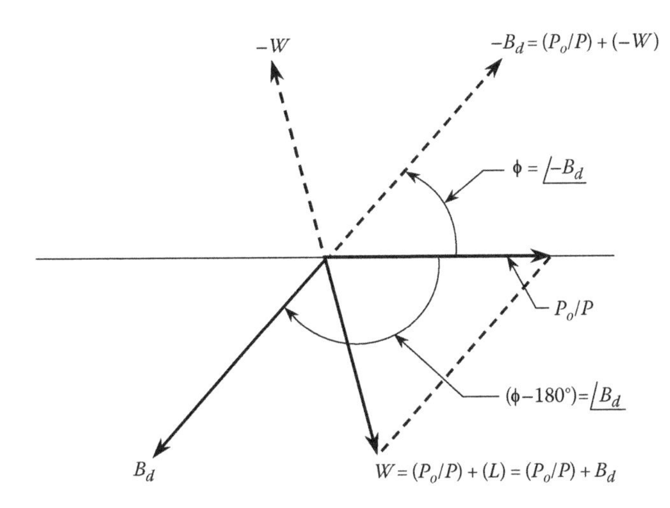

FIGURE 3.14
Phasor relationship of Equation 3.42, which is Equation 3.38 with $L_o = B_d$.

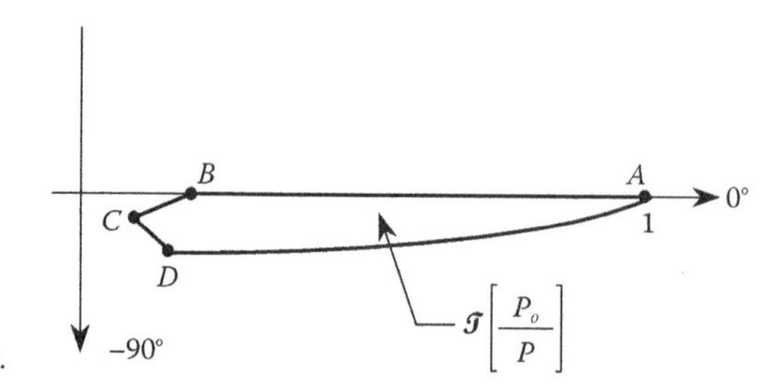

FIGURE 3.15
Template in polar coordinates.

b. Based upon Figure 3.14 and the location of the phasor $-W(j\omega)$, the solution for $-B_d$ is obtained from

$$-B_d = \Im\left[\frac{P_o}{P}\right] - W \tag{3.43}$$

For one value of P_o/P shown in Figure 3.16, the value of $-W$ is plotted and $-B_d = |-B_d| \angle\phi$ is obtained. This graphical evaluation of $B_d = |B_d|\angle(\phi - 180°)$ is performed for various points around the perimeter of $\Im[P_o/P]$ in Figure 3.15. A simple graphical evaluation yields a more restrictive bound (the worst case). Use a compass to mark off arcs with a radius equal to the distance $|W(j\omega_i)|$ at a number of points on the perimeter of $\Im[P_o(jw_i)/P(jw_i)]$. Draw a curve that is tangent to these arcs to form the first quadrant portion of the Q-contour shown in Figure 3.16. Depending on the plant type desired for L, it may be necessary to extend this contour into the second and fourth quadrants.

4. Based upon Equation 3.43 and Figure 3.15, the phasor from the origin of Figure 3.16 to the Q-contour represents $-B_d(j\omega_i)$. This contour includes the plant uncertainty as represented by $\Im[P_o(jw_i)/P(jw_i)]$. In the frequency range $\omega_x \leq \omega_i \leq \omega_h$, if $|W(j\omega_i)| \gg |P_o(j\omega_i)/P(j\omega_i)|$, then the Q-contour is essentially a circle about the origin with radius $|W(j\omega_i)| \equiv |B_d(j\omega_i)|$.
5. Assuming the partial Q-contour of Figure 3.16 is sufficient, measure from the graph the length $B_d(j\omega_i)$ for every 10° of $B_d(j\omega_i)$ and create Table 3.3 for each value of ω_i.
6. Plot the values of $B_D(j\omega_i)$ from Table 3.3 for each value of ω_i, on the same NC as $B_R(j\omega_i)$.

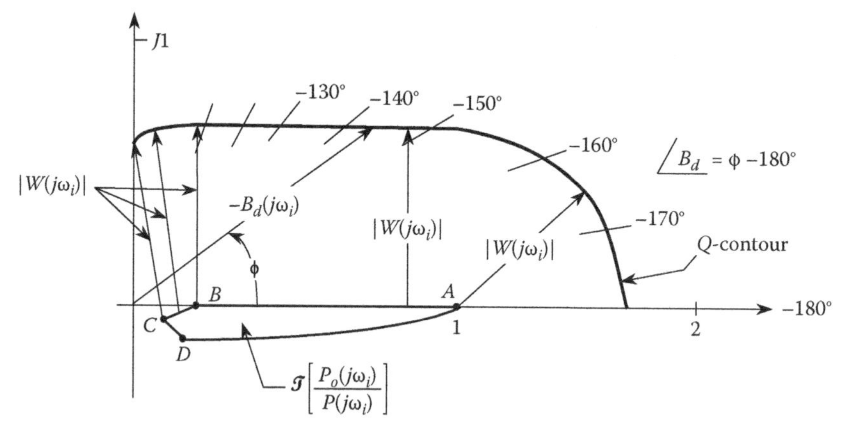

FIGURE 3.16
Graphical evaluation of $B_d(j\omega_i)$.

TABLE 3.3

Data Points for the Boundary $B_D(j\omega)$

$\angle B_d(j\omega_i)$	$\lvert B_d(j\omega_i)\rvert$	$B_D(j\omega_i) = Lm\ B_d(j\omega_i)$
−180°		
−170°		
−160°		

3.11 Composite Boundary $B_o(j\omega_i)$

The composite bound $B_o(j\omega_i)$ that is used to synthesize the desired loop transmission transfer function $L_o(s)$ is obtained in the manner shown in Figure 3.17. The composite bound $B_o(j\omega_i)$, for each value of ω_i, is composed of those portions of each respective bound $B_R(j\omega_i)$ and $B_D(j\omega_i)$ that are the most restrictive. For the case shown in Figure 3.17a, the bound $B_o(j\omega_i)$ is composed of those portions of each respective bound $B_R(j\omega_i)$ and $B_D(j\omega_i)$ that have the largest values. For the situation of Figure 3.17b, the outermost of the two boundaries $B_R(j\omega_i)$ and $B_D(j\omega_i)$ becomes the perimeter of $B_o(j\omega_i)$. The situations of Figure 3.17 occur when the two bounds have one or more intersections. If there are no intersections, then the bound with the largest value or with the outermost boundary dominates. The synthesized $L_o(j\omega_i)$, for the situation of Figure 3.17a, must lie on or just above the bound $B_o(j\omega_i)$. For the situation of Figure 3.17b the synthesized $L_o(j\omega_i)$ *must not lie in the interior* of the $B_o(j\omega_i)$ contour. For complement information about the bounds see Appendices B and C and Refs. 42 through 63.

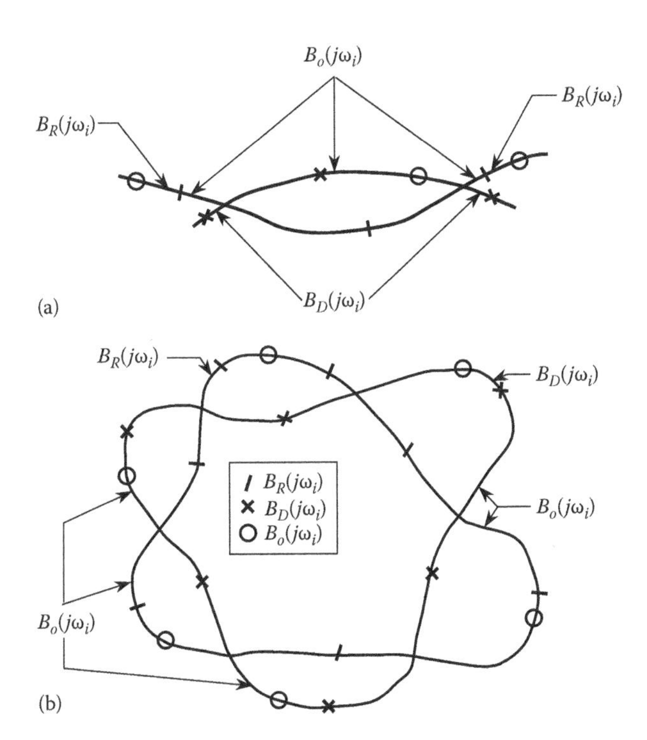

FIGURE 3.17
Composite $B_o(j\omega_i)$. (a) Determination of B_o: largest value case and (b) determination of B_o: outermost case.

3.12 Shaping of $L_o(j\omega)$

A realistic definition of optimum[126] in an LTI system is the minimization of the high-frequency loop gain K while satisfying the performance bounds. This gain affects the high-frequency response since $\lim_{\omega\to\infty}[L(j\omega)] = K(j\omega)^{-\lambda}$ where λ is the excess of poles over zeros assigned to $L(j\omega)$. Thus, only the gain K has a significant effect on the high-frequency response, and the effect of the other parameter uncertainty is negligible. Also, the importance of minimizing the high-frequency loop gain is to minimize the effect of sensor noise whose spectrum, in general, lies in the high-frequency range (see Chapters 6 and 9 of Ref. 6). It has been shown that the optimum $L_o(j\omega)$ exists; it lies on the boundary $B_o(j\omega_i)$ at all ω_i, and it is unique.[32,116] Note the bounds can be also calculated using an iterative algorithm that computes them through quadratic inequalities (see Appendix B).

Previous sections have described how tolerances on the closed-loop system frequency response, in combination with plant uncertainty templates, are translated into bounds on a nominal loop transmission function $L(j\omega)$. In Figure 3.18, the template $\Im P(j\omega_i)$ is located

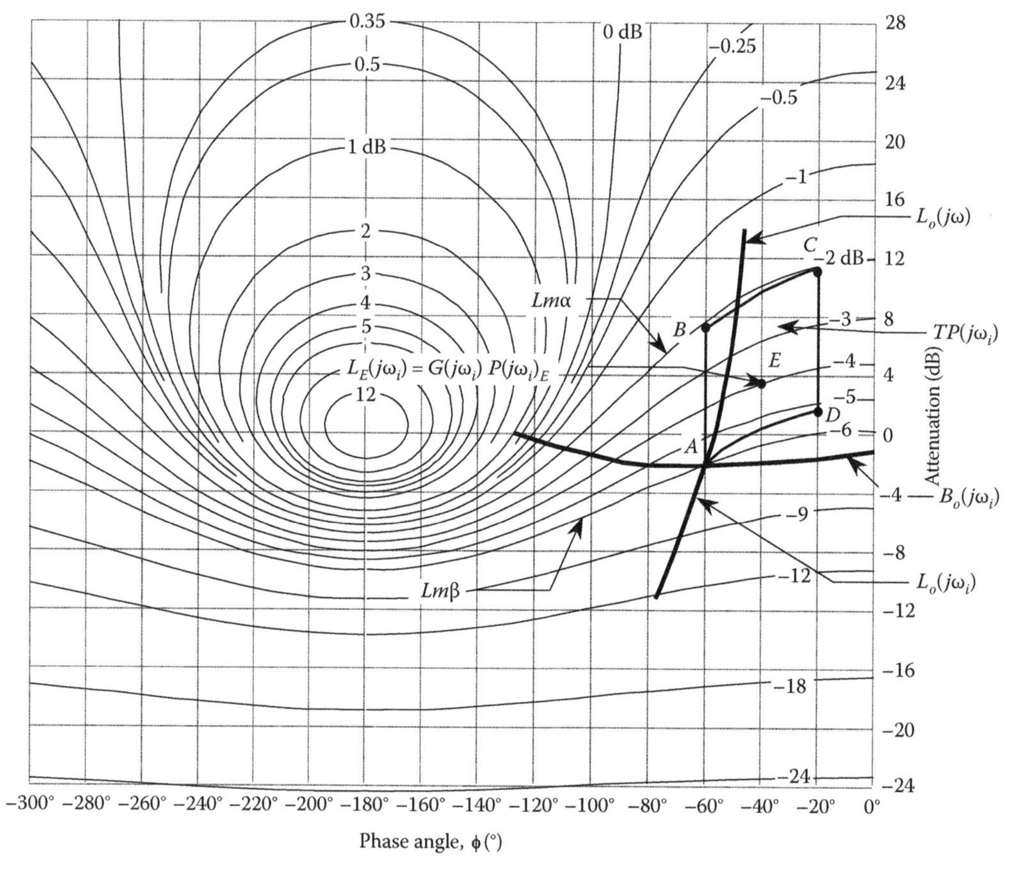

FIGURE 3.18
Graphical determination of $Lm\ T_1(j\omega_i)$ for $P \in \mathcal{P}$.

on the corresponding bound $B_o(j\omega_i)$ where point A is on the constant M-curve $Lm\beta$ and point C on the constant M-curve $Lm\alpha$ such that

$$\delta_R(\omega_i) = Lm\alpha - Lm\beta = 4\,\text{dB} \tag{3.44}$$

and where $\alpha = T_{max}$, $\beta = T_{min}$, and $\delta_R(\omega_i) = 4\,\text{dB}$.

For the set of plant parameters that correspond to point E within $\Im P(jw_i)$ and for a synthesized L_o, as shown in Figure 3.18, the open-loop transfer function is $L_E = P_E G$. The control ratio

$$T_E = \frac{L_E}{1 + L_E} \tag{3.45}$$

obtained from the constant M-contours on the NC has the value $LmT_E = -4\,\text{dB}$. Thus, any value of P within $\Im P(j\omega_i)$ yields a value of LmT ($=LmT_\iota$) of Equation 3.5 between points A and C in Figure 3.18. Therefore, any value of P that lies within the region of uncertainty (see Figure 3.6 for the example of this chapter) yields a maximum variation in $T(j\omega_i)$ that satisfies the requirement

$$LmT_{max} - LmT_{min} \le \delta_R(\omega_i) \tag{3.46}$$

Thus, proper design of the prefilter F (see Figure 3.1) yields a tracking control ratio T_R that lies between LmT_U and LmT_L in Figure 3.4b.

For the plant of Equation 2.7, the shaping of $L_o(j\omega)$ is shown by the dashed curve in Figure 3.19. A point such as $LmL_o(j2)$ must be on or above the curve labeled $B_o(j2)$. Further, in order to satisfy the specifications, $L_o(j\omega)$ cannot violate the U-contour. In this example, a reasonable $L_o(j\omega)$ closely follows the U-contour up to $\omega = 40\,\text{rad/s}$ and stays below it above $\omega = 40$ as shown in Figure 3.19.

Additional specifications are $\lambda = 4$, that is, there are four poles in excess of zeros, and that it also must be Type 1 (one pole at the origin).

A representative procedure for choosing a rational function $L_o(s)$ that satisfies the aforementioned specifications is now described (see also Appendix D). It involves building up the function

$$L_o(j\omega) = L_{ok}(j\omega) = P_o(j\omega)\prod_{k=0}^{w}[K_k G_k(j\omega)] \tag{3.47}$$

where for $k = 0$, $G_o = 1\ \angle 0°$, and

$$K = \prod_{k=0}^{w} K_k$$

In order to minimize the order of the compensator, a good starting point for "building up" the loop transmission function is to initially assume that $L_{o0}(j\omega) = P_o(j\omega)$ as indicated in Equation 3.47. $L_o(j\omega)$ is built up term by term in order to stay just outside the U-contour in the NC of Figure 3.19. The first step is to find the $B_o(j\omega_i)$, which *dominates* $L_o(j\omega)$.

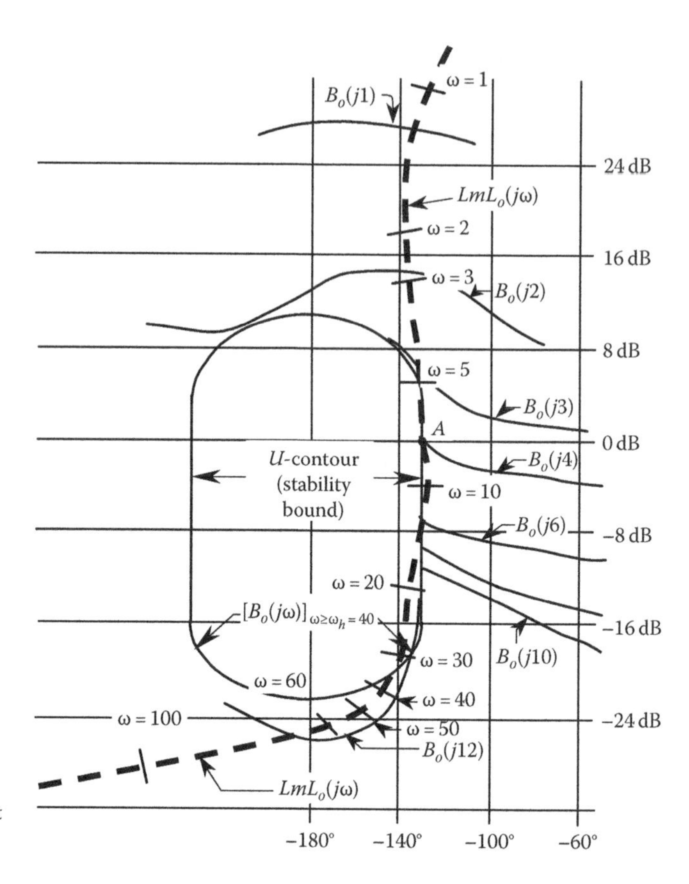

FIGURE 3.19
Shaping of $L_o(j\omega)$ on the NC for the plant of Equation 2.7.

For example, suppose that $L_{o0}(j4) = K_o P_o'(j4) = 0\,\text{dB}\ \angle{-135°}$ (point A in Figure 3.19) and that at $\omega = 1$ the required $LmL_{o0}(j1)$ is approximately 27 dB. In order for $LmL_o(j\omega)$ to decrease from 27 to about 0 dB in two octaves, the slope of $L_o(j\omega)$ must be about −14 dB/octave, with $\angle L_o < -180°$ (see Figure 3.20). Since the M_L stability contour must be satisfied for all J plants, it is required that $L_o(j\omega)$ have a phase-margin angle γ of 45° over the entire frequency range for which $L_o(j\omega)$ follows the vertical right-hand side of the U-contour and not just at the 0-dB crossover. Hence $B_o(j1)$ dominates $L_o(j\omega)$ more than does $B_o(j4)$. In the same way it is seen that $B_o(j1)$ dominates all other $B_o(j\omega)$ in Figure 3.19.

The $B_o(j\omega)$ curves for $\omega < 1$ are not shown in Figure 3.19 because it is assumed that a slope of −6 dB/octave for $\omega < 1$ suffices (additional values are 33 dB at $\omega = 0.5$, 39 dB at $\omega = 0.25$, etc.). By selecting $L_{o0}(s) = kP_o'(s) = k/[s(s+1)]$, the first denominator factor of L_{o0} has a corner frequency at $\omega = 1$ (i.e., a pole at −1), which maintains −135° for $\omega \geq 1$. Thus, the value Lm $L_o(j1)$ on the straight-line approximation is selected at 30 dB (to allow for the −3 dB correction at the corner frequency). The function $L_o(s)$ determined so far is

$$L_{o0}(s) = 31.6/[s(s+1)]$$

whose angle $\angle L_{o0}(j\omega)$ is sketched in Figure 3.20.

$L_{o0}(j\omega)$ violates the −135° bound at $\omega \geq 1.2$, hence a numerator term $(1 + j\omega T_2)$ must be added. At $\omega = 5$, $\angle L_{o0}(j5) = -169°$ (see Figure 3.20), and therefore a lead angle of 29° is needed at this frequency. Since a second denominator term $(1 + j\omega T_3)$ will be needed, allow an

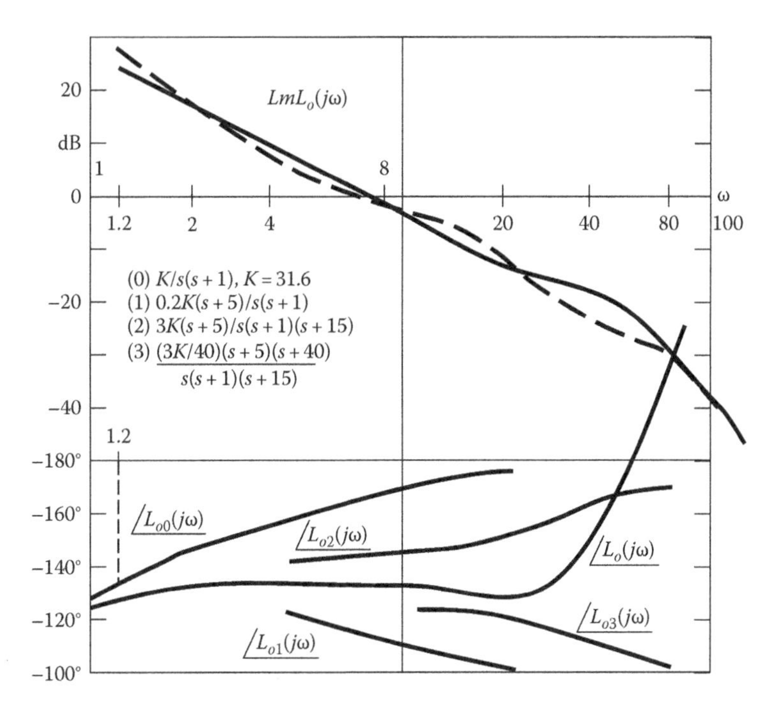

FIGURE 3.20
Shaping of $L_o(j\omega)$ on the Bode plot.

additional 15° for this factor, giving a total of 15° + 29° = 45° lead angle required at $\omega = 5$. This is achieved by selecting $T_2 = 0.2$, that is, a zero at −5. This results in the composite value

$$L_{o1}(s) = \frac{31.6(1+s/5)}{s(1+s)}$$

whose phase angle $\angle L_{o1}(j\omega)$ is sketched in Figure 3.20. In the NC of Figure 3.19, $\omega \geq 5$ is the region where the maximum phase *lag* allowed is −135° (i.e., $\angle L_o(j\omega)$ must be ≥−135°). At $\omega = 10$, since $\angle L_{o1}(j10) = -112°$, 135° −112° = 23° more lag angle is needed and is provided with an additional denominator term $(1 + j\omega T_3)$. However, this is to be followed by an additional numerator term $(1 + j\omega T_4)$, so allow about 10° for it, giving 23° + 10° = 33° more lag allowable from $(1 + j\omega T_3)$. This requires selecting the corner frequency at $\omega = 15$ (or $T_3 = 1/15$), giving

$$L_{o2}(s) = \frac{31.6(1+s/5)}{s(1+s)(1+s/15)}$$

A sketch of $\angle L_{o2}(j\omega)$ is shown in Figure 3.20.

Looking ahead at $\omega = 40$, $LmL_{o2}(j40) = -20$ dB, thus $L_o(j\omega)$ can make its asymptotic left turn *under* the *U*-contour. The plan is to add two more numerator factors, $(1 + jT_4)$ and $(1 + jT_5)$, and finally two complex-pole pairs, in order to have an excess $\lambda = 4$ of poles over zeros and to minimize the *BW*. A zero is assigned at $\omega = 40$ with $T_4 = 1/40$, giving

$$L_{o3}(s) = \frac{31.6(1+s/5)(1+s/40)}{s(1+s)(1+s/15)}$$

A sketch of $\angle L_{o3}(j\omega)$ is shown in Figure 3.20.

In order to achieve (an asymptotic) horizontal segment for $LmL_o(j\omega)$, before the final −24 dB/octave slope is achieved, a final zero obtained from $(1 + j\omega T_5)$ is needed. Since the bottom of the *U*-contour (see Figure 3.19) is at −22.5 dB, allow a 2 dB safety margin, a 3 dB correction due to $(1 + j\omega T_5)$, and a 1.5 dB for the effect of $(1 + j\omega T_4)$, giving a total of $-(22.5 + 2 + 3 + 1.5) = -29$ dB. A damping ratio of $\zeta = 0.5$ is selected for the two complex-pole pairs, so no correction is needed for them. Thus, the final corner frequency for the straight-line curve of $LmL_{o3}(j\omega)$ is at −29 dB. Since $LmL_{o3}(j\omega)$ achieves this at $\omega = 60$, the last corner frequency is at $\omega = 60$, that is, $T_5 = 1/60$. The resulting phase angle, due to L_{o3} and $(1 + j\omega/60)$, is −66. An angle of −180° could be selected at this point, but an additional 15° margin is allowed (this is a matter of judgment that depends on the problem, which may include the presence of higher-order modes, etc.). This means that 100° phase lag is permitted: 50° due to each complex-pole pair (180° − 66° − 15° ≈ 100°). Thus, a different value of damping ratio, with the appropriate dB correction applied to the log magnitude plot, is chosen, that is, $\zeta = 0.6$. This locates the corner frequency at $\omega = 100$. Thus

$$L_o(s) = \frac{31.6(1+s/5)(1+s/40)(1+s/60)}{s(1+s)(1+s/15)(1+1.2s/100+s^2/10^4)^2}$$

$$= KP_o(s)\prod_{k=1}^{4} G_k(s) \tag{3.48}$$

The optimal loop transfer function $L_o(j\omega)$ is sketched in Figure 3.19 and is shown by solid curves in Figure 3.20. A well-designed, that is, an "economical" $L_o(j\omega)$ is close to the $B_o(j\omega)$ boundary at each ω_i. The vertical line at −140° in Figure 3.19 is the dominating vertical boundary for $L_o(j\omega)$ for $\omega < 5 = \omega_x$, and the right side of the *U*-contour line at −135° is the vertical boundary effectively for $\omega_x \approx 5 < \omega < 30 \approx \omega_y$. The final $L_o(j\omega)$ is good in this respect since it is close to these boundaries. Although there is a "slight" infringement of the *U*-contour, for $\omega > 15$, because of the inherent overdesign, no reshaping of $L_o(s)$ is done unless the simulation reveals that the specifications are not met.

There is a trade-off between the complexity of $L_o(s)$ (the number of its poles and zeros) and its final cutoff corner frequency, which is $\omega = 100$, and the phase-margin frequency $\omega_\phi = 7$. There is some phase to spare between $L_o(j\omega)$ and the boundaries. Use of more poles and zeros in $L_o(s)$ permits this cutoff frequency to be reduced a bit below 100, but not by much. On the other hand, if it is desired to reduce the number of poles and zeros of $L_o(s)$, then the price in achieving this is a larger cutoff frequency. It is possible to economize significantly by allowing more phase lag in the low-frequency range. If −180° is permitted at $\omega = 1$, then a decrease of $LmL_o(j\omega)$ at a rate of 12 dB/octave can be achieved, for example, with $LmL_o(j1) = 25$ dB, then it will be 13 dB at $\omega = 2$ (instead of the present 18 dB). Even with no more savings, this 5 dB difference allows a cutoff frequency of about $\omega = 70$ instead of 100.

Figure 3.19 reveals immediately, without any reshaping of $L_o(j\omega)$ required, that reduction (i.e., easing) of the specifications at $\omega = 1$ to about 21 dB (instead of about 26 dB) has the same effect as mentioned earlier. How badly the specifications are compromised by such easing can easily be checked. The design technique is thus highly "transparent" in revealing the trade-offs between performance and uncertainty tolerances, complexity of the compensation, stability margins, and the "cost of feedback" in BW.

3.13 Guidelines for Shaping $L_o(j\omega)$

Some general guidelines for the shaping of $L_o(j\omega)$ are as follows:

1. For the *beginner* it is best not to use a CAD program. Use the straight-line approximations on the Bode diagram for the log magnitude at the start of the design problem.
2. On the graph paper for the Bode diagram, plot the points representing $LmB_o(j\omega_i)$ and the angles corresponding to the right side of the *U*-contour (the desired phase-margin angle γ) for the frequency range of $\omega_x \le \omega \le \omega_y$. For the example illustrated in Figure 3.19 the phase-margin angle $\gamma = -45°$ must be maintained for the frequency range $\omega_x = 5 \le \omega \le \omega_y \approx 30$.
3. Do the shaping of $LmL_o(j\omega)$ on the Bode plot using straight-line approximations for $LmL_o(j\omega)$, with the plotted information of Step 2, and employ the shaping discussion that follows Equation 3.48 for the frequency range $\omega < \omega_x$ as guidelines in achieving Equation 3.48.
4. Use frequencies an octave above and below and a decade above and below a corner frequency for both first- and second-order terms,[2,206] while maintaining the phase-margin angle corresponding to the right side of the *U*-contour in shaping $L_o(j\omega)$.
5. The last two poles that are added to $L_o(j\omega)$ are generally a complex pair (the nominal range is $0.5 < \zeta < 0.7$), which tends to minimize the *BW*.
6. Once $L_o(j\omega)$ has been shaped, determine $F(s)$ (see Section 3.14) and then verify that $L_o(j\omega)$ does meet the design objectives by using a CAD program (see Appendix F). If the synthesized $L_o(j\omega)$ yields the desired performance, then the required compensator is given by

$$G(s) = \frac{L_o(s)}{P_o(s)}$$

Specific guidelines for shaping $L_o(j\omega)$ are as follows:

1. An optimum design of $L_o(j\omega)$ requires that $L_o(j\omega_i)$ be on the corresponding bound. In practice, place $L_o(j\omega_i)$ as close as possible to the bound $B_o(j\omega_i)$, but above it, in order to keep the *BW* of $L_o/[1 + L_o]$ to a minimum.[32]
2. Since exact cancellation of a pole by a zero is rarely possible, any right-hand-plane (RHP) poles and/or zeros of $P_o(s)$ should be included in $L_o(s)$. A good starting $L_o(s)$ is $L_{o0}(s) = K_0 P_o(s)$. If it is desired that $y(\infty) = 0$ for $d(t) = u_{-1}(t)$, it is necessary to insure that $T_D(s)$ has a zero at the origin. For this situation a possible starting point is $L_{o0}(s) = K_0 P_o(s)/s$.
3. If $P(s)$ has an excess of poles over zeros, $\#p's - \#z's = \lambda$, which is denoted *by* e^{λ}, then, in general, the *final form* of $L_o(s)$ must have an excess of poles over zeros of at least $e^{\lambda+\mu}$ where $\mu \ge 1$. If the *BW* is too large, then increase the value of *i*. Experience shows that a value of $\lambda + i$ of 3 or more for $L(s)$ yields satisfactory results.[31,32]

4. Generally, the *BW* of $L_o(s)/[1 + L_o(s)]$ is larger than required for an acceptable rise time t_R for the tracking of $r(t)$ by $y(t)$. An acceptable rise time can be achieved by the proper design of the prefilter $F(s)$ (see Section 3.14).
5. Generally, it is desirable to first find the bounds $B_D(j\omega_i)$ and then the bounds $B_R(j\omega_i)$. After finding the first $B_R(j\omega_i)$, it may be evident that all or some of the $B_D(j\omega_i)$ are completely dominant compared to $B_R(j\omega_i)$. In that case the $B_D(j\omega_i)$ boundaries are the optimal boundaries, that is, $B_D(j\omega_i) = B_o(j\omega_i)$.
6. If $\delta_R(\omega_i)$ is not continuously increasing as ω_i increases, then it will be necessary to utilize complex poles and/or zeros in $G(s)$ in order to achieve an optimal $L_o(s)$. This assumes that the tracking model is not designed to yield this increasing characteristic for $\delta_R(\omega_i)$.
7. The ability to shape the nominal loop transmission $L_o(s)$ is an art developed by the designer only after much practice and patience.

The success of the compensator design strongly depends on the experience of the designer. The aforementioned rules and general guidelines for shaping give a good way to obtain the compensator from the engineering point of view. On the other hand, in the last few years some automatic loop-shaping procedures to help in the QFT compensator design have been developed (see Refs. 66–68, 71–73). Although very often they do not reach the optimum solution, they could contribute to show some tracks that help in the compensator design. For complement information about the loop-shaping see Appendix D and Refs. 64 through 73.

3.14 Design of the Prefilter *F(s)*

Design of a proper $L_o(s)$ guarantees only that the variation in $|T_R(j\omega)|$, that is, ΔT_R, is less than or equal to that allowed. The purpose of the prefilter is to position $LmT(j\omega)$ within the frequency-domain specifications. For the example in this chapter, the magnitude of the frequency response must lay within the bounds B_U and B_L shown in Figure 3.4b, which is redrawn in Figure 3.21. A method for determining the bounds on $F(s)$ is as follows[6]: Place

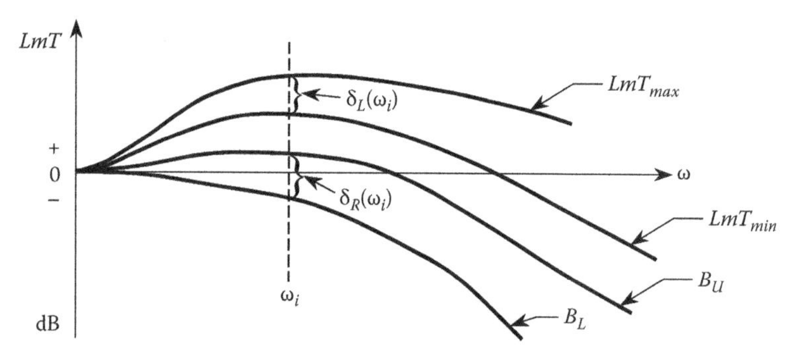

FIGURE 3.21
Requirements on $F(s)$.

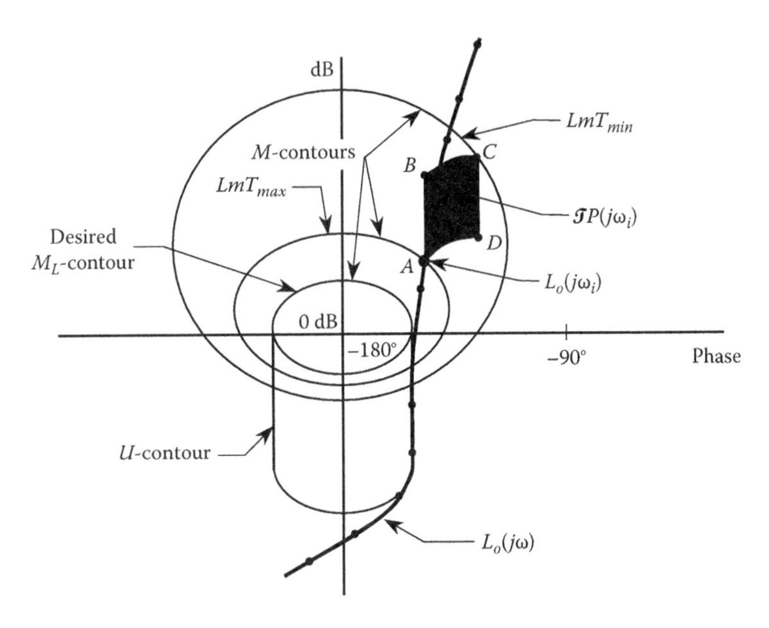

FIGURE 3.22
Prefilter determination.

the nominal point A of the ω_i plant template on the $L_o(j\omega_i)$ point of the $L_o(j\omega)$ curve on the NC (see Figure 3.22). Traversing the template, determine the maximum LmT_{max} and minimum LmT_{min} values of

$$LmT(j\omega_i) = \frac{L(j\omega_i)}{1+L(j\omega_i)} \tag{3.49}$$

obtained from the M-contours. These values are plotted as shown in Figure 3.21. The tracking control ratio is $T_R = FL/[1 + L]$ and

$$LmT_R(j\omega_i) = LmF(j\omega_i) + LmT(j\omega_i) \tag{3.50}$$

The variations in Equations 3.49 and 3.50 are due to the variation in P; thus

$$\delta_L(\omega_i) = LmT_{max} - LmT_{min} \le \delta_R = B_U - B_L \tag{3.51}$$

If values of $L_o(j\omega_i)$, for each value of ω_i, lie exactly on the tracking bounds $B_R(j\omega_i)$, then $\delta_L = \delta_R$. Therefore, based upon Equation 3.50, it is necessary to determine the range in dB by which $LmT(j\omega_i)$ must be raised or lowered to fit within the bounds of the specifications by using the prefilter $F(j\omega_i)$. The process is repeated for each frequency corresponding to the templates used in the design of $L_o(j\omega)$. Therefore, in Figure 3.23 the difference between the $LmT_{R_U} - LmT_{max}$ and the $LmT_{R_L} - LmT_{min}$ curves yields the requirement for $LmF(j\omega)$, that is, from Equation 3.50.

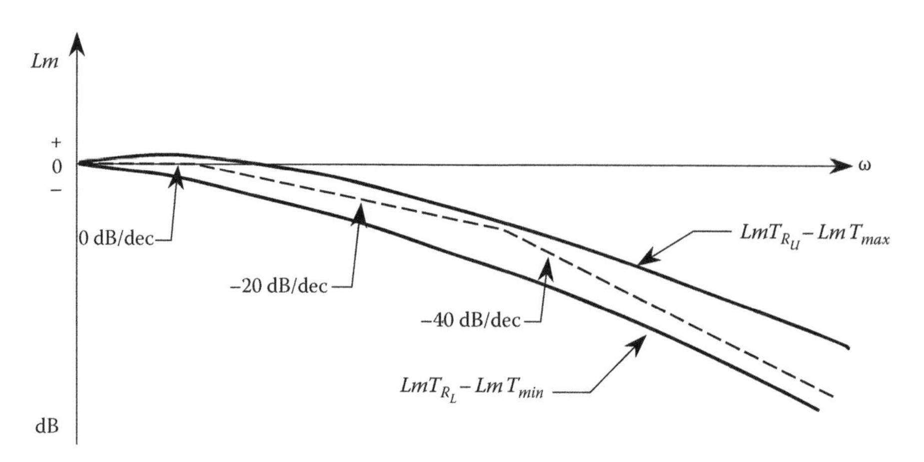

FIGURE 3.23
Frequency bounds on the prefilter $F(s)$.

The procedure for designing $F(s)$ is summarized as follows:

1. Use templates in conjunction with the $L_o(j\omega)$ plot on the NC to determine T_{max} and T_{min} for each ω_i. This is done by placing $\Im P(jw_i)$ with its nominal point on the point $Lm\ L_o(j\omega_i)$. Then use the M-contours to determine $T_{max}(j\omega_i)$ and $T_{min}(j\omega_i)$ (see Figure 3.22).
2. Obtain the values of LmT_{R_U} and LmT_{R_L} for various values of ω_i from Figure 3.4b. From the values obtained in Steps 1 and 2, plot

$$[LmT_{R_U} - LmT_{max}] \text{ and } [LmT_{R_L} - LmT_{min}] \text{ versus } \omega \tag{3.52}$$

as shown in Figure 3.23.

Use straight-line approximations to synthesize an $F(s)$ so that $LmF(j\omega)$ lies within the plots of Step 3. For step forcing functions the resulting $F(s)$ must satisfy

$$\lim_{s\to 0}[F(s)] = 1 \tag{3.53}$$

3.15 Basic Design Procedure for a MISO System

The basic concepts of the QFT technique are explained by means of a design example. The system configuration shown in Figure 3.1 contains four inputs. Parameter uncertainty for the plant of Equation 2.7 is shown in Figure 3.6. The first objectives are to track a step input $r(t) = u_{-1}(t)$ with no steady-state error and to satisfy the performance specifications of Figure 3.3. Additional objectives are to attenuate the system response caused by external step disturbance inputs $d_1(t)$, $d_2(t)$, and $d(t)$, as described in Section 3.4.2, and to minimize the effect of noise. An outline of the basic design procedure for the QFT technique, as applied to an *m.p.* plant, is as follows (Note that the CAD Package, *QFTCT,* can be used in all the steps of the methodology. See Appendices F and G and Refs. 149 and 150, and visit the websites http://cesc.case.edu or http://www.crcpress.com/product/isbn/9781439821794.):

1. Synthesize the tracking model control ratio $T_R(s)$ in the manner described in Section 3.4, based upon the desired tracking specifications (see Figures 3.3 and 3.4b).
2. Synthesize the disturbance-rejection model control ratios $T_D(s)$ in the manner described in Section 3.10, based upon the disturbance-rejection specifications.
3. Obtain templates of $P(j\omega_i)$ that pictorially describe the plant uncertainty on the NC for the desired pass-band frequency range (see also Appendix A).
4. Select a nominal plant from the set of Equation 3.1 and denote it as $P_o(s)$.
5. Determine the U-contour based upon the specified values of $\delta_R(\omega_i)$ for tracking, M_L for disturbance rejection, and V for the UHFB B_h in conjunction with Steps 6–8 (see Appendices B and C).
6. Use the data of Steps 2 and 3 and the values of $\delta_D(\omega_i)$ (see Figure 3.5) to determine the disturbance bound $B_D(j\omega_i)$ on the loop transmission $L(j\omega_i) = P(j\omega_i)G(j\omega_i)$. For *m.p.* systems, this requires that the synthesized loop transmission $LmL(j\omega_i)$ must be on or above the curve for $LmB_D(j\omega_i)$ on the NC (see Figure 3.19 assuming $B_D = B_o$) (see also Appendices B and C).
7. Determine the tracking bound $B_R(j\omega_i)$ on the nominal transmission $L_o(j\omega_i) = P_o(j\omega_i)G(j\omega_i)$, using the tracking model (Step 1), the templates $P(j\omega_i)$ (Step 3), the values of $\delta_R(\omega_i)$ (see Figure 3.4b), and M_L (see Equation 3.19). For *m.p.* systems this requires that the synthesized loop transmission satisfy the requirement that $LmL_o(j\omega_i)$ is on or above the curve for $LmB_R(j\omega_i)$ on the NC (see Appendices B and C).
8. Plot curves of $LmB_R(j\omega_i)$ versus $\phi_R = \angle B_R(j\omega_i)$ and $LmB_D(j\omega_i)$ versus $\phi_D = \angle B_D(j\omega_i)$ on the same NC. For a given value of ω_i at various values of the angle ϕ, select the value of $LmB_D(j\omega_i)$ or $LmB_R(j\omega_i)$, whichever is the largest value (termed the "worst" or "most severe" boundary). Draw a curve through these points. The resulting plot defines the overall boundary $LmB_o(j\omega_i)$ versus ϕ. Repeat this procedure for sufficient values of ω_i (see Appendices B and C).
9. Design $L_o(j\omega_i)$ to be as close as possible to the boundary value $B_o(j\omega_i)$ by selecting an appropriate compensator transfer function $G(j\omega)$ (see Appendix D). Synthesize an $L_o(j\omega) = G(j\omega)P_o(j\omega)$ using the $LmB_o(j\omega_i)$ boundaries and U-contour so that $LmL_o(j\omega_i)$ is on or above the curve for $LmB_o(j\omega_i)$ on the NC. This procedure achieves the lowest possible value of the loop transmission frequency (phase-margin frequency $\omega\phi$). Note that $|L_o(j\omega_i)| \geq |B_o(j\omega_i)|$ represents the loop transfer function that satisfies the most severe boundaries B_R and B_D.
10. Based upon the information available from Steps 1 and 9, synthesize an $F(s)$ that results in a LmT_R (Equation 3.5) versus ω that lies between B_U and B_L of Figure 3.4 (see Appendix D).
11. Obtain the time-response data for $y(t)$: (a) with $d(t) = u_{-1}(t)$ and $r(t) = 0$ and (b) with $r(t) = u_{-1}(t)$ and $d(t) = 0$ for sufficient points around the parameter space describing the plant uncertainty (see Figure 3.6).

Simulation in time domain is always recommended to check the final fulfillment of the initial control specifications. Moreover, if the original specifications are given in time domain, although with the designed compensator the corresponding frequency requirements are accomplished, the final time responses could show some small differences from the initial specifications. This is to be expected since the translation into the frequency domain is good and usually enough of an approximation, but it is just a design approximation. In fact, it does not exist as a formal and exact translation between these two domains. Thus, if the designer finds such small differences, a second iteration could be useful to improve the controller design.

For the $L_o(j\omega)$ obtained, the plot of $Lm\{L(j\omega)/[1 + L(j\omega)]\}$ may be larger or smaller than LmT_{R_U} or LmT_{R_L} of Figure 3.4b, but $\delta_R(\omega_i)$ is satisfied. Because of the proper design of the input filter $F(s)$, $LmT_R = Lm\{FL/[1 + L]\}$ will lie within the bounds of LmT_{R_U} and LmT_{R_L}.

In problems with very large uncertainty and in disturbance rejection requiring a very large $|L(j\omega)|$ over a "large" BW,

$$\left|\frac{L}{1+L}\right| \approx 1 \tag{3.54}$$

and $T \approx F$. For these situations, design $F(s)$ in the same manner as for the tracking models of Section 3.4 and Ref. 218.

A situation may occur in which it may be impossible to satisfy all the desired performance specifications and a design trade-off decision needs to be made (see Appendices B and C). For example, the gain that is required to satisfy the dominating $B_D(j\omega_i)$ bounds, that is, $B_o = B_D$, may be too high resulting in saturation and/or sensor noise effects. The analysis of the equation

$$Y(j\omega) = \left[\frac{P(j\omega)}{1 + P(j\omega)G(j\omega)}\right] D_1(j\omega)$$

for the condition $|P(j\omega)G(j\omega)| \gg 1$, due to the gain in $G(j\omega)$, over the desired BW results in

$$Y(j\omega) \approx \frac{D_1(j\omega)}{G(j\omega)}$$

Thus, by ignoring the disturbance-rejection specification, synthesize the loop transmission function $L_o(s)$ by satisfying the bounds $B_o(j\omega_i) = B_R(j\omega_i)$. The possibility exists that the gain required in $G(s)$ in order to satisfy the tracking performance specifications and the overdesign characteristic of the QFT technique may result in satisfying the performance requirement:

$$|Y(j\omega)| = \left|\frac{D_1(j\omega)}{G(j\omega)}\right| \le \alpha_p$$

The simple plant illustrated by Equation 2.7 is used in the following sections to illustrate the details in applying this QFT design procedure. The CAD package *QFTCT* included with this book (see Appendix F) is used to obtain the data and to execute the design procedures.

3.16 Design Example 1

This design example is for the control system of Figure 3.1 with $r(t) = d_2(t) = u_{-1}(t)$ and $d_1(t) = 0$. The plant transfer function

$$P(s) = \frac{Ka}{s(s+a)} \quad 1 \le K \le 10 \quad 1 \le a \le 10 \tag{3.55}$$

for the nominal values $a = K = 1$.

Step 1: Modeling the tracking control ratio $T_R(s) = Y(s)/R(s)$:

$$T_{R_U}: M_p = 1.2,\ t_s = 2\,\text{s}$$

$$T_{R_L}: \text{Overdamped},\ t_s = 2\,\text{s}$$

1. A tracking model for the upper bound, based upon the given desired performance specifications, is tentatively identified by

$$T_{R_U}(s) = \frac{19.753}{s + 2 \pm j3.969} \tag{3.56}$$

   A zero is inserted in Equation 3.56, which does not affect the desired performance specifications but widens δ_{hf} between T_{R_U} and T_{R_L} in the high-frequency range. Thus

$$T_{R_U}(s) = \frac{0.6584(s+30)}{s + 2 \pm j3.969} \tag{3.57}$$

   The FOM for this transfer function with a step input are $M_p = 1.2078$, $t_R = 0.342\,\text{s}$, $t_p = 0.766\,\text{s}$, $t_s = 1.84\,\text{s}$, $Lm\ M_m = 1.95\,\text{dB}$, and $\omega_m = 3.4\,\text{rad/s}$.

2. A tracking model (see Section 3.4) for the lower bound, based upon the desired performance specifications, is tentatively identified by

$$T_{R_L}(s) = \frac{120}{(s+3)(s+4)(s+10)} \tag{3.58}$$

   and is modified with the addition of a pole to yield

$$T_{R_L}(s) = \frac{3520}{(s+4)(s+4)(s+4.4)(s+50)} \tag{3.59}$$

   The pole at $s = -50$ is inserted in Equation 3.59 to widen δ_{hf} further between T_{R_U} and T_{R_L} at high frequencies. The FOM for Equation 3.59 are as follows: $M_p = 1$, $t_R = 1.02\,\text{s}$, and $t_s = 1.844\,\text{s}$. In practice, these synthesized models should yield FOM within 1% of the specified values.

3. *Determination of* $\delta_R(\omega_i)$. From the data for the log magnitude plots of Equations 3.57 and 3.59, the values obtained for $\delta_R(\omega_i)$ are listed in Table 3.4.

TABLE 3.4

Magnitude Data for $\delta_R(\omega_i)$

ω	$\delta_R(\omega_i)$, dB
0.5	0.851
1	1.01
2	3.749
5	11.538
7	11.735
9	12.194
10	12.594
15	15.224
100	46.81

As seen from Table 3.4, $\delta_R(\omega_i)$ has the desired characteristic of an increasing value with increased frequency. Sufficient values of ω up to ω_h should be obtained in order to ensure that this desired characteristic has been achieved.

Step 2: Modeling the disturbance control ratio $T_{Du}(s) = Y(s)/D_2(s)$.
Since $y(0) = 1$, it is desired that $y(t)$ decays "as fast as possible," that is, $|y(t)| \le 0.01$ for $t \ge t_x = 60$ ms. Let the model disturbance control ratio be of the form

$$T_{Du} = \frac{Y(s)}{D_2(s)} = \frac{s(s+g)}{(s+g)^2 + h^2} \tag{3.60}$$

For $d_2(t) = u_{-1}(t)$ the output response has the form $y(t) = e^{-gt}\cos(ht)$. The desired specifications (see Figure 3.5) are satisfied by choosing

$$T_{Du}(s) = \frac{s(s+70)}{s+70 \pm j18} = \frac{1}{1+L_D} \tag{3.61}$$

For Equation 3.61, the output decays rapidly so that $|y(t)| \le 0.01$ at $t_x \approx 0.0565$ s.

Steps 3 and 4: Forming the templates of $P(j\omega_i)$.
Analysis of $P(s)$ is based upon the nominal values $K = a = 1$ (point A in Figure 3.24). The template for each frequency is based on the data listed in Table 3.5, which is computed using Equation 3.55. The values in Table 3.5 are used to draw the templates $\Im P(jw_i)$ shown in Figure 3.25.

Step 5: Determination of the *U*-contour.

1. Determination of V (see Equation 3.21) for the plant of Equation 3.55 yields

$$\lim_{s\to\infty}[P(s)] = \left[\frac{Ka}{s^2}\right] \quad \text{and} \quad \Delta Lm[P]_{max} = Lm[Ka]_{max} = 40 \text{ dB} \tag{3.62}$$

To determine the value of ω_h, where $\Delta LmP \approx V = 40$ dB and $\omega \approx 100$ rad/s, select various values of ω_i as shown in Table 3.6.

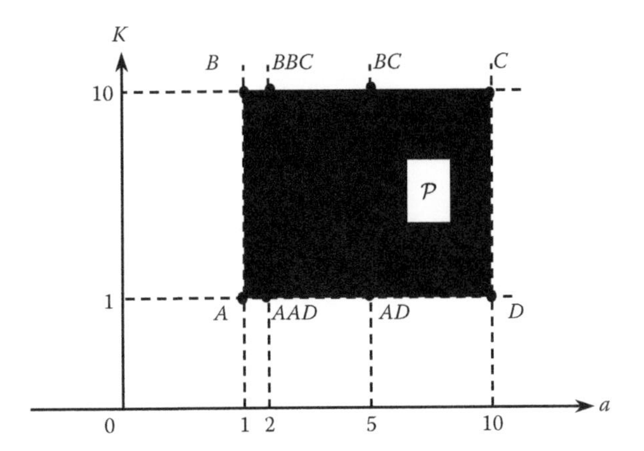

FIGURE 3.24
Values for the parameters with uncertainty.

TABLE 3.5
Data Points for $\Im P(jw_i)$

Point	ω_i	0.5	1	2	5	10	15
A	*LmP*	5	−3	−13	−28.1	−40	−47
	∠*P*	−116.6°	−135°	−153.5°	−168.7°	−174.3°	−176.2°
B	*LmP*	25	17	7	−8.1	−20	−27
	∠*P*	−116.6°	−135°	−153.5°	−168.7°	−174.3°	−176.2°
BBC	*LmP*	25.8	19	11	−2.6	−14.2	−21.1
	∠*P*	−104.1°	−116.6°	−135°	−158.2°	−168.7°	−171.4°
BC	*LmP*	26	19.8	13.3	3	−7	−13.5
	∠*P*	−95.7°	−101.3°	−111.8°	−135°	−153.5°	−161.6°
C	*LmP*	26	20	13.8	5	−3	−8.64
	∠*P*	−93°	−95.7°	−101.3°	−116.6°	−13.5°	−146.4°
D	*LmP*	6	−0.04	−6.2	−14.95	−23	−28.6
	∠*P*	−93°	−95.7°	−101.3°	−116.6°	−135°	−146.4°
AD	*LmP*	6	−0.17	6.67	−17	−27	−33.5
	∠*P*	−95.7°	−101.3°	−111.8°	−135°	−153.5°	−116.6°
ADD	*LmP*	5.65	0.97	−9	−22.6	−34.2	−41.1
	∠*P*	−104.1°	−116.6°	−135°	−158.2°	−168.7°	−172.4°

Note: Values of *Lm* are in dB.

From the data shown in Table 3.6, it is seen that the maximum value of $\Delta LmP(j\omega_i)$ occurs between points *A* and C. For other values of ω_i the maximum values of ΔLm $P(j\omega_i)$ are as shown in Table 3.7. Therefore it can be seen, for this example, that *V* is achieved essentially at $\omega_h \approx 40$.

2. *Determination of the* B_h *boundary*. Select the *M*-contour that represents the desired value of M_L for *T* (see Section 3.8). At various points around the lower half of this contour measure down 40 dB and draw the B_h boundary in the manner shown in Figure 3.8.

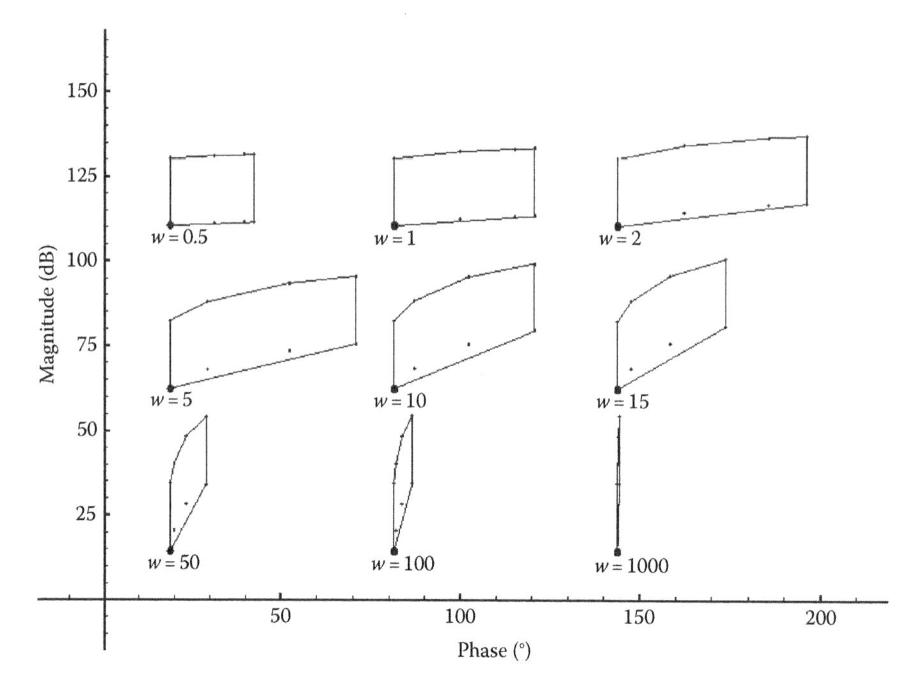

FIGURE 3.25
Construction of the plant templates.

TABLE 3.6

Data Points for $LmP(j\omega_i)$, $\omega \approx 100$ rad/s

Point	$LmP(j\omega_i)$
A	−80
B	−60
C	−40
D	−60

TABLE 3.7

Other Maximum Values of $\Delta LmP(j\omega_i)$

Frequency ω_i	max $\Delta LmP(j\omega_i)$, dB
15	38.4
16	38.6
20	39.0
40	39.65

Step 6: Determination of bounds $B_{D2}(j\omega_i)$ for

$$T_{D_U} = \frac{Y}{D_2} = \frac{1}{1+L} = \frac{\ell}{1+\ell} \tag{3.63}$$

where $\ell = 1/L = 1/PG$ and $Y = D_2 T_{D_U}$. For the disturbance-rejection case,

$$\Delta LmY = \Delta Lm\left[\frac{\ell}{1+\ell}\right] = \Delta LmT_{D_U} \tag{3.64}$$

Note that there are no lower bounds on the response $y(t)$. Thus, choose $L \geq B_{D2}(j\omega_i)$ to satisfy

$$Lm\left[\frac{Y}{D_2}\right] = Lm\left[\frac{\ell}{1+\ell}\right] \leq Lm[T_{D_U}] \tag{3.65}$$

for each frequency ω_i of $T_D(j\omega_i)$ over the frequency range $0 \leq \omega \leq 15$ rad/s. The data in Table 3.8 are obtained from Equation 3.61. Point A of the templates of Step 3 yields the maximum value for $\ell = 1/L$. For $0 \leq \omega \leq 15$ the contours of $LmT_D(j\omega_i)$ (see Figure 3.26) become the contours for B_D, that is, $LmT_D = LmB_D$. For $\omega \geq 20$, the values of LmL_D in Table 3.8, obtained by using Equation 3.61, are points of $B_D(j\omega_i)$ on and in the vicinity of the U-contour. For this example, as shown in the next step, the B_D bounds completely dominate over the B_R bounds. Thus, $B_D = B_R$ as shown in Figure 3.26.

Step 7: Determination of the bounds $B_R(j\omega_i)$.
By using the templates, the values of $\delta_R(\omega_i)$ given in Step 2, the M_L-contour, and the B_h boundary, the bounds $B_R(j\omega_i)$ and the U-contour are determined and are drawn in Figure 3.27. From this figure it can be seen that $\gamma \approx 58°$.

TABLE 3.8

Frequency Data for Equation 3.61

ω	LmT_D, dB	$\angle T_D$,°	LmL_D
0.5	−43.5	89.65	43.5
1	−37.5	89.3	37.5
2	−31.44	88.6	31.4
5	−23.5	86.4	23.5
10	−17.52	82.85	17.5
15	−14.1	79.31	14.1
20	−11.67	75.81	11.4
50	−4.84	56.8	4.46
100	−1.67	36.17	−2.91
200	−0.453	19.55	−9.05
500	−0.0737	7.99	−17.1
1000	−0.01845	4.007	−21.6
2000	−0.0029954	2.005	−31.0
4000	−0.001154	1.003	−35.2

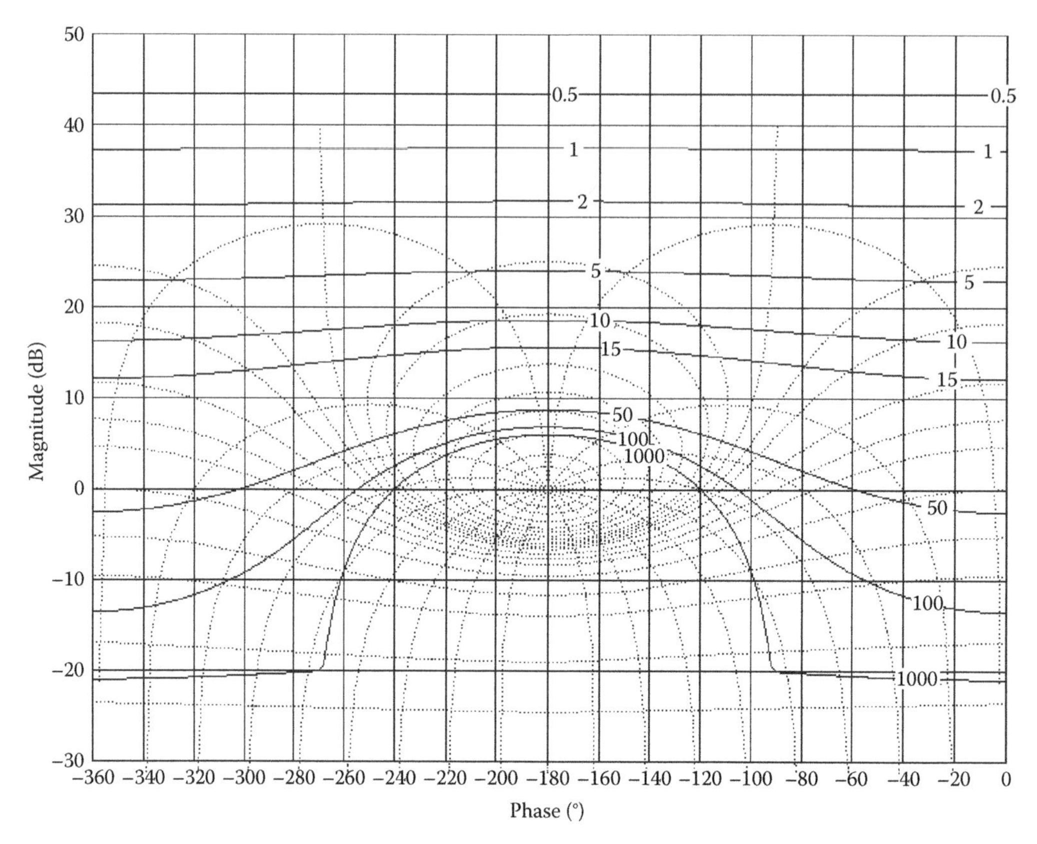

FIGURE 3.26
Construction of the $B_D(j\omega)$ bounds.

Step 8: Determination of the composite bounds $B_o(j\omega_i)$.
An analysis of Figures 3.26 and 3.27 reveals that the bounds $B_D(j\omega_i)$ for this example all lie above the tracking contours $B_R(j\omega_i)$. Thus, the B_D contours of Figure 3.26 are more severe and become the $B_o(j\omega_i)$ contours for the overall system of Figure 3.28.

Step 9: Synthesizing or shaping of $L_o(s)$.

1. *Minimum structure.* Since $P(s)$ is Type 1, $L_0(s)$ must be at least a Type 1 in order to maintain the Type 1 tracking characteristic for $L/(1 + L)$. Thus, the initial $L_o(s)$ has the form

$$L_{o0}(s) = \frac{K_{o1}}{s^m} \tag{3.66}$$

where $m = 1$ in this example. (*Note*: In general, select $L_{o0}(s) = P(s)$, so that the desired system type is achieved.) For the case where $T_D = P/(1 + L)$, it is desired that T_D have a zero at the origin so that $y(\infty) = 0$ for $d_1(t) = u_{-1}(t)$. Therefore $G(s)$ must have a pole at the origin. This requirement may place a severe restriction on trying to synthesize an L_o.

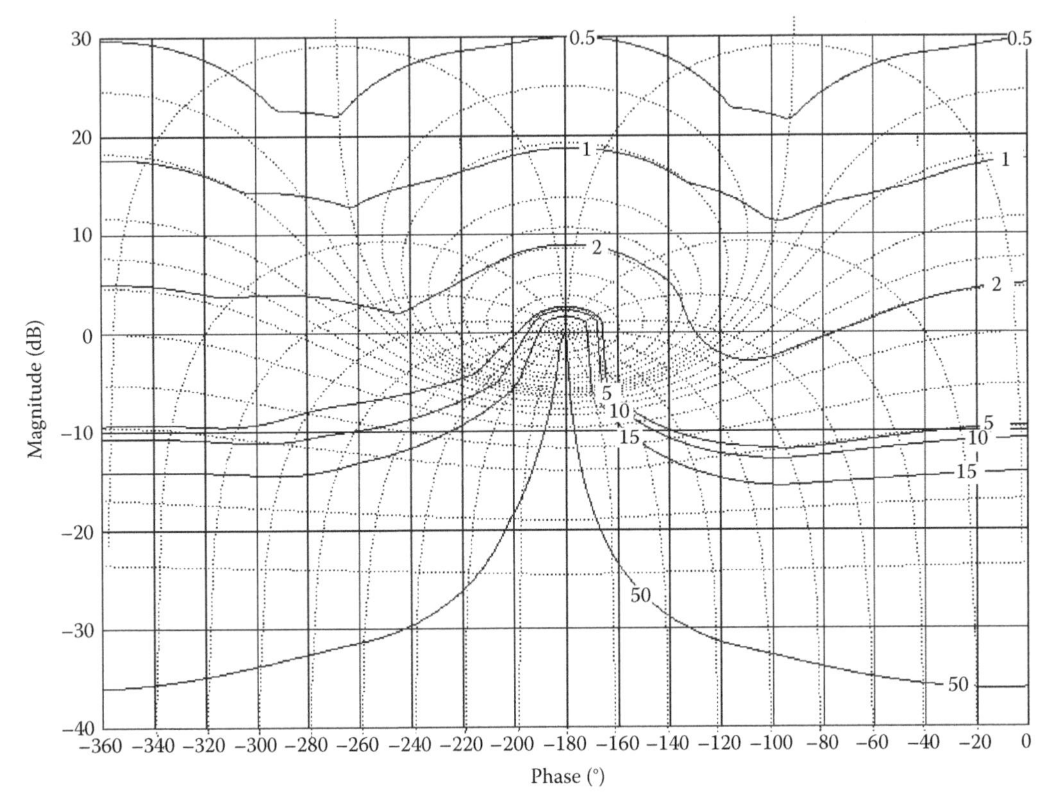

FIGURE 3.27
Construction of the $B_R(j\omega_i)$ contours.

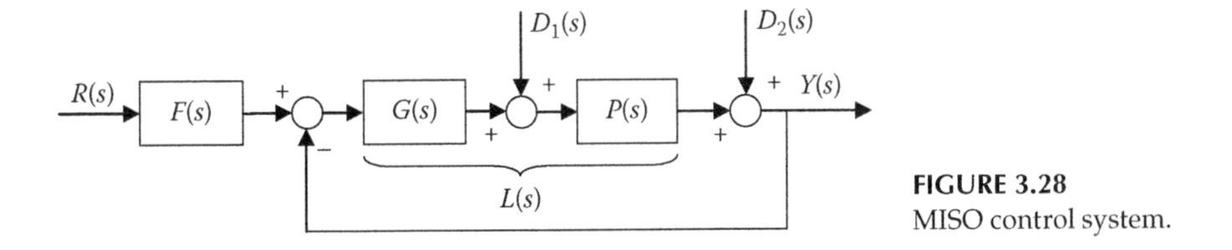

FIGURE 3.28
MISO control system.

2. *Synthesis of* $L_o(s)$. Using four-cycle semi-log graph paper and the data in Table 3.9, construct or shape $L_o(s)$ so that it comes as close as possible to the *U*-contour. Then $L_o(j\omega_i)$ must be as close as possible to $B_o(j\omega_i)$ but never below it. For the approximate range $\omega_x \approx 15 \leq \omega_i \leq \omega_y$, the angle $\angle L_o(j\omega_i) \geq -123°$ *must* be satisfied. For this example the frequency ω_y is the value of $L_o(j\omega_y)$, which results in $LmL_o(j\omega_i) \approx -36$ dB. These restrictions can be satisfied by assuming the format of the optimal transfer function as

$$L_o(j\omega) = L_{o0}(j\omega)\left[\frac{(j\omega - z_1)(j\omega - z_2)\cdots}{(j\omega - p_3)(j\omega - p_4)\cdots\left[\left(1-\left(\omega^2/\omega_n^2\right)\right)+j2\varsigma(\omega/\omega_n)\right]}\right] \tag{3.67}$$

TABLE 3.9

Data for Loop Shaping

		Angle Contribution of Poles (p) and Zero(s) (z) in Degrees				
ω	$\angle L_{o1}$	z_1	P_2	z_2	$p_{3,4}$	$-L_o$
1	−180°	32.0	−2.20	1.27	0	−149.0°
5	−180°	72.26	−10.89	6.34	−0.40	−112.7°
10	−180°	80.90	−21.04	12.53	−0.80	−108.4°
15	−180°	83.91	−29.98	18.43	−1.20	−108.8°
50	−180°	88.17	−62.53	48.01	−4.01	−110.4°
100	−180°	89.08	−75.43	65.77	−16.26	−108.6°
200	−180°	89.54	−82.59	77.32	−90.0	−111.99°
1000	−180°	89.90	−88.51	87.42		−181.2°
2000	−180°	89.95	−89.26	88.71	−136.98	227.6°

which has w real zeros, $n = w + \lambda - 2$ real poles, and a pair of complex-conjugate poles. For this example a table having a similar format shown in Table 3.9 is used to assist in obtaining the desired $L_o(s)$. The resulting loop transfer function is

$$L_o(s) = \frac{4.11\times 10^{10}(s+0.6)^2(s+176)}{s^2(s+1)(s+200)(s+14,700\pm j15,000)} \tag{3.68}$$

where $\zeta = 0.7$, $\omega_n = 1000\,\text{rad/s}$, and $\omega_\phi = 200\,\text{rad/s}$. $L_o(j\omega)$ is drawn on the Bode plot of Figure 3.29. It is also plotted on the NC of Figure 3.33. Since L_o is Type 2, $T_D = 1/(1 + L)$ has two zeros at the origin.

3. *Determination of* $G(s)$. Use $L_o(s)$ and $P_o(s)$, where $P_o(s) = 1/s(s + 1)$ is the nominal plant, to obtain the transfer function $G(s) = L_o(s)/P_o(s)$.

Step 10: The input filter $F(s)$ is synthesized to yield the desired tracking of the input by the output $y(t)$ in the manner described in Section 3.14. The curves of B_U, B_L, LmT_{min}, and LmT_{max} are plotted in Figure 3.30. Since, for this example $|L_o(j\omega)| \gg 1$ over the desired BW, $F(s) = T_R(s)$. Thus, the following $F(s)$ is synthesized to lie between B_U and B_L as shown in Figure 3.30:

$$F(s) = \frac{18.85}{(s+2.9)(s+6.5)} \tag{3.69}$$

The output responses for this example with a unit step input are all identical, as shown in Figure 3.31 for the entire region of plant parameter uncertainty where $M_p = 1.002$ and $t_s = 1.606\,\text{s}$.

Step 11: Time responses for a disturbance input is $d(t) = u_{-1}(t)$. For each point of Figure 3.24, substitute the corresponding data into

$$Y_D(s) = \frac{1}{1+P(s)G(s)}D(s) \tag{3.70}$$

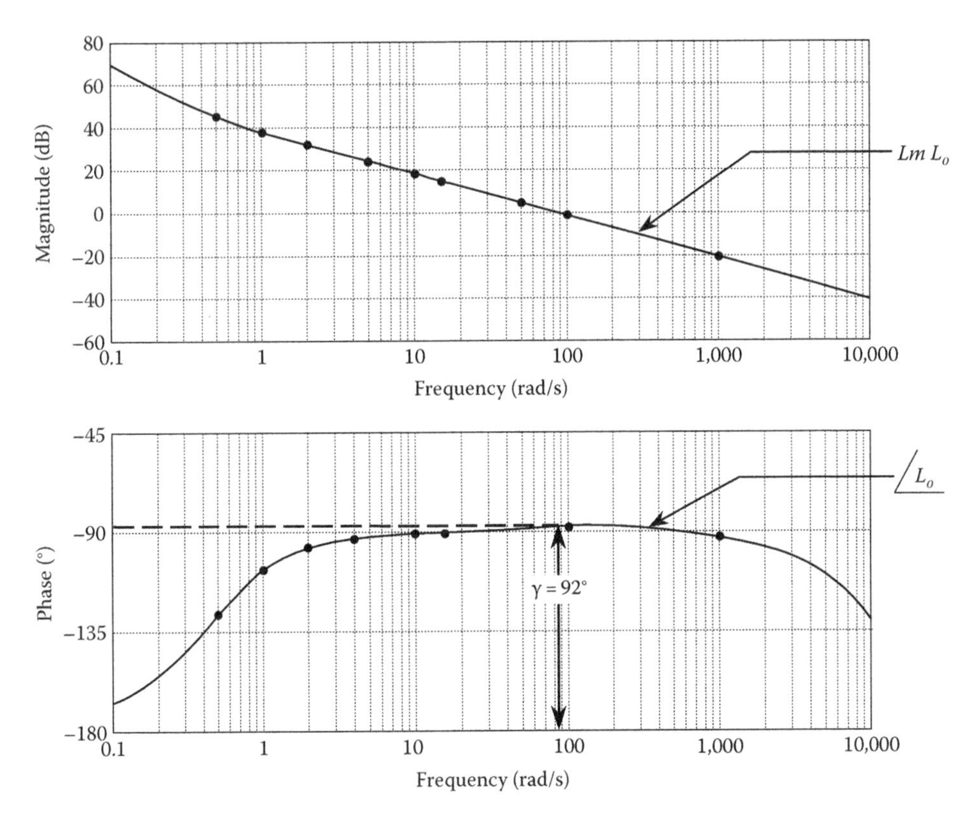

FIGURE 3.29
Bode plot of $L_o(j\omega)$ of Equation 3.68.

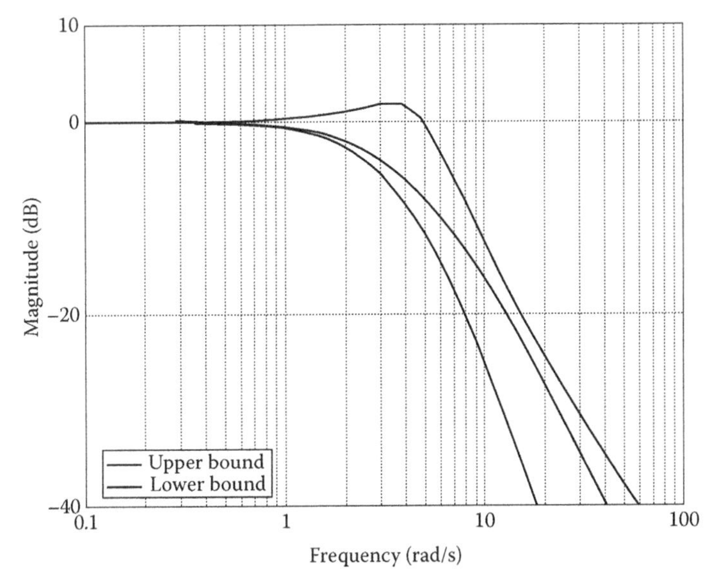

FIGURE 3.30
Requirements on $F(s)$ for Design Example 1.

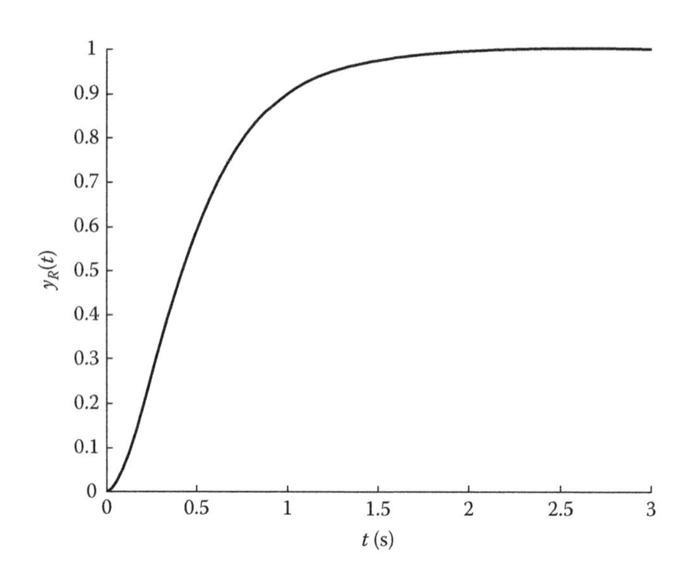

FIGURE 3.31
Time responses to unit step input for all J plants.

TABLE 3.10
Time-Response Characteristics for Equation 3.70

Point	$y_D(t)$	Specified t_x	Actual t_x for $\lvert y_D(t) \rvert = 0.01$	$y(\infty)$
A	0.00998	0.06	0.058	0
B	0.0090	0.06	0.0074	0
C	0.00004	0.06	0.00226	0
D	0.000399	0.06	0.007	0

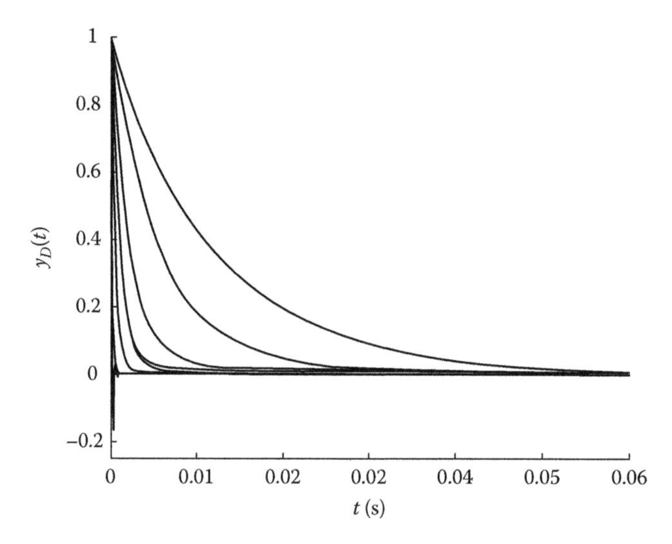

FIGURE 3.32
Disturbance-response plots for all J plants.

and determine $y_D(t)$. As indicated in Table 3.10, the disturbance time-response characteristics for points A, B, C, and D satisfy the specification that $|y(t_x)| \leq 0.01$ for $t_x \geq 60$ ms. The disturbance-response plots, for the eight cases, are shown in Figure 3.32. As shown in the figure, although the initial peak overshoots exceed the magnitude of 0.01, the value of $|y_D(t)| \leq 0.01$ for $t_x \geq 60$ ms is achieved.

3.17 Design Example 2

The plant and tracking specifications of Design Example 1 are used for this example, with $r(t) = d_1(t) = u_{-1}(t)$, $d_2(t) = 0$, and $\alpha_p = 0.1$. Steps 1, 3–5, and 7 are the same as for example 1. For Step 2, $M_p \approx M_m = 0.1$; thus $LmT_{D_1} = LmM_{mD} = -20$ dB. A QFT CAD is used to obtain Figure 3.33, which shows the U-contour, B_o bounds, and $LmL_o(j\omega)$ given by Equation 3.71. The synthesized $L_o(s)$ is

$$L_o(s) = \frac{9.93\times10^6(s+1.3)(s+1.6)(s+45)}{s^2(s+1)(s+26)(s+900\pm j975)} \tag{3.71}$$

Equations 3.57 and 3.59 and Figure 3.33 are used to obtain the data for plotting B_U, B_L, LmT_{min}, and LmT_{max} in Figure 3.34. Since the tracking specifications and models are identical for both examples, the prefilter for this example is the same as given by Equation 3.69 and is plotted in Figure 3.34. The simulation results are shown in Figures 3.35 and 3.36. These figures reveal that all disturbance responses are below 0.03 and thus satisfy the specification $|y_D(t)_{max}| \leq 0.1$.

Also, all the tracking responses lie in the range of $1 < M_p < 1.022$ and $t_s \leq 1.62$ s. Thus, the desired performance specifications have all been met and a robust design has been achieved.

3.18 Template Generation for Unstable Plants

In the generation of templates for unstable plants, proper care must be exercised in analyzing the angular variation of $P(j\omega)$ as the frequency is varied from zero to infinity. As an example, consider the plant

$$P(j\omega) = \frac{K(j\omega - z_1)}{j\omega(j\omega - p_1)(j\omega - p_2)(j\omega - p_3)} \tag{3.72}$$

where p_1 and p_2 are a complex-conjugate pair and

$$\angle \lim_{\omega\to\infty} P(j\omega) = \frac{\angle K}{(j\omega)^3} = -270^\circ \tag{3.73}$$

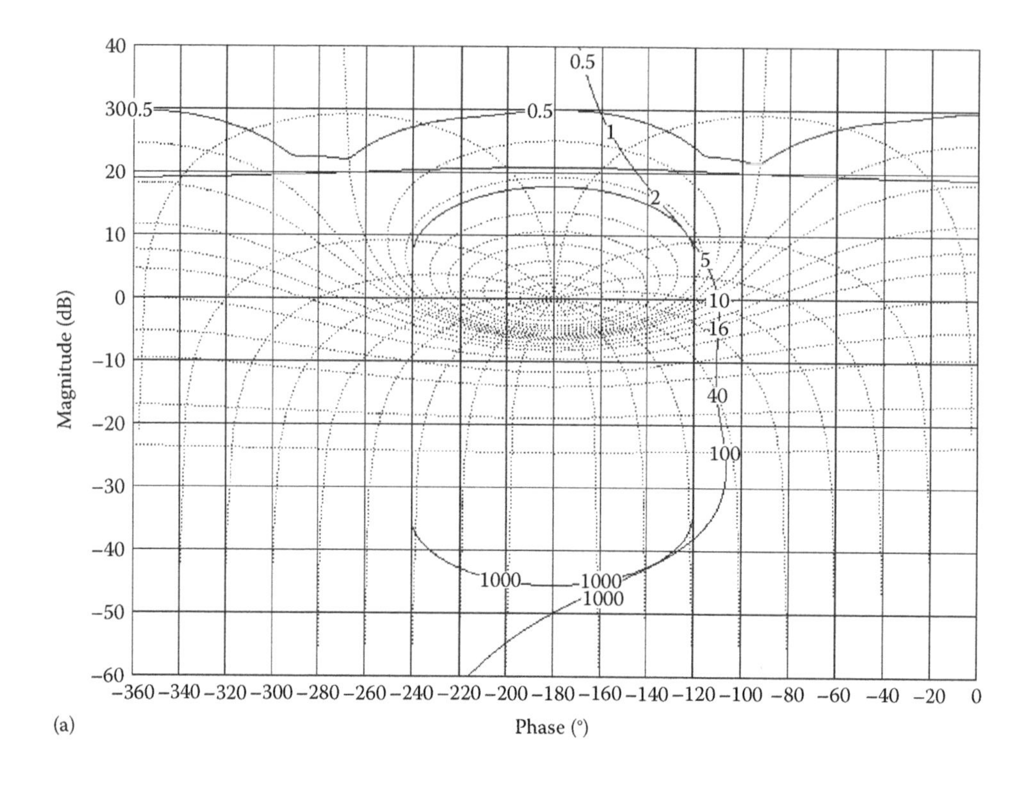

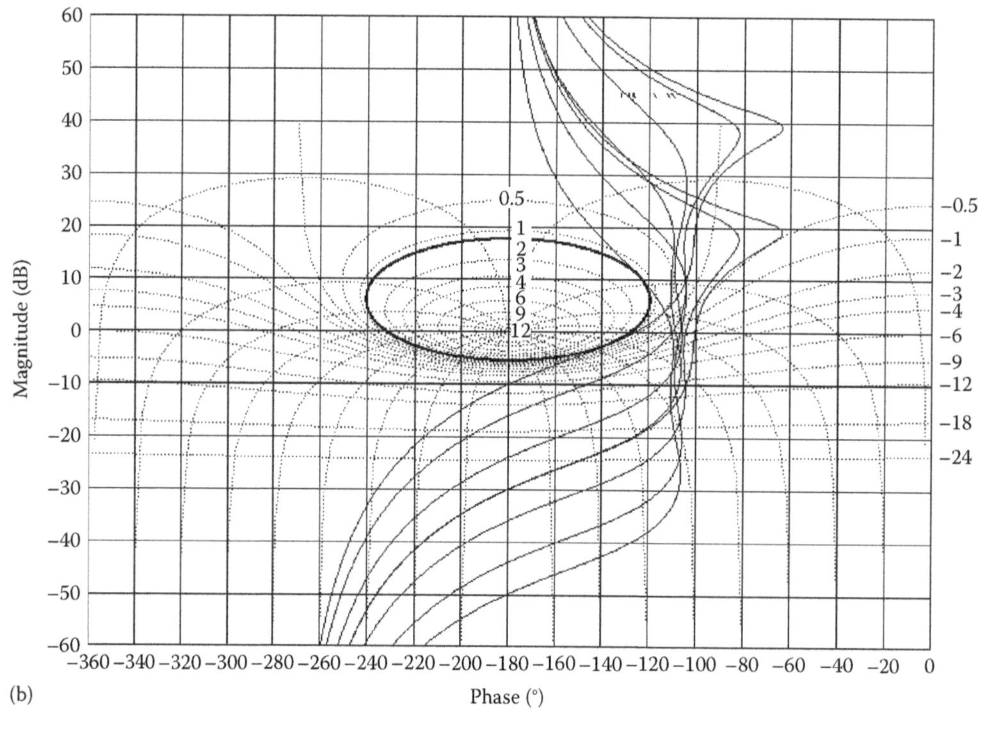

FIGURE 3.33
(a) *U*-contour, B_o bounds, and $Lm\ L_o(j\omega_i)$ for obtaining $L_o(s)$. (b) All of the $L_i(j\omega)$ plotted on one Nichols plot along with the stability margin.

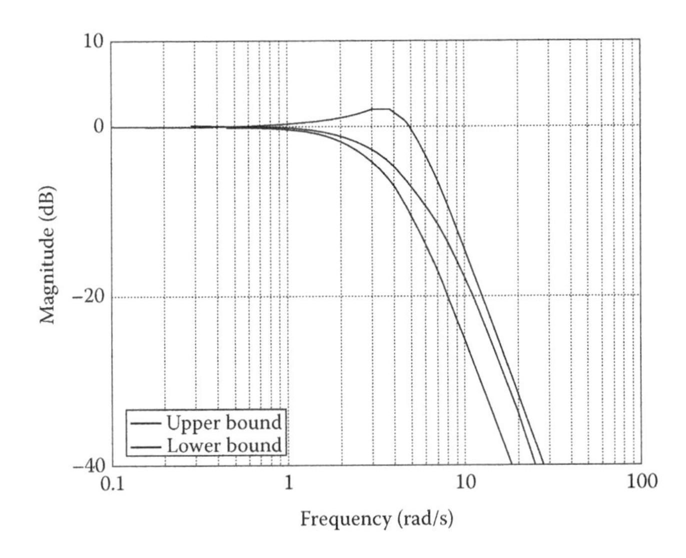

FIGURE 3.34
Requirements and resulting prefilter for design examples.

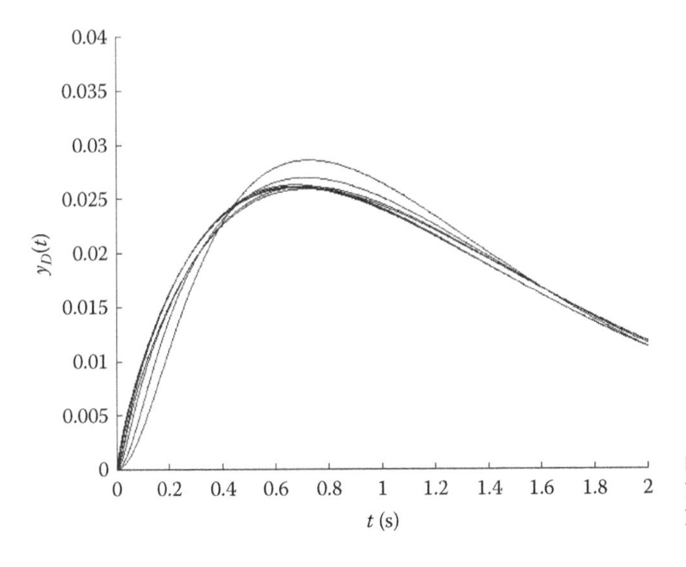

FIGURE 3.35
Disturbance responses for Design Example 2.

In Equation 3.73, the convention used is that, in the limit as $\omega \rightarrow \infty$, each $j\omega$ term contributes an angle of +90°. Two cases are analyzed for Equation 3.72 as follows:

Case 1. The poles and zeros of Equation 3.72 for a stable plant are plotted in Figure 3.37a. In Equation 3.72 the angle of $(j\omega_i - p_1)$ is negative clockwise for $\omega_i < \omega_{d1}$ and is positive counterclockwise (ccw) for $\omega_i > \omega_{d1}$. All other angles are positive (ccw) for all $0 \leq \omega \leq \infty$ as illustrated in Figure 3.37a, and in Table 3.11.

Case 2. The poles and zeros of Equation 3.72 for an unstable plant are plotted in Figure 3.37b. For Equation 3.72, all angles of first-order factors $(j\omega_i - p_i)$ are positive (ccw), as illustrated in Figure 3.37b and in Table 3.11.

Note that the difference in the angle of $P(j0)$ between the two cases is 360° and is necessary to account for the RHP poles. Thus, because all angles are measured ccw, the angle of $P(j\omega)$ varies continuously in a given direction as the frequency is varied between zero

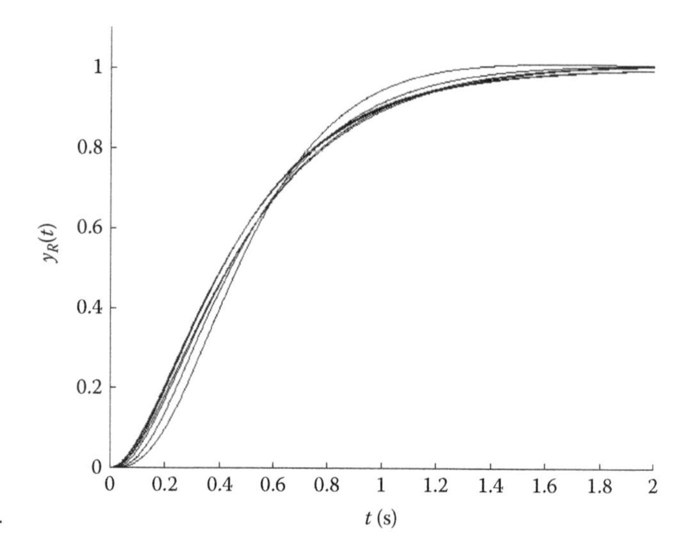

FIGURE 3.36
Tracking responses for Design Example 2.

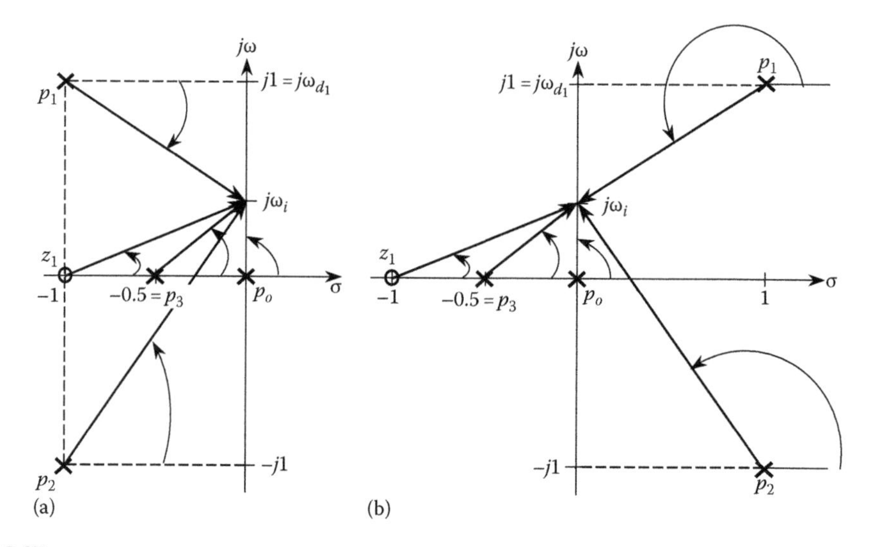

FIGURE 3.37
Poles and zeros in the s-plane for (a) a stable plant and (b) an unstable plant.

TABLE 3.11

Angular Variations of the Minimum-Phase Stable and Unstable Plants $P(j\omega)$ (See Equation 3.70) Shown in Figure 3.37

Case	ω_i	Angle in Degrees					
		$\angle p_0$	$\angle p_1$	$\angle p_2$	$\angle p_3$	$\angle z_1$	$\angle P(j\omega_i)$
1	0	+90	−45	+45	0	0	−90
	1	+90	0	+63.5	+63.5	+45	−172
	∞	+90	+90	+90	+90	+90	−270
2	0	+90	+225	+135	0	0	−450
	1	+90	+180	+116.5	+63.5	+45	−405
	∞	+90	+90	+90	+90	+90	−270

and infinity. This feature is very important in obtaining a template where P contains both stable and unstable plants. The following guidelines should be used in the angular determination of $\angle P(j\omega_i)$:

1. For stable plants, all angular directions are taken so that their values always lie within the range of −90° to +90°.
2. For plants with RHP poles and zeros: (a) the angular directions as for the stable plant left-hand-plane (LHP) poles and zeros of $P(s)$ are taken in the same manner; (b) the angular directions of all RHP poles and zeros are all taken ccw.

For complement information about the loop-shaping of unstable systems see also Appendix D.

3.19 Summary

A general introduction to the MISO QFT technique was presented in this chapter. It is a frequency-response design method applied to the design of a MISO control system with an uncertain MISO plant P. As shown in Chapters 5 and 6, a MIMO system can be represented by an equivalent set of MISO systems. Four QFT MIMO system design methods are available in which the equivalent MISO loops are designed according to the MISO design method presented in this chapter. This design is based upon the following:

1. Specifying the tolerance in the ω-domain by means of the sets of plant transfer functions and closed-loop control ratios, $\mathcal{P}(jw) = \{P(jw)\}$ and $\mathcal{T}(j\omega) = \{T(j\omega)\}$, respectively
2. Finding the resulting bounds on the loop transfer functions L_i and input filter transfer functions F of Figure 3.27

The robust design technique presented in this chapter permits the design of an analog control system that satisfies the desired performance specifications within the specified range of structured plant parameter variation. Since the QFT technique is based upon the design of the loop transmission function $L(s)$ of the MISO control system of Figure 3.1, it is referred to as a MISO design technique. Although this chapter dealt with *m.p.* plants, the QFT technique can be applied to *n.m.p.* plants and to digital control systems (see Chapter 4 and Refs. 116 through 121). The CAD Package, *QFTCT*, can be used in all the steps of the methodology. See Appendices F and G and Refs. 149 and 150, and visit the websites http://cesc.case.edu or http://www.crcpress.com/product/isbn/9781439821794.

4

Discrete Quantitative Feedback Technique

4.1 Introduction

A digital control system (see Figure 4.1) utilizes a digital computer or a microprocessor (μP) to compute the control law or algorithm, an analog-to-digital converter (A/D C) to sample the analog sensor signals and convert them into digits or numbers for the algorithm, and a digital-to-analog converter (D/A C) to convert the result of the control algorithm (numbers) into an analog control signal for the actuators.[104,206] This chapter focuses on the application of the quantitative feedback theory (QFT) technique to multiple-input-single-output (MISO)-sampled data (digital) control systems.[104]

The QFT sampled-data (S-D) system design process is tuned to the bounds of uncertainty, the performance tolerances, and the sampling time T (or sampling frequency $\omega_s = 2\pi/T$). The QFT technique requires, as discussed in Section 4.5, the determination of the minimum sampling frequency $(\omega_s)_{min}$ bandwidth (BW) that is needed for a satisfactory design. The larger the plant uncertainty and the narrower the system performance tolerances, the larger must be the value of $(\omega_s)_{min}$. The use of the z- to the w'-domain bilinear transformation[206] permits the analysis and design of S-D systems by the use of the *digitization (DIG) technique*.[206] That is, the w'-plane-detailed QFT design procedure essentially parallels very closely that for continuous-time systems of Chapter 3, the difference being that the design must take into account the right-half-plane (RHP) zero(s) that results in the w' plant transfer function due to the bilinear transformation. Note that when a plant $P(s)$ is minimum phase (*m.p.*), it becomes a non-minimum-phase (*n.m.p.*) plant when transformed into the w'-domain. Thus, proper care must be exercised in satisfying the stability bounds prescribed for the QFT design.

The *pseudo-continuous*-time (PCT) approach, which is discussed in Section 4.8, is another DIG technique that allows the QFT design of the $D(z)$ controller to be done in the s-domain.

Once the s-domain controller has been synthesized, in the manner described in Chapter 3, it is transformed into the z-domain by using the *Tustin transformation*, a bilinear transformation, to obtain $D(z)$. The advantage of this approach, when the plant is *m.p.*, is that it eliminates dealing with an *n.m.p.* plant and the problem associated with satisfying the stability bounds. Thus, the transformation either into the w'- or s-domain of the S-D MISO or MIMO control system enables the use of the MISO QFT analog design technique to be readily used, with minor exceptions, to perform the QFT design for the controller $D(w')$ or $D(s)$. If the w'- or s-domain simulations satisfy the desired performance specifications, then by using the bilinear transformation the z-domain controller $G(z)$ is obtained. With this z-domain controller, a discrete time-domain simulation is obtained to verify the goodness of the design. For complement information about digital QFT see Refs. 104 through 107 and 206.

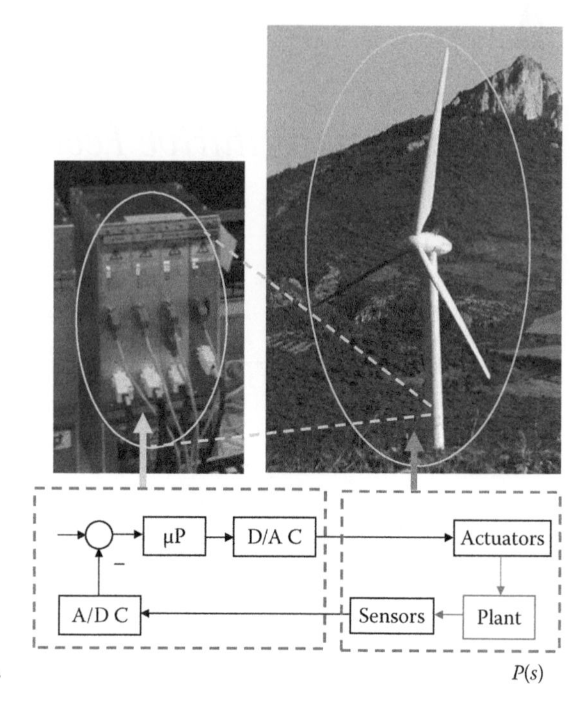

FIGURE 4.1
A digital control system of a wind turbine.

4.2 Bilinear Transformations

Each of the two DIG techniques, w'- or s-domain design, requires the use of a bilinear transformation.[206] This section presents the z- to the w'-domain to the z-domain and the s- to the z-domain transformations.

4.2.1 w- and w'-Domain Transformations

The QFT design of the S-D control system requires the use of the bilinear transformations

$$z = \frac{w+1}{-w+1} \tag{4.1}$$

$$w = \frac{z-1}{z+1} \tag{4.2}$$

where $z = e^{sT}$. Note that the w-domain lacks the desirable property that as the sampling time T approaches zero (continuous case), w should approach s; that is, using the Taylor polynomial expansion,

$$w\big|_{T\to 0} = \lim_{T\to 0}\left[\frac{z-1}{z+1} = \frac{e^{sT}-1}{e^{sT}+1} = \frac{sT+(sT)^2/2!+\cdots}{2+sT+(sT)^2/2!+\cdots}\right] = 0 \tag{4.3}$$

and where

$$z\,|_{T\to 0} = \lim_{T\to 0}[z = e^{sT}] = 1$$

This situation is overcome by defining

$$w' \equiv \frac{2}{T}w \equiv \frac{2}{T}\frac{sT + (sT)^2/2! + \cdots}{2 + sT + (sT)^2/2! + \cdots} \tag{4.4}$$

Thus, in the w'-plane the desirable property that $w' \to s$ as $T \to 0$ is achieved. This w'-plane property establishes the conceptual basis for defining a quantity in the w'-domain, which is analogous to a quantity in the s-domain. Substituting $w = Tw'/2$ into Equations 4.1 and 4.2 yields, respectively, the z- to w'-plane and the w'- to z-plane transformations as follows:

$$w' = \frac{2}{T}\frac{z-1}{z+1} \tag{4.5}$$

$$z = \frac{Tw'+2}{-Tw'+2} \tag{4.6}$$

Equation 4.6 represents an approximation of $z = e^{sT}$. The mapping of the z-plane via Equation 4.5 into the w'-plane is shown in Figure 4.2. From now on, *the w' transformation is used throughout the chapter and the prime designator is omitted* (i.e., $w' = w$) except where noted in Table 4.1.

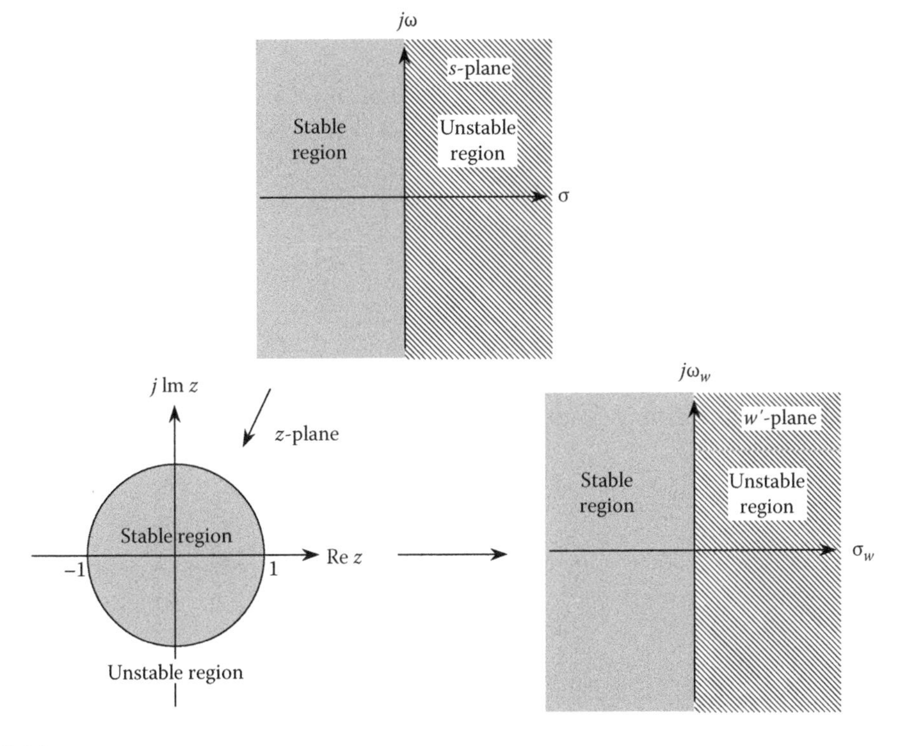

FIGURE 4.2
Mapping of the s-plane into the z-plane by means of $z = e^{sT}$, and of the z-plane into the w-plane by means of Equation 4.5.

TABLE 4.1

w- and w'-Domain Values of v for ω_s = 120 and 240 rad/s

	w-Plane v Values		w'-Plane v Values		
s-Plane ω Values	$\omega_s = 120$	$\omega_s = 240$	$\omega_s = 120$	$\omega_s = 240$	Valid Approximation Range
0.5	0.013	0.00655	0.5000	0.5000	↓
1.0	0.02618	0.01309	1.0002	1.0000	↓
2.0	0.0524	0.02619	2.0018	2.0004	↓
5.0	0.13165	0.06554	5.0287	5.0071	↓
10.0	0.26795	0.13165	10.235	10.057	↓
20.0	0.57735	0.26795	22.053	20.47	—
30.0	1.0	0.41420	38.2	31.63	—
40.0	1.7320	0.57735	66.15	44.11	—
50.0	3.73205	0.76733	142.55	58.619	—
60.0	—	1.0	—	76.394	—
70.0	—	1.3032	—	99.559	—
100.0	—	3.732	—	285.108	
120.0	—	—	—	—	

4.2.2 s-Plane and w-Plane Relationship

The relationship between the s- and w-plane can be found by examination of Equation 4.6 in terms of $z = e^{sT}$, that is,

$$w = u + jv = \sigma_{wp} + j\omega_{wp}$$

$$= \frac{2}{T}\left[\frac{e^{ST}-1}{e^{ST}+1}\right]\frac{e^{-ST/2}}{e^{-ST/2}} = \frac{2}{T}\left[\frac{e^{ST/2}-e^{-ST/2}}{e^{ST/2}+e^{-ST/2}}\right] \tag{4.7}$$

which yields

$$w = \frac{2\tanh(sT/2)}{T} \tag{4.8}$$

For the s-plane imaginary axis, substitute $s = j\omega_{sp}$ into Equation 4.8 to obtain

$$w = \sigma_{wp} + j\omega_{wp} = \frac{2\tanh(j\omega_{sp}T/2)}{T} = \frac{j2\tan(\omega_{sp}T/2)}{T} \tag{4.9}$$

or

$$v = \omega_{wp} = \frac{2\tan(\omega_{sp}T/2)}{T} \tag{4.10}$$

Thus, the imaginary axis in the primary strip of the s-plane is mapped onto the entire imaginary axis of the w-plane (see Figure 4.2). If $\omega_{sp}T/2$ is small ($\omega_{sp}T/2 \le 0.297$), then from

Equation 4.10 we obtain $v = \omega_{wp} \approx \omega_{sp}$. For the real axis, substitute $s = \sigma_{sp}$ into Equation 4.8 to obtain

$$w = \frac{2\tanh(\sigma_{sp}T/2)}{T} = \sigma_{wp} = u \tag{4.11}$$

By letting $\alpha = \sigma_{sp}T/2$, $\tanh \alpha$ can be expressed in the expanded form as

$$\tanh\alpha = \frac{\alpha + \alpha^2/3! + \cdots}{1 + \alpha^2/2! + \cdots} \tag{4.12}$$

For $\alpha^2 \ll 2$, from Equations 4.11 and 4.12,

$$\sigma_{sp} \approx \sigma_{wp} \tag{4.13}$$

Thus, when the approximations are valid,

$$w = u + jv = \sigma_{wp} + j\omega_{wp} \approx \sigma_{sp} + j\omega_{sp} = s \tag{4.14}$$

If the approximations are not valid, then Equations 4.10 and 4.11 must be used to locate the s-plane poles and zeros properly in the w-plane. The mapping of the s-plane poles and zeros into the w-plane by using these equations is referred to as *pre-warping of the s-plane poles and zeros.* The relationships of Equation 4.14 are the basis of a QFT design method (DIG technique) in the w-plane. Table 4.1 illustrates the frequency range for which the degree of accuracy of the relationships of Equation 4.14 is very good and how it is dependent upon the sampling time T.

A characteristic of a bilinear transformation is that, in general, it transforms an unequal-order z-domain transfer function, one whose order of the numerator on_z and order of its denominator od_z are unequal, into one for which the order of the numerator (on_w) is equal to the order of its denominator (od_w). That is, $on_z \neq od_z$ in the z-domain and $on_w = od_w$ in the w-domain. Further, note that in transforming $G(z) = K_xG'(z)$ to $G(w) = K_wG'(w)$ by means of Equation 4.5, the value of the gain constant K_w may be positive or negative.[206] The sign of K_w is determined by the coefficients of $G(z)$, which in turn are functions of T. This characteristic must be kept in mind when synthesizing $G(w)$ and $F(w)$.

4.2.3 *s*- to *z*-Plane Transformation: Tustin Transformation

One of the most popular s- to z-plane transformations is the *Tustin algorithm.*[201,296] The function of z that is substituted for s^q in implementing the Tustin transformation is

$$s^q = \left(\frac{2}{T}\frac{1 - z^{-1}}{1 + z^{-1}}\right)^q \tag{4.15}$$

where q is an integer number. An advantage of the Tustin algorithm is that it is comparatively easy to implement. Also, the accuracy of the response of the Tustin z-domain transfer function is good compared with the response of the exact z-domain transfer function; that is, the accuracy increases as the frequency increases. The Tustin transformation for $q = 1$ is defined as

$$s \equiv \frac{2}{T}\frac{1-z^{-1}}{1+z^{-1}} = \frac{2}{T}\frac{z-1}{z+1} \tag{4.16}$$

which is a bilinear transformation and can be equated to the trapezoidal integration (s^{-1}) method. Note, for $q = 1$, Equation 4.16 is identical to the w- to z-domain transformation of Equation 4.5. Thus, both equations are from now on referred to as *Tustin transformation*. Also, Equation 4.16 can be derived by approximating $z = e^{sT}$ as a finite series. The operator equation (4.16) is as applicable to matrix equations as it is to scalar differential equations. Therefore, the following discussion is useful in understanding the mapping result for the vector model.

To represent functionally the s- to z-plane mapping, Equation 4.16 is rearranged to yield

$$z = \frac{1+sT/2}{1-sT/2} \tag{4.17}$$

Utilizing the mathematical expression

$$\frac{1+a}{1-a} = e^{j2\tan^{-1}a}$$

and letting $s = j\hat{\omega}_{sp}$ in Equation 4.17, the following expression is obtained:[206]

$$z = \frac{1+j\hat{\omega}_{sp}T/2}{1-j\hat{\omega}_{sp}T/2} = e^{j2\tan^{-1}\hat{\omega}_{sp}T/2} \tag{4.18}$$

The exact $\mathcal{Z}$-transform yields $z = e^{j\omega_{sp}T}$, where ω_{sp} is an equivalent s-plane frequency. Thus, the following equation is obtained from Equation 4.18:

$$e^{j\omega_{sp}T} = e^{j2\tan^{-1}\left(\frac{\hat{\omega}_{sp}T}{2}\right)} \tag{4.19}$$

Equating the exponents yields

$$\frac{\omega_{sp}T}{2} = \tan^{-1}\frac{\hat{\omega}_{sp}T}{2} \tag{4.20}$$

or

$$\tan\frac{\omega_{sp}T}{2} = \frac{\hat{\omega}_{sp}T}{2} \tag{4.21}$$

When $\omega_{sp}T/2 < 17°$, or ≈ 0.30 rad, then

$$\omega_{sp} \approx \hat{\omega}_{sp} \tag{4.22}$$

which means that in the frequency domain the Tustin approximation is good for small values of $\omega_{sp}T/2$.

Returning to Equation 4.19, it is easy to realize that the s-plane imaginary axis is mapped into the *unit circle* (U.C.) in the z-plane as shown in Figure 4.2. The left-half (LH) s-plane is mapped into the inside of the U.C. The same stability regions exist for the exact $\mathcal{Z}$-transform and the Tustin approximation.

Also, in this approximation, the entire s-plane imaginary axis is mapped only once onto the U.C. The Tustin approximation prevents pole and zero aliasing since the folding phenomenon does not occur with this method. However, there is again a warping penalty.[206] Compensation can be accomplished by using Equation 4.21, which is depicted in Figure 4.3. To compensate for the warping, prewarping of ω_{sp} by using Equation 4.21 generates $\hat{\omega}_{sp}$. The continuous controller is mapped onto the z-plane by means of Equation 4.16 using the prewarped frequency $\hat{\omega}_{sp}$. The digital compensator (controller) must be tuned (i.e., its numerical coefficients adjusted) to finalize the design since approximations have been employed. As seen from Figure 4.3, Equation 4.22 is a good approximation when $\omega_{sp}T/2$ and $\hat{\omega}_{sp}T/2$ are both less than 0.3 rad.

The prewarping approach for the Tustin approximation takes the s-plane imaginary axis and folds it back to $\pi/2$ to $-\pi/2$ as seen from Figure 4.3. The spectrum of the input must also be taken into consideration when selecting an approximation procedure with or without prewarping. It should be noted that in the previous discussion only the frequency has been prewarped due to the interest in the controller frequency response. The real part of the s-plane pole influences such parameters as rise time, overshoot, and settling time. Thus, consideration of the warping of the real pole component is now analyzed as a

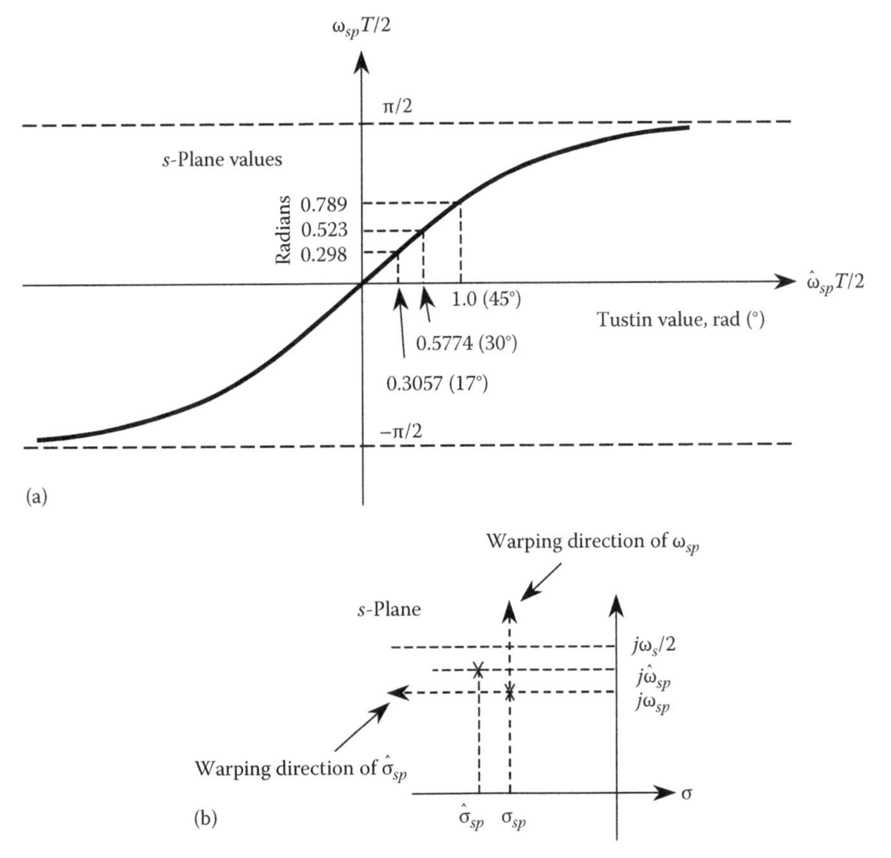

FIGURE 4.3
$\hat{\omega} = 2(\tan \omega_{sp}T/2)/T$ map. (a) Equation 4.19 plot and (b) warping effect.

fine-tuning approach. Proceeding in the same manner as used in deriving Equation 4.22, substitute $z = e^{\sigma_{sp}T}$ and $s = \hat{\sigma}_{sp}$ into Equation 4.17 to yield

$$e^{\sigma_{sp}T} = \frac{1+\hat{\sigma}_{sp}T/2}{1-\hat{\sigma}_{sp}T/2} \tag{4.23}$$

Replacing $e^{\sigma_{sp}T}$ by its exponential series and dividing the numerator by the denominator in Equation 4.23 results in the expression

$$1+\sigma_{sp}T+\frac{(\sigma_{sp}T)^2}{2}+\cdots = 1+\frac{\hat{\sigma}_{sp}T}{1-\hat{\sigma}_{sp}T/2} \tag{4.24}$$

If $|\sigma_{sp}T| \gg (\sigma_{sp}T)^2/2$ (or $1 \gg |\sigma_{sp}T/2|$) and $1 \gg |\hat{\sigma}_{sp}T/2|$, then

$$|\hat{\sigma}_{sp}| \approx |\sigma_{sp}| \ll \frac{2}{T} \tag{4.25}$$

Thus, with Equations 4.22 and 4.25 satisfied, the Tustin approximation in the s-domain is good for small magnitudes of the real and imaginary components of the variable s. The shaded area in Figure 4.4 represents the allowable location of the poles and zeros in the s-plane for a good Tustin approximation.

Because of the mapping properties and its ease of use, the Tustin transformation is employed for the *DIG* technique in this chapter. Figure 4.3b illustrates the warping effect of a pole (or zero) when the approximations are not satisfied.

A *matched $\mathcal{Z}$-transform* can be defined as a direct mapping of each s-domain root to a z-domain root; that is, $s + a \rightarrow 1 - e^{-aT}z^{-1}$. The poles of $G(z)$ using this approach are identical to those resulting from the exact $\mathcal{Z}$-transformation of a given $G(s)$. However, the zeros and the dc gain are usually different!

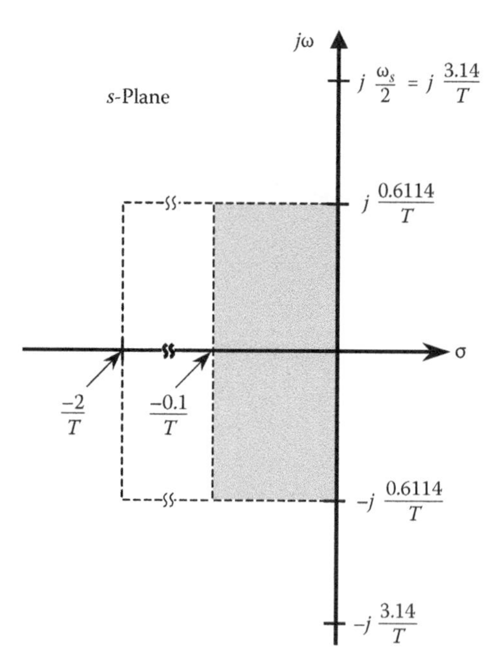

FIGURE 4.4
Allowable location (shaded area) of dominant poles and zeros in the s-plane for a good Tustin approximation.

A characteristic of a bilinear transformation, as pointed out in Section 4.2.2, is that, in general, it transforms an unequal-order transfer function ($on_s \neq od_s$) in the *s*-domain into one for which the order of the numerator is equal to the order of its denominator ($on_z = od_z$) in the *z*-domain. This characteristic must be kept in mind when synthesizing *G*(*s*) and *F*(*s*).

4.3 Non-Minimum-Phase Analog Plant

The analog QFT design technique for *m.p.* plants can be modified when the plant is *n.m.p.* For complement information about *n.m.p.* QFT, see Refs. 116 through 121. The case where the *n.m.p.* plant has only one RHP zero is considered in this section. For this case,

$$L(s) \equiv (1-\tau s)L'_m(s) \tag{4.26}$$

where $L'_m(s)$ is an *m.p.* transfer function whose gain constant has a positive value. Equation 4.26 is modified to

$$L(s) = \left[\frac{1-\tau s}{1+\tau s}\right][(1+\tau s)L'_m(s)] = A'(s)L_m(s) \tag{4.27}$$

where

$$A'(s) \equiv \frac{1-\tau s}{1+\tau s} \tag{4.28}$$

is *n.m.p.* and

$$L_m(s) = (1+\tau s)L'_m(s) \tag{4.29}$$

is *m.p.* The frequency-domain characteristics of *A*′(*s*) are determined by substituting $s = j\omega$ into Equation 4.28. Thus,

$$A'(j\omega) = |A'|\angle A'(j\omega) \tag{4.30}$$

where

$$|A'| = \frac{\sqrt{1+(\tau\omega)^2}}{\sqrt{1+(\tau\omega)^2}} = 1 \tag{4.31}$$

$$\phi_{A'} = \angle A'(j\omega) = \tan^{-1}\left(\frac{-\omega\tau}{1}\right) - \tan^{-1}\left(\frac{\omega\tau}{1}\right) = \phi_N - \phi_D \tag{4.32}$$

and $\phi_N = -\phi_D < 0°$ (see Figure 4.5). Thus, Equations 4.31 and 4.32 reveal that *A*′(*s*) is an *all-pass filter* (*a.p.f.*), where its angular contribution

$$\phi_{A'} = \phi_N - \phi_D = 2\phi_N < 0 \tag{4.33}$$

contributes an "extra" phase lag to the *m.p.* function $L_m(s)$ as shown in Figure 4.6.

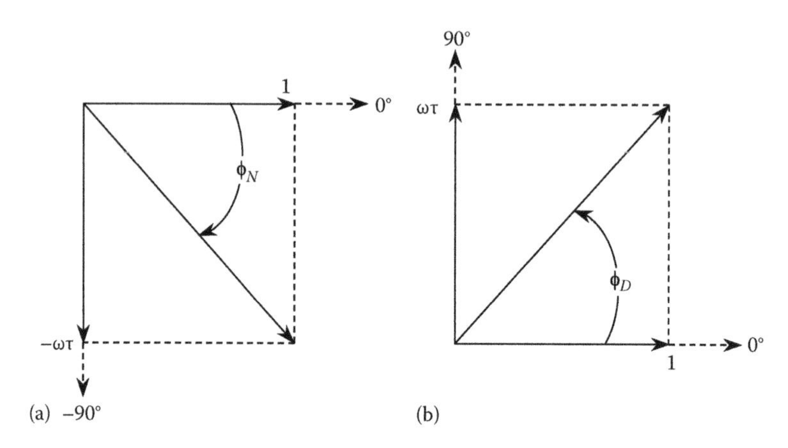

FIGURE 4.5
Analysis of the angular contribution of Equation 4.31: (a) numerator contribution $\phi_N < 0°$ and (b) denominator contribution $\phi_D > 0°$.

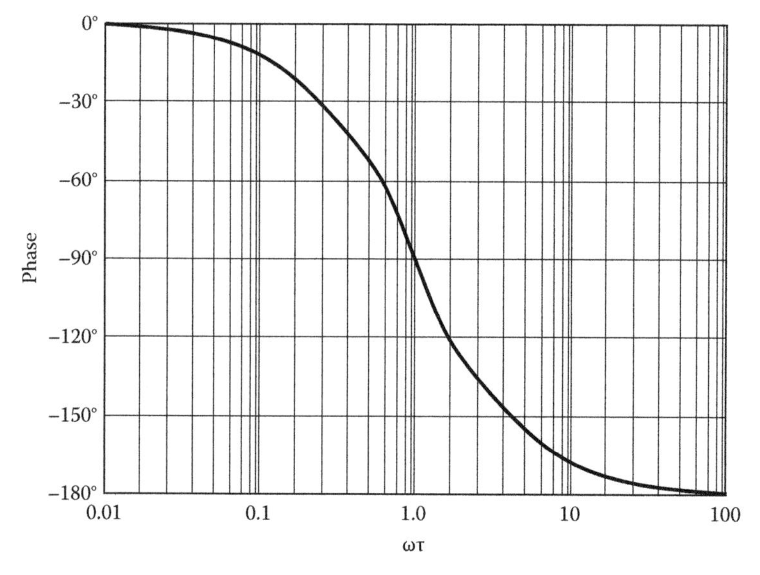

FIGURE 4.6
An analog *all-pass filter* angle characteristic.

In the low-frequency range, $0 \le \omega_i \le \omega_L$, where the magnitude of the angle of Equation 4.33 is very small, the *a.p.f.* lag characteristic has essentially no effect on the determination of the optimal loop transmission function $L_o(s)$ or $L_{mo}(s)$. Whether the *a.p.f.*'s angular contribution is detrimental in achieving a satisfactory $L_{mo}(s)$ can be determined by the following approach: if the QFT design requires a phase-margin angle $\gamma = 40°$ at $\omega_i = \omega_\phi$, for which $\angle L_{mo}(j\omega_\phi) = -90°$, then

$$\angle L_{mo}(j\omega_\phi) - \gamma = -130°$$

This results in the restriction that $\phi_{A'} \ge -180° - (-130°) = -50°$. If at $\omega = \omega_\phi$, the angle $\phi_{A'} < -50°$, then the actual phase-margin angle will be less than 40°, that is, $\gamma < 40°$ at $\omega = \omega_\phi$, resulting in a larger value of M_m (or M_p). The limit on an achievable ω_ϕ for an *n.m.p.* plant having a RHP zero at z_i is $\omega_\phi = 0.5|z_i|$.[22]

4.3.1 Analog QFT Design Procedure for an *n.m.p.* Plant

The analog design procedure of Chapter 3 is modified for an *n.m.p.* analog plant as follows:

1. For the nominal plant P_o, from Equation 4.27, the nominal loop transmission function is denoted as

$$L_o(s) \equiv A'(s)L_{mo}(s) \tag{4.34}$$

   where, as shown in Equation 4.27,
   $L_{mo}(s)$ is an *m.p.* transfer function whose gain constant has a positive value
   $A'(s)$ is an *all-pass filter*

2. Obtain the *U*-contour (B'_h-contour) and the B_D, B_R, and the B'_o bounds corresponding to the template frequencies ω_i for $L_o(s)$ in the same manner as for the analog plant. Note that the B_h-contour and B_o bounds of Figure 3.19 are now relabeled as B'_h-contours and B'_o bounds, respectively. Figure 4.7 shows the B'_h-contour and the $B'_o(j\omega_1)$ bound. It is necessary to modify the angle characteristic of $L_{mo}(j\omega)$ in order to compensate for $\phi_{A'} = \angle A'(j\omega) < 0°$. This is accomplished as follows: Equation 4.34 is rearranged to

$$L_{mo}(s) = \frac{L_o(s)}{A'(s)} \tag{4.35}$$

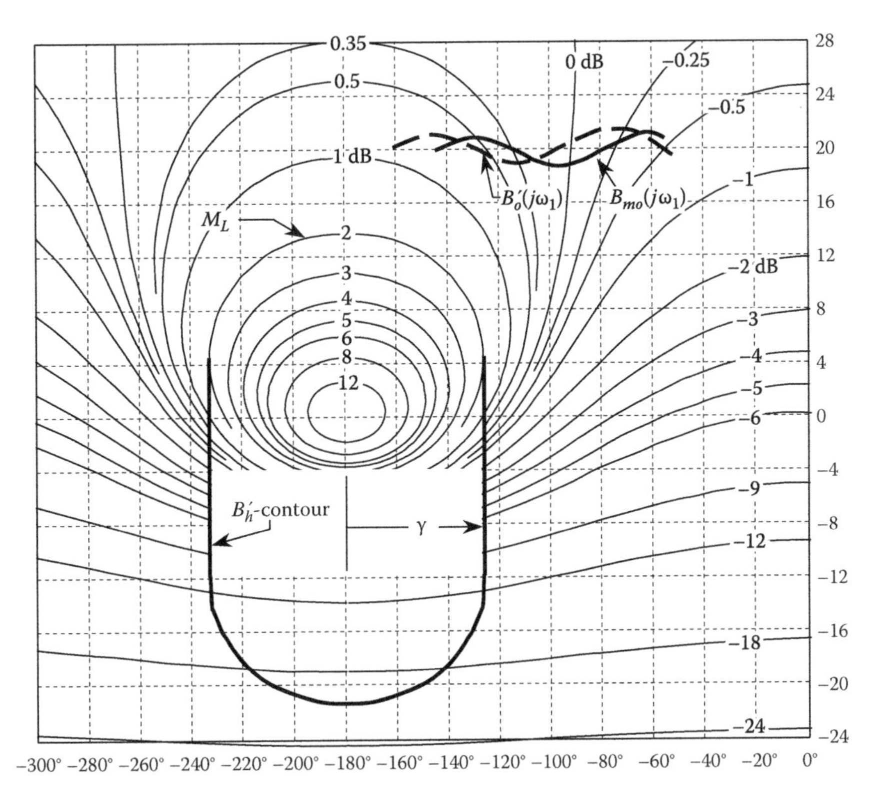

FIGURE 4.7
Nichols chart characteristic for an *n.m.p.* $L(s)$.

which yields

$$\angle L_{mo}(j\omega) = \angle L_o(j\omega) - \phi_{A'}(j\omega) = \angle L_o(j\omega) + [-\phi_{A'}(j\omega)] \tag{4.36}$$

Since $\phi_{A'}(j\omega) < 0°$, Equation 4.36 implies that the B_o' bounds and the *U*-contour, for each value of frequency corresponding to the template frequencies, are shifted to the right by $|\phi_{A'}|$ to yield the corresponding B_{mo} bounds, as shown in Figure 4.7, and the corresponding B_h-contours, as shown in Figure 4.8.

3. Before proceeding with the design, the question that needs to be addressed is: Can a realizable $L_o(s)$ $[L_{mo}(s)]$ be synthesized that satisfies all the B_h'-bounds (B_{mo} bounds) and yet be able to go from a dB value of 20 $\log_{10}[L_{mo}(j\omega_i)]$ to 20 $\log_{10}[L_{mo}(j\omega_K)]$? For example, let

$$\Delta dB = \{|B(j\omega_i)| = 16\,\text{dB}\} - \{|B(j\omega_{i+j})| = -50\,\text{dB}\} = 66\,\text{dB}$$

where
$\omega_i = 3$
$\omega_{i+j} = 100\,\text{rad/s}$

Thus, in order to achieve this decrease, $\Delta dB = 66\,\text{dB}$, in one plus decade, the slope of 20 $\log_{10}[L_{mo}(j\omega)]$ must be about 60 dB/dec. This value for the slope results in a large "phase lag" characteristic, which can force 20 $\log_{10}[L_{mo}(j\omega)]$ to decrease so fast that it cannot satisfy all the $B_{mo}(j\omega_{i+l})$ bounds let alone yield a stable system. If this

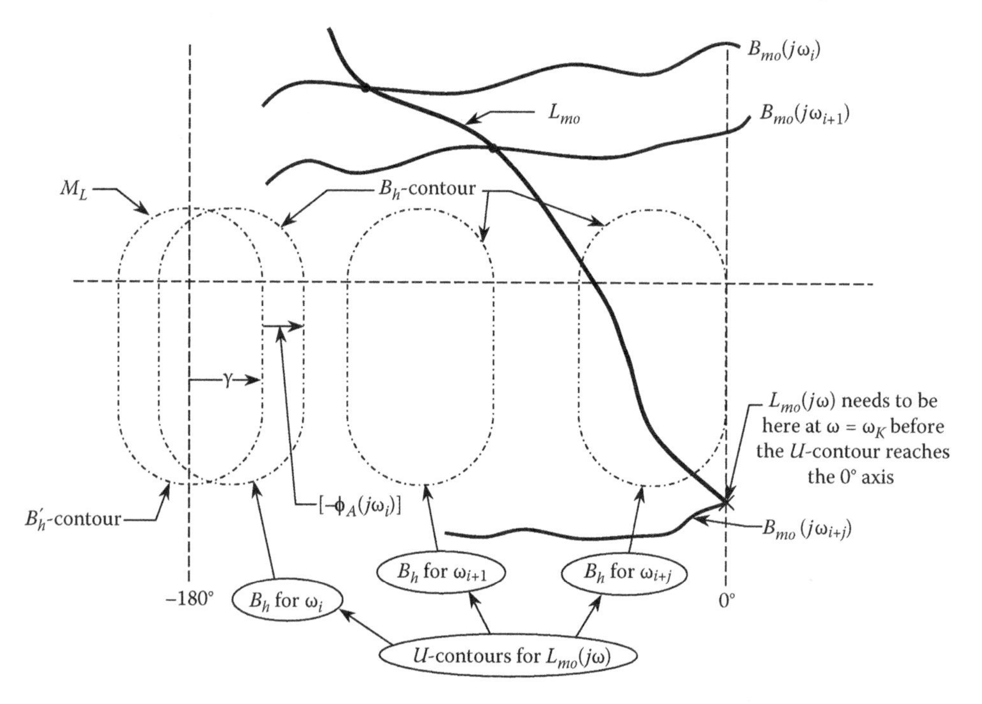

FIGURE 4.8
The shifted *U*-contours for an *n.m.p.* $L_o(s)$.

occurs, then a QFT design can be achieved to only meet the stability requirement, that is, satisfy only the M_L specification. Accepting a larger value for M_L results in a "shrunken" U-contour and may allow a QFT design that satisfies all the specifications. That is, the synthesized $L_{mo}(s)$ lies on or above each $B_{mo}(j\omega_i)$ bound and is to the "right" or to the "right bottom" of the shifted U-contour (B_h-contour).

4. In synthesizing $L_{mo}(s)$, for the case where only the stability M_L requirement is satisfied, one needs to exercise judgment as to where to try to locate the initial zero(s) that is (are) inserted into the synthesized function. In order to minimize the value of t_s, locate the zero(s) as far left from the imaginary axis as possible.
5. Form $L_o(s) = A'(s)L_{mo}(s)$ in order to obtain

$$G(s) = \frac{L_o(s)}{P_o(s)} \tag{4.37}$$

6. Design the filter and perform a simulation in the same manner as described in Chapter 3.

4.4 Discrete MISO Model with Plant Uncertainty

As the state of the art of digital computers makes great strides, digital control systems such as digital flight control systems are playing a more important and greater role today. Thus, it is important to extend the QFT continuous-time system design technique to an S-D control system as represented by Figure 4.9. In this figure the ZOH unit represents a *zero-order-hold* (ZOH) device whose transfer function is $G_{zo}(s) = (1 - e^{-sT})/s = (1 - z^{-1})/s$. Also shown in this figure are switches representing an ideal sampler. The basic equations that describe the MISO system of Figure 4.9 are as follows:

$$Y(s) = P(s)W(s) \qquad W(s) = M(s) + D(s) \qquad M(s) = G_{zo}(s)X^*(s)$$

$$X^*(s) = G_1^*(s)E^*(s) \qquad E^*(s) = V^*(s) - Y^*(s) \qquad V^*(s) = F^*(s)R^*(s)$$

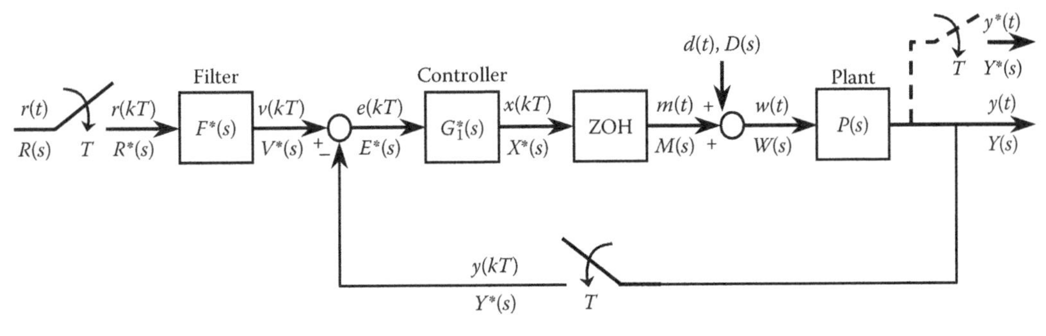

FIGURE 4.9
A MISO S-D control system.

Note that the starred functions represent impulse transfer functions; for example, $Y^*(s)$ represents the ideal impulse transform of $Y(s)$. These equations are manipulated and transformed to yield the input/output z-domain relationships:

$$Y(z) = \frac{L(z)F(z)}{1+L(z)}R(z) + \frac{PD(z)}{1+L(z)} = Y_R(z) + Y_D(z) \tag{4.38}$$

where

$$L(z) = G_{zo}P(z)G_1(z) \tag{4.39}$$

$$P_z(z) = G_{zo}P(z) = (1-z^{-1})\mathcal{Z}\left[\frac{P(s)}{s}\right] = (1-z^{-1})P_e(z) \tag{4.40}$$

$$P_e(z) \equiv \mathcal{Z}\left[\frac{P(s)}{s}\right] = \mathcal{Z}\left[P_e(s)\right] \tag{4.41}$$

$$P_e(s) \equiv \frac{P(s)}{s} \tag{4.42}$$

$$PD(z) = \mathcal{Z}[P(s)D(s)] \tag{4.43}$$

$$T_R(z) = \frac{F(z)L(z)}{1+L(z)} = F(z)T_R'(z) \tag{4.44}$$

$$T_R'(z) = \frac{L(z)}{1+L(z)} \tag{4.45}$$

$$Y_D(z) = \frac{PD(z)}{1+L(z)} \tag{4.46}$$

Substituting Equation 4.40 into Equation 4.39 yields

$$L(z) = G_1(z)(1-z^{-1})P_e(z) = G_1(z)P_z(z) \tag{4.47}$$

Note, for a unit-step disturbance input function, $D(s) = 1/s$,

$$P_e(s) = P(s)D(s) = \frac{P(s)}{s} \tag{4.48}$$

and thus

$$P_e(z) = PD(z) \tag{4.49}$$

The QFT design is based upon the uncertain plant being defined by Equation 4.41. Once $L(z)$ has been synthesized then the controller $G_1(z)$, which is to be implemented, is readily determined. Analyzing and designing the S-D control system in the z-domain is referred to as the *direct* (*DIR*) design technique.[206]

4.5 QFT w-Domain DIG Design

As discussed in Section 4.1, the design techniques that have been highly developed for the s-domain can readily be applied in the w-domain. This feature enables the QFT design technique for analog systems to be utilized for the design of discrete systems in the w-domain if certain conditions related to T, the sampling time, hold. The pertinent s-, z-, and w-plane relationships from Section 4.2 are repeated in the following:

$$s = \sigma_{sp} + j\omega_{sp} = \sigma + j\omega \tag{4.50}$$

$$z = \frac{Tw + 2}{-Tw + 2} \tag{4.51}$$

$$w = \sigma_{wp} + j\omega_{wp} = u + jv = \left(\frac{2}{T}\right)\left[\frac{z-1}{z+1}\right] \tag{4.52}$$

$$v = \left(\frac{2}{T}\right)\tan\left(\frac{\omega T}{2}\right) = \frac{\omega_S}{\pi}\tan\left(\frac{\omega\pi}{\omega_S}\right) \tag{4.53}$$

$$\omega_S = \frac{2\pi}{T} \tag{4.54}$$

$$z = e^{sT} = e^{\sigma T}\angle\omega T = |z|\angle\omega T \tag{4.55}$$

If

$$\alpha^2 = \left(\frac{\sigma_{sp}T}{2}\right)^2 \ll 2 \quad \text{and} \quad \frac{\omega_{sp}T}{2} \le 0.297 \tag{4.56}$$

are both satisfied, then $s \approx w$. If, for the range of parameter uncertainty and for a specified value of T, both the conditions of Equation 4.56 are satisfied in the low-frequency range, then the approximation

$$\Im P(s) \approx \Im P(w) \tag{4.57}$$

is valid in the low-frequency range. The w-domain QFT design procedure presented in this chapter is based upon a stable uncertain plant.

Example 4.1 For the plant

$$P(s) = \frac{K}{s(s+a)} \quad \text{where } 1 \le K \le 10 \quad \text{and} \quad 1 \le a \le 4$$

let $T = 0.05\,\text{s}$. Consider the following two cases: (a) $0 \le a \le 20$ and (b) $0 \le a \le 10$. The range of values for which Equation 4.57 is valid is determined by utilizing Equation 4.56, which yields

$$\text{(a)} \quad \alpha^2 = \left(\frac{\sigma_{sp}T}{2}\right)^2 = \left(\frac{-aT}{2}\right)^2 = 0.25 \ll 2 \rightarrow 1 \ll 8$$

$$\text{(b)} \quad \alpha^2 = 0.0625 \ll 2 \rightarrow 1 \ll 32$$

For both ranges of a: $\omega_{sp}T/2 \le 0.297 \rightarrow \omega_{sp} \le 11.88$. Thus, for $0 \le a \le 10$ and $0 \le \omega_{sp} \le 11.88$ Equation 4.57 is valid. As the value of a approaches 20, from the low side, then Equation 4.57 is a "fair" approximation with some warping.

Example 4.2 Given the plant

$$P(s) = \left[\frac{K_x}{s(s+2)}\right]\left[\frac{K_y(s+a)}{s+b}\right] = P_x P_y \tag{4.58}$$

obtained by using either $P(s)$ or $P_y(s)$ will be identical. Remember the $1/s$ where P_x is the LTI portion and P_y is the plant uncertainty portion of $P(s)$. Note that the "s" in the denominator of Equation 4.41 comes from the ZOH unit and is not part of the plant $P(s)$.

Note, for simplification purposes, the double subscripts in Equation 4.50 are dropped for the s-plane representation, and the $w = u + jv$ notation is used for the w-plane representation.

If ω_s is fixed a priori, then the QFT design can proceed in the w-domain precisely in the same manner as for the continuous system.[104,218] The design can be successful only if ω_s is large enough for the specific plant uncertainty. Thus, for S-D control systems for which the minimum value of sampling time has not been specified, it is best to assume initially that ω_s is not known a priori and the derivation of the minimum ω_s needed becomes one of the principal design problems.[104] *For an achievable QFT design, it is required that in the*

$$\lim_{v \to \infty}[P_{z_\iota}(jv)] \tag{4.59}$$

the resulting values of gain for all J plants, where $\iota = 1, 2, \ldots, J$, *do not change sign.*[120,129]

4.5.1 Closed-Loop System Specifications

Figure 3.4b represents the upper and lower bounds for the desired tracking responses and Figure 3.5 represents the upper bound for the desired disturbance response. The $\delta_R(\omega_i)$ specification suffices to control the time-domain response for *m.p.* systems and those systems for which the RHP zeros are known.[2] The latter case applies for the w-domain QFT design for a system whose plant $P(s)$ is *m.p.* As derived in Ref. 206, for ideal impulse sampling,

$$H(j\omega) \approx TH^*(j\omega) \tag{4.60}$$

where

$H(j\omega)$ is the analog transfer function

$H^*(j\omega)$ is its corresponding impulse sampled function

$\omega \ll \omega_s$

Equation 4.60 is valid for $\omega \leq 0.1\omega_s$, where 0.1 is a good engineering rule-of-thumb value. Thus, where for the analog QFT design it is desired that

$$|B_L(j\omega)| \leq 20 \log_{10}|T_R(j\omega)| \leq |B_U(j\omega)| \tag{4.61}$$

and

$$\delta_R(\omega) = |B_U(j\omega)| - |B_L(j\omega)| \geq |\Delta T_R(j\omega)| \tag{4.62}$$

for all P in $\mathcal{P}$, then for a discrete QFT designed system, Equations 4.61 and 4.62 are modified to

$$|B_L(j\omega)| \leq 20 \log_{10}|T_R^*(j\omega)| \leq |B_U(j\omega)| \tag{4.63}$$

$$\delta_R^*(j\omega) = |B_U^*(j\omega)| - |B_L^*(j\omega)| \geq |\Delta T_R^*(j\omega)| \tag{4.64}$$

$$|B_L^*(j\omega)| = \left|\frac{B_L(j\omega)}{T}\right| \leq 20 \log_{10}|T_R^*(j\omega)| \leq \left|\frac{B_U(j\omega)}{T}\right| = |B_U^*(j\omega)| \tag{4.65}$$

which is valid for a "low-enough" part of the frequency spectrum.

The specified B_L and B_U bounds[6] for the design example to be used in this chapter is shown in Figure 4.10. For the QFT technique it suffices if Equations 4.63 through 4.65 are valid up to where B_L is approximately $20 \log_{10}(\alpha_h) \equiv -24\,\text{dB}$, with $\omega_h \equiv 10\,\text{rad/s}$ in Figure 4.10. The reason for this is that for $\omega > 10$, B_L decreases rapidly, so the allowable $|T_R(j\omega)|$ uncertainty (i.e., $B_U - B_L$) soon exceeds the plant uncertainty.

The requirement for a reasonable stability margin independent of B_L and B_U is to maximize the value of ω at which the $-24\,\text{dB}$ value occurs on the B_L curve. For example, is the range 0–10 rad/s "low enough"? In general, in feedback systems, the price paid for the benefits of feedback is in the BW of the loop transmission L of Equations 4.44 and 4.45.[1] This tends to force ω_s to be significantly larger than what is required by the sampling theorem. Thus, it is very reasonable to simply use Equations 4.63 through 4.65 for the sampled model, which is based upon the continuous model.

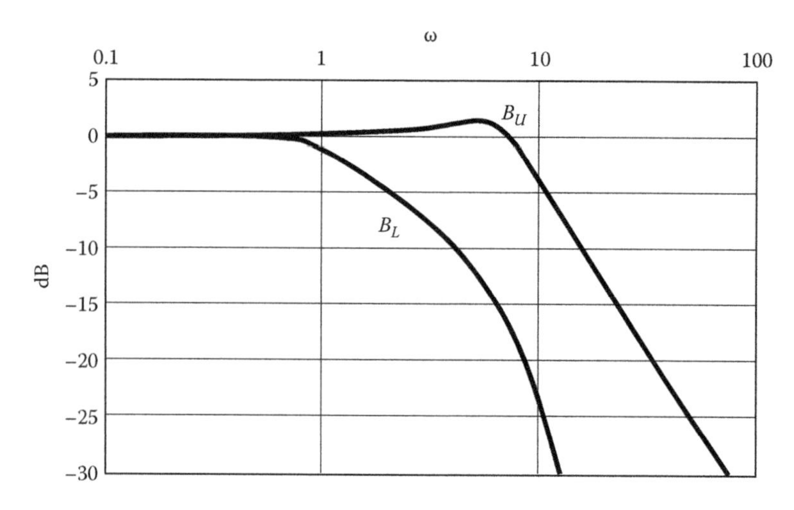

FIGURE 4.10
Bounds on $20\ \log_{10}|T_R(j\omega)|$ used for $T_R(j\omega)$ up to the −24dB point. *Note*: Curves are drawn approximately to the scale.

A loop stability requirement similar to that done for the continuous system must be added to the disturbance-response performance requirements, as indicated in Figure 3.5.[6] This requirement (a damping constraint) is

$$\left|\frac{L^*}{1+L^*}\right| < M_L \tag{4.66}$$

which is a constant for all ω and for all P_e in $\mathcal{P}_e$. For the example in this chapter a value of $20\ \log_{10}(M_L) = 6\,\text{dB}$ is used. If conditionally stability is not allowed, then the following additional requirement is imposed:

$$\angle L^* \geq -180^\circ + \gamma \tag{4.67}$$

where $\gamma > 0$ for $\omega < \omega_c$, ω_c is the crossover frequency $|L^*(j\omega_c)| = 1$, and $|L^*(j\omega)| < 1$ for $\omega > \omega_c$.

Thus, in order to proceed with a QFT design, it is necessary at the outset to determine the bounds and the values of M_L and γ. The change in T_R due to parameter variation can be obtained (see Chapter 3) from Equations 4.44 and 4.45 in the logarithmic domain and is expressed as follows:

$$\Delta(20\ \log_{10}[T_R]) = 20\ \log_{10}[T_R] - 20\ \log_{10}[F] = 20\ \log_{10}\left[\frac{L}{1+L}\right] = 20\ \log_{10}[T'_R] \tag{4.68}$$

As it is seen from Equation 4.68, $\Delta\{20\ \log_{10}[T_R(z)]\}$ can be made arbitrarily small by choosing $L(z)$ sufficiently large. For the QFT technique, the designer finds the minimum $L(z)$ [or $L(jv)$] to satisfy Equations 4.63 through 4.67.

4.5.2 Plant Templates

As is required for continuous-time control systems, it is also necessary for S-D control systems to determine the plant templates $\Im P_e(j\omega_i) = \{P_e(j\omega_i)\}$ at a "sufficient" number of ω_i values. The number of plants J within the $\mathcal{P} = \{P_{e_\iota}\}$, where ι = 1, 2,…, J is chosen to

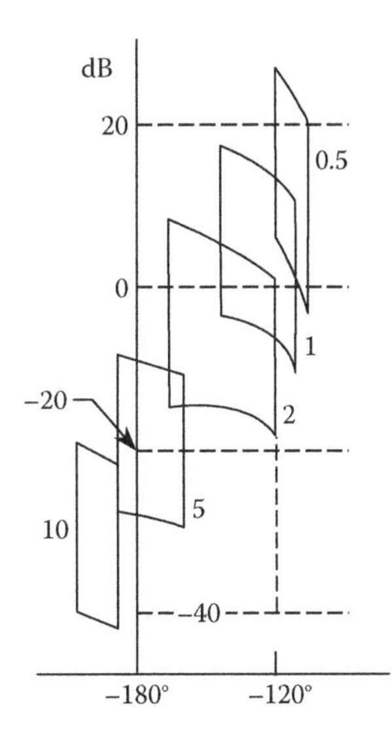

FIGURE 4.11
Plant templates at ω_i = 0.5, 1, 2, 5, and 10 rad/s calculated for ω_s = 60 rad/s, but almost identical for ω_s = 120 and 240 rad/s and for the continuous-time system. (From Horowitz, I.M. and Liao, Y., *Int. J. Control*, 44, 665, 1986.)

adequately describe the contour of the template, which represents the region of plant uncertainty for each ω_i. For S-D systems, these "plant templates" are functions of the sampling frequency ω_s, but this becomes apparent only for $\omega > 0.25\omega_s \equiv \omega_{0.25}(v_i > v_{0.25})$ (an approximate engineering value, see Section 4.5.6). For the values of $\omega_i \leq \omega_{0.25}$, the plant templates $\Im P(j\omega_i) = \{P(j\omega_i)\}$ for the continuous system are essentially the same as the plant templates $\Im P_e(jv_i) = \{P_e(jv_i)\}$ for the S-D system.

Since at this point of the design process the desired value of ω_s is yet to be determined, it is therefore reasonable to tentatively use the continuous $\Im P(j\omega_i)$ templates in order to proceed with the design and eventually to be able to determine the required value of ω_s. A number of plant templates (for $v_i = (2/T)\tan(\omega_i T/2)$) are shown in Figure 4.11 for the plant of Example 4.1 calculated at ω_s = 60 rad/s. These templates are very similar to the templates for the continuous-time system and those for ω_s = 120 and 240 rad/s. To accentuate the value of doing a w-domain design (via Equation 4.2) versus a w'-domain design (via Equation 4.5), a correlation is made between the values of

$$w = j\tan\left(\frac{\omega\pi}{\omega_s}\right) \quad \text{and} \quad w' = jv = j\frac{\omega_s}{\pi}\tan\left(\frac{\omega\pi}{\omega_s}\right)$$

for various values of ω_i in the frequency range $0 < \omega_i \leq \omega_s/2$, in Table 4.1. Note, henceforth, once again the prime designator is omitted (i.e., $w' = w$).

4.5.3 Bounds $B(jv)$ on $L_o(jv)$

Recall that the jv-axis of the w-plane, where $0 \leq v \leq \infty$, corresponds to the upper half of the U.C. in the z-plane and to the primary strip $0 \leq j\omega_s/2$ in the s-plane (see Section 4.2). Since QFT requires shaping of L to satisfy certain bounds, such a shaping is difficult to do within the U.C. in the z-domain. Therefore, it is best to design F and G_1 in Figure 4.9 in the $w = jv$ domain. As a consequence, the same design techniques that are used to perform loop shaping in the $s = j\omega$ domain can be used. The appropriate values of v_i (see Sections 4.5.5 and 4.5.6) to be used for the templates are available from Equations 4.50 through 4.54, only after ω_s is chosen. Hence, the tentative value ω_s = 120 rad/s is selected to initiate the design process. The choice is intentionally made low in order to stress the relationship between $(\omega_s)_{min}$, the plant uncertainty, and the performance tolerances. After an individual achieves a sufficient understanding of the QFT technique, a good estimate can be made a priori for the value of $(\omega_s)_{min}$.

In order to determine the bounds, a nominal plant $P_{eo}(w)$ must be selected. Thus, $K = 1$ and $a = 1$ are selected as the nominal plant parameter values. The template $\Im P_e(jv_i)$, with its corresponding $\delta_R(jv_i)$ value, is shifted on the NC to determine the bounds $B(jv_i)$ for each value of v_i. Based upon Equations 4.40 and 4.47 the corresponding nominal $L_o(w)$ and $P_{zo}(w)$ are, respectively,

$$L_o(w) = G_1(w)P_{zo}(w) = G_1(w)\left[\frac{2w}{w+2/T}\right]P_{eo}(w) \tag{4.69}$$

$$P_{zo}(w) = \left[\frac{2w}{w+2/T}\right]P_{eo}(w) \tag{4.70}$$

where

$$P_{eo}(w) = \frac{-2.99\times 10^{-6}(w-38.2)(w+4377)(w+38.2)}{w^2(w+0.9998)} \tag{4.71}$$

Thus, Equation 4.65 imposes the requirement that

$$|\Delta T_R(jv)| \le |B_U(jv)| - |B_L(jv)| \tag{4.72}$$

for all P_e in $\mathcal{P}_e$. The constraint imposed on 20 $\log_{10}|L_o(jv)|$ by Equation 4.72 is reflected on the NC by the bounds $B_o(jv_i)$ on 20 $\log_{10}|L_o(jv_i)|$ as discussed in Chapter 3. The NC is convenient because it contains loci of constant $|L/(1 + L)|$. For example, in Figure 4.12, 20 $\log_{10}[L_o(j0.5)]$, although not plotted, must be on or above $B_{mo}(j0.5)$ (B_o shifted to the right as discussed in Section 4.5.4) in order to satisfy Equation 4.72 for all P_e in $\mathcal{P}_e$.

A good rule in selecting the nominal plant P_{eo} from $\mathcal{P}_e$ is to select the P_{e_1} in $\mathcal{P}_e$ that has the smallest magnitude and the largest phase lag over all other sP_e in $\mathcal{P}_e$, for all values of v_i, as P_{eo}. This selection criterion for $P_{eo}(P_{zo})$ minimizes the number of plant templates needed to determine the bounds.

4.5.4 Non-Minimum-Phase $L_o(w)$

It is important to note that in the w-domain any practical $L(w)$ is non-minimum phase (*n.m.p.*) with a zero at $2/T$ (see Equations 4.69 and 4.70). (Note the zero in the "true" w-domain is at 1 [Equation 4.1]). This result is due to the fact that any practical $L(z)$ has an excess of at least one pole over zeros, which leads via Equation 4.52 to the zero at $2/T$. Thus, the design technique *for a stable uncertain plant* is modified[31] to incorporate the *a.p.f.*

$$A'(w) = \frac{2/T - w}{2/T + w} \tag{4.73}$$

as follows: let the nominal loop transmission be defined as

$$L_o(w) \equiv -L_{mo}(w)A(w) = L_{mo}A'(w) \tag{4.74}$$

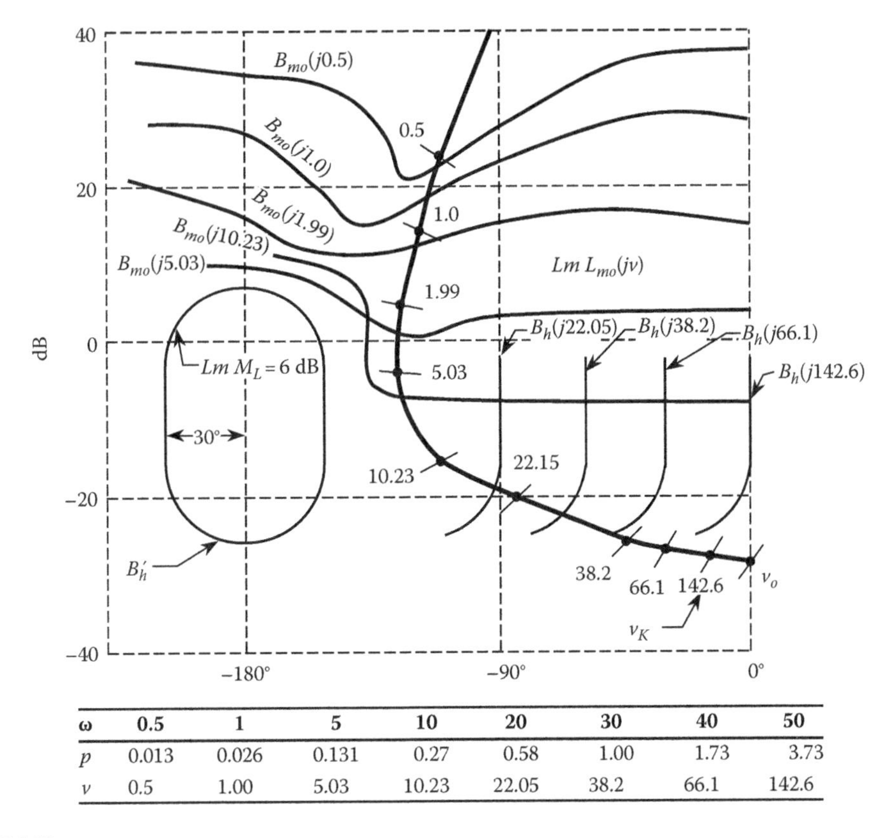

ω	**0.5**	**1**	**5**	**10**	**20**	**30**	**40**	**50**
p	0.013	0.026	0.131	0.27	0.58	1.00	1.73	3.73
v	0.5	1.00	5.03	10.23	22.05	38.2	66.1	142.6

FIGURE 4.12
Unsuccessful design at $\omega_s = 120$ rad/s. *Note*: Curves are drawn approximately to the scale. (From Horowitz, I.M. and Liao, Y., *Int. J. Control*, 44, 665, 1986.)

where

$$A(w) \equiv \frac{w - 2/T}{w + 2/T} = -A'(w) \tag{4.75}$$

and where $|A'(jv)| = 1$ for $0 \le v \le \infty$ and L_{mo} is an *m.p.* function considered to have a positive value of gain. Since $|L_o(jv) = |L_{mo}(jv)|$ and

$$\angle L_o(jv) = \angle L_{mo}(jv) + \angle A'(jv) \tag{4.76}$$

or

$$\angle L_{mo}(jv) = \angle L_o(jv) - \angle A'(jv) \tag{4.77}$$

a bound $B'(jv)$ on $L_o(jv)$ becomes the bound $B(jv)$ on $L_{mo}(jv)$ by shifting $B'(jv)$ positively (to the right on the *NC*) by the angle

$$-\angle A'(jv) = 2\tan^{-1}\left(\frac{vT}{2}\right) = 2\tan^{-1} p > 0 \tag{4.78}$$

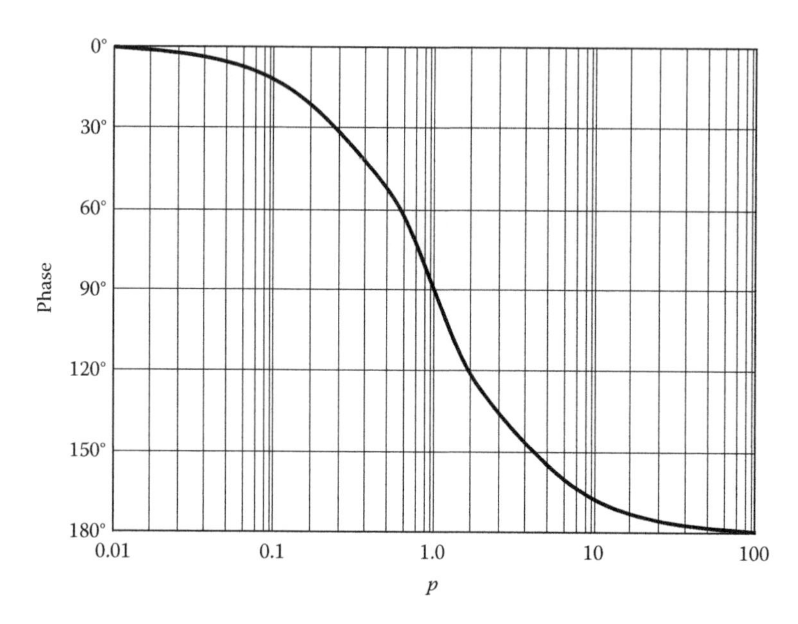

FIGURE 4.13
Angle characteristic of Equation 4.78.

A nondimensionalized $p = vT/2$ plot of Equation 4.78 is shown in Figure 4.13. The amount of shift is 90° at $p = 1$ ($v = 2/T$), 151.93° at $p = 4$ ($v = 8/T$), and it approaches 180° as $p(v)$ approaches infinity.

The inherent limitations on the feedback capabilities of an *n.m.p.* system are readily revealed as follows. At larger values of v_i (or ω_i), the templates tend to be vertical lines (due to the bilinear transformation characteristic) and Equations 4.63 through 4.65 tend to dominate, leading in *m.p.* systems to a single universal high-frequency contour (bound) B_h' basically effective for all ω (or v) greater than some ω_h (or v_h). For example, in Figure 4.12, B_h' is shown with a width of 2(30°) for $M_L = 6\,\text{dB}$. However, in the *n.m.p.* system, the B_h'-contour is shifted in the NC by $2\tan^{-1} vT/2$, to become $B_h(jp)$, a function of p whose right boundary shifts to the vertical line $\angle L_{mo}(jv) = 0°$ at the value $p = p_K$. Thus, for Figure 4.12, where $\gamma = 30°$,

$$2\tan^{-1} p_K = 180° - 30° = 150° \tag{4.79}$$

results in $p_K = 3.73$ ($v_K = 142.55$ for $\omega_s = 120$) for this example. In general, although not done in this chapter, the B_h contours are denoted as a function of p in order for these contours to be independent of the sampling frequency. The Nyquist stability criterion dictates that the $L_{mo}(jv)$ plot is on the "right side" or the "bottom right side" of the $B_h(jp)$ (or $B_h(jv)$) contours for the frequency range of $0 \le v_i \le v_K$. It has been shown that[1,2,31]

1. $L_{mo}(jv)$ must reach the *right-hand bottom* of $B_h(jv_K)$, that is, approximately point K in Figure 4.12, at a value of $v \le v_K$
2. $\angle L_{mo}(jv_K) < 0°$ in order that there exists a practical L_{mo} that satisfies the bounds $B(jv)$ and provides the required stability

Requirement (2) is necessary in order to obtain an $L_{mo}(jv)$ whose angle is always negative for all frequencies, a necessary requirement for a stable system. These requirements are very useful for determining the minimum ω_s needed for any specific uncertainty problem as discussed in Section 4.5.5.

For the situation where the uncertain plant parameter characteristics can yield an unstable plant(s), amongst the J plants being considered, if the nominal plant chosen is one of these unstable plants, *then the a.p.f. to be used in the QFT design must include all RHP zeros of* P_{zo}. This situation is not discussed in this text.

4.5.5 Synthesizing $L_{mo}(w)$

Based upon the analysis made in the previous sections, it is possible to divide the frequency spectrum $0 < v_i < \infty$ into four general regions, as shown in Figure 4.14, in order to formalize the design procedure. These regions are as follows:

Region I. For the frequency range $0 < v_i < v_{0.25}$, where Equation 4.56 is essentially satisfied, use the analog templates, that is, $\Im P(j\omega_i) \approx \Im P_e(jv_i)$. Note, depending on the value of T, v_h may be less than $v_{0.25}$.

Region II. For the frequency range $v_{0.25} < v_i \le v_h$, use the w-domain templates $\Im P_e(jv_i)$ to satisfy both the B_{mo} bounds and the B'_h- and B_h-contours, assuming T is small enough. If T is not small enough and cannot be decreased, use these templates to satisfy the stability requirement, that is, satisfy only the B'_h- and B_h-contours. Depending on the value of T, the value of v_h may be less than that of $v_{0.25}$.

Region III. For the frequency range $v_h < v_i \le v_K$, satisfy only the B_h-contours (stability bounds).

Region IV. For the frequency range $v_i > v_K$, use the w-domain templates to satisfy only the M_L-contours and obtain the corresponding B_S-contours.

A good engineering rule of thumb, based upon the definition of *Region I*, is to let $L_{mo1} \approx P_o|_{s=w}$ be the starting loop transmission function, with the appropriate gain adjustment, upon which the final synthesized $L_{mo}(w)$ is constructed. *Only the w-domain location of the s-domain poles of P_o should be used in setting $L_{mo1}(w) = P_o(s)|_{s=w}$.* Thus, from Equation 4.69,

$$G_1(w) = \frac{L_o(w)}{P_{zo}(w)} \tag{4.80}$$

assuming perfect cancellation, results in a minimal-order controller. It should be noted that for *a tracking control system, $L_{mo}(w)$ must be a Type 1 or higher transfer function.* For example, from Equation 4.68, the poles of $P_o(s) = 1/[s(s + 1)]$ are used for the starting L_{mo1}, that is,

$$L_{mo1} = \frac{K_{o1}}{w(w + 0.9998)} \tag{4.81}$$

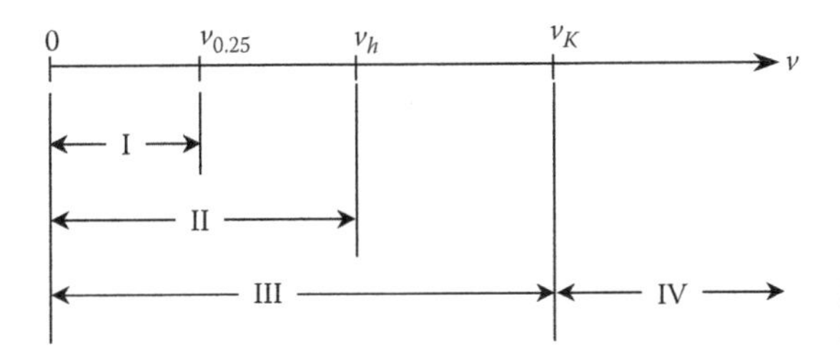

FIGURE 4.14
Regions of the frequency spectrum for synthesizing $L_{mo}(jv)$.

Note: When starting with a Type n $P_o(s)$, in obtaining $L_{mo1}(w)$ by using a CAD package, the CAD data may not result in n poles of $L_{mo1}(w)$ being *exactly at the origin*. If this occurs, the value of these n poles must be corrected in the CAD program to reflect their true value being *exactly at the origin*.

It should be noted that the discussion dealing with the discrete QFT design procedure is based upon the gain value of $P_{eo}(w)$ being a negative quantity. This results in the gain of $G_1(w)$ being a positive quantity. The sign of the gain value of $P_{eo}(w)$ is a function of the sampling time.[206] If the gain of $P_{eo}(w)$ is a positive quantity, the same design procedure is followed, the only difference being that the gain of the resulting $G_1(w)$ is a negative quantity.

4.5.6 ω_s = 120 Is Too Small

The bounds on $L_{mo}(jv)$ are shown in Figure 4.12. Is there a practical $L_{mo}(jv)$ that can satisfy these bounds? In other words, is there an $L_{mo}(jv)$ that lies on or outside the $B_{mo}(jv)$ bounds at small values of v but then decreases fast enough versus v on the right side of the $B_h(jv)$ contour in order to reach the "bottom" at $v < v_K$? The difficulty is that $d\{20 \log_{10}|L_{mo}(jv)|\}/d\{20 \log_{10}[v]\}$ (being d the derivative) is a function of $\angle L_{mo}(jv)$ resulting in an average of 20α dB/dec decrease for $\angle L_{mo}(jv) = -90°\alpha$ ($\alpha = 1, 2, \ldots$). Thus, the more phase lag available for L_{mo}, the faster $|L_{mo}(jv)|$ can decrease versus v. However, the $B_h(jv)$ contour steadily moves to the right versus v, because of the *n.m.p.* $A(w)$, until at v_O there is no more phase lag for a decrease of $|L_{mo}(jv)|$. Hence, it is necessary that $L_{mo}(jv)$ reaches the bottom of this moving barrier (contour) before the value of v reaches the value of v_O.

The following numerical procedure is convenient for determining the minimum ω_s that is required. Work backward, that is, assign approximately −10° to $\angle L_{mo}(jv_K)$, where $v_K < v_O$, and the corresponding right-side phase values of $B_h(jv)$ to $L_{mo}(jv)$ for $v < v_K$: see Figure 4.12 for −30° at v = 66.1, −45° at v = 38.2, and −90° at 22.05. For v < 22.05, in Figure 4.12, "reasonable" values of $-\angle A'(jv)$ (see Equation 4.78), based upon loop-shaping experience, are tried; for example, the right-side phase angles of $B_h(jv)$ of −120° occurs at v = 10.3, of −135.2° occurs at v = 4.97, and similarly the corresponding angles for v = 1.986, 0.993, and 0.497 can be determined. As is well known,[218] $L_{mo}(jv)$ determines $|L_{mo}(jv)|$ within an arbitrary multiplier, and an algorithm due to Bode,[1] used in Gera and Horowitz,[64] is especially efficient in calculating numerically $|L_{mo}(jv)|$ for values obtained from Equation 4.77 for specified values of v_i.

The resulting $20 \log_{10}|L_{mo}(jv)|$, so obtained, is shifted vertically on the NC in Figure 4.12 such that $20 \log_{10}|L_{mo}(jv_K)|$ is at point K and checked as to whether it satisfies the other $B_h(jv_i)$ bounds. It is impossible to do so in this example, as seen in Figure 4.12, since the bounds in the range of

$$0.497 < v < 22.15 \quad (0.0065 < p < 0.58)$$

are violated. If $|L_{mo}(jv)|$ is shifted upward so that it satisfies all the bounds, then it is above the critical point K at v_K = 142.6 (p = 3.73). Then for $v > v_K$, $\angle L_{mo} > 0$ in order to satisfy the $B_h(jv)$ contours for $v > v_K$, so $d|L_{mo}|/dv > 0$ and $|L_{mo}|$ must monotonically increase, which is impractical. That is, since the zeros of L_{mo} dominate, they yield a positive angle and a positive dB value for L_{mo}. Hence, ω_s = 120 and, in turn, ω_s = 200 are too small a sampling frequency for this uncertainty problem.

If the original plant is *m.p.*, it is guaranteed that a satisfactory ω_s exists, because the sampled system approaches the continuous system as ω_s approaches infinity (T approaching zero).

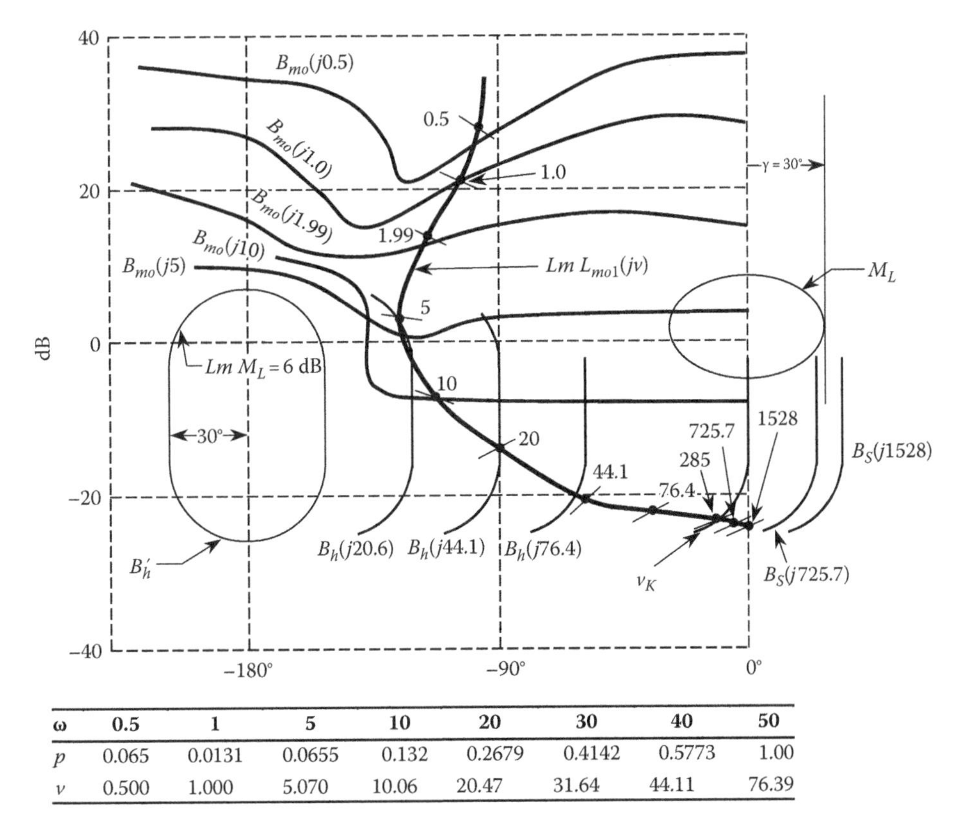

ω	0.5	1	5	10	20	30	40	50
p	0.065	0.0131	0.0655	0.132	0.2679	0.4142	0.5773	1.00
v	0.500	1.000	5.070	10.06	20.47	31.64	44.11	76.39

FIGURE 4.15
A satisfactory design: L_{mo1} at $\omega_s = 240$. *Note*: Curves are drawn approximately to the scale.

The challenge is to use the minimum ω_s value that is achievable; the value $\omega_s = 240$ is satisfactory for this example, as seen in Figure 4.15, where the plot of $20 \log_{10}|L_{mo1}(jv)|$ satisfies all the bounds. The mechanism whereby a larger value of ω_s does this is seen in Figures 4.12 and 4.15.

The $B_h(jv_i)$ high-frequency contours in these figures, for $v_i \geq 20$, are identical in shape and differ only in the frequency value associated for a given contour; that is, the corresponding contours are related as follows:

$$B_h(jv_i)_{\omega_s=120} = B_h(j2v_i)_{\omega_s=240} \tag{4.82}$$

The bounds $B_{mo}(jv_i)$, for ω_s = 120 and 240, are the same for $v_i \leq 5.03$ because the low-frequency (ω) bounds are basically invariant with respect to ω. An analysis of these two figures reveals that the same decrease in log magnitude from about 20 dB to −30 dB is achieved in the range $0.497 \leq v_i \leq 285$ in Figure 4.15 instead of in the range $0.497 \leq v_i \leq 142.3$ in Figure 4.12; that is, an extra octave is available to achieve this decrease in log magnitude. In the event a preliminary analog QFT design is accomplished for the plant $P(s)$, then a good rule of thumb in selecting a suitable sampling frequency is as follows: Select the value of ω_s to be two to three times the value of ω_c that results in $\angle L_o(j\omega_c) = -180°$. In order to use this rule it is assumed that the value of the analog gain-margin frequency ω_c is a practical or realistic value. For this example, the Bode algorithm[1] provides a satisfactory $20 \log_{10}|L_{mo}(jv)|$, labeled $Lm\ L_{mo1}$ in Figure 4.15.

In Figure 4.15 the synthesized $L_{mo}(s)$ has to satisfy all the bounds $B_{mo}(j\omega_i)$ through $B_{mo}(j\omega_{i+j})$ and yet achieve the necessary dB decrease (ΔdB) in $Lm\ L_{mo}(j\omega)$ in the frequency range ω_i to ω_K. An S-D control system satisfying both the requirements may not be obtainable for a given value of M_L and/or T. Increasing the value of M_L and/or decreasing the value of T widens the frequency BW from v_i up to v_K. Examples of this situation are illustrated in Figures 4.12 and 4.15. In order to determine if all requirements can be satisfied for a specified frequency BW, let

$$2^d = \frac{v_K}{v_i} \tag{4.83}$$

where d represents the number of octaves between v_i and v_K. One needs to estimate and select the dominant low-frequency bound $B_{mo}(jv_i)$ and a starting point for $Lm\ L_{mo}(jv_i)$ on this bound. Assuming an average phase change of $\theta_{av} = 75°$, the average negative slope is given by

$$\left[\frac{\theta_{av}}{90^\circ}\right]\left[\frac{6\,\text{dB}}{Oct}\right] \tag{4.84}$$

Therefore the estimated decrease in dB ($est.\ dB$) is given by

$$est.\ dB = \left[\frac{\theta_{av}}{90^\circ}\right]\left[\frac{6\ \text{dB}}{Oct}\right]d \tag{4.85}$$

If the $est.\ dB > \Delta dB$ then the value of ω_s that has been chosen is satisfactory. Applying Equation 4.83 to Figure 4.12 results in

$$2^d = \left[\frac{v_K}{v_i}\right] = \left[\frac{142.6}{0.5}\right]$$

which yields $d = 8.153$ and an $est.\ dB = 40.76 < 43\,\text{dB} = \Delta dB$. Thus, the value of $\omega_s = 120$ is too small. If ω_s is increased to 240, it results in $d = 9.153$ and an $est.\ dB = 45.76$ ($>43\,\text{dB} = \Delta dB$), then from Figure 4.15,

$$2^d = \left[\frac{v_K}{v_i}\right] = \left[\frac{285.1}{0.5}\right]$$

which indicates that the value of $\omega_s = 240$ is satisfactory.

The numerator and denominator of the synthesized $L_{mo1}(jv)$ w-plane transfer function must be of the same degree. The function numerically obtained in this manner can be achieved as accurately as desired by a rational function $L_{mo}(w)$,[1] which can however, be of high degree. Hence, there is some trade-off between ω_s and the degree of $L_{mo}(w)$, which can be made at this numerical stage by allowing some overdesign. For the uncertain plant of Section 2.3.2.1, $K \in \{1,10\}$ and $a \in \{1,4\}$, the synthesized loop transmission is

$$\begin{aligned} L_{o1} &= \frac{-0.059559(w+12.987)(w+45.84)(w-76.39)}{w(w+3.056)(w+76.39)} \\ &= -L_{mo1}(w)A(w) = L_{mo1}(w)[A'(w)] \end{aligned} \tag{4.86}$$

which is sketched in Figure 4.15. Stability is assured when loop shaping for the *m.p.* $L_{mo}(w)$ is performed on the NC. When the *n.m.p.* function $L_o(w)$ is formed the proper sign on its gain term to assure a stable system *can also be ascertained from a polar plot analysis.* That is, the Nyquist stability criterion is applied to $L_o(w)$ in the polar plot domain.

If the disturbance-response specification, for a unit-step disturbance input $d(t)$, is $y(\infty) = 0$, then $L_{mo}(w)$ and, in-turn, $L_o(w)$ loop transmission functions must have at least two poles at the origin. *This ensures, for a nominal Type 1 P_{eo} plant, a Type 1 system characteristic for T_R and that $Y_D(w)$ will not have a pole at the origin.* Thus, the resulting *pseudo control ratio*

$$\left[\frac{Y_D(w)}{D(w)}\right]_P \tag{4.87}$$

has a zero at the origin. Equation 4.87 is obtained by dividing the resulting $Y_D(w)$ by the bilinear transformation of the $\mathcal{Z}$-transform for a unit-step forcing function. For a Type n P_{eo}, then $L_o(w)$ must have $n + 1$ poles *precisely* at the origin.

4.5.7 Error in the Design

There is always the problem of choosing sufficient ω_i values, for the analog system, or v_i values, for the discrete system, especially at large values of ω or v. The poles and zeros of $P(s)$ or $P_e(w)$ are important, because the plant templates do not remain at almost vertical lines (giving the B_h type of bounds for all $\omega >$ some ω_h), until a frequency several times larger than the largest pole or zero of P_e. It is usually easy to detect the largest pole or zero in the continuous plant, as well as the largest pole in $P_e(w)$, because the poles of $P(s)$ map directly onto $P_e(w)$. However, there may be large shifts in the zeros of $P_e(w)$, as was the case for this example, with a far-off zero at $w = -4377$ for the nominal plant $P_{eo}(w)$ of Equation 4.70.

The result is almost vertical templates and hence B_h-type of bounds from, approximately, $v = v_h = 20/0.9998 = 20/$(the pole -0.9998 of Equation 4.71) to $v = 4377/20 = 219 \approx v_K$ (the zero -4377 of Equation 4.71) (allowing a factor of 20 for the effect of a pole or zero on the phase). Note that 0.9998 is the smallest magnitude and 4377 is the largest magnitude of a pole or zero in Equation 4.71. Since the templates $\Im P_e(jv_i)$ broaden out again for $v > 219$, as shown in Figure 4.16a, it is necessary to obtain the more stringent bounds B_S in Figure 4.16b for $v = 1{,}528$, $15{,}280$, $1{,}528{,}000, \ldots$. These bounds, for $v_i > v_K$, are obtained by shifting the M_L-contour to the right corresponding to the template frequencies, up to its maximum shift of 180°. That is, the maximum shift occurs when $-\angle A'(jv) = 180°$ as shown in Figure 4.13. For $v > v_K$, the templates are translated to the right so that they are always tangent to their corresponding M_L-contour in order to obtain sufficient points to determine the corresponding B_S bounds. Note that the templates are measured from the shifted M_L-contours, not the shifted B_h-contour. These high-frequency M_L bounds are overlooked for the first trial design, with the "universal" type B_h assumed in Figure 4.15 for all $v > 20.6$. The first design stops at point W in Figure 4.16b, definitely violating the bounds for $v \geq 15{,}280$. The bound is moderately violated at $v = 15{,}280$ and violated by $\approx$12 dB at $v = 152{,}800$, enough to cause instability. The L_{mo1} is augmented by a high-frequency pole-zero pair to correct the situation. To determine the required pole-zero pair(s), the very-high-frequency (*v.h.f.*) templates of Figure 4.16a are utilized to determine the B_S bounds shown in Figure 4.16b. As illustrated in Figure 4.17, these templates are shifted around the corresponding M_L-contours to obtain the corresponding B_S-contours.

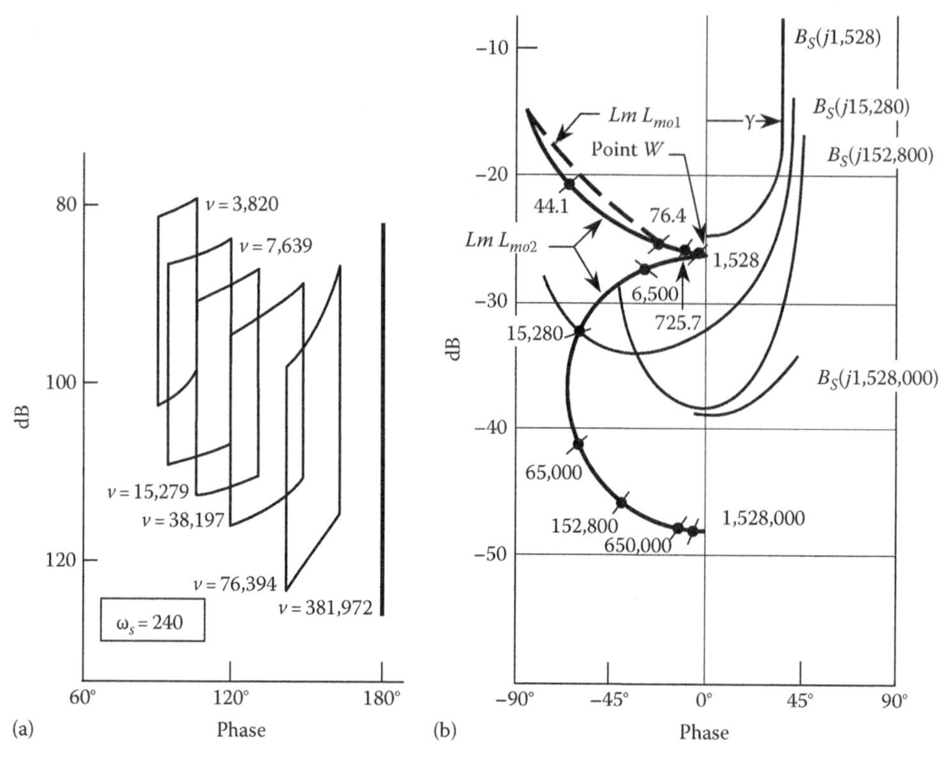

FIGURE 4.16
(a) Neglected plant templates at higher v. (b) Reason for unstable L_{mo1} design and successful L_{mo2}.

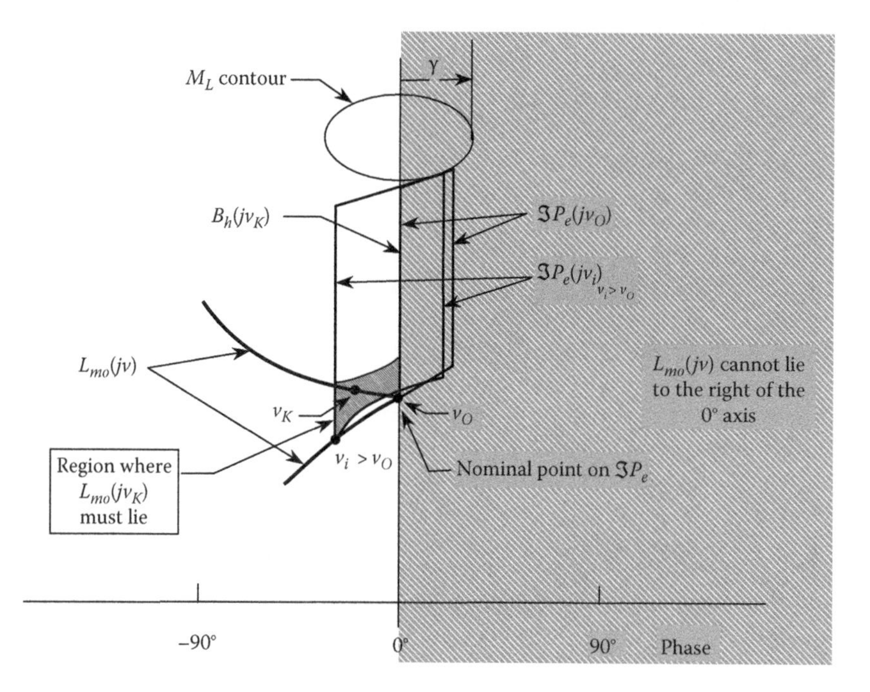

FIGURE 4.17
Determination of the $v.h.f.$ B_S bounds.

The pole-zero pair(s) that are selected must result in an $L_{mo2}(w)$ whose corresponding *v.h.f.* points lie on or below the corresponding B_S bounds. Thus

$$L_{mo2}(w) = \frac{1+w/140566}{1+w/8785.4} L_{mo1}(w) \tag{4.88}$$

which satisfies the $B_S(jv)$ boundaries, as seen in Figure 4.16b. Thus the final loop transmission function, in the w-plane, is

$$\begin{aligned} L_{o2}(w) &= -L_{mo2}(w)A(w) = L_{mo2}(w)A'(w) \\ &= G_1(w)P_{zo}(w) \end{aligned} \tag{4.89}$$

Once $L_o(w)$ is determined, the w-plane controller transfer function is determined from Equation 4.89, that is,

$$G_1(w) = \frac{L_{o2}(w)}{P_{zo}(w)} \tag{4.90}$$

where for $\omega_s = 240$,

$$P_{zo}(w) = \frac{-0.7476\times 10^{-6}(w-76.39)(w+17510)}{w(w+0.9999)} \tag{4.91}$$

utilizing Equations 4.86, 4.88 through 4.90 the controller transfer function is

$$G_1(w) = \frac{4,979.19(w+0.9999)(w+12.987)(w+45.84)(w+140,566)}{(w+3.056)(w+76.39)(w+8,785.4)(w+17,510)} \tag{4.92}$$

Substituting Equation 4.51 into Equation 4.88 yields

$$L_{mo2}(z) = \frac{0.10634(z-0.7094)(z-0.25)(z+0.999891)}{(z-1)(z-0.9231)(z+0.98276)} \tag{4.93}$$

The resulting loop-shaping function is

$$\begin{aligned} L_o(z) &= A(z)L_{mo2}(z) \\ &= \frac{0.10634(z-0.7094)(z-0.25)(z+0.99891)}{z(z-1)(z-0.9231)(z+0.98276)} \end{aligned} \tag{4.94}$$

where $A(z) = 1/z$.

As a general rule of thumb, let the highest frequency of concern for v_i be 10 times the largest magnitude of a pole or zero of $P_e(w) \in \mathcal{P}$.

4.5.8 Design of the Prefilter $F(w)$

The design of $F(w)$ is done exactly in the same manner as for the continuous systems (see Section 3.14). The loop-shaping function $L_o(w) = A(w)L_{mo2}(w)$ only guarantees that the variation $\delta_R(v_i)$ does not exceed that permitted by Equations 4.63 through 4.65. For example, if at some v_i, $B_L(jv_i) = -6\,\text{dB}$ and $B_U(jv_i) = 0\,\text{dB}$ then the permitted $\delta_R(v_i) = 6\,\text{dB}$. An acceptable $L_o(jv)$ design may result in $|T_R'(jv_i)|$, where $T_R = FT_R'$, extremes of −4 and 1 dB with $\delta_R(v_i) =$ 5 dB. It is then necessary that

$$-2 \le |F(jv_i)| \le -1\ \text{dB}$$

The 1 dB overdesign of $|T_R'(jv_i)|$ allows 1 dB tolerances for $|F(jv_i)|$. $F(w)$ is chosen to satisfy the resulting tolerances on $|F(jv_i)|$. For this example, the satisfactory prefilter

$$F(w) = \frac{0.033329(w+229.2)}{w+7.639} \tag{4.95}$$

is obtained. Note that the prefilter $F(w) = 7.639/(w + 7.639)$ satisfies the w-domain but not the z-domain performance specifications. The reason for this is that applying the bilinear transformation in order to transform this *simple* $F(w)$ to the z-domain results in an $F(z)$ whose numerator and denominator orders are the same. In other words, due to the transformation the zero of $F(z)$ will alter the desired QFT prefilter frequency-domain characteristics. In order to prevent this from occurring, it is necessary after the *simpler* $F(w)$ is synthesized, according to the QFT design procedure, that enough *nondominant w-domain zeros are inserted into the simple $F(w)$ in order to create a prefilter transfer function whose numerator and denominator are of the same order.* This was done in obtaining Equation 4.95 whose zero at −229.2 is nondominant. By using Equation 4.51, the prefilter in the z-domain is

$$F(z) = \frac{0.1212(z+0.5)}{z-0.8182} \tag{4.96}$$

Utilizing Equations 4.47 and 4.94, where

$$P_{eo} = P_{e1}(z) = \frac{3.397\times 10^{-4} z(z+0.99131)}{(z-1)^2(z-0.97416)} \tag{4.97}$$

yields

$$G_1(z) = \frac{313.05(z-0.7094)(z-0.25)(z+0.99891)(z-0.97416)}{z(z-0.9231)(z+0.98276)(z+0.99131)} \tag{4.98}$$

Since the zero at −0.99891 and the pole at −0.99131 are very close to one another, on the negative real axis, Equation 4.98 can be approximated as

$$G_1(z) = \frac{313.05(z-0.7094)(z-0.25)(z-0.97416)}{z(z-0.9231)(z+0.98276)} \tag{4.99}$$

The poles of $G_1(z)$ at −0.98276 and −0.99131, since they are close to the unit ($U.C.$), can present a stability problem when $L_\iota(z) = G_1(z)P_{z_\iota}(z)$ is formed for all $\iota = 1, 2, \ldots, J$ cases. The same situation can occur when the controller of Equation 4.98 is utilized. The characteristic equation, $Q_\iota(z) = 1 + L_\iota(z)$ should be obtained in order to ascertain that a stable system exists for all J cases. This is illustrated in the following section. One must also ascertain that the Dahlin effect, as discussed in Ref. 206, is not a problem.

4.6 Simulation

As stressed in the previous sections, the essence of the QFT technique is the synthesis or loop shaping of the optimal loop transmission function. Applying the bilinear transformation to $G_1(w)$, to obtain $G_1(z)$, can result in significant warping of some of the poles and zeros of G_1. This "warping" is due to some pole and zeros of $G_1(w)$ lying outside the good Tustin (bilinear transformation) region of Figure 4.4. This warping may result in a significant degradation of the desired loop-shaping characteristics, that is, the violation of the B_o, B_h, and B_S bounds in the z-domain. Whether warping is significant or not can be determined by obtaining the Bode plots of $G_1(w)$ and $G_1(z)$. If the Bode plots within the BW determined by where the −24 dB value occurs on B_L in Figure 4.10 lie essentially on top of one another, then warping is insignificant with respect to satisfying the loop-shaping requirements in the z-domain. The controller Bode plots of $G_1(w)$ and $G_1(z)$, for this example, within the BW of $0 \le \omega \le 10$ lie essentially on top of one another indicating insignificant warping. There is an excellent correlation between these plots up to ≈40 rad/s.

A computer synthesis and simulation flowchart is presented in Appendix E of Ref. 49. The pertinent w-domain transfer functions for the $\iota = 1, 2, \ldots, J$ plants necessary to obtain the tracking control ratio $T_{R_\iota}(w)$, based upon Equations 4.39 through 4.49, and 4.87, are

$$P_{e_\iota}(s) = P_\iota(s)D(s) \tag{4.100}$$

$$P_{e_\iota}(w) = P_\iota D(w) \tag{4.101}$$

$$P_{z_\iota}(w) = \frac{2wP_{e_\iota}(w)}{w + 2/T} \tag{4.102}$$

$$G(w) = \frac{2wG_1(w)}{w + 2/T} \tag{4.103}$$

$$L_\iota(w) = G(w)P_{e_\iota}(w) = P_{z_\iota}(w)G_1(w) \tag{4.104}$$

$$T_{R_\iota}(w) = \frac{F(w)L_\iota(w)}{1 + L_\iota(w)} \tag{4.105}$$

$$T_{R_\iota}(w) = \frac{F(w)L_\iota(w)}{1 + L_\iota(w)} \tag{4.106}$$

Substituting Equation 4.104 into Equation 4.105 yields

$$T_{R_\iota}(w) = F(w)\left[\frac{G_1(w)P_{z_\iota}(w)}{1 + G_1P_{z_\iota}(w)}\right] \tag{4.107}$$

The pseudo-disturbance control ratio $T_{D_\iota}(w)$, Equation 4.87, is obtained based upon the *unit-step disturbance function* $d(t) = u_{-1}(t)$ and the following additional w-domain transfer functions:

$$Y_D(w) = \frac{P_\iota D(w)}{1 + L_\iota(w)} = \frac{P_\iota D(w)}{1 + P_{z_\iota}(w)G_1(w)} \tag{4.108}$$

$$D(w) = \frac{w + 2/T}{2w} \tag{4.109}$$

Note that Equation 4.109 is the bilinear transformation of the $\mathcal{Z}$-transfer function for a unit-step function, $z/(z-1)$. Thus, based upon these equations the pseudo-disturbance control ratio is

$$T_{D_\iota}(w) = \frac{Y_{D_\iota}(w)}{[D(w)]_P} = \frac{1}{D(w)}\left[\frac{P_\iota D(w)}{1 + P_{z_\iota}(w)G_1(w)}\right] \tag{4.110}$$

Since from Equations 4.102 and 4.109,

$$\left[\frac{P_\iota D(w)}{D(w)}\right] = \left[\frac{2w}{w + 2/T}\right] P_\iota D(w) = P_{z_\iota}(w) \tag{4.111}$$

Equation 4.109 can be expressed as follows:

$$T_{D_\iota}(w) = \left[\frac{P_{z_\iota}(w)}{1 + P_{z_\iota}(w)G_1(w)}\right] \tag{4.112}$$

From Equations 4.44 and 4.47, the z-domain tracking control ratio can be expressed as

$$T_{R_\iota}(z) = \left[\frac{F(z)L_\iota(z)}{1 + L_\iota(z)}\right] = \left[\frac{F(z)G_1(z)P_{z_\iota}(z)}{1 + P_{z_\iota}G_1(z)}\right] \tag{4.113}$$

The z-domain pseudo-disturbance control ratio, based upon Equations 4.40, 4.46, and 4.47, and $D(z) = z/(z-1)$, can be expressed as

$$T_{D_\iota}(z) = \left[\frac{Y_{D_\iota}(z)}{D(z)}\right]_P = \left[\frac{z-1}{z}\right]\left[\frac{P_\iota D(z)}{1 + P_{z_\iota}(z)G_1(z)}\right]$$

$$= \left[\frac{[(z-1)/z]P_{e_\iota}(z)}{1 + P_{z_\iota}(z)G_1(z)}\right] = \left[\frac{P_{z_\iota}(z)}{1 + P_{z_\iota}(z)G_1(z)}\right] \tag{4.114}$$

where

$$P_{z_\iota}(z) = \frac{z-1}{z} P_{e_\iota}(z)$$

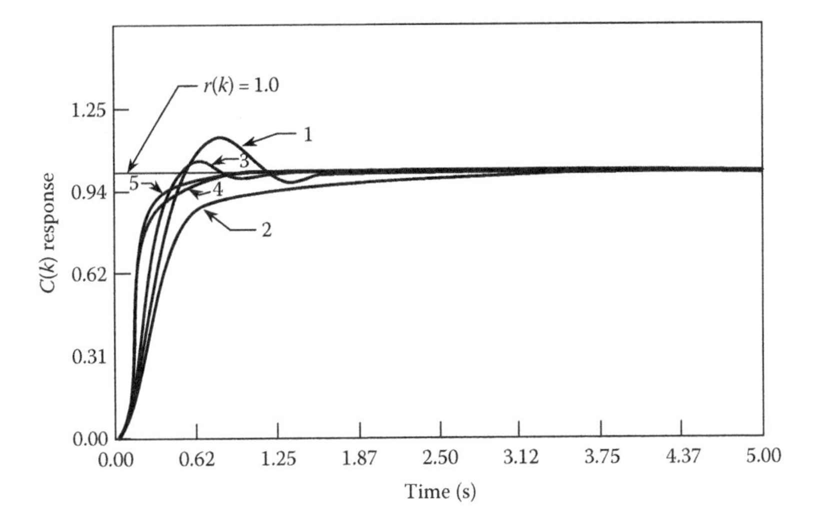

FIGURE 4.18

Step responses for:

Case	1	2	3	4	5
a	1	4	2	2	4
k	1	1	3	10	10

The w-domain tracking step responses for five extreme cases are shown in Figure 4.18. The w-domain figures of merit (FOM) shown in Tables 4.2 and 4.3 satisfy the specifications as represented by the boundaries in Figure 4.15. The Bode plots for these cases all lie within the area bounded by the B_U and B_L bounds of Figure 4.10. Since the specifications are met, the bilinear transformation of Equation 4.51 is applied to Equation 4.92 to obtain the z-domain controller transfer function of Equation 4.98. The poles and zeros of $L_\iota(z) = P_{z_\iota}(z)G_1(z)$ for all five cases are shown in Table 4.4. A plot of the poles and zeros of $L_\iota(s)$ is shown in Figure 4.19.

TABLE 4.2

Tracking *FOM*

Case (Plants)	Domain	Controller	M_p	t_p (s)	t_s (s)
1	w	$G_1(w)$	1.139	0.664	1.38
	z	$G_1(z)$	1.14	0.6280	0.97
		$G_1(z)_m$	1.14	0.6283	1.335
2	w	$G_1(w)$	1.000	—	2.893
	z	$G_1(z)$	1.000	—	2.906
		$G_1(z)_m$	1.000	—	2.801
3	w	$G_1(w)$	1.016	0.451	0.353
	z	$G_1(z)$	1.017	0.445	0.340
		$G_1(z)_m$	1.013	0.44506	0.3403
4	w	$G_1(w)$	1.000	—	0.478
	z	$G_1(z)$	1.000	—	0.471
		$G_1(z)_m$	1.000	—	0.4711
5	w	$G_1(w)$	1.000	—	0.609
	z	$G_1(z)$	Unstable	—	—
		$G_1(z)_m$	1.000	—	0.6021

TABLE 4.3
Disturbance *FOM*[a]

Domain	Controller	Case (Plant) 1	2	3	4	5
w	$G_1(w)$	0.08619	0.08618	0.08665	0.08619	0.08618
z	$G_1(z)$	0.08648	0.08648	0.8648	0.08648	Unstable
	$G_1(z)_m$	0.0820	0.0820	0.0820	0.0820	0.0820

[a] All responses are overdamped; only final values are shown.

TABLE 4.4
Poles and Zeros (All Real) of $L_1(z)$

Item	Case ι 1	2	3	4	5
Zeros	0.99131	0.9657	−0.9827	−0.99131	−0.9657
			0.7094		
			0.2500		
			0.97416		
Poles	0.97416	0.9006	0.9490	0.97416	0.9006
			0		
			1		
			0.9231		
			−0.98276		

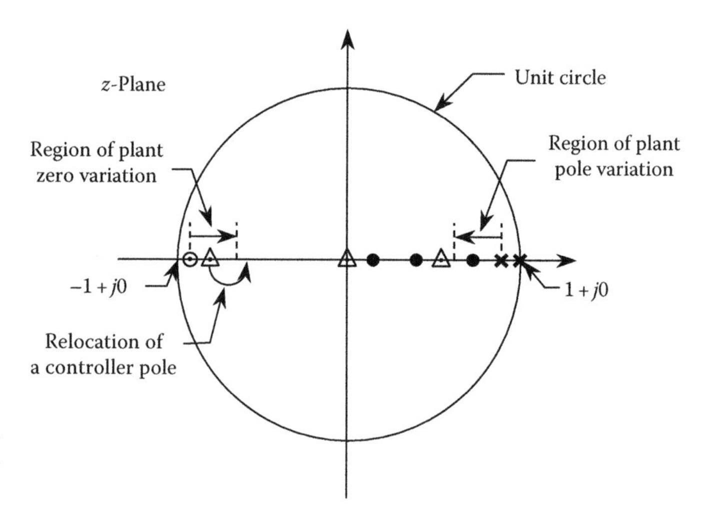

FIGURE 4.19
A root-locus pole-zero location resulting in an unstable system pole **x**, plant poles; △, controller poles ⊙, plant zeros; ● controller zeros.

Based upon the root-locus analysis of $L_1(z) = -1$, see Figure 4.20, for all five cases, and the resulting characteristic equations $Q_1(z)$, it is determined for case 5 that one of the closed-loop poles is at −1.008. This unstable pole, see Figure 4.20c, is a result of the $G_1(z)$ controller pole p_{g_v} at −0.98276 and the plant zero $z_{z_5}(z)$ at −0.9657. By trial and error, this controller pole is moved by an incremental amount toward the origin until all factors of $Q_1(z)$ are

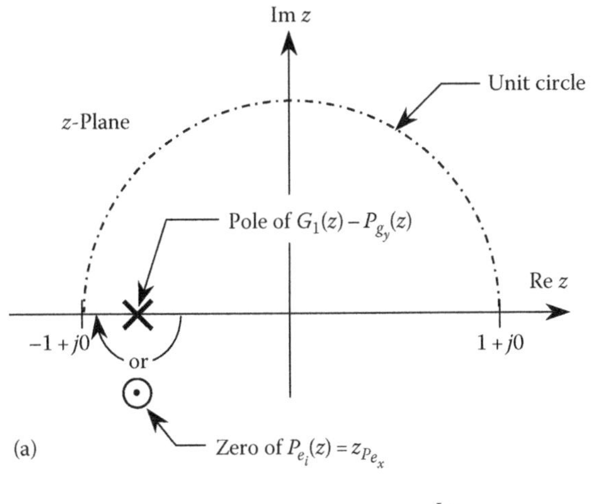

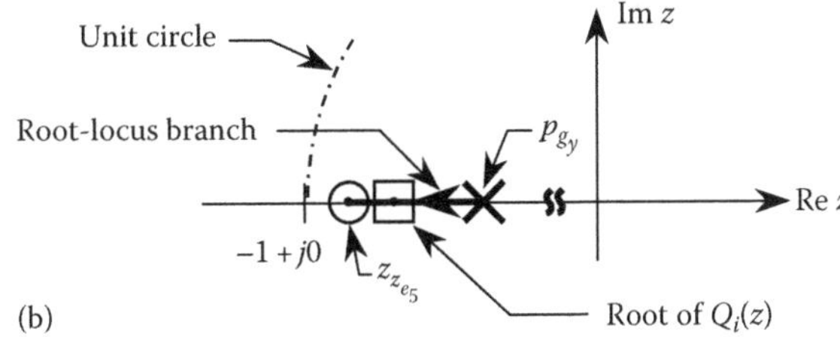

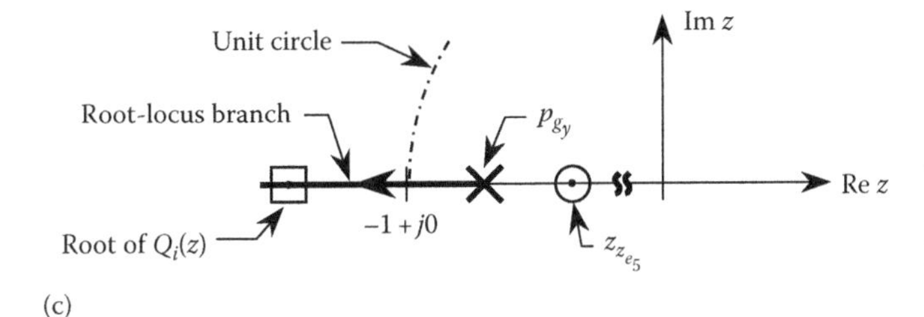

FIGURE 4.20
Analysis of root-locus plot of $L(z) = G_1(z)P_{e_1}(z) = -1$: (a) pole-zero locations determining system stability; location yielding: (b) a stable system and (c) an unstable system.

inside the *U.C.* For this example, this controller pole is replaced by a pole at −0.88 to yield the modified controller

$$G_1(z)_m = \frac{313.05(z-0.7094)(z-0.25)(z-0.97416)}{z(z-0.9231)(z+0.88)} \tag{4.115}$$

As Tables 4.2 and 4.3 indicate, the w-domain simulation results are essentially duplicated by the z-domain simulation utilizing the modified controller $G_1(z)_m$.

4.7 Basic Design Procedure for a MISO S-D Control System

The design procedure of Chapter 3 is repeated in this section but amplified and modified, where necessary, to be applicable to S-D MISO systems. In order to minimize the warping effect, as discussed in the previous section, the smallest value of sampling time allowable should be selected. The following step numbers, in general, correspond to those of Chapter 3.

1. Synthesize the T_{R_U} and the T_{R_L} models in the s-domain.
   a. The $T_{R_U}(s)$ model usually contains one zero and two poles; thus, it has an excess of poles over zeros of one. Since no ZOH device is involved, the $\mathcal{Z}$-transform of $T_{R_U}(s)$ cannot be taken. If Equation 4.14 is applicable, then the w-domain model is obtained as follows:

$$T_{R_U}(w) = T_{R_U}(s)\Big|_{s=w} \tag{4.116}$$

$$T_{R_L}(w) = T_{R_L}(s)\Big|_{s=w} \tag{4.117}$$

   In general, for the value of T that is chosen and the model that is synthesized, Equation 4.14 will be satisfied. The reader must ascertain that $\alpha^2 \ll 2$ and $\omega_{sp}T/2 \ll 0.297$ in order to utilize Equations 4.116 and 4.117.
   b. If Equation 4.16 is not applicable, one can (1) select a lower value of T, if this is permissible, or (2) synthesize $T_{R_U}(s)$ that has an excess of poles over zeros of two in order to be able to obtain the $\mathcal{Z}$-transform of $T_{R_U}(s)$. In doing the latter, it is necessary to modify $T_{R_L}(s)$ to have an excess of four poles over zeros in order to maintain the requirement that as ω_i increases in value $\delta_R(j\omega_i)$ also continually increases. Since the $T_{R_L}(s)$ model, by the QFT-imposed requirement, has at least an excess of two poles over zeros, its $\mathcal{Z}$-transform $T_{R_L}(z)$ can readily be obtained. Since the $\mathcal{Z}$-transform of the modified $T_{R_U}(s)$ and $T_{R_L}(s)$ can be obtained, $T_{R_U}(w)$ and $T_{R_L}(w)$ can be obtained using Equation 4.5.
2. The $T_D(w)$ model can be obtained from $T_D(s)$ as follows:
   a. For the s-domain model that has only an excess of poles over zeros of one, the procedure outlined in Step 1a can be repeated to obtain $T_D(w)$.
   b. If Equation 4.56 is not satisfied, then $T_D(s)$ can be modified in the same manner as done for $T_{R_U}(s)$ in Step 1b.
   c. For a discrete QFT design, it is best to select $T_D(s)$ = a constant value.
3. The frequency intervals for a QFT design may be divided (see Section 4.5) as follows:
   Region I. Where Equation 4.56 is satisfied for all J plants, the analog QFT templates $\Im P_e(j\omega)$ can be used to determine the appropriate bounds.
   Region II. For the frequency range of 0 to v_h, assuming the value of T is small enough, all performance specifications (B_{mo}, B_h, B'_h, and B_l) can be satisfied.
   Region III. For the frequency range of $v_h < v_i \le v_K$, only the specifications associated with the B_h-contours need to be satisfied.
   Region IV. For the frequency range of $v_K < v_i$, only the specifications associated with the B_S (M_L) stability contours need to be satisfied.
4. The number of templates that are necessary for the QFT design can be determined as follows:
   a. For the frequency range $0 < v_i < v_h$,
      i. If an analog QFT design has been accomplished for a given system, and if the *low-frequency range* ($0 < v_i < v_{0.25}$) Equation 4.56 is applicable, then $\Im P_e(j\omega_i) \approx \Im P_e(jv_i)$.

ii. Whether or not Equation 4.56 is applicable, the data for the templates can be determined in the w-domain in the same manner as is done for the analog design.

b. For the frequency range $v_h < v_i < v_K$, the templates in the w-domain have the same characteristic as that in the s-domain, that is, they approach a straight vertical line of V dB height.

c. For the frequency range $v_K < v_i$, the templates (see Figure 4.16a) broaden out again due to the "far out" pole or zero of $P_e(w)$ and then again approach a straight line. Sufficient B_S-contours need to be obtained in order to ensure a stable system. As a general rule, templates below v_K should be approximately an octave apart. Above v_K, the templates should be between two octaves to a decade apart. This can be modified depending on the evolving shaping characteristic of L_{mo}. See Section 4.5.6 for some additional guidelines.

5. Select the plant from the set of plants P_{e_ι} that has the smallest dB value and the largest (most negative) phase lag characteristics as the nominal plant P_{eo}.
6. Once the B'_h-contour has been constructed, based upon the M_L and v_i values, construct the $B_h(jv_i)$-contours corresponding to the v_i values of the templates.
7. *Through item 9.* The optimal bounds B_{mo} can be determined as follows:
   a. For $0 < v_i <$ 20/(smallest magnitude of a pole or zero; other than for the origin of $P_{eo}) = v_p$, determine the B_R, B_D, and B_{mo} bounds in the same manner of the analog QFT design.
   b. For frequencies greater than v_p, the $B_h(jv_i)$-contours are determined by the appropriate angular shift to the right due to the *a.p.f.* characteristic for $v_p < v_i < v_K$.
8. For $v_i > v_K$, the stability bounds B_S are determined by using the corresponding templates and applying the appropriate shift to the right of the M_L-contour due to the *a.p.f.* characteristic. The template is translated right or left while it is kept tangent to the corresponding M_L-contour. For the selected tangent points, the nominal point on the template yields points for the corresponding B_S bound.
9. Synthesize $L_{mo}(w)$ in the same manner as done for the analog QFT design.
   a. *First trial design*: Synthesize $L_{mo1} = P_o(s)|_{s=w}$ for the frequency range $0 < v_i < v_K$.
   b. *Second trial design*: Synthesize

$$L_{mo2} = [L_{mo1}]\left[\frac{(\,)\ldots(\,)}{(\,)\ldots(\,)}\right] \tag{4.118}$$

to satisfy the high-frequency ($v_i > v_K$) B_S-contour bounds on their left or bottom sides (see Figure 4.16a). Once a satisfactory $L_{mo}(w)$ has been synthesized, the controller transfer function is obtained from Equation 4.69, that is,

$$G_1(w) = \left(w + \frac{2}{T}\right)\frac{L_o(w)}{2wP_{eo}(w)} \tag{4.119}$$

or

$$G(w) = \frac{L_o(w)}{P_{zo}(w)} = \frac{-L_{mo}(w)A(w)}{P_{eo}(w)} \tag{4.120}$$

where $A(w) = -A'(w)$.

10. Synthesize $F(w)$ in the same manner as for the analog QFT design employing the templates and the plot of $L_o(jv)$. *Remember $F(w)$ must be equal order over equal order.* i.e., the order of the numerator and the order of the denominator of $F(w)$ must be the same.
11. Simulation is first accomplished in the w-domain in order to ensure that all system performance specifications have been satisfied. If the w-domain design is not satisfactory then do another w-domain design to try to achieve the desired performance. Once a satisfactory w-domain design has been achieved, transform $G(w)$ into the z-domain by using the bilinear transformation of Equation 4.6 to obtain $G(z)$. In order to validate that the desired loop shaping that has been achieved in the w-domain has been maintained in the z-domain, it is necessary to validate that the Bode plots $Lm\ G(jv_i)$ and $\angle G(jv_i)$ versus v_i are essentially the same as the Bode plots of $Lm\ G(z_i)$ and $\angle G(z_i)$ versus z_i in the frequency range of $0 < \omega_i < \omega_s/2$ where $z = e^{j\omega_i T}$. If these plots are not "reasonably" close to one another, then warping has been sufficient enough to degrade the desired z-domain loop-shaping characteristics. If this occurs, it will be necessary to modify $G(z)$ until the z-domain Bode plots of $G(z)$ are essentially the same as those for $G(w)$. Once a satisfactory $G(z)$ has been achieved, a discrete-time domain simulation is performed in order to validate that the desired S-D control system performance specifications have been achieved. Before proceeding with the z-domain simulation, the factors of the characteristic equation $Q_\iota(z)$ should be determined to ascertain that stable responses are achievable for all of the J cases as discussed in Section 4.6.

4.8 QFT Technique Applied to the PCT System

As noted in the preceding sections, the *n.m.p.* characteristic of a w-domain plant transfer function requires the use of an *a.p.f.* in order to apply the QFT technique[206]. The resulting modification of the *m.p.* analog QFT design procedure of Chapter 3, in order to take into account the use of this filter, results in a more involved w-domain QFT design procedure. When the requirements in Section 4.2.3 for a *PCT* representation of an S-D system are satisfied and $P(s)$ is *m.p.*, the simpler analog QFT design procedure of Chapter 3 can be applied to the *m.p.* PCT system. This PCT design approach is also referred to as a DIG design method. Using the minimum allowable practical sampling time enhances the possibility for a given *m.p.* plant to satisfy these requirements.

4.8.1 Introduction to PCT System DIG Technique

The DIG method of designing an S-D system, in the complex-frequency s-plane, requires a satisfactory PCT model of the S-D system. In other words, for the S-D system of Figure 4.21, the sampler and the ZOH units must be approximated by a linear continuous-time unit $G_A(s)$, as shown in Figure 4.22c. The DIG method requires that the dominant poles and zeros of the PCT model lie in the shaded area of Figure 4.4 for a high level of correlation with the S-D system. To determine $G_A(s)$, first consider the frequency component $E^*(j\omega)$ representing the continuous-time signal $E(j\omega)$, where all its sidebands are multiplied by $1/T$ (see Equation 6.9 of Ref. 206). Because of the low-pass filtering characteristics of an S-D system, only the primary component needs to be considered in the analysis of the system. Therefore, the PCT approximation of the sampler of Figure 4.21 is shown in Figure 4.22b.

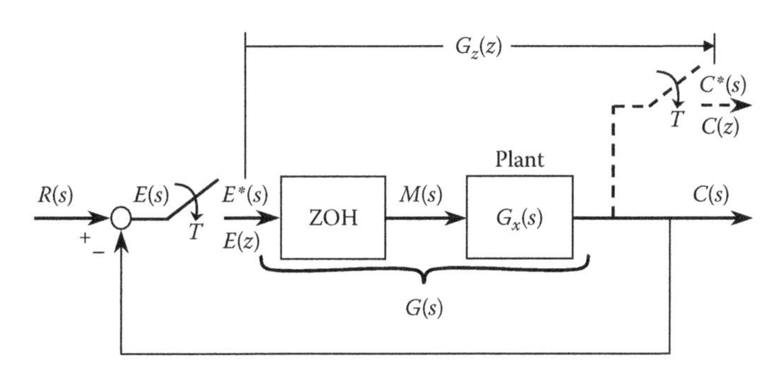

FIGURE 4.21
The uncompensated S-D control system.

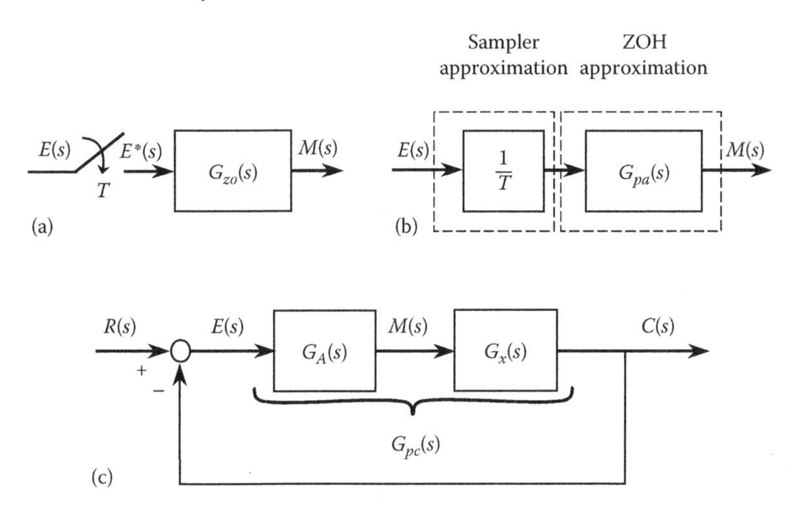

FIGURE 4.22
(a) Sampler and ZOH. (b) Approximations of the sampler and ZOH. (c) The approximate continuous-time control system equivalent of Figure 4.21.

Using the first-order Padé approximation (see Ref. 13), the transfer function of the ZOH device, when the value of T is small enough, is approximated as follows:

$$G_{zo}(s) = \frac{1-e^{-Ts}}{s} \approx \frac{2T}{Ts+2} = G_{pa}(s) \tag{4.121}$$

Thus, the Padé approximation $G_{pa}(s)$ is used to replace $G_{zo}(s)$ as shown in Figure 4.22a. This approximation is good for $\omega_c \leq \omega_s/10$, whereas the second-order approximation is good for $\omega_c \leq \omega_s/3$ (Ref. 297). Therefore, the sampler and the ZOH units of an S-D system are approximated in the PCT system of Figure 4.22c by the transfer function

$$G_A(s) = \frac{1}{T}G_{pa}(s) = \frac{2}{Ts+2} \tag{4.122}$$

Since Equation 4.122 satisfies the condition $\lim_{T\to 0} G_A(s) = 1$, it is an accurate PCT representation of the sampler and ZOH units, because it satisfies the requirement that as $T \to 0$ the output of $G_A(s)$ must equal its input. Further, note that in the frequency domain as $\omega_s \to \infty$ ($T \to 0$), the primary strip becomes the entire frequency-spectrum domain, which is the representation for the continuous-time system.[206]

TABLE 4.5

Analysis of a PCT System Representing a S-D Control System for $\zeta = 0.45$

Method	T (s)	Domain	K_x	M_p	t_p (s)	t_s (s)
DIR	0.01	z	4.147	1.202	4.16	9.53
DIG		s	4.215	1.206	4.11	9.478
DIR	0.1	z	3.892	1.202	4.2–4.3	9.8+
DIG		s	3.906	1.203	4.33–	9.90+
DIR	1	z	2.4393	1.199	6	13–14
DIG		s	2.496	1.200	6.18	13.76

Note that in obtaining PCT systems for an S-D system, the factor $1/T$ replaces only the sampler that is sampling the continuous-time signal. This multiplier of $1/T$ attenuates the fundamental frequency of the sampled signal, and all its harmonics are attenuated. To illustrate the effect of the value of T on the validity of the results obtained by the DIG method, consider the S-D closed-loop control system of Figure 4.21 where

$$G_x(s) = \frac{K_x}{s(s+1)(s+5)} \tag{4.123}$$

The closed-loop system performance for three values of T are determined in both the s- and z-domains, that is, the DIG technique and the DIR *technique* (the z-analysis), respectively. Table 4.5 presents the required value of K_x and the time-response characteristics for each value of T. Note that for $T \leq 0.1$ s there is a high level of correlation between the DIG and DIR models. For $T \leq 1$ s there is still a relatively good correlation. (The designer needs to specify, for a given application, what is considered to be "good correlation.")

4.8.2 Simple PCT Example

Figure 4.21 represents a basic or uncompensated S-D control system where

Case 1

$$G_x(s) = \frac{K_x}{s(s+1)} \tag{4.124}$$

is used to illustrate the approaches for improving the performance of a basic system. The root-locus plot $G_x(s) = -1$ shown in Figure 4.23a yields, for $\zeta = 0.7071$, $K_x = 0.4767$.

One approach for designing an S-D unity-feedback control system is to first obtain a suitable closed-loop model $[C(s)/R(s)]_T$ for the PCT unity-feedback control system of Figure 4.22, utilizing the plant of the S-D control system. This model is then used as a guide for selecting an acceptable $C(z)/R(z)$. Thus, for the plant of Equation 4.124,

$$G_{PC}(s) = G_A(s)G_x(s) = \frac{2K_x/T}{s(s+1)(s+2/T)} \tag{4.125}$$

and for $T = 0.1$ s,

$$\left[\frac{C(s)}{R(s)}\right]_T = \frac{G_{PC}(s)}{1+G_{PC}(s)} = \frac{20K_x}{s^3+21s^2+20s+20K_x} \tag{4.126}$$

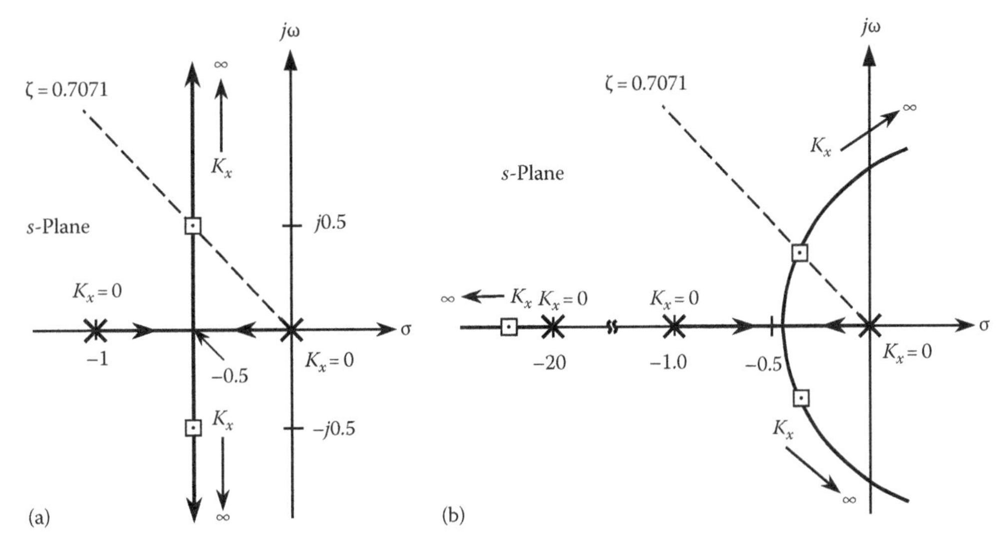

FIGURE 4.23
Root locus for (a) case 1, Equation 4.124 and (b) case 2, Equation 4.125.

The root locus for $G_{PC}(s) = -1$ is shown in Figure 4.23b. For comparison purposes, the root-locus plot for $G_x(s) = -1$ is shown in Figure 4.23a. These figures illustrate the effect of inserting a lag network in cascade in a feedback control system; that is, the lag characteristic of $G_{zo}(s)$ reduces the degree of system stability, as illustrated by Figure 4.23, which transformed a completely stable system into a conditionally stable system. Thus, for a given value of ζ, the values of t_p and t_s (and T_s) are increased. Therefore, as stated previously, the ZOH unit degrades the degree of system stability.

For the model it is assumed that the desired value of the damping ratio ζ for the dominant roots is 0.7071. Thus, for a unit-step input,

Case 2

$$[C(s)]_T = \frac{9.534}{s(s^3 + 21s^2 + 20s + 9.534)} = \frac{9.534}{s(s + 0.4875 \pm j0.4883)(s + 20.03)} \quad (4.127)$$

where $K_x = 0.4767$. The real and imaginary parts of the desired roots of Equation 4.127, for $T = 0.1$ s, lie in the acceptable region of Figure 4.4 for a good Tustin approximation. Note that the pole at −20.03 is due to the Padé approximation for $G_{zo}(s)$ (see Equation 4.121).

4.8.3 S-D Control System Example

The determination of the time-domain performance of the S-D control system of Figure 4.21 may be achieved by either obtaining the exact expression for $C(z)$ or applying the Tustin transformation to Equation 4.126 to obtain the approximate expression for $C(z)$.

Proceeding with the exact approach first requires the $\mathcal{Z}$-transfer function of the forward loop of Figure 4.21. For the plant transfer function of Equation 4.124,

$$G_z(z) = \mathcal{Z}\left[\frac{K_x(1-e^{-sT})}{s^2(s+1)}\right] = (1-z^{-1})\mathcal{Z}\left[\frac{K_x}{s^2(s+1)}\right]$$

$$= \frac{K_x[(T-1+e^{-T})z+(1-Te^{-T}-e^{-T})]}{z^2-(1+e^{-T}+K_x-TK_x-K_xe^{-T})z+e^{-T}+K_x-K_x(T+1)e^{-T}} \quad (4.128)$$

Thus, for $T = 0.1$ s and $K_x = 0.4767$,

Case 3

$$G_z(z) = \frac{0.002306(z+0.9672)}{(z-1)(z-0.9048)} \quad (4.129)$$

or

$$\frac{C(z)}{R(z)} = \frac{0.002306(z+0.9672)}{(z-0.9513 \pm j0.04649)} \quad (4.130)$$

The DIG technique requires that the s-domain model control ratio be transformed into a z-domain model. Applying the Tustin transformation to

$$\left[\frac{C(s)}{R(s)}\right]_T = \frac{G_{PC}(s)}{1+G_{PC}(s)} = F_T(s) \quad (4.131)$$

yields

$$\frac{[C(z)]_{TU}}{[R(z)]_{TU}} = [F(z)]_{TU} \quad (4.132)$$

This equation is rearranged to

$$[C(z)]_{TU} = [F(z)]_{TU}[R(z)]_{TU} \quad (4.133)$$

As stated in Section 4.8.1

$$R(z) = \mathcal{Z}[r^*(t)] = \frac{1}{T}[R(z)]_{TU} \quad (4.134)$$

$$C(z) = \mathcal{Z}[c^*(t)] = \frac{1}{T}[C(z)]_{TU} \quad (4.135)$$

Substituting from Equations 4.134 and 4.135 into Equation 4.132 yields

$$\frac{C(z)}{R(z)} = [F(z)]_{TU} \quad (4.136)$$

Substituting from Equation 4.135 into Equation 4.133 and rearranging yields

$$C(z) = \frac{1}{T}[F(z)]_{TU}[R(z)]_{TU} = \frac{1}{T}[\text{Tustin of } F_T(s)R(s)] \tag{4.137}$$

Thus, based upon Equation 4.136, the Tustin transformation of Equation 4.126, with $K_x = 0.4767$, results in a Tustin model of the control ratio as follows:

Case 4

$$\frac{C(z)}{R(z)} = \left[\frac{C(z)}{R(z)}\right]_{TU}$$

$$= \frac{5.672\times10^{-4}(z+1)^3}{(z-0.9513\pm j0.04651)(z+6.252\times10^{-4})} \tag{4.138}$$

Note that the dominant poles of Equation 4.138 are essentially the same as those of Equation 4.130 due to the value of T used, which resulted in the dominant roots lying in the good Tustin region of Figure 4.4. The nondominant pole is due to the Padé approximation of $G_{zo}(s)$. In using the exact $\mathcal{Z}$-transform, the order of the numerator polynomial of $C(z)/R(z)$, Equation 4.130, is one less than the order of its denominator polynomial. When using the Tustin transformation, the order of the numerator polynomial of the resulting $[C(z)/R(z)]_{TU}$ is in general equal to the order of its corresponding denominator polynomial (see Equation 4.138). Thus, $[C(z)]_{TU}$ results in a value of $c^*(t) \neq 0$ at $t = 0$, which is in error based upon zero initial conditions. Table 4.6 illustrates the effect of this characteristic of the Tustin transformation on the time response due to a unit-step forcing function. The degradation of the time response by the use of the Tustin transformation is minimal; that is, the resulting values of the FOM are in close agreement to those obtained by exact $\mathcal{Z}$-transform. Therefore, the Tustin transformation is a valid design tool when the dominant zeros and poles of $[C(s)/R(s)]_{TU}$ lie in the acceptable Tustin region of Figure 4.4.

Table 4.7 summarizes the time-response FOM for a unit-step forcing function of (1) the continuous-time system of Figure 4.22 for the two cases of $G(s) = G_x(s)$ (with $G_A(s)$ removed) and $G(s) = G_A(s)G_x(s)$ and (2) the S-D system of Figure 4.21 based upon the exact and Tustin expressions for $C(z)$. Note that the value of M_p occurs between 6.4 and 6.5 s and the value of t_s occurs between 8.6 and 8.7 s. The table reveals the following:

1. In converting a continuous-time system into an S-D system, the time-response characteristics are degraded.
2. The time-response characteristics of the S-D system, using the values of gain obtained from the continuous-time model, agree favorably with those of the continuous-time model. As may be expected, there is some variation in the values obtained when utilizing the exact $C(z)$ and $[C(z)]_{TU}$.

4.8.4 PCT System of Figure 4.9

The PCT system representing the MISO digital control system of Figure 4.9 is shown in Figure 4.24a. Note that *the sampler sampling the forcing function and the sampler in the system's output y(t) are replaced by a factor of* $1/T$. This diagram is simplified to the one shown in Figure 4.24b and is the structure that is used for the QFT design.

TABLE 4.6

Comparison of Time Responses between $C(z)$ and $[C(z)]_{TU}$ for a Unit-Step Input and $T = 0.1$ s

	c(kT)	
k	Case 3 (Exact), $C(z)$	Case 4 (Tustin), $[C(z)]_{TU}$
0	0.	0.5672E–03
2	0.8924E–02	0.9823E–02
4	0.3340E–01	0.3403E–01
6	0.7024E–01	0.7064E–01
8	0.1166	0.1168
10	0.1701	0.1701
12	0.2284	0.2283
14	0.2897	0.2894
18	0.4153	0.4148
22	0.5370	0.5364
26	0.6485	0.6478
30	0.7461	0.7453
34	0.8281	0.8273
38	0.8944	0.8936
42	0.9460	0.9452
46	0.9844	0.9836
50	1.011	1.011
54	1.029	1.028
58	1.039	1.038
60	1.042	1.041
61	1.042	1.042
62	1.043	1.043
63	1.043	1.043
64	1.043	1.043
65	1.043	1.043
66	1.043	1.043
67	1.043	1.043
68	1.042	1.042
85	1.022	1.022
86	1.021	1.021
87	1.019	1.019

TABLE 4.7

Time-Response Characteristics of the Uncompensated System

	M_p	t_p (s)	t_s (s)	K_1 (s^{-1})	Case
Continuous-time system					
$G(s) = G_x(s)$	1.04821	6.3	8.40–8.45	0.4767	1
$G_M(s) = G_A(s)G_x(s)$	1.04342	6.48	8.69		2
Sampled-data system					
$C(z)$	1.043	6.45	8.75	0.4765	3
$[C(z)]_{TU}$	1.043	6.45	8.75		4

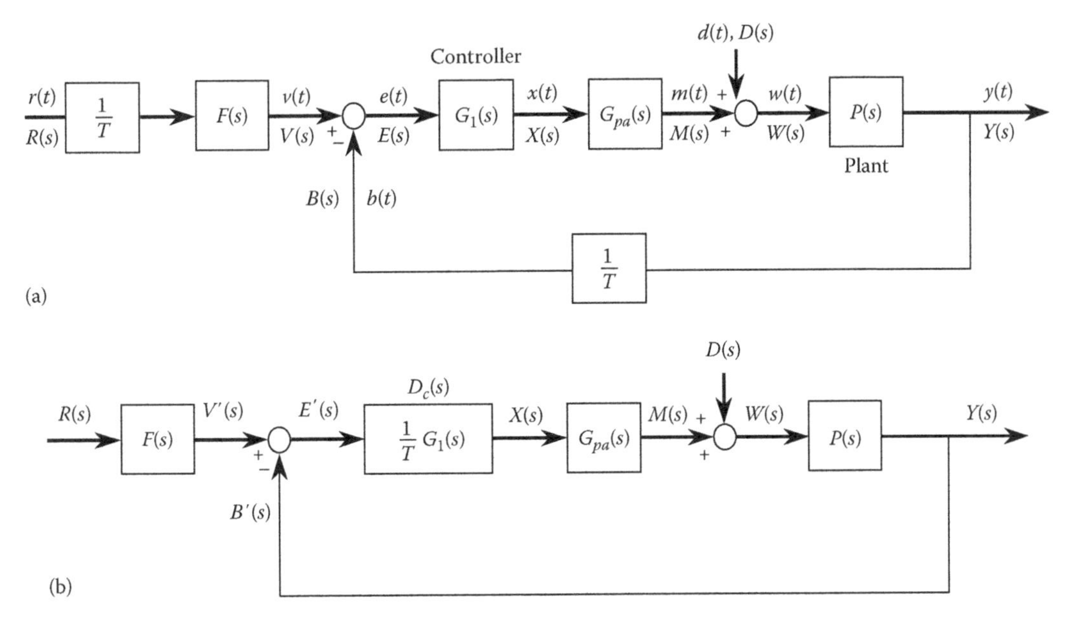

FIGURE 4.24
(a) PCT equivalent of Figure 4.9, and (b) diagram mathematical equivalent of Figure 4.24(a).

4.8.5 PCT Design Summary

Once a satisfactory controller $D_c(s)$ (see Figure 4.24b), whose orders of the numerator and denominator are, respectively, on_s and od_s, has been achieved, (1) if od_s is greater than $on_s + 2$, use the exact $\mathcal{Z}$-transform to obtain the discrete controller $D_c(z)$, or (2) if $on_s < od_s < on_s + 2$, use the Tustin transformation to obtain $[D_c(z)]_{TU}$.[206]

A final check should be made before simulating the discrete design; that is, the Bode plots of $D_c(s)|_{s=j\omega}$ and $D_c(z)|_{z=\exp(j\omega T)}$ should be essentially the same within the desired BW in order to ascertain that the discrete-time system response characteristics will be essentially the same as those obtained for the analog PCT system response. If the plots differ appreciably, it implies that warping has occurred and the desired discrete-time response characteristics may not be achieved (depending on the degree of the warping).

4.9 Applicability of Design Technique to Other Plants

As stated in Chapter 3, the QFT design techniques can be applied to an $n \times n$ MIMO control system by transforming this system into n^2 equivalent MISO systems (see Chapters 5 and 6). Each MISO system has two inputs, one desired (tracking) and one unwanted (cross-coupling effects) inputs, and one output. The solution of these MISO systems is guaranteed to satisfy the original MIMO problem. Thus, the MISO loop QFT design techniques of Chapters 3 and 4 can be applied to the QFT design of the n^2 equivalent MISO systems as discussed in Chapters 5 and 6.

4.10 Designing $L(w)$ Directly

It is possible to perform the loop shaping of the *n.m.p.* $L(w)$ by the method described in Chapter 3 with the aid of the *QFTCT* CAD package of Appendix F. Since the QFT design is done on the NC, it is necessary to verify all the J characteristic equations in order to determine if the sign of the controller gain is correct for achieving a stable system. This approach is simpler than using the approach of Section 4.5.

4.11 Summary

4.11.1 Minimum-Phase, Non-Minimum-Phase, and Unstable $P(s)$

As pointed out earlier, the minimum required sampling frequency ω_s is very strongly a function of the uncertainty problem, that is, to the extent of the plant uncertainty and of the system's performance tolerances. If the continuous plant $P(s)$ is *m.p.*, then the sampled loop transmission function $L(w)$ needs to have only one RHP zero at $w = 2/T$. This can be asserted even though the zeros of $P(s)$, unlike the poles, do not map directly onto the zeros of $P_e(w)$. The reason is that the range of the s-plane, over which the mapping is almost directly so, can be made as large as desired by increasing ω_s. The larger the ω_s, the smaller the importance of the surplus phase lags of the *a.p.f.* $A(w) = (-w + 2/T)/(w + 2/T)$, see Figure 4.13, because the BW of importance recedes into lower v values (compare the low v bounds of Figures 4.12 and 4.13).

However, if $P(s)$ is *n.m.p.*, then it may be impossible to achieve the desired performance for all P in $\mathcal{P}$, precisely as for the continuous systems.[31] One should then first check that the problem is solvable as a QFT continuous system design. If so, then a QFT discrete system design is achievable by using a large enough value for ω_s. If it is barely solvable in the former, ω_s will have to be very large. If it is not solvable for a continuous system design, it is certainly not solvable as a discrete system design. Unstable but *m.p.* plants pose no problem, even though the sampled $L(w)$ has RHP poles and the RHP zero at $2/T$. The reason is that with large enough ω_s, the RHP poles are arbitrarily close to the origin in the w-plane.[104] Unstable *n.m.p.* $P(s)$ do pose a great problem,[120] but it has been shown how time-varying compensation may be used for such plants to achieve stability (but not small system sensitivity) over large plant parameter uncertainity.[104]

4.11.2 Digital Controller Implementation

This chapter cannot be concluded without mentioning the importance of the implementation issue stressed in Figure 1.6. In designing a digital controller to achieve the desired control system performance requirements, the engineer must be aware of the factors that play an important role in the implementation of the controller. For example, how is maximum computational accuracy achieved? This particular aspect is discussed in Chapter 9 in Ref. 6. The reader is referred to the technical literature that discuss the other factors involved in controller implementation (e.g., Ref. 206).

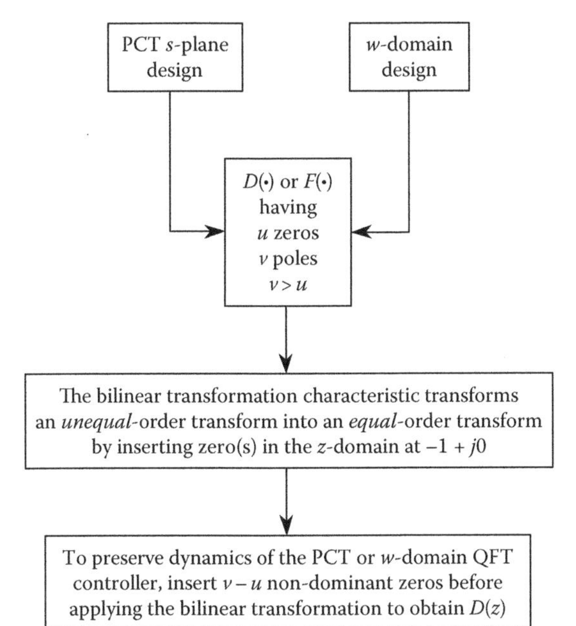

FIGURE 4.25
Bilinear transformation characteristics.

4.11.3 Conclusions

The basic feedback concepts and quantitative design procedures are the same in the w-domain for the S-D systems as in the s-domain for the continuous systems, if the design is executed in the v-domain ($w = jv$) for the former. The price paid is in the loop BW, that is, the desired BW may be achievable in the former by making ω_s large enough. Any practical S-D system is *n.m.p.* in the w-domain, but the harmful effect of the RHP zero at $2/T$ can be made arbitrarily small by making ω_s large enough. The main difference in quantitative design is the need for some experimentation to find $(\omega_s)_{min}$. When applicable, as determined by the plant characteristics and the minimum practical sampling that can be used, the QFT technique can be applied to the PCT system representation of the S-D control system. For an *m.p.* plant, the resulting PCT system is also *m.p.*, eliminating the need for the use of an *a.p.f.* The QFT design procedure of Chapter 3 can be applied to the PCT system, and the resulting $G_1(s) = D_c(s)$, by means of the Tustin transformation, is transformed into the z-domain to yield $D_c(z)$. Some of the characteristics involved in the use of a bilinear transformation, which must be kept in mind by a control system designer, are stressed in Figure 4.25. As pointed out in Section 4.10, with the aid of the *QFTCT* CAD package of Appendix F, it is possible to apply the method of Chapter 3 directly to obtain the *n.m.p.* $L(w)$ with the requirement to verify all J characteristic equations to assure that the sign of the controller gain is correct for a stable robust system design.

5

Diagonal MIMO QFT

5.1 Introduction

Control of multiple-input-multiple-output (MIMO) systems with model uncertainty is still one of the most difficult problems the control engineer has to face in real-world applications. (Now it is considered $\boldsymbol{Y}(s) = \boldsymbol{P}(s)\ \boldsymbol{U}(s)$, where $\boldsymbol{P}$ is an $n \times n$ transfer matrix, $\boldsymbol{P} \in \mathcal{P}$, and $\mathcal{P}$ is the set of possible plants due to uncertainty [see Figure 5.1].) Two of the main characteristics that define a MIMO system are the input and output directionality (different possible vectors to actuate $\boldsymbol{U}$ and different possible vectors to measure $\boldsymbol{Y}$) and the coupling among control loops (each input u_i can affect some outputs y_i, and each output can be affected by one or several inputs). This problem, which is known as interaction or coupling, makes the control system design less intuitive since any change in one loop interferes with the rest of the plant loops.

The previous chapters presented the design procedures of the quantitative feedback theory (QFT) synthesis technique for multiple-input-single-output (MISO) control systems. This chapter extends the QFT MISO design technique to enable the design of MIMO QFT control systems.[77,78,81] Section 5.2 introduces a MIMO plant example to illustrate and motivate the study. Section 5.3 presents an overview of the main control challenges and dynamic characteristics of a MIMO system. Section 5.4 describes the principal existing QFT techniques to design MIMO robust controllers ($\boldsymbol{G}(s)$). Sections 5.5 through 5.8 develop two diagonal MIMO QFT control design techniques ($\boldsymbol{G}(s)$, with $g_{ij} = 0$): the nonsequential technique (*Method 1*) and the sequential technique (*Method 2*). Both techniques consider the $n \times n$ MIMO control system as n equivalent single-loop MISO systems and n^2 prefilter/cross-coupling problems, which are each designed as MISO systems, as outlined in Chapter 3.[79]

5.2 Examples and Motivation

5.2.1 MIMO Plant Example

The multivariable plant, the $\boldsymbol{P}$ matrix, may be formed from the system linear differential equations, the system State Space matrix representation, or the transfer matrix description. Those *linear time-invariant* (LTI) MIMO system representations are as follows:

$$\begin{aligned} \dot{\boldsymbol{X}}(t) &= \boldsymbol{A}\boldsymbol{X}(t) + \boldsymbol{B}\boldsymbol{U}(t) \\ \boldsymbol{Y}(t) &= \boldsymbol{C}\boldsymbol{X}(t) + \boldsymbol{D}\boldsymbol{U}(t) \end{aligned} \tag{5.1}$$

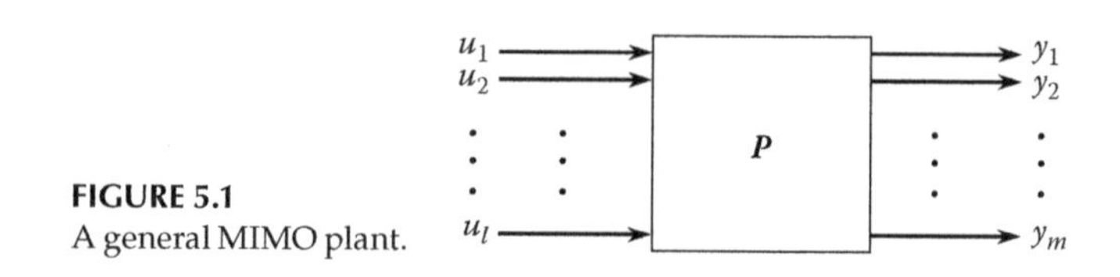

FIGURE 5.1
A general MIMO plant.

where
A, *B*, *C*, and *D* are constant matrices
X, *Y*, and *U* are *n*-dimensional vectors

Considering $D = 0$, the plant transfer function matrix (TFM) $P(s)$ is evaluated as

$$P(s) = C[sI - A]^{-1}B; \quad Y(t) = P(s)U(s) \tag{5.2}$$

If the plant model consists of *n* coupled *LTI* differential equations, the general plant model for a MIMO system with two inputs and two outputs has the form

$$\begin{aligned} a(s)y_1(s) + b(s)y_2(s) &= f(s)u_1(s) + g(s)u_2(s) \\ c(s)y_1(s) + d(s)y_2(s) &= h(s)u_1(s) + k(s)u_2(s) \end{aligned} \tag{5.3}$$

where
$a(s)$ through $k(s)$ are polynomials in *s*
$y_1(s)$ and $y_2(s)$ are the outputs
$u_1(s)$ and $u_2(s)$ are the inputs

In matrix notation the system is represented by

$$\begin{bmatrix} a(s) & b(s) \\ c(s) & d(s) \end{bmatrix} Y(s) = \begin{bmatrix} f(s) & g(s) \\ h(s) & k(s) \end{bmatrix} U(s) \tag{5.4}$$

This is defined as a 2×2 system. In the general case, with *n* inputs and *n* outputs, the system is defined as $n \times n$. Let the matrix premultiplying the output vector $Y(s)$ be $\alpha(s)$ and the matrix premultiplying the input vector $U(s)$ be $\beta(s)$. Thus, Equation 5.5 may then be written as

$$\alpha(s)Y(s) = \beta(s)U(s) \tag{5.5}$$

The solution of Equation 5.6 for $Y(s)$, where $\alpha(s)$ must be nonsingular, yields

$$Y(s) = \alpha^{-1}(s)\beta(s)U(s) = P(s)U(s) \tag{5.6}$$

Thus, the $n \times n$ plant TFM, $P(s)$, is

$$P(s) = \alpha^{-1}(s)\beta(s) \tag{5.7}$$

This plant matrix $P(s) = [p_{ij}(s)]$ is a member of the set $\mathcal{P} = \{P(s)\}$ of possible plant matrices, which are functions of the unstructured and/or structured uncertainty in the plant parameters. In practice, only finite set of $\boldsymbol{P}$ matrices are formed, representing the extreme boundaries of the projection on the Nichols chart (NC) of the plant uncertainty under varying conditions. Only LTI systems are considered in this chapter. By obtaining these LTI plants that describe the extremities of that projection, the QFT technique can also achieve, for many real-world nonlinear problems, a satisfactory control system design based on linear techniques.

Example 5.1

The Kirchoff's voltage law (loop method) is applied to the electrical network of Figure 5.2 to yield the following differential equations (the "D" operator notation for "d/dt" is used):

$$e_1(t) = (R_1 + R_4 + DL)i_1(t) - (R_4 + DL)i_2(t) \tag{5.8}$$

$$e_2(t) = -(R_4 + DL)i_1(t) + (R_2 + R_3 + R_4 + DL)i_2(t) \tag{5.9}$$

The Laplace transform of these equations are

$$(R_1 + R_4 + Ls)I_1(s) - (R_4 + Ls)I_2(s) = E_1(s) \tag{5.10}$$

$$-(R_4 + Ls)I_1(s) + (R_2 + R_3 + R_4 + Ls)I_2(s) = E_2(s) \tag{5.11}$$

assuming zero initial conditions. Let

$$\boldsymbol{Y} = \underbrace{\begin{bmatrix} y_1 \\ y_2 \end{bmatrix} = \begin{bmatrix} i_1 \\ i_2 \end{bmatrix}}_{n \text{ Outputs}} \quad \text{and} \quad \boldsymbol{U} = \underbrace{\begin{bmatrix} u_1 \\ u_2 \end{bmatrix} = \begin{bmatrix} e_1 \\ e_2 \end{bmatrix}}_{n \text{ Inputs}}$$

Thus, Equations 5.10 and 5.11 are of the form

$$d_{11}(s)y_1(s) + d_{12}(s)y_2(s) = m_{11}(s)u_1(s) + m_{12}(s)u_2(s) \tag{5.12}$$

$$d_{21}(s)y_1(s) + d_{22}(s)y_2(s) = m_{21}(s)u_1(s) + m_{22}(s)u_2(s) \tag{5.13}$$

where, for this example, $m_{12} = m_{21} = 0$. These equations are of the general form

$$d_{i1}(s)y_1(s) + \cdots + d_{in}(s)y_n(s) = m_{i1}(s)u_1(s) + \cdots + m_{in}(s)u_n(s) \tag{5.14}$$

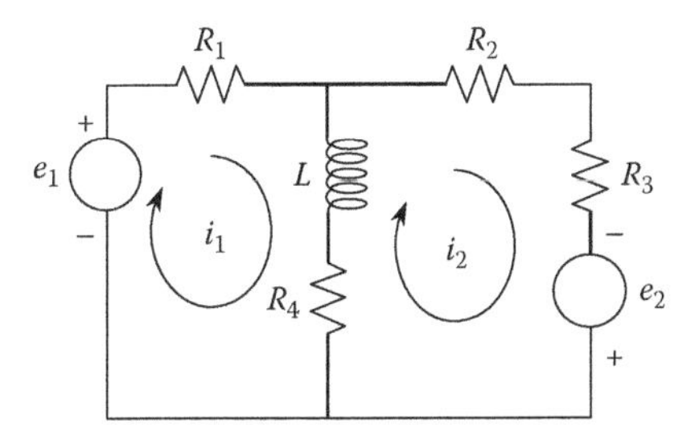

FIGURE 5.2
An electrical network.

where $i = 1, 2, \ldots, n$

$$D = \begin{bmatrix} d_{11} & d_{12} & \ldots & d_{1n} \\ d_{21} & d_{22} & \ldots & d_{2n} \\ \vdots & \vdots & & \vdots \\ d_{ni} & d_{n2} & \ldots & d_{nn} \end{bmatrix} \tag{5.15}$$

$$M = \begin{bmatrix} m_{11} & m_{12} & \ldots & m_{1n} \\ m_{21} & m_{22} & \ldots & m_{2n} \\ \vdots & \vdots & & \vdots \\ m_{n1} & m_{n2} & \ldots & m_{nn} \end{bmatrix} \tag{5.16}$$

Thus, Equations 5.12 and 5.13 can be expressed as follows:

$$\underset{n\times n}{D} \underset{n\times 1}{\begin{bmatrix} y_1 \\ y_2 \\ \vdots \\ y_n \end{bmatrix}} = \underset{n\times n}{M} \underset{n\times 1}{\begin{bmatrix} u_1 \\ u_2 \\ \vdots \\ u_n \end{bmatrix}} \tag{5.17}$$

which is of the form of Equation 5.5, with $\boldsymbol{\alpha} = \boldsymbol{D}$, $\boldsymbol{\beta} = \boldsymbol{M}$. For this example, based on Equation 5.6 and where $n = 2$, the expression for $\boldsymbol{Y}(s)$ is

$$\boldsymbol{Y}(s) = \begin{bmatrix} y_1(s) \\ y_2(s) \end{bmatrix} = \begin{bmatrix} p_{11}(s) & p_{12}(s) \\ p_{21}(s) & p_{22}(s) \end{bmatrix} \begin{bmatrix} u_1(s) \\ u_2(s) \end{bmatrix} = \begin{bmatrix} p_{11}(s)u_1(s) & p_{12}(s)u_2(s) \\ p_{21}(s)u_1(s) & p_{22}(s)u_2(s) \end{bmatrix} \tag{5.18}$$

Suppose that $p_{11}\, p_{22} = p_{12}\, p_{21}$, resulting in

$$|\boldsymbol{P}| = \begin{vmatrix} p_{11} & p_{12} \\ p_{21} & p_{22} \end{vmatrix} = p_{11}p_{22} - p_{12}p_{21} = 0 \tag{5.19}$$

Thus, for this example, $|\boldsymbol{P}|$ is singular, and Equation 5.18 yields

$$\begin{aligned} y_2(s) &= p_{21}(s)u_1(s) + p_{22}(s)u_2(s) = p_{21}(s)u_1(s) + \left(\frac{p_{12}(s)p_{21}(s)}{p_{11}(s)} \right) u_2(s) \\ &= \frac{p_{21}(s)[p_{11}(s)u_1(s) + p_{12}(s)u_2(s)]}{p_{11}(s)} = \left(\frac{p_{12}(s)}{p_{11}(s)} \right) y_1(s) \end{aligned} \tag{5.20}$$

Equation 5.20 reveals that y_1 and y_2 are not independent of each other. Thus, they cannot be controlled independently, that is, it is an uncontrollable system. Therefore, *when* ***P*** *is singular the system is uncontrollable.*

From Equation 5.18 the signal flow graph (SFG) of Figure 5.3a is obtained, which represents a plant with structured/unstructured plant parameter uncertainty with no cross-coupling effects. Figure 5.3b shows the SFG of the compensated MIMO closed-loop control system where the compensator and prefilter matrices are, respectively,

$$\boldsymbol{G}(s) = \begin{bmatrix} g_{11}(s) & g_{12}(s) \\ g_{21}(s) & g_{22}(s) \end{bmatrix}; \quad \boldsymbol{F}(s) = \begin{bmatrix} f_{11}(s) & f_{12}(s) \\ f_{21}(s) & f_{22}(s) \end{bmatrix} \tag{5.21}$$

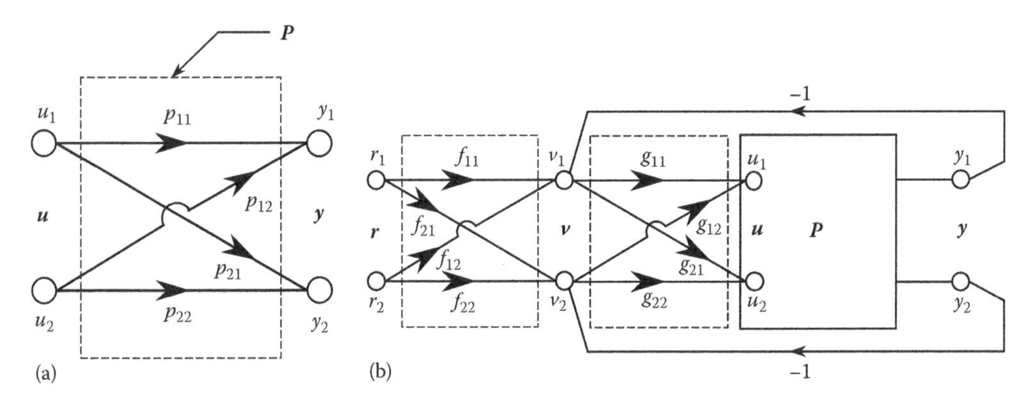

FIGURE 5.3
The SFG of (a) Equation 5.14 and (b) the compensated MIMO control system.

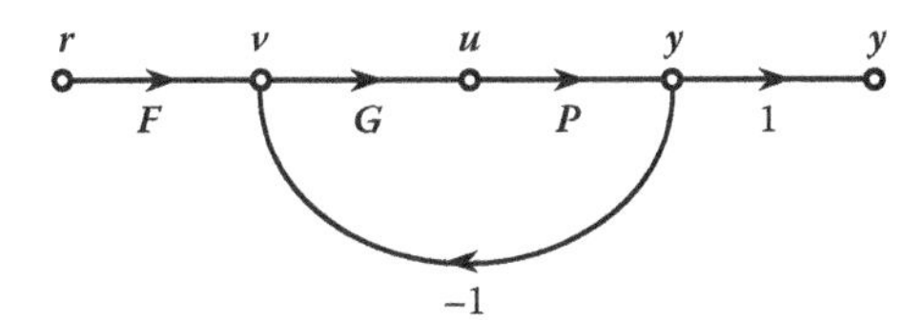

FIGURE 5.4
The simplified SFG of Figure 5.3b.

The control ratio matrix $\boldsymbol{T}$ is

$$\boldsymbol{T}(s) = \begin{bmatrix} t_{11}(s) & t_{12}(s) \\ t_{21}(s) & t_{22}(s) \end{bmatrix} \tag{5.22}$$

where the closed-loop system control ratio $t_{ij} = y_i/r_j$ relates the *ith output to the jth input*, and the *tolerance (specifications) matrix* is given by

$$\Im(s) = \begin{bmatrix} \tau_{11}(s) & \dots & \tau_{1n}(s) \\ \vdots & \ddots & \vdots \\ \tau_{n1}(s) & \dots & \tau_{nn}(s) \end{bmatrix} \tag{5.23}$$

and where the elements of the $\mathcal{T}(s)$ matrix are given by $\tau_{ij}(s) \rightarrow a_{ij} \leq t_{ij}(s) \leq b_{ij}$. Figure 5.3b may be represented by the simplified SFG of Figure 5.4.

5.2.2 Introduction to MIMO Compensation

Figure 5.3b has the $n \times n$ closed-loop MIMO feedback control structure of Figure 5.5 in which $\boldsymbol{F}$, $\boldsymbol{G}$, $\boldsymbol{P}$, $\boldsymbol{T}$ are each $n \times n$ matrices, and $\mathcal{P} = \{\boldsymbol{P}\}$ is a set of transfer matrices with plant uncertainty. There are n^2 closed-loop system transfer functions (transmissions) $t_{ij}(s)$ relating the outputs $y_i(s)$ to the inputs $r_j(s)$, that is, $y_i(s) = t_{ij}(s)r_j(s)$. In a quantitative problem statement, there are *tolerance (specifications) bounds* on each $t_{ij}(s)$, giving n^2 sets of acceptable regions $\tau_{ij}(s)$, which are to be specified in the design; thus, $t_{ij}(s) \in \tau_{ij}(s)$ and $\mathcal{T}(s) = \{t_{ij}(s)\}$. The application of QFT to 2×2 and 3×3 systems has been highly developed and is illustrated in later sections.

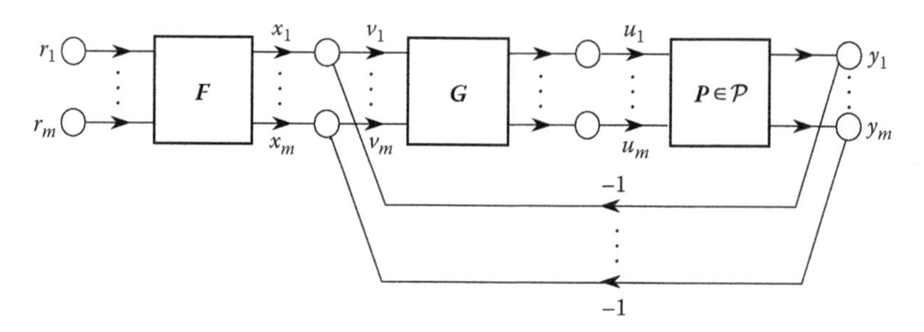

FIGURE 5.5
MIMO feedback structure.

In Figure 5.5, the compensator **G** may be characterized by either a diagonal or a non-diagonal matrix. The use of a fully populated (non-diagonal) **G** matrix allows the control system designer more design flexibility to achieve the desired performance specifications in some cases (see Chapter 6).[93,97,102] In other words, the use of non-diagonal elements of **G** may ease the diagonal compensator design problem. This chapter is devoted to the use of a diagonal **G**. Chapter 6 discusses the use of a non-diagonal compensator for a MIMO QFT design. From Figure 5.5, the following equations can be written as

$$\boldsymbol{y} = \boldsymbol{P}\boldsymbol{u}; \quad \boldsymbol{u} = \boldsymbol{G}\boldsymbol{v}; \quad \boldsymbol{v} = \boldsymbol{x} - \boldsymbol{y}; \quad \boldsymbol{x} = \boldsymbol{F}\boldsymbol{r} \tag{5.24}$$

In these equations, $\boldsymbol{P}(s) = [p_{ij}(s)]$ is the matrix of plant transfer functions, $\boldsymbol{G}(s)$ is the matrix of compensator transfer functions and is often simplified so that it is diagonal, that is, $\boldsymbol{G}(s) = \text{diag}\{g_i(s)\}$, and $\boldsymbol{F}(s) = \{f_{ij}(s)\}$ is the matrix of prefilter transfer functions, which may also be a diagonal matrix. The first two expressions yield

$$\boldsymbol{y} = \boldsymbol{P}\,\boldsymbol{G}\,\boldsymbol{v} \tag{5.25}$$

which is utilized with the remaining two expressions to obtain

$$\boldsymbol{y} = \boldsymbol{P}\,\boldsymbol{G}\,[\boldsymbol{x} - \boldsymbol{y}] = \boldsymbol{P}\,\boldsymbol{G}[\boldsymbol{F}\boldsymbol{r} - \boldsymbol{y}] \tag{5.26}$$

This equation is rearranged to yield

$$\boldsymbol{y} = [\boldsymbol{I} + \boldsymbol{P}\,\boldsymbol{G}]^{-1}\,\boldsymbol{P}\,\boldsymbol{G}\,\boldsymbol{F}\,\boldsymbol{r} \tag{5.27}$$

where the system control ratio $\boldsymbol{T}$ relating $\boldsymbol{r}$ to $\boldsymbol{y}$ is

$$\boldsymbol{T} = [\boldsymbol{I} + \boldsymbol{P}\,\boldsymbol{G}]^{-1}\,\boldsymbol{P}\,\boldsymbol{G}\,\boldsymbol{F} \tag{5.28}$$

To appreciate the difficulty of the design problem, note the very complex expression for t_{11} given by Equation 5.29, which is the first element of Equation 5.22 for the case $n = 3$ and with a diagonal **G** matrix. However, the QFT design procedure systematizes and simplifies the manner of achieving a satisfactory system design:

$$
\begin{aligned}
t_{11} = \{&(p_{11}f_{11}g_1 + p_{12}f_{21}g_2 + p_{13}f_{31}g_3)(1+p_{22}g_2)(1+p_{33}g_3 - p_{23}p_{32}g_2g_3) \\
&-(p_{21}f_{11}g_1 + p_{22}f_{21}g_2 + p_{23}f_{31}g_3)\,[p_{12}g_2(1+p_{33}g_3) - p_{32}p_{13}g_2g_3] \\
&+(p_{31}f_{11}g_1 + p_{32}f_{21}g_2 + p_{33}f_{31}g_3)[p_{23}p_{12}g_2g_3 - (1+p_{22}g_2)p_{13}g_3]\}/ \\
&\{(1+p_{11}g_1)[(1+p_{22}g_2)(1+p_{33}g_3) - p_{23}p_{32}g_2g_3] \\
&-p_{21}g_1[p_{12}g_2(1+p_{33}g_3) - p_{32}p_{13}g_2g_3] + p_{31}g_1[p_{12}p_{23}g_2g_3 - p_{13}g_3(1+p_{22}g_2)]\} \qquad (5.29)
\end{aligned}
$$

There are $n^2 = 9$ transfer functions such that the $t_{ij}(s)$ expressions all have the same denominator, and there may be considerable uncertainty in the nine plant transfer functions $p_{ij}(s)$. The design objective is a system, which behaves as desired for the entire range of uncertainty. This requires finding nine $f_{ij}(s)$ and three $g_i(s)$ such that each $t_{ij}(s)$ stays within its acceptable region $\tau_{ij}(s)$, no matter how the $p_{ij}(s)$ may vary. Clearly, this is a very difficult problem. Even the stability problem alone, ensuring that the characteristic polynomial (the denominator of Equation 5.28) has no factors in the right-half plane (RHP) for all possible $p_{ij}(s)$, is extremely difficult. Most design approaches treat stability for fixed parameter set, neglecting uncertainty and attempting to cope with the plant uncertainty by trying to design the system to have conservative stability margins. Two highly developed diagonal MIMO QFT design techniques, *Method 1* and *Method 2*, exist for the design of such systems and are presented in this chapter. In both approaches the MIMO system is converted into an equivalent set of single-loop systems. *Methods 1* and *2* utilize the MISO design method of Chapters 3 and 4. *Method 2*, "the improved method," is an outgrowth of *Method 1* in which the designed components of the previously designed loop are used in the design of the succeeding loops.

5.2.3 MIMO Compensation

The basic MIMO compensation structure for a 2 × 2 MIMO system is shown in Figure 5.6. The structure for a 3 × 3 MIMO system is shown in Figure 5.7. These consist of the uncertain plant matrix **P**, the diagonal compensation matrix **G**, and the prefilter matrix **F**.

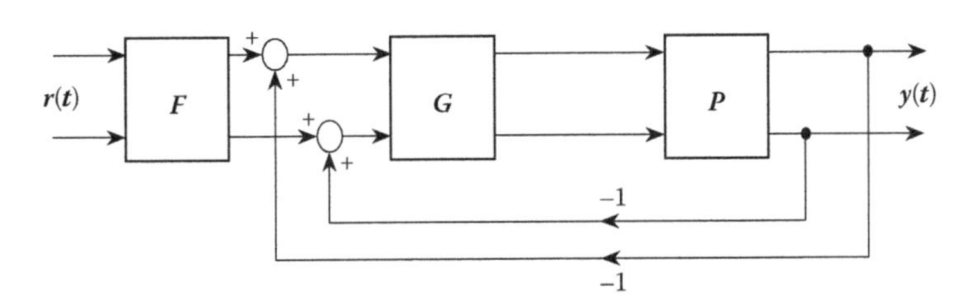

FIGURE 5.6
MIMO control structure 2 × 2 system.

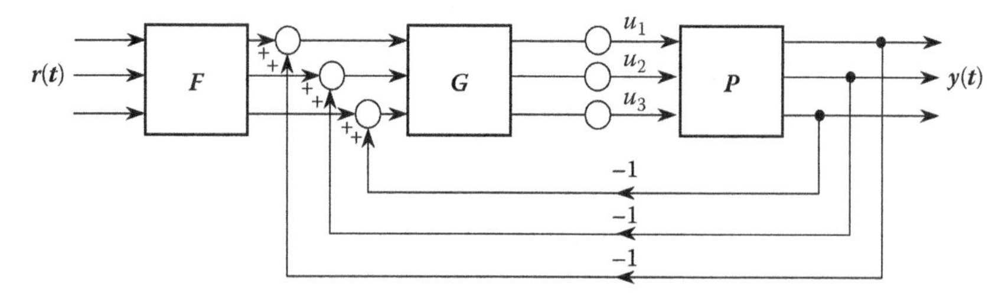

FIGURE 5.7
MIMO control structure 3 × 3 system.

This chapter considers only a diagonal **G** matrix, though a non-diagonal **G** matrix (see Chapter 6) allows the designer much more design flexibility.[69,93,97,102] These matrices are defined in Equation 5.30.

The dashes in Equation 5.30 denote the **G**, **F**, and **P** matrices for a 2 × 2 system. Substituting these matrices into Equation 5.28 yields the $t_{ij}(s)$ control ratios relating the ith output to

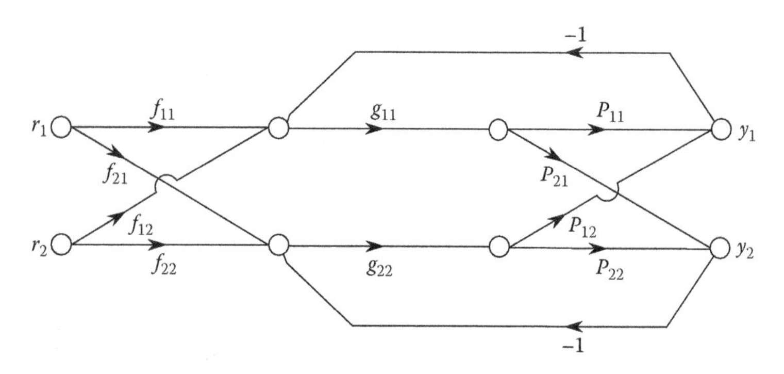

FIGURE 5.8
2 × 2 MIMO signal flow graph.

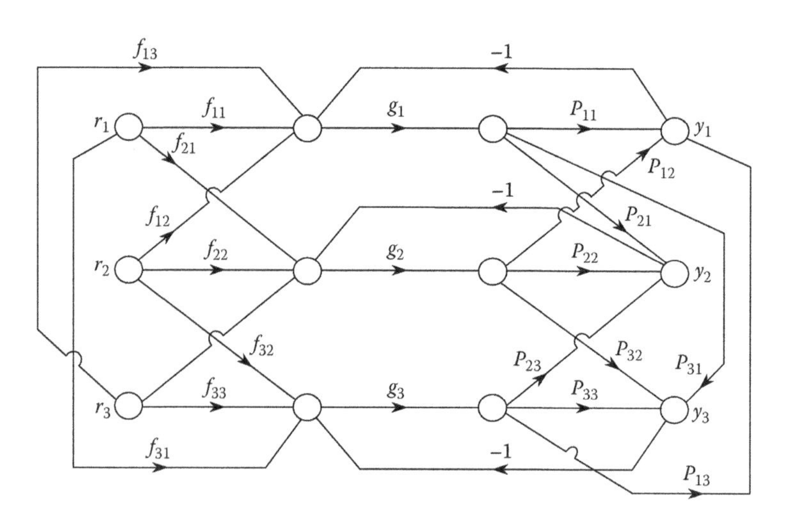

FIGURE 5.9
3 × 3 MIMO signal flow graph.

the jth input. From these $t_{ij}(s)$ expressions the SFG of Figure 5.8 is obtained. The SFG of Figure 5.9 for a 3 × 3 system is obtained in a similar manner:

$$G = \begin{bmatrix} g_1 & 0 & \dots & 0 \\ 0 & g_2 & \dots & 0 \\ \vdots & \vdots & \ddots & \vdots \\ 0 & 0 & \dots & g_n \end{bmatrix}; \quad F = \begin{bmatrix} f_{11} & f_{12} & \dots & f_{1n} \\ f_{21} & f_{22} & \dots & f_{2n} \\ \vdots & \vdots & \ddots & \vdots \\ f_{n1} & f_{n2} & \dots & f_{nn} \end{bmatrix}; \quad P = \begin{bmatrix} p_{11} & p_{12} & \dots & p_{1n} \\ p_{21} & p_{22} & \dots & p_{2n} \\ \vdots & \vdots & \ddots & \vdots \\ p_{n1} & p_{n2} & \dots & p_{nn} \end{bmatrix} \tag{5.30}$$

Division I: General Overview of MIMO Systems

Division I of this chapter, Sections 5.3 and 5.4, presents a general overview of MIMO systems and a review of MIMO QFT control design techniques respectively.

5.3 MIMO Systems—Characteristics and Overview[187]

The systems considered from now on are supposed to be linearizable, at least within a range of operating conditions, as is the case with most physical real problems. This type of systems can be described by means of an $n \times m$ matrix of transfer functions $\boldsymbol{P}(s) = [p_{ij}(s)]$, also called as the plant TFM, which relates the m input variables (manipulated variables, $u_j(s)$ with $j = 1, \dots, m$) with the n output variables (controlled variables, $y_i(s)$ with $i = 1, \dots, n$), so that $[y_i(s)] = \boldsymbol{P}(s)\,[u_j(s)]$.

In general, the MIMO TFM $\boldsymbol{P}(s)$ can be rectangular. However, most of the related literature deals with square systems, that is, with the same number of inputs and outputs. If it is not the case for the plant under study, there exist different procedures that can be followed, such as using weighting matrices that reduce the system to a square effective plant matrix,[6] leaving some outputs (inputs) uncontrolled (not used), or looking for independent extra inputs or outputs, depending on which one is in excess.[211]

Multivariable systems have aroused great interest within the control community and many design techniques have been developed. This is not only because of their mathematical and computational challenges (derived from the matrix representation) but also due to their inherent features that do not appear in single-input-single-output (SISO) systems. This particular nature of MIMO systems poses additional difficulties to control design such as directionality, coupling, and transmission zeros and all with the intrinsic uncertainty of real-world applications.

5.3.1 Loops Coupling and Controller Structure

The most distinctive aspect of MIMO plants is the existence of coupling among the different control loops. Thus, one input (manipulated variable) can affect various outputs (controlled variables), and the other way around; that is, an output can be affected by one or several inputs. Consequently, applying a control signal to one of the plant inputs causes the responses at more than one output, which hampers the controller design. Then, it becomes hard to predict the type and amount of control action simultaneously needed at several inputs in order to get outputs to behave as desired.

The first and easiest way that comes to mind for dealing with a MIMO system is to reduce it to a set of SISO problems ignoring the system interactions, which is the so-called decentralized control.[219] Then, each input is responsible for only one output and the resulting controller is diagonal. Finding a suitable input–output pairing therefore becomes essential for decentralized control. However, this approach is only valid provided the coupling among variables is not important, which unfortunately is not the case for many real applications. In other approaches the goal is to remove, or at least greatly reduce, the effects of the interaction before performing a decentralized control of the somehow decoupled plant as if they are independent input–output pairs.

In any case, it is necessary to quantify the amount of coupling present in the system. Many of the MIMO design techniques, particularly the sequential ones, strongly depend on the correct selection and pairing of inputs and outputs at the beginning of the design procedure. Determining the controller structure is also crucial. This means deciding whether the multivariable system can be divided into several SISO or smaller MIMO subsystems and establishing the off-diagonal compensators needed if a populated matrix controller is to be designed, avoiding non-required extra controllers. This issue becomes extremely complex in the presence of large coupling and has generated great interest within the control community, as shown by the numerous related works such as Refs. 205, 253, 256–258, 264, 268, and 270. Nevertheless, too often, only the extreme controller structures (the fully centralized [fully populated matrix] and the fully decentralized [set of SISO loops]) are discussed.[261]

5.3.1.1 Interaction Analysis

An extensive amount of work on the way of quantifying the system interaction can be found in the literature.[204,219] One of the most popular techniques is the *relative gain array* (RGA) defined by Bristol as a matrix of relative gains Λ based on the steady-state gains of the plant:[224]

$$\mathbf{\Lambda} = \begin{bmatrix} \lambda_{11} & \lambda_{12} & \cdots & \lambda_{1n} \\ \lambda_{21} & \lambda_{22} & \cdots & \lambda_{2n} \\ \vdots & \vdots & \cdots & \vdots \\ \lambda_{n1} & \lambda_{n2} & \cdots & \lambda_{nn} \end{bmatrix} \quad (5.31)$$

The elements λ_{ij} that constitute this matrix are dimensionless and represent the relation between the following gains of the system:

$$\lambda_{ij} = \frac{K_{OFF}}{K_{ON}} \quad (5.32)$$

where

K_{OFF} is the open-loop gain between the output i and the input j when the rest of loops are open

K_{ON} is the open-loop gain between the same output i and input j when the remaining loops are working in automatic mode, that is, they are closed

Another way of computing the RGA is through the following matrix expression:

$$\Lambda = P_0 \otimes (P_0^{-1})^T \tag{5.33}$$

where P_0 is an $n \times n$ matrix representing the steady-state ($\omega = 0$) process. Its elements are determined by applying the final value theorem to the transfer functions describing the system dynamics. The operator [$\otimes$] denotes element-by-element multiplication (Hadamard or Schur product).

The RGA provides a scaling-independent measure of the coupling among loops and useful information on how to achieve the best possible pairing of variables.[198] Its elements λ_{ij} are closely related to the interaction among the different control loops. Several possible values and their interpretations are as follows:

- $\lambda_{ij} = 1 \Rightarrow$ The closure of the rest of loops does not change the influence of the input j on the output i. Hence the ij loop is decoupled from the rest of the system and can be treated as a SISO subsystem.
- $\lambda_{ij} = 0 \Rightarrow$ There is no influence of the manipulated variable j over the control variable i.
- $0 < \lambda_{ij} < 1 \Rightarrow$ When the rest of the loops are closed, the gain between the input j and the output i increases, that is, $K_{ON} > K_{OFF}$.
- $\lambda_{ij} < 0 \Rightarrow$ At the closure of the remaining loops, the system gain changes its sign. Providing a controller with negative gain for the normal situation (all the loops closed and working), the system reacts in the opposite direction if some of the remaining loops are open for any reason. Then, integrity is lost.
- $\lambda_{ij} > 1 \Rightarrow$ When all the loops are closed, higher gains are required. The interaction reduces the gain in the ij control loop: $K_{OFF} > K_{ON}$.
- $\lambda_{ij} > 10 \Rightarrow$ Pairings of variables with large RGA values are undesirable. They are sensitive to modeling errors and to small variations in the loop gain.

Given its importance, the RGA method has been the subject of multiple revisions and research. For instance, although originally defined for the steady-state gain, the RGA was extended to a frequency-dependent definition and used to assess the interaction at frequencies other than zero.[198,219,238,259] In most cases, it is the value of RGA at frequencies close to crossover that is the most important one, and both the gain and the phase are to be taken into account. For a detailed analysis of the plant, RGA is considered as a function of frequency

$$RGA(j\omega) = P(j\omega) \otimes (P^{-1}(j\omega))^T \tag{5.34}$$

where $P(j\omega)$ is a frequency-dependent matrix.

According to the meaning of the RGA elements outlined earlier, it is desired to pair variables so that λ_{ij} is positive and close to 1, because this means that the gain from the input u_j to the output y_i is not very much affected by closing the other loops. On the other hand, a pairing corresponding to $0 < \lambda_{ij} < 1$ values means that the other loops reinforce the gain of the given loop. A pairing corresponding to $1 < \lambda_{ij}$ values means that the other loops reduce the gain of the given loop. And negative values of λ_{ij} are undesirable because it means that the steady-state gain in the given loop changes sign when the other loops are closed.

As a conclusion, to avoid instability caused by interactions, in the crossover region one should prefer pairings for which the RGA matrix in this frequency range is close to identity. In the same way, to avoid instability caused by poor integrity, one should avoid pairings with negative steady-state RGA elements.

Further information on how to perform the pairing is available in Ref. 198, and different properties of the RGA can be consulted in Refs. 219, 224, 253, 262, 263, 267, and 271.

Other measures of interaction that exist in the literature are the *block relative gain*;[257,260,265] the *relative disturbance gain*;[254,255,262,263] and the *generalized relative disturbance gain*.[255]

5.3.2 Multivariable Poles and Zeros

Due to the aforementioned interaction among loops, the poles and zeros of a multivariable system may differ from what could be deduced from observation of the elements of the plant TFM.[204] In fact, the pole positions can be inferred from the matrix elements $p_{ij}(s)$, but not their multiplicity, which is of great importance when applying Nyquist-like stability theorems in the presence of RHP poles. Regarding the multivariable zeros (also known as transmission zeros), neither the position nor the multiplicity can be derived from the direct observation of $p_{ij}(s)$. These multivariable zeros present a transmission-blocking property, since they provoke the loss of rank of the plant TFM.

Thus, it is necessary to determine the effective poles and zeros of a MIMO system, for example, by using the so-called Smith–McMillan form,[223] as Rosenbrock first suggested.[188,190,229,233] Alternative definitions for transmission zeros can be found in Refs. 191, 231, 232, 234, and 240. Further information on this issue is available in Refs. 188, 196, 204, and 225.

5.3.3 Directionality

Among the main reasons why SISO analysis and design tools are difficult to translate to the MIMO case is the existence of directionality, which is one of the most important differences between MIMO and SISO plants.[202,219] A given direction is a combination of input signal values: for instance, $[u_1, u_2, u_3] = [4\ 1\ 3]$ has the same direction as $[u_1, u_2, u_3] = [8\ 2\ 6]$, which is $2 \times [4\ 1\ 3]$. Inherently, MIMO systems present spatial (directional) and frequency dependency. Basically, such systems respond differently to input signals lying in distinct directions. As a result, the relationship between the open-loop and closed-loop properties of the feedback system is less obvious. This directionality is completely in accordance with the TFM representation for MIMO systems.

5.3.3.1 Gain and Phase

The concept of gain of a system is somehow easy to translate to MIMO plants through the *singular value decomposition* (SVD) of the TFM,[203,219,239,241] which provides the plant gain at each particular frequency with respect to the main directions (determined by the corresponding singular vectors).

However, the extension of the notion of phase, as understood in scalar systems, is not so straightforward. Several attempts have been made to define a multivariable phase, such as can be seen in Refs. 197, 202, 246, and 248. On the other hand, transmission zeros contribute with extra phase lag in some directions, but not in others.[244] Generally speaking, the change imposed by a MIMO system upon a vector signal can be observed in the magnitude, the direction, and the phase.[202]

5.3.3.2 Effect of Poles and Zeros

The effect of multivariable poles and zeros strongly depends on directionality as well. That is, their nature is only perceptible for particular directions. So, the TFM transmittance gets unbounded when the matrix is evaluated at a pole, but only in the directions determined by the residue matrix at the pole. Likewise, transmission zeros exert their blocking influence provided the TFM is evaluated at the zero, and the input signal lies in the corresponding null space.[202]

5.3.3.3 Disturbance and Noise Signals

Because of directionality, disturbance and noise signals generally do not equally affect all the loops. In general, they have more influence on some loops than on others. Depending on the disturbance direction, that is, the direction of the system output vector resulting from a specific disturbance, some disturbances may be easily rejected, while others may not. The disturbance direction can have an influence in two ways: (1) through the magnitude of the manipulated variables needed to cancel the effect of the disturbance at steady state, independent of the designed controllers, and (2) through its effect on closed-loop performance of the controlled outputs. To address this issue, Skogestad and Morari defined the *disturbance condition number.*[262,263] It measures the magnitude of the manipulated variables needed to counteract a disturbance acting in a particular direction relative to the "best" possible direction.

5.3.4 Uncertainty

Uncertainty, present in all real-world systems, adds a bigger complexity to MIMO systems, especially in the crossover frequency region. Indeed, uncertainty is one of the reasons (together with the presence of disturbances and the original instability of the plant, if that is the case) why feedback is necessary in control systems.

There exist multiple sources of uncertainty (model/plant mismatch) such as the following:

- The model is known only approximately or is inaccurately identified.
- The model varies because of a change in the operating conditions (experimental models are accurate for a limited range of operating conditions), wear of components, nonlinearities, etc.
- Measurement devices are not perfect and their resolution range may be limited.
- The structure or order of the system is unknown at high frequencies.
- The plant model is sometimes simplified to carry out the controller design, the neglected dynamics being considered as uncertainty.
- Other events such as sensor and actuator failures, changes in the control objectives, the switch from automatic to manual (or the other way around) in any loop, inaccuracy in the implementation of the control laws, etc.

The uncertainty can be characterized as *unstructured* when the only available knowledge is the loop location, the stability, and a frequency-dependent magnitude of the uncertainty. The weights used for that magnitude (or bound) are generally stable and minimum phase to avoid additional problems, and multiplicative weights are usually preferred. This description is useful for representing unmodeled dynamics, particularly in the high-frequency range, and small nonlinearities. Different ways of mathematically expressing the unstructured uncertainty and its corresponding properties are available in the book by Skogestad and Postlethwaite.[219]

Nevertheless, unstructured uncertainty is often a poor assumption for MIMO plants. It can sometimes lead to highly conservative designs since the controller has to face events that, in fact, are not likely to exist. On the one hand, errors on particular model parameters, such as mode shapes, natural frequencies, damping values, etc., are highly structured. This is the so-called *parametric uncertainty*. Likewise, parameter errors arising in linearized models are correlated, that is, they are not independent. On the other hand, uncertainty that is unstructured at a component level becomes structured when analyzed at a system level.

Thus, in all those cases, it is more convenient to use *structured uncertainty*. Several approaches can be followed to represent this type of uncertainty. For example, a diagonal block can be utilized[249,250] or a straightforward and accurate representation of the uncertain elements can be performed by means of the plant templates (which are particularly useful for parametric uncertainty). As seen in Chapters 2 and 3, the templates describe the set of possible frequency responses of a plant at each frequency. Indeed, the QFT robust control theory can quantitatively handle both types of uncertainty, structured and unstructured.

Alternative approaches for describing uncertainty are also available, but so far its practicality is somehow limited for controller design. An example is the assumption of a probabilistic distribution (e.g., normal, uniform) for parametric uncertainty.

As for the rest of system features, uncertainty in MIMO systems also displays directionality properties. One loop may contain substantially more uncertainty due to unmodeled dynamics or parameter variations than do other loops. Added to this, and again because of directionality, uncertainty at the plant input or output has a different effect. Primarily, input uncertainty is usually a diagonal perturbation, since, in principle, there is no reason to assume that the perturbations in the manipulated variables are correlated. This uncertainty represents errors on the change rather than on the absolute value.[262,263]

5.3.5 Stability

Stability of MIMO systems is also a crucial point in the design process. In the literature, and depending on the design methodology applied, there exist different ways of assessing the feedback system stability.

One of the main approaches is the *generalized Nyquist stability criterion*, in its direct and inverse versions.[188,237] It places an encirclement condition on the Nyquist plot of the determinant of the return difference matrix.[190] However, it is necessary to get a diagonally dominant system for this criterion to be practical because of loop interaction. This is achieved by means of precompensation. The designer is helped in this task by the Gershgorin and Ostrowski bands (see Refs. 188, 190, and 204) or by Mees' theorem (see Ref. 247). This stability criterion is mainly used in nonsequential classical methodologies (e.g., the inverse Nyquist array[226] and the direct Nyquist array).[188,190] By contrast, sequential classical techniques do not make a direct use of it. Proofs of the multivariable Nyquist stability criterion have been given from different viewpoints. See, for instance, Refs. 194, 230, 236, and 243.

An alternative way of checking stability is by means of the *Smith–McMillan poles*.[223] This approach is applied in classical sequential methodologies through stability conditions such as those defined by De Bedout and Franchek for non-diagonal sequential techniques.[92]

A completely different strategy is adopted by synthesis techniques, which make use of stability robustness results such as the *small-gain theorem*.[192] This states that a feedback loop composed of stable operators will remain stable if the product of all the operator gains is smaller than unity. The theorem is applied to systems with unstructured uncertainty. When the phases of perturbations, rather than their gains, can be bounded, the *small-phase theorem* can be used.[248] However, the main drawback of this approach is the highly conservative results it may provide. In the presence of structured uncertainty, results based on the *structured singular value* (SSV) can be used instead.[250]

5.4 MIMO QFT Control—Overview[187]

As indicated in previous chapters, QFT is an engineering control design methodology, which explicitly emphasizes the use of feedback to simultaneously reduce the effects of plant uncertainty and satisfy performance specifications. It is deeply rooted in classical frequency response analysis involving Bode diagrams, template manipulations, and NCs. It relies on the observation that the feedback is needed principally when the plant presents model uncertainty or when there are uncertain disturbances acting on the plant.

Model uncertainty, frequency domain specifications, and desired time-domain responses translated into frequency domain tolerances, lead to the so-called Horowitz–Sidi bounds (or constraints). These bounds serve as a guide for shaping the nominal loop transfer function $L(s) = P(s)G(s)$, which involves the selection of gain, poles, and zeros to design the appropriate controller $G(s)$. On the whole, the main objective of the QFT is to synthesize (loop shape) a simple, low-order controller with minimum bandwidth (BW), which satisfies the desired performance specifications for all the possible plants due to the model uncertainty. The use of CAD tools have made the QFT controller design much simpler (see, for instance, the QFT Control Toolbox for MATLAB® included in this book [see Appendix F and visit the websites http://cesc.case.edu or http://www.crcpress.com/product/isbn/9781439821794.], developed by Garcia-Sanz, Mauch, and Philippe for the European Space Agency;[149,150] the popular QFT Control Design MATLAB Toolbox developed by Borghesani et al.;[144] the pioneer AFIT CAD tool developed by Sating and Houpis;[6,142,143] and the Qsyn CAD tool developed by Gutman[145]).

The first proposal for MIMO QFT design was made by Horowitz in his first book in 1963,[2] where he pointed out the possibility of using diagonal controllers for quantitative design. This was divided into different frequency ranges: for the low-frequency interval, the controller gain generally needs to be high and is easily determined. As for the medium- and high-frequency bands, he suggested the progressive tuning loop by loop sorted in increasing order. A more systematic and precise approach was later introduced by Shaked et al. in 1976.[235] However, no proof of convergence to a solution was provided.

The first rigorous MIMO QFT methodology was developed by Horowitz in 1979.[77] This nonsequential technique, labeled as *Method 1* in this book (Section 5.5), translates the original $n \times n$ MIMO problem with uncertainty into n MISO systems with uncertainty, disturbances, and specifications derived from the initial problem. The coupling is then treated as a disturbance at the plant input, and the individual solutions guarantee the whole multivariable solution. This is assured by the application of the Schauder's fixed point theorem. This theory maps the desired fixed point on the basis of unit impulse functions.

As before, there exist differentiated frequency ranges in the design procedure. Loops are designed as basically noninteracting (BINC) at low frequency (see Section 5.7), whereas in the middle- and high-frequency ranges attention must be paid to the effect of the noise at the plant input, especially in problems with significant uncertainty.

On the whole, Horowitz's method is a direct technique oriented toward MIMO plants with uncertainty. It also allows the trade-off among loops in the ranges of frequency. Nevertheless, the type of plant that can be dealt with is constrained in several ways, and the method places necessary conditions depending on the system size, which hampers its application to high-order systems. In addition, it presents potential overdesign and may generate highly conservative designs. Additional references on this methodology and its applications are available in Refs. 78–80 and 152.

An improvement of the preceding technique was also provided by Horowitz with a sequential procedure,[81] labeled as *Method 2* here (see Section 5.6). There exist some similarities between this technique and the SRD method by Mayne,[228,242] such as the fact that the resulting controller is diagonal or that they proceed as if each input–output pair was a standard SISO system with loop interaction behaving as an external disturbance. Besides, both methods incorporate the effects of each loop once it is designed into the subsequent loop designs.

Nevertheless, the main difference is that Horowitz's methodology relies on a factorization of the return difference matrix, which is based on the inverse of the plant TFM. By using the inverse plant, a much simpler relationship between the closed-loop and the open-loop TFMs is obtained. One of Horowitz's major contributions with this technique is that he dealt with the problem of robust stability by considering parametric uncertainty.

The stability proof for Horowitz's *Method* 2 was provided in the work by Yaniv and Horowitz,[83] and De Bedout and Franchek.[92] By and large, the method constituted a great step forward in MIMO QFT design techniques. First, as mentioned earlier, parametric uncertainty was considered. Second, the Schauder's fixed point theorem was no longer needed. Third, the limitation related to the system size from the first method was avoided. Finally, it reduced the conservativeness of the former method by using the concept of equivalent plant (which takes into account the controllers that were previously designed). All in all, the second method is a much more powerful technique (although obviously more complicated than other classical approaches), and the physical sense is kept all along the procedure.

Different authors, such as Nwokah, Yaniv, and Horowitz, again made some improvements of these first two MIMO QFT design methods in subsequent works.[82–84] A detailed compilation of the aforementioned techniques is presented in the book by Houpis et al.[6]

An alternative approach to MIMO QFT methodologies was presented by Park, Chait, and Steinbuch in 1994, who developed a direct technique. In other words, the inversion of the plant matrix was not required anymore, which therefore simplified the design process to some extent.[85]

The methodologies outlined so far only deal with the problem of designing a diagonal controller. Nevertheless, there exist potential benefits in the use of full-matrix compensators. Horowitz already commented that the use of diagonal controllers was established just to simplify the theoretical development, but that in practice it could be convenient to consider the off-diagonal elements as well. These terms could then be used to reduce the level of coupling in open loop and, therefore, reduce the amount of feedback needed in the diagonal compensators to fulfill the required specifications.[77]

Furthermore, as Franchek et al. demonstrated,[272] non-diagonal compensators can be used for ensuring that no SISO loop introduces extra unstable poles into the subsequent loops in sequential procedures based on the inverse plant domain, for example, *Method 2* in Section 5.6 (accordingly, this is not possible in Mayne's,[228,242] or Park's[85] framework). As a result, it can be reduced to the minimum crossover frequency needed to achieve closed-loop stability in these succeeding loops. In other words, the actuation BW requirements can be relaxed. Additionally, specific integrity objectives can be achieved, allowing the design of fault-tolerant MIMO systems. In the case of Horowitz's diagonal sequential method (*Method 2*), however, it is not possible to remove the unstable poles originally present in those subsequent loops, but a more general design technique can be developed for that purpose.[92] On the other hand, diagonal compensators are limited for the correction of the plant directionality when needed. There even exist cases where a diagonal or triangular controller cannot stabilize the system.[92]

On balance, the designer has greater flexibility to design the MIMO feedback control system when using fully populated controllers. But the introduction of such non-diagonal controllers poses two main issues: the way of determining the off-diagonal compensators and the need for suitable stability conditions. In systems controlled by a full-matrix compensator, the property of diagonal dominance is not assured. The Gershgorin circles become too conservative in that case and the stability test gets more complicated. As a result, different stability results are needed. Sufficient stability conditions for non-diagonal sequential procedures have been defined by De Bedout and Franchek.[92]

Regarding the determination of the needed off-diagonal compensators, different techniques are available. The first attempt in non-diagonal MIMO QFT was proposed by Horowitz and coworkers,[22,80] who suggested the premultiplication of the plant by a full matrix. Yaniv[87] presented a procedure where a non-diagonal decoupler is applied as a precompensator and a classical diagonal controller is designed afterward. Therein, the main objective becomes the improvement of the system BW.

A different approach was adopted by Boje and Nwokah.[191,238] They used the Perron–Frobenius root as a measure of interaction and of the level of triangularization of the uncertain plant. The full-matrix precompensator is accordingly designed to reduce the coupling before designing a diagonal QFT controller.

On the other hand, Franchek et al.[269,272] introduced a non-diagonal sequential procedure. They made use of the Gauss elimination technique[252] to introduce the effects of the controllers previously designed by means of a recursive expression. Integrity considerations are also included. The controller is then divided into three parts with differentiated roles in the design process. The technique achieves the reduction of the required BW with respect to previous classical sequential techniques. Additionally, De Bedout and Franchek established sufficient stability conditions for non-diagonal sequential procedures.[92]

Another important sequential technique to be considered is the one presented by Garcia-Sanz and collaborators.[6,93,97,98,178,180,181,183] Following Horowitz's ideas, they extended Horowitz's sequential methodology[81] to the design of fully populated MIMO controllers. The role of the non-diagonal terms is simultaneously analyzed for the fundamental cases of reference tracking, disturbance rejection at plant input, and disturbance rejection at plant output. The compensators are aimed at the reduction of the coupling on the basis of defined coupling matrices, which are accordingly minimized. This method has been proven by Garcia-Sanz and collaborators to be a convincing design tool in real applications from different fields, such as heat exchangers,[177] robotics,[97] vehicles,[179] industrial furnaces,[178] waste water treatment plants,[180,184] spacecraft flying in formation (NASA-JPL),[181] spacecraft with flexible appendages (ESA-ESTEC),[183] and wind turbines,[174,379] among others.

In 2009, Garcia-Sanz and Eguinoa introduced a reformulation of the full-matrix QFT robust control methodology for MIMO plants with uncertainty.[102] This methodology includes a generalization of their previous non-diagonal MIMO QFT techniques. It avoids former hypotheses of diagonal dominance and simplifies the calculations for the off-diagonal elements and the method itself. It also reformulates the classical matrix definition of MIMO specifications by designing a new set of loop-by-loop QFT bounds on the

NC, with necessary and sufficient conditions, giving explicit expressions to share the load among the loops of the MIMO system to achieve the matrix specifications and also for stability, reference tracking, disturbance rejection at plant input and output, and noise attenuation problems. The new methodology was also applied to the design of a MIMO controller for a spacecraft flying in formation in a low Earth orbit.

Regarding the field of nonsequential MIMO QFT techniques, the approach by Kerr, Jayasuriya and coworkers is remarkable.[95,96,100,176,182] Stability conditions have also been established within this framework.[95,99]

Other approaches have also been introduced for particular types of MIMO systems. For example, there are results on *n.m.p.* MIMO plants.[120] It is noted that not all the $n \times n$ transfer functions have to suffer the limitations imposed by the *n.m.p.* behavior.[118] The MIMO system has the capacity to relocate the RHP zeros in those outputs, which are not so determining, while the critical outputs are kept as minimum-phase loops. Likewise, some research has been done for unstable and strongly *n.m.p.* MIMO systems, for example, the X-29 aircraft.[119,153,182] One interesting suggestion is the *singular-G method*,[119,153] which makes use of a singular compensator (i.e., with a determinant equal to zero, which implies that one output is dependent from the rest of outputs). In this way, the technique allows easing the *n.m.p.* problem and the instability in the MIMO system, simultaneously achieving good performance.

Division II: Method 1

The second division of this chapter, Section 5.5, introduces *Method 1*, a nonsequential technique to design diagonal matrix compensators for MIMO systems with model uncertainty.

5.5 Nonsequential Diagonal MIMO QFT (*Method 1*)

From Equation 5.28 obtain

$$T = \frac{[\text{adj}(\boldsymbol{I} + \boldsymbol{PG})]\,\boldsymbol{PGF}}{\det|\boldsymbol{I} + \boldsymbol{PG}|} \tag{5.35}$$

For a 3×3 plant and a diagonal $\boldsymbol{G}$ matrix, the denominator of this equation becomes

$$\det|\boldsymbol{I} + \boldsymbol{PG}| = \begin{vmatrix} 1 + p_{11}g_1 & p_{12}g_2 & p_{13}g_3 \\ p_{21}g_1 & 1 + p_{22}g_2 & p_{23}g_3 \\ p_{31}g_1 & p_{32}g_2 & 1 + p_{33}g_3 \end{vmatrix} \tag{5.36}$$

which is a very "messy" equation for the purpose of analysis and synthesis in achieving a satisfactory design of the control system. In analyzing Equation 5.28, it is noted that if $|\boldsymbol{P}(j\omega)\boldsymbol{G}(j\omega)| \gg \boldsymbol{I}$, then $\boldsymbol{T} \approx \boldsymbol{F}$ and the system becomes insensitive to the parameter variations in $\boldsymbol{P}$.

5.5.1 Effective MISO Equivalents

The objective of this section is to find a suitable mapping that permits the analysis and synthesis of a MIMO control system by a set of equivalent MISO control systems. This mapping results in n^2 equivalent systems, each with two inputs and one output. One input is designated as a "desired" input and the other as an "unwanted" input (cross-coupling effects and/or external system disturbances). First, Equation 5.28 is premultiplied by $[\boldsymbol{I} + \boldsymbol{PG}]$ to obtain

$$[\boldsymbol{I}+\boldsymbol{PG}]\,\boldsymbol{T}=\boldsymbol{PGF} \tag{5.37}$$

When $\boldsymbol{P}$ is nonsingular, premultiplying both sides of Equation 5.37 by $\boldsymbol{P}^{-1}$ yields

$$[\boldsymbol{P}^{-1}+\boldsymbol{G}]\,\boldsymbol{T}=\boldsymbol{GF} \tag{5.38}$$

which puts the constrained or the structured parametric uncertainty in one place in the equation. Let

$$\boldsymbol{P}^{-1}=\boldsymbol{P}^{*}=[p_{ij}^{*}]=\begin{bmatrix} p_{11}^{*} & p_{12}^{*} & \cdots & p_{1n}^{*} \\ p_{21}^{*} & p_{22}^{*} & \cdots & p_{2n}^{*} \\ \vdots & \vdots & \ddots & \vdots \\ p_{n1}^{*} & p_{n2}^{*} & \cdots & p_{nn}^{*} \end{bmatrix} \tag{5.39}$$

The n^2 effective plant transfer functions are based upon defining

$$q_{ij}\equiv\frac{1}{p_{ij}^{*}}=\frac{\det[\boldsymbol{P}]}{adj_{ij}[\boldsymbol{P}]} \tag{5.40}$$

The $\boldsymbol{Q}$ matrix is then formed as

$$\boldsymbol{Q}=\begin{bmatrix} q_{11} & q_{12} & \cdots & q_{1n} \\ q_{21} & q_{22} & \cdots & q_{2n} \\ \cdots & \cdots & \cdots & \cdots \\ q_{n1} & q_{n2} & \cdots & q_{nn} \end{bmatrix}=\begin{bmatrix} 1/p_{11}^{*} & 1/p_{12}^{*} & \cdots & 1/p_{1n}^{*} \\ 1/p_{21}^{*} & 1/p_{22}^{*} & \cdots & 1/p_{2n}^{*} \\ \cdots & \cdots & \cdots & \cdots \\ 1/p_{n1}^{*} & 1/p_{n2}^{*} & \cdots & 1/p_{nn}^{*} \end{bmatrix} \tag{5.41}$$

where

$$\boldsymbol{P}=[p_{ij}];\quad \boldsymbol{P}^{-1}=[p_{ij}^{*}]=[1/q_{ij}];\quad \boldsymbol{Q}=[q_{ij}]=[1/p_{ij}^{*}]$$

The matrix $\boldsymbol{P}^{-1}$ is partitioned to the form

$$\boldsymbol{P}^{-1}=[p_{ij}^{*}]=[1/q_{ij}]=\boldsymbol{\Lambda}+\boldsymbol{B} \tag{5.42}$$

where
- $\boldsymbol{\Lambda}$ is the diagonal part
- $\boldsymbol{B}$ is the balance of $\boldsymbol{P}^{-1}$

thus $\lambda_{ii} = 1/q_{ii} = p^*_{ii}$, $b_{ii} = 0$, and $b_{ij} = 1/q_{ij} = p^*_{ij}$ for $i \neq j$. Next, rewrite Equation 5.38 using Equation 5.42 and with $\mathbf{G}$ diagonal. This yields $[\mathbf{\Lambda} + \mathbf{G}]\mathbf{T} = \mathbf{GF} - \mathbf{BT}$, which produces

$$\mathbf{T} = [\mathbf{\Lambda} + \mathbf{G}]^{-1}[\mathbf{GF} - \mathbf{BT}] \tag{5.43}$$

This expression is used to define the desired fixed point mapping, based upon *unit impulse functions*, where each of the n^2 matrix elements on the right side of Equation 5.43 can be interpreted as a MISO problem. Proof of the fact that design of the individual MISO feedback loops will yield a satisfactory MIMO design is based on the Schauder fixed point theorem.[77] This theorem is described by defining a mapping on $\Im$ as follows:

$$\mathbf{Y}(\mathbf{T}_i) \equiv [\mathbf{\Lambda} + \mathbf{G}]^{-1}[\mathbf{GF} - \mathbf{BT}_i] \equiv \mathbf{T}_j \tag{5.44}$$

where each $\mathbf{T}_i$ and $\mathbf{T}_j$ is from the acceptable set $\Im$. If this mapping has a fixed point, that is, $\mathbf{T}_i$, $\mathbf{T}_j \in \Im$ such that $\mathbf{Y}(\mathbf{T}_i) = \mathbf{T}_j$ (see Figure 5.10), then a solution to the robust control problem has been achieved yielding a solution in the acceptable set $\Im$. Recalling that $\mathbf{\Lambda}$ and $\mathbf{G}$ in Equation 5.43 are both diagonal, the (1,1) element on the right side of Equation 5.44 for the 3 × 3 case, for a *unit impulse input*, yields the output

$$y_{11} = \frac{q_{11}}{1 + g_1 q_{11}}\left[g_1 f_{11} - \left(\frac{t_{21}}{q_{12}} + \frac{t_{31}}{q_{13}}\right)\right] \tag{5.45}$$

This corresponds precisely to the first structure in Figure 5.11 and is the *control ratio* that relates the *i*th output to the *j*th input, where $i = j = 1$ in Equation 5.45. Similarly, each of the nine structures in Figure 5.11 corresponds to one of the elements of $\mathbf{Y}(\mathbf{T})$ of Equation 5.43.

The general transformation result of n^2 MISO system loops is shown in Figure 5.12. Figure 5.11 shows the four effective MISO loops (in the boxed area) resulting from a 2 × 2 system and the nine effective MISO loops resulting from a 3 × 3 system.[79] The control ratios, for unit impulse inputs, for the $n \times n$ system of Figure 5.12 obtained from Equation 5.44 have the form

$$y_{ij} = w_{ii}(v_{ij} + c_{ij}) \tag{5.46}$$

where

$$w_{ii} = \frac{q_{ii}}{(1 + g_i q_{ii})} \tag{5.47}$$

$$v_{ij} = g_i f_{ij} \tag{5.48}$$

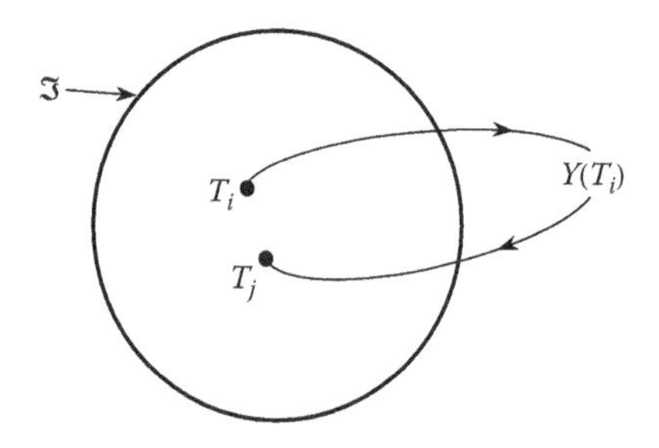

FIGURE 5.10
Schauder fixed point mapping.

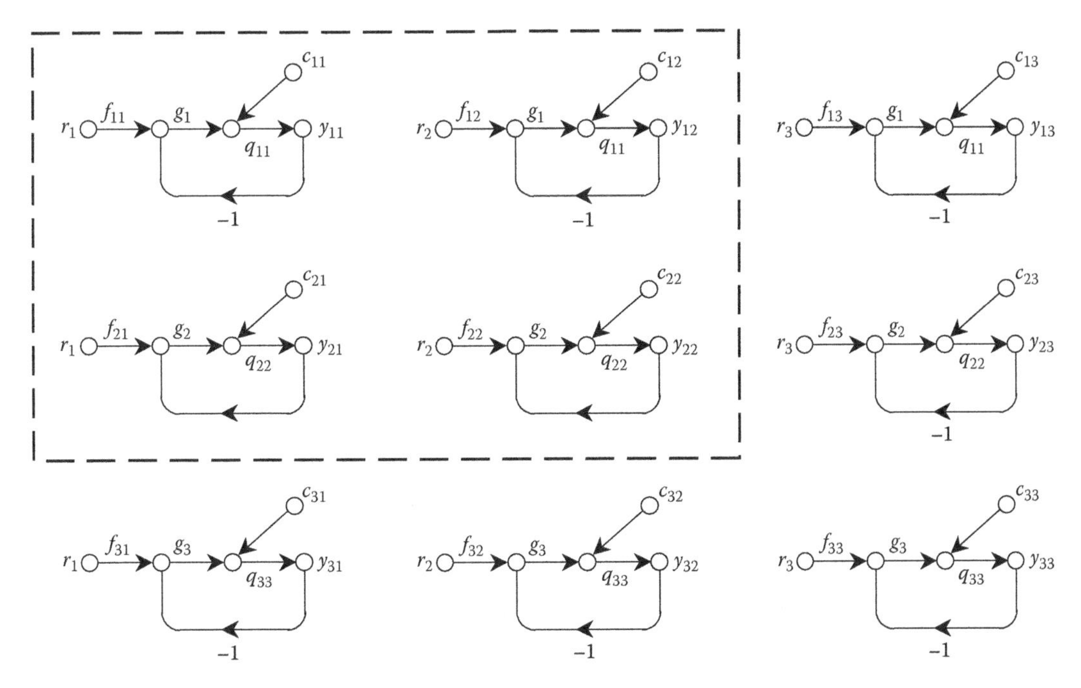

FIGURE 5.11
Effective MISO loops 2 × 2 (boxed in loops) and 3 × 3 (all nine loops).

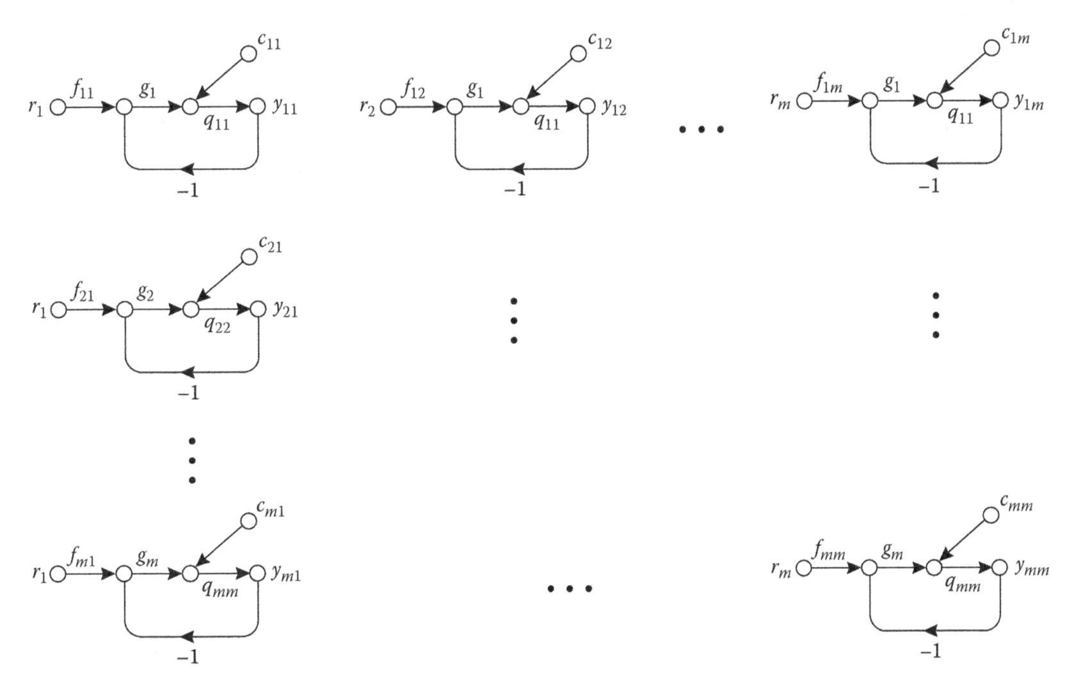

FIGURE 5.12
Effective MISO loops (in general).

and

$$c_{ij} = -\sum_{k \neq i} \left[\frac{t_{kj}}{q_{ik}} \right], \quad k = 1, 2, \ldots, n \tag{5.49}$$

Equation 5.49 represents the interaction (cross-coupling) between the loops.

Thus, Equation 5.46 represents the control ratio of the ith MISO loop where the transfer function $w_{ii}v_{ij}$ relates the ith output to ith "desired" input r_i and the transfer function $w_{ii}c_{ij}$ relates the ith output to the jth "cross-coupling effect" input c_{ij}. The transfer function of Equation 5.46 can thus be expressed, with $r_i(s) = 1$ (*a unit impulse function*), as

$$y_{ij} = (y_{ij})_{r_j} + (y_{ij})_{c_{ij}} = y_{r_j} + y_{c_{ij}} \tag{5.50}$$

or

$$t_{ij} = t_{r_{ij}} + t_{c_{ij}} \tag{5.51}$$

where

$$t_{r_{ij}} = y_{r_j} = w_{ii}v_{ij} \tag{5.52}$$

$$t_{c_{ij}} = y_{c_{ij}} = w_{ii}c_{ij} \tag{5.53}$$

and now the upper bound, in the low frequency range, is expressed as

$$b_{ij} = b'_{ij} + \tau_{c_{ij}} \tag{5.54}$$

Thus

$$\tau_{c_{ij}} = b_{ij} - b'_{ij} \tag{5.55}$$

represents the maximum portion of b_{ij} allocated toward cross-coupling effect rejection and b'_{ij} represents the upper bound for the tracking portion of t_{ij}.

For any particular loop there is a cross-coupling effect input, which is a function of all the other loop outputs. The object of the design is to have each loop track its desired input while minimizing the outputs due to the cross-coupling effects.

In each of the nine structures of Figure 5.11, the control ratio $t_{ij}(s)$ must be a member of the acceptable set $t_{ij} \in \Im_{ij}(s)$ (see Figure 5.10). All the $g_i(s)$, $f_{ij}(s)$ must be chosen to ensure this condition is satisfied, thus constituting nine MISO design problems. If all of these MISO problems are solved, there exists a fixed point, and then $y_{ij}(s)$ on the left side of Equation 5.44 may be replaced by a t_{ij} and all the elements of $\boldsymbol{T}$ on the right side of the same equation by t_{kj}. This means that there exist nine t_{ik} and t_{kj}, each in its acceptable set, which is a solution to Figure 5.5. If each element is 1:1, then this solution must be unique. A more formal and detailed treatment is given in Ref. 77. Note that if the plant has transmission zeros in the RHP, it only indicates that q_{ij} may be non-minimum phase (*n.m.p.*) or the det $\boldsymbol{P}$ may have zeros in the RHP (see Section 6.2 of Ref. 6). For a controllable and observable plant, the transmission zeros can be computed from the determinant of the *system matrix*, which is defined by

$$\begin{vmatrix} s\boldsymbol{I} - \boldsymbol{A} & \boldsymbol{B} \\ -\boldsymbol{C} & \boldsymbol{0} \end{vmatrix} = 0$$

where $\boldsymbol{A}$, $\boldsymbol{B}$, $\boldsymbol{C}$, and $\boldsymbol{I}$ are the $n \times n$ matrices of the State Space description. The number of transmission zeros[118] for a system having the same number of inputs as outputs is equal to d, where d is the rank deficiency of the matrix product of $\boldsymbol{CB}$.

5.5.2 Effective MISO Loops of the MIMO System

There are two design methods for designing diagonal MIMO QFT controllers. In the first method each MISO loop in Figures 5.11 and 5.12 is treated as an individual MISO design problem, which is solved using the procedures explained in Chapters 3 and 4. The $f_{ij}(s)$ and $g_i(s)$ are the compensator elements of the $\boldsymbol{F}(s)$ and $\boldsymbol{G}(s)$ matrices described previously.

The cross-coupling effect $c_{ij}(s)$, expressed by Equation 5.49, represents the interaction between the loops and can be bounded (worst case of interaction) by replacing t_{kj} by the specification b_{kj}, that is,

$$c_{ij} = -\sum_{k \neq i} \left[\frac{b_{kj}}{q_{ik}} \right], \quad k = 1, 2, \ldots, n \tag{5.56}$$

where the numerator b_{kj} is the upper response bound, T_{R_U} or T_D in Figure 3.4b or 3.5, for the respective output–input relationship. These are obtained from the design specifications.[79] The first subscript k refers to the output variable and the second subscript j refers to the input variable. Therefore, b_{kj} is a function of the response requirements on the output y_k due to the input r_j. The lower bound a_{kj} needs defining only when there is a command input. It should be noted that if the phase margin frequencies ω_ϕ of the loops are not widely separated, there is considerable interaction between the loops.[22]

5.5.3 Example: The 2 × 2 Plant

For this example, a diagonal $\boldsymbol{G}$ matrix is utilized. The use of a diagonal matrix results in restricting the design freedom available to achieve the desired performance specifications. This is offset by the resulting simplified design process. The elements of a diagonal $\boldsymbol{G}$ are denoted with a single subscript, that is, g_i. The $\boldsymbol{P}$ and $\boldsymbol{P}^{-1}$ matrices are, respectively,

$$\boldsymbol{P} = \begin{bmatrix} p_{11} & p_{12} \\ p_{21} & p_{22} \end{bmatrix} \tag{5.57}$$

$$\boldsymbol{P}^{-1} = \begin{bmatrix} p_{11}^* & p_{12}^* \\ p_{21}^* & p_{22}^* \end{bmatrix} = \frac{1}{\Delta} \begin{bmatrix} p_{22} & -p_{12} \\ -p_{21} & p_{11} \end{bmatrix} \tag{5.58}$$

where $\Delta = p_{11}p_{22} - p_{12}p_{21}$.

From Equation 5.40,

$$\boldsymbol{P}^{-1} = \begin{bmatrix} p_{11}^* & p_{12}^* \\ p_{21}^* & p_{22}^* \end{bmatrix} = \begin{bmatrix} 1/q_{11} & 1/q_{12} \\ 1/q_{21} & 1/q_{22} \end{bmatrix} \tag{5.59}$$

where

$$q_{11} = \frac{\Delta}{p_{22}}, \quad q_{12} = \frac{-\Delta}{p_{12}}$$
$$q_{21} = \frac{-\Delta}{p_{21}}, \quad q_{22} = \frac{\Delta}{p_{11}} \tag{5.60}$$

Substituting Equation 5.59 into Equation 5.38 yields

$$\begin{bmatrix} \frac{1}{q_{11}} + g_1 & \frac{1}{q_{12}} \\ \frac{1}{q_{21}} & \frac{1}{q_{22}} + g_2 \end{bmatrix} \begin{bmatrix} t_{11} & t_{12} \\ t_{21} & t_{22} \end{bmatrix} = \begin{bmatrix} g_1 f_{11} & g_1 f_{12} \\ g_2 f_{21} & g_2 f_{22} \end{bmatrix} \tag{5.61}$$

The responses due to *input* 1, obtained from Equation 5.61, are as follows:

$$\left(\frac{1}{q_{11}} + g_1\right) t_{11} + \frac{t_{21}}{q_{12}} = g_1 f_{11}$$
$$\frac{t_{11}}{q_{21}} + \left(\frac{1}{q_{22}} + g_2\right) t_{21} = g_2 f_{21} \tag{5.62}$$

The responses caused by *input* 2, obtained from Equation 5.61, are as follows:

$$\left(\frac{1}{q_{22}} + g_1\right) t_{12} + \frac{t_{22}}{q_{12}} = g_1 f_{12}$$
$$\frac{t_{12}}{q_{21}} + \left(\frac{1}{q_{22}} + g_2\right) t_{22} = g_2 f_{22} \tag{5.63}$$

These equations are rearranged into a format that readily permits the synthesis of the g_is and the f_is that will result in the MIMO control system achieving the desired system performance specifications. Equations 5.62 and 5.63 are manipulated to achieve the following format:

$$\text{For input } r_1: \quad t_{11} = \frac{g_1 f_{11} - (t_{21}/q_{12})}{(1/q_{11}) + g_1} \qquad t_{21} = \frac{g_2 f_{21} - (t_{11}/q_{21})}{(1/q_{22}) + g_2}$$
$$\text{For input } r_2: \quad t_{12} = \frac{g_1 f_{12} - (t_{22}/q_{12})}{(1/q_{11}) + g_1} \qquad t_{22} = \frac{g_2 f_{22} - (t_{12}/q_{21})}{(1/q_{22}) + g_2} \tag{5.64}$$

Multiplying the t_{11} and t_{12} equations by q_{11}, and the t_{21} and t_{22} equations by q_{22} in Equation 5.64, respectively, yields the equations shown in Figure 5.13. Associated with each equation in this figure is its corresponding SFG.

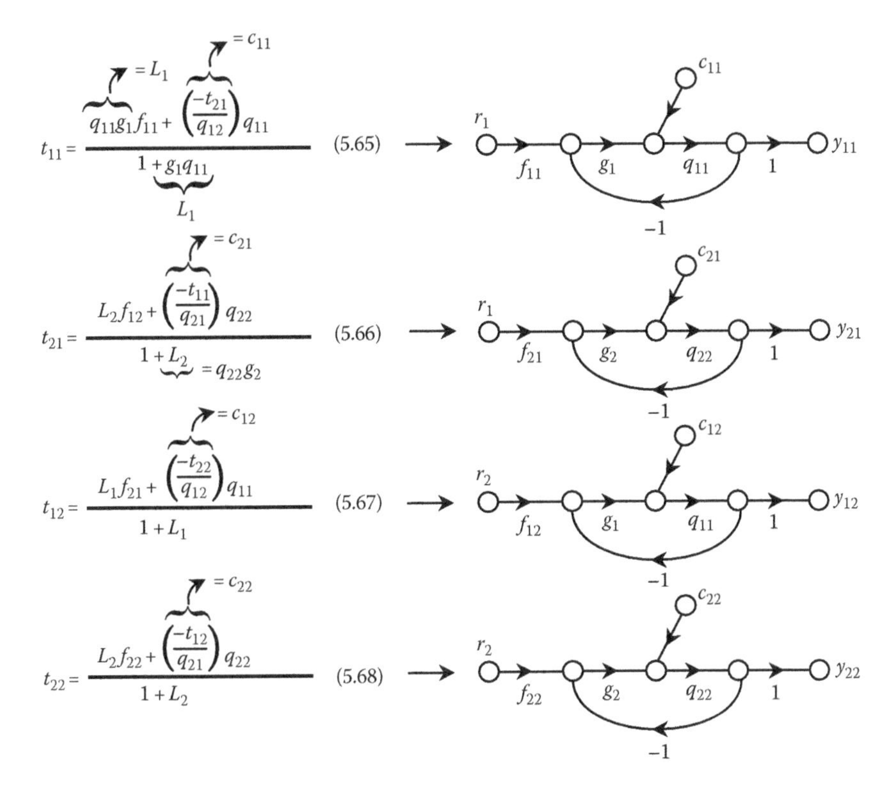

FIGURE 5.13
2 × 2 MISO structures and their respective t_{ij} equations.

Equations 5.65 through 5.68 are of the format of Equations 5.46 through 5.55. Note that

1. t_{22} and t_{21} can automatically be obtained from the expressions for t_{11} and t_{12} by interchanging $1 \to 2$ and $2 \to 1$ in the equations for t_{11} and t_{12}.
2. The c_{ij} terms in these equations represent the cross-coupling effect from the other loops. These terms are functions of the other t_{kj}'s and the structured parameter uncertainty of the plant.
3. Theoretically, by making $|L_i(j\omega)|$ "large enough," so that $c_{ij} \approx 0$ a "decoupled system" is achieved.

In a similar fashion, the t_{ij} expressions and their corresponding SFG may be obtained for any $n \times n$ control system.

5.5.4 Performance Bounds

Based upon *unit impulse inputs,* from Equation 5.51 obtain

$$t_{ij} = t_{r_{ij}} + t_{c_{ij}} \tag{5.69}$$

Let

$$\phi_{ij} \text{ be the actual value of } t_{ij}$$
$$\tau_{r_{ij}} \text{ be the actual value of } t_{r_{ij}}$$
$$\tau_{c_{ij}} \text{ be the actual value of } t_{c_{ij}}$$

A 2 × 2 control system is used to illustrate the concept of performance bounds. The "actual value" expression corresponding to the t_{11} expression of Equation 5.65 is

$$\phi_{11} = \tau_{r_{11}} + \tau_{c_{11}}$$

where

the τ_r term represents the transmission due to the command input r_1
the τ_c term represents the transmission due to the cross-coupling effects

For LTI system, the linear superposition theorem is utilized in the development of the performance bounds.

For all $\boldsymbol{P} \in \mathcal{P} = \{\boldsymbol{P}\}$ and $t_{21} \in \tau_{21}$, and $c_{11} = -t_{21}/q_{12}$, the output ϕ_{11} must satisfy the performance specifications on t_{11}. Thus, it is necessary to specify a priori the closed-loop transfer functions t_{ij}. Consider the specifications b_{11} and a_{11} on t_{11}, as illustrated in Figure 5.14 (see Equations 5.54 and 5.55) for a minimum-phase (*m.p.*) system. Only magnitudes need to be considered for the *m.p.* system since the magnitude determines the phase in these systems. Thus, in terms of the actual values,

$$\begin{array}{ccc} a_{11}(\omega) \leq & |\phi_{11}(j\omega)| & \leq b_{11}(\omega) \\ \leq & |\tau_{r_{11}} + \tau_{c_{11}}| & \leq \\ \Uparrow & & \Uparrow \\ \text{lower bound} & & \text{upper bound} \end{array} \quad (5.70)$$

Since the relative phases of the τ's are not known and not required for *m.p.* systems, to ensure the achievement of the desired performance specifications, Equation 5.70 is expressed as follows:

$$a_{11}(\omega) \leq \left|\,|\tau_{r_{11}}| - |\tau_{c_{11}}|\,\right| \quad \text{the smallest bound} \quad (5.71)$$

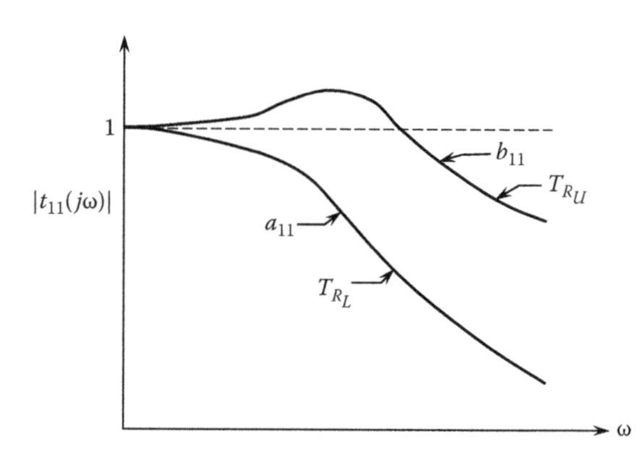

FIGURE 5.14
Upper and lower tracking bounds for t_{11}.

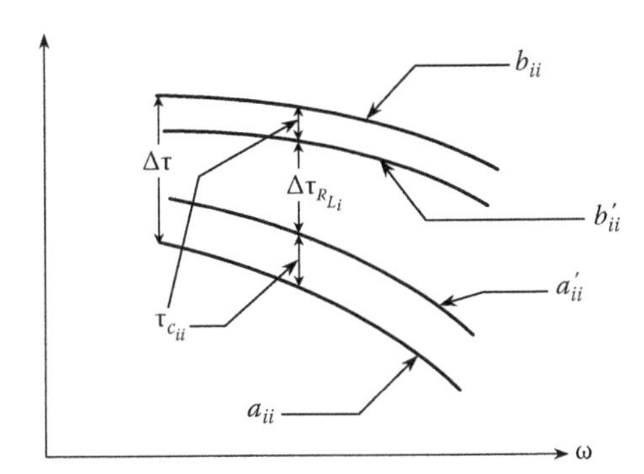

FIGURE 5.15
Allocation for tracking and cross-coupling specifications.

and

$$(|\tau_{r_{11}}| + |\tau_{c_{11}}|) \leq b_{11}(\omega) \quad \text{the largest bound} \tag{5.72}$$

These represent an overdesign since they result in a more restrictive performance. A pictorial representation of Equations 5.71 and 5.72, for $\omega = \omega_i$, is shown in Figure 5.15. Note that the τs in this figure and in the remaining discussion in this section represent only magnitudes. From this figure the following expression is obtained:

$$\Delta\tau = \Delta\tau_{r_{11}} + 2\tau_{c_{11}} = b_{11} - a_{11}; \quad b'_{11} = b_{11} - \tau_{c_{11}}; \quad a'_{11} = a_{11} + \tau_{c_{11}} \tag{5.73}$$

In the "low-frequency range" the BW of concern, $0 < \omega < \omega_h$, $\Delta\tau$ is split up based upon the desired performance specifications. As is discussed later, for the high-frequency range, only an upper bound is of concern; thus, there is no need to be concerned with a "split."

Example 5.2

Consider the bound determinations b_{ii}, b'_{ii}, a_{ii}, a'_{ii}, and b_{cii} on $L_1 = g_1 q_{11}$, for a 2 × 2 system. Thus, b_{r11}, based upon

$$a'_{ii} = (\tau_{r_{11}})_L \leq |t_{r_{11}}| \leq (\tau_{r_{11}})_U = b'_{ii} \tag{5.74}$$

and Figure 5.15, for t_{11} can be determined in the same manner as for a MISO system (see Chapter 3). In Figure 5.15, for illustrative purposes only, $\tau_{c_{11}} = 0.05$ and $\Delta\tau_{r_{11}} = 0.2$ at $\omega = \omega_i$. Referring to Figure 5.13 and to Equation 5.65, the bound on the cross-coupling effect $b_{c_{ij}}$, is determined as follows:

$$|t_{c_{11}}| = \left|\frac{c_{11} q_{11}}{1 + L_1}\right| \leq \tau_{c_{ij}} = \tau_{c_{11}} = 0.05 \tag{5.75}$$

where for the cross-coupling effect, the *upper bound* for $t_{c_{11}}$ is given by

$$|t_{c_{ij}}| = |t_{c_{11}}| \leq b_{c_{ij}} = b_{c_{11}}, \quad \text{where } i = j = 1$$

Substituting the expression for c_{11} (see Equation 5.56) into Equation 5.75 and then manipulating this equation, the following constraint on $L_1(j\omega_i)$, where b_{21} is the upper bound on t_{21}, is obtained:

$$|1+L_1| \geq \frac{|c_{11}q_{11}|}{\tau_{c_{11}} = 0.05} = 20\left|\frac{t_{21}q_{11}}{q_{12}}\right| = 20\left|\frac{b_{21}q_{11}}{q_{12}}\right| \tag{5.76}$$

where the upper bound b_{21} is inserted for t_{21}. It is necessary to manipulate this equation in a manner that permits the utilization of the NC. This is accomplished by substituting $L_1 = 1/\eta_1$ into this equation. Thus,

$$\left|1+\frac{1}{\eta_1}\right| = \left|\frac{1+\eta_1}{\eta_1}\right| \geq 20\left|\frac{b_{21}q_{11}}{q_{12}}\right| \tag{5.77}$$

Inverting this equation yields

$$\left|\frac{\eta_1}{1+\eta_1}\right| \leq \frac{1}{20}\left|\frac{q_{12}}{b_{21}q_{11}}\right| \tag{5.78}$$

which is of the mathematical format that allows the use of the NC for the graphical determination of the cross-coupling bound $b_{c_{11}}$ (see Section 3.10). Equations 5.77 and 5.78 are, respectively, of the following mathematical format:

$$\begin{aligned} |A| &\geq |B| \quad \text{(a)} \\ |C| &\leq |D| \quad \text{(b)} \end{aligned} \tag{5.79}$$

In determining these bounds, it is necessary to insert the actual plant parameters into these equations. That is, for each of the J plant models, $\mathbf{P}_\iota$ ($\iota = 1, 2,\ldots, J$), insert the corresponding plant parameters into these equations and determine the magnitudes A_ι, B_ι, C_ι, and D_ι. The following magnitudes, for each value of ω_i, are used from these J sets of values to determine the cross-coupling bound:

$$\begin{aligned} |A_\iota|_{min} &\geq |B_\iota|_{max} \\ |C_\iota|_{max} &\leq |D_\iota|_{min} \end{aligned} \tag{5.80}$$

"High bounds" on the NC require high gain (*h.g.*); thus, in order to *minimize* the required compensator gain, the *optimum* choice of the

$$\Delta\tau_{\eta_{11}}, \tau_{c_{11}} \tag{5.81}$$

specifications is those that result in achieving essentially the same tracking and cross-coupling bounds, that is,

$$b_{c_{11}} \approx b_{\eta_{11}} \tag{5.82}$$

A recommended method for determining an appropriate set of constraints (specifications) on Equation 5.81 is to create a design initially based on the following assumption:

$$\Delta\tau_{\eta_{11}} = b'_{11} - a'_{11} \tag{5.83}$$

With this design, determine how big $\tau_{c_{11}}$ is and then, by trial and error, if a CAD package is not available, adjust $\tau_{c_{11}}$ until the condition

$$\Delta\tau_{r_{11}} \approx 2\tau_{c_{11}} \tag{5.84}$$

is satisfied.

The *QFTCT* CAD included in this book, and discussed in the appendices, can help with this calculation. This procedure is illustrated in Figure 5.16. Depending on the starting quantities, decreasing (increasing) one quantity and increasing (decreasing) the other quantity expedites achieving the condition of Equation 5.82. By the procedure just described, the optimal bound $b_{o_{11}}$ is given by Equation 5.82.

For the cross-coupling effect rejections problem, the responses must be less than some bound, that is,

$$(y_{ij})_{c_{ij}} \leq \tau_{c_{ij}} = b_{c_{ij}} \tag{5.85}$$

Thus, according to Equations 5.46 and 5.50, the loop equations become

$$\tau_{ij} \geq (y_{ij})_{c_{ij}} = \frac{c_{ij}\,|\,q_{ii}\,|}{|\,1+g_i q_{ii}\,|} \tag{5.86}$$

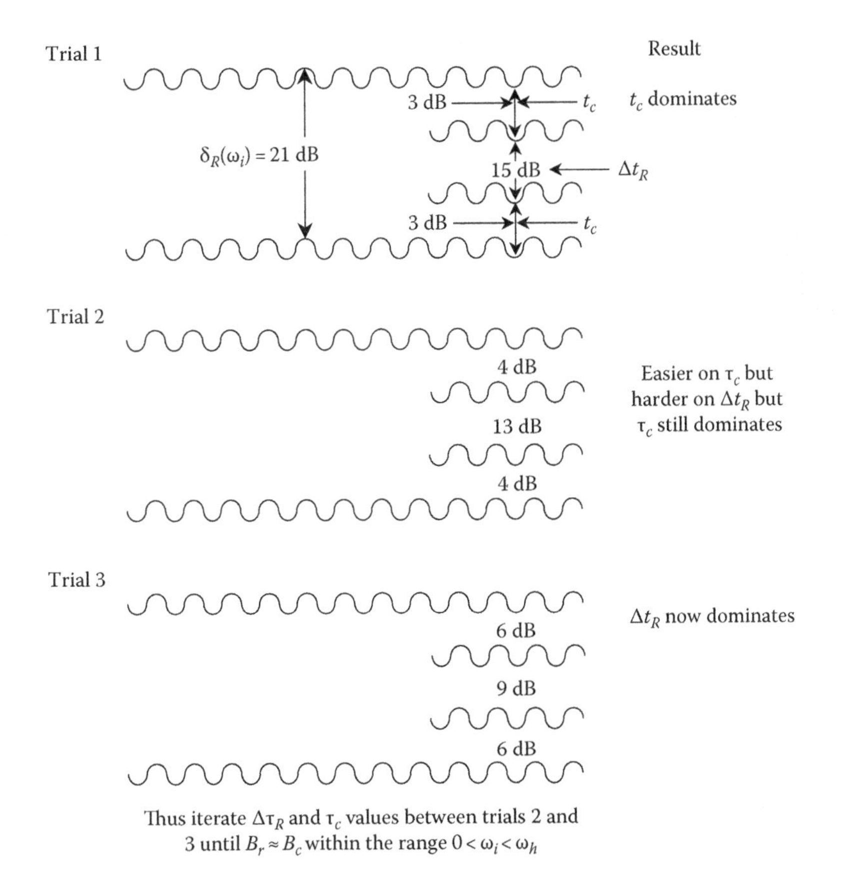

FIGURE 5.16
Procedure pictorial representation to achieve $B_r \approx B_c$.

where $g_i\, q_{ii} = L_i$. Equation 5.86 is manipulated to yield

$$|1 + L_i| \geq \frac{c_{ij}\, |q_{ii}|}{\tau_{ij}} \tag{5.87}$$

Substituting Equation 5.56 into Equation 5.87 yields

$$|1 + L_i| \geq \left| -\sum_{k \neq i} \frac{b_{kj}}{q_{ik}} \right| \frac{|q_{ii}|}{\tau_{ij}} \tag{5.88}$$

For example, in a 3 × 3 system for the first loop, L_1, where $i = 1, j = 2$, and in the first term $k = 2$ and in the second term $k = 3$, Equation 5.88 becomes

$$|1 + L_1| \geq \frac{\left|(b_{22}/|q_{12}|) + (b_{32}/|q_{13}|)\right| |q_{11}|}{\tau_{c12}} \tag{5.89}$$

Remember to use only the magnitude in the cross-coupling calculations. This assumes the worst case.

5.5.5 Constraints on the Plant Matrix[137]

In order to use the diagonal MIMO QFT techniques, the following critical conditions must be satisfied.

Condition 1: $\boldsymbol{P}$ must be nonsingular for any combination of possible plant parameters to ensure that $\boldsymbol{P}^{-1}$ exists.

In the high-frequency range, $y_{r_i}(j\omega)$ approaches zero as $\omega \to \infty$. Therefore, Equations 5.50 and 5.51 can be approximated by

$$t_{ij} \approx y_{c_{ij}} \tag{5.90}$$

Thus from Equations 5.46, 5.47, and 5.56, with impulse input functions, Equation 5.90 can be expressed as

$$y_{c_{ij}} = \frac{c_{ij} q_{ii}}{1 + g_i q_{ii}} \tag{5.91}$$

Consider first the 2 × 2 plant, that is, $m = 2$. Specifying that $|y_{c_{11}}| \leq b_{ii}$ (the given upper bound in the high-frequency range), Equations 5.56 and 5.91 yield, for $i = j = 1$,

$$b_{11} \geq \left| \frac{-b_{21}}{q_{12}} \right| \left| \frac{q_{11}}{1 + L_1} \right| \tag{5.92}$$

For $i = 2, j = 1$, where it is specified that $(y_c)_{21} \leq b_{21}$ (the given cross-coupling upper bound). Henceforth, the cross-coupling bound notation $b_{c_{ij}}$ is simplified to b_{ij} for $i \neq j$; thus, Equations 5.56 and 5.92 yield

$$b_{21} \geq \left| \frac{-b_{11}}{q_{21}} \right| \left| \frac{q_{22}}{1 + L_2} \right| \tag{5.93}$$

Equations 5.92 and 5.93 are rearranged to

$$|1+L_1| \geq \frac{b_{21}}{b_{11}}\left|\frac{q_{11}}{q_{12}}\right| \tag{5.94}$$

$$|1+L_2| \geq \frac{b_{11}}{b_{21}}\left|\frac{q_{22}}{q_{21}}\right| \tag{5.95}$$

Multiplying Equation 5.94 by Equation 5.95, where $L_1 = L_2 \approx 0$ in the high-frequency range, results in

$$1 \geq \left|\frac{q_{11}q_{22}}{q_{12}q_{21}}\right| \tag{5.96}$$

Substituting Equations 5.60 into Equation 5.96 yields Condition 2.

Condition 2: (2 × 2 plant).[190]
As $\omega \to \infty$

$$|p_{11}p_{22}| > |p_{12}p_{21}| \quad \text{(a)} \quad \text{or} \quad |p_{11}p_{22}| - |p_{12}p_{21}| > 0 \quad \text{(b)} \tag{5.97}$$

Since p_{11} and p_{22} are elements of the diagonal of $\mathbf{P}$, Equation 5.97 is the *diagonal dominance condition* for the 2 × 2 plant (also known as the Rosenbrock's diagonal dominance condition).[190] Equation 5.97 must also be satisfied for all frequencies. This condition is obtained considering only the left column of the MISO loops for the 2 × 2 plant of Figure 5.11. This condition may also be obtained by using the right column of the MISO loops in Figure 5.11 since the loop transmissions L_1 and L_2 are again involved in the derivation.

Next, consider the 3 × 3 plant, that is, $m = 3$. For $i = j = 1$ (the left column of Figure 5.11), where $L_1 = L_2 = L_3 \approx 0$ as $\omega \to \infty$ and where it is specified that

$$|y_{c_{11}}| \leq b_{11}, \qquad |y_{c_{21}}| \leq b_{21}, \qquad |y_{c_{31}}| \leq b_{31} \tag{5.98}$$

Equations 5.56 and 5.91 yield, for $i = 1, 2, 3$, respectively,

$$1 \geq \left|\frac{q_{11}}{b_{11}}\right|\left[\left|\frac{b_{21}}{q_{12}}\right| + \left|\frac{b_{31}}{q_{13}}\right|\right] \tag{5.99}$$

$$1 \geq \left|\frac{q_{22}}{b_{21}}\right|\left[\left|\frac{b_{11}}{q_{21}}\right| + \left|\frac{b_{31}}{q_{23}}\right|\right] \tag{5.100}$$

$$1 \geq \left|\frac{q_{33}}{b_{31}}\right|\left[\left|\frac{b_{11}}{q_{31}}\right| + \left|\frac{b_{21}}{q_{32}}\right|\right] \tag{5.101}$$

Letting

$$\lambda_1 = \frac{b_{21}}{b_{11}}, \quad \lambda_2 = \frac{b_{31}}{b_{11}} \tag{5.102}$$

Equations 5.99 through 5.101 become, respectively,

$$1 \geq |q_{11}| \left[\left| \frac{\lambda_1}{q_{12}} \right| + \left| \frac{\lambda_2}{q_{13}} \right| \right] \tag{5.103}$$

$$\lambda_1 \geq |q_{22}| \left[\left| \frac{1}{q_{21}} \right| + \left| \frac{\lambda_2}{q_{23}} \right| \right] \tag{5.104}$$

$$\lambda_2 \geq |q_{33}| \left[\left| \frac{1}{q_{31}} \right| + \left| \frac{\lambda_2}{q_{32}} \right| \right] \tag{5.105}$$

From Equations 5.104 and 5.105 and using

$$\gamma_{ij} = \left[\frac{q_{ii}q_{jj}}{q_{ij}q_{ji}} \right]_{\substack{i=2\\j=3}} = \gamma_{23} = \frac{q_{22}q_{33}}{q_{23}q_{32}} \tag{5.106}$$

the following expressions for λ_1 and λ_2 are obtained:

$$\lambda_1 \geq \frac{|(q_{22}/q_{21})| + |(q_{22}q_{33}/q_{23}q_{31})|}{|1-\gamma_{23}|} \tag{5.107}$$

$$\lambda_2 \geq \frac{|(q_{33}/q_{31})| + |(q_{22}q_{33}/q_{21}q_{32})|}{|1-\gamma_{23}|} \tag{5.108}$$

Substitute Equations 5.107 and 5.108 into Equation 5.103 to obtain

$$\left| \frac{q_{11}}{q_{12}} \right| \left[\left| \frac{q_{22}}{q_{21}} \right| + \left| \frac{q_{22}q_{33}}{q_{23}q_{31}} \right| \right] + \left| \frac{q_{11}}{q_{13}} \right| \left[\left| \frac{q_{33}}{q_{31}} \right| + \left| \frac{q_{33}q_{22}}{q_{21}q_{32}} \right| \right] \leq |1-\gamma_{23}| \tag{5.109}$$

Substituting $q_{ij} = 1/p^*_{ij}$ into Equation 5.109 yields

$$\left| \frac{p^*_{12}}{p^*_{11}} \right| \left[\left| \frac{p^*_{21}}{p^*_{22}} \right| + \left| \frac{p^*_{23}p^*_{31}}{p^*_{22}p^*_{33}} \right| \right] + \left| \frac{p^*_{13}}{p^*_{11}} \right| \left[\left| \frac{p^*_{31}}{p^*_{33}} \right| + \left| \frac{p^*_{21}p^*_{32}}{p^*_{22}p^*_{33}} \right| \right] \leq \left| 1 - \frac{p^*_{23}p^*_{32}}{p^*_{22}p^*_{33}} \right| \tag{5.110}$$

Multiplying Equation 5.110 by $p^*_{11}p^*_{22}p^*_{33} = p^*_{123}$ yields Condition 3.

Condition 3: (3 × 3 plant).[190]
As $\omega \to \infty$

$$|p^*_{123}| \geq |p^*_{12}|[|p^*_{21}p^*_{33}| + |p^*_{23}p^*_{31}|] + |p^*_{13}|[|p^*_{22}p^*_{31}| + |p^*_{21}p^*_{32}|] + |p^*_{11}p^*_{23}p^*_{32}| \tag{5.111}$$

See Ref. 77 for higher order plants. This condition is necessary only if Equation 5.7 is used to generate the plant.

Condition 1 ensures controllability of the plant since the inverse of $\boldsymbol{P}$ produces the effective transfer functions used in the design. If the $\boldsymbol{P}$ matrix, resulting from the original ordering of the elements of the input and output vectors, does not satisfy Condition 2 (or 3), then a reordering of the input and output vectors may result in satisfying these conditions.

Division III: Method 2

The third division of this chapter, Section 5.6, introduces *Method 2*, a sequential technique to design diagonal matrix compensators for MIMO systems with model uncertainty.

5.6 Sequential Diagonal MIMO QFT (*Method 2*)[81]

The previous diagonal MIMO QFT technique, *Method 1*, inherently involves some overdesign, as seen from Equations 5.49 and 5.56, in which $t_{21}, t_{31}, \ldots,$ can be any members of their acceptable sets $\tau_{21}, \tau_{31}, \ldots,$ and $q_{12}, q_{13}, \ldots,$ any members of their uncertainty sets. As noted in Equation 5.56, it is therefore necessary to use the worst case values ($t_{kj} = b_{kj}$, where b_{kj} is the upper response bound, T_{R_U} or T_D), which leads to an overdesign. Actually, in real systems, there is usually a correlation between the $t_{21}, t_{31}, \ldots,$ and the $q_{11}, q_{13}, \ldots$. For example, in Equation 5.65 it is possible that q_{12} is large when t_{21} is large in the expression for c_{11}. Such a correlation can only help make c_{11} smaller. Thus, for *Method 1*, it is not possible to use this correlation, and so one must take the largest t_{21}, the smallest q_{12}, etc. This is the price paid for converting the MIMO problem into the much simpler MISO problems and avoiding having to work with the very complicated denominator in Equation 5.29.

Method 1 was improved by the so-called *Method 2* or sequential method,[81] which considers successive steps of an iterative method with an equivalent plant that also takes into account the controllers designed in the previous steps. The QFT design *Method 2*, for many problems, may yield a better control system design. This method is an improvement over *Method 1*. It utilizes the resulting designed g_is and f_{ij}s of the first MISO equivalent loop that is designed in the design of the succeeding loops, etc. This feature of *Method 2* reduces the overdesign in the early part of the design process. *Method 2* may involve a trade-off in the design parameters. The final MISO equivalent loop to be designed uses the exact transfer functions of the previously designed g_is and f_{ij}s; thus, this loop has the least amount of overdesign.[79] Using Equation 5.39, Equation 5.38 becomes

$$\begin{bmatrix} p^*_{11} + g_1 & p^*_{12} & \cdots & p^*_{1m} \\ p^*_{21} & p^*_{22} + g_2 & \cdots & p^*_{2m} \\ \cdots & \cdots & \cdots & \cdots \\ p^*_{m1} & p^*_{m2} & \cdots & p^*_{mm} + g_m \end{bmatrix} \boldsymbol{T} = \boldsymbol{GF} \tag{5.112}$$

Now, in order to make $[\boldsymbol{P}^* + \boldsymbol{G}]$ in Equation 5.112 a triangular matrix (which allows explicit expressions for each loop problem), the Gauss elimination method is applied by premultiplying both sides by $\boldsymbol{G}_2, \boldsymbol{G}_3, \ldots, \boldsymbol{G}_n$ so that $\boldsymbol{G}_n \ldots \boldsymbol{G}_3\boldsymbol{G}_2[\boldsymbol{P}^* + \boldsymbol{G}]\boldsymbol{T} = \boldsymbol{G}_n \ldots \boldsymbol{G}_3\boldsymbol{G}_2\boldsymbol{GF}$, with

$$G_2 = \begin{bmatrix} 1 & 0 & \dots & 0 \\ \dfrac{-p_{21}^{*1}}{p_{11}^{*1} + g_1} & 1 & \dots & 0 \\ \dots & \dots & \dots & \dots \\ \dfrac{-p_{n1}^{*1}}{p_{11}^{*1} + g_1} & 0 & \dots & 1 \end{bmatrix}; \quad G_3 = \begin{bmatrix} 1 & 0 & 0 & \dots & 0 \\ 0 & 1 & 0 & \dots & 0 \\ 0 & \dfrac{-p_{32}^{*2}}{p_{22}^{*2} + g_2} & 1 & \dots & 0 \\ \dots & \dots & \dots & \dots & \dots \\ 0 & \dfrac{-p_{n2}^{*2}}{p_{22}^{*2} + g_2} & 0 & \dots & 1 \end{bmatrix}$$

$$G_n = \begin{bmatrix} 1 & \dots & 0 & \dots & \dots & 0 \\ \dots & \dots & \dots & \dots & \dots & \dots \\ 0 & \dots & 1 & \dots & \dots & 0 \\ 0 & \dots & \dfrac{-p_{km}^{*m}}{-p_{mm}^{*m} + g_m} & 1 & \dots & 0 \\ \dots & \dots & \dots & \dots & \dots & \dots \\ 0 & \dots & \dfrac{-p_{mm}^{*m}}{p_{mm}^{*m} + g_m} & 0 & \dots & 1 \end{bmatrix} \tag{5.113}$$

and with the iterative expressions

$$\begin{aligned} p_{ij}^{*1} &= [\boldsymbol{P}]_{11} \\ p_{ij}^{*k+1} &= p_{ij}^{*k} - \frac{p_{ik}^{*k} p_{kj}^{*k}}{p_{kk}^{*k} + g_k}; \quad i = k+1, 2, \dots, n; \; j = 1, 2, \dots, n \end{aligned} \tag{5.114}$$

Then Equation 5.112 becomes

$$\begin{bmatrix} p_{11}^{*1} + g_1 & p_{12}^{*1} & \dots & \dots & \dots & p_{1n}^{*1} \\ 0 & p_{22}^{*2} + g_2 & \dots & p_{2k}^{*2} & \dots & p_{2n}^{*2} \\ \dots & \dots & \dots & \dots & \dots & 0 \\ 0 & 0 & \dots & p_{kk}^{*k} + g_k & \dots & p_{kn}^{*k} \\ \dots & \dots & \dots & \dots & \dots & \dots \\ 0 & 0 & \dots & 0 & \dots & p_{nn}^{*n} + g_n \end{bmatrix} \begin{bmatrix} t_{11} & \dots & t_{1n} \\ \dots & \dots & \dots \\ t_{n1} & \dots & t_{nn} \end{bmatrix}$$

$$= \begin{bmatrix} g_1 & 0 & \dots & 0 & \dots & 0 \\ -p_{21}^{*2} & g_2 & \dots & 0 & \dots & 0 \\ \dots & \dots & \dots & \dots & \dots & 0 \\ -p_{k1}^{*k} & -p_{k2}^{*k} & \dots & g_k & \dots & 0 \\ \dots & \dots & \dots & \dots & \dots & \dots \\ -p_{n1}^{*n} & -p_{n2}^{*n} & \dots & -p_{nk}^{*n} & \dots & g_n \end{bmatrix} \begin{bmatrix} f_{11} & \dots & f_{1n} \\ \dots & \dots & \dots \\ f_{n1} & \dots & f_{nn} \end{bmatrix} \tag{5.115}$$

where, the t_{mj} element for the $n \times n$ case, is

$$t_{mj} = \frac{1}{\overset{*m}{p}_{mm} + g_m} \left[g_m f_{mj} - \sum_{u=m+1}^{n} \overset{*m}{p}_{mu} t_{uj} - \sum_{u=1}^{m-1} \overset{*m}{p}_{mu} f_{uj} \right], \quad j = 1,2,\ldots,n \tag{5.116}$$

For instance, t_{11} for the 3×3 case is $t_{11} = (\overset{*1}{p}_{11} + g_1)^{-1}(g_1 f_{11} - \overset{*1}{p}_{12} t_{21} - \overset{*1}{p}_{13} t_{31})$

Now, the transfer functions for the $n \times n$ system obtained from Equation 5.116 have the form

$$t_{mj} = w_{mm}(v_{mj} + c_{mj}) \tag{5.117}$$

where

$$w_{mm} = \frac{1}{\overset{*m}{p}_{mm} + g_m} = \frac{\overset{m}{q}_{mm}}{1 + \overset{m}{q}_{mm} g_m}, \quad \overset{m}{q}_{mm} = \left(\frac{1}{\overset{*m}{p}_{mm}} \right) \tag{5.118}$$

$$v_{mj} = g_m f_{mj} \tag{5.119}$$

$$c_{mj} = -\sum_{u=m+1}^{n} \overset{*m}{p}_{mu} t_{uj} - \sum_{u=1}^{m-1} \overset{*m}{p}_{mu} f_{uj}, \quad j = 1,2,\ldots,n \tag{5.120}$$

For example, the equations of *Method 2* for a 2×2 system are the following:

Loop 1

$$\begin{aligned} t_{1j} = y_{1j} &= \frac{L_1}{1+L_1} f_{1j} + \frac{q_{11} c_{1j}}{1+L_1}; \quad L_1 = g_1 q_{11} \\ c_{1j} &= \frac{-t_{2j}}{q_{12}}; \quad t_{2j} \in \tau_{2j}; \quad \mathbf{P}^{-1} = [\overset{*}{p}_{ji}] = [1/q_{ji}] \end{aligned} \tag{5.121}$$

Loop 2

$$\begin{aligned} t_{2j} &= \frac{L_{2e}}{1+L_{2e}} f_{2j} + \frac{c_{2j}}{1+L_{2e}}; \quad L_{2e} = g_2 q_{22e} = g_2 \frac{q_{22}(1+L_1)}{1-\gamma_{12}+L_1} \\ c_{2j} &= \frac{g_1 f_{1j} p_{21}(1-\gamma_{12})}{1-\gamma_{12}+L_1}; \quad \gamma_{12} = \frac{p_{21} p_{12}}{p_{22} p_{11}}; \quad \mathbf{P} = [p_{ji}] \end{aligned} \tag{5.122}$$

which can also be calculated from Equation 5.116. In the same manner, the equations of *Method 2* for a 3×3 system are the following:

Loop 1

$$\begin{aligned} t_{1j} = y_{1j} &= \frac{L_1}{1+L_1} f_{1j} + \frac{q_{11} c_{1j}}{1+L_1}; \quad L_1 = g_1 q_{11} \\ c_{1j} &= \frac{-t_{2j}}{q_{12}}; \quad t_{2j} \in \tau_{2j}; \quad \mathbf{P}^{-1} = [\overset{*}{p}_{ji}] = [1/q_{ji}] \end{aligned} \tag{5.123}$$

Loop 2

$$t_{2j} = y_{2j} = \frac{f_{2j}L_{2e} + c_{2j}}{1 + L_{2e}}; \quad L_{2e} = \frac{g_2 q_{22}}{1 - (\gamma_{12}/(1 + L_1))}$$

$$\gamma_{ji} = \frac{p_{ij}p_{ji}}{p_{ii}p_{jj}}; \quad p_{21}^{e} = \frac{L_1 p_{21}^{*}}{1 + L_1}; \quad L_{2e} = g_2 q_{22e}$$

$$c_{2j} = \frac{L_{2e}}{g_2}\left(t_{3j}\left[\frac{p_{21}^{*}p_{13}^{*}}{p_{11}^{*}(1 + L_1)} - p_{23}^{*}\right] - f_{1j}p_{21}^{*}\right) \tag{5.124}$$

$$P^{-1} = \left[p_{ij}^{*}\right]$$

Loop 3

$$t_{3j} = \frac{f_{3j}L_{3e}}{1 + L_{3e}}; \quad L_{3e} = \frac{L_3\varsigma}{\varsigma - \Lambda}; \quad L_3 = g_3 q_{33}$$

$$\varsigma = (1 + L_1)(1 + L_{2e}) - \gamma_{12}$$

$$\Lambda = \gamma_{23}(1 + L_1) + \gamma_{13}(1 + L_{2e}) - (\gamma_{12}\mu_2 + \gamma_{13}\mu_3)$$

$$\mu_2 = \frac{p_{23}^{*}p_{31}^{*}}{p_{21}^{*}p_{33}^{*}}; \quad \mu_3 = \frac{p_{32}^{*}p_{21}^{*}}{p_{31}^{*}p_{22}^{*}}$$

$$c_{3j} = \frac{f_{1j}L_1 q_{33}\eta_1 + f_{2j}L_2 q_{33}\eta_2}{\varsigma - \Lambda} \tag{5.125}$$

$$\eta_1 = q_{22}p_{21}^{*}p_{32}^{*} - p_{31}^{*}(1 +_{2e})$$

$$\eta_2 = q_{11}p_{12}^{*}p_{31}^{*} - p_{32}^{*}(1 + L_1)$$

which can be also calculated from Equation 5.116.

The order in which the MISO loops are designed with *Method 2* is important. Any order may be chosen but some orders may produce less overdesign (lower BW) than others. The general rule in the choice of the design order of the loops is *that the most constrained loop is chosen as the starting loop.* That is, the performance specifications (P.S.) to be satisfied has placed a *constraint* on which loop is to be chosen as the starting loop. For example, the BW specifications for each loop, ω_{hii}, to be satisfied places the constraint on which loop is to be chosen as the starting loop. Thus, choose the "starting" loop i for which it is most important to minimize the BW requirements. Some of the factors involved in determining the BW requirements are as follows:

1. Sensor noise.
2. Loop i has *severe* bending mode problems that other loops do not have.
3. The "high-frequency gain" (*h.f.g.*) uncertainty of q_{ii} may be very much greater than for the other loop(s). The *h.f.g.* uncertainty may affect the size of the templates.

Professor Horowitz provides the following insight:

- Analyze the various q_{ii} templates over a reasonable range of frequencies: almost vertical at low and high frequencies for the analog case and for the discrete case up to the frequency range in which the templates first narrow, before widening again for $\omega_i > \omega_s/2$ (ω_s sampling frequency), rather than the final frequency range ($\omega_i > \omega_s/2$), in which the templates approach a vertical line (see Chapter 4).
- If the feedback requirements per loop are roughly the same, which indicates the t_{ii} specifications are about the same, and the t_{ij}, $i \neq j$, are also roughly similar for the different loops, then the loop with the smallest q_{ii} templates should be the "starting loop." That is, the loop with the smallest amount of feedback should be chosen as the starting loop. The reason for this choice is that there is a tendency toward BW propagation as the design proceeds from the first loop to the final loop that is designed. Therefore, for this case, the loop with the smallest BW requirement becomes the most constrained loop due to the BW propagation effect. Thus, it is the loop to be designed first and the second loop to be designed should be the one having the second smallest feedback requirement, etc. Also, see the discussion in Chapter 9 of Ref. 6.

All of these factors emerge from the *transparency* of QFT, which helps to reduce the trial and error that is involved in achieving a satisfactory design. If choosing the loop with the most severe BW limitation causes a design problem in the succeeding designed loops, then try a different starting loop selection. This is based on the knowledge, that in general, the BW of the succeeding designed loops is higher than the BW of the previously designed loops.

If *Method 2* cannot satisfy the BW requirements for all loops, then *Method 1* must be used. If *Method 1* cannot be used, then the non-diagonal *Method 3* or *Method 4* could be helpful (see Chapter 6).[93,97,102] If there are still some problems, then reevaluate the performance specifications, and so on. Chapters 6 and 7 of Ref. 6 provide more insight into the methods.

5.7 Basically Noninteracting Loops

A basically noninteracting (BNIC) loop[79] is one in which the output $y_k(s)$ due to the input $r_j(s)$ is ideally zero. Plant uncertainty and loop interaction (cross-coupling) makes the ideal response unachievable. Thus, the system performance specifications describe a range of acceptable responses for the commanded output and a maximum tolerable response for the uncommanded outputs. The uncommanded outputs are treated as cross-coupling effects (akin to disturbances).

For an LTI plant, having no parameter uncertainty, it is possible to essentially achieve zero cross-coupling effects, that is, the output $y_k \approx 0$ due to c_{ij}. This desired result can be achieved by postmultiplying $\boldsymbol{P}$ by a matrix $\boldsymbol{W}$ to yield $\boldsymbol{P}_n = \boldsymbol{PW} = [p_{ijn}]$, where $p_{ijn} = 0$ for $i \neq j$, resulting in a diagonal $\boldsymbol{P}_n$ matrix for $\boldsymbol{P}$ representing the nominal plant case in the set $\mathcal{P}$. With plant uncertainty, the off-diagonal terms of $\boldsymbol{P}_n$ will not be zero but "very small" in comparison to $\boldsymbol{P}$ for the non-nominal plant cases in $\boldsymbol{P} \in \mathcal{P} = \{\boldsymbol{P}\}$. In some design problems it may be necessary or desired to determine a $\boldsymbol{P}_n$ upon which the QFT design is accomplished. Doing this minimizes the effort required to achieve the desired BW and

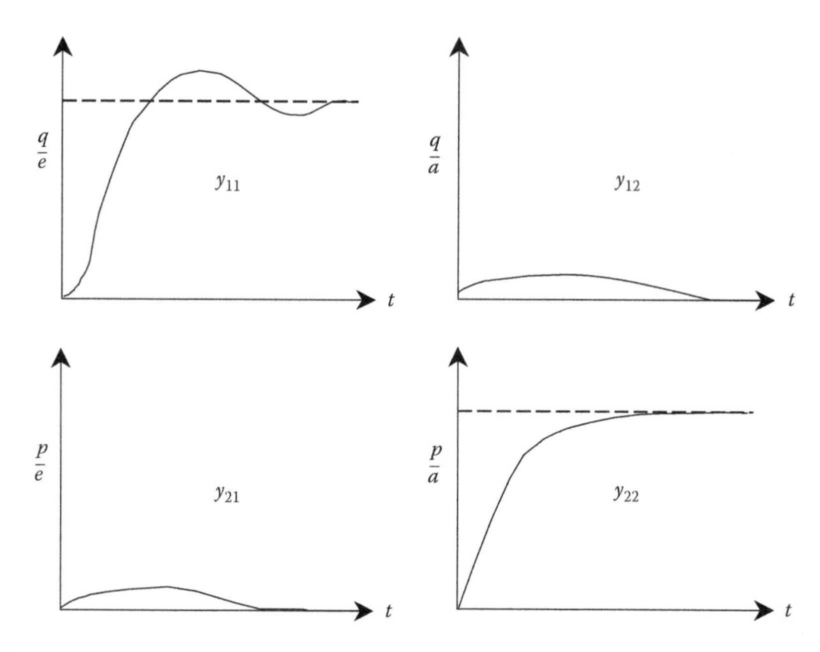

FIGURE 5.17
Output time-response sketch for 2 × 2 plant: *a*, aileron deflection; *q*, pitch rate; *e*, elevator deflection; *p*, roll rate.

minimizes the cross-coupling effects. Since $|t_{ij}(j\omega)| \leq b_{ij}(\omega)$, $i \neq j$, for all ω, it is clearly best to let $f_{ij} = 0$, for $i \neq j$. Thus,

$$t_{ij} = t_{c_{ij}} = \frac{c_{ij}q_{ii}}{1+L_i} \quad \text{for all } i \neq j$$

This was done on an AFTI-16 design by Horowitz[152] as shown in Figure 5.17. In general, an upper bound b_{ij}, $i \neq j$, is specified in order to achieve the performance specification:

$$|\tau_{c_{ij}}| \leq b_{ij} \quad \text{for all } P \in \mathcal{P}$$

A system designed to this specification is called BNIC.

5.8 MIMO QFT with External (Input) Disturbances

Previous discussions of diagonal MIMO QFT did not consider external input disturbance in the calculation of cross-coupling rejection bounds. The following development quantifies external uncertain disturbances. Figure 5.18 represents an $n \times n$ MIMO closed-loop system in which $\mathbf{G}(s)$, $\mathbf{P}(s)$, and $\mathbf{P}_d(s)$ are $n \times n$ matrices. $\mathcal{P}(s) = \{\mathbf{P}(s)\}$ and $\mathcal{P}_d(s) = \{\mathbf{P}_d(s)\}$ are sets of matrices due to plant and disturbance uncertainties, respectively. The objective is to find a suitable mapping that permits the analysis and synthesis of a diagonal MIMO control system by a set of equivalent MISO control systems. The following equations are derived from Figure 5.18:

$$\mathbf{y}(s) = \mathbf{P}(s)\mathbf{u}(s) + \mathbf{P}_d(s)\mathbf{d}_{ext}(s); \quad \mathbf{u}(s) = \mathbf{G}(s)\mathbf{v}(s); \quad \mathbf{v}(s) = \mathbf{r}(s) - \mathbf{y}(s) \tag{5.126}$$

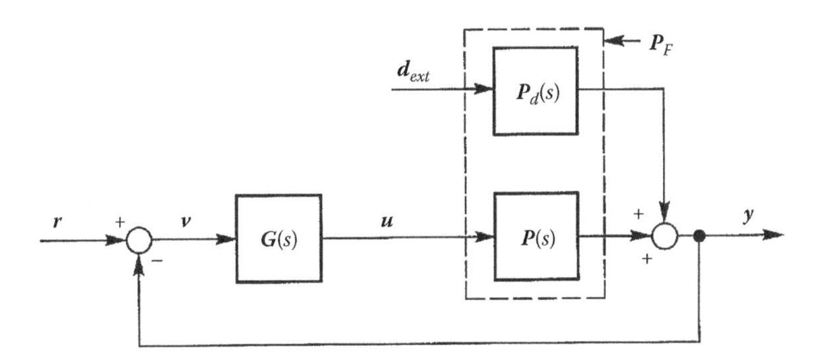

FIGURE 5.18
QFT compensator with output external disturbance.

where for the regulator case with zero tracking input,

$$r(t) = [0, 0, 0]^T \tag{5.127}$$

From Equations 5.126 and 5.127, where henceforth the (s) is dropped in the continuing development

$$v = -y; \quad u = -Gy \tag{5.128}$$

which yields

$$y = -PGy + P_d d_{ext} \tag{5.129}$$

Equation 5.129 is rearranged to yield

$$y = [I + PG]^{-1} P_d d_{ext} \tag{5.130}$$

Based upon unit impulse disturbance inputs for d_{ext}, the system control ratio relating d_{ext} to y is

$$T_d = [I + PG]^{-1} P_d \tag{5.131}$$

Premultiplying Equation 5.131 by $[I + PG]$ yields

$$[I + PG]\, T_d = P_d \tag{5.132}$$

Premultiplying both sides of Equation 5.132 by P^{-1} results in

$$[P^{-1} + G] T_d = P^{-1} P_d \tag{5.133}$$

Using again the definition of P^{-1} (Equation 5.39), q_{ij} (Equation 5.40), and Q (Equation 5.41), where $P = [p_{ij}]$, $P^{-1} = [p^*_{ij}] = [1/q_{ij}]$ and $Q = [q_{ij}] = [1/p^*_{ij}]$. The P^{-1} matrix is again partitioned as $P^{-1} = [p^*_{ij}] = [1/q_{ij}] = \Lambda + B$, (Equation 5.42), where Λ is the diagonal part of P^{-1} and B is the balance of $\boldsymbol{P}^{-1}$. Thus $l_{ii} = 1/q_{ii} = p_{ii}^*$, $b_{ii} = 0$, and $b_{ij} = 1/q_{ij} = p_{ij}^*$ for $i \neq j$. Substituting this result into Equation 5.133 with $\boldsymbol{G}$ diagonal results in

$$[\boldsymbol{\Lambda}+\boldsymbol{B}+\boldsymbol{G}]\boldsymbol{T}_d = [\boldsymbol{\Lambda}+\boldsymbol{B}]\boldsymbol{P}_d \tag{5.134}$$

Rearranging Equation 5.134 produces

$$\boldsymbol{T}_d = [\boldsymbol{\Lambda}+\boldsymbol{G}]^{-1}[\boldsymbol{\Lambda}\boldsymbol{P}_d + \boldsymbol{B}\boldsymbol{P}_d - \boldsymbol{B}\boldsymbol{P}_d] = \{t_{d_{ij}}\} \tag{5.135}$$

Rearranging, for a unit impulse external disturbance input, Equation 5.135 yields

$$\boldsymbol{Y}(\boldsymbol{T}_d) = [\boldsymbol{\Lambda}+\boldsymbol{G}]^{-1}\left[\boldsymbol{\Lambda}\boldsymbol{P}_d + \boldsymbol{B}\boldsymbol{P}_d - \boldsymbol{B}\boldsymbol{T}_d\right] \tag{5.136}$$

This equation defines the desired fixed point mapping, where each of the n^2 matrix elements on the right side of Equation 5.135 is interpreted as a MISO problem. Proof of the fact that the design of each MISO system yields a satisfactory MIMO design is based on the Schauder fixed point theorem.[77] The theorem defines a mapping $\boldsymbol{Y}(\boldsymbol{T}_d)$ where each member of $\boldsymbol{T}_d$ is from the acceptable set $\Im_d$ (see Figure 5.10). If this mapping has a fixed point, that is, $\boldsymbol{T}_d \in \Im_d$, then this $\boldsymbol{T}_d$ is a solution of Equation 5.135.

Figure 5.19 shows the effective MISO loops resulting from a 3 × 3 system. Since $\boldsymbol{\Lambda}$ and $\boldsymbol{G}$ in Equation 5.135 are diagonal, the (1,1) element on the right side of Equation 5.136 for the 3 × 3 case, for a *unit impulse input*, provides the output

$$y_{d_{11}} = \frac{q_{11}}{1+g_1 q_{11}}\left[\frac{p_{d_{11}}}{q_{11}} + \frac{p_{d_{21}}}{q_{12}} + \frac{p_{d_{31}}}{q_{13}} - \left(\frac{t_{c_{21}}}{q_{12}} + \frac{t_{c_{31}}}{q_{13}}\right)\right] \tag{5.137}$$

Equation 5.137 corresponds precisely to the first structure in Figure 5.19. Similarly, each of the nine structures in this figure corresponds to one of the elements of $\boldsymbol{Y}(\boldsymbol{T}_d)$ of Equation 5.135.

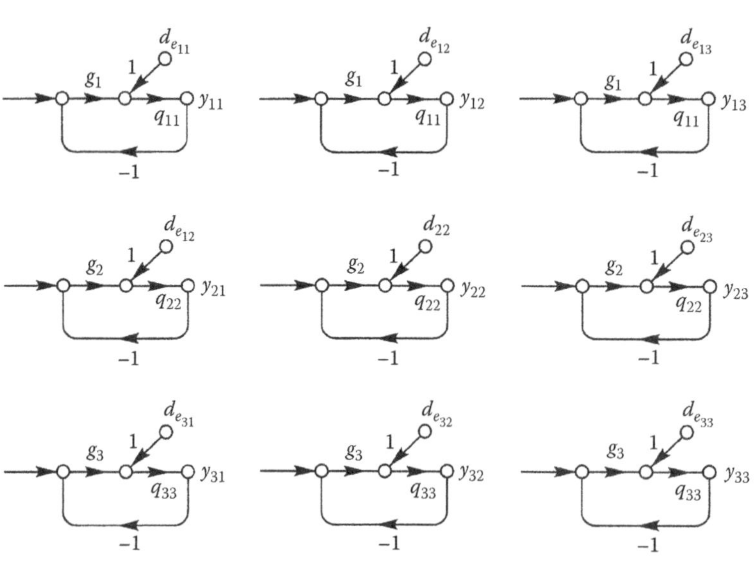

FIGURE 5.19
3 × 3 MISO equivalent loops for external output disturbance.

The control ratios for the external disturbance inputs $d_{e_{ij}}$ and the corresponding outputs y_i for each feedback loop of Equation 5.136 have the form

$$y_{ii} = w_{ii}(d_{e_{ij}}) \tag{5.138}$$

where $w_{ii} = q_{ii}/(1 + g_i q_i)$ and

$$d_{e_{ij}} = (d_{ext})_{ij} - c_{ij} = \sum_{k=1}^{x} \frac{p_{d_{kj}}}{q_{ik}} - \sum_{k \neq 1}^{m} \frac{t_{c_{kj}}}{q_{ik}} \tag{5.139}$$

where
  x is the number of disturbance inputs
  m is the dimension of square MIMO system

Thus, the *interaction term,* Equation 5.139, not only contains the cross-coupling interaction but also the external disturbances, that is,

$$\begin{aligned} (d_{ext})_{ij} &= \sum_{k=1}^{x} \frac{p_{d_{kj}}}{q_{ik}} \\ c_{ij} &= \sum_{k \neq 1}^{m} \frac{t_{c_{kj}}}{q_{ik}} \end{aligned} \tag{5.140}$$

where
  $(d_{ext})_{ij}$ represents the external disturbance effects
  c_{ij} represents the cross-coupling effects

Additional equations, quantifying both the external disturbance $(d_{ext})_{ij}$ and the internal cross-coupling effects c_{ij}, are derived to utilize the improved diagonal MIMO QFT design technique (*Method* 2). These equations are used to define the disturbance bounds for subsequent loops based on the completed design of the first loop. For this development, the equations for the case of a 2 × 2 MIMO system are presented.

From Equation 5.138, for the 1–2 loop case, which is the output of *loop* 1 due to *disturbance input* 2, including the cross-coupling terms from *loop* 2 yields, for *unit impulse inputs,* the following control ratio:

$$t_{d_{12}} = y_{12} = w_{11}(d_{e_{12}}) = \frac{q_{11}}{1+L_1}\left[\frac{p_{d_{12}}}{q_{11}} + \frac{p_{d_{22}}}{q_{12}} + \frac{t_{d_{22}}}{q_{12}}\right] \tag{5.141}$$

Substituting for $t_{d_{22}}$ yields

$$t_{d_{12}} = \frac{q_{11}}{1+L_1}\left[\frac{p_{d_{12}}}{q_{11}} + \frac{p_{d_{22}}}{q_{12}} + \frac{q_{22}d_{e_{22}}}{(1+L_2)q_{12}}\right] \tag{5.142}$$

$$t_{d_{12}} = \frac{q_{11}}{1+L_1}\left[\frac{p_{d_{12}}(1+L_2)q_{12} + p_{d_{22}}(1+L_2)q_{11} - q_{22}q_{11}d_{e_{22}}}{(1+L_2)q_{11}q_{12}}\right] \tag{5.143}$$

$$t_{d_{12}} = \frac{(1+L_2)(p_{d_{12}}q_{12}+p_{d_{22}}q_{11}) - q_{22}q_{11}\left(\dfrac{p_{d_{12}}}{q_{21}}+\dfrac{p_{d_{22}}}{q_{22}}+\dfrac{t_{c_{12}}}{q_{21}}\right)}{(1+L_1)(1+L_2)q_{12}} \quad (5.144)$$

Substituting for $d_{e_{22}}$ and rearranging yields

$$t_{d_{12}} = \frac{(1+L_2)(p_{d_{12}}q_{12}+p_{d_{22}}q_{11}) - (q_{22}q_{11}p_{d_{12}}/q_{21}) - q_{11}p_{d_{22}} + (q_{22}q_{11}t_{d_{12}}/q_{21})}{(1+L_1)(1+L_2)q_{12}} \quad (5.145)$$

$$t_{d_{12}} = \frac{(1+L_1)(p_{d_{12}}q_{12}+p_{d_{22}}q_{11})q_{21} - q_{22}q_{11}p_{d_{12}} - q_{11}q_{21}p_{d_{22}} + q_{22}q_{11}t_{d_{12}}}{(1+L_1)(1+L_2)q_{12}q_{21}} \quad (5.146)$$

Doing $\gamma_{12} = q_{11}q_{22}/q_{21}q_{12}$ and solving for $t_{d_{12}}$ yields

$$t_{d12}(1+L_1)(1+L_2) = \frac{(1+L_2)(p_{d_{12}}q_{12}+p_{d_{22}}q_{11})}{q_{12}} - \frac{q_{11}p_{d_{22}}}{q_{12}} - \gamma\gamma_{12}p_{d_{12}}t_{d_{12}} \quad (5.147)$$

$$t_{d_{12}}[(1+L_1)(1+L_2)-\gamma_{12}] = (1+L_2)\frac{q_{11}}{q_{12}}p_{d_{22}} + (1+L_2)p_{d_{12}} - \frac{q_{11}}{q_{12}}p_{d_{22}} - \gamma_{12}p_{d_{12}} \quad (5.148)$$

$$t_{d_{12}} = \frac{L_2(q_{11}/q_{12})p_{d_{22}} + (1+L_2-\gamma_{12})p_{d_{12}}}{L_1(1+L_2)+1+L_2-\gamma_{12}} \quad (5.149)$$

Equation 5.149 is rearranged as follows:

$$t_{d_{12}} = \frac{((L_2(q_{11}/q_{12})p_{d_{22}})/(1+L_2-\gamma_{12})) + p_{d_{12}}}{1+((L_1(1+L_2))/(1+L_2-\gamma_{12}))} \quad (5.150)$$

From Equation 5.150, the effective plant is defined as

$$q_{11_e} \equiv \frac{q_{11}(1+L_2)}{1+L_2-\gamma_{12}} \quad (5.151)$$

Substituting Equation 5.151 into Equation 5.150 yields

$$t_{d_{12}} = \frac{(L_2p_{d_{22}}q_{11_e}/q_{12}(1+L_2)) + p_{d_{12}}}{1+g_1q_{11_e}} \quad (5.152)$$

Thus, in general, for the 2 × 2 case, the *Method 2* control ratio of the *j*th interaction input to the *i*th system output is

$$t_{d_{ij}} = \frac{(L_kp_{d_{kj}}q_{ii_e}/q_{ik}(1+L_k)) + p_{d_{ij}}}{1+g_iq_{ii_e}} \quad \text{where } i = 1,2 \text{ and } k \neq i \quad (5.153)$$

The interaction bounds (the optimal bounds for a pure regulator control system), representing the cross-coupling and the external disturbance effects, are calculated at a given frequency to satisfy

$$(B_{d_e})_{ij} \geq \left|t_{d_{ij}}\right| = \frac{\left|(L_k p_{d_{kj}} q_{ii_e} / q_{ik}(1+L_k)) + p_{d_{ij}}\right|}{\left|1+g_i q_{ii_e}\right|}, \quad \text{where } i = 1,2 \text{ and } k \neq i \tag{5.154}$$

or

$$\left|1+g_i q_{ii_e}\right| \geq \frac{1}{(B_{d_e})_{ij}} \left|\frac{L_k p_{d_{kj}} q_{ii_e}}{q_{ik}(1+L_k)} + p_{d_{ij}}\right| \quad \text{where } i = 1,2 \text{ and } k \neq i \tag{5.155}$$

Method 2 uses these equations to reduce the overdesign inherent in the original design process. As pointed out previously, the order in which the loops are designed is important. Any order can be used, but some orders produce less overdesign (less BW) than others. The last loop designed has the least amount of overdesign, therefore, the most constrained loop is done first by *Method 1*. Then, the design is continued through the remaining loops by using *Method 2*. As a second iteration, the first loop is then redesigned using *Method 2*.

At this point it is important to point out that when the interaction term specification is considered, the designer must decide how much is to be allocated for the cross-coupling effects and how much for the external disturbance effects. In other words, the designer can "tune" the external disturbance rejection specification depending on the nature of the interaction term for a particular loop. For example, if one loop is only affected by external disturbance, the interaction term specification would consider external disturbance effects only. But if the loop interaction term is a mix of cross-coupling and external disturbance, the designer must then "tune" the interaction term specification accordingly. Since each loop may not exhibit the same interaction characteristics, interaction term specification tuning provides flexibility in the QFT design process.

5.9 Summary

This chapter extended the MISO QFT design technique to enable the design of diagonal MIMO QFT control systems. The first division of this chapter presented a general overview of MIMO systems, the second division presented the nonsequential technique (*Method 1*), and the third division the sequential technique (*Method 2*). Both the techniques represent the MIMO control system by n^2 *MISO equivalent control systems*. Reference tracking, external disturbance rejection, and coupling attenuation problems were also presented in this chapter.

6

Non-Diagonal MIMO QFT[93,97,102]

6.1 Introduction

The design of non-diagonal (full-matrix) controllers for MIMO systems with model uncertainty is still one of the most difficult challenges faced in control engineering. Problems like the compensator/controller matrix structure, sensor–actuator pairings, input–output directions, loops coupling, transmission zeros, etc., become much more difficult when the plant presents model uncertainty. In the last few decades, a significant amount of work on MIMO systems has been published.

A non-diagonal (fully populated) matrix compensator allows the designer much more design flexibility to regulate MIMO systems than the classical diagonal controller structure. Two methodologies (*Method 3* and *Method 4*) to design non-diagonal matrix compensators (fully populated $\mathbf{G}$) for MIMO systems with model uncertainty are presented in this chapter. They extend the classical diagonal QFT compensator design of *Method 1* and *Method 2* of Chapter 5 to new systems with more loop coupling and/or more demanding specifications.

Division I of this chapter introduces *Method 3*. Three classical cases are studied under this division: reference tracking, external disturbance rejection at plant input, and external disturbance rejection at plant output. The method analyses the role played by the non-diagonal compensator elements g_{ij} $(i \neq j)$ by means of three coupling matrices ($\mathbf{C}_1$, $\mathbf{C}_2$, $\mathbf{C}_3$) and a quality function η_{ij}. It quantifies the amount of interaction among the control loops and proposes a sequential design methodology for the fully populated matrix compensator that yields n equivalent tracking SISO systems and n equivalent disturbance rejection SISO systems.[93,97,98,178,180,181,183,184] As a result, the off-diagonal elements g_{ij} $(i \neq j)$ of the compensator matrix reduce (or cancel if there is no uncertainty) the level of coupling between loops, and the diagonal elements g_{kk} regulate the system with less bandwidth requirements than the diagonal elements of the previous diagonal $\mathbf{G}$ methods.

Division II of this chapter introduces *Method 4*. It is a formal reformulation of *Method 3* that includes a generalization of previous techniques. It avoids former hypotheses of diagonal dominance and simplifies the calculations for the off-diagonal elements and the method itself. The method reformulates the classical matrix definition of MIMO specifications by designing a new set of loop-by-loop QFT bounds on the Nichols chart with necessary and sufficient conditions. It also gives explicit expressions to share the load among the loops of the MIMO system to achieve the matrix specifications and also for stability, reference tracking, disturbance rejection at plant input and output, and noise attenuation problems.[102]

Division I: *Method* 3[93,97,98,178,180,181,183,184]

The first division of this chapter, Sections 6.2 through 6.8, introduces *Method 3*, a technique to design non-diagonal (fully populated) matrix compensators for MIMO systems with model uncertainty.

6.2 Non-Diagonal MIMO QFT: A Coupling Minimization Technique (*Method* 3)

Coupling among the control loops is one of the main challenges in MIMO systems. This section defines an index (the coupling matrix) that allows one to quantify the loop interaction in MIMO systems. Following some Horowitz's ideas introduced in the previous chapters, consider an $n \times n$ linear multivariable system (see Figure 6.1), composed of a plant $\boldsymbol{P}$, a fully populated matrix compensator $\boldsymbol{G}$, and a prefilter $\boldsymbol{F}$. These matrices are defined as follows:

$$\boldsymbol{P} = \begin{bmatrix} p_{11} & p_{12} & \cdots & p_{1n} \\ p_{21} & p_{22} & \cdots & p_{2n} \\ \vdots & \vdots & \ddots & \vdots \\ p_{n1} & p_{n2} & \cdots & p_{nn} \end{bmatrix}; \quad \boldsymbol{G} = \begin{bmatrix} g_{11} & g_{12} & \cdots & g_{1n} \\ g_{21} & g_{22} & \cdots & g_{2n} \\ \vdots & \vdots & \ddots & \vdots \\ g_{n1} & g_{n2} & \cdots & g_{nn} \end{bmatrix};$$
$$\boldsymbol{F} = \begin{bmatrix} f_{11} & f_{12} & \cdots & f_{1n} \\ f_{21} & f_{22} & \cdots & f_{2n} \\ \vdots & \vdots & \ddots & \vdots \\ f_{n1} & f_{n2} & \cdots & f_{nn} \end{bmatrix} \tag{6.1}$$

Figure 6.1 also shows the plant input disturbance transfer function $\boldsymbol{P}_{di}$ and the plant output disturbance transfer function $\boldsymbol{P}_{do}$, where $\boldsymbol{P} \in \mathcal{P}$, and $\mathcal{P}$ is the set of possible plants due to uncertainty. The reference vector $\boldsymbol{r}'$ and the external disturbance vectors at plant input $\boldsymbol{d}_i'$ and plant output $\boldsymbol{d}_o'$ are the inputs of the system. The output vector $\boldsymbol{y}$

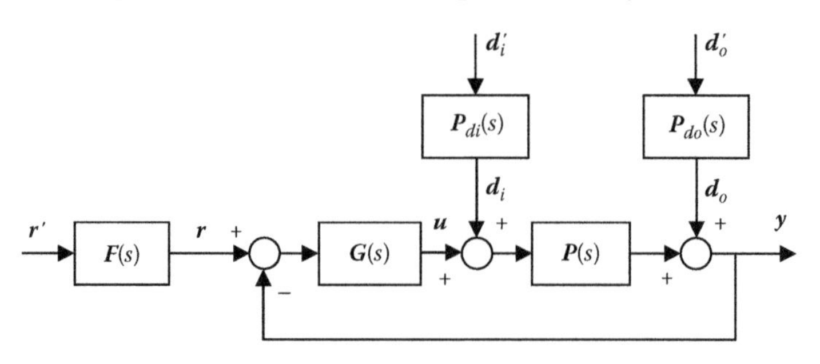

FIGURE 6.1
Structure of a 2DOF (two-degree-of-freedom) MIMO system.

represents the variables to be controlled. The plant inverse P^{-1}, denoted by P^* in this chapter, is presented in the following format:

$$P^{-1} = P^* = [p_{ij}^*] = \Lambda + B = \begin{bmatrix} p_{11}^* & 0 & 0 \\ 0 & \cdots & 0 \\ 0 & 0 & p_{nn}^* \end{bmatrix} + \begin{bmatrix} 0 & \cdots & p_{1n}^* \\ \cdots & 0 & \cdots \\ p_{n1}^* & \cdots & 0 \end{bmatrix} \quad (6.2)$$

and the compensator matrix is broken up into two parts as follows:

$$G = G_d + G_b = \begin{bmatrix} g_{11} & 0 & 0 \\ 0 & \cdots & 0 \\ 0 & 0 & g_{nn} \end{bmatrix} + \begin{bmatrix} 0 & \cdots & g_{1n} \\ \cdots & 0 & \cdots \\ g_{n1} & \cdots & 0 \end{bmatrix} \quad (6.3)$$

Note that Λ is the diagonal part and B is the balance of P^* and that G_d is the diagonal part and G_b is the balance of G.

The following three sections introduce a measurement index to quantify the loop interaction in the three classical cases: reference tracking, external disturbances at the plant input, and the external disturbances at the plant output. In this chapter the measurement index is called the *coupling matrix* C and, depending on the case, shows three different notations C_1, C_2, C_3, respectively. The use of these coupling matrices enables the achievement of essentially n equivalent tracking SISO systems and n equivalent disturbance rejection SISO systems.

6.2.1 Tracking

The transfer function matrix of the control system for the reference tracking problem (Figure 6.1), without any external disturbance ($d_i' = 0$, $d_o' = 0$), is written as shown in the following equation:

$$y = (I + PG)^{-1} PGr = T_{y/r}\, r = T_{y/r}\, Fr' \quad (6.4)$$

Using Equations 6.2 and 6.3, Equation 6.4 is rewritten as

$$T_{y/r}r = (I + \Lambda^{-1}G_d)^{-1}\Lambda^{-1}G_d r + (I + \Lambda^{-1}G_d)^{-1}\Lambda^{-1}\left[G_b r - (B + G_b)T_{y/r}r\right] \quad (6.5)$$

which is another expression to represent the same idea introduced in Chapter 5, as

$$y_{ij} = t_{ij}^{y/r} r_j = (t_{r_{ii}} + t_{c_{ij}}) r_j, \quad i, j = 1, 2, \ldots, n$$

where

$$t_{r_{ii}} = w_{ii} g_{ii}; \quad t_{c_{ij}} = w_{ii} c_{ij}$$

$$w_{ii} = \frac{1}{p_{ii}^* + g_{ii}}$$

$$c_{ij} = -\sum_{k \neq i} t_{kj} p_{ik}^*, \quad k = 1, 2, \ldots, n$$

An analysis of Equation 6.5, the closed-loop transfer function matrix, reveals that it can be broken up into two parts as follows:

1. A diagonal term T_{y/r_d} is given by

$$T_{y/r\_d} = (I + \Lambda^{-1}G_d)^{-1}\Lambda^{-1}G_d \tag{6.6}$$

which represents a pure diagonal structure. Note that it does not depend on the non-diagonal part of the plant inverse B or on the non-diagonal part of the compensator G_b.

It is equivalent to n reference tracking SISO systems formed by plants equal to the elements of Λ^{-1} when the n corresponding parts of a diagonal G_d control them, as shown in Figure 6.2a. In this figure $t_i(s)$ represents the closed-loop control ratio.

2. A non-diagonal term T_{y/r_b} is given by

$$T_{y/r\_b} = (I + \Lambda^{-1}G_d)^{-1}\Lambda^{-1}[G_b - (B + G_b)T_{y/r}] = (I + \Lambda^{-1}G_d)^{-1}\Lambda^{-1}C_1 \tag{6.7}$$

which represents a non-diagonal structure. It is equivalent to the same n previous systems with cross-coupling (internal) disturbances $c_{1ij}r_j$ at the plant input and to n disturbance rejection SISO systems (Figure 6.2b).

In Equation 6.7, the matrix C_1 is the only part that depends on the non-diagonal parts of both the plant inverse B and the compensator G_b. Hence, this matrix comprises the coupling, and from now on C_1 represents the *coupling matrix* C of the equivalent system for reference tracking problems. The bracketed term in Equation 6.7 represents C_1, that is,

$$C_1 = G_b - (B + G_b)T_{y/r} = \begin{bmatrix} 0 & c_{112} & \cdots & c_{11m} \\ c_{121} & 0 & \cdots & c_{12m} \\ \vdots & \vdots & \ddots & \vdots \\ c_{1m1} & c_{1m2} & \cdots & 0 \end{bmatrix} \tag{6.8}$$

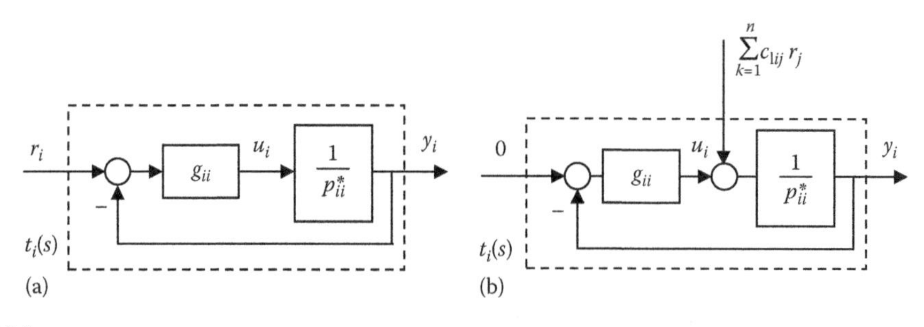

FIGURE 6.2
ith equivalent decoupled SISO systems with cross coupling (internal) disturbances: (a) diagonal term, Equation 6.6 and (b) non-diagonal term, Equation 6.7.

Each element $c_{1_{ij}}$ of this matrix obeys

$$c_{1ij} = g_{ij}(1-\delta_{ij}) - \sum_{k=1}^{n} (p_{ik}^{*} + g_{ik})t_{kj}(1-\delta_{ik}) \tag{6.9}$$

where δ_{ik} is the Kronecker delta that is defined as

$$\delta_{ik} = \begin{cases} \delta_{ik} = 1 \Leftrightarrow k = i \\ \delta_{ik} = 0 \Leftrightarrow k \neq i \end{cases} \tag{6.10}$$

and which is an extension of the cross-coupling c_{ij} elements introduced previously in Chapter 5.

6.2.2 Disturbance Rejection at Plant Input

The transfer matrix from the external disturbance $\boldsymbol{d}_i'$, at the plant input, to the plant output $\boldsymbol{y}$ (Figure 6.1) is written as shown in the following equation:

$$\boldsymbol{y} = (\boldsymbol{I}+\boldsymbol{PG})^{-1}\boldsymbol{P}\boldsymbol{d}_i = \boldsymbol{T}_{y/di}\boldsymbol{d}_i = \boldsymbol{T}_{y/di}\boldsymbol{P}_{di}\,\boldsymbol{d}_i' \tag{6.11}$$

Using Equations 6.2 and 6.3, Equation 6.11 is rewritten as

$$\boldsymbol{T}_{y/di}\boldsymbol{d}_i = (\boldsymbol{I}+\boldsymbol{\Lambda}^{-1}\boldsymbol{G}_d)^{-1}\boldsymbol{\Lambda}^{-1}\boldsymbol{d}_i - (\boldsymbol{I}+\boldsymbol{\Lambda}^{-1}\boldsymbol{G}_d)^{-1}\boldsymbol{\Lambda}^{-1}[(\boldsymbol{B}+\boldsymbol{G}_b)\boldsymbol{T}_{y/di}]\boldsymbol{d}_i \tag{6.12}$$

From Equation 6.12 it is possible to define two different terms as follows:

1. A diagonal term $\boldsymbol{T}_{y/di\_d}$ is given by

$$\boldsymbol{T}_{y/di\_d} = (\boldsymbol{I}+\boldsymbol{\Lambda}^{-1}\boldsymbol{G}_d)^{-1}\boldsymbol{\Lambda}^{-1} \tag{6.13}$$

   Again, Equation 6.13 is equivalent to n regulator SISO systems, as shown in Figure 6.3a.
2. A non-diagonal term $\boldsymbol{T}_{y/di\_b}$ is given by

$$\boldsymbol{T}_{y/di\_b} = (\boldsymbol{I}+\boldsymbol{\Lambda}^{-1}\boldsymbol{G}_d)^{-1}\boldsymbol{\Lambda}^{-1}(\boldsymbol{B}+\boldsymbol{G}_b)\boldsymbol{T}_{y/di} = (\boldsymbol{I}+\boldsymbol{\Lambda}^{-1}\boldsymbol{G}_d)^{-1}\boldsymbol{\Lambda}^{-1}\boldsymbol{C}_2 \tag{6.14}$$

   which represents a non-diagonal structure, equivalent to the same n previous systems with external disturbances $c_{2ij}\, di_j$ at the plant input, as shown in Figure 6.3b.

In Equation 6.14, the matrix $\boldsymbol{C}_2$ comprises the coupling. Thus, from now on $\boldsymbol{C}_2$ represents the *coupling matrix* of the equivalent system for external disturbance rejection at the plant input problems and is given by

$$\boldsymbol{C}_2 = (\boldsymbol{B}+\boldsymbol{G}_b)\boldsymbol{T}_{y/di} \tag{6.15}$$

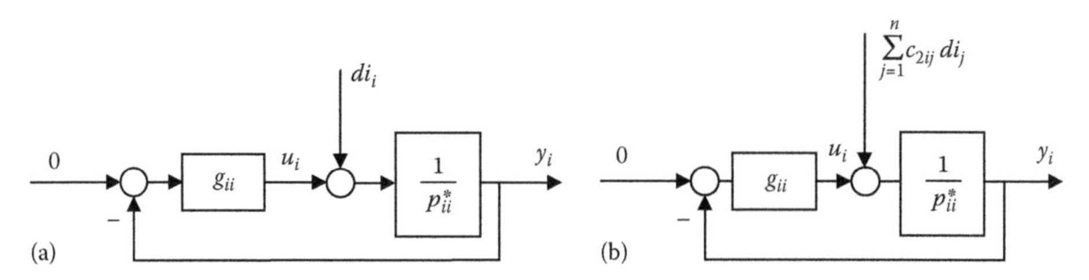

FIGURE 6.3
ith equivalent decoupled SISO systems with external disturbances at the plant input: (a) diagonal term, Equation 6.13 and (b) non-diagonal term, Equation 6.14.

Each element $c_{2_{ij}}$ of this matrix obeys

$$c_{2_{ij}} = \sum_{k=1}^{n} \left(p^*_{ik} + g_{ik}\right) t_{kj} (1 - \delta_{ik}) \tag{6.16}$$

where δ_{ik} is the Kronecker delta defined in Equation 6.10.

6.2.3 Disturbance Rejection at Plant Output

The transfer matrix from the external disturbance $\boldsymbol{d}'_o$ at the plant output to the output $\boldsymbol{y}$ (Figure 6.1) is written as shown in Equation 6.17:

$$\boldsymbol{y} = (\boldsymbol{I} + \boldsymbol{PG})^{-1}\boldsymbol{d}_o = \boldsymbol{T}_{y/do}\boldsymbol{d}_o = \boldsymbol{T}_{y/do}\boldsymbol{P}_{do}\boldsymbol{d}'_o \tag{6.17}$$

Using Equations 6.2 and 6.3, repeating the procedure of the previous sections, Equation 6.17 is rewritten as

$$\boldsymbol{T}_{y/do}\boldsymbol{d}_o = (\boldsymbol{I} + \boldsymbol{\Lambda}^{-1}\boldsymbol{G}_d)^{-1}\boldsymbol{d}_o + (\boldsymbol{I} + \boldsymbol{\Lambda}^{-1}\boldsymbol{G}_d)^{-1}\boldsymbol{\Lambda}^{-1}[\boldsymbol{B} - (\boldsymbol{B} + \boldsymbol{G}_b)\boldsymbol{T}_{y/do}]\boldsymbol{d}_o \tag{6.18}$$

From Equation 6.18 it is also possible to define the following two terms:

1. A diagonal term $\boldsymbol{T}_{y/do\_d}$ given by

$$\boldsymbol{T}_{y/do\_d} = (\boldsymbol{I} + \boldsymbol{\Lambda}^{-1}\boldsymbol{G}_d)^{-1} \tag{6.19}$$

   Once again, Equation 6.19 is equivalent to the n regulator SISO systems shown in Figure 6.4a.

2. A non-diagonal term $\boldsymbol{T}_{y/do\_b}$ is given by

$$\boldsymbol{T}_{y/do\_b} = (\boldsymbol{I} + \boldsymbol{\Lambda}^{-1}\boldsymbol{G}_d)^{-1}\boldsymbol{\Lambda}^{-1}[\boldsymbol{B} - (\boldsymbol{B} + \boldsymbol{G}_b)\boldsymbol{T}_{y/do}] = (\boldsymbol{I} + \boldsymbol{\Lambda}^{-1}\boldsymbol{G}_d)^{-1}\boldsymbol{\Lambda}^{-1}\boldsymbol{C}_3 \tag{6.20}$$

   which represents a non-diagonal structure. It is equivalent to the same n previous systems with external disturbances $c_{3_{ij}}do_j$ at the plant input, as shown Figure 6.4b.

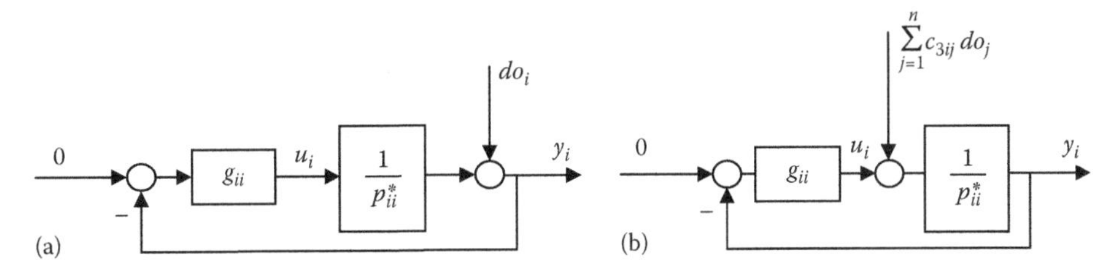

FIGURE 6.4
*i*th equivalent decoupled SISO systems with external disturbances at the plant output: (a) diagonal term, Equation 6.19 and (b) non-diagonal term, Equation 6.20.

In Equation 6.20, the matrix C_3 comprises the coupling. Thus, from now on it represents the *coupling matrix* of the equivalent system for external disturbance rejection for the plant output problems and is given by

$$C_3 = B - (B + G_b)T_{y/do} \tag{6.21}$$

Each element of the coupling matrix, c_{3ij}, obeys

$$c_{3ij} = p^*_{ij}(1-\delta_{ij}) - \sum_{k=1}^{n}(p^*_{ik} + g_{ik})t_{kj}(1-\delta_{ik}) \tag{6.22}$$

where δ_{ik} is the Kronecker delta as defined in Equation 6.10.

6.3 Coupling Elements

In order to design a MIMO compensator with a low coupling level, it is necessary to study the influence of every non-diagonal element g_{ij} on the coupling elements c_{1ij}, c_{2ij}, and c_{3ij} as defined by Equations 6.9, 6.16, and 6.22, respectively.

To easily quantify the main coupling effects, these elements are simplified by applying the following hypothesis.

Hypothesis H1: Suppose that in Equations 6.9, 6.16, and 6.22,

$$\left|\left(p^*_{ij} + g_{ij}\right)t_{jj}\right| \gg \left|\left(p^*_{ik} + g_{ik}\right)t_{kj}\right| \quad \text{for } k \neq j \text{ and in the bandwidth of } t_{jj} \tag{6.23}$$

Note that it is not a too bold hypothesis, considering that it is desirable for the diagonal elements t_{jj} to be much larger than the non-diagonal elements t_{kj}, when once the pairing of the most convenient variables have been applied. Thus,

$$\left|t_{jj}\right| \gg \left|t_{kj}\right| \quad \text{for } k \neq j \text{ and in the bandwidth of } t_{jj} \tag{6.24}$$

Now, two simplifications are applied to facilitate the quantification of the coupling effects c_{1ij}, c_{2ij}, c_{3ij}.

Simplification S1: Using hypothesis H1, Equations 6.9, 6.16, and 6.22, which describe the coupling elements in the tracking problem, the disturbance rejection at the plant input and the disturbance rejection at the plant output, respectively, are rewritten as

$$c_{1ij} = g_{ij} - t_{jj}(p_{ij}^* + g_{ij}), \quad i \neq j \tag{6.25}$$

$$c_{2ij} = t_{jj}(p_{ij}^* + g_{ij}), \quad i \neq j \tag{6.26}$$

$$c_{3ij} = p_{ij}^* - t_{jj}(p_{ij}^* + g_{ij}), \quad i \neq j \tag{6.27}$$

Simplification S2: The elements t_{jj} are, respectively, computed for each case from the equivalent system derived from Equations 6.6, 6.13, and 6.19, so that

$$t_{jj} = \frac{g_{jj}(p_{ij}^*)^{-1}}{1 + g_{jj}(p_{ij}^*)^{-1}} \tag{6.28}$$

$$t_{jj} = \frac{(p_{ij}^*)^{-1}}{1 + g_{jj}(p_{ij}^*)^{-1}} \tag{6.29}$$

$$t_{jj} = \frac{1}{1 + g_{jj}(p_{ij}^*)^{-1}} \tag{6.30}$$

Due to simplifications S1 and S2, the coupling effects c_{1ij}, c_{2ij}, and c_{3ij} are computed as

$$c_{1ij} = g_{ij} - \frac{g_{jj}(p_{ij}^* + g_{ij})}{(p_{jj}^* + g_{jj})}, \quad i \neq j \tag{6.31}$$

$$c_{2ij} = \frac{(p_{ij}^* + g_{ij})}{(p_{jj}^* + g_{jj})}, \quad i \neq j \tag{6.32}$$

$$c_{3ij} = p_{ij}^* - \frac{p_{jj}^*(p_{ij}^* + g_{ij})}{(p_{ij}^* + g_{jj})}, \quad i \neq j \tag{6.33}$$

6.4 Optimum Non-Diagonal Compensator

As stated previously, the purpose of non-diagonal compensators is to reduce the coupling effect in addition to achieving the desired loop performance specifications. The optimum non-diagonal compensators for the three cases (tracking and disturbance rejection at the plant input and output) are obtained making the loop interaction of Equations 6.31 through 6.33 equal to zero.

Note that both elements p^*_{ij} and p^*_{jj} of these three equations are uncertain elements of $\boldsymbol{P}^*$. In general, every uncertain plant p^*_{ij} can be any plant represented by the family

$$\{p^*_{ij}\} = p^{*N}_{ij}(1+\Delta_{ij}), \quad 0 \le |\Delta_{ij}| \le \Delta p^*_{ij} \quad \text{for } i,j = 1,\ldots,n \tag{6.34}$$

where

p^{*N}_{ij} is the selected nominal plant for the non-diagonal controller expression

Δp^*_{ij} is the maximum of the nonparametric uncertainty radii $|\Delta_{ij}|$ (see Figure 6.5b)

Note: Δp^*_{ij} depends on the selection of p^{*N}_{ij}.

The selected plants p^{*N}_{ij} and p^{*N}_{jj} that are chosen for the optimum non-diagonal compensator must comply with the following rules:

1. If the uncertain parameters of the plants show a uniform probability distribution (see Figure 6.5a), which is typical in the QFT methodology, then the elements p^*_{ij} and p^*_{jj} for the optimum non-diagonal compensator are the plants p^{*N}_{ij} and p^{*N}_{jj}. These plants minimize the maximum of the nonparametric uncertainty radii Δp^*_{ij} and Δp^*_{jj} that comprise the plant templates (see Figure 6.5b).

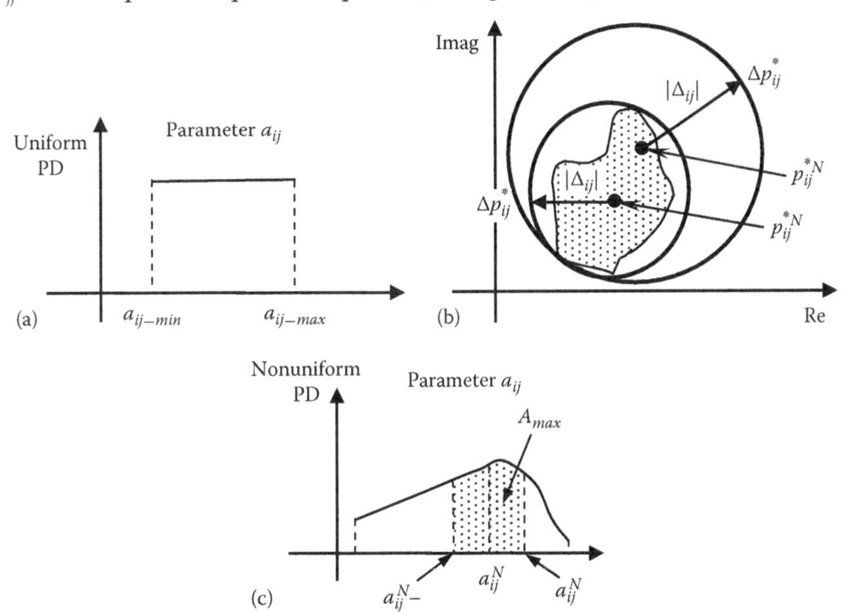

FIGURE 6.5
(a, c) Probability distribution of the parameter a_{ij}; (b) two possibilities of the maximum nonparametric uncertainty radii Δp^*_{ij} that comprise the plant templates.

2. If the uncertain parameters of the plants show a nonuniform probability distribution (see Figure 6.5c), then the elements p_{ij}^* and p_{jj}^* for the optimum non-diagonal compensator are the plants p_{ij}^{*N} and p_{jj}^{*N}, whose set of parameters maximize the area of the probability distribution in the regions $[a_{ij} - \varepsilon, a_{ij} + \varepsilon]$ and $[a_{jj} - \varepsilon, a_{jj} + \varepsilon]$ ($\forall$ parameter $a_{ij}, b_{ij}, \ldots, a_{jj}, b_{jj}, \ldots$), respectively.

These rules of selection are analyzed again in Section 6.5, where the coupling effects with the optimum non-diagonal compensator are computed. By setting Equations 6.31 through 6.33 equal to zero and using Equation 6.34, the optimum non-diagonal compensator for each of the following cases (Equations 6.35 through 6.37) is obtained.

6.4.1 Tracking

$$g_{ij}^{opt} = F_{pd}\left(g_{jj} \frac{p_{ij}^{*N}}{p_{jj}^{*N}} \right) \quad \text{for } i \neq j \tag{6.35}$$

6.4.2 Disturbance Rejection at Plant Input

$$g_{ij}^{opt} = F_{pd}\left(-p_{ij}^{*N} \right) \quad \text{for } i \neq j \tag{6.36}$$

6.4.3 Disturbance Rejection at Plant Output

$$g_{ij}^{opt} = F_{pd}\left(g_{jj} \frac{p_{ij}^{*N}}{p_{jj}^{*N}} \right) \quad \text{for } i \neq j \tag{6.37}$$

where the function $F_{pd}(A)$ means in every case a proper stable and minimum-phase function made from the dominant poles and zeros of the expression A.

6.5 Coupling Effects

The rules of Section 6.4 are utilized for choosing the plants p_{ij}^{*N} and p_{jj}^{*N}. These plants are inserted into Equations 6.35 through 6.37 in order to obtain the respective g_{ij}^{opt}, which are in turn utilized for determining the minimum achievable coupling effects given by Equations 6.38, 6.40, and 6.42. In a similar manner, the maximum coupling effects for the diagonal compensator matrix case, given by Equations 6.39, 6.41, and 6.43, are computed by substituting $g_{ij} = 0$ $(i \neq j)$ into the coupling expressions of Equations 6.31 through 6.33, respectively.

6.5.1 Tracking

$$\left| c_{1ij} \right|_{g_{ij}=g_{ij}^{opt}} = \left| \psi_{ij} \left(\Delta_{jj} - \Delta_{ij} \right) g_{jj} \right| \tag{6.38}$$

$$\left| c_{1ij} \right|_{g_{ij}=0} = \left| \psi_{ij} (1 + \Delta_{ij}) g_{jj} \right| \tag{6.39}$$

6.5.2 Disturbance Rejection at Plant Input

$$\left|c_{2ij}\right|_{g_{ij}=g_{ij}^{opt}} = \left|\psi_{ij}\Delta_{ij}\right| \tag{6.40}$$

$$\left|c_{2ij}\right|_{g_{ij}=0} = \left|\psi_{ij}(1+\Delta_{ij})\right| \tag{6.41}$$

6.5.3 Disturbance Rejection at Plant Output

$$\left|c_{3ij}\right|_{g_{ij}=g_{ij}^{opt}} = \left|\psi_{ij}(\Delta_{ij}-\Delta_{jj})g_{jj}\right| \tag{6.42}$$

$$\left|c_{3ij}\right|_{g_{ij}=0} = \left|\psi_{ij}(1+\Delta_{ij})g_{jj}\right| \tag{6.43}$$

where

$$\psi_{ij} = \frac{p_{ij}^{*N}}{(1+\Delta_{jj})p_{jj}^{*N}+g_{jj}} \tag{6.44}$$

and where the uncertainty is given by

$$0 \le \left|\Delta_{ij}\right| \le \Delta p_{ij}^{*}, \quad 0 \le \left|\Delta_{jj}\right| \le \Delta p_{jj}^{*} \quad \text{for } i,j=1,\ldots,n$$

The coupling effects, calculated for the pure diagonal compensator cases, result in three expressions (6.39), (6.41), and (6.43) that still present a non-zero value, even if (p_{ij}^{*N}, p_{jj}^{*N}) is selected so that the actual plant mismatching disappears ($\Delta_{ij}=0$ and $\Delta_{jj}=0$). However, the coupling effects obtained with the optimum non-diagonal compensators (see Equations 6.38, 6.40, and 6.42) tend to zero when the mismatching disappears ($\Delta_{ij}=0$ and $\Delta_{jj}=0$).

6.6 Quality Function of the Designed Compensator

Figure 6.6 shows the appearance of three different coupling bands for a common system. The maximum $|c_{ij}|_{g_{ij}=0}$, computed from Equations 6.39, 6.41, and 6.43, and the minimum coupling effects without any non-diagonal compensator g_{ij} limit the first one: the top cross-hatched pair.

The second pair of curves (dashed lines) is bounded by the maximum (the upper dashed curve) and the minimum (the lower dashed curve) coupling effects with a nonoptimum decoupling element g_{ij}.

Finally, the minimum coupling effect $|c_{ij}|_{g_{ij}=g_{ij}^{opt}}$ with the optimum decoupling element g_{ij}^{opt} (the bottom solid curve) presents a maximum value, computed from Equations 6.38, 6.40, and 6.42.

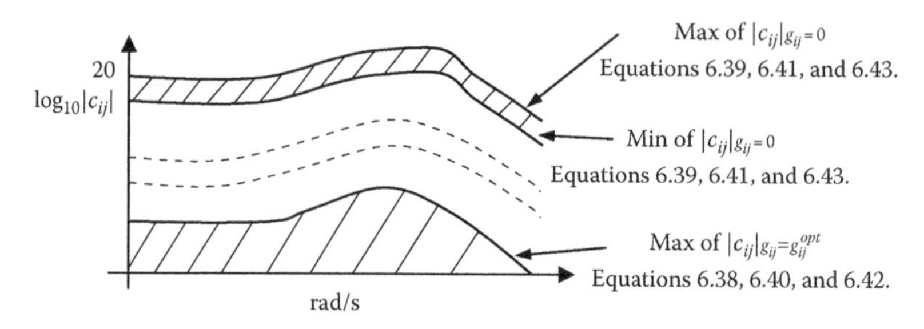

FIGURE 6.6
Coupling effect bands with different non-diagonal compensators.

From this analysis a quality function η_{ij} is defined for a non-diagonal compensator g_{ij} $(i \neq j)$ so that

$$\eta_{ij}(\%) = 100\left\{\frac{\log_{10}\left[\max\left\{|c_{ij}|_{g_{ij}=0}\right\}\Big/\max\left\{|c_{ij}|_{g_{ij}=g_{ij}}\right\}\right]}{\log_{10}\left[\max\left\{|c_{ij}|_{g_{ij}=0}\right\}\Big/\max\left\{|c_{ij}|^{opt}_{g_{ij}=g_{ij}}\right\}\right]}\right\} \tag{6.45}$$

The quality function becomes a proximity measure of the coupling effect c_{ij} to the minimum achievable coupling effect. Thus, the function is useful to quantify the amount of loop interaction and to design the non-diagonal compensators.

If η_{ij} is closed to 100%, then the coupling effect is a minimum and the g_{ij} compensator tends to be the optimum one. A suitable non-diagonal compensator maximizes the quality function of Equation 6.45.

6.7 Design Methodology

The compensator design method is a sequential procedure by closing the loops that only needs to fulfill Hypothesis H1 (Equations 6.23 and 6.24).

6.7.1 Methodology

Step A: Input/output pairing and loop ordering. First, the methodology begins paring the plant inputs and outputs with the relative gain analysis (RGA) technique,[183,219,224] where RGA$(j\omega)$ = $\boldsymbol{P}(j\omega) \otimes (\boldsymbol{P}^{-1}(j\omega))^T$, and where $\otimes$ denotes element-by-element multiplication (Schur product) (see Section 5.3).

This is followed by arranging the matrix $\boldsymbol{P}^*$ so that $(p_{11}^*)^{-1}$ has the smallest phase margin frequency, $(p_{22}^*)^{-1}$ has the next smallest phase margin frequency, and so on. The sequential compensator design technique (as described in Figure 6.7), composed of n stages (n loops), utilizing the following steps (B and C) is repeated for every column $k = 1$ to n.

Step B: Design of the diagonal compensator elements g_{kk}. This design of the element g_{kk} is calculated using the standard QFT loop-shaping technique for the inverse of the equivalent

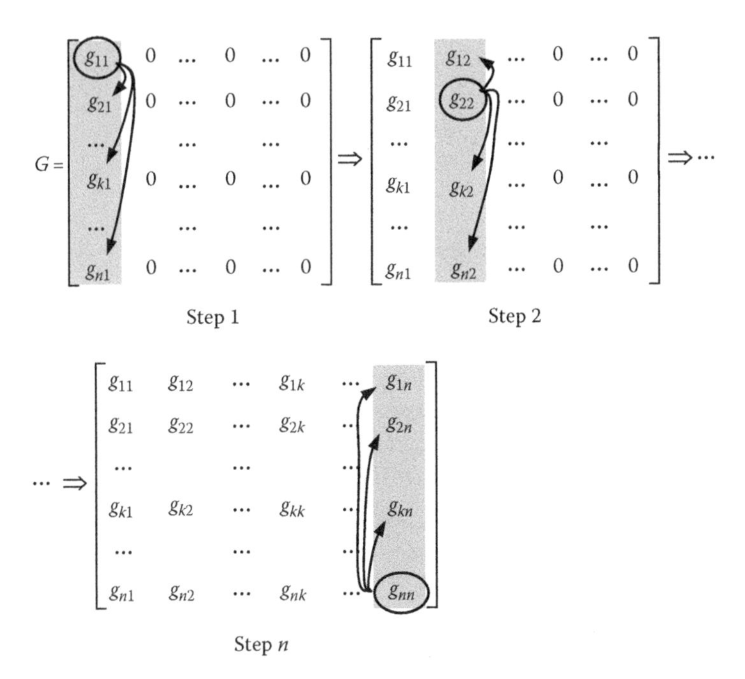

FIGURE 6.7
n stages of the sequential compensator design technique.

plant $\left(p_{kk}^{*e}\right)^{-1}$ in order to achieve robust stability and robust performance specifications. The equivalent plant satisfies the recursive relationship[272] given by Equation 6.46:

$$\left[p_{ii}^{*e}\right]_k = \left[p_{ii}^{*}\right]_{k-1} - \frac{\left(\left[p_{i(i-1)}^{*}\right]_{k-1} + \left[g_{i(i-1)}\right]_{k-1}\right)\left(\left[p_{(i-1)i}^{*}\right]_{k-1} + \left[g_{(i-1)i}\right]_{k-1}\right)}{\left[p_{(i-1)(i-1)}^{*}\right]_{k-1} + \left[g_{(i-1)(i-1)}\right]_{k-1}}, \quad i \geq k, \; [\mathbf{P}^*]_{k=1} = \mathbf{P}^* \tag{6.46}$$

This equation is an extension of that for the non-diagonal case of the recursive expression proposed by Horowitz[81] as the *improved design technique,* also called *Method 2* in Chapter 5.

At this point, the design should also fulfill the following two stability conditions[92]: (a) $L_i(s) = g_{ii}(s)(p_{ii}^{*e})^{-1}$ has to satisfy the Nyquist encirclement condition and (b) no right-half-plane (RHP) pole-zero cancellations have to occur between $g_{ii}(s)$ and $(p_{ii}^{*e})^{-1}$. If the system requires the tracking specifications as

$$a_{ii}(\omega) \leq \left|t_{ii}^{y/r}(j\omega)\right| \leq b_{ii}(\omega)$$

and since

$$t_{ii}^{y/r} = t_{r_{ii}} + t_{c_{ii}}$$

the tracking bounds b_{ii} and a_{ii} are corrected to take into account the cross-coupling specification $\tau_{c_{ii}}$, so that

$$b_{ii}^{c} = b_{ii} - \tau_{cii}, \quad a_{ii}^{c} = a_{ii} + \tau_{cii} \tag{6.47}$$

$$t_{c_{ii}} = w_{ii}c_{ii} \leq \tau_{c_{ii}} \tag{6.48}$$

$$a_{ii}^{c}(\omega) \leq \left| t_{r_{ii}}(j\omega) \right| \leq b_{ii}^{c}(\omega) \tag{6.49}$$

These are the same corrections proposed by Horowitz[77,81] for the original MIMO QFT *Methods 1* and *2* (Chapter 5).

However, for the non-diagonal compensator, these corrections are less demanding. The coupling expression $t_{c_{ii}} = w_{ii}c_{ii}$ is now minor as compared to the previous diagonal methods (for instance, compare Equations 6.38 and 6.39). That is, the off-diagonal elements g_{ij} $(i \neq j)$ of the matrix compensator attenuate or cancel the cross-coupling effects. This results in the diagonal elements g_{kk} of the non-diagonal method requiring a smaller bandwidth than the diagonal elements of the diagonal compensator methods.

Step C: Design of the non-diagonal compensator elements g_{ij}. The $(n-1)$ non-diagonal elements g_{ik} $(i \neq k, i = 1,2,\ldots,n)$ of the kth compensator column are designed to minimize the cross-coupling terms c_{ik} given by Equations 6.31 through 6.33. The optimum compensator elements (see Equations 6.35 through 6.37) are utilized in order to achieve this goal.

Once the design of $\mathbf{G}(s)$ is finished, the design has to fulfill two more stability conditions also[92]: (c) no Smith–McMillan pole-zero cancellations have to occur between $\mathbf{P}(s)$ and $\mathbf{G}(s)$ and (d) no Smith–McMillan pole-zero cancellations have to occur in $|\mathbf{P}^*(s) + \mathbf{G}(s)|$.

Remark 6.1

Although it is very remote, theoretically, there exists the possibility of introducing RHP transmission zeros due to the controller design. This undesirable situation cannot be detected until the whole multivariable system design is completed. To avoid this problem, the proposed methodology (Steps A, B, and C) is introduced in the next procedure.[98]

- *Stage 1: Design of the controller matrix.* First of all, the matrix compensator $\mathbf{G}(s)$ is designed conforming to the methodology described in Steps A, B, and C. At the end of this stage, the transmission zeros of the whole multivariable system are able to be evaluated.
- *Stage 2: Calculation of transmission zeros.* The multivariable zeros of $\mathbf{P}(s)$ $\mathbf{G}(s)$ are determined using the Smith–McMillan form over the set of possible plants $\mathcal{P}$ due to the uncertainty. If there are no new RHP zeros apart from those that might already be present in $\mathbf{P}(s)$, the method concludes. Otherwise, proceed to Stage 3.
- *Stage 3: Modification of the RHP transmission zero positions.* Once we have observed that there exist RHP transmission zeros introduced by the matrix compensator elements, we proceed to rectify this undesirable situation by modifying the non-diagonal elements placed in the last column of the matrix $\mathbf{G}(s)$, according to the Smith–McMillan expressions.

Step D: Prefilter. The design of a prefilter $\mathbf{F}(s)$ is necessary in case of reference tracking specifications. Once the full-matrix compensator $\mathbf{G}(s)$ has been designed the prefilter does not present any difficulty, because the final $\mathbf{T}_{y/r}$ function shows less loop interaction. Therefore, the prefilter $\mathbf{F}(s)$ can be matrix diagonal.

6.8 Some Practical Issues

The sequential non-diagonal MIMO QFT technique introduced in this chapter arrives at a robust stable closed-loop system if, for each $\boldsymbol{P} \in \mathcal{P}$,[92]

1. Each $L_i(s) = g_{ii}(s)\left(p_{ii}^{*e}\right)^{-1}$, $i = 1,\ldots, n$, satisfies the Nyquist encirclement condition
2. No RHP pole-zero cancellations occur between $g_{ii}(s)$ and $\left(p_{ii}^{*e}\right)^{-1}$, $i = 1, \ldots, n$
3. No Smith–McMillan pole-zero cancellations occur between $\boldsymbol{P}(s)$ and $\boldsymbol{G}(s)$
4. No Smith–McMillan pole-zero cancellations occur in $|\boldsymbol{P}^*(s) + \boldsymbol{G}(s)|$

On the other hand, the resulting matrix $\boldsymbol{PG}$ should be checked in every step of the methodology to ensure that RHP transmission zeros or unstable modes have not been introduced by the new compensator elements g_{kk} or g_{ik}, which would obviously cause an unnecessary loss of control performance. If these non-minimum-phase (*n.m.p.*) zeros appear due to the designed compensator elements, supplementary constraints in the determinant of $\boldsymbol{PG}$ should be imposed to recalculate the compensator. There again, if the plant elements, p_{kk} or p_{ik}, are the cause of the introduction of *n.m.p.* elements in the equivalent plant $\left(p_{kk}^{*e}\right)^{-1}$, the theory proposed for *n.m.p.* MISO feedback systems in Chapter 4 can be applied to properly design the compensators in the loop-shaping step.

Incidentally, arbitrarily picking the wrong order of the loops to be designed can result in the nonexistence of a solution. This may occur if the solution process is based on satisfying an upper limit of the phase margin frequency ω_ϕ for each loop. To avoid that potential problem, as it has been introduced in Step A of the methodology, loop *i* having the smallest phase margin frequency has to be chosen as the first loop to be designed. The loop that has the next smallest phase margin frequency is designed next, and so on.[6]

Finally, it is important to notice that the calculation of the equivalent plant $\left(p_{kk}^{*e}\right)^{-1}$ usually introduces some exact pole-zero cancellations. This operation can be precisely done by using symbolic mathematic tools, but could be erroneously done when using numerical calculus due to the typical computer round errors.

Division II: *Method 4*[102]

The second division of this chapter, Sections 6.9 through 6.11, introduces *Method 4*, a reformulation to design non-diagonal (fully populated) matrix QFT compensators for MIMO systems with model uncertainty.[102]

6.9 Non-Diagonal MIMO QFT: A Generalized Technique (*Method 4*)

The MIMO QFT reformulation introduced in the following sections, also known as *Method 4*, is a generalization of previous non-diagonal MIMO QFT techniques found in the references and of *Method 3*, in particular. It avoids former hypotheses of diagonal dominance

(Equations 6.23 and 6.24). The method simplifies the calculations of the off-diagonal elements and the method explained in Section 6.7 and in Figure 6.7. It reformulates the classical matrix definition of MIMO specifications by designing a new set of loop-by-loop QFT bounds on the Nichols chart with necessary and sufficient conditions. It also gives explicit expressions to share the load among the loops of the MIMO system to achieve the matrix specifications and also for stability, reference tracking, disturbance rejection at plant input and output, and noise attenuation problems.

One of the most significant differences observed between classical MIMO control techniques (such as H-infinity, μ-synthesis, and others) and QFT techniques is the way of defining the performance specifications. While classical techniques deal with a matrix definition of specifications (e.g., $\|T(j\omega)\|_\infty < \delta_1$ or $\|S(j\omega)\|_\infty < \delta_2$), QFT describes the specifications in terms of a loop-by-loop definition (e.g., $|t_{ii}(j\omega)| < \delta_r(\omega)$ or $|s_{ii}(j\omega)| < \delta_k(\omega)$). This fact is analyzed and solved in Section 6.11, where the classical matrix MIMO specifications are formally translated into a set of loop-by-loop QFT bounds on the Nichols chart with necessary and sufficient conditions.

Sections 6.10 and 6.11 introduce the reformulation of the full-matrix MIMO QFT technique. In particular, Section 6.10 presents the new methodology itself, and Section 6.11 presents the technique to translate the classical matrix performance specifications into the QFT methodology.

6.10 Reformulation

The first objective of the non-diagonal elements (g_{ij}, $i \neq j$) of the full-matrix controller **G** is to minimize the loop interactions (coupling). They act as feed-forward functions, canceling some of the dynamics of the non-diagonal elements (p_{ij}, $i \neq j$) of the plant matrix **P**. As a consequence, they are sensitive to model uncertainty. But their inclusion relieves the design of the diagonal compensators by reducing the amount of feedback (bandwidth) necessary to fulfill the required specifications. They are also required to achieve internal stability in some systems where diagonal or triangular controllers are not enough.[92] By contrast, the diagonal elements of the controller (g_{ii}) act as feedback functions, minimizing the effect of the uncertainty (sensitivity) and fulfilling the performance and stability specifications. They also have to overcome the uncertainty of the non-diagonal plant elements (p_{ij}, $i \neq j$), minimizing the residual coupling that the non-diagonal elements of the controller (g_{ij}, $i \neq j$) have not cancelled.

Taking these considerations as a starting point, the reformulation presented in this section for the full-matrix controller design methodology adopts a four-step procedure (A through D). The dominant characteristic of the system will determine the way of designing the off-diagonal compensators: reference tracking or disturbance rejection at plant output (Case 1, Section 6.10.1) or disturbance rejection at plant input (Case 2, Section 6.10.2). Of course, independent of which case is chosen, any type of specification (reference tracking, disturbance rejection at plant input and output, noise attenuation, etc.) can be introduced if required when it comes to designing the diagonal controller elements, as is usual within the QFT framework. The distinction is therefore just based on the role assigned to the off-diagonal compensators.

6.10.1 Case 1: Reference Tracking and Disturbance Rejection at Plant Output

Consider an $n \times n$ linear multivariable system (Figure 6.1, $\boldsymbol{d}_i' = 0$), composed of a plant $\boldsymbol{P}$, a fully populated matrix controller $\boldsymbol{G} = \boldsymbol{G}_\alpha \boldsymbol{G}_\beta$, a prefilter $\boldsymbol{F}$, and a plant output disturbance transfer function $\boldsymbol{P}_{do}$, where $\boldsymbol{P} \in \mathcal{P}$, and $\mathcal{P}$ is the set of possible plants due to uncertainty, and

$$\boldsymbol{P} = \begin{bmatrix} p_{11} & p_{12} & \cdots & p_{1n} \\ p_{21} & p_{22} & \cdots & p_{2n} \\ \cdots & \cdots & \cdots & \cdots \\ p_{n1} & p_{n2} & \cdots & p_{nn} \end{bmatrix}; \quad \boldsymbol{F} = \begin{bmatrix} f_{11} & f_{12} & \cdots & f_{1n} \\ f_{21} & f_{22} & \cdots & f_{2n} \\ \cdots & \cdots & \cdots & \cdots \\ f_{n1} & f_{n2} & \cdots & f_{nn} \end{bmatrix} \tag{6.50}$$

$$\begin{aligned} \boldsymbol{G} &= \begin{bmatrix} g_{11} & g_{12} & \cdots & g_{1n} \\ g_{21} & g_{22} & \cdots & g_{2n} \\ \cdots & \cdots & \cdots & \cdots \\ g_{n1} & g_{n2} & \cdots & g_{nn} \end{bmatrix} = \boldsymbol{G}_\alpha \boldsymbol{G}_\beta; \\ \boldsymbol{G}_\alpha &= \begin{bmatrix} g_{11}^\alpha & g_{12}^\alpha & \cdots & g_{1n}^\alpha \\ g_{21}^\alpha & g_{22}^\alpha & \cdots & g_{2n}^\alpha \\ \cdots & \cdots & \cdots & \cdots \\ g_{n1}^\alpha & g_{n2}^\alpha & \cdots & g_{nn}^\alpha \end{bmatrix}; \quad \boldsymbol{G}_\beta = \begin{bmatrix} g_{11}^\beta & 0 & \cdots & 0 \\ 0 & g_{22}^\beta & \cdots & 0 \\ \cdots & \cdots & \cdots & \cdots \\ 0 & 0 & \cdots & g_{nn}^\beta \end{bmatrix} \end{aligned} \tag{6.51}$$

Note that for the sake of clarity the dependence on (s) is omitted all along the rest of this chapter. The reference vector $\boldsymbol{r}'$ and the external disturbance vector at plant output $\boldsymbol{d}_o'$ are the inputs of the system. The output vector $\boldsymbol{y}$ is the variable to be controlled. The plant inverse $\boldsymbol{P}^*$ is given by

$$\boldsymbol{P}^{-1} = \boldsymbol{P}^* = [p_{ij}^*] = \begin{bmatrix} p_{11}^* & p_{12}^* & \cdots & p_{1n}^* \\ p_{21}^* & p_{22}^* & \cdots & p_{2n}^* \\ \vdots & \vdots & \ddots & \vdots \\ p_{n1}^* & p_{n2}^* & \cdots & p_{nn}^* \end{bmatrix} \tag{6.52}$$

6.10.1.1 Methodology

Step A: Controller structure. In the same way as in *Method 3* (Section 6.7, Step A), by using the RGA[224] over the frequencies of interest (see Equation 6.53)[219] and taking into account the requirement of minimum complexity for the controller, the method identifies the input–output pairing and the most appropriate structure for the matrix compensator: the required off-diagonal and diagonal elements, which correspond with the significant values of the RGA matrix:[183]

$$\text{RGA}(j\omega) = \boldsymbol{P}(j\omega) \otimes (\boldsymbol{P}^{-1}(j\omega))^T \tag{6.53}$$

where $\otimes$ denotes element-by-element multiplication (see Section 5.3).

Step B: Design of G_α. The fully populated matrix controller G is composed of two matrices: $G = G_\alpha G_\beta$ (see Equation 6.51). The main objective of the pre-compensator G_α is to diagonalize the plant P as much as possible. The initial expression used to determine G_α is

$$G_\alpha = \left[g_{ij}^{\alpha}\right] = \hat{P}^{-1}\hat{P}_{diag} = \left[\frac{\hat{p}_{jj}\hat{\Delta}_{ji}}{\hat{\Delta}}\right]_{ij} \tag{6.54}$$

where
- $\hat{P}$ is a plant matrix selected within the uncertainty
- $\hat{P}_{diag}$ is its diagonal part
- $\hat{\Delta}$ is the determinant of the $\hat{P}$ matrix
- $\hat{\Delta}_{ji}$ is the jith cofactor of the $\hat{P}$ matrix

The plant matrix $\hat{P}$ is selected so that the expression of the extended matrix in Equation 6.55 presents the closest form to a diagonal matrix, nulling the off-diagonal terms as much as possible:

$$P^x = PG_\alpha = \left[p_{ij}^x\right] \tag{6.55}$$

The starting point for the design of G_α is the well-known ideal decoupling method.[227] A further discussion on this subject is given in Remark 6.2. However, instead of performing it directly, and to avoid some of its drawbacks, the interest is focused on approximating the frequency response of the matrix in Equation 6.54 over the frequencies of interest and the uncertainty. Thus, the $[g_{ij}^{\alpha}]$ compensators are shaped following the mean value in magnitude and phase at every frequency of the region plotted by Equation 6.54 within the uncertainty. Due to this uncertainty, no exact cancellation is achieved. But the residual coupling, or error committed, is then managed through the design of the diagonal feedback controller G_β. Note that since the feedback compensators in G_β are designed by robust MIMO QFT, and not as a decentralized control system, the role of the controller G_α is not to achieve an exact decoupling but to ease the design of G_β. That is, to reduce the amount of feedback needed to achieve the robust performance specifications.

Besides, this approach allows modifying, when necessary, the final form of the controller G_α in Equation 6.54 so that

- No RHP or imaginary-axis pole-zero cancellation occurs between P and G_α or its elements
- No RHP transmission elements (Smith–McMillan) are introduced by the controller G_α
- The relative difference of the number of poles and zeros in each element of the G_α matrix is the same as in Equation 6.54 in order to ease the design of the G_β controller
- The RGA of the system is improved, looking for positive and close-to-one diagonal elements λ_{ii} of the RGA matrix. That is, the pre-compensator G_α decouples the system to some extent, which is its main goal

Step C: Design of G_β. After determining G_α, the method proceeds with the design of a diagonal matrix G_β that fulfills the desired robust stability and robust performance specifications for the extended plant $P^x = PG_\alpha$. Its inverse, the P^{x*} matrix in Equation 6.56, is reorganized so that $(p_{11}^{x*})^{-1}$ has the smallest bandwidth, $(p_{22}^{x*})^{-1}$ the next smallest bandwidth, and so on. This eases the existence of a solution and avoids unnecessary overdesign related to the order in which loops are designed:[6]

$$P^{x*} = (P^x)^{-1} = \left[p_{ij}^{x*}\right] = \begin{bmatrix} p_{11}^{x*} & p_{12}^{x*} & \cdots & p_{1n}^{x*} \\ p_{21}^{x*} & p_{22}^{x*} & \cdots & p_{2n}^{x*} \\ \vdots & \vdots & \ddots & \vdots \\ p_{n1}^{x*} & p_{n2}^{x*} & \cdots & p_{nn}^{x*} \end{bmatrix} \tag{6.56}$$

The compensators g_{kk}^{β} (from $k = 1$ to n) are calculated by using a sequential (loop-by-loop) standard QFT loop-shaping methodology[6] for the inverse of the equivalent extended plant $\left[p_{ii}^{x*e}\right]_i^{-1}$. It satisfies the recursive relationship of Equation 6.57 (see also Ref. 272 with $g_{ij} = 0$, $i \neq j$) and takes into account the compensator elements previously designed (this is the so-called *sequential method*):

$$\left[p_{ij}^{x*e}\right]_k = \left[p_{ij}^{x*e}\right]_{k-1} - \frac{\left[p_{i(k-1)}^{x*e}\right]_{k-1}\left[p_{(k-1)j}^{x*e}\right]_{k-1}}{\left[p_{(k-1)(k-1)}^{x*e}\right]_{k-1} + g_{(k-1)(k-1)}^{\beta}}, \quad i,j \geq k, \ [P^{x*e}]_{k=1} = P^{x*} \tag{6.57}$$

The presence of model uncertainty ($P \in \mathcal{P}$) reduces the diagonalization effect of the pre-compensator G_α over P. This diagonalization is a real cancellation only when the plant P is exactly at $\hat{P}$ (working point): $PG_\alpha = \hat{P}\hat{P}^{-1}\hat{P}_{diag} = \hat{P}_{diag}$. If the plant P is working on a different point, the extended plant that the compensator G_β sees includes off-diagonal elements as well. Consequently, the performance specifications (disturbance rejection) used to design the elements g_{kk}^{β} have to be demanding enough in order to avoid the residual coupling, according to the classical methodology that takes into account the coupling effect loop by loop.[6]

Step D: Design of the prefilter. Finally, when tracking specifications are required, the design of the prefilter F does not present any difficulty whenever the complementary sensitivity function shows a low level of loop interaction. Therefore, the prefilter F can be diagonal.

Example 6.1: ***Case* 2 × 2**

By using Equations 6.50 through 6.52 and 6.54, the expression of the final controller G for a 2 × 2 MIMO plant is

$$G = G_\alpha G_\beta = \begin{bmatrix} g_{11} & g_{12} \\ g_{21} & g_{22} \end{bmatrix} = \begin{bmatrix} \dfrac{\hat{p}_{11}\hat{p}_{22}}{\hat{p}_{11}\hat{p}_{22} - \hat{p}_{12}\hat{p}_{21}} g_{11}^{\beta} & \dfrac{-\hat{p}_{12}}{\hat{p}_{11} - (\hat{p}_{12}\hat{p}_{21}/\hat{p}_{22})} g_{22}^{\beta} \\ \dfrac{-\hat{p}_{21}}{\hat{p}_{22} - (\hat{p}_{12}\hat{p}_{21}/\hat{p}_{11})} g_{11}^{\beta} & \dfrac{\hat{p}_{11}\hat{p}_{22}}{\hat{p}_{11}\hat{p}_{22} - \hat{p}_{12}\hat{p}_{21}} g_{22}^{\beta} \end{bmatrix} \tag{6.58}$$

For tracking or disturbance rejection at plant output, the expression for 2 × 2 MIMO plants of the previous non-diagonal MIMO QFT methodology (see *Method 3*, Sections 6.2 through 6.8) is

$$G = \begin{bmatrix} g_{11} & g_{12} \\ g_{21} & g_{22} \end{bmatrix} = \begin{bmatrix} g_{11} & \dfrac{-\hat{p}_{12}}{\hat{p}_{11}} g_{22} \\ \dfrac{-\hat{p}_{21}}{\hat{p}_{22}} g_{11} & g_{22} \end{bmatrix} \tag{6.59}$$

Method 4 is more general than *Method 3*, because with *Method 4* the hypothesis described in Equations 6.23 and 6.24 is not needed. Furthermore, the diagonal compensators in *Method 4* are directly involved in the reduction of coupling ($\mathbf{G}_\alpha$ is a full matrix). In addition, the resulting off-diagonal compensators, given by Equation 6.58, are the same as those of *Method 3*, given by Equation 6.59, plus some new terms. Precisely, these new terms in Equation 6.58 disappear and the diagonal compensators reduce to the feedback ones if ($|\hat{p}_{11}\hat{p}_{22}| >> |\hat{p}_{12}\hat{p}_{21}|$) is applied, which is the diagonal dominance hypothesis of the *Method 3* (see Equations 6.23 and 6.24). Thus, *Method 4* generalizes and broadens the scope of the preceding *Method 3*. Further discussion on the comparison of both techniques is developed in Section 6.12.

Remark 6.2

Ideal decoupling belongs to the more general group of inverse-based controllers and has been extensively discussed, especially in the chemical engineering literature and in the context of distillation columns.[198,227,245,251] The works mentioned, which are solely based upon observation (simulation or theoretical study) of the dual composition control of distillation towers,[273] encountered some drawbacks for this inverse-based approach, such as the sensitivity of decouplers to modeling errors or the possibly complicated expressions for the decouplers.

However, *Method 4* uses ideal decoupling just as a starting expression for Equation 6.54, whose frequency response is approximated in the range of frequencies of interest and within the uncertainty. This avoids the use of excessively complicated expressions, potential problems of realization and implementation, and the cancellation of RHP elements (see Section 6.10.1, Step B). Moreover, the subsequent $\mathbf{G}_\beta$ feedback controller is designed through MIMO QFT, by taking into account the residual coupling, and not through decentralized control, which has been the approach traditionally used in conjunction with decoupling. This fact explains some of problems encountered in the application of decoupling.

On the other hand, there exists high sensitivity of inverse-based controllers to input uncertainty when the plant has large RGA elements.[219,262] If the input uncertainty is critical for the system under study, then the designer should consider the off-diagonal compensators as a means of fighting against this disturbance at the plant input and, consequently, select Case 2 (Section 6.10.2) for his/her design. Likewise, for extreme cases, one-way or triangular decouplers could be applied instead, since they are much less sensitive to input uncertainty.

6.10.2 Case 2: Disturbance Rejection at Plant Input

In case the designer wants to use the off-diagonal compensators to mainly deal with disturbance rejection at plant input (see Figure 6.1, $\mathbf{r}' = 0$, $\mathbf{d}'_o = 0$), the controller design methodology is based on the one defined in Refs. 6, 93, 97, 98, 178, 181, and 183, with some additional modifications and remarks. Now the full-matrix controller is given by $\mathbf{G} = \mathbf{G}_d + \mathbf{G}_b$, where

$$\mathbf{G} = \begin{bmatrix} g_{11} & g_{12} & \cdots & g_{1n} \\ g_{21} & g_{22} & \cdots & g_{2n} \\ \cdots & \cdots & \cdots & \cdots \\ g_{n1} & g_{n2} & \cdots & g_{nn} \end{bmatrix} = \mathbf{G}_d + \mathbf{G}_b;$$
$$\mathbf{G}_d = \begin{bmatrix} g_{11} & 0 & \cdots & 0 \\ 0 & g_{22} & \cdots & 0 \\ \cdots & \cdots & \cdots & \cdots \\ 0 & 0 & \cdots & g_{nn} \end{bmatrix}; \quad \mathbf{G}_b = \begin{bmatrix} 0 & g_{12} & \cdots & g_{1n} \\ g_{21} & 0 & \cdots & g_{2n} \\ \cdots & \cdots & \cdots & \cdots \\ g_{n1} & g_{n2} & \cdots & 0 \end{bmatrix} \tag{6.60}$$

and the transfer function between the disturbance $\mathbf{d}_i$ and the output $\mathbf{y}$ is

$$\mathbf{y} = [\mathbf{I} + \mathbf{P}\mathbf{G}]^{-1}\mathbf{P}\mathbf{d}_i = [\mathbf{P}^* + \mathbf{G}]^{-1}\mathbf{d}_i \tag{6.61}$$

In this particular case, among the objectives of the controller are to reject external disturbances at plant input $\mathbf{d}_i$ and to reduce the loop interaction. In other words, the disturbance rejection specification could be defined as a diagonal matrix so that $\mathbf{T}_D = [t_{dii}] = \mathbf{y}/\mathbf{d}_i$, and using Equation 6.61

$$\mathbf{T}_D = (\mathbf{P}^* + \mathbf{G})^{-1} \tag{6.62}$$

or

$$\mathbf{G} = \mathbf{T}_D^{-1} - \mathbf{P}^{-1} \tag{6.63}$$

Step B: Design of $\mathbf{G}_b$. The Step A for controller structure design is performed as before (Section 6.10.1, Step A). Then, based on Equation 6.63, the off-diagonal elements g_{ij} $(i \neq j)$ of the controller ($\mathbf{G}_b$ part) are calculated first, so that

$$g_{ij}(i \neq j) = -\hat{p}_{ij}^* \tag{6.64}$$

where ($\wedge$) denotes the plant that minimizes the maximum of the nonparametric uncertainty radii comprising the plant templates on the Nichols chart.[6] That is, to say, the compensator $g_{ij}(s)$ $(i \neq j)$ is shaped following the mean value at every frequency $\omega \in [\omega_{min}, \omega_{max}]$ of the region plotted by the uncertain plants $\left[-p_{ij}^*(s)\right]$ in magnitude and phase.

Step C: Design of $\mathbf{G}_d$. The diagonal elements $g_{kk}(s)$ are calculated through standard QFT loop shaping,[6] but with off-diagonal controller elements, for the inverse of the equivalent plant $\left[p_{kk}^{*e}(s)\right]_k^{-1}$ in order to achieve robust stability and robust performance specifications.[6,93,97,98,178,181,183] The equivalent plant satisfies the recursive relationship Equation 6.65[272]:

$$\left[p_{ij}^{*e}(s)\right]_k = \left[p_{ij}^{*e}(s)\right]_{k-1} - \frac{\left(\left[p_{i(k-1)}^{*e}(s)\right]_{k-1} + \left[g_{i(k-1)}(s)\right]_{k-1}\right)\left(\left[p_{(k-1)j}^{*e}(s)\right]_{k-1} + \left[g_{(k-1)j}(s)\right]_{k-1}\right)}{\left[p_{(k-1)(k-1)}^{*e}(s)\right]_{k-1} + \left[g_{(k-1)(k-1)}(s)\right]_{k-1}};$$

$$i, j \geq k; \quad \left[P^{*e}(s)\right]_{k=1} = P^{*}(s) \tag{6.65}$$

Based on Equation 6.63, a starting expression for the diagonal elements $g_{kk}(s)$ in the loop-shaping stage could be $g_{ii} = t_{dii}^{*} - \hat{p}_{ii}^{*}$.

6.10.3 Stability Conditions and Final Implementation

6.10.3.1 *Stability Conditions*

The closed-loop stability of the MIMO system with the full-matrix controller $\mathbf{G} = [g_{ij}]$ is guaranteed by the following sufficient conditions, previously introduced in Section 6.8[92]:

1. Each $L_i(s) = a_{ii}(s)b_{ii}(s)$, $i = 1,\ldots, n$, satisfies the Nyquist encirclement condition, where $a_{ii}(s) = g_{ii}^{b}(s)$, $b_{ii}(s) = \left[p_{ii}^{x*e}(s)\right]_i^{-1}$ for Case 1 of the method (Section 6.10.1) and $a_{ii}(s) = g_{ii}(s)$, $b_{ii}(s) = \left[p_{ii}^{*e}(s)\right]_i^{-1}$ for Case 2 (Section 6.10.2)
2. No RHP pole-zero cancellations occur between $a_{ii}(s)$ and $b_{ii}(s)$, $i = 1,\ldots,n$, $a_{ii}(s)$ and $b_{ii}(s)$ being those defined in the previous point
3. No Smith–McMillan pole-zero cancellations occur between $\mathbf{P}(s)$ and $\mathbf{G}(s)$
4. No Smith–McMillan pole-zero cancellations occur in $|\mathbf{P}^{*}(s) + \mathbf{G}(s)|$

Conditions 1 and 2 are checked loop by loop when the compensators $a_{ii}(s)$ are calculated in *Step C*, where $a_{ii}(s) = g_{ii}^{b}(s)$ and $a_{ii}(s) = g_{ii}(s)$, for Case 1 and Case 2, respectively. Conditions 3 and 4 are checked after *Step C* is finished.

6.10.3.2 *Final Implementation*

The final implementation of the full-matrix controller $\mathbf{G}$ (either $\mathbf{G}_\alpha \mathbf{G}_\beta$ or $\mathbf{G}_d + \mathbf{G}_b$) will take the form

$$F_{pd}(\mathbf{G}) = \begin{bmatrix} F_{pd}(g_{11}) & F_{pd}(g_{12}) & \cdots & F_{pd}(g_{1n}) \\ F_{pd}(g_{21}) & F_{pd}(g_{22}) & \cdots & F_{pd}(g_{2n}) \\ \cdots & \cdots & \vdots & \cdots \\ F_{pd}(g_{n1}) & F_{pd}(g_{n2}) & \cdots & F_{pd}(g_{nn}) \end{bmatrix} \tag{6.66}$$

where the function $F_{pd}(A)$ means in every case a proper and causal function. If it is necessary to make any modification to get a proper function, it will have to be made always preserving the dominant poles and zeros (low and medium frequency) of the expression A.

Remark 6.3

Once the design is completed, the Smith–McMillan form of the matrix compensator $G(s)$ must be analyzed to check whether it has introduced additional RHP transmission zeros. If so, it is necessary to apply the procedure introduced previously in *Remark 6.1*.[98]

6.11 Translating Matrix Performance Specifications[102]

One of the main differences between MIMO QFT techniques and the classical MIMO control methods (such as H-infinity, μ-synthesis, and others) is the way of defining the specifications. While the latter are able to define them in terms of matrices (singular value decomposition, $\|T(j\omega)\|_\infty < \delta_1$ or $\|S(j\omega)\|_\infty < \delta_2$, etc.), QFT usually does the work loop by loop. Horowitz already pointed out the existence of interaction between the performance that can be achieved for each loop.[79]

Once the equilibrium among loops is attained, this interaction becomes a trade-off, and the burden on any loop cannot be reduced without increasing it on some other loops. However, Horowitz did not formalize it from the viewpoint of matrix specifications or for the $n \times n$ case. In this section an approach to deal with these classical matrix specifications within QFT is introduced, solving the mentioned problem.

6.11.1 Case $n \times n$

Consider an $n \times n$ linear MIMO system (Figure 6.1), composed of a set of uncertain plants, $\mathcal{P}(j\omega_i) = \{P(j\omega_i),\ \omega_i \in \cup\Omega_k\}$, with a full-matrix controller G and a prefilter F, both to be designed. The vectors $\boldsymbol{r}$, $\boldsymbol{n}$, $\boldsymbol{d}_i$, $\boldsymbol{d}_o$, and $\boldsymbol{y}$, represent, respectively, the reference input, sensor noise input, external disturbances at plant input and output, and the plant output.

6.11.1.1 *Tracking*

Looking at Figure 6.1, the reference tracking problem gives the following matrix equations:

$$y = T\,r \tag{6.67}$$

$$y = [P^* + G]^{-1} G\,F\,r \tag{6.68}$$

$$[P^* + G]T = G\,F \tag{6.69}$$

6.11.1.2 *Disturbance Rejection at Plant Output*

Analogously, looking at Figure 6.1, the disturbance rejection at plant output problem gives the following matrix equations:

$$y = S_o\, d_o \tag{6.70}$$

$$y = [P^* + G]^{-1} P^*\, d_o \tag{6.71}$$

$$[P^* + G]S_o = P^* \tag{6.72}$$

6.11.1.3 Disturbance Rejection at Plant Input

In the same way, the disturbance rejection at plant input problem gives the following matrix equations:

$$y = S_i d_i \tag{6.73}$$

$$y = [P^* + G]^{-1} d_i \tag{6.74}$$

$$[P^* + G] S_i = I \tag{6.75}$$

6.11.1.4 Noise Attenuation

Finally, the noise attenuation problem yields the following equations:

$$y = S_n n \tag{6.76}$$

$$y = [P^* + G]^{-1}(-G) n \tag{6.77}$$

$$[P^* + G] S_n = -G \tag{6.78}$$

6.11.1.5 General Expression

The four previous problems (Equations 6.69, 6.72, 6.75, and 6.78) can now be written in a compact form, so that

$$[P^* + G]\boldsymbol{\alpha} = \boldsymbol{\beta} \tag{6.79}$$

where, depending on the case, the matrices $\boldsymbol{\alpha}$ and $\boldsymbol{\beta}$ are defined so that

- *Tracking*: $\boldsymbol{\alpha} = T$; $\boldsymbol{\beta} = GF$
- *Disturbance rejection at plant output:* $\boldsymbol{\alpha} = S_o$; $\boldsymbol{\beta} = P^*$
- *Disturbance rejection at plant input:* $\boldsymbol{\alpha} = S_i$; $\boldsymbol{\beta} = I$
- *Noise attenuation:* $\boldsymbol{\alpha} = S_n$; $\boldsymbol{\beta} = -G$

Now, in order to make $(P^* + G)$ in Equation 6.79 a triangular matrix (which allows explicit expressions for each loop problem), and following the Gauss elimination method, the equation is pre-multiplied by $(M_{n-1} M_{n-2} M_{n-3} \cdots M_1)$ so that

$$M_{n-1} M_{n-2} M_{n-3} \cdots M_1 [P^* + G]\boldsymbol{\alpha} = M_{n-1} M_{n-2} M_{n-3} \cdots M_1 \boldsymbol{\beta} \tag{6.80}$$

where

$$M_k = \begin{bmatrix} & & & Column(k) & & & \\ 1 & 0 & \cdots & 0 & \cdots & 0 & \\ 0 & 1 & \cdots & 0 & \cdots & 0 & \\ \cdots & \cdots & \cdots & \cdots & \cdots & \cdots & \\ 0 & 0 & \cdots & 1 & \cdots & 0 & \\ 0 & 0 & \cdots & \dfrac{-\left(p_{k+1,k}^{*k} + g_{k+1,k}\right)}{p_{k,k}^{*k} + g_{k,k}} & \cdots & 0 & row(k+1) \\ \cdots & \cdots & \cdots & \cdots & \cdots & \cdots & \\ 0 & 0 & \cdots & \dfrac{-\left(p_{n,k}^{*k} + g_{n,k}\right)}{p_{k,k}^{*k} + g_{k,k}} & \cdots & 1 & \end{bmatrix} \tag{6.81}$$

$$p_{ij}^{*k} = p_{ij}^{*k-1} - \frac{\left(p_{i,k-1}^{*k-1} + g_{i,k-1}\right)\left(p_{k-1,j}^{*k-1} + g_{k-1,j}\right)}{p_{k-1,k-1}^{*k-1} + g_{k-1,k-1}} \tag{6.82}$$

$$\text{for } i = k\ldots n; \quad j = k\ldots n; \quad k = 1\ldots n; \quad \text{and} \quad \left[p_{ij}^{*1}\right] = [\boldsymbol{P}]^{-1}$$

After some calculations, the first part of Equation 6.80 is

$$\boldsymbol{M}_{n-1}\boldsymbol{M}_{n-2}\boldsymbol{M}_{n-3}\ldots\boldsymbol{M}_1[\boldsymbol{P}^* + \boldsymbol{G}]\boldsymbol{\alpha}$$

$$= \begin{bmatrix} \sum_{k=1}^{n}\left(p_{1k}^{*1} + g_{1k}\right)\alpha_{k1} & \sum_{k=1}^{n}\left(p_{1k}^{*1} + g_{1k}\right)\alpha_{k2} & \cdots & \sum_{k=1}^{n}\left(p_{1k}^{*1} + g_{1k}\right)\alpha_{kn} \\ \sum_{k=2}^{n}\left(p_{2k}^{*2} + g_{2k}\right)\alpha_{k1} & \sum_{k=2}^{n}\left(p_{2k}^{*2} + g_{2k}\right)\alpha_{k2} & \cdots & \sum_{k=2}^{n}\left(p_{2k}^{*2} + g_{2k}\right)\alpha_{kn} \\ \cdots & \cdots & \cdots & \cdots \\ \sum_{k=n}^{n}\left(p_{nk}^{*n} + g_{nk}\right)\alpha_{k1} & \sum_{k=n}^{n}\left(p_{nk}^{*n} + g_{nk}\right)\alpha_{k2} & \cdots & \sum_{k=n}^{n}\left(p_{nk}^{*n} + g_{nk}\right)\alpha_{kn} \end{bmatrix} \tag{6.83}$$

In other words, the element ij of the matrix is

$$\boldsymbol{M}_{n-1}\boldsymbol{M}_{n-2}\boldsymbol{M}_{n-3}\ldots\boldsymbol{M}_1[\boldsymbol{P}^* + \boldsymbol{G}]\boldsymbol{\alpha} = \left[\sum_{k=i}^{n}\left(p_{ik}^{*i} + g_{ik}\right)\alpha_{kj}\right]_{ij} \tag{6.84}$$

$$\text{for } i = 1\ldots n; \quad j = 1\ldots n$$

with

$$p_{ik}^{*i} = p_{ik}^{*i-1} - \frac{\left(p_{i,i-1}^{*i-1} + g_{i,i-1}\right)\left(p_{i-1,k}^{*i-1} + g_{i-1,k}\right)}{p_{i-1,i-1}^{*i-1} + g_{i-1,i-1}} \tag{6.85}$$

$$\text{for } i = 1\ldots n; \quad j = 1\ldots n; \quad \text{and} \quad \left[p_{ik}^{*1}\right] = [\boldsymbol{P}]^{-1}$$

Making

$$q_{ii}^{i} = \frac{1}{p_{ii}^{*i}} \quad \text{and} \quad L_i^i = q_{ii}^i g_{ii} \tag{6.86}$$

Equation 6.84 becomes

$$\boldsymbol{M}_{n-1}\boldsymbol{M}_{n-2}\boldsymbol{M}_{n-3}\ldots\boldsymbol{M}_1[\boldsymbol{P}^* + \boldsymbol{G}]\boldsymbol{\alpha} = \left[\frac{1+L_i^i}{q_{ii}^i}\alpha_{ij} + \sum_{k=i+1}^{n}\left(p_{ik}^{*i} + g_{ik}\right)\alpha_{kj}\right]_{ij} \tag{6.87}$$

$$\text{for } i = 1\ldots n; \quad j = 1\ldots n$$

$$[\alpha_{ij}] = \begin{cases} \boldsymbol{T} \text{ (for tracking)} \\ \boldsymbol{S}_o \text{ (for disturbance rejection at plant output)} \\ \boldsymbol{S}_i \text{ (for disturbance rejection at plant input)} \\ \boldsymbol{S}_n \text{ (for noise attenuation)} \end{cases} \tag{6.88}$$

Analogously, after some calculations, the second part of Equation 6.80 is

$$\boldsymbol{M}_{n-1}\boldsymbol{M}_{n-2}\boldsymbol{M}_{n-3}\ldots\boldsymbol{M}_1\boldsymbol{\beta} = \left[\left(\sum_{x=1}^{i-1} a_{ix}\beta_{xj}\right) + \beta_{ij}\right]_{ij} \tag{6.89}$$

where

$$a_{ix} = 1 \text{ (if } i = x); \quad a_{ix} = 0 \text{ (if } i < x);$$

$$a_{ix} = \sum_{\substack{\text{from } m=1 \\ \text{to } m=i-x}}^{\binom{i-x-1}{m-1}\text{elements}} \prod_{m \text{ elements}} \frac{-\left(p_{ab}^{*b} + g_{ab}\right)}{\left(p_{bb}^{*b} + g_{bb}\right)} \quad \text{(if } i > x) \tag{6.90}$$

and where there exist all the possible combinations of couples ab, from $a = i$ to $b = x$, without repetition of couples ab and with $a \neq b$, with

$$\binom{i-x-1}{m-1} = \frac{(i-x-1)!}{(m-1)!(i-x-m)!}$$

$$0 \le (m-1) \le (i-x-1); \quad 1 \le m \le (i-x)$$

and where

$$[\beta_{ij}] = \begin{cases} \boldsymbol{GF} \text{ (for tracking)} \\ \boldsymbol{P}^* \text{ (for disturbance rejection at plant output)} \\ \boldsymbol{I} \text{ (for disturbance rejection at plant input)} \\ -\boldsymbol{G} \text{ (for noise attenuation)} \end{cases} \tag{6.91}$$

Combining Equations 6.87 through 6.91 and substituting in Equation 6.80, the explicit expression of α_{ij} is

$$\alpha_{ij} = \frac{q_{ii}^i}{1+L_i^i}\left[\left(\sum_{x=1}^{i-1} a_{ix}\beta_{xj}\right) + \beta_{ij} - \sum_{k=i+1}^{n}\left(p_{ik}^{*i} + g_{ik}\right)\alpha_{kj}\right]_{ij} \tag{6.92}$$

Now, (1) defining the desired matrix performance specification and its elements $\boldsymbol{\tau} = [\tau_{ij}]$ for the matrices $\boldsymbol{\alpha} = [\alpha_{ij}]$, where $\boldsymbol{\alpha} = \boldsymbol{T}$ or $\boldsymbol{\alpha} = \boldsymbol{S}_o$ or $\boldsymbol{\alpha} = \boldsymbol{S}_i$ or $\boldsymbol{\alpha} = \boldsymbol{S}_n$ (depending on the case); (2) comparing the particular element α_{ij} of Equation 6.92 with its specification, $\alpha_{ij} \le \tau_{ij}$; and (3) substituting the remaining α_{kj} ($k = i + 1$ to n), which are still unknown, with their corresponding τ_{kj}, the general expression for the performance specifications becomes

$$\left|\frac{q_{ii}^i}{1+L_i^i}\right| \le \frac{\tau_{ij}}{\left|\left\{\left(\sum_{x=1}^{i-1} a_{ix}\beta_{xj}\right) + \beta_{ij} - \sum_{k=i+1}^{n}\left(p_{ik}^{*i} + g_{ik}\right)\tau_{kj}\right\}_{ij}\right|_{max}} \tag{6.93}$$

where

$$a_{ix} = 1 \text{ (if } i = x\text{)}; \quad a_{ix} = 0 \text{ (if } i < x\text{)}$$

$$a_{ix} = \sum_{\substack{\text{from } m=1 \\ \text{to } m=i-x}}^{\binom{i-x-1}{m-1}\text{elements}} \prod_{m \text{ elements}} \frac{-\left(p_{ab}^{*b} + g_{ab}\right)}{\left(p_{bb}^{*b} + g_{bb}\right)} \quad \text{(if } i > x\text{)}$$

and where there exist all the possible combinations of couples ab, from $a = i$ to $b = x$, without repetition of couples ab and with $a \ne b$, with

$$\binom{i-x-1}{m-1} = \frac{(i-x-1)!}{(m-1)!(i-x-m)!}$$

$$0 \le (m-1) \le (i-x-1); \quad 1 \le m \le (i-x)$$

for the element ij, and where the elements of α_{ij} have been replaced with their specification τ_{ij}, and where

- *Tracking:* $[\tau_{ij}] = spec(\boldsymbol{T})$; $[\beta_{ij}] = \boldsymbol{GF}$
- *Disturbance rejection at plant output:* $[\tau_{ij}] = spec(\boldsymbol{S}_o)$; $[\beta_{ij}] = \boldsymbol{P}^*$
- *Disturbance rejection at plant input:* $[\tau_{ij}] = spec(\boldsymbol{S}_i)$; $[\beta_{ij}] = \boldsymbol{I}$
- *Noise attenuation:* $[\tau_{ij}] = spec(\boldsymbol{S}_n)$; $[\beta_{ij}] = -\boldsymbol{G}$

The notation $|\bullet|_{max}$ in Equation 6.93 means the maximum absolute value within the plant uncertainty. This expression describes the sufficient condition (C_{suf}) to fulfill the specification. It is also possible to define a necessary condition (C_{nec}) to fulfill the specification with the notation $|\bullet|_{min}$, which means the minimum absolute value within the plant uncertainty. In other words,

$$\left|\frac{q_{ii}^{i}}{1+L_{i}^{i}}\right| \leq C_{suf} \leq C_{nec} \tag{6.94}$$

where the sufficient condition C_{suf} is

$$C_{suf} = \frac{\tau_{ij}}{\left|\left\{\left(\sum_{x=1}^{i-1} a_{ix}\beta_{xj}\right)+\beta_{ij}-\sum_{k=i+1}^{n}\left(p_{ik}^{*i}+g_{ik}\right)\tau_{kj}\right\}_{ij}\right|_{max}} \tag{6.95}$$

and the necessary condition C_{nec} is

$$C_{nec} = \frac{\tau_{ij}}{\left|\left\{\left(\sum_{x=1}^{i-1} a_{ix}\beta_{xj}\right)+\beta_{ij}-\sum_{k=i+1}^{n}\left(p_{ik}^{*i}+g_{ik}\right)\tau_{kj}\right\}_{ij}\right|_{min}} \tag{6.96}$$

These last equations generate two different bounds on the Nichols chart for the loop-by-loop controller design method: the sufficient bound and the necessary bound for every kind of matrix specifications and every frequency (see Figure 6.8). The controller design

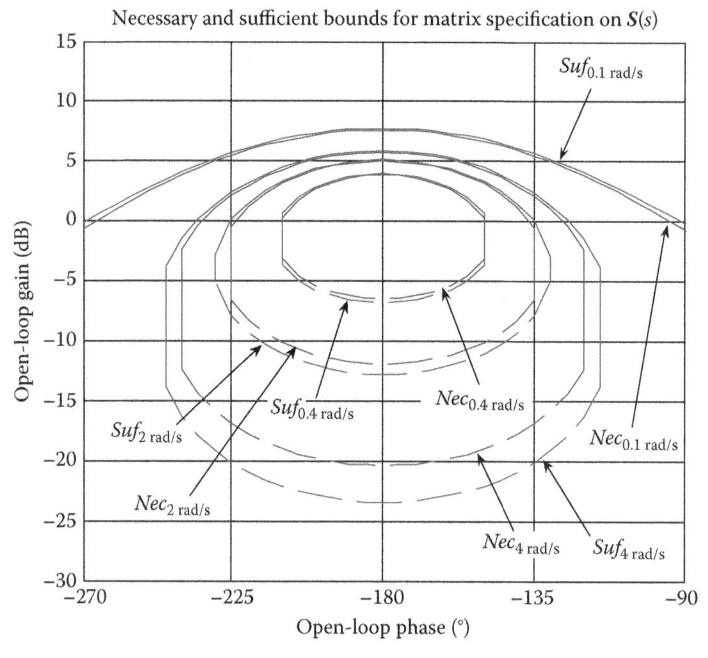

FIGURE 6.8
Loop 1: Necessary and sufficient bounds for matrix specification on $S(s)$.

stage (loop shaping) will consider both, depending on the level of conservativeness/ robustness needed for the design.

Remark 6.4

It is through the corresponding matrix $[\tau_{ij}]$ for each case (tracking, disturbance rejection at plant output and plant input, and noise attenuation) that global matrix specifications can be taken into account. The designer has at his/her disposal a tool of great versatility. On the one hand, the control engineer can deliberately decide how much burden is allocated to each loop on the basis of the constraints that a particular application presents (restrictions on bandwidth, presence of flexible modes, noise, capacity of actuators, precision needed for a particular output, etc.), which are the real engineering requirements. In this decision on how to share the load among loops in a realistic way (i.e., not demanding simultaneously unachievable loop performance levels), the explicit expressions on interaction among loop specifications, developed for the case 2 × 2 in the next section, are used as a guideline. The specification matrix elements τ_{ij} are then accordingly shaped.

But at the same time and prior to the design, the designer can make sure that these specifications fulfill a required H-infinity norm (or even the desired frequency response for the maximum singular value of the matrix). This is an additional step that is not typically considered in MIMO QFT techniques, where the H-infinity norm behavior has to be assessed after the design is performed. The approach introduced in this section allows the QFT designer to make sure that the defined specifications fulfill the matrix requirements as well.

Of course, it leads to choosing one within the infinite possibilities, but this is a degree of freedom explicitly and deliberately used by the control engineer prior to the controller design and on the basis of the rest of the engineering requirements.

On the other hand, if the focus is on optimizing the H-infinity norm instead of predetermining the directionality of the specification matrix $[\tau_{ij}]$, an H-infinity design could be previously performed. The resulting τ_{ij}, optimum from the H-infinity point of view, could then be introduced as specifications for this procedure. In both situations, the methodology assures the translation of constraints on all matrix elements to requirements on the design of each feedback loop. Besides, these translated specifications can be combined with classical QFT loop-by-loop specifications.

6.11.2 Case 2 × 2

In this section, Equation 6.93 is particularized for the case 2 × 2, and for the four classical problems: tracking, disturbance rejection at plant output and input, and noise attenuation.

6.11.2.1 Tracking

Consider the following specifications:

$$\begin{bmatrix} y_1 \\ y_2 \end{bmatrix} = \begin{bmatrix} a_{11} \le t_{11} \le b_{11} & 0 \\ 0 & a_{22} \le t_{22} \le b_{22} \end{bmatrix} \begin{bmatrix} r_1 \\ r_2 \end{bmatrix} \quad \text{(for tracking)} \tag{6.97}$$

$$\begin{bmatrix} y_1 \\ y_2 \end{bmatrix} = \begin{bmatrix} \tau_{c11} & \tau_{c12} \\ \tau_{c21} & \tau_{c22} \end{bmatrix} \begin{bmatrix} r_1 \\ r_2 \end{bmatrix} \quad \text{(for coupling attenuation)} \tag{6.98}$$

Now, Equation 6.93 yields the following sufficient conditions:

- *Loop 1*

$$a'_{11} \leq \left| \frac{L_1^1}{1+L_1^1} \right| \leq b'_{11} \quad \text{with } a'_{11} = a_{11} + \tau_{c11};\ b'_{11} = b_{11} - \tau_{c11} \tag{6.99}$$

$$\left| \frac{q_{11}^1}{1+L_1^1} \right| \leq \frac{\tau_{c11}}{\left| \left(\left(1/q_{12}^1\right) + g_{12} \right) \tau_{c21} \right|_{max}} \tag{6.100}$$

$$\left| \frac{q_{11}^1}{1+L_1^1} \right| \leq \frac{\tau_{c12}}{\left| g_{12} - \left(\left(1/q_{12}^1\right) + g_{12} \right) b_{22} \right|_{max}} \tag{6.101}$$

- *Loop 2*

$$a'_{22} \leq \left| \frac{L_2^2}{1+L_2^2} \right| \leq b'_{22} \quad \text{with } a'_{22} = a_{22} + \tau_{c22};\ b'_{22} = b_{22} - \tau_{c22} \tag{6.102}$$

$$\left| \frac{q_{22}^2}{1+L_2^2} \right| \leq \frac{\tau_{c22}}{\left| \left(q_{11}^1/\left(1+L_1^1\right) \right) \left(\left(1/q_{21}^1\right) + g_{21} \right) g_{12} \right|_{max}} \tag{6.103}$$

$$\left| \frac{q_{22}^2}{1+L_2^2} \right| \leq \frac{\tau_{c21}}{\left| g_{21} - \left(q_{11}^1/\left(1+L_1^1\right) \right) \left(\left(1/q_{21}^1\right) + g_{21} \right) g_{11} \right|_{max}} \tag{6.104}$$

Note that there is a specific interaction between Equation 6.100 and Equation 6.104, through the definition of the specification τ_{c21}, and between Equation 6.101 and Equation 6.102, through the definition of the specification b_{22}. If τ_{c21} is higher, then Equation 6.100 is more demanding and Equation 6.104 is less demanding. Analogously, if b_{22} is higher, then Equation 6.101 is more demanding and Equation 6.102 is less demanding. So, the proper definition of the specifications τ_{c21} and b_{22} establishes a load sharing between the loops 1 and 2. Observe that this is not the typical trade-off between performance and coupling concerning specifications from the same loop, but a trade-off of design burden among performance specifications from different loops. This load-sharing problem shows that, when it comes to define the specifications, the designer cannot demand too much of a loop without affecting the achievable performance of rest of the loops, and therefore it can be used as a guideline.

6.11.2.2 Disturbance Rejection at Plant Output

Consider the following specifications:

$$\begin{bmatrix} y_1 \\ y_2 \end{bmatrix} = \begin{bmatrix} \tau_{so11} & \tau_{so12} \\ \tau_{so21} & \tau_{so22} \end{bmatrix} \begin{bmatrix} d_{o1} \\ d_{o2} \end{bmatrix} \tag{6.105}$$

Now, Equation 6.93 yields the following sufficient conditions:

- *Loop 1*

$$\left|\frac{q_{11}^{1}}{1+L_1^1}\right| \leq \frac{\tau_{so11}}{\left|(1/q_{11}^{1})-((1/q_{12}^{1})+g_{12})\tau_{so21}\right|_{max}} \tag{6.106}$$

$$\left|\frac{q_{11}^{1}}{1+L_1^1}\right| \leq \frac{\tau_{so12}}{\left|(1/q_{12}^{1})-((1/q_{12}^{1})+g_{12})\tau_{so22}\right|_{max}} \tag{6.107}$$

- *Loop 2*

$$\left|\frac{q_{22}^{2}}{1+L_2^2}\right| \leq \frac{\tau_{so22}}{\left|(1/q_{22}^{1})-(q_{11}^{1}/(1+L_1^1))((1/q_{21}^{1})+g_{21})(1/q_{12}^{1})\right|_{max}} \tag{6.108}$$

$$\left|\frac{q_{22}^{2}}{1+L_2^2}\right| \leq \frac{\tau_{so21}}{\left|(1/q_{21}^{1})-(1/(1+L_1^1))((1/q_{21}^{1})+g_{21})\right|_{max}} \tag{6.109}$$

As for reference tracking, there exists a load sharing between loops 1 and 2 through the specifications τ_{so21} and τ_{so22}.

6.11.2.3 Disturbance Rejection at Plant Input

Consider the following specifications:

$$\begin{bmatrix} y_1 \\ y_2 \end{bmatrix} = \begin{bmatrix} \tau_{si11} & \tau_{si12} \\ \tau_{si21} & \tau_{si22} \end{bmatrix} \begin{bmatrix} d_{i1} \\ d_{i2} \end{bmatrix} \tag{6.110}$$

Now, Equation 6.93 yields the following sufficient conditions:

- *Loop 1*

$$\left|\frac{q_{11}^{1}}{1+L_1^1}\right| \leq \frac{\tau_{si11}}{\left|(1-(1/q_{12}^{1})+g_{12})\tau_{si21}\right|_{max}} \tag{6.111}$$

$$\left|\frac{q_{11}^{1}}{1+L_1^1}\right| \leq \frac{\tau_{si12}}{\left|((1/q_{12}^{1})+g_{12})\tau_{si22}\right|_{max}} \tag{6.112}$$

- *Loop 2*

$$\left|\frac{q_{22}^{2}}{1+L_2^2}\right| \leq \tau_{si22} \tag{6.113}$$

$$\left|\frac{q_{22}^2}{1+L_2^2}\right| \le \frac{\tau_{si21}}{\left|\left(q_{11}^1/\left(1+L_1^1\right)\right)\left(\left(1/q_{21}^1\right)+g_{21}\right)\right|_{max}} \tag{6.114}$$

Again a load sharing appears between loops 1 and 2 through specification elements τ_{si21} and τ_{si22}.

6.11.2.4 Noise Attenuation

Consider the following specifications:

$$\begin{bmatrix} y_1 \\ y_2 \end{bmatrix} = \begin{bmatrix} b_{n11} & \tau_{sn12} \\ \tau_{sn21} & b_{n22} \end{bmatrix} \begin{bmatrix} n_1 \\ n_2 \end{bmatrix} \quad \text{(for noise attenuation)} \tag{6.115}$$

$$\begin{bmatrix} y_1 \\ y_2 \end{bmatrix} = \begin{bmatrix} \tau_{nc11} & 0 \\ 0 & \tau_{nc22} \end{bmatrix} \begin{bmatrix} n_1 \\ n_2 \end{bmatrix} \quad \text{(for coupling attenuation)} \tag{6.116}$$

Now, Equation 6.93 yields the following sufficient conditions:

- *Loop 1*

$$\left|\frac{L_1^1}{1+L_1^1}\right| \le b'_{n11} \quad \text{with } b'_{n11} = b_{n11} - \tau_{nc11} \tag{6.117}$$

$$\left|\frac{q_{11}^1}{1+L_1^1}\right| \le \frac{\tau_{nc11}}{\left|\left(\left(1/q_{12}^1\right)+g_{12}\right)\tau_{sn21}\right|_{max}} \tag{6.118}$$

$$\left|\frac{q_{11}^1}{1+L_1^1}\right| \le \frac{\tau_{sn12}}{\left|-g_{12}-\left(\left(1/q_{12}^1\right)+g_{12}\right)b_{n22}\right|_{max}} \tag{6.119}$$

- *Loop 2*

$$\left|\frac{L_2^2}{1+L_2^2}\right| \le b'_{n22} \quad \text{with } b'_{n22} = b_{n22} - \tau_{nc22} \tag{6.120}$$

$$\left|\frac{q_{22}^2}{1+L_2^2}\right| \le \frac{\tau_{nc22}}{\left|\left(q_{11}^1/\left(1+L_1^1\right)\right)\left(\left(1/q_{21}^1\right)+g_{21}\right)g_{12}\right|_{max}} \tag{6.121}$$

$$\left|\frac{q_{22}^2}{1+L_2^2}\right| \le \frac{\tau_{sn21}}{\left|-g_{21}-\left(q_{11}^1/\left(1+L_1^1\right)\right)\left(\left(1/q_{21}^1\right)+g_{21}\right)g_{11}\right|_{max}} \tag{6.122}$$

As shown earlier, specific interaction appears between loop 1 and 2 specifications through the τ_{sn21} and b_{n22} elements.

6.12 Comparison of *Methods* 3 and 4

This section makes a comparison between the two methodologies introduced in this chapter: *Method 3* and *Method 4*. Both techniques are based on the same principle: the off-diagonal compensators play the role of feed-forward functions that reduce the coupling and facilitate the design of the diagonal compensators, which are feedback functions in charge of fulfilling the robust stability and robust performance specifications. Additionally, for the *Method 4*, and particularly for its Case 1 (Section 6.10.1), the diagonal elements are also directly involved in coping with coupling (the pre-compensator G_α is a full matrix).

The existence of plant uncertainty causes the decoupling (feed-forward) to be inexact. This residual coupling, however, is taken into account through a proper design of the diagonal compensators (feedback). Beyond these similarities, the way in which off-diagonal compensators are designed and the procedure itself are different for the two methodologies.

Method 3 requires a diagonal dominance hypothesis on the plant, limiting the application of the technique. This hypothesis is a fundamental requirement on the plant itself and is essential to obtain explicit expressions of the coupling matrices, which are defined therein to quantitatively compute the coupling. Besides, it allows two simplifications leading to the explicit optimal expressions for the off-diagonal compensators. By contrast, *Method 4* no longer needs such a hypothesis or any simplification. As a result, this removal extends the application of the method to a larger scope of plants. Furthermore, as is shown in Section 6.10.1, the expressions obtained for the off-diagonal compensators in Equation 6.58 are the same as those of the former method (Equation 6.59) plus some new terms, which disappear if the aforementioned diagonal dominance hypothesis is applied. So, *Method 4* generalizes *Method 3* and extends its field of application.

Additionally, the procedure has been simplified. In *Method 3*, the approach is performed by columns, designing first the diagonal compensator and then the off-diagonal elements of each column, which is somehow a complicated procedure. In the case of needing a redesign in any diagonal compensator, it implies the redesign of all the elements, diagonal and off-diagonal, from the subsequent loops. However, *Method 4* performs separately the design of the off-diagonal and the diagonal parts. Regardless of the case chosen (Case 1, Section 6.10.1 or Case 2, Section 6.10.2), when it comes to designing the feedback diagonal compensators, all the feed-forward off-diagonal elements have already been determined.

Finally, both methods allow the introduction not only of classical QFT specifications but also of classical matrix specifications and provide the guideline to perform a load sharing among loops, distributing the burden as needed.

6.13 Summary

Two methodologies (*Method 3* and *Method 4*) to design non-diagonal matrix compensators (fully populated G) for MIMO systems with model uncertainty were presented in this chapter. They extend the classical diagonal QFT compensator design of *Method 1* and *Method 2* of Chapter 5 to new systems with more loop coupling and/or more demanding specifications.

The first division of this chapter introduced *Method 3*, studying three classical cases: reference tracking, external disturbance rejection at plant input, and external disturbance rejection at plant output. The method analyses the role played by the non-diagonal compensator elements g_{ij} $(i \neq j)$ by means of three coupling matrices ($\mathbf{C}_1$, $\mathbf{C}_2$, $\mathbf{C}_3$) and a quality function η_{ij}. It quantifies the amount of interaction among the control loops and proposes a sequential design methodology for the fully populated matrix compensator that yields n equivalent tracking SISO systems and n equivalent disturbance rejection SISO systems. As a result, the off-diagonal elements g_{ij} $(i \neq j)$ of the compensator matrix reduce (or cancel if there is no uncertainty) the level of coupling between loops, and the diagonal elements g_{kk} regulate the system with less bandwidth requirements than the diagonal elements of the previous diagonal $\mathbf{G}$ methods.[6,93,97,98,178,181,183]

The second division of this chapter introduced *Method 4*. It is a formal reformulation of *Method 3* that includes a generalization of previous techniques. It avoids former hypotheses of diagonal dominance and simplifies the calculations for the off-diagonal elements and the method itself. It also reformulates the classical matrix definition of MIMO specifications by designing a new set of loop-by-loop QFT bounds on the Nichols chart with necessary and sufficient conditions. It gives explicit expressions to share the load among the loops of the MIMO system to achieve the matrix specifications and also for stability, reference tracking, disturbance rejection at plant input and output, and noise attenuation problems.[102]

7

QFT for Distributed Parameter Systems[115]

7.1 Introduction

This chapter introduces a practical quantitative feedback theory (QFT) robust control technique to design one-point feedback controllers for distributed parameter systems (DPS) with uncertainty. The method considers (1) the spatial distribution of the relevant points where the inputs and outputs of the control system are applied (actuators, sensors, disturbances, and control objectives) and (2) a new set of transfer functions (TFs) that describe the relationships between those distributed inputs and outputs.

Based on the definition of such distributed TFs, the classical robust stability and robust performance specifications are extended to the DPS case, developing a new set of quadratic inequalities to define the QFT bounds. The method can deal with uncertainty in both the model and the spatial distribution of the inputs and the outputs. This chapter also includes some examples to illustrate the use and simplicity of the proposed methodology.

7.2 Background

More than a decade after developing the QFT fundamentals, in the early 1980s Horowitz proposed to Kannai an interesting mathematical task: the analysis of the causality and stability of linear systems that are described by partial differential operators.[298] As a consequence of these preliminary results, Horowitz and Azor[108,109] developed the very first approach of QFT to DPS. In that approach, the uncertain plant P was described by partial differential equations (PDEs) with two independent variables: time t and space x. The approach was based upon the performance specifications on the system output $y(t,x)$ in response to the command input $r(t,x)$ over the defined (t and x) intervals for all the set of plants. Therefore, a frequency-response method with two complex variables s_1 and s_2 was used to synthesize the system. The method was a two-variable generalization of the QFT design technique developed for lumped uncertain plants.

From the graphical and mathematical point of view, the aforementioned approach requires a three-dimensional Bode diagram and a double Laplace transform for the one-dimensional space; a four-dimensional Bode diagram and a triple Laplace transform for the two-dimensional space; and a five-dimensional Bode diagram and a quadruple Laplace transform for the three-dimensional space, etc. The development also requires the implementation of a nonrealistic continuous feedback loop at every single point of the spatial distribution.[112]

Some years later, and as a result of new solutions achieved by Horowitz et al.,[112] a one-point feedback approach to distributed linear systems was presented. The proposed controller design methodology was more realistic, dealing with only a single output point at a time. However, no guidelines or tutorials to solve the practical system difficulties were included.

Meanwhile, Chait et al.[110] presented a new Nyquist stability criterion for DPS, as well as studied the "spillover" effects of the truncation of DPS models, and Hedge and Nataraj[114] developed a two-point feedback approach.

In this context, in October 2000, at the Public University of Navarra (Spain), and during one of his last public presentations, Professor Horowitz challenged Professor Mario Garcia-Sanz and his PhD students to explore new ideas related to DPS and QFT.[115] This chapter is the result of these original conversations with Professor Horowitz and explains the resulting methodology to design a one-point feedback QFT controller for DPS with uncertainty. Section 7.3 describes the proposed control structure, Section 7.4 introduces the extension of the QFT to DPS, Section 7.5 details some of the modeling difficulties to be met and some tools to solve them, and Section 7.6 includes two examples presented by Kelemen et al.[112] to compare previous results with the ones obtained with the new methodology. Finally, Section 7.7 sums up the main ideas and conclusions of the chapter.

7.3 Generalized DPS Control System Structure

As seen in the previous chapters, the classical 2 degree of freedom (2DOF) general feedback structure, with a loop compensator G and a prefilter F in cascade with the feedback loop, is one of the mostly used control topologies (Figure 7.1). From that structure the following well-known equations are obtained:

$$Y = \frac{1}{1+PGH} D_2 + \frac{P}{1+PGH} D_1 + \frac{PG}{1+PGH}(W+FR) - \frac{PGH}{1+PGH} N \tag{7.1}$$

$$U = \frac{G}{1+PGH}(W+FR) - \frac{GH}{1+PGH}(N + D_2 + PD_1) \tag{7.2}$$

$$E = -\frac{H}{1+PGH} D_2 + \frac{PH}{1+PGH} D_1 + \frac{PGH}{1+PGH} W + \frac{1}{1+PGH} FR - \frac{H}{1+PGH} N \tag{7.3}$$

where

the Laplace variable s is omitted for simplicity

P denotes the plant with uncertainty

R, W, N, D_1, D_2 represent the inputs (reference signal for tracking, error disturbances, sensor noise, and external disturbances at the plant input and output, respectively)

To deal with DPS, this chapter proposes a generalization of the previous classical 2DOF feedback structure shown in Figure 7.1. The characteristic input and output points for the new generalized structure are the actuator, disturbances, sensor and control objectives, and are located at x_a, x_d, x_s, and x_o, respectively, as shown in Figure 7.2. Between these points of interest, the corresponding PDE TFs are defined. These PDE TFs vary with the relative

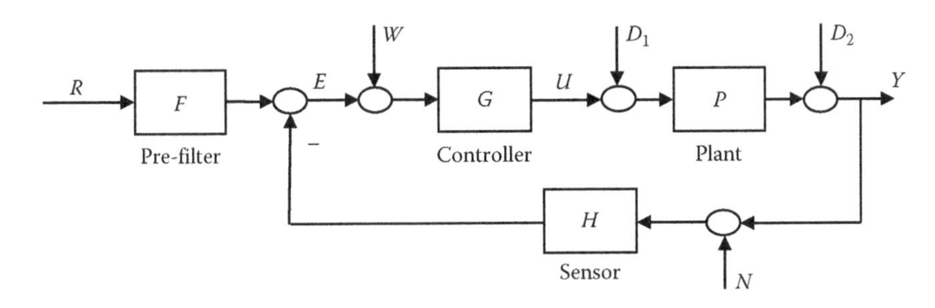

FIGURE 7.1
Classical 2DOF feedback structure.

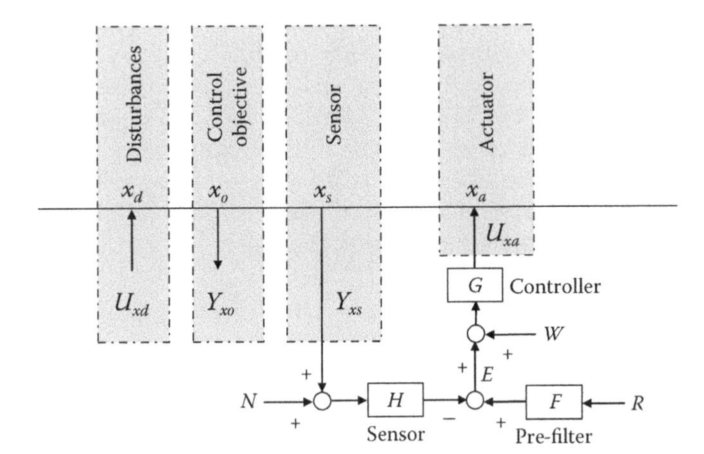

FIGURE 7.2
Generalized 2DOF DPS control system structure.

position, topology, and required spatial distribution. In a similar way to Equations 7.1 through 7.3, the dynamics of the DPS system is now explained by

$$Y_{xo} = P_{xo\,xd}U_{xd} + P_{xo\,xa}U_{xa} \tag{7.4}$$

$$Y_{xs} = P_{xs\,xd}U_{xd} + P_{xs\,xa}U_{xa} \tag{7.5}$$

$$U_{xa} = G[FR + W - H(N + Y_{xs})] \tag{7.6}$$

$$E = Y_{xo} - R \tag{7.7}$$

where

P_{x2x1} denote the Laplace Transform between an input x_1 and an output x_2

U_{xd}, U_{xa}, E, Y_{xo}, and Y_{xs} represent the external disturbances, the actuator, the error, the control objective, and the sensor output signals, respectively

Substituting Equation 7.5 into Equation 7.6 and the result into Equations 7.4 through 7.7, the expressions that explain the DPS 2DOF feedback structure of Figure 7.2 are

$$Y_{xo} = \frac{GP_{xo\,xa}}{1+GHP_{xs\,xa}}(FR+W) + \left[P_{xo\,xd} - \frac{GHP_{xo\,xa}P_{xs\,xd}}{1+GHP_{xs\,xa}}\right]U_{xd} - \frac{GHP_{xo\,xa}}{1+GHP_{xs\,xa}}N \tag{7.8}$$

$$U_{xa} = \frac{G}{1+GHP_{xs\,xa}}(FR+W) - \frac{GHP_{xs\,xd}}{1+GHP_{xs\,xa}}U_{xd} - \frac{GH}{1+GHP_{xs\,xa}}N \tag{7.9}$$

$$E = \frac{1}{1+GHP_{xs\,xa}}FR - \frac{GHP_{xs\,xa}}{1+GHP_{xs\,xa}}W - \frac{HP_{xs\,xd}}{1+GHP_{xs\,xa}}U_{xd} - \frac{H}{1+GHP_{xs\,xa}}N \tag{7.10}$$

$$Y_{xs} = \frac{GP_{xs\,xa}}{1+GHP_{xs\,xa}}(FR+W) + \frac{P_{xs\,xd}}{1+GHP_{xs\,xa}}U_{xd} - \frac{GHP_{xs\,xa}}{1+GHP_{xs\,xa}}N \tag{7.11}$$

where the TFs depend on the compensator G and the prefilter F, the sensor dynamics H and the spatial distribution, and the topology through the four TFs $P_{xs\ xa}$, $P_{xo\ xa}$, $P_{xs\ xd}$, and $P_{xo\ xd}$.

The structure defines the DPS through a set of TFs $\{P_{xs\ xa}, P_{xo\ xa}, P_{xs\ xd}, P_{xo\ xd}\}$, describing the relationships between the outputs and the inputs for the points of significance. These representations are simplified structures of the PDEs that still preserve the spatial configuration of the problem. Furthermore, the implementation of a rational form allows the designer to use classical control theory tools related to rational TFs.

Remark 7.1

One of the most powerful characteristics of the proposed methodology is the fact that the points of interest may not be fixed at a particular location. They can slowly move along in space or else their current location is uncertain. In other words, the control objective, the disturbance, the sensor, and the actuator may not be defined as a single point but as a distribution of points.

Remark 7.2

One of the difficulties to be met in designing a controller for a DPS is the definition of the TFs $\{P_{xs\ xa}, P_{xo\ xa}, P_{xs\ xd}, P_{xo\ xd}\}$ that being rational must comply with the distributed nature of the system. Mostly, PDE models are defined by approximating the solution by an irrational form with infinite sums and/or products, where a limited but high enough number of elements must be taken into account. See Section 7.5 for details.

Remark 7.3

It is also easy to see that the structure proposed is just a generalization of the classical 2DOF. Consider for instance the case where $x_d = x_a$ and $x_o = x_s$ (see Figure 7.3). The four TFs turn out to be the same: $P_{xs\ xa} = P_{xo\ xa} = P_{xo\ xd} = P_{xs\ xd}$. Hence, substituting this equivalence in Equation 7.8, the control objective Y_{xo} expression becomes

$$Y_{xo} = \frac{GP_{xs\,xa}}{1+GHP_{xs\,xa}}(FR+W) + \frac{P_{xs\,xa}}{1+GHP_{xs\,xa}}U_{xd} - \frac{GHP_{xs\,xa}}{1+GHP_{xs\,xa}}N \tag{7.12}$$

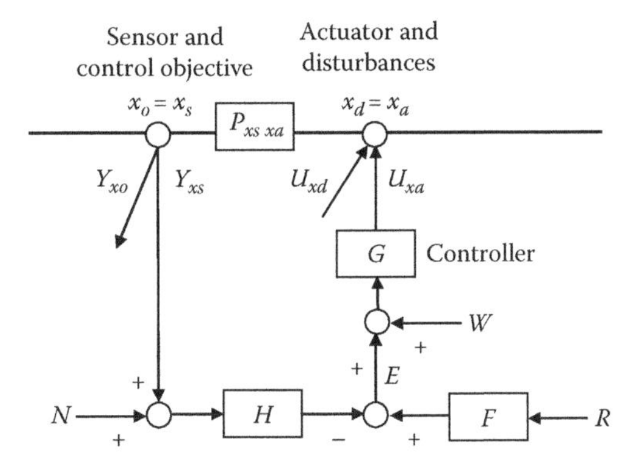

FIGURE 7.3
DPS structure when, $x_d = x_a$ and $x_o = x_s$.

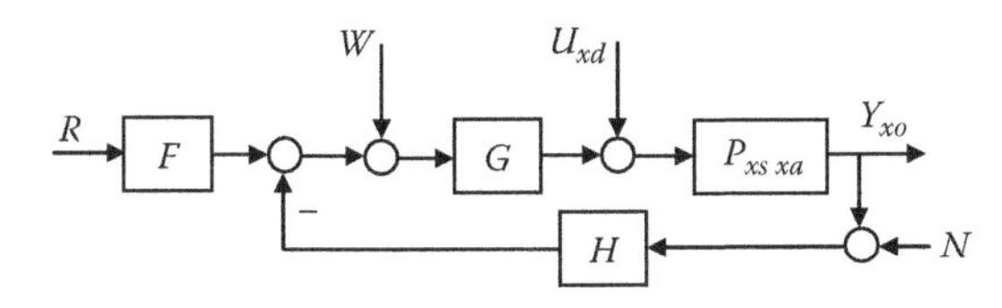

FIGURE 7.4
Lumped 2DOF case. Disturbance at plant input.

which matches with the classical lumped system ($P = P_{xs\,xa}$) with a disturbance at the plant input ($D_1 = U_{xd}$) (Figure 7.4). Further topologies can be explored checking other lumped configurations; for instance, let $x_1 = x_2$, which results in $P_{x2\,x1} = 1$.

7.4 Extension of Quantitative Feedback Theory to DPS

7.4.1 Classical QFT for Lumped Systems

As developed by Horowitz (see Chapter 3), QFT uses different TFs between the input and output signals of the classical 2DOF feedback structure (see Figure 7.1 and Equations 7.1 through 7.3) to define stability and performance specifications in terms of inequalities in the frequency domain (see Equations 7.13 through 7.17 in Table 7.1), where $\{|T_k(j\omega_i)| \leq \delta_k(\omega_i),\ \forall\omega_i \in \Omega_k,\ k = 1, 2, \ldots; i = 1, 2, \ldots\}$, $T_k(j\omega_i)$ are the TFs, and $\delta_k(\omega_i)$ is the magnitude of the specifications in the frequency domain.

The standard QFT defines the plant model taking into account the parameter uncertainty at every frequency of interest ω_i. It deals with a set of plants $P(j\omega_i) = \{P(j\omega_i)_m,\ \omega_i \in \Omega_k$, where $m = 1, 2, \ldots, J$ plants$\}$ that are evaluated and represented on the Nichols chart (NC) at every frequency of interest (*the templates*). Then, for a nominal plant $P_0 \in P(j\omega_i)$, the QFT methodology converts the system specifications $\delta_k(\omega_i)$ and the model plant uncertainty $P(j\omega_i)$ into a set of bounds for every frequency ω_i. Such a great integration of information in a set of simple curves (*the bounds*) allows one to design the controller using only a single plant, the nominal plant P_0, and shaping the resulting open-loop function $L_0 = P_0G$ (see Chapter 3).

Therefore, for every frequency ω_i the feedback specifications $\{|T_k(j\omega_i)| \leq \delta_k(\omega_i), k = 1, \ldots, 5\}$ from Table 7.1 (see Equations 7.13 through 7.17) are translated into quadratic inequalities in Table 7.2 (see Equations 7.18 through 7.22), where each plant within the model uncertainty is expressed as $P_r(j\omega_i) = p(\omega_i)e^{j\theta(\omega_i)} = p\angle\theta$ and the compensator as $G(j\omega_i) = g(\omega_i)e^{j\phi(\omega_i)} = g\angle\phi$. The compensator phase ϕ varies from -2π to 0.

TABLE 7.1

Feedback Specifications for Lumped Systems when $H = 1$

k	Inequalities	Equation
1	$\lvert T_1(j\omega)\rvert = \left\lvert \frac{Y(j\omega)}{R(j\omega)F(j\omega)} \right\rvert = \left\lvert \frac{U(j\omega)}{D_1(j\omega)} \right\rvert = \left\lvert \frac{Y(j\omega)}{N(j\omega)} \right\rvert = \left\lvert \frac{P(j\omega)G(j\omega)}{1+P(j\omega)G(j\omega)} \right\rvert \le \delta_1(\omega), \quad \omega \in \Omega_1$	(7.13)
2	$\lvert T_2(j\omega)\rvert = \left\lvert \frac{Y(j\omega)}{D_2(j\omega)} \right\rvert = \left\lvert \frac{1}{1+P(j\omega)G(j\omega)} \right\rvert \le \delta_2(\omega), \quad \omega \in \Omega_2$	(7.14)
3	$\lvert T_3(j\omega)\rvert = \left\lvert \frac{Y(j\omega)}{D_1(j\omega)} \right\rvert = \left\lvert \frac{P(j\omega)}{1+P(j\omega)G(j\omega)} \right\rvert \le \delta_3(\omega), \quad \omega \in \Omega_3$	(7.15)
4	$\lvert T_4(j\omega)\rvert = \left\lvert \frac{U(j\omega)}{D_2(j\omega)} \right\rvert = \left\lvert \frac{U(j\omega)}{N(j\omega)} \right\rvert = \left\lvert \frac{U(j\omega)}{R(j\omega)F(j\omega)} \right\rvert = \left\lvert \frac{G(j\omega)}{1+P(j\omega)G(j\omega)} \right\rvert \le \delta_4(\omega), \quad \omega \in \Omega_4$	(7.16)
5	$\delta_{5inf}(\omega) < \lvert T_5(j\omega)\rvert = \left\lvert \frac{Y(j\omega)}{R(j\omega)} \right\rvert = \left\lvert F(j\omega)\frac{P(j\omega)G(j\omega)}{1+P(j\omega)G(j\omega)} \right\rvert \le \delta_{5sup}(\omega), \quad \omega \in \Omega_5$ $\frac{\lvert G(j\omega)P_d(j\omega)\rvert}{\lvert G(j\omega)P_e(j\omega)\rvert}\frac{\lvert 1+G(j\omega)P_e(j\omega)\rvert}{\lvert 1+G(j\omega)P_d(j\omega)\rvert} \le \delta_5(\omega) = \frac{\delta_{5sup}(\omega)}{\delta_{5inf}(\omega)}, \quad \omega \in \Omega_5$	(7.17)

TABLE 7.2

Quadratic Inequalities for Lumped Systems with $H = 1$

k	Inequality	Equation
1	$p^2\left(1-\frac{1}{\delta_1^2}\right)g^2 + 2p\cos(\phi+\theta)g + 1 \ge 0$	(7.18)
2	$p^2g^2 + 2p\cos(\phi+\theta)g + \left(1-\frac{1}{\delta_2^2}\right) \ge 0$	(7.19)
3	$p^2g^2 + 2p\cos(\phi+\theta)g + \left(1-\frac{p^2}{\delta_3^2}\right) \ge 0$	(7.20)
4	$\left(p^2-\frac{1}{\delta_4^2}\right)g^2 + 2p\cos(\phi+\theta)g + 1 \ge 0$	(7.21)
5	$p_e^2p_d^2\left(1-\frac{1}{\delta_5^2}\right)g^2$ $+2p_ep_d\left(p_e\cos(\phi+\theta_d) - \frac{p_d}{\delta_5^2}\cos(\phi+\theta_e)\right)g$ $+\left(p_e^2-\frac{p_d^2}{\delta_5^2}\right) \ge 0$	(7.22)

Note that the format of each quadratic expression is

$$I_{\omega_i}^k(p,\theta,\delta_k,\phi) = ag^2 + bg + c \ge 0 \tag{7.23}$$

where a, b, c depend on p, θ, ϕ, and δ_k. Chait and Yaniv[48] developed an algorithm to compute the bounds based on these quadratic inequalities, simplifying much of the work on

TABLE 7.3

Feedback Specifications for DPS when $H = 1$

k	Inequalities	Equation
1	$\lvert T_1(j\omega)\rvert = \left\lvert\frac{Y_o(j\omega)}{R(j\omega)F(j\omega)}\right\rvert = \left\lvert\frac{Y_{xo}(j\omega)}{N(j\omega)}\right\rvert = \left\lvert\frac{Y_{xo}(j\omega)}{W(j\omega)}\right\rvert = \left\lvert\frac{G(j\omega)P_{xo\,xa}(j\omega)}{1+G(j\omega)P_{xs\,xa}(j\omega)}\right\rvert \le \delta_1(\omega), \quad \omega \in \Omega_1$	(7.24)
2	$\lvert T_2(j\omega)\rvert = \left\lvert\frac{Y_{xo}(j\omega)}{U_{xd}(j\omega)}\right\rvert = \left\lvert P_{xd\,xo}(j\omega) - \frac{G(j\omega)P_{xo\,xa}(j\omega)P_{xs\,xd}(j\omega)}{1+G(j\omega)P_{xs\,xa}(j\omega)}\right\rvert \le \delta_2(\omega), \quad \omega \in \Omega_2$	(7.25)
3	$\lvert T_3(j\omega)\rvert = \left\lvert\frac{U_{xa}(j\omega)}{R(j\omega)F(j\omega)}\right\rvert = \left\lvert\frac{U_{xa}(j\omega)}{W(j\omega)}\right\rvert = \left\lvert\frac{U_{xa}(j\omega)}{N(j\omega)}\right\rvert = \left\lvert\frac{G(j\omega)}{1+G(j\omega)P_{xs\,xa}(j\omega)}\right\rvert \le \delta_3(\omega), \quad \omega \in \Omega_3$	(7.26)
4	$\lvert T_4(j\omega)\rvert = \left\lvert\frac{U_{xa}(j\omega)}{U_{xd}(j\omega)}\right\rvert = \left\lvert\frac{G(j\omega)P_{xs\,xd}(j\omega)}{1+G(j\omega)P_{xs\,xa}(j\omega)}\right\rvert \le \delta_4(\omega), \quad \omega \in \Omega_4$	(7.27)
5	$\delta_{5inf}(\omega) \le \lvert T_5(j\omega)\rvert = \left\lvert\frac{Y_{xo}(j\omega)}{R(j\omega)}\right\rvert = \left\lvert F(j\omega)\frac{G(j\omega)P_{xo\,xa}(j\omega)}{1+G(j\omega)P_{xs\,xa}(j\omega)}\right\rvert \le \delta_{5sup}(\omega), \quad \omega \in \Omega_5$	(7.28)
	$\frac{\lvert G(j\omega)P_{xo\,xa\,d}(j\omega)\rvert}{\lvert G(j\omega)P_{xo\,xa\,e}(j\omega)\rvert}\frac{\lvert 1+G(j\omega)P_{xs\,xa\,e}(j\omega)\rvert}{\lvert 1+G(j\omega)P_{xs\,xa\,d}(j\omega)\rvert} \le \delta_5(\omega) = \frac{\delta_{5sup}(\omega)}{\delta_{5inf}(\omega)}, \quad \omega \in \Omega_5$	
6	$\lvert T_6(j\omega)\rvert = \left\lvert\frac{E(j\omega)}{R(j\omega)F(j\omega)}\right\rvert = \left\lvert\frac{E(j\omega)}{N(j\omega)}\right\rvert = \left\lvert\frac{1}{1+G(j\omega)P_{xs\,xa}(j\omega)}\right\rvert \le \delta_6(\omega), \quad \omega \in \Omega_6$	(7.29)
7	$\lvert T_7(j\omega)\rvert = \left\lvert\frac{E(j\omega)}{U_{xd}(j\omega)}\right\rvert = \left\lvert\frac{Y_{xs}(j\omega)}{U_{xd}(j\omega)}\right\rvert = \left\lvert\frac{P_{xs\,xd}(j\omega)}{1+G(j\omega)P_{xs\,xa}(j\omega)}\right\rvert \le \delta_7(\omega), \quad \omega \in \Omega_7$	(7.30)
8	$\lvert T_8(j\omega)\rvert = \left\lvert\frac{E(j\omega)}{W(j\omega)}\right\rvert = \left\lvert\frac{Y_{xs}(j\omega)}{R(j\omega)F(j\omega)}\right\rvert = \left\lvert\frac{Y_{xs}(j\omega)}{N(j\omega)}\right\rvert = \left\lvert\frac{Y_{xs}(j\omega)}{W(j\omega)}\right\rvert = \left\lvert\frac{G(j\omega)P_{xs\,xa}(j\omega)}{1+G(j\omega)P_{xs\,xa}(j\omega)}\right\rvert \le \delta_8(\omega), \quad \omega \in \Omega_8$	(7.31)

traditional manual bound computation (see Appendix B). From these inequalities, it is possible to compute the bounds on the NC and to loop shape the compensators afterward. The compensator G in the 2DOF feedback structure reduces the sensitivity to the uncertainty and fulfils the stability and performance specifications, whereas the prefilter F makes the system meet the desired tracking specifications.

7.4.2 QFT for Distributed Parameter Systems

Likewise, for the lumped approach, the DPS generalized TFs $T_k(j\omega_i)$ extracted from Equations 7.8 through 7.11 are shown in Table 7.3 (see Equations 7.24 through 7.31) as inequality performance specifications. In turn, they are translated into quadratic inequalities in Table 7.4 (see Equations 7.32 through 7.39).

When extending this methodology to the DPS case, the main difference encountered is the fact that the set of plants $\mathcal{P}_{x2\,x1}(j\omega_i) = \{P_{x2\,x1}(j\omega_i), \omega_i \in \Omega_k\}$ not only includes parametric and nonparametric uncertainty related to physical characteristics of the model, but also uncertainty related to the location of the inputs and outputs (x_s, x_o, x_a, and x_d). In addition, now, the nominal plant corresponds to $P_{xs\,xa0} \in \mathcal{P}_{xs\,xa}$, and the templates and QFT bounds consider uncertainty related to the whole DPS model with the TFs $\{P_{xs\,xa}, P_{xo\,xa}, P_{xo\,xd}, P_{xs\,xd}\}$.

TABLE 7.4

Quadratic Inequalities for DPS with $H = 1$

k	Inequalities	Equation
1	$\left(p_{xsxa}^2 - \frac{p_{xoxa}^2}{\delta_1^2}\right)g^2 + 2p_{xsxa}\cos(\phi + \theta_{xsxa})g + 1 \geq 0$	(7.32)
2	$\left(p_{xsxa}^2 - \frac{(p_{xsxa}\, p_{xoxd})^2 + (p_{xoxa}\, p_{xsxd})^2 - 2p_{xsxa}p_{xoxd}p_{xoxa}p_{xsxd}\cos(-\theta_{xsxa} - \theta_{xoxd} + \theta_{xoxa} + \theta_{xsxd})}{\delta_2^2}\right)g^2$ $+\left(\frac{2p_{xsxa}\cos(\phi + \theta_{xsxa}) - 2p_{xoxd}\left[p_{xsxa}\, p_{xoxd}\cos(\phi + \theta_{xsxa}) - p_{xoxa}\, p_{xsxd}\cos(-\phi - \theta_{xsxd} + \theta_{xoxd} - \theta_{xoxa})\right]}{\delta_2^2}\right)g$ $+\left(1 - \frac{p_{xoxd}^2}{\delta_2^2}\right) \geq 0$	(7.33)
3	$\left(p_{xsxa}^2 - \frac{1}{\delta_3^2}\right)g^2 + 2p_{xsxa}\cos(\phi + \theta_{xsxa})g + 1 \geq 0$	(7.34)
4	$\left(p_{xsxa}^2 - \frac{p_{xsxd}^2}{\delta_4^2}\right)g^2 + 2p_{xsxa}\cos(\phi + \theta_{xsxa})g + 1 \geq 0$	(7.35)
5	$\left(p_{xoxa_e}^2 p_{xsxa_d}^2 - \frac{p_{xoxa_d}^2 p_{xsxa_e}^2}{\delta_5^2}\right)g^2$ $+2\left(p_{xoxa_e}^2 p_{xsxa_d}\cos(\phi + \theta_{xsxa_d}) - \frac{p_{xoxa_d}^2 p_{xsxa_e}\cos(\phi + \theta_{xsxa_e})}{\delta_5^2}\right)g$ $+\left(p_{xoxa_e}^2 - \frac{p_{xoxa_d}^2}{\delta_5^2}\right) \geq 0$	(7.36)
6	$p_{xsxa}^2 g^2 + 2p_{xsxa}\cos(\phi + \theta_{xsxa})g + \left(1 - \frac{1}{\delta_6^2}\right) \geq 0$	(7.37)
7	$p_{xsxa}^2 g^2 + 2p_{xsxa}\cos(\phi + \theta_{xsxa})g + \left(1 - \frac{p_{xsxd}^2}{\delta_7^2}\right) \geq 0$	(7.38)
8	$p_{xsxa}^2\left(1 - \frac{1}{\delta_8^2}\right)g^2 + 2p_{xsxa}\cos(\phi + \theta_{xsxa})g + 1 \geq 0$	(7.39)

Remark 7.4

Note that the proposed methodology can deal with uncertainty at the location of the input and output points (x_s, x_o, x_a, and x_d) extending the templates of $P_{x1\,x2}$ to new very interesting cases. See also Remark 7.1.

Remark 7.5

When controlling any DPS by a one-point feedback structure, some limitations due to the spatial distribution arise. From Equation 7.8, it follows that the compensator G cannot totally change the disturbance effects at the output. In particular, there is no solution to the DPS problem if $P_{xo\,xd}$ has poles in the right-half plane (RHP). In the same way, $P_{xo\,xa}$ and $P_{xs\,xd}$ must be stable to allow the closed-loop system meet the performance requirements. This is easily concluded from Equations 7.8 through 7.11.

7.5 Modeling Approaches for PDE

According to the classical book by Farlow,[319] a general second-order linear PDE in two variables is of the form

$$AU_{xx} + BU_{xy} + CU_{yy} + DU_x + EU_y + FU = G \tag{7.40}$$

where
$U_{xx} = \partial^2 U/\partial x^2$
$U_y = \partial U/\partial y$, etc.

This equation is written here with the two independent variables x and y. However, in many problems, one of the two variables stands for time and hence it is written in terms of x and t.

Following Equation 7.40, there is a basic classification to linear PDE involving three types of equations. They are (1) *parabolic equations*, which satisfy the property $B^2 - 4AC = 0$ (as for heat flow and diffusion processes); (2) *hyperbolic equations*, which satisfy the property $B^2 - 4AC > 0$ (as for heat flow and diffusion processes); and (3) *elliptic equations*, which satisfy the property $B^2 - 4AC < 0$ (steady-state phenomena).

Some very well-known examples of these three types are as follows: the heat exchange equation for parabolic type $\{\partial T/\partial t = K\partial^2 T/\partial x^2\}$; the one-dimensional wave equation for hyperbolic type $\{\partial^2 T/\partial t^2 = K\partial^2 T/\partial x^2\}$; and the Euler–Lagrange equation for the elliptic type $\{\partial^2 T/\partial t^2 + \partial^2 T/\partial x^2 = 0\}$.

In general, for parabolic systems, an easy approach is the equivalent electrical model (see Figure 7.5 and example in Section 7.6.2). Generally speaking, the set of differential equations describing the dynamic performance of this kind of DPS can be formulated in terms of *through variables* (I) and *across variables* (U) (see Refs. 189 and 222). The through variable (I) can represent an electrical current, a mechanical force or torque, a heat flow rate, etc. The across variable (V) can represent an electrical voltage difference, a mechanical linear or angular velocity difference, or a temperature difference, etc.

The equivalent electrical model P_{x2x1} (Figure 7.5) is defined with enough Π (Pi) elements in series between the points of interest X_1 and X_2. The more Π elements introduced, the larger

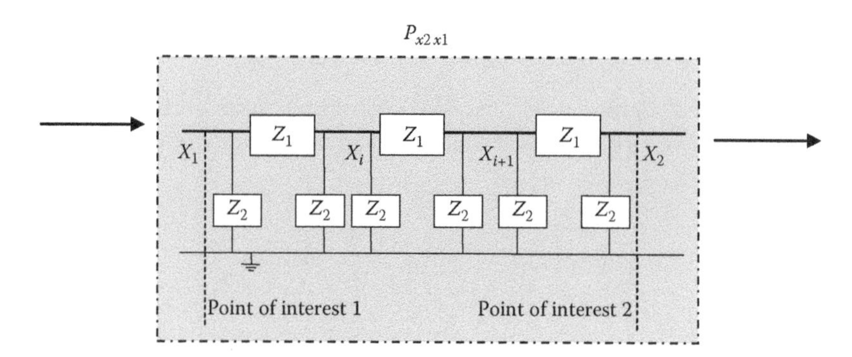

FIGURE 7.5
Π equivalent electrical model.

the bandwidth of the model, which is accurate enough to be used. The minimum amount of elements to be considered can be studied following the Anderson–Parks's criteria.[299]

For the three types of PDEs, another general approach of P_{x2x1} can be obtained by getting the solution of the PDE. This is not a trivial task, and often it leads to either an irrational solution (see, for instance, Equation 7.49) or a very complex TF. In these cases, it is convenient to transform the irrational solutions into a series of sums or products by using some mathematical approaches such as the Taylor series, Fourier series, or Weirstrass factorization (see, e.g., Equation 7.42). As before, the exact amount of elements can be studied following the Anderson–Parks's criteria.[299]

One way of solving the system equations is by separating the variables and defining the solution as Fourier series. With the defined solution, the problem is solved for each term individually, or even better, for the general term by evaluating it into the system equation. In any case, enough number of elements must be taken and used for modeling purposes. See the following section for illustrative examples.

7.6 Examples

In this section, the proposed methodology is illustrated by means of two well-known PDE plants, also explored by Kelemen et al.[112] in his preliminary work. The examples are the Bernoulli–Euler beam and the heat exchange equation.

7.6.1 Bernoulli–Euler Beam

7.6.1.1 Definition

Consider the simply supported beam shown in Figure 7.6 and the equation

$$\begin{aligned}
&\frac{\partial y}{\partial t} - 2\varepsilon\frac{\partial^3 y}{\partial t\,\partial x^2} + \frac{\partial^4 y}{\partial x^4} = \vartheta(x,t),\\
&\vartheta(x,t) = (\delta_{t=0})(\delta_{x=xi}), \quad x \in (0,\pi), t \geq 0, \varepsilon = 0.1\\
&y(x,0) = \frac{\partial y}{\partial t}(\pi,t) = 0, \quad x \in [0,\pi]\\
&y(0,t) = y(\pi,t) = \frac{\partial^2 y}{\partial x^2}(0,t) = \frac{\partial^2 y}{\partial x^2}(0,t) = 0
\end{aligned} \tag{7.41}$$

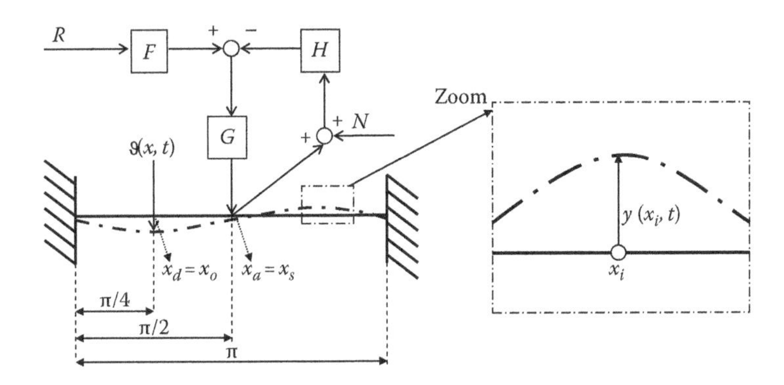

FIGURE 7.6
Bernoulli–Euler beam diagram.

where $y(x, t)$ and $\vartheta(x, t)$ are the distributed transversal displacement and the input, respectively. The sensor and the actuator are located at the same point, $x_s = x_a = \pi/2$. The disturbance is applied at the objective point $x_o = x_d = \pi/4$ (see Figure 7.6).

7.6.1.2 Modeling

As explained in Section 7.5, the PDE modeled by Equation 7.41 is hard to handle. In this case, a direct rational approach is suggested by using a Fourier series approach, assuming that the solution of the PDE is a sum of the series terms, so that

$$P_{x_2x_1}(s) = \frac{2}{\pi}\sum_{k}^{\infty} \frac{\sin(kx_1)\sin(kx_2)}{s^2 + 2\varepsilon ks + k^4} \tag{7.42}$$

where
x_2 stands for the control objective and sensor locations
x_1 stands for the actuator and disturbance locations

For notation purposes, when only the first k elements are taken $P_{x2x1}(s)$ is written as $P_{x2x1}k(s)$ or $Pk(s)$ for simplicity. Figure 7.7 shows the Bode diagram of Equation 7.42 for $P_{xs\,xa}(s)$, with one term P1(s), or two terms P2(s), or three terms P3(s), etc. Similar diagrams can be obtained for $P_{xo\,xa}(s)$, $P_{xs\,xd}(s)$, or $P_{xo\,xd}(s)$. The larger the number of elements, the higher the frequencies that are described by the model. In this case, three elements are enough to describe the model up to the frequency crossover. This can be deduced by the Anderson's criteria.

7.6.1.3 Control Specifications

The system performance specifications are selected according to the following requirements:

a. *Stability*. It is defined by the most restrictive condition of the following two expressions:

$$\left|\frac{GP_{xo\,xa}}{1+GP_{xs\,xa}}\right| \le 1.2 \quad \forall\omega \tag{7.43}$$

$$\left|\frac{GP_{xs\,xa}}{1+GP_{xs\,xa}}\right| \le 1.2 \quad \forall\omega \tag{7.44}$$

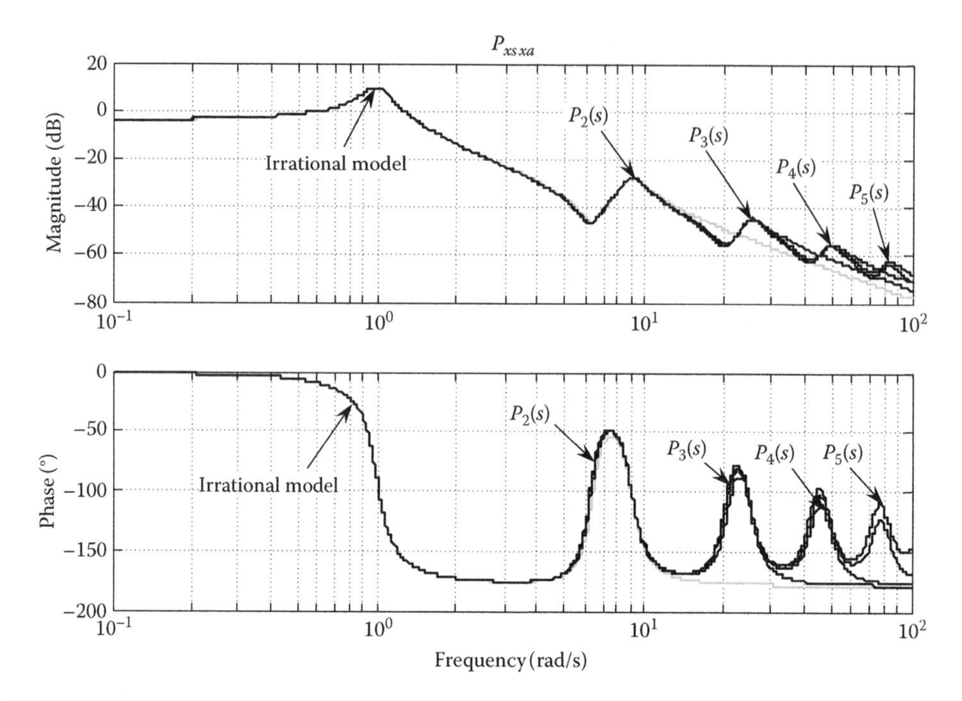

FIGURE 7.7
Bode diagram for P_{xsxa} according to the irrational model and the rational Fourier approach given by Equation 7.42 for 2, 3, 4, and 5 elements.

b. *Disturbance rejection*

$$\left|\frac{Y_{xo}}{U_{xd}}\right| = \left|P_{xoxd} - \frac{GP_{xoxa}P_{xsxd}}{1+GP_{xsxa}}\right| \leq 0.3\left|P_{xoxd}\right|, \quad \omega < 1.1\,\text{rad/s} \tag{7.45}$$

7.6.1.4 Compensator Design

The templates are calculated from the equivalent TFs $\{P_{xsxa}, P_{xoxa}, P_{xsxd}, P_{xoxd}\}$, and the robust stability and disturbance rejection bounds $B(j\omega)$ are obtained from the quadratic inequalities corresponding to Equations 7.32, 7.33, and 7.39 of Table 7.4. The nominal open-loop expression is $L_1 = G_1P_{xsxa0}$. The G_1 compensator (see Equation 7.46) is obtained by using a standard loop-shaping QFT technique with the bounds calculated with the DPS technique introduced in this chapter:

$$G_1(s) = \frac{11.6(s^2 + 0.2s + 1)}{s(s/60+1)(s/60+1)} \tag{7.46}$$

Figure 7.8 shows on the NC such QFT bounds, together with the loop TF $L_1 = G_1P_{xsxa0}$. Note that the loops shown on the NC are the resonance peaks represented in Figure 7.7.

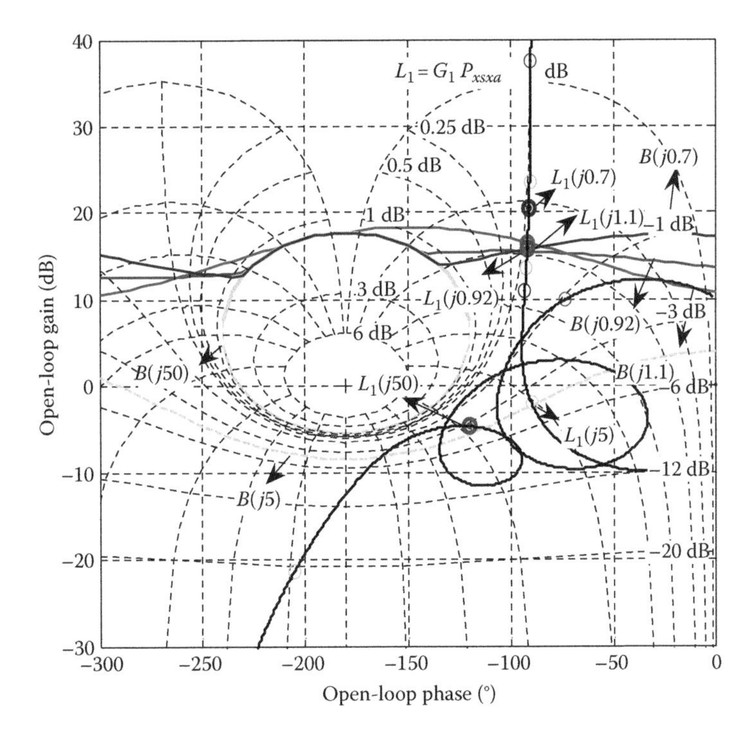

FIGURE 7.8
Loop shaping for the compensator. Bernoulli–Euler beam open-loop TF: $L_1 = P_{xsxa0}\, G_1$.

Equation 7.47 also shows the compensator calculated by Kelemen et al.[112] for this example using his previous methodology:

$$G_2(s) = \frac{0.1(s/0.3+1)}{(s/0.5+1)^2} \tag{7.47}$$

7.6.1.5 Simulations

Both compensators, $G_1(s)$ and $G_2(s)$, are implemented and tested in the DPS simulator. The closed-loop time responses of the desired output Y_{xo} and the control signal U_{xa}, when external disturbances U_{xd} are introduced, appear in Figure 7.9 for both designs. Note that the prefilter $F(s)$, dependent on the spatial configuration of the one-point feedback DPS, is fixed at a constant value $F(s) = 2$ in both cases. The desired output Y_{xo} shows a more damped behavior and a smaller tracking error when using the $G_1(s)$ compensator designed with the methodology introduced in this chapter. In addition, the new technique is simpler, it can deal with the model uncertainty in the structure, parameters, and location of the relevant input–output points, and it is able to work with distributed specifications.

Remark 7.6

This example has been studied without uncertainty at the location of the relevant points to make a strict comparison with Kelemen's former work. However, the new methodology is more general and can also introduce uncertainty at the structure, parameters of the model, and the location of the relevant points (inputs and outputs).

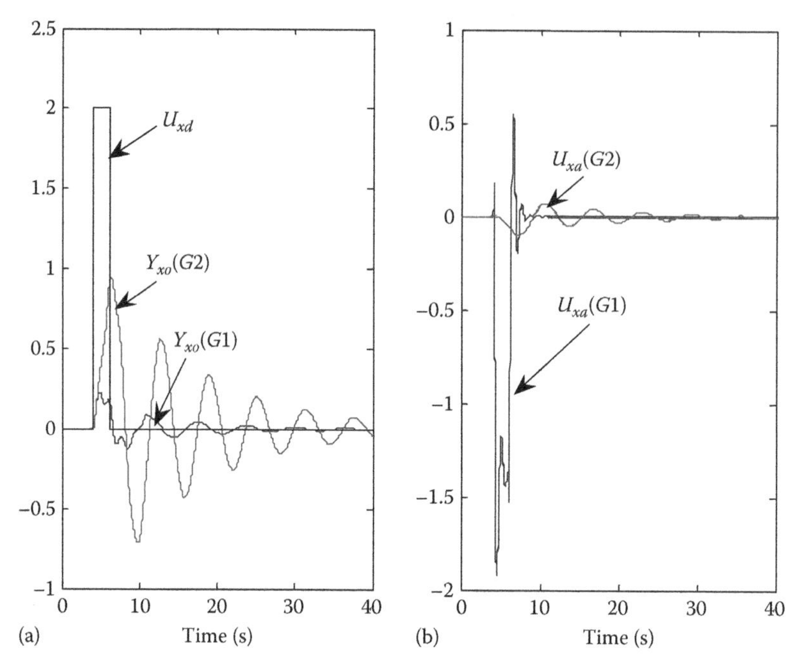

FIGURE 7.9
(a) Disturbance U_{xd} and output Y_{xo} at $x = x_d = x_o$. (b) Control signal U_{xa} at $x = x_a = x_s$.

7.6.2 Heat Equation Problem

7.6.2.1 Definition

Consider the following model of the temperature-distributed prismatic plant of Figure 7.10:

$$\begin{aligned} &\frac{\partial T}{\partial t} - \frac{\partial^2 T}{\partial x^2} = \vartheta(x,t), \quad x \in (0,\pi), t \geq 0 \\ &T(x,0) = 0, \quad x \in [0,\pi] \\ &T(0,t) = T(\pi,t) = 0; \quad \vartheta(x,t) = (\delta_{t=0})(\delta_{x=xi}) \end{aligned} \tag{7.48}$$

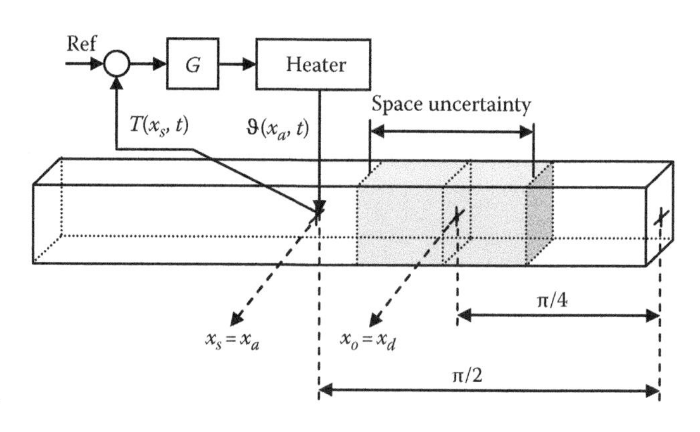

FIGURE 7.10
Temperature-distributed prismatic plant.

where $T(x,t)$ and $\vartheta(x,t)$ are the temperature and heat input, respectively. It is assumed that the sensor and the actuator point match up, $x_s = x_a = \pi/2$, and a disturbance is applied at x_d. The example shows the design of one feedback loop to meet the desired performance specifications at $x_o = x_d = \pi/4$.

7.6.2.2 Modeling

The original paper by Kelemen et al.[112] shows an irrational TF, Equation 7.49, which represents the behavior of the temperature at the distributed prismatic plant.

$$P_{x\,xi}(s) = \frac{\left(e^{\sqrt{s}(\pi-x)} - e^{-\sqrt{s}(\pi-x)}\right)\left(e^{\sqrt{s}x_i} - e^{-\sqrt{s}\,x_i}\right)}{2\sqrt{s}\left(e^{\sqrt{s}\,\pi} - e^{-\sqrt{s}\,\pi}\right)} \tag{7.49}$$

As detailed in Section 7.5, the irrational solution calculated for the problem can be either transformed into a sum of terms or into a product of terms. In this case, since it is a parabolic equation, an equivalent electrical model of the plant is chosen (see Figure 7.11). Solving the resultant system for the 3, 4, and 8 Π elements cases, the respective groups of rational TFs P_{xsxa}, P_{xoxa}, P_{xoxd}, P_{xsxd} are obtained. For this case, the parameters used are the distances $k_1 = \pi/4$, $k_2 = \pi/4$, and $k_3 = \pi/2$; the resistance per meter $r = 1/kA$; the capacitance per meter $c = \rho CpA$, where ρ is the density (kg/m^3), the thermal conductivity k (W/m °C), the thermal capacitance Cp (J/kg °C), and the section area A (m^2). As in Kelemen et al.,[112] all ρ, k, Cp, and A are equal to unity.

The electric circuit of Figure 7.11 can be easily solved by classical circuit techniques, resulting in the following system:

$$\begin{bmatrix} Y_{xo} \\ Y_{xs} \end{bmatrix} = \begin{bmatrix} \dfrac{1}{k_1 r} + \dfrac{1}{\dfrac{2}{k_1 c\, s}} + \dfrac{1}{k_2 r} + \dfrac{1}{\dfrac{2}{k_2 c\, s}} & -\dfrac{1}{k_2 r} \\ -\dfrac{1}{k_2 r} & \dfrac{1}{k_2 r} + \dfrac{1}{\dfrac{2}{k_2 c\, s}} + \dfrac{1}{k_3 r} + \dfrac{1}{\dfrac{2}{k_3 c\, s}} \end{bmatrix}^{-1} \begin{bmatrix} U_{xd} + \dfrac{T_c}{k_1 r} \\ U_{xa} + \dfrac{T_c}{k_3 r} \end{bmatrix} \tag{7.50}$$

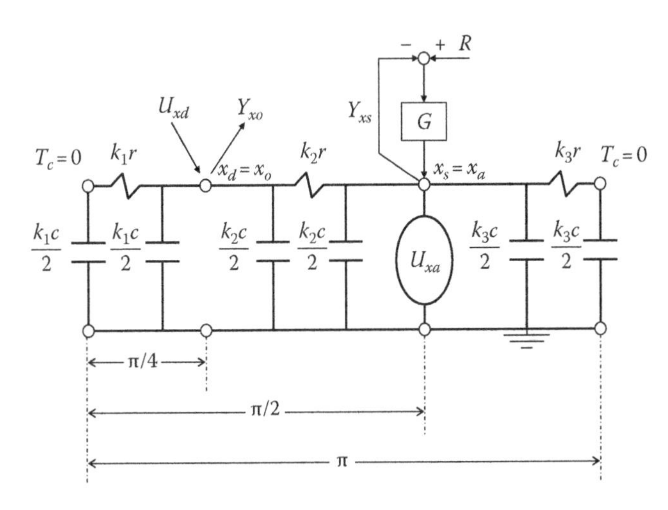

FIGURE 7.11
Π equivalent electrical model.

Further development of the system equations ($T_c = 0$, $r = 1$; $c = 1$) yields

$$\begin{bmatrix} Y_{xo} \\ Y_{xs} \end{bmatrix} = \begin{bmatrix} P_{xo\,xd} & P_{xo\,xa} \\ P_{xs\,xd} & P_{xs\,xa} \end{bmatrix} \begin{bmatrix} U_{xd} \\ U_{xa} \end{bmatrix} \tag{7.51}$$

where

$$\begin{bmatrix} P_{xo\,xd} & P_{xo\,xa} \\ P_{xs\,xd} & P_{xs\,xa} \end{bmatrix} = Q \begin{bmatrix} \frac{1}{k_2} + \frac{1}{k_3} + \frac{(k_2 + k_3)}{2} s & \frac{1}{k_2} \\ \frac{1}{k_2} & \frac{1}{k_1} + \frac{1}{k_2} + \frac{(k_1 + k_2)}{2} s \end{bmatrix} \tag{7.52}$$

$$Q = \frac{1}{\left[\left(\frac{1}{k_1} + \frac{1}{k_2} + \frac{(k_1 + k_2)}{2} s \right) \left(\frac{1}{k_2} + \frac{1}{k_3} + \frac{(k_2 + k_3)}{2} s \right) - \frac{1}{k_2^2} \right]} \tag{7.53}$$

Now the TFs extracted from the system, with $k_1 = \pi/4$, $k_2 = \pi/4$, and $k_3 = \pi/2$, are

$$P_{xo\,xd} = \frac{12\pi(\pi^2 s + 16)}{3\pi^4 s^2 + 144s + 1024} \tag{7.54}$$

$$P_{xs\,xd} = P_{xo\,xa} = \frac{128\pi}{3\pi^4 s^2 + 144s + 1024} \tag{7.55}$$

$$P_{xs\,xa} = \frac{8\pi(\pi^2 s + 32)}{3\pi^4 s^2 + 144s + 1024} \tag{7.56}$$

These TFs can represent a single Π element plant. The same procedure can be used to obtain the 3, 4, and 8 Π elements model, just by introducing more equivalent Π electrical elements inside the equations. Figure 7.12 shows how the equivalent electrical models (for 3, 4, and 8 Π elements) approximate the irrational equation of Equation 7.49.

The 8 Π elements equivalent model fits well with the original DPS. According to Anderson and Parks,[299] the approach is valid for $\omega < 2/RC = 2/(\pi/8)^2 = 12.97$ rad/s, where R and C are the resistance and capacitance per Π element length, respectively.

7.6.2.3 Control Specifications

As before the desired performance specifications are as follows:

a. *Stability*. The most restrictive expression of the previous expression equations (7.42) and (7.43).
b. *Disturbance rejection*

$$\left| \frac{Y_{xo}}{U_{xd}} \right| = \left| P_{xo\,xd} - \frac{G P_{xo\,xa} P_{xs\,xd}}{1 + G P_{xs\,xa}} \right| \le 0.9 \left| P_{xo\,xd} \right|, \quad \omega < 1.5 \text{ rad/s} \tag{7.57}$$

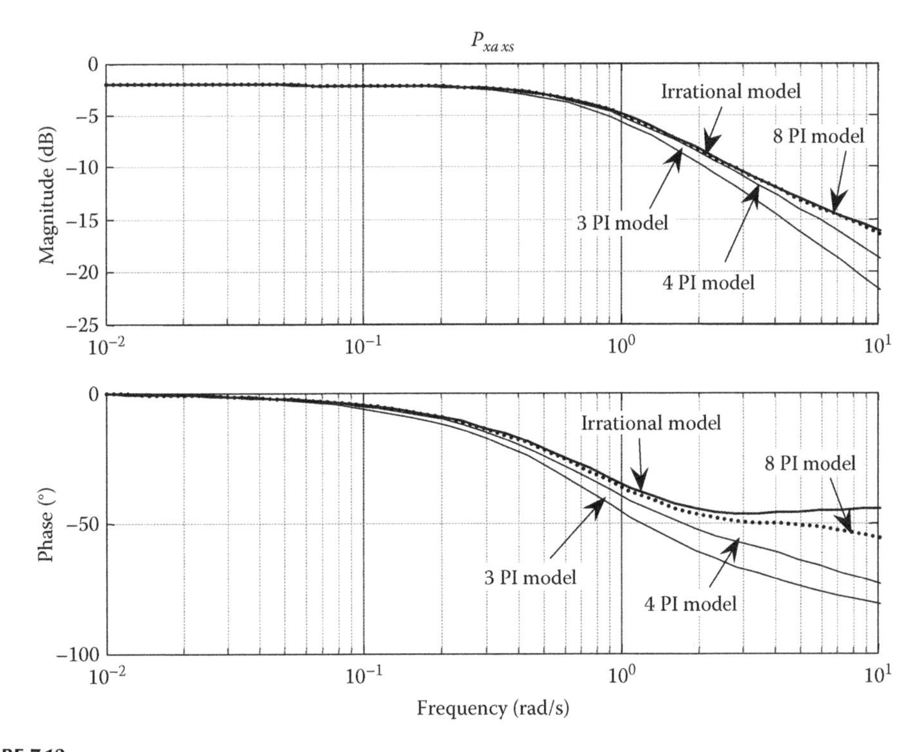

FIGURE 7.12
Bode diagram for P_{xsxa} according to the irrational model equation (7.49), and the rational electric circuit approach given by 3, 4, and 8 Π elements.

7.6.2.4 Compensator Design

The templates are calculated from the equivalent TFs $P_{xs\,xa}$, $P_{xo\,xa}$, $P_{xo\,xd}$, and $P_{xs\,xd}$ of the 8 Π elements case. The robust stability and disturbance rejection bounds $B(j\omega)$ are obtained from the quadratic inequalities given by Equations 7.32 and 7.33, respectively.

The nominal open-loop expression is $L_1 = G_1 P_{xs\,xa0}$. By using a standard loop-shaping QFT technique, the G_1 compensator (see Equation 7.58) is designed. Figure 7.13 shows the NC for both loop TFs (L_1 and L_2), with the proposed compensator G_1 and the Kelemen compensator G_2 (see Equation 7.59), respectively,

$$G_1(s) = \frac{11.5\,(s/2.9+1)}{s(s/5+1)} \tag{7.58}$$

$$G_2(s) = \frac{60(s/1.5+1)(s/2.5+1)}{(s/0.4+1)^2} \tag{7.59}$$

7.6.2.5 Simulations

Figure 7.14 compares the time responses obtained by using both compensators G_1 and G_2. The results show a similar performance of both approaches. However, as before, the proposed methodology is simpler, deals with model uncertainty, and is able to work with distributed location of the relevant points and specifications.

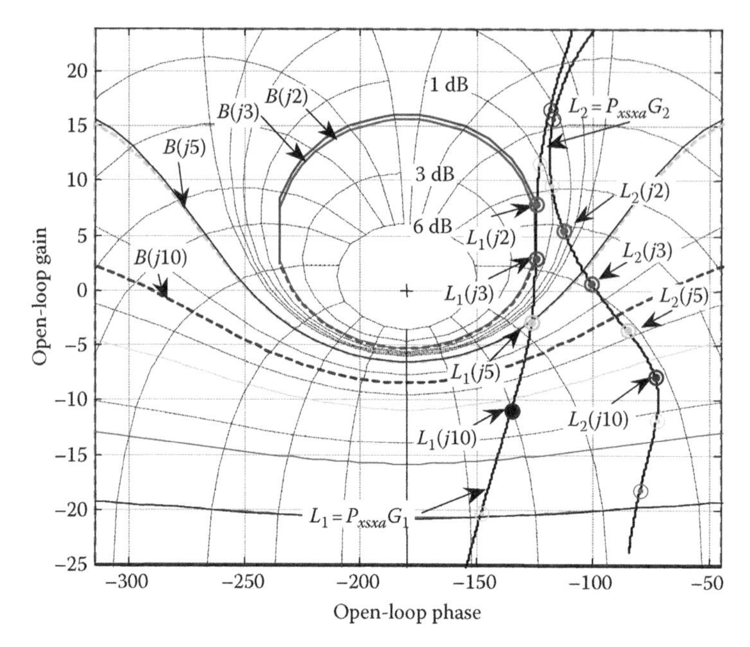

FIGURE 7.13
Loop shaping for the compensators. The temperature-distributed plant problem with the open-loop TFs: $L_1 = P_{xsxa0}G_1$ and $L_2 = P_{xsxa0}G_2$.

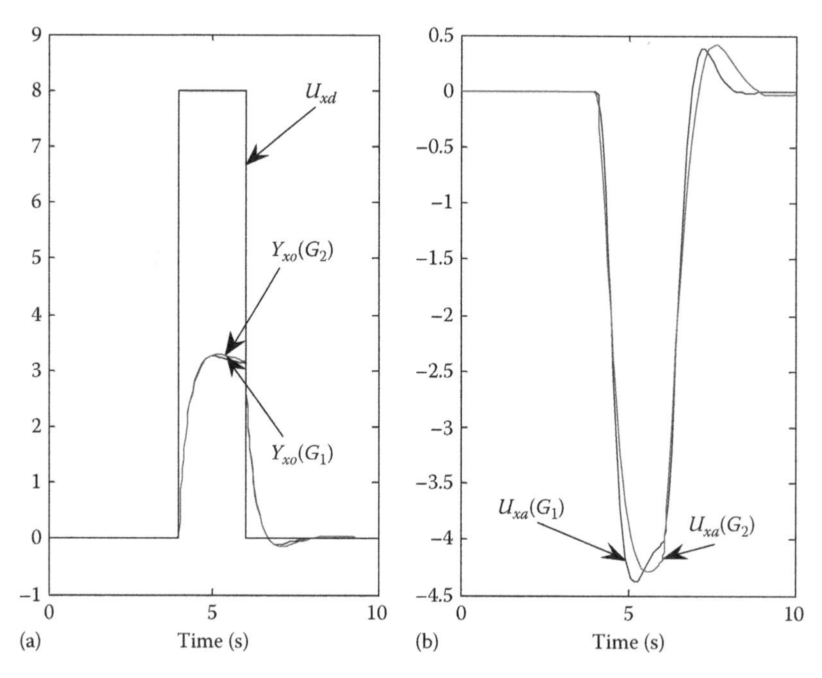

FIGURE 7.14
(a) Disturbance U_{xd} and output signals Y_{xo} at $x = x_d = x_o$. (b) Control signal U_{xa} at $x = x_a = x_s$. $F = 2$.

7.7 Summary

There is no need for very complex methodologies to design controllers for DPS. On the contrary, by using the practical approach presented in this chapter, classical QFT methods based on a new bound definition can be used to solve DPS control problems. A spatial distribution of the location of the relevant points was considered, allowing the designer to place the actuator, disturbances, sensor, and control objectives in distributed locations with uncertainty. From this topology, new stability and performance specifications, TFs, and quadratic inequalities for DPS were introduced. The simplicity and excellent practical results of the methodology were illustrated by comparison with two previously published examples.

8

Nonlinear Switching Control Techniques[140]

8.1 Introduction

This chapter introduces a hybrid methodology to design nonlinear robust control systems that are able to go beyond the classical linear limitations.[140] Combining robust compensator/controller designs and stable switching, the new system optimizes the time response of the plant by fast adaptation of the controller parameters during the transient response according to certain rules based on the amplitude of the error. The methodology is based on both a graphical frequency-domain stability criterion for switching linear systems and the use of the robust quantitative feedback technique (QFT) control system design technique.

Switching control has demonstrated to be an efficient tool in achieving tight performance specifications in control systems.[288,293] This enhancement can be reached by designing various parallel controllers with different characteristics, and continuously selecting among them the optimum one that controls the system best (Figure 8.1). Performance specifications that are not achievable by a simple linear controller, as the limitation theory predicts,[2,289] can be attained through suitable switching rules.

One of the main issues in switching control techniques is that the system stability is not assured a priori, even if the switching is made between stable controllers. This is the reason why most of the current literature about switching systems is still devoted to stability issues. See Refs. 291, 292, and 295 for general results about stability criteria applied to some practical cases.

This chapter introduces a nonlinear switching robust controller design methodology, including a graphical frequency-domain criterion to ensure the system stability and several illustrative examples. In Chapter 14, the methodology is also applied to the design of a pitch control system, which is one of the most critical issues in wind turbines, where a tight combination of high reliability (robustness), minimum mechanical fatigue, and performance optimization is required.

8.2 System Stability under Switching

Figure 8.1 shows the general scheme to be used with the switching control system. A set of controllers are designed and a supervisor selects the most suitable one, depending on the state of the system and/or some external parameters.

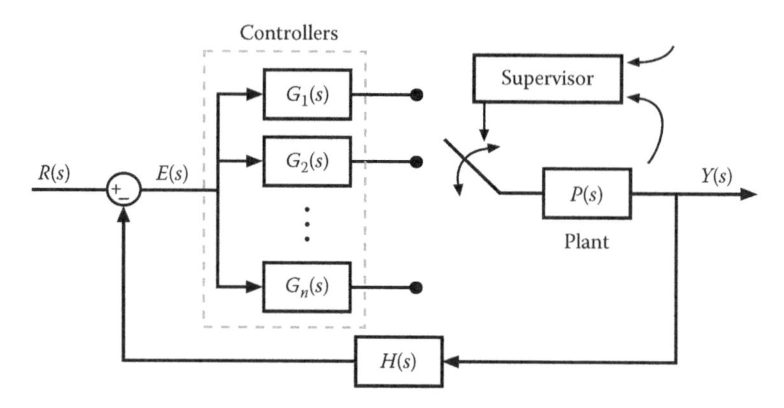

FIGURE 8.1
Switching control scheme.

One of the main difficulties found when switching techniques are applied is that, in general, the system stability is not assured, even if switching is made between stable controllers. Some extra conditions must be met to assure stability. In particular, it is proven that a system represented by[293]

$$\dot{x}(t) = A(t)x(t), \quad A(t) \in \mathcal{A} = \{A_1, \ldots, A_m\}, \quad A_i \text{ Hurwitz} \tag{8.1}$$

with arbitrary switching within the set of matrices $\boldsymbol{A}$ is exponentially stable if and only if there exists a common Lyapunov function (*CLF*) for all A_i in the set $\boldsymbol{A}$.[287]

It has also been proven that the existence of a common quadratic Lyapunov function (*CQLF*) is a sufficient condition for exponential stability.[290] In this context, the main issue in linear switching systems is finding conditions under which the existence of a *CQLF* is assured. In particular, the circle criterion provides the necessary and sufficient conditions for the existence of a *CQLF* for two systems in companion form,[285,286,294] that is, the systems

$$\dot{x}(t) = Ax(t) \tag{8.2}$$

$$\dot{x}(t) = (A - g\Delta^T)x(t) \tag{8.3}$$

with

$$A = \begin{bmatrix} 0 & 1 & 0 & \cdots & 0 & 0 \\ 0 & 0 & 1 & \cdots & 0 & 0 \\ 0 & 0 & 0 & \cdots & 0 & 0 \\ \vdots & \vdots & \vdots & \ddots & \vdots & \vdots \\ 0 & 0 & 0 & \cdots & 0 & 1 \\ -e_0 & -e_1 & -e_2 & \cdots & -e_{n-2} & -e_{n-1} \end{bmatrix}, \quad g = \begin{bmatrix} 0 \\ 0 \\ 0 \\ \vdots \\ 0 \\ 1 \end{bmatrix}, \quad \Delta = \begin{bmatrix} \Delta e_0 \\ \Delta e_1 \\ \Delta e_2 \\ \vdots \\ \Delta e_{n-2} \\ \Delta e_{n-1} \end{bmatrix} \tag{8.4}$$

have a *CQLF* if and only if

$$1 + \text{Re}\{\Delta^T (sI - A)^{-1} g\} > 0, \quad s = j\omega, \text{ for all frequency } \omega \tag{8.5}$$

This chapter considers stability for arbitrary switching between two closed-loop systems with transfer functions $T_1(s) = L_1(s)/[1 + L_1(s)]$ and $T_2(s) = L_2(s)/[1 + L_2(s)]$, both stable. For these two transfer functions, $L_1(s) = P(s)G_1(s)$ and $L_2(s) = P(s)G_2(s)$ are proper. Additionally, $L_1(s)$ and $L_2(s)$ have the same number of poles and the same number of zeros. The switching takes place by changing the gain and the position of the poles and zeros. The open-loop transfer functions for both systems are

$$L_1(s) = \frac{b_{n-1}s^{n-1} + \cdots + b_0}{s^n + a_{n-1}s^{n-1} + \cdots + a_0} = \frac{N(s)}{D(s)} \tag{8.6}$$

and

$$L_2(s) = \frac{(b_{n-1} + \Delta b_{n-1})s^{n-1} + \cdots + (b_0 + \Delta b_0)}{s^n + (a_{n-1} + \Delta a_{n-1})s^{n-1} + \cdots + (a_0 + \Delta a_0)} = \frac{N(s) + \Delta N(s)}{D(s) + \Delta D(s)} \tag{8.7}$$

For the sake of clarity, and without losing generality, in this analysis a general expression where the order of the numerator is 1 less than that of the denominator is used. If it is not the case, then the coefficients b_{n-1}, etc., will be zero. The closed-loop transfer functions are

$$T_1(s) = \frac{L_1(s)}{1 + L_1(s)} = \frac{N(s)}{D(s) + N(s)} \tag{8.8}$$

and

$$T_2(s) = \frac{L_2(s)}{1 + L_2(s)} = \frac{N(s) + \Delta N(s)}{D(s) + \Delta D(s) + N(s) + \Delta N(s)} \tag{8.9}$$

where the characteristic equations are

$$D(s) + N(s) = s^n + e_{n-1}s^{n-1} + \cdots + e_1 s + e_0 \tag{8.10}$$

and

$$D(s) + \Delta D(s) + N(s) + \Delta N(s) = s^n + (e_{n-1} + \Delta e_{n-1})s^{n-1} + \cdots + (e_0 + \Delta e_0) \tag{8.11}$$

with $e_i = a_i + b_i$ and $\Delta e_i = \Delta a_i + \Delta b_i$.

Using these expressions for the coefficients e_i and Δe_i, the matrices $\boldsymbol{A}$, $\boldsymbol{g}$, and $\boldsymbol{\Delta}$ are defined. Now the circle criterion is applied to guarantee stability under arbitrary switching as

$$(s\boldsymbol{I} - \boldsymbol{A})^{-1} = s^n + e_{n-1}s^{n-1} + \cdots + e_1 s + e_0$$

and

$$\boldsymbol{\Delta}^T \boldsymbol{g} = \Delta e_{n-1}s^{n-1} + \cdots + \Delta e_1 s + \Delta e_0$$

Equation 8.5 becomes

$$1+\text{Re}\left\{\frac{\Delta e_{n-1}s^{n-1}+\cdots+\Delta e_1 s+\Delta e_0}{s^n+e_{n-1}s^{n-1}+\cdots+e_1 s+e_0}\right\}>0, \quad s=j\omega \tag{8.12}$$

for all frequency ω.

After some simple manipulation,

$$\begin{aligned}
&\text{Re}\left\{1+\frac{\Delta e_{n-1}s^{n-1}+\cdots+\Delta e_1 s+\Delta e_0}{s^n+e_{n-1}s^{n-1}+\cdots+e_1 s+e_0}\right\} \\
&=\text{Re}\left\{\frac{N(s)+\Delta N(s)+D(s)+\Delta D(s)}{N(s)+D(s)}\right\} \\
&=\text{Re}\left\{\frac{\left((N(s)+\Delta N(s)+D(s)+\Delta D(s))/(N(s)+\Delta N(s))\right)}{\left((N(s)+D(s))/N(s)\right)}\left(\frac{N(s)+\Delta N(s)}{N(s)}\right)\right\} \\
&=\text{Re}\left\{\frac{1+L_2(s)}{1+L_1(s)}\left(\frac{D(s)+\Delta D(s)}{D(s)}\right)\right\}
\end{aligned} \tag{8.13}$$

The condition can be expressed in the following form:

$$\text{Re}\left\{\frac{1+L_2(s)}{1+L_1(s)}\left(\frac{D(s)+\Delta D(s)}{D(s)}\right)\right\}>0, \quad s=j\omega \tag{8.14}$$

for all frequency ω.

As this formulation of the circle criterion is applied to open-loop transfer functions, due to symmetry, it is sufficient to check it at only positive frequencies. The condition in Equation 8.14 is then equivalent to

$$\left|\arg\{1+L_2(j\omega)\}-\arg\{1+L_1(j\omega)\}+\arg\left\{\frac{D(j\omega)+\Delta D(j\omega)}{D(j\omega)}\right\}\right|<\frac{\pi}{2} \tag{8.15}$$

for all $\omega \geq 0$.

By denoting the equations

$$\phi_{12}(\omega)^\circ=\left|\arg\{1+L_2(j\omega)\}-\arg\{1+L_1(j\omega)\}\right| \tag{8.16}$$

$$\alpha(\omega)^\circ=\left|\arg\left\{\frac{D(j\omega)+\Delta D(j\omega)}{D(j\omega)}\right\}\right| \tag{8.17}$$

and using the triangle inequality, a sufficient condition for Equation 8.15 is

$$\phi_{12}(\omega)^{\circ} < 90^{\circ} - \alpha(\omega)^{\circ} \tag{8.18}$$

for all $\omega \geq 0$.

With this result, the criterion can be applied graphically to both the complex plane and the Nichols diagram. In the first case, the criterion is expressed by requiring $L_1(j\omega)$ and $L_2(j\omega)$ to be inside of an arc of $[90^{\circ} - \alpha(\omega)^{\circ}]$ degrees around the critical point (−1,0) at each frequency.

Similarly, in the Nichols diagram the condition based upon angles is easily checked by plotting $[1 + L_1(j\omega)]$ and $[1 + L_2(j\omega)]$ and the distance of $\phi_{12}(\omega)$ on the horizontal axis at each frequency. To assure stability, this distance should be less than $[90^{\circ}-\alpha(\omega)^{\circ}]$ degrees.

It is noted that the function $\alpha(\omega)$ can also be expressed according to Equation 8.19, so that it may be considered as a measurement of the change of the controller poles, as shown in Figure 8.2. Consequently, the larger the movement made by the poles, the larger the conservativeness introduced by the triangle inequality of Equation 8.18. Note that for each frequency ω_i there is a different angle $\alpha(\omega_i)^{\circ}$ given by

$$\alpha(\omega)^{\circ} = \left| \arg\left\{ \frac{D(j\omega)+\Delta D(j\omega)}{D(j\omega)} \right\} \right|$$

$$= \left| \arg\left\{ \frac{\prod_{j=1}^{n} (j\omega + p_j + \Delta p_j)}{\prod_{j=1}^{n} (j\omega + p_j)} \right\} \right| = \left| \sum_{j=1}^{n} \arg\{j\omega + p_j + \Delta p_j\} - \arg\{j\omega + p_j\} \right| \tag{8.19}$$

At this point two issues arise. Firstly, the criterion presented earlier is applied to switching between two isolated controllers with the same structure and the same plant model. However, it is possible that the designer may desire to do switching among more than two systems, or even among an infinite number of systems, which can also be considered as a *linear parameter-varying* (*LPV*) system, where the controller varies continuously. Secondly, real systems present uncertainty, so the criterion must be modified in some manner to take the uncertainty into account. The next discussion undertakes both issues.

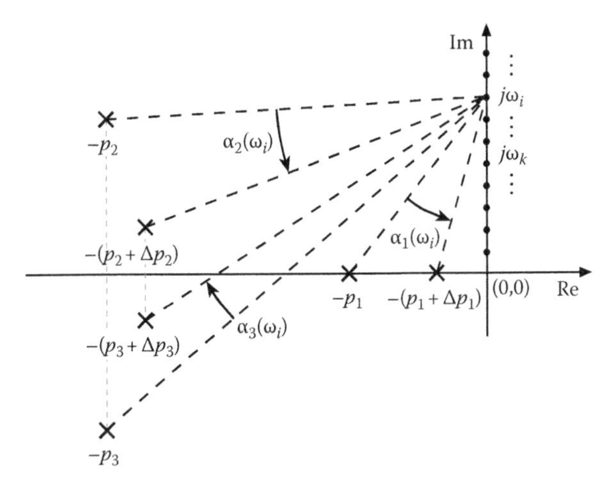

FIGURE 8.2
Complex plane. $\alpha(s)$ for a system with three switching poles: $\alpha(\omega_i)^{\circ} = \alpha_1(\omega_i)^{\circ} + \alpha_2(\omega_i)^{\circ} + \alpha_3(\omega_i)^{\circ}$.

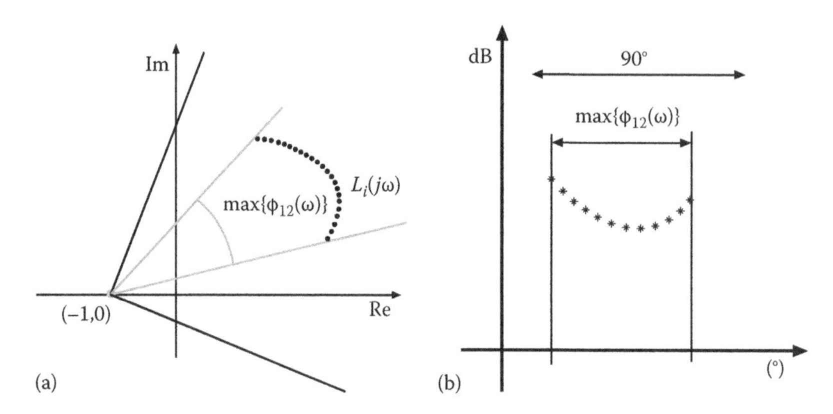

FIGURE 8.3
Criterion for continuous switching: (a) complex plane, plotting $L_i(j\omega)$; (b) Nichols plot, plotting $[1 + L_i(j\omega)]$.

If the switching is made among a set of several controllers, the criterion must be satisfied for each pair of controllers. Checking this condition may be an impossible task if there is more than one pole moving, because the angle α is different for each pair of controllers. For this reason, consider the case in which a controller whose parameters change continuously with the error, and allow the gain and zeros to be the variables. Then, the angle $\alpha(\omega)^\circ$ is null for every frequency, and the only condition to be satisfied is that the angle between any two possible systems $L_i(j\omega)$ and the critical point (−1,0) is less than 90°. Moreover, under this premise the conservativeness introduced in Equation 8.18 vanishes. The condition can be checked graphically with a grid of the possible open-loop systems that the controller variation can generate, as shown in Figure 8.3a. The maximum angle $\phi_{12}(\omega)^\circ$ must be contained in a 90° arc from (−1,0). In the Nichols plot, the way to apply the criterion is to draw the grid of a possible $1 + L_i(j\omega)$ systems, and check that the maximum horizontal distance is less than 90°. Figure 8.3b illustrates the criterion on the Nichols plot.

Using similar arguments, it is also easy to deal with uncertainty. For an uncertain system, the *template* T $P(j\omega)$ is the area of the possible plants within the uncertainty at the frequency ω_i. If the system is governed by a switching controller, each point of the template can change its position due to the switching. From this point of view, switching can be considered as a mechanism that modifies the position and the shape of the templates of $[1 + L_i(j\omega)]$. To be sure that the switching is stable, the aforementioned criterion must be applied to the whole template.

It has been traditionally considered in robust control theory that uncertainty changes the plant slowly in comparison with the system dynamics. If the switching laws depend on the state of the system, then the switching is much faster than changes due to uncertainty. Consequently, for each point of the departure template there is only one corresponding point in the arrival template. Furthermore, it is assumed that uncertainty does not affect the angle α.

Then the Nichols chart is a very clear diagram to test the stability of the uncertain switching system. Figure 8.4 shows the templates of $[1 + L_i(j\omega)]$ and the application of the method. If, during the displacement of each point of the first template to its corresponding point of the second one the maximum horizontal distance between any two points of this path is less that 90°, the stability condition is satisfied at that particular frequency. Although it is a laborious task, usually it is not necessary to check each point of the template at each frequency, because the whole set of templates is much closer together than the critical distance.

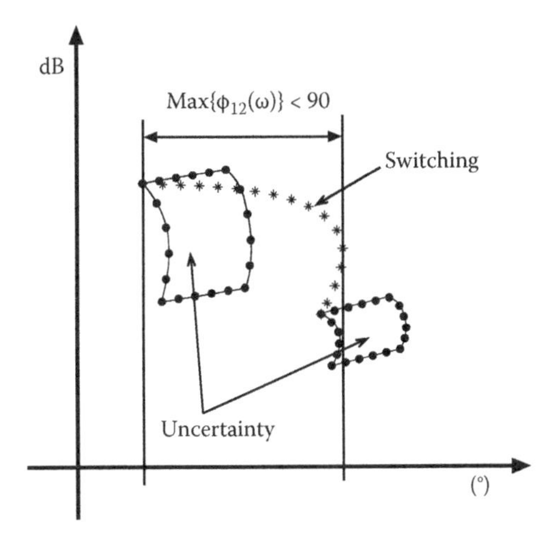

FIGURE 8.4
Templates of $[1 + L_i(j\omega)]$ and the stability criterion on the Nichols chart.

8.3 Methodology

QFT[6] has been demonstrated to be an excellent tool to deal with several, often conflicting, control specifications. As it is shown in the previous chapters, it is a transparent design technique that allows the designer to consider all the specifications simultaneously and in the same plot. The QFT philosophy permits to design controllers that satisfy all the required performance specifications for every plant within the uncertainty. It quantifies and minimizes the amount of feedback (controller magnitude) needed at each frequency of interest.

To design very high performance systems, going beyond the classical performance limitations imposed by any linear system, the hybrid methodology is used. It combines the QFT robust compensator/controller design method introduced in the previous chapters and the stable switching technique presented in the previous section.

In particular, the error amplitude is used as the switching signal, so that, when the output is far from the reference (large error), the system needs to be more stable, and also faster, but precision is not so necessary. Conversely, when there is a small error, some amount of stability margin can be sacrificed in order to increase the low-frequency gain and therefore to increase the precision and the disturbance rejection. These ideas, combined with the QFT method, lead to the design procedures, Method 1 and 2, which are presented as follows

Method 1: $L_i(s)$, $i = 1, 2, \ldots n$, *with gain and zeros variation, constant poles*
For switching among many controllers (two or more) and for systems with uncertainty

Step 1: Obtain a preliminary linear controller for the system by applying the QFT robust control methodology, dealing with parametric and/or nonparametric uncertainty (templates), time-domain and frequency-domain specifications (QFT bounds), and applying loop-shaping techniques, as shown in the previous chapters.

Step 2: Use the preliminary QFT controller as the starting point to design two extreme controllers with the same structure, where the gain and the zeros can vary freely but the poles remain unchanged. The characteristics of these two extreme controllers can be related with the error amplitude. As Boris Lurie[215] explains, when the error is large, the bandwidth must increase to obtain a fast response, but the loop gain does not need to be high. Additionally, when the error is small, the bandwidth is reduced to avoid the effects of noise, while the low-frequencies gain is increased to minimize the jitter and the reference tracking error. In terms of loop shaping, these rules are as follows: (1) for small errors, increase the low-frequency gain and bring nearer the zeros and (2) for high errors, decrease the low-frequency gain and move further away the zeros. Apart from this, reasonable stability margins must be maintained, although they can be considerably reduced for the small error situation if necessary.

Step 3: The robustness of the extreme designs guarantees that both linear systems are stable for every plant within the uncertainty. However, it is necessary to apply the criterion presented in the previous section (see Equation 8.18 with $\alpha = 0$) to assure that the switching between both controllers is also stable. One advantage of this graphical criterion is that it gives information about the frequencies where conditions are not satisfied, so that the designer can go back to Step 2 and change the extreme controllers in this region if necessary.

Step 4: The switching function relates the error amplitude with the position of the controller parameters. Select the most appropriate switching function from the time-domain simulation with each controller (see Example 8.3 in the next section or Chapter 14 for more information).

Method 2: $L_i(s)$, $i = 1, 2$, *with variation in gain, poles, and zeros*

For switching among two controllers and for systems without uncertainty

The stability is checked graphically, Equations 8.18 and 8.19, in both the complex plane and the Nichols diagram. In the first case, the criterion is expressed by requiring $L_1(j\omega)$ and $L_2(j\omega)$ to be inside of an arc of $[90° - \alpha(\omega)°]$ degrees around the point (–1,0) at each frequency (Figure 8.3a); in the second case, plotting $[1 + L_1(j\omega)]$ and $[1 + L_2(j\omega)]$, the distance $\phi_{12}(\omega)$ on the horizontal axis at each frequency should be less than $[90° - \alpha(\omega)°]$ degrees (Figure 8.3b).

8.4 Examples

To validate the theory introduced in the previous sections, three illustrative switching examples are presented here: two based on Shorten et al.[294] and one based on a benchmark problem presented at the American Control Conference in 1992 (the classical two-mass-spring system).[300,301]

Example 8.1

There does not exist a *CQLF*.

According to the methodology presented by Shorten et al.,[294] the stability under switching between the two systems given by the matrices

$$A_1 = \begin{bmatrix} 0 & 1 & 0 \\ 0 & 0 & 1 \\ -1 & -2 & -3 \end{bmatrix}, \quad A_2 = \begin{bmatrix} 0 & 1 & 0 \\ 0 & 0 & 1 \\ -2 & -3 & -1 \end{bmatrix} \tag{8.20}$$

cannot be assured because A_1A_2 has two real negative eigenvalues.

This result, based on the A_1A_2 eigenvalues,[294] can be also found by applying the methodology introduced in Sections 8.2 and 8.3.

It is assumed that A_1 and A_2 represent two closed-loop systems. What the designer should do depends on the open-loop transfer functions $L_i(s)$ that defined the closed-loop systems A_1 and A_2.

Two possible cases to define $L_1(s)$ and $L_2(s)$ are presented here. The first one with gain and poles variation and constant zeros, and the second one with gain and zeros variation and constant poles.

Case 1

$L_i(s)$ with gain and poles variation, constant zeros: Method 2

The systems A_1 and A_2 could come, for instance, from a switching between the following two $L_i(j\omega)$ transfer functions:

$$L_1(s) = \frac{(s+1)}{s(s+2.618)(s+0.382)} \tag{8.21}$$

$$L_2(s) = \frac{2(s+1)}{s(s+0.5 \pm 0.8666j)} \tag{8.22}$$

The MATLAB® code used in this case is as follows:

```
A1=[0 1 0;0 0 1;-1 -2 -3], B1=[-1 0 0]', C1=[0 1 1], D1=[0]
[numT1,denT1] = ss2tf(A1,B1,C1,D1);
numL1=numT1; denL1=denT1-numT1; L1=tf(numL1,denL1);

A2=[0 1 0;0 0 1;-2 -3 -1], B2=[-1 0 0]', C2=[0 1 1], D2=[0]
[numT2,denT2] = ss2tf(A2,B2,C2,D2);
numL2=numT2; denL2=denT2-numT2; L2=tf(numL2,denL2);

w=[0.01:0.01:10]; % frequency vector

FR_1L1 = freqresp( (1+L1),w);
An_FR_1L1 = angle(FR_1L1(1,:));

FR_1L2 = freqresp( (1+L2),w);
An_FR_1L2 = angle(FR_1L2(1,:));

Phi12 = abs((An_FR_1L2 - An_FR_1L1))*180/pi; % Phi12

FR_alpha = freqresp(tf(denL2,denL1),w);
alpha = abs(angle(FR_alpha(1,:)))*180/pi; % alpha

figure, plot(w,Phi12+alpha,w,90+w*0);
xlabel ('w (rad/s)'); ylabel('Phi12+alpha & 90 degrees');
```

The switching produces a gain increase and a change of two real poles by a pair of complex conjugate ones (Method 2, Section 8.3). Figure 8.5 shows the results of Equation 8.18 at every frequency. The condition described in Equation 8.18 fails in the range of frequencies $1 \le \omega \le 1.67$ rad/s.

A simple solution can be to reduce the second controller gain. It has been checked that multiplying the second transfer function $L_2(s)$ by 0.59 the stability of the switching system is assured (see also Figure 8.5).

Case 2

$L_i(s)$ with gain and zeros variation, constant poles: Method 1

On the contrary, if the systems A_1 and A_2 can come from a switching between the following two $L_i(j\omega)$ transfer functions,

$$L_1(s) = \frac{3s(s+0.6667)}{(s+1)(s-0.5 \pm 0.8660j)} \tag{8.23}$$

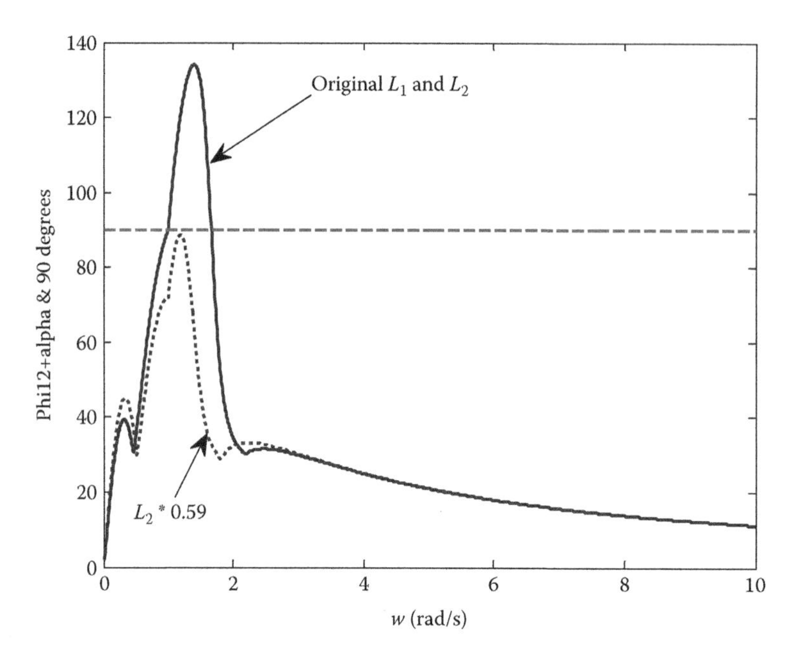

FIGURE 8.5
Example 8.1, Case 1. Equation 8.18, original L_1 and L_2, and with 0.59 L_2.

$$L_2(s) = \frac{(s+2.618)(s+0.382)}{(s+1)(s-0.5\pm 0.8660j)} \tag{8.24}$$

then the changes are only the zeros and the gain, so $\alpha(\omega)$ is null for all frequencies (Method 1, Section 8.3). In this case, as is shown in Figure 8.6, the condition of Equation 8.18 fails in the range of frequencies $0.9 \le \omega \le 1.52$ rad/s.

The MATLAB code used in this case is as follows:

```
A1=[-3 -2 -1; 1 0 0; 0 1 0], B1=[1 0 0]', C1=[3 2 0], D1=[0]
[numT1,denT1] = ss2tf(A1,B1,C1,D1);
numL1=numT1; denL1=denT1-numT1; L1=tf(numL1,denL1);

A2=[-1 -3 -2; 1 0 0; 0 1 0], B2=[1 0 0]', C2=[1 3 1], D2=[0]
[numT2,denT2] = ss2tf(A2,B2,C2,D2);
numL2=numT2; denL2=denT2-numT2; L2=tf(numL2,denL2);

w=[0.01:0.01:10]; % frequency vector

FR_1L1 = freqresp((1+L1),w);
An_FR_1L1 = angle(FR_1L1(1,:));

FR_1L2 = freqresp((1+L2),w);
An_FR_1L2 = angle(FR_1L2(1,:));

Phi12 = abs((An_FR_1L2 - An_FR_1L1))*180/pi; % Phi12

figure, plot(w,Phi12,w,90+w*0);
xlabel ('w (rad/s)'); ylabel('Phi12 & 90 degrees,(alpha=0)');
```

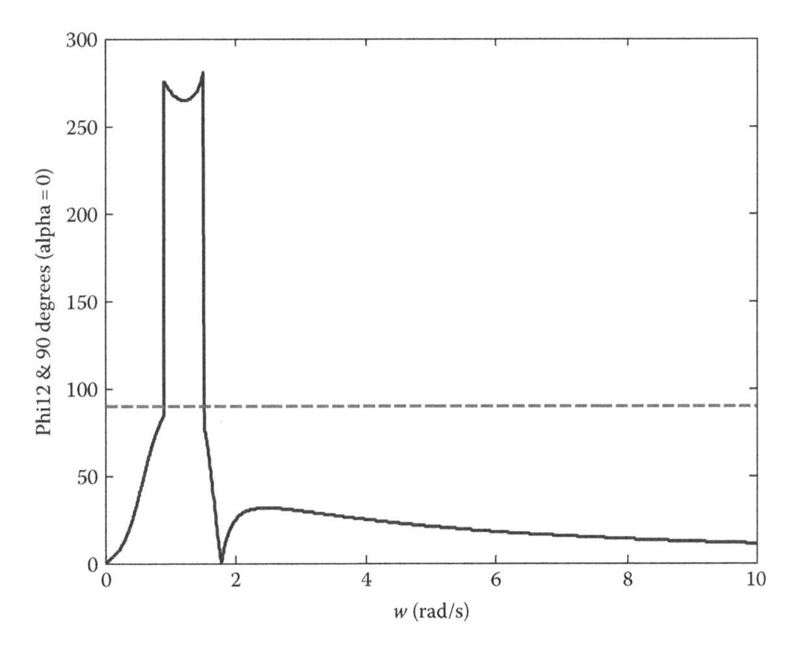

FIGURE 8.6
Example 8.1, Case 2. Equation 8.18.

Example 8.2

There exist a *CQLF*.

Again, according to the methodology presented by Shorten et al.,[294] the stability under switching between the two systems given by the matrices

$$A_1 = \begin{bmatrix} 0 & 1 & 0 \\ 0 & 0 & 1 \\ -1 & -2 & -3 \end{bmatrix}, \quad A_2 = \begin{bmatrix} 0 & 1 & 0 \\ 0 & 0 & 1 \\ -1 & -1 & -3 \end{bmatrix} \tag{8.25}$$

is assured because A_1A_2 has no real negative eigenvalues.

This result, based on the A_1A_2 eigenvalues,[294] can also be found by applying the methodology introduced in Sections 8.2 and 8.3. Again it is assumed that these matrices represent two closed-loop systems and that the switching that produces these matrices is done by the two open-loop transfer functions

$$L_1(s) = \frac{1}{s(s+2)(s+1)} \tag{8.26}$$

$$L_2(s) = \frac{1}{s(s+2.618)(s+0.382)} \tag{8.27}$$

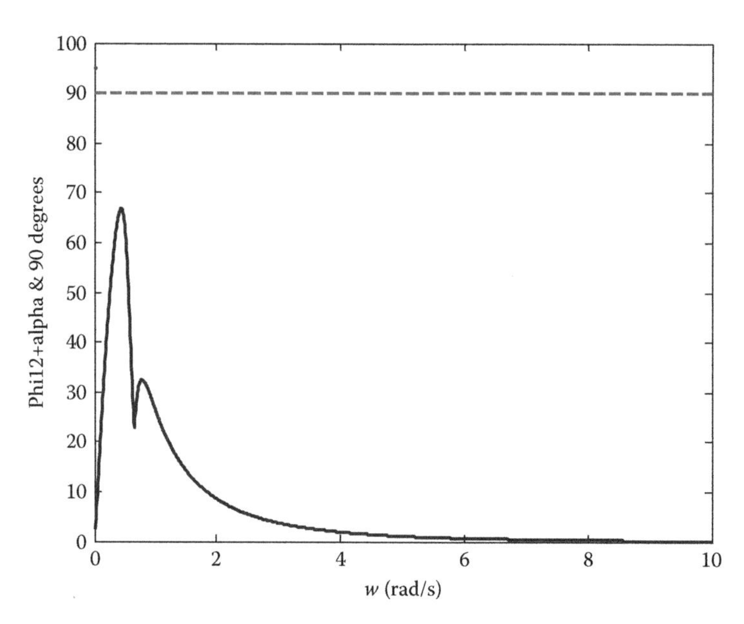

FIGURE 8.7
Example 8.2, Equation 8.18.

Now, as it is shown in Figure 8.7, the condition of Equation 8.18 is fulfilled for every frequency of interest. The MATLAB code used in this case is as follows:

```
A1=[-3 -2 -1;1 0 0;0 1 0], B1=[1 0 0]', C1=[0 0 1], D1=[0]
[numT1,denT1] = ss2tf(A1,B1,C1,D1);
numL1=numT1; denL1=denT1-numT1; L1=tf(numL1,denL1);

A2=[-3 -1 -1;1 0 0;0 1 0], B2=[1 0 0]', C2=[0 0 1], D2=[0]
[numT2,denT2] = ss2tf(A2,B2,C2,D2);
numL2=numT2; denL2=denT2-numT2; L2=tf(numL2,denL2);

w=[0.01:0.01:10]; % frequency vector
FR_1L1 = freqresp((1+L1),w);
An_FR_1L1 = angle(FR_1L1(1,:));
FR_1L2 = freqresp((1+L2),w);
An_FR_1L2 = angle(FR_1L2(1,:));
Phi12 = abs((An_FR_1L2 - An_FR_1L1))*180/pi; % Phi12
FR_alpha = freqresp(tf(denL2,denL1),w);
alpha = abs(angle(FR_alpha(1,:)))*180/pi; % alpha
figure, plot(w,Phi12+alpha,w,90+w*0);
xlabel ('w (rad/s)'); ylabel('Phi12+alpha & 90 degrees');
```

Example 8.3

A new solution for the ACC'92 benchmark problem.

This example presents and improves one of the numerous solutions published for the classical two-mass-spring system (Figure 8.8), also known as the ACC'92 benchmark problem.[301] By applying the switching strategy introduced in this chapter (Section 8.3, Method 1), it is shown how easily the results obtained at the benchmark can be improved.[300]

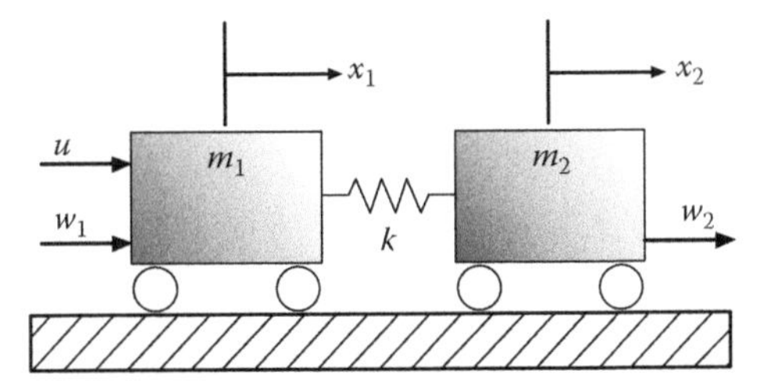

FIGURE 8.8
Two-mass-spring system ($m_1 = m_2 = k = 1$).

The linear controller to be modified is the one presented in Ref. 300:

$$G(s) = \frac{2246.3(s+0.237)(s^2-0.681s+1.132)}{(s+33.19)(s+11.79)(s^2+4.95s+7.563)} \tag{8.28}$$

For the small error regime, one pole is placed nearer the origin, while for the large error regime this is done with a zero. The threshold value in the error is selected based on simulations with the original controller. Doing so, the new controller is

$$G(s) = \begin{cases} \dfrac{2246.3(s+0.237)(s^2-0.681s+1.132)}{(s+33.19)(s+10.5)(s^2+4.95s+7.563)} & \text{for } |e(t)| \le 0.5 \\ \dfrac{2246.3(s+0.13)(s^2-0.681s+1.132)}{(s+33.19)(s+11.79)(s^2+4.95s+7.563)} & \text{for } |e(t)| > 0.5 \end{cases} \tag{8.29}$$

Stability of the design is assured by applying the technique presented in this chapter. Figures 8.9 through 8.11 show the improvement with respect to the design introduced in Ref. 300. The transient response of the system to disturbances w_1 on mass m_1 is shown in Figure 8.9. The transient response of the system to disturbances w_2 on mass m_2 is shown in Figure 8.10. Also, the transient response of the system to a change in the reference signal is shown in Figure 8.11.

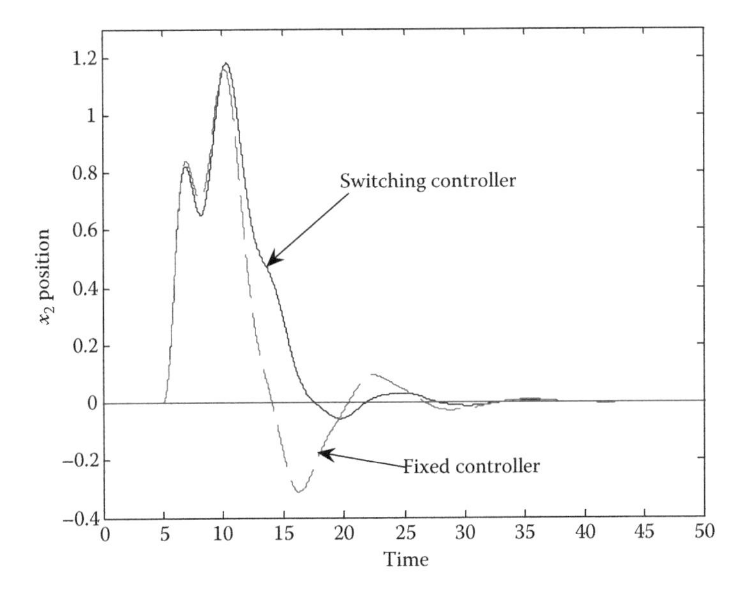

FIGURE 8.9
Response to unit impulse disturbance w_1 on m_1 at $t = 5$ s.

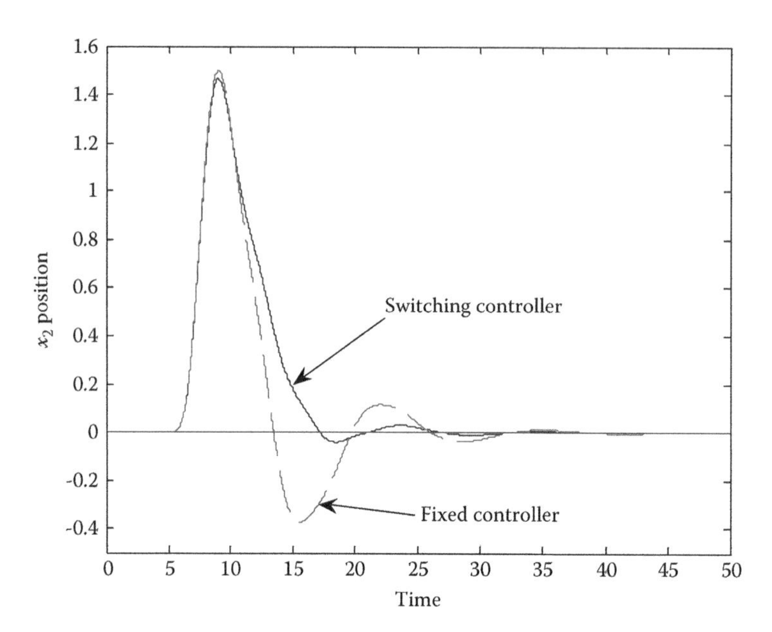

FIGURE 8.10
Response to unit impulse disturbance w_2 on m_2 at $t = 5\,\mathrm{s}$.

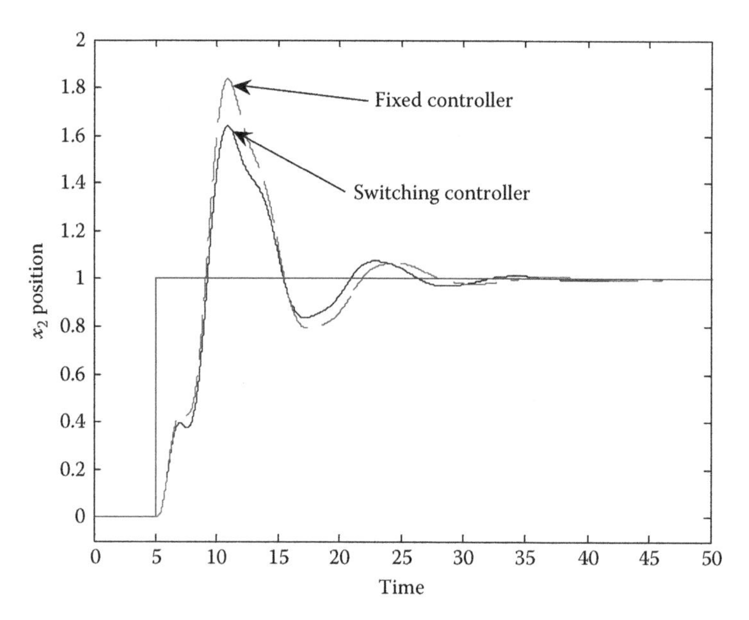

FIGURE 8.11
Response to a unit-step command tracking.

8.5 Summary

A practical methodology to design robust controllers that work under a switching mechanism was presented in this chapter. The method is capable of optimizing performance and stability simultaneously, going beyond the classical linear limitations and giving a solution for the well-known robustness-performance trade-off. Based on the frequency-domain approach, the method combines a new graphical stability criterion for switching linear systems and the robust QFT technique. The new formulation was applied here to three illustrative examples. In addition, the technique is also applied in Chapter 14 to design an advanced pitch control system for a doubly fed induction-generator wind turbine.

Part II

Wind Turbine Control

9

Introduction to Wind Energy Systems

9.1 Introduction[315,318,338,339]

Harvesting energy on a large scale is, without a doubt, one of the main challenges of this century. Future energy sustainability strongly depends on how the renewable energy problem is addressed in the next few decades. Considering only 13% of the Earth's surface and using conventional onshore wind turbines (WTs), with 80 m towers, the estimated wind power that could be commercially viable is 72 TW. That is almost five times the global power consumption in all forms, which currently is about 15 TW on average. With a capacity that has tripled in the last 5 years, wind energy is the fastest growing energy source in the world. WTs are used to collect kinetic energy and to convert it into electricity. The average power output of a WT unit has increased significantly in the last few years. Most major manufacturers have developed large turbines that produce 1.5–3.5 MW of power, reaching even 5–6 MW per turbine in some remarkable cases. Grouped together to create wind farms, the global collective capacity generates 340 TWh of energy, equaling 2% of the global electricity consumption, with 200 GW of wind-powered generators worldwide by the end of 2010.

At the same time, and based on significant governmental subsidies, several countries have achieved relatively high levels of wind power "penetration," which is the fraction of energy produced by wind compared with the total available generation capacity: about 24% in Denmark, 14.5% in Spain, 14.5% in Portugal, 10.1% in Ireland, and 9.3% in Germany.

While wind energy is a clean and renewable source of electrical power, there are many challenges to address. Because WTs are complex systems, with large flexible structures working under very turbulent and unpredictable environmental conditions, and subject to a variable and demanding electrical grid, their efficiency and reliability strongly depend on the applied control strategy. As wind energy penetration in the grid increases, additional challenges are being revealed, such as response to grid voltage dips, active power control and frequency regulation, reactive power control and voltage regulation, restoration of grid services after power outages, wind prediction, etc.

This chapter analyses how the results achieved in the control engineering field are playing a central role in the success of the wind energy development. Section 9.2 introduces a piece of the history of the birth of the modern WTs. Section 9.3 describes current and future market sizes and investments in the wind energy field. This section is followed by Section 9.4, which analyses future challenges and opportunities, as well as new control engineering solutions needed to open virgin global markets.

9.2 Birth of Modern Wind Turbines[302,304,311]

It is commonly accepted that Charles F. Brush designed and erected the world's first automatically operating WT for electricity generation (see Figure 9.1). The turbine was installed in Cleveland, Ohio, in 1887. It operated for 20 years, delivering 12 kW of power to Brush's home on 37th Euclid Avenue. The WT was a giant, the world's largest, with a rotor diameter of 17 m, and 144 blades made of cedar wood. The tower was rectangular in form and about 18 m high and 36,300 kg in weight. The tail was 18 m long and 6 m wide. A 6 m long and 16.5 cm diameter shaft, inside the tower, turned pulleys, belts, and a step-up gearbox (50:1) in order to turn a vertical direct current generator at its required operational speed, which at top performance, was 500 revolutions per minute (rpm). The dynamo was connected to 408 batteries in Brush's basement. The dry cells illuminated 350 incandescent lamps and operated three electric motors and two arc lights. The whole production was 12 kW at its peak. An automatic control system was arranged so that the dynamo's operation went into effective action at 330 rpm, and the dc voltage was kept between 70 and 90 V. Charles F. Brush was one of the founders of the American electrical industry, with 50 significant patents in the electrical arena. Born in Euclid, Ohio, in 1849, he was one of the founders of the Case School of Applied Science and a trustee of Western Reserve University. Brush died in Cleveland on June 15, 1929.

Another very remarkable project in the early days of wind energy development was the 1.25 MW WT developed by Palmer Putnam in the United States, in 1939–1945 (see Figure 9.2). It was a giant WT of 53 m (175 ft) diameter and had two blades with a hydraulic pitch control system. The Smith–Putnam turbine was developed for the S. Morgan Smith Co., a manufacturer of hydroelectric turbines. Although it was very successful at the beginning, after a blade failure, it was dismantled in 1945. The project became a technological major breakthrough, showing the way for ulterior downwind designs of large WTs.

The years following 1970 are typically considered as the birth of the modern wind-driven electrical generators. At that time, the average power output of a WT unit was about 50 kW, with blades that are 8 m long. Since then, the size of the machines has increased dramatically.

FIGURE 9.1
Charles F. Brush' WT. 1887, Cleveland, Ohio. The world's first *automatically operating* WT for the *generation of electricity.*

FIGURE 9.2
Smith–Putnam's WT project, 1939–1945, United States. Power: 1.25 MW.

Nowadays, the typical values of power output of the modern turbines deployed around the world are about 1.5–3.0 MW, with blades that are more than 40 m long for onshore and 60 m for offshore applications. Simultaneously, the cost per kW has decreased significantly, and the efficiency, reliability, and availability of the machines have definitely improved.

9.3 Market Sizes and Investments[315,318,338,339]

As it is addressed in the introduction, the potential onshore wind power at 80 m of the Earth surface, covering only 13% of the land, is five times greater (72 TW) than the current all forms of world energy consumption (15 TW). On top of that, one can also add the tremendous potential of the newly available offshore locations, both at shallow and deep waters, and the new promising high-altitude airborne WT projects, where the wind energy density is much higher (e.g., at 300 m the wind energy density can be about three times more than near the surface). All in all, the wind energy available with the onshore, offshore, and airborne technologies is far larger than the practical energy needs worldwide. Providing a good technology, the limit of the market size and its exploitation will be set just by economic, environmental, and political factors.

With many thousands of WTs operating, the total worldwide installed capacity is currently about 200 GW, and according to the World Wind Energy Association, it is expected that a net growth rate of more than 21% per year will be achieved. The top five countries, United States, Germany, Spain, China, and India, currently share about 73% of the world's capacity.

A very effective mechanism adopted by many countries to activate the wind energy activities is the renewable portfolio standard (RPS). It is a regulation that requires increasing the production of energy from renewable sources. In particular, the RPS generally places an obligation on utilities to produce a specified fraction of their electricity from renewable (mainly wind) energy. The amount of global RPS share that will be achieved by the European Union is 20% by 2020 and by the United States is 20% by 2030, with different targets and years depending on the State (e.g., Arizona's 15% by 2025 or Colorado's 20% by 2020).

The cost of utility-scale WTs has dropped by more than 80% over the last 20 years, reaching values of about 2.2 and 3.5–5.5 million dollars per MW for onshore and offshore applications, respectively, in 2011.

According to the Department of Energy (DOE) of the United States, there is still room to decrease the capital cost of onshore applications by about 10% over the next two decades. In addition, several countries have adopted special programs to subsidize and promote the development of wind energy. Among them, the most successful ones are the feed-in-tariff (FiT) programs and the production tax credit (PCT) programs.

The FiT programs have been adopted by more than 60 countries all over the world, including some of the top-producing countries like Germany, Spain, Canada, Denmark, etc., and a number of states within the United States. The program typically includes (1) a guaranteed grid access to the wind farm, (2) long-term contracts to sell the electricity produced by the WTs, and (3) good purchase prices that eventually will tend toward the grid parity.

The PCT program has been adopted in the United States. It is a federal incentive that provides a credit of some cents per kWh (currently 2.1 cents). Since its establishment in 1992, the PCT has had an "on-again/off-again" status, which has contributed to boom–bust cycles of the wind energy industry in the United States.

9.4 Future Challenges and Opportunities[392]

The enormous and unique worldwide possibilities for large-scale wind energy development in the next few decades strongly depend on how some critical technology challenges (TCs) are addressed. New ideas and control engineering solutions are needed to open virgin global markets. Among others that need to be emphasized are the three applications (App) and the four TCs shown in Table 9.1.

9.4.1 Offshore Wind Turbine Applications (Application 1)[318,410–432]

Offshore wind power is a very promising technology with an enormous energy potential. The leading suppliers of offshore WTs are currently developing machines within a power

TABLE 9.1

Wind Energy Challenges

App. 1	Offshore WT applications
App. 2	Extreme weather conditions
App. 3	Airborne WTs
TC 1	Cost reduction for a zero-incentive situation
TC 2	Efficiency maximization
TC 3	Mechanical load attenuation
TC 4	Large-scale grid integration and penetration

range of 2.5–5 MW and 90–120 m rotor diameter. With less logistic constraints than the onshore applications, over the next few years the offshore turbines will reach a typical size of 5–8 MW, and a rotor diameter of more than 150 m, adopting blade-tip speeds slightly higher than those of onshore turbines, which currently have speeds of 75–80 m/s. The offshore tower height is generally lower than the onshore ones because wind shear offshore profiles are less steep. The offshore foundation system depends on the water depth. Most of the projects installed so far have been in water less than 22 m deep with a demonstration project in Scotland at a depth of 45 m. Shallow-water technology currently uses monopiles for a depth of about 20 m. Very deep water applications, with floating platforms, still need reliable solutions, including advanced control systems to deal with wind, ocean waves, tides, and water currents simultaneously.

Apart from solutions for floating platforms, research opportunities for offshore applications include the following: (1) new ideas to reduce the cost; (2) remote, intelligent, turbine condition monitoring, and self-diagnostic systems; (3) dedicated deployment vessels; (4) analytical models to characterize wind, ocean currents, tides, and ocean waves; (5) high-reliable systems; (6) predictive maintenance techniques; (7) smart blade systems; and (8) identifying the root causes of component failures, understanding the frequency and cost of each event, and appropriately implementing design improvements, etc. See Chapter 18 for more information.

9.4.2 Extreme Weather Conditions (Application 2)[315]

Extreme cold and humid weather conditions can force the WTs to stop working during winter months, due to ice formation on the blades in quantities that would degrade the turbine performance and cause blade imbalance. By integrating ice protection systems in the blades and managing them from an appropriate control system, the WTs can produce a greater amount of power during winter. This solution opens the wind energy to new larger geographical and still virgin areas, including north latitudes and many offshore locations such as the fresh water Great Lakes in the United States and Canada. The ice protection system will also serve to minimize wear and tear on the turbine itself thereby reducing maintenance costs and turbine downtime.

9.4.3 Airborne Wind Turbines (Application 3)[350,499–517]

An airborne WT system is a WT that is supported in the air without a tower. Two technologies have been proposed: (1) Ground generator systems and (2) aloft generator systems. In both cases the WTs have the advantage of an almost constant and high-speed wind and a low-cost structure without the expense of tower construction. Advanced multivariable robust control strategies, for attitude and position control of the flying structure, and reliable control algorithms to govern the system under bad weather conditions, such as lightning or thunderstorms, are critical strategies that need to be developed. No commercial airborne WTs are in regular operation yet. See Chapter 19 for more information.

9.4.4 Cost Reduction for Zero Incentive (TC 1)[315,318,518–521]

Although the cost of utility-scale wind farms has dropped by more than 80% over the last 20 years, most of the wind energy systems, including all the offshore applications, still need a significant government support to be feasible. This subsidy cannot be sustained globally in the long term by any country, because it eventually would affect the national economy producing a significant energy cost increase. As a consequence, long-term economic sustainability of wind energy imperatively requires improving the wind energy business model so that the cost is similar to conventional power generation.

Wind projects are calculated in terms of their initial *installed capital cost* (*ICC*), as well as their life-cycle cost, known as the *levelized cost of energy* (*LCOE*) or simply life-cycle cost of energy (*CoE*).

The *ICC* includes the planning, equipment purchase, construction and installation costs for a turn-key wind system, ready to operate. It is usually referred as ICC/P_r in million dollars per MW of power (P_r is the rated power of the WT). The *ICC* is typically broken down into two main aspects: the *turbine capital cost* (*TCC*) and the *balance of plant* (*BoP*) construction contract ($ICC = TCC + BoP$). The *BoP* contract covers the civil and electrical works. It includes turbine foundations, access roads, crane pads and laydown areas, turbine transformers (although these are sometimes included in the turbine supply contract), electrical lines, substation and meteorological mast.

Table 9.2 presents the WT cost breakdown of two existing machines in the market: an onshore 1.5 MW unit and an offshore shallow-water near-shoreline 3.0 MW unit. The numbers in both columns are in %; the total cost of the machine being 100% in 20 years. Note that in 2011 baseline cost, 1% in the first column represents 40,640 dollars (1.5 MW onshore WT) and 1% in the second column represents 180,542 dollars (3.0 MW offshore WT), all in 2011 U.S. dollars. Thus, by multiplying the first column by 40,640 dollars and the second one by 180,542 dollars the percentage numbers are translated into 2011 U.S. dollars.

According to Equation 9.1, the *LCOE* includes the addition of the *ICC*, cost of capital (financing), cost of land lease, operation and maintenance costs (*O&M*), subsystems replacement costs (*LRC*), and insurance, warrantees, administration, and decommissioning costs, all divided by the amount of energy to be generated. *LCOE* is usually calculated as a unit of currency per unit of energy, typically cents of dollar per kWh or dollar per MWh:

$$LCOE = \frac{(DRF\ ICC) + (LRC + O\&M + Fees)}{AEP_{net}} \tag{9.1}$$

where

LCOE is the levelized cost of energy (\$/kWh) (constant \$)
DRF is the discount rate factor (1/year)
ICC is the installed capital cost (\$)
LRC is the levelized replacement/overhaul cost per year (\$)
O&M is the levelized operation and maintenance cost per year (\$)
Fees is the land lease, insurance, warrantees, administration costs per year (\$)
AOE is the annual operating expenses = (*LRC* + *O&M* + *Fees*) (\$)
AEP_{net} is the net annual energy production = $8760\ P_r\ [CF\ (1 - EL)\ Av]$ (kWh/year)
P_r is the rated power of the WT (kW)
Av is the availability: time the machine is available for generation (per unit)
EL represents the losses from the machine to the grid wind farm connection. (1 − *EL*) is the associated efficiency (per unit)
CF is the capacity factor, see Section 13.5.1

The AEP_{net} is calculated based on the projected energy output of the turbine based on a given annual average wind speed. The *DFR* is the annual amount per dollar of initial capital cost needed to cover the capital cost. It includes financing fees, return on debt and

TABLE 9.2

WT Cost Breakdown: Onshore/Offshore (in %)

		Onshore	Offshore Shallow, Near
Power (Pr)		1.5 MW	3.0 MW
Rotor diameter (r_b)		70 m	90 m
Hub height (h)		65 m	80 m
		%	%
A.1 Rotor		**13.4**	**6.5**
	Blades (3)	8.6	4.4
	Hub	2.4	0.9
	Pitch mechanism and bearings	2.4	1.2
A.2 Drive-train, nacelle		**34.9**	**19.5**
	Low-speed shaft	1.2	0.8
	Bearings	0.7	0.4
	Gearbox	8.6	5.6
	Mechanical brake	0.2	0.1
	Generator	5.5	2.9
	Power electronics	6.7	3.6
	Yaw drive and bearings	1.1	0.6
	Main frame	5.2	2.3
	Electrical connections	3.4	2.1
	Hydraulic, cooling systems	1.0	0.6
	Nacelle cover	1.2	0.5
A.3 Control, safety system, condition monitoring		**2.0**	**0.8**
A.4 Tower		**8.3**	**5.7**
A.	*TCC*	**58.5**	**32.6**
B.1 Foundation		**2.6**	**15.3**
B.2 Transportation		**2.8**	**3.8**
B.3 Roads, civil work		**4.5**	**0.0**
B.4 Assembly and installation		**2.1**	**6.1**
B.5 Electrical lines and connections		**6.9**	**12.7**
B.6 Engineering and permits		**1.8**	**2.5**
B.7 Decommissioning, Scour protect.		**2.0**	**5.2**
B.	*BoP*	**22.7**	**45.6**
C.1 Levelized replacement cost (20 years)	($LRC \times 20$ years, $CF = 0.3$)	**6.1**	**5.1**
C.2 Regular maintenance cost (20 years)	($O\&M \times 20$ years, $CF = 0.3$)	**8.1**	**14.1**
C.3 Fees (20 years)	($Fees \times 20$ years)	**4.5**	**2.6**
C.	*Operating expenses* (20 years)	**18.7**	**21.8**
$A + B + C = TCC + BoP + Operation\ expenses\ (\%) =$		**100.0**	**100.0**
$CF\ (1 - EL)\ Av$ (considering capacity factor, losses and availability) =		0.3	0.3
AEP_{net} = net annual energy production = $8670\ P_r\ [CF\ (1 - EL)\ Av]$ (kWh/year)		3942000	7884000
$LCOE = (DRF\ ICC + LCR + O\&M + Fees)/AEP_{net}$ ($/KWh)		**0.1018**	**0.2219**
2011 baseline. Onshore: 1% = $40640, Offshore: 1% = $180542 aprox. (U.S. $)		DFR =	0.11
Onshore, aprox.: $ICC/P_r = (TCC + BoP)/P_r = (40{,}640 * 81.3)/1.5 =$ **2.2 M$/MW**		Tax reduc. C.1, C.2 of 40%	
Offshore, aprox.: $ICC/P_r = (TCC + BoP)/P_r = (180{,}542 * 78.2)/3 =$ **4.7 M$/MW**		=0.6 cost	

Sources: European Wind Energy Association, *Wind Energy—The Facts*, Earthscan, London, U.K., 2009; Fingersh, L. et al., Wind turbine design cost and scaling model, Technical Report, NREL/TP-500-40566, 2006.

equity, depreciation, income tax, property tax, and insurance. *DFR* is calculated according to Equation 9.2:

$$DFR = \frac{d}{1-\left(1/(1+d)^N\right)} \frac{(1-T\ PVDEP)}{1-T} \tag{9.2}$$

where

d is the discount rate per period (the rate of return that could be earned on an investment in the financial markets with similar risk)

N is the number of periods

T is the marginal income tax rate (the amount of tax paid on an additional dollar of income; as income rises, so does the tax rate)

PVDEP is the present value of depreciation

The power purchase agreement, *PPA*, is the contract between two parties, one who generates electricity for the purpose of sale (the seller) and one who purchases that electricity (the buyer). The *PPA* is measured in cents of dollar per kWh of energy and depends on the *LCOE*.

Although it is difficult to give precise numbers, due to global market and regulatory uncertainties, it can be estimated that in 2011 the ICC/P_r for onshore projects is about 2.2 million dollars per MW of installed power and for offshore projects between 3.5 and 5.5 million dollars per MW (depending on water depth and distance to shoreline). At the same time, the *LCOE* for onshore is about 10 cents of dollar per kWh of energy produced and for offshore projects about 22 cents of dollar per kWh. The objective of the U.S. DOE is to reduce the cost of offshore to a value between 7 and 9 cents of dollar per kWh by 2030, with a medium-term objective of 13 cents of dollar per kWh by 2020.[318,341]

The WT control system plays a central role in the business model, affecting directly the costs and the efficiency (see Table 9.2 and Equation 9.1). In particular, it affects the *control, safety system, and condition monitoring* costs (A.3) and also the costs of *rotor* (A.1), *drive-train* (A.2), *tower* (A.4), and *foundation* (B.1), as well as the costs of *operation and maintenance* (C.2) and *replacement* (C.1). In addition, the control system can improve the energy efficiency, reliability, and availability of the WT, which finally improves AEP_{net} as well.

An advanced control system can (1) significantly reduce the mechanical fatigue of the WT, lowering the required weight of major components, and thus the costs; (2) improve the efficiency, reliability, and availability of the machines, and thus AEP_{net}; and (3) improve the *O&M* by implementing remote, intelligent, turbine condition monitoring and self-diagnostic systems.

Reducing the cost and improving the energy production requires a *concurrent engineering* way of thinking and working, with the leadership of a multidisciplinary engineer, which is required for the control engineer, to be able to achieve an optimum design, taking into account all the aspects of the project (see also Chapters 1, 15, and 18).

9.4.5 Efficiency Maximization (TC 2)[392,486,495]

Efficiency maximization means to generate more energy over the low-medium operating wind spectrum (Regions 1 and 2 shown in Figure 9.3). Achieving efficiency maximization requires research opportunities in the areas of (1) high-efficiency airfoils and rotor configurations; (2) smart blade systems; (3) optimum and robust extremum-seeking control

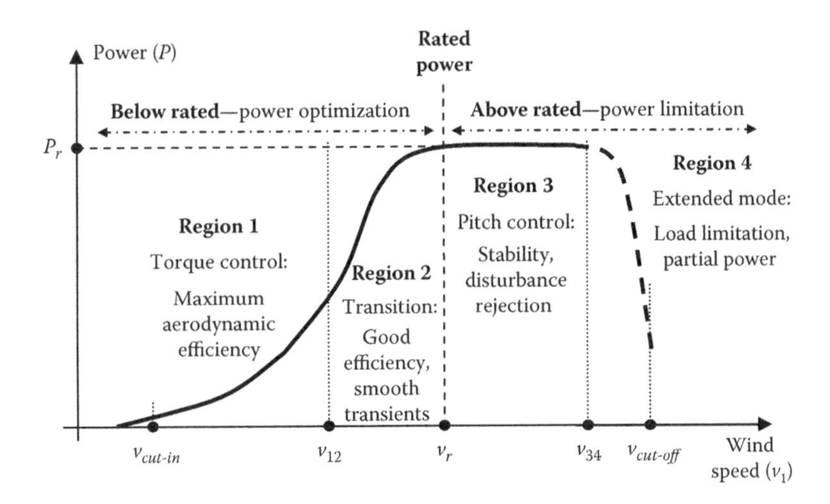

FIGURE 9.3
Power curve of a WT.

strategies to achieve maximum aerodynamic efficiency; (4) variable-diameter rotors, which could significantly increase the efficiency of the turbine by presenting a large area to capture more energy in low winds and have a reduced area to protect the system in high winds; and (5) turbines with taller towers to capture more energy in regions with high wind shear.

In all cases, advanced control strategies to damp out the tower motion by using blade pitch and generator torque control are critical. See Chapters 11 and 17 for more information.

9.4.6 Mechanical Load Attenuation (TC 3)[392,486,495]

Attenuating loads on the WT rotor offers great reduction to the total cost of WTs. This fact is particularly important when dealing with offshore WTs, which demand larger blades, and as a consequence deal with bigger aerodynamic and mechanical loads. Every load reduction on the WT rotor can produce a multiple chain-effect benefit in the machine, simplifying the blade structure, drive-train, tower design, foundation, and the required deployment vessel for the offshore installation, as well as increasing the life span and system reliability, and reducing the maintenance costs.

The aerodynamic and structural loads applied to a WT during operation are aeroelastically coupled and randomly variable. This depends on the following several factors: (1) the horizontal and vertical wind shear, (2) the wind turbulence and rotational variation, (3) the tower shadow, (4) the yaw and tilt misalignments, (5) the gravity forces, (6) the wakes of other turbines, etc.

Design loads on WTs are typically divided into extreme loads and fatigue loads. Fatigue loads are a key factor for the WT design. Reducing fatigue loads can result in a significant reduction in cost, increasing life span, simplifying required materials and maintenance costs, and improving system reliability.

Smart blade systems are capable of considerably reducing extreme and fatigue loads on WTs by means of two main methods: passive and active control.

Passive load control counteracts changes in wind speed through passively adapting aeroelastic response of the rotor blades. Some advanced solutions like tension–torsion coupling, bend–twist coupling, and sweep–twist coupling are still under investigation.

In general, all of them are inherently open loop, being effective only over limited operating conditions.

With active control, the blade is adapted by adjusting its aerodynamic properties (angle of attack or lift coefficient) based on specific real-time measurements. These techniques are inherently closed-loop control systems. They utilize distributed sensors, which measure the current state of the rotor along with dynamic models, and real-time control strategies in microprocessors, to alter the blades' response through innovative distributed actuators that change the aerodynamic surfaces of the blades. As a consequence, active control techniques offer significantly more flexibility, especially when dealing with unsteady changes in a flow state. Individual pitch control, with a pitch angle adjustment per blade instead of a collective one, is one of the most advanced smart blade active control techniques applied nowadays. See Chapter 17 for more information.

9.4.7 Large-Scale Grid Penetration (TC 4)[308,312,315,397–409]

On a large-scale wind energy scenario, the wind farms will have to support the grid by (1) providing fault-ride-through capability, (2) providing primary frequency control, (3) providing voltage regulation, (4) providing reactive power control, (5) avoiding power flickers, (6) providing low harmonics content, and (7) carrying a share of power control capability for the grid. In that manner, the WTs, and the power electronics and control systems in particular, will be designed so that they help the grid operator by carrying out some of the traditional duties of conventional power plants. This requires operating below the maximum power most of the times in order to offer some energy capture for grid ancillary services.

At the same time, wind energy "penetration," which refers to the fraction of energy produced by wind compared with the total available generation capacity, will increase to very significant levels in the near future. There is no generally accepted "maximum" level of wind penetration nowadays. The limit for a particular grid will depend on the existing generating plants, WT technology, WT control systems, grid demand management, pricing mechanisms, capacity and type of storage, etc. High levels of wind penetration have been achieved in some European countries, with more than 33% in Spain and Denmark on very windy days. At the same time, some advanced studies have shown that the limit could be much higher than that, depending on the grid configuration and the existing backup generators. See Chapters 13 and 15 for more information.

TABLE 9.3

References: Main Topics Part II

Area	References
Books related to wind energy	302–318
Books related to engineering	319–336
Associations, agencies, labs, and industry related to wind energy	337–350
Papers related to WTs: modeling, design, and control	351–392
Papers related to electrical modeling of WTs	393–396
Papers related to grid integration and energy storage of WTs	397–409
Papers related to off-shore wind energy: modeling and control	410–432
Papers related to smart WT blades	433–498
Papers related to airborne WTs	499–517
Papers related to wind energy economics	518–521

9.5 Summary

The success of the future large-scale wind energy development strongly depends on some new critical engineering solutions, including advanced control systems, which will play a central role in many cases. After a description of the past successful wind energy applications and an analysis of the market size and the economic investments required for the near future, this chapter addressed the current challenges and research opportunities in the field, including the following three applications (App.) and four technology challenges (TC): App. 1, offshore wind energy applications; App. 2, extreme cold weather areas; App. 3, airborne WTs; TC 1, cost reduction for a zero-incentive situation; TC 2, efficiency maximization; TC 3, mechanical load attenuation; and TC 4, large-scale grid integration and penetration. Table 9.3 shows the references for the main topics of Part II of this book.

10

Standards and Certification for Wind Turbines

10.1 Introduction

This chapter presents an introduction of technical standards and certification processes for wind turbines (WTs). Section 10.2 defines the technical standard formal concept and describes its strategic value. Section 10.3 presents the structure and development procedures for new standards. Section 10.4 introduces the fundamentals of the certification of WTs, as well as the impact of the controller design in the process. Some key definitions are given in Section 10.5.

10.2 Standards: Definition and Strategic Value

A *technical standard* is a formal document that establishes uniform engineering or technical criteria, methods, processes, and practices about technical systems. Standards have existed for thousands of years (e.g., roads, water networks, and buildings in the Roman Empire). Standards are among the most important building blocks for all national economies and international trade. They define necessary measurements to improve product performance, quality, safety, uniformity, functionality, etc., as well as procedures to lower product costs. It is estimated that more than 0.5 million standards exist in the world today to support every technical field and the global marketplace. In 1999, the Organization for Economic Cooperation and Development estimated the value of technical standards at more than 4 trillion U.S. dollars, affecting directly more than 80% of the world trade market (www.thinkstandards.net/benefits.html).

The development process of an international technical standard is a multidisciplinary procedure to find a balance of interests among many disciplines, such as engineering, technology, science, economics, trade, public policy, law, etc. Analysis and preparation are necessary to ensure the successful development of an international standard. That typically includes (a) reviewing existing consensus standards; (b) reconciliation of such standards; (c) reviewing existing policies, laws, and regulations; (d) reviewing health, safety, environment, trade, and competition issues; (e) establishing a realistic development schedule; (f) providing sufficient resources necessary to ensure the success of the project, etc.

Broadly speaking, an international standards organization develops international technical criteria, methods, processes, and practices. There are many international standards organizations. The International Electrotechnical Commission (IEC)[337] is the world's leading organization that prepares and publishes international standards for all electrical, electronic, and related technologies (for more details see http://www.iec.ch/index.htm).

10.3 Standards: Structure and Development

Although the structure of a new standard may vary according to circumstances, the typical contents include the following:

1. *Scope.* It specifies products or issues the standard will apply to.
2. *Normative references.* These refer to other existing standards.
3. *Definitions.* These describe all words or terms unique to the standard.
4. *Symbols and units.* These are the universal language and measurement.
5. *Test procedures.* These describe how a technology is evaluated.
6. *Acceptance criteria.* These provide the acceptable level of design and performance of the product.
7. *Effective date.* It is the date that the consensus standard takes effect.

Reliable, fair, open, transparent, and well-defined development procedures are essential for the creation of new international consensus standards. The procedures are typically published for public review with an opportunity for comment and usually include the following stages:

1. *Preliminary*: planning of future work, usually driven by market needs.
2. *Proposal*: proposals are typically recommended by a national committee.
3. *Preparatory*: preparation of a working draft.
4. *Committee*: submission of the working draft to all national committees.
5. *Enquiry*: all national committees vote.
6. *Approval*: revised version sent to standards central office.
7. *Publication*: after final approval.

IEC standards development procedures may be reviewed at http://www.iec.ch/ourwork/stages-e.htm

10.4 Certification of Wind Turbines

According to the European Standard EN-45020, *certification* is "the confirmation of compliance of a product or a service with defined requirements" (e.g., codes, guidelines, and standards). In the wind energy field, the scope of certification consists of the examination of structural integrity, safety, reliability, strength, fatigue, and compliance of the complete WT and/or its components, such as rotor blades, gearbox, tower, foundation, etc., with the standards (e.g., IEC,[337] GL,[347] etc.).

TABLE 10.1

IEC Normative Documents for WTs

IEC 61400-1	*Wind Turbine Safety and Design*
IEC 61400-2	*Small Wind Turbine Safety*
IEC 61400-3	*Design Requirements for Offshore Wind Turbines*
IEC 61400-11	*Noise Measurement*
IEC 61400-12	*Power Performance*
IEC 61400-13	*Mechanical Load Measurements*
IEC 61400-21	*Power Quality*
IEC 61400-23	*Blade Structural Testing*
IEC 61400-24	*Lightning Protection*
IEC 61400-25	*Communication Standards for Control and Monitoring*
IEC 60050-415	*International Electro-technical Vocabulary—Part 415: Wind Turbine Generator Systems*
IEC WT-01	*IEC System for Conformity Testing and Certification of Wind Turbines, Rules and Procedures, 2001-04*

Source: International Electrotechnical Commission, http://www.iec.ch/

The rapid expansion of the wind energy industry and the increasing size of turbines themselves are enforcing financing banks, insurance companies, manufacturers, grid operators, and authorities to require reliability and safety evaluations of the wind energy projects. These assessments are performed following different certification processes for individual turbines or wind farms, for either onshore or offshore applications.

International standards for WTs have been developed by the Technical Committee 88 (TC-88) of the IEC.[337] Since 1988, TC-88 has been working in standards and technical specifications for the safety of WTs, rotor blade tests, power curve, noise, load measurements, and power quality under the scope of the IEC 61400 series with a significant number of working groups, project teams, and maintenance teams. The list of normative documents compiled by IEC in the area of wind energy so far is summarized in Table 10.1 (see also www.iec.ch).

The American National Standards Institute (ANSI) has designated the American Wind Energy Association (AWEA)[338] as the lead organization for wind energy standards in the United States. AWEA is represented on the IEC TC-88 subcommittee (see www.awea.org/standards).

In Europe, there exist additional normative documents for WTs, such as the EN standards, the German DIN standards, and the popular Germanischer Lloyd (GL) WindEnergie GmbH normative[347] (Table 10.2).

The certification of a new WT is usually obtained by the designer or the manufacturer. It is an important step to demonstrate that the new system meets the international standards

TABLE 10.2

GL and DIN EN Normative Documents for WTs

GL WindEnergie GmbH	*Guideline for the Certification of Wind Turbines, Edition 2010*
GL WindEnergie GmbH	*Guideline for the Certification of Offshore Wind Turbines, Edition 2005, Reprint 2007*
DIN EN 50308	*Labour Safety*
DIN EN 50373	*Electromagnetic Compatibility*
DIN EN 50376	*Declaration of Sound Power Level and Tonality Values of Wind Turbines*

Source: Germanischer Lloyd Certification GmbH, http://www.gl-group.com/

and to prove that it is ready for investment and commercialization. Although the certification is different from the manufacturer's warranty, it is the base to prove that the machine has been tested by an official accredited entity and has passed the key requirements for structural integrity, mechanical loads and fatigue, acoustic emissions, safety, control, power performance, power quality, and other characteristics. According to the GL and IEC standards, the certification of a new WT requires the completion of the requirements summarized in Figure 10.1. This includes the following three main topics:

a. *Design evaluation*, which includes three steps: (a1) *C-design assessment*, (a2) *B-design assessment*, and (a3) *A-design assessment*.
b. *Type certificate*, which includes three steps: (b1) *design evaluation*—point (a), (b2) *prototype testing*, and (b3) *manufacturing evaluation*.
c. *Project certificate*, which includes (c1) *type certificate*—point (b), (c2) *site-specific assessment*, (c3) *surveillance during production*, (c4) *surveillance during transport and erection*, (c5) *surveillance during commissioning*, and requires (c6) *periodic monitoring*.

The control system design, control optimization, and control analysis play a central role in the certification process of the WT, as we can see in Figure 10.1, in the following stages: *design evaluation, prototype testing, site-specific assessment, surveillance during production*, and *periodic monitoring*. A more detailed explanation of each stage is presented in the following.

10.4.1 Design Evaluation

Design evaluation is the first step in the certification process (see Figure 10.1). It is based on design documents and is divided in two main parts:

1. *Safety aspects, control design, and load calculations of the complete system*

   The load calculations are made for the entire WT, taking into account the interaction among all the subsystems (blades, nacelle, tower, foundation, control system, electrical system, mechanical structures, aerodynamics surfaces, pitch system, yaw system, power electronics, generator, etc.; see Figure 10.2). The analysis is usually based on aeroelastic codes, stochastic wind fields, modal or finite element analysis techniques, electrical models, and the specific control system methodology. In the case of offshore WTs, the fluid–structure interaction of the submerged part is also included. There are several CAD packages available for this analysis such as *Bladed*, developed by Garrad Hasssan,[348] and *FAST*, developed by NREL.[342]

   Figure 10.2 explains graphically this first step of the certification. As shown in the figure, it is necessary to develop the dynamic models of each main component of the WT and the dynamic interaction among all of them, and then to design and integrate the control system (pitch control, yaw control, torque control, etc.) in the WT simulator (which obviously depends on the WT design itself).

   In addition, and following the standards, it is critical to identify and prepare the set of worst-case scenarios (wind speed variation, changes in wind direction, faults in the grid, potential faults in the WT, a combination of them according to reasonable understanding of realistic simultaneousness, etc.) where the WT is to be tested with the simulator (see the first column of Figure 10.2).

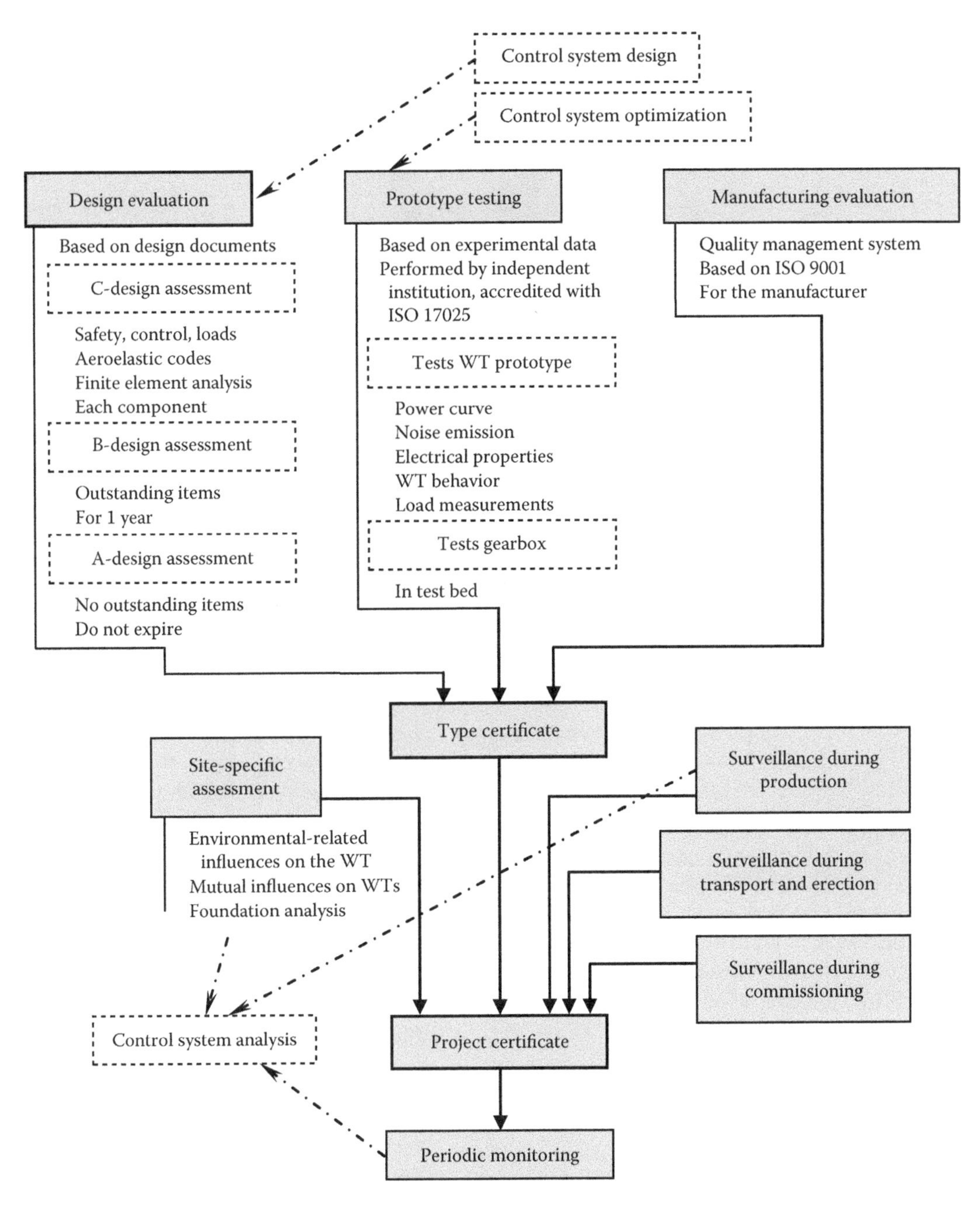

FIGURE 10.1
Certification process: interaction with control design/analysis.

The output of the simulation is the mechanical loads supported by each component, at every point of the structure, including the blades, nacelle, tower, foundation, etc. (see the third column of Figure 10.2), and the evaluation of that design, which might require a new redesign and iteration.

Note that this procedure is kind of "chicken and egg" problem. To design the components of the WT (structure, materials, strength, reinforcement, etc.)

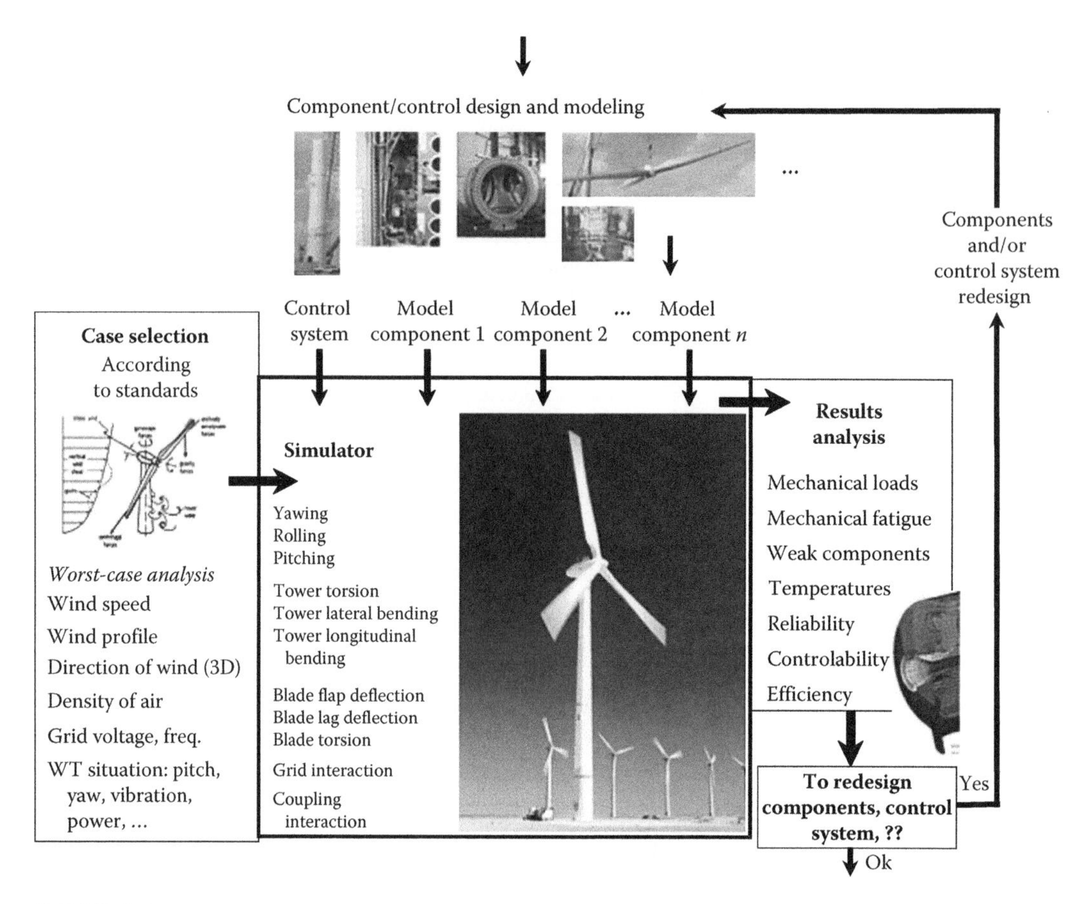

FIGURE 10.2
Design evaluation, *C-design assessment* (see Figure 10.1).

a previous control system, which defines the final dynamic behavior, that is, the mechanical fatigue, maximum mechanical loads, etc., is necessary. At the same time, to design the control system, the components of the WTs are required to be known, because they define the inertia, stiffness, damping, and all the dynamics of the plant to be controlled.

Usually, this difficult "chicken and egg" problem is solved by several iterations among all the engineers that work in the design (see also Figure 15.1 for this *concurrent engineering* concept).

Standards used at this point are, among others, IEC 61400-1, IEC 61400-13, IEC 61400-23, IEC 61400-24, and GL WindEnergie GmbH (*Guideline for the Certification of Wind Turbines, Edition 2010*) (see Tables 10.1 and 10.2).

2. *Analysis of every single component*

It is based on the previously approved loads and the existing standards and guidelines (blades, nacelle, tower, foundation, control system, electrical system, mechanical structures, aerodynamics surfaces, pitch system, yaw system, power electronics, generator, etc.) (See the third column of Figure 10.2.)

Rotor blade testing, lightning protection, and testing of the electrical equipment are also part of this first step of the certification process. The design evaluation ends with the fundamental manuals and procedures for manufacturing, transport, erection, start-up, commissioning, operation and maintenance, as well as personnel safety.

The design evaluation process is composed of three blocks: C-design assessment, B-design assessment, and A-design assessment (Figure 10.1).

The C-design assessment checks the design documentation for the WT prototype, reviewing the calculations, components, and control system introduced previously. The output of this study is the *Statement of Compliance for the C-Design Assessment*, which is valid for the installation of the prototype and the operation of a maximum of 2 years or 4000 equivalent hours at full power.

The B-design assessment checks some additional items that are outstanding, if they are not directly safety relevant. It is valid for a 1 year period. The A-design assessment checks some additional items that are not outstanding. It does not expire unless the design is modified.

10.4.2 Prototype Testing

This is the second step in the certification process (see Figure 10.1). It is based on experimental data. It must be performed by an independent institution, accredited according to the International Organization for Standardization (ISO) 17025. In the WT prototype, it requires the experimental measurement of (a) the power curve, (b) the noise emission, (c) the electrical properties, (d) the turbine behavior, (e) the mechanical loads, etc., and (f) the experimental testing of the gearbox in an independent test bench.

10.4.3 Manufacturing Evaluation

This is the third step in the certification process (see Figure 10.1). It consists of the evaluation of the quality management system of the manufacturer. It is based on the standards ISO 9001 and is required to confirm the quality of the product.

10.4.4 Type Certificate

The *type certificate* (see Figure 10.1) requires the completion of the *design evaluation* (Section 10.4.1), the *prototype testing* (Section 10.4.2), and the *manufacturing evaluation* (Section 10.4.3). As an option, the *type certificate* can also include a *foundation design evaluation* and a *type characteristic measurement*.

Standards used at this point are, among others, IEC 61400-12 for the power curve measurement, IEC 61400-11 for the noise emissions, IEC 61400-13 for mechanical load measurements, IEC 61400-21 for electrical characteristics, IEC 61400-23 for blade tests, and the GL WindEnergie GmbH (*Guideline for the Certification of Wind Turbines, Edition 2010*) for safety and function tests (see Tables 10.1 and 10.2).

10.4.5 Project Certificate

This is the last step in the certification process (see Figure 10.1). It requires the completion of the *type certificate* (10.4.4), and the *site-specific assessment*, the *surveillance during production*, the *surveillance during transport and erection*, and the *surveillance during commissioning*. It also requires *periodic monitoring*.

The *project certificate* is intended for projects covering more than one single WT, such as wind farms. It is valid for a period of up to 5 years. During that period annual reports are to be sent to the certification agency, describing any deviation in operating experience and minor modifications. In case of major modifications, a recertification is required.

The *project certificate* does not expire as long as *periodic monitoring* is carried out. Major modifications or repairs affect the validity of the certificate, and require a new certification process.

The *site-specific assessment* requires the study of environmental-related influences on the WT (hot/cold climate, ice formation, earthquakes, corrosion, electrical network conditions, soil conditions, etc.), the mutual influences among the WTs in the wind farm, and an analysis of the foundation. This study demonstrates the integrity of the design. In some cases, it might require some modifications of the WT design for some specific sites, with the corresponding recertification from the very beginning, etc.

The *surveillance during production*, the *surveillance during transport and erection*, and the *surveillance during commissioning* are also subject to approved procedures and operation and safety issues, which must also be inspected and tested.

The *periodic monitoring* is related to safety, maintenance, and operation inspection, and must be carried out by a third party every 2–4 years.

10.5 General Concepts

10.5.1 Turbulence Intensity and Wind Classes

The IEC and GL standards define the longitudinal *turbulence intensity* (I_{long}) of the wind speed as[304,337,347]

$$I_{long} = \frac{\sigma_{v,long}}{\mu_{v,long}} \tag{10.1}$$

where

$\sigma_{v,long}$ is the annual standard deviation of the longitudinal wind speed at hub height

$\mu_{v,long}$ is the annual average of the 10 minutes mean of the longitudinal wind speed at hub height

and

$$\sigma_{v,long} = \frac{I_{15}(15 + \lambda\mu_{v,long})}{(\lambda + 1)} \tag{10.2}$$

where I_{15} is the measured turbulence intensity for a longitudinal wind speed of 15 m/s. For $I_{15} = 0.18$ or more (higher turbulence sites) the standards take $\lambda = 2$, and for $I_{15} = 0.16$ or less (lower turbulence sites) the standards take $\lambda = 3$ (see Table 10.3).

For lateral (I_{lat}) and vertical (I_{ver}) turbulence intensities, the standards allow to take either $I_{long} = I_{lat} = I_{ver}$ (isotropic model) or $I_{lat} = 0.8\, I_{long}$, $I_{ver} = 0.5\, I_{long}$.

TABLE 10.3

Basic Parameters for Wind Classes (IEC and GL)

		WT Classes				
		I	II	III	IV	S
v_{ref} (m/s)		50	42.5	37.5	30	Values to be specified by the manufacturer
v_{ave} (m/s)		10	8.5	7.5	6	
a	I_{15}	0.18	0.18	0.18	0.18	
	λ	2	2	2	2	
b	I_{15}	0.16	0.16	0.16	0.16	
	λ	3	3	3	3	

Source: International Electrotechnical Commission, http://www.iec.ch/; Germanischer Lloyd Certification GmbH, http://www.gl-group.com/

According to IEC and GL, the classification of the wind speed classes is calculated by measuring the extreme wind speed, the average wind speed, and the turbulence intensity (see Table 10.3). The manufacturers design different machines according to that classification (see Chapter 15, Tables 15.1 through 15.4 as an example).

In Table 10.3, v_{ref} is the reference wind speed (10 min mean of the extreme wind speed with a recurrence period of 50 years at the hub height), v_{ave} is the annual average wind speed at hub height, a is the category for higher turbulence sites, b is the category for lower turbulence sites, λ is the slope parameter, and I_{15} is the characteristic value of the turbulence intensity at 15 m/s in Equation 10.2.

10.5.2 Wind Speed Distribution

The wind speed distribution at the site is important information to determine the frequency of occurrence of loads in the WT. The cumulative probability function of the wind speed at the hub height can be calculated as a Weibull or as a Rayleigh distribution so that

$$P_{Weibull}(v) = 1 - e^{-(v/C)^k} \tag{10.3}$$

or

$$P_{Rayleigh}(v) = 1 - e^{-\pi[v/(2v_{ave})]^2} \tag{10.4}$$

where

v is the wind speed at the hub height
k is the shape parameter of the Weibull distribution ($k = 2$ in the standard WT class)
C is the scale parameter (m/s) of the Weibull function

C and k are derived from real data. The Rayleigh distribution is identical to the Weibull distribution if $k = 2$ and C and v_{ave} satisfy the equation $v_{ave} = C$ sqrt(π)/2.

The distribution functions specify the cumulative probability when the wind speed is less than v. On the other hand, the difference $[P(v_1) - P(v_2)]$ indicates the time that the wind speed v is between v_1 and v_2 (see Figure 10.3 as an example).

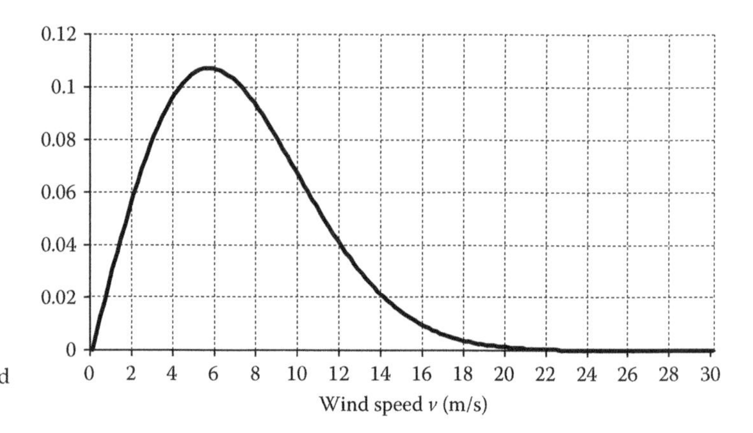

FIGURE 10.3
Weibull distribution for wind speed v, with $k = 2$, $C = 8$.

10.5.3 Wind Speed Profile

According to the IEC and GL standards, the wind speed profile with height, or wind shear, can be calculated as[337,347]

$$v_z(z) = v_h\left(\frac{z}{h}\right)^{\alpha} \tag{10.5}$$

where

z and h are heights above the ground (m)
$v_z(z)$ is the unknown wind speed (m/s) at z meters above the ground
$v_h(z)$ is the wind speed (m/s) measured at h meters above the ground
α is the shear exponent (typically $\alpha = 0.11$ or 0.16)

10.5.4 Frequency Analysis: Campbell Diagram

A specific frequency-domain analysis required by the international standards (IEC, GL, etc.)[337,347] is the Campbell diagram. In particular, the *design evaluation* step in the certification process requires a Campbell diagram containing the natural frequencies (modes) of the rotor blades, drive-train, and tower, including the relevant excitations, such as the rotor speed 1P, 3P, 6P, etc. (excitation of the WT through the rotor speed multiplied by the factor 1, 3, 6 [due to the blades' rotation in three-blade rotors]; 1P = per 1 revolution, 3P = per 3 revolutions, etc.).

A Campbell diagram is basically the representation of the resonance frequencies (modes, Ω_j) of the system as functions of rotor speed (ω). Figure 10.4 shows an example of a three-blade rotor WT. The diagram allows the designer to determine if some resonant frequencies can be excited during the normal operation of the WT. Thus, for this example the required changes can be determined in order to avoid exciting resonant frequencies.

Typically, some resonance frequencies (modes) of the WTs increase quite strongly with rotor speed (ω) because the stiffness of some components (like the blades) strongly increases with rotor speed.

The Campbell diagram does not show the amplitude of the resonant frequencies but only the potential problems (critical speed locations ω_n) in the intersections of the lines in the diagram. In the authors' experience, and as it is also pointed out by Adams,[334] it is

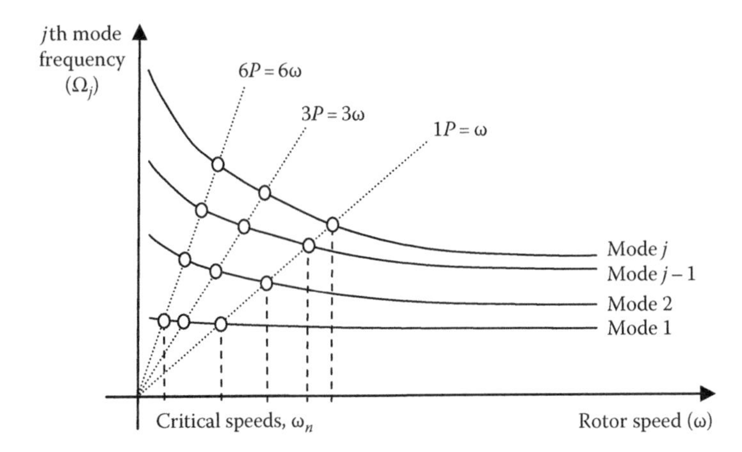

FIGURE 10.4
Typical Campbell diagram for a three-blade rotor turbine.

a good practice to combine the Campbell diagram with the fast Fourier transform (FFT) analysis, to see the amplitudes (X_n) at the critical speeds (ω_n) at the same time.

The information derived from the Campbell diagram is very important to design a reliable control system, which must avoid exciting the WT at the resonant frequencies. This engineering analysis can be easily introduced within the frequency-domain QFT methodology.

10.6 Summary

This chapter briefly introduced the international technical standards and the certification procedures for WTs, including formal concepts, strategic value, fundamentals and procedures, as well as the impact of the controller design in the certification process and final product.

11

Wind Turbine Control Objectives and Strategies

11.1 Introduction

This chapter introduces the main characteristics, objectives, and strategies of the control system of a wind turbine (WT). Section 11.2 describes the principal objectives and specifications to be taken into account in the design of a WT controller. Section 11.3 describes the typical control strategies and methodologies used to fulfill such objectives, and Section 11.4 analyzes the classical control loops and components to regulate the WT.[351–392]

11.2 Control Objectives

The interest in the development of wind energy has widely grown throughout the world in the past two decades. Thus, due to the significant increase in the number of WTs connected to the grid, new technologies aimed to assure power system quality and stability have become not only an important area for research and development but a necessity.[305,309] In particular, the design of advanced control techniques has emerged as a crucial issue for wind energy to penetrate at a large scale into the electrical market. On the other hand, the need for more reliable control systems has become even more crucial with the increasing size of the machines and the related shift in their natural frequencies.[367,369,389] As WTs become larger and more flexible, the design of innovative control strategies with load reduction as an explicit objective turns out to be the most critical.[376,462]

As a first approach, some of the main control objectives to be taken into account in the design of a WT controller are as follows:

- *Reliability and availability*: Advanced control techniques need to be used to assure enough reliability of the WTs under extreme external conditions and internal changes, long-term operation, and high availability.[377] These features are even more relevant for offshore turbines due to the limited accessibility and high maintenance costs.[374]
- *Energy capture maximization*: One of the first requirements for the turbine's control algorithms is to maximize the energy capture over the low-medium operating wind spectrum (see Figure 11.5, Regions 1 and 2). This requires optimizing the conversion efficiency below rated wind speed conditions and limiting the energy production above the rated in order to minimize the mechanical fatigue and extreme loads of the turbine structure.[364]

- *Mechanical load and fatigue attenuation*: As turbines increase in size, alleviating structural loads plays a vital role in the design of controllers.[376,466] Thus, one of the main purposes of the control system is to keep both the maximum mechanical loads on the turbine structure and the mechanical fatigue within the design limits. The latter is usually the most demanding one in the WT design, because WTs are normally designed for a lifetime of 20 years.
- *Provide damping*: The increase in the size of WTs leads to the requirement of bigger blades and lighter and more flexible structures. As a consequence, the tower shifts its structural modes toward lower frequencies and the blade itself provides less damping. The natural flap (out-of-plane) frequency of the blades also decreases as the size of the blades increases. The control system must ensure that the dynamics of the WT are sufficiently damped to avoid resonant modes in the closed-loop system that may be excited by external disturbances.
- *Energy quality and grid stability*: In the past decade, wind power quality demand has become essential. The increase in the number of wind farms in some European countries and some states in the U.S. (see Chapter 9) has brought a significant impact on grid stability and operation.[305,309] Thus, the WT control system must assure the quality of the power supplied, including the ability to (1) react to sudden voltage dips, (2) avoid power flicker, (3) correct the power factor, (4) contribute to the grid frequency stability, (5) regulate the voltage, and (6) provide low harmonics content.

11.3 Control Strategies

Roughly speaking, there are essentially two types of WTs: constant-speed and variable-speed machines. Until the end of the 1990s, the constant-speed concept dominated the market. It represents a significant part of the operating WTs even today, but due to newer grid requirements it has led to the emergence of variable-speed designs.

Alternatively, there are three main strategies for regulating the amount of power captured by the rotor: passive stall control or fixed pitch,[362,382] variable pitch control,[174,379] and active stall control.[329] So far, over the entire range of WT sizes, none of these strategies have taken the lead over the others. However, as machines get larger and the loads become greater, there is a trend toward pitch control and active stall control.

11.3.1 Constant-Speed Wind Turbines

The configuration of a constant-speed WT is based on a gearbox and an asynchronous generator, which is usually a squirrel cage induction generator (SCIG) to reduce costs (e.g., NEG Micon NM72/2000, Bonus 1300). The gearbox speeds up the rotational shaft speed from the rotor to a fixed generator speed (see Figure 11.1). The generator produces electricity through a direct grid connection with a set of capacitors to compensate reactive power. Due to the lack of a frequency converter, the generator speed is dictated by the grid frequency.

One of the main drawbacks of the constant-speed operation is the poor aerodynamic efficiency, particularly at partial load operation (see Figure 11.5, Regions 1 and 2). From the

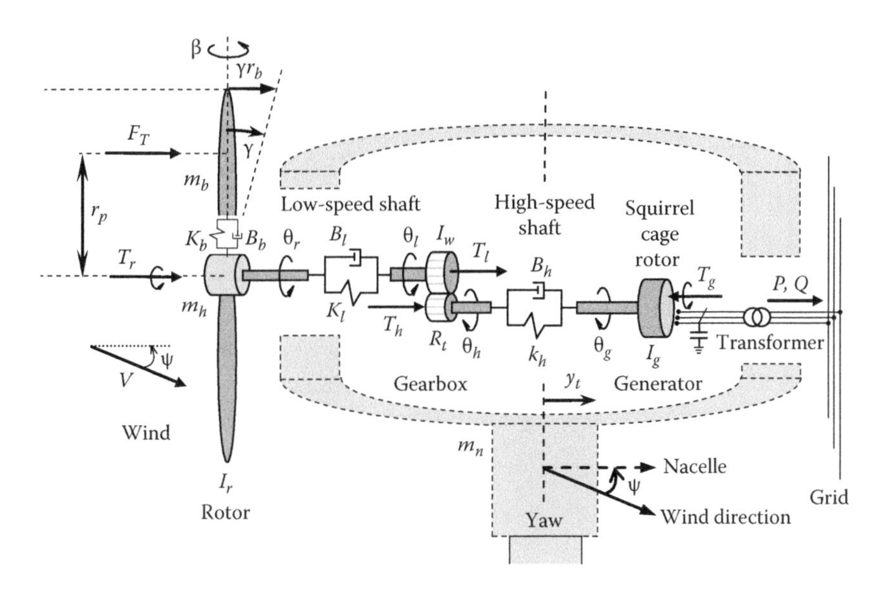

FIGURE 11.1
A classical constant-speed SCIG WT.

electrical system's point of view, another disadvantage is that this type of operation gives only a limited power quality because asynchronous generators demand reactive energy from the grid. This is the reason why they have to be connected through a set of capacitors. On the other hand, wind gusts provoke higher mechanical load variations, and these wind fluctuations also lead to fluctuations of the generated electrical power (flicker).

11.3.2 Variable-Speed Wind Turbines

Many different techniques, such as the following, have been developed to achieve some degree of speed variation[357]: (1) dual-speed generators with pole switching, (2) high-slip asynchronous generators for a low range of variable speed, (3) doubly fed induction generators (DFIG) for a moderate range of variable speed, and finally (4) direct-drive systems for a wide range of variable speed. These techniques reduce noise, increase energy capture at low winds and improve power quality at high winds.

Dual speed means, by changing the number of pole pairs, the rotor spins at two different speeds: using a lower speed at low winds, which improves the performance and reduces the noise emission, and a higher speed (by using a lower number of poles) at high winds (e.g., Vestas V82-1.65, NEG Micon NM900/52).

Another alternative is the so-called slip control method, which adjusts the slip continuously (e.g., Vestas V80-1.8, V90-2.0). In this case, a wound rotor is connected to some variable resistors through slip rings. By changing the electrical resistance of the rotor it is possible to compensate small changes in the rotational speed variation of about 10% above the synchronous speed, without varying the generator output frequency. The main drawback of this design is that the slip rings wear easily due to the movement. Some companies overcome this difficulty by using an optical fiber to avoid physical contact.[365]

Another approach that has a considerable dominance in the current market is the DFIG, also called wound rotor induction generator (see Figure 11.2).[380] In this machine, the stator windings are directly connected to the grid, and a partial scaled frequency converter links the standard wound rotor and the grid (e.g., Gamesa G80-2.0, General Electric GE-1.5/77,

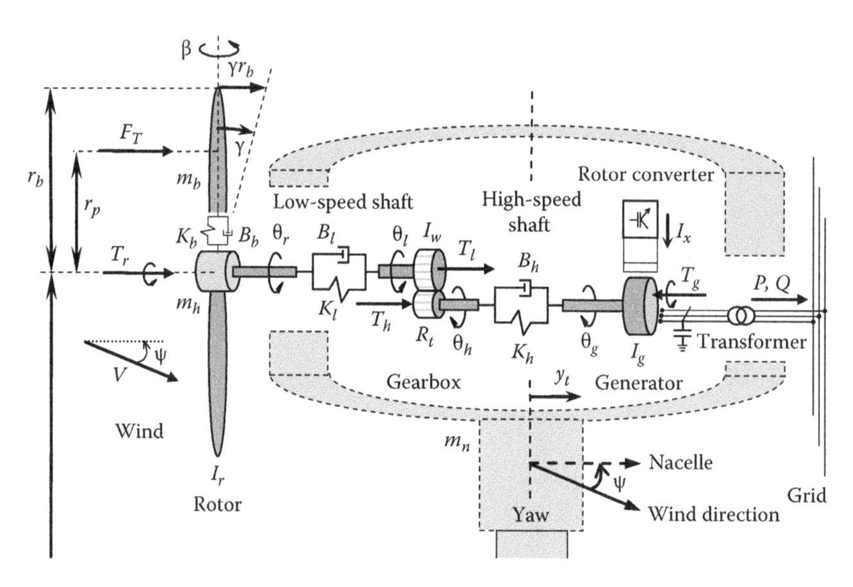

FIGURE 11.2
A classical variable-speed DFIG WT.

Dewind-D70-1500). This configuration allows the machine to control the slip in the generator, and so the rotor speed can vary moderately, achieving a better aerodynamic efficiency. Besides, as the converter controls the rotor voltage in magnitude and phase angle, a partial active and reactive power control is also possible. As power electronics are not rated for full power capacity, simpler and less expensive converters are employed. The main drawbacks of this design are the difficulty to work under voltage dips, the limitation to control effectively the grid variables with such a small converter, and the high maintenance costs of the gearbox.

In some designs a synchronous generator is indirectly connected to the grid through a full power electronic converter, which handles all the energy produced (see Figures 11.3 and 11.4).

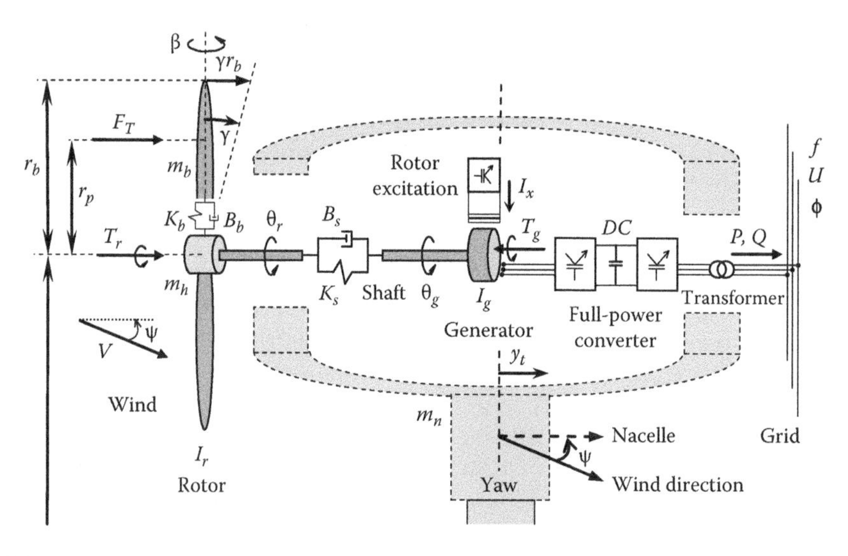

FIGURE 11.3
A direct-drive multipole synchronous generator WT.

FIGURE 11.4
Direct-drive variable-speed pitch-controlled WT. (Courtesy of MTorres, Navarra, Spain.)

This concept, also called direct-drive machine (DD), takes advantage of the wide-range speed operation allowed by the full-scale frequency converter.[174,379,358] The generator can operate at any rotational speed of the rotor, and the rotor operates at the optimal speed for each wind condition. Among its main advantages are the low maintenance costs and high reliability due to the omission of the gearbox, the improved aerodynamic efficiency, and the ability to assist the grid stability and performance. Both wound rotor (e.g., MTorres TWT-1.65/82, Enercon E66-1.8, E112-4500, Lagerwey LW72/1500) and permanent magnet rotor (e.g., Eozen-Vensys 77-1.5) designs are used in this approach.

On the whole, the power electronics used to accomplish variable-speed operation are a little more expensive. However, variable-speed machines are able to reduce mechanical loading and drive-train fatigue and to capture more energy over a wider range of wind speeds, something that increases remarkably the annual production. Another important advantage of some of the variable-speed WTs is the ability to control independently their active and reactive power and thus to assist the power system.

11.3.3 Passive Stall Control

For passive stall-controlled WTs, the generator reaction torque regulates rotor speed below rated operation (see Figure 11.5, Regions 1 and 2) to maximize energy capture. Whereas above rated operation (see Figure 11.5, Region 3), rotor speed is controlled by deliberately inducing stall over a specific wind speed (e.g., Ecotecnia 62-1.3, Made AE61-1.3, Bonus MkIV). In this manner, the power delivered by the rotor is limited at high winds thanks to a particular design of the blades that provokes a loss of efficiency. For such stalling to take place, it is essential to hold the rotor speed constant, and this is usually accomplished with an asynchronous generator connected to the electrical network. Additionally, in this type of control the pitch angle is fixed and the tip brakes are the only part of the blade that can rotate with the aim of spilling off the spare energy to shut down the WT.

11.3.4 Variable Pitch Control

In pitch control, the blades regulate the power delivered by the rotor, either by pitching the blades toward the wind to maximize energy capture or by pitching to feather to discard the excess of power and ensure the mechanical limitations are not exceeded (e.g., MTorres TWT-1.65/77, Nordex S77-1.5). At rated operation (see Figure 11.5, Region 3), the aim is to

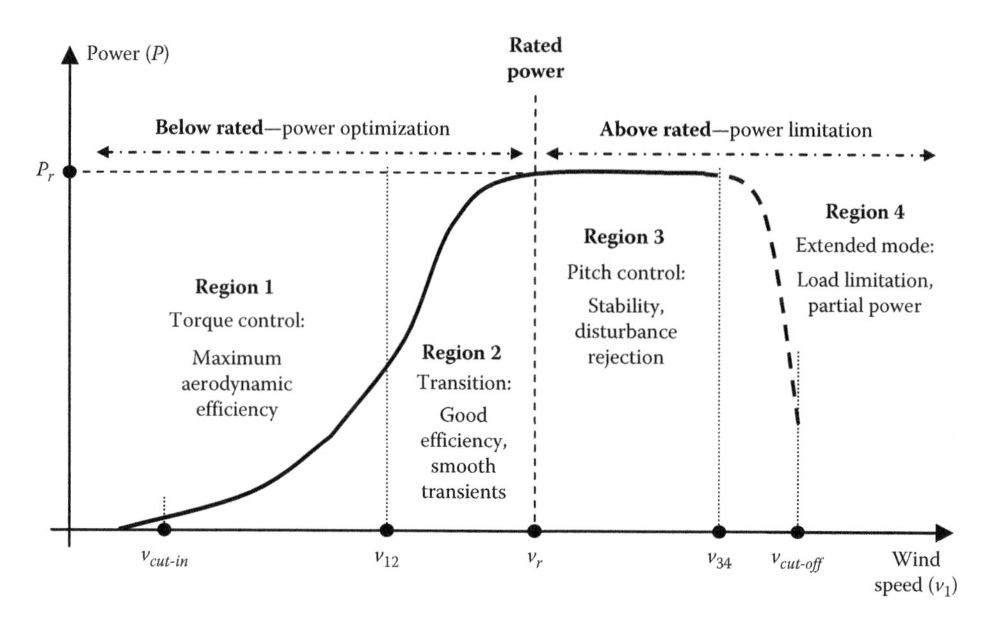

FIGURE 11.5
Power curve of a WT.

maintain power and rotor speed to their rated value. To achieve this, the torque is held constant and the pitch is continually changed following the demands of a closed-loop rotor speed controller that limits energy capture and follows wind-speed variations. In contrast, below rated operation (see Figure 11.5, Regions 1 and 2) there is no pitch control, and the blade is set to a fine pitch position to yield higher values of power capture (see Figure 11.6) while the generator torque itself regulates the rotor speed (see Section 11.4.5.1).

In this turbine the controller turns off the machine simply by moving the blades (pitch angle) toward the feathered position, which is the position perpendicular to the wind direction. Consequently, pitch control requires variable-speed operation, allowing slight accelerations and decelerations of the WT's rotor at wind gusts and lulls.[174,379]

11.3.5 Active Stall Control

This technique is a combination of stall and pitch control. Active-stall was the name used by NEG-Micon, now Vestas (e.g., NM72C/1500, NM82/1650), while a similar concept is

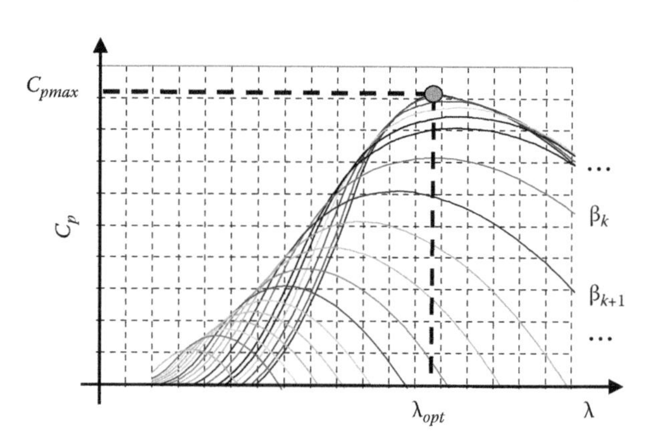

FIGURE 11.6
Aerodynamic power coefficient C_p as a function of λ, β ($\beta_k < \beta_{k+1}$).

named Combi-stall by Bonus–Siemens (e.g., Bonus 1.3MW/62, Siemens SWT-2.3-82). Active-stall/Combi-stall control has the same regulation possibilities as the pitch-regulated turbine, but using the stall properties of the blades. The blades are designed in a similar manner as the stall control blades but the entire blade is turned 90° to adjust its pitch. Thus, the tip of the blade brakes used in passive stall are not required. In this technique, the blades are rotated only by small amounts and less frequently than in the pitch control machines.[382] The idea is to pitch the blade gradually with a negative pitch angle, "pitch to stall," in order to optimize the performance of the blades over the range of wind speeds, especially with low wind speeds. The power control method adjusts the amount of blade that is at stall to compensate the reductions of power output. It allows a more accurate control at high wind speeds while resulting in a more constant power value at rated power.

As a summary, stall-controlled WTs are still attractive because they are relatively simple to construct and operate and therefore much cheaper. Nevertheless, during the last few years the market interest for the fixed-speed operation has decreased slightly, and the trends have pointed to variable-speed WTs. One of the main reasons is that the latter improves the power quality and produces more energy than stall machines.[373] Alternatively, among all the possible configurations, and apart from the fixed speed with passive stall control, the variable speed together with pitch control techniques is one of the most demanded approaches. Recently, some manufactures have also been using active stall control in conjunction with variable-speed operation because it is not as complex and expensive as pitch control. In contrast, other configurations like fixed speed with variable pitch control, although very popular in the past, are nowadays generally rejected because they produce very large transients in the power output when controlling power.

11.4 Control System

The control system of a WT consists of a hardware/software configuration with appropriate sensors, actuators, and microprocessors. It utilizes all the necessary real-time information to regulate the WT (see Section 11.3) and achieve the overall objectives described in Section 11.2, with special emphasis on reliability, performance, and availability.

A generic qualitative power curve for a variable-speed pitch-controlled WT is shown in Figure 11.5. Four regions and two areas are indicated in this figure. The power P of the WT, that is, the actual power supplied to the grid, which is the incoming wind power minus losses, separates the graph into two main areas: below and above rated power P_r. Below rated power (below rated wind speed, $v_1 < v_r$) the WT produces only a fraction of its total design power, and, therefore, an optimization strategy to capture the maximum amount of energy at every wind speed needs to be performed. On the other hand, above rated power (above rated wind speed, $v_r < v_1$), where the wind speed has more power than the rated power P_r, a limitation strategy to generate only the rated power is required.

The four regions of the power curve present the following characteristics:

- *Region 1. Torque control*: Lowest wind, typically between $v_{cut\text{-}in}$ = 3–4 m/s and v_{12} = 7–8 m/s. The objective in this region is to obtain the maximum aerodynamic efficiency. This is usually done by means of manipulating the electrical torque T_g in order to get a particular and pre-fixed ratio (optimum tip speed ratio, λ_{opt}) between wind speed and rotor speed. In this manner, the maximum aerodynamic power coefficient C_P is achieved (see Figure 11.6 and Equation 11.2).

- *Region 2. Transition*: Medium wind, typically between v_{12} = 7–8 m/s and v_r = 11–13 m/s. In this region it is not possible to obtain the desired tip speed ratio because the rotor speed is near to its maximum value. The torque has to be increased until the turbine reaches its rated power. This is usually done by following a high slope torque/rotor-speed reference or by implementing a closed-loop controller.
- *Region 3. Pitch control*: High wind, typically between v_r = 11–13 m/s and v_{34} = 20–25 m/s. The blades have to move the pitch angle β in order to limit the incoming power, control the rotor speed Ω_r, and minimize the mechanical loads at the same time. An important cost reduction and life-operating augmentation can be achieved if load attenuation is targeted as a primary objective when designing the controller or if a blade-independent control strategy to reduce loads is performed.[462]
- *Region 4. Extended mode*: Very high wind, typically between v_{34} = 20–25 m/s and $v_{cut\text{-}off}$ = 25–30 m/s. Although not very common, an extended mode in very high winds can be obtained by means of varying the pitch closed-loop performance. Through a rotor speed Ω_r limitation, the extreme loads can be reduced.

To fulfill the control objectives detailed in the previous sections, a complex hardware/software system needs to be implemented. A certain number of sensors, which provide information to the main control unit or controller, are needed. This controller commands the actuators with appropriate outputs. A complementary and independent safety system is still needed to achieve redundancy and guarantee the security of the WT.[337,347]

11.4.1 Sensors

Some of the main measurements required to control effectively the WT are the following:

- *Meteorological measurements*: anemometer, wind vane, and temperature. Wind speed is usually measured with an anemometer at the nacelle. The accuracy of this measurement is not sufficient enough to be used in the controller. This is so because of the existence of some wind acceleration due to the nacelle shape, some wind turbulences due to the rotor movement, and due to the fact that a single measurement is not representative enough of the mean wind speed that crosses the very large rotor surface. The use of wind speed estimators is sometimes proposed to improve control algorithms.[356]
- *Angular measurements*: rotor speed, yaw angle, pitch speed and angle. Sometimes, an azimuth or rotor position sensor is also used to perform independent blade control.
- *Electrical measurements*: power, voltage, and current of the generator, converters and grid, power factors, total harmonics distortion (*THD*), etc.
- *Mechanical measurements*: torques, temperatures, tower vibration, nacelle acceleration, blade efforts, etc.

11.4.2 Actuators

The following four classical actuators are important features of concern in the operation of the WT:

- *Pitch*: The pitch system is a mechanical, electrical, or hydraulic device, acting on the blade root to turn it in both directions. The pitch system has a crucial role on the turbine, not only because it is the main actuator but also because of the safety of the machine. An incorrect operation of the pitch system causes the turbine to suffer dramatic and irreversible damage. To avoid this situation, in modern and large WTs, an autonomous and independent pitch system has to be implemented in each blade to achieve redundancy and reliability; if some error or malfunction occurs, the other two blades go to the safety position, stopping the machine. An energy storage device is also required to be able to move the blades during a grid loss.
- *Torque*: The generator torque is the other main actuator of a WT. It is manipulated by means of a power drive or electrical converter. Its action response is fast, but smaller than the one of the pitch system. Generator torque is used below rated with the aim of maximizing the aerodynamic efficiency, as explained below.
- *Yaw*: The yaw system is in charge of tracking the wind direction. It usually turns at a constant speed. Thus, a very simple configuration consisting of several motors working together is enough to position the nacelle.
- *Brakes*: The shaft brakes consist of some hydraulic devices that are able to actuate over the high-speed shaft. Modern direct-drive machines usually do not need a shaft brake, and the rotor is stopped only by means of a highly reliable pitch system.

11.4.3 Controller

The controller is usually a real-time control industrial system with a dedicated microprocessor that runs the main control software with the specific control algorithms. The principal functions of the controller are as follows:

- Data acquisition input/output system. Inputs are signals coming from the sensors whereas outputs are signals going to the actuators. All these devices have an electronic interface to sample and translate physical signals to digital variables for the WT controller.
- Signal processing.
- Evaluation of the state of the system, execution of control laws, and supervision.
- Handling data acquisition for the *supervisory control and data acquisition* (SCADA) system (see Section 11.4.7) for storage, maintenance, and remote-operating purposes.

11.4.4 Safety System

The safety system is a redundant hardwired security chain with autonomy to shut down the WT in a secure mode under all circumstances. The safety system is capable of driving back the blades to the feather position, to switch off every electrical device, to provide an emergency digital signal to the controller, and to generate an appropriate alarm for the maintenance staff. Physically, it consists of a long path of contacts in series, beginning by a power supply, which provides a reference voltage, and ending in a safety relay. The opening of any contact causes the loss of that reference voltage. Thus, the safety relay breaks the "OK" hardwired signal of every subsystem (converter, pitch, yaw, brakes, etc.) causing the turbine to stop in a safe manner.

11.4.5 Main Control Loops

The main control loops of the WT are described in the following sections.

11.4.5.1 Torque

The power P generated by the WT and injected to the grid is

$$P = \frac{1}{2}\rho A_r C_p(t) v_1(t)^3 \eta = T_r(t)\Omega_r(t)\eta \tag{11.1}$$

where

T_r is the mechanical torque at the shaft due to the wind
Ω_r is the rotor speed
ρ is the air density
A_r is the rotor effective surface
v_1 is the undisturbed upstream wind speed
η is the mechanical and electrical efficiency
C_p is the aerodynamic power coefficient according to Figure 11.6

and

$$C_p(t) = f\left[\lambda(t), \beta(t)\right] \tag{11.2}$$

where λ is the tip speed ratio, defined as a function of the rotor radius r_b, rotor speed Ω_r, and wind speed v_1, so that

$$\lambda(t) = \frac{r_b \Omega_r(t)}{v_1(t)} \tag{11.3}$$

The electrical torque T_g is manipulated by the control system below the rated power (see Figure 11.5, Regions 1 and 2), to achieve the maximum aerodynamic efficiency C_{pmax}. This strategy aims to keep the relationship between wind speed v_1 and rotor speed Ω_r optimal as long as possible (see λ_{opt} in Figure 11.6 and Equation 11.3). It modifies the rotor speed by changing the electrical torque T_g against the mechanical torque T_r, producing an acceleration torque to follow the wind speed changes and keep $\lambda = \lambda_{opt}$.

From Equation 11.1, the mechanical torque T_r at the shaft of the rotor, due to the wind speed, is

$$T_r(t) = \frac{0.5\rho A_r C_p(t) v_1(t)^3}{\Omega_r(t)} \tag{11.4}$$

Hence, introducing in Equation 11.4, $C_p = C_{pmax}$ and $[r_b\ \Omega_r(t)/v_1(t)] = \lambda = \lambda_{opt}$, the resulting torque for maximum power capture at every wind speed is the demanded electrical torque T_g, as given by the following two equations:

$$T_g(t) = \left(\frac{0.5\rho A_r r_b^3 C_{pmax}}{\lambda_{opt}^3}\right)\Omega_r(t)^2 = K_a\Omega_r(t)^2 \tag{11.5}$$

$$K_a = \frac{0.5\rho A_r r_b^3 C_{pmax}}{\lambda_{opt}^3} \tag{11.6}$$

where C_{pmax} is the maximum power coefficient obtained at λ_{opt} (see Figure 11.6).

Equations 11.5 and 11.6 show a very simple and useful expression to set up the torque at the below rated region. The expression is based on the C_p/λ curves provided by the blade manufacturer (Figure 11.6), which usually only gives a first estimate for steady-state and laminar flow conditions. For a more complete approach some improvement can be done by slightly changing Equations 11.5 and 11.6, taking into account some dynamic conditions. Due to the erosion and dirtiness of the blades and due to the variation of the air density value ρ at different weather conditions, some parameters such as C_p become time variant. A reduction of the value of the constant K_a, or the application of adaptive techniques to estimate it, can optimize the energy capture.

The electrical torque T_g obtained from Equation 11.5 is not usually enough to cover the entire zone below the rated value (Regions 1 and 2), because the rated rotor speed is reached before reaching the maximum torque. Hence, a different strategy needs to be performed to increase the torque until it reaches its maximum (rated) value, usually following a high slope torque/rotor-speed reference or by implementing a closed-loop controller (at Region 2) (see also Refs. 381 and 387).

11.4.5.2 Pitch

The main purpose of the pitch system is to control the rotor speed Ω_r. This can be accomplished by limiting the incoming aerodynamic torque by moving the pitch angle β. There are two possible strategies to accomplish this: pitch to feather and pitch to stall. The methodology to design the pitch control loop is based on the dynamic model of the aerodynamic torque (see Chapter 12 for more details). As indicated earlier, T_r can be expressed as a nonlinear function that depends on the rotor speed Ω_r, the wind speed v_1, and the pitch angle β. A simple linear proportional-integral-derivative (PID) controller[208] is a good starting point, although further considerations need to be observed.[6] As wind speed increases, the $\partial\Omega/\partial\beta$ ratio becomes higher (see Chapter 12). The controller will take this effect into account in order to keep the mechanical fatigue within the design limits and to guarantee the system stability. The design is carried out by using robust and/or adaptive control strategies.[6] Additionally, a good controller should filter the resonance modes of the blades, tower, drive-train, gears, etc., as well as the frequencies related to the rotor revolution (1P), blade crossing in front of the tower (3P), and some multiple (6P). The increasing size of WTs and the use of larger and more flexible structures result in a more demanding design of the pitch control system (see Chapters 14 and 15, Section 15.3).

11.4.5.3 Yaw Angle

The main objective of the yaw controller is to follow the wind direction, maximizing the energy capture, unwinding the cable when necessary, and avoiding any kind of dangerous gyroscopic effects or oscillatory behavior. The potential loss of energy due to yaw errors is a function of the projection of the rotor area A_r on the perpendicular plane of the wind speed; for example, a function of cos(ψ) (see Figures 11.1 through 11.3). As an example, a yaw error of 20° means a reduction of 6% of power. A closed control loop is needed to compensate this effect. A wind vane in the nacelle supplies the reference angle to the controller, which sends a signal to the yaw actuators to move the nacelle toward the wind direction.

11.4.6 External Grid

As the number of wind farms increases, more advanced control techniques are required by the grid operator. Active-power/frequency and reactive-power/voltage loops can be performed to contribute to the stability, quality, and reliability of the overall electrical system. The modern direct-drive gearless turbines with full power converters are in a good position to contribute to such innovative trends (see Chapter 15, Sections 15.4 and 15.5).

11.4.7 Supervisory Control and Data Acquisition

The so-called SCADA *system* is a software tool that is able to start up and shut down each turbine operation, supervise them, and coordinate their usage in the wind farm (see Chapter 15, Figure 15.10). It is usually running in a PC, which is independent of the one with the main real-time turbine control system. It has the following functions:

- Handles warnings and alarms for preventive and corrective maintenances
- Visualizes the status of specific variables of the WT for operating purposes
- Develops reports, including performance, faults, events, production, etc.
- Provides an interface for external communication and remote operation tasks
- Coordinates the turbines of the wind farm

11.5 Summary

This chapter introduced, as a first approach, the main characteristics, objectives, and strategies of the control system of a WT. Section 11.2 described the principal objectives and specifications, which are taken into account in the design of a WT controller, including stability, captured energy maximization, mechanical load reduction, fatigue limitation, power-train damping, reliability optimization, availability aspects, etc. Section 11.3 described the typical control strategies and methodologies used to fulfill such objectives, and Section 11.4 analyzed the classical control loops and components to regulate the WT.

12

Aerodynamics and Mechanical Modeling of Wind Turbines

12.1 Introduction

This chapter introduces the aerodynamics and the mechanical models that describe the dominant characteristics of a wind turbine (WT), as a first step to understand its dynamical behavior and to design the controllers.

Section 12.2 derives the aerodynamic models for the ideal (Section 12.2.1) and the real conditions (Sections 12.2.2 through 12.2.5) of the machine. By applying the Euler–Lagrange method (energy-based approach), Section 12.3 introduces the general mechanical State Space WT models for both the direct-drive synchronous generator (DD) (Section 12.3.2) and the doubly fed induction generator (DFIG) (Section 12.3.3). Finally, Section 12.3.4 develops the transfer matrix representation of a general (DD and DFIG) WT and Section 12.3.5 the rotor speed transfer functions.

12.2 Aerodynamic Models

12.2.1 Maximum Aerodynamic Efficiency: Betz Limit

The main objective of a WT is to transform the kinetic energy contained in the moving air into mechanical/electrical energy. Applying elementary physical laws, and according to the analysis made by Albert Betz between 1922 and 1925, the extraction of mechanical energy from an air stream passing through a given cross-sectional area is restricted to a certain maximum value of the incoming wind. This is the so-called *Betz limit*.[304,306] This section derives the Betz limit of a WT. It is assumed that the WT is working with the following ideal hypothesis:

- Homogeneous, incompressible, steady-state fluid flow
- No frictional drag
- An infinite number of blades
- Uniform thrust over the rotor area
- A nonrotating wake
- The static pressure far upstream and far downstream is equal to the undisturbed ambient static pressure

The kinetic energy E_k of an air mass m moving at a constant velocity v is

$$E_k = \frac{1}{2}mv^2 \tag{12.1}$$

The volume flow of air $\dot{V}$ and the mass flow $\dot{m}$ with an air density ρ flowing through a cross-sectional area A during a certain time unit are

$$\dot{V} = vA \tag{12.2}$$

$$\dot{m} = \rho vA \tag{12.3}$$

The amount of energy passing through a cross-sectional area A per unit time, that is, the power P, can be easily derived from the derivative of the kinetic energy equation (12.1) with respect to time and from the mass flow equation (12.3) resulting in

$$P = \frac{1}{2}\rho Av^3 \tag{12.4}$$

Consider a free stream air flow passing through a WT rotor with a cross-sectional area A_r and an air velocity v_r (see Figure 12.1). Consider also a far upstream air velocity v_1 with

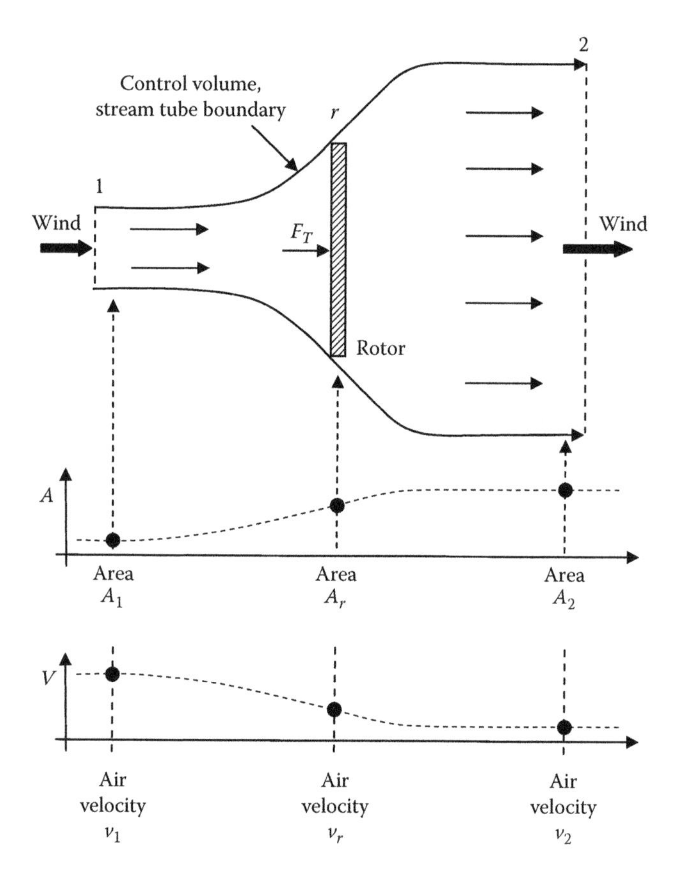

FIGURE 12.1
WT air stream flow diagram.

a cross-sectional area A_1 and a downstream air velocity v_2 with a cross-sectional area A_2. The power P extracted by the WT rotor can be expressed as

$$P = \frac{1}{2}\rho A_1 v_1^3 - \frac{1}{2}\rho A_2 v_2^3 = \frac{1}{2}\rho\left(A_1 v_1^3 - A_2 v_2^3\right) \tag{12.5}$$

The mass flow continuity equation is

$$\dot{m} = \rho v_1 A_1 = \rho v_2 A_2 = \rho v_r A_r = \text{constant} \tag{12.6}$$

Thus,

$$P = \frac{1}{2}\rho A_1 v_1\left(v_1^2 - v_2^2\right) = \frac{1}{2}\dot{m}\left(v_1^2 - v_2^2\right) \tag{12.7}$$

Using the law of conservation of linear momentum for a one-dimensional, incompressible, time-invariant flow, the force F_T, so-called thrust, which the air exerts on the WT rotor, is equal and opposite to the change in momentum of air stream in the control volume (see Figure 12.1). It is given by the following equation:

$$\text{Rate of change of momentum} = F_T = \dot{m}(v_1 - v_2) \tag{12.8}$$

The power extracted by the WT rotor can now be expressed as

$$P = F_T v_r = \dot{m}(v_1 - v_2)v_r \tag{12.9}$$

Now, comparing the last two expressions of power, Equations 12.7 and 12.9, yields

$$\frac{1}{2}\dot{m}(v_1^2 - v_2^2) = \dot{m}(v_1 - v_2)v_r \tag{12.10}$$

Then, the rotor air velocity v_r is equal to the arithmetic mean of v_1 and v_2, as given by

$$v_r = \frac{1}{2}(v_1 + v_2) \tag{12.11}$$

Let $v = v_r$ and $A = A_r$ in Equation 12.3 where the subscript "r" refers to the rotor. Using that equation and Equation 12.11, the mass flow can be expressed as

$$\dot{m} = \rho v_r A_r = \frac{1}{2}\rho A_r(v_1 + v_2) \tag{12.12}$$

Using Equations 12.7 and 12.12, the mechanical power P extracted by the WT rotor is

$$P = \frac{1}{4}\rho A_r\left(v_1^2 - v_2^2\right)(v_1 + v_2) \tag{12.13}$$

According to Equation 12.4, the power $P = P_0$ of the free-air stream that flows through the same cross-sectional area of the rotor $A = A_r$ with a velocity $v = v_1$, and without any mechanical power being extracted from it, is

$$P_0 = \frac{1}{2}\rho A_r v_1^3 \tag{12.14}$$

The ratio between the mechanical power extracted by the WT, Equation 12.13, and the original power of the free-air stream that flows through the same cross-sectional area, Equation 12.14, is called the power coefficient C_P, which is given by

$$C_P = \frac{P}{P_0} = \frac{(1/4)\rho A_r \left(v_1^2 - v_2^2\right)(v_1 + v_2)}{(1/2)\rho A_r v_1^3} = 4a(1-a)^2 \tag{12.15}$$

where a is the so-called axial induction factor given by

$$a = \frac{1}{2}\left(1 - \frac{v_2}{v_1}\right) \tag{12.16}$$

or, in other words,

$$v_2 = (1-2a)v_1 \tag{12.17}$$

Combining this expression with Equation 12.11 yields

$$v_r = (1-a)v_1 \tag{12.18}$$

Figure 12.2 represents the power coefficient C_P where the axial induction factor a varies from 0 to 1. The maximum achievable value of C_P occurs when

$$\frac{dC_P}{da} = 4(1-a)(1-3a) = 0 \tag{12.19}$$

which gives the value of $a = 1/3$. Hence,

$$C_{Pmax} = \frac{16}{27} = 0.593 \tag{12.20}$$

This theoretical maximum for an ideal WT is known as the *Betz limit*, and it represents the maximum power efficiency of the rotor, considering the ideal assumptions enumerated in this section. In this ideal situation, $a = 1/3$, and using Equation 12.18, the air velocity v_r that flows through the cross-sectional area of the rotor is

$$v_r = (1-a)v_1 = \frac{2}{3}v_1 \tag{12.21}$$

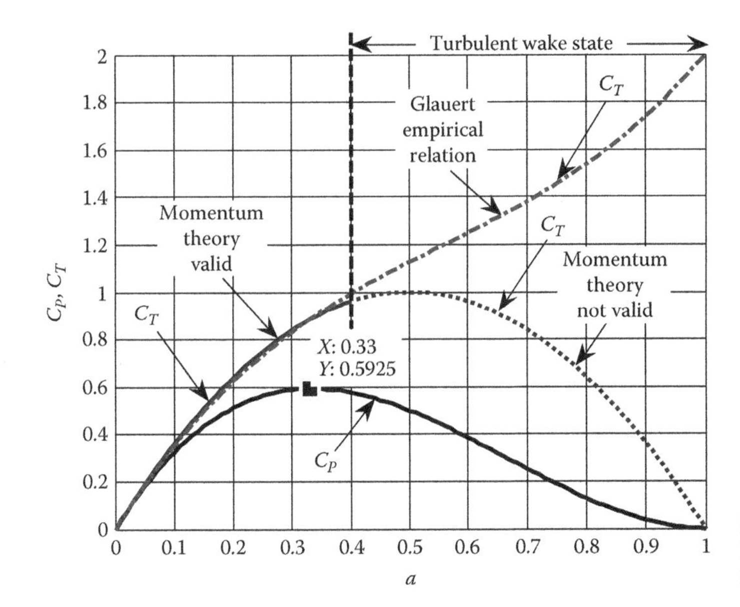

FIGURE 12.2
Plot of C_P and C_T as a varies from 0 to 1.

With Equation 12.17, the reduced velocity v_2, behind the WT, is given by

$$v_2 = (1-2a)v_1 = \frac{1}{3}v_1 \tag{12.22}$$

Utilizing Equations 12.14 and 12.15 the mechanical power extracted by the WT is given by

$$P = P_0 C_P = \frac{1}{2}\rho A_r v_1^3 C_P \tag{12.23}$$

Substituting Equation 12.15 into Equation 12.23 yields

$$P = \frac{1}{2}\rho A_r v_1^3 C_P = \frac{1}{2}\rho A_r v_1^3 [4a(1-a)^2] \tag{12.24}$$

Thus, the force F_T, or thrust, that the air exerts on the WT rotor becomes

$$F_T = \frac{P}{v_r} = \frac{(1/2)\rho A_r v_1^3 [4a(1-a)^2]}{v_r} \tag{12.25}$$

which, with Equation 12.18, becomes

$$F_T = \frac{1}{2}\rho A_r v_1^2 [4a(1-a)] = \frac{1}{2}\rho A_r v_1^2 C_T \tag{12.26}$$

where C_T is the thrust coefficient (see Figure 12.2), so that

$$C_T = 4a(1-a) = \frac{C_P}{(1-a)} \tag{12.27}$$

In the ideal situation, $a = 1/3$, the maximum power coefficient (Betz limit) is $C_{Pmax} = 16/27$, and the thrust coefficient $C_T = 8/9$ (see Figure 12.2).

12.2.2 Wake Rotation Effect

In the previous analysis, it is assumed that no rotation is imparted to the air flow behind the rotor. However, in the case of a rotating WT rotor, the flow behind the WT rotates in the opposite direction to the rotor, in reaction to the torque exerted by the flow on the rotor, according to the conservation of angular momentum (Figure 12.3).

The previous section is based on the classical momentum theory. The equations derived there are valid only for an axial induction factor a of less than approximately 0.4. For greater values, the free shear layer at the edge of the wake becomes unstable. This situation is called the turbulent-wake state. For these conditions, different empirical relations for the thrust C_T coefficients have been derived. These relations, developed by Glauert and Hansen,[317,351] are given in Equation 12.28. Figure 12.2 also describes the empirical Glauert's relation for C_T:

$$\begin{cases} C_T = 4a(1-a)f_1 \quad \text{for } a < 0.33 \\ C_T = 4a\left[1 - \frac{1}{4}(5-3a)a\right]f_1 \quad \text{for } a > 0.33 \end{cases} \tag{12.28}$$

$$f_1 = (2/\pi)\cos^{-1}\left(e^{-N(r_p - r)/2r\sin(\phi)}\right)$$

where

f_1 is the Prandtl's correction factor that takes into account a finite number of blades (see also Section 12.2.4) and the losses due to the air circulation at the blade tip

λ is the tip speed ratio, with $\lambda = \Omega_r r_b / v_1$

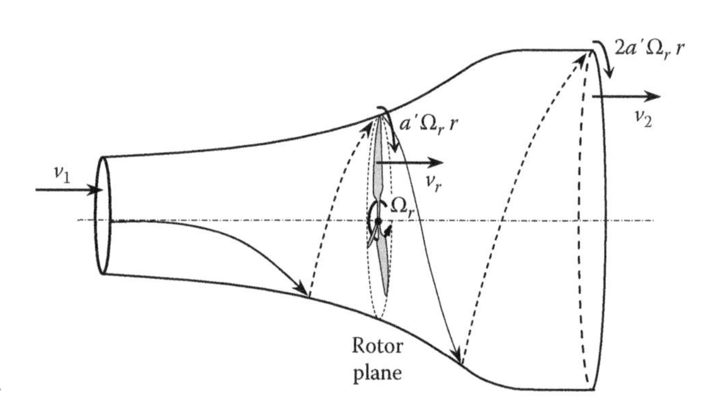

FIGURE 12.3
The wake rotation effect.

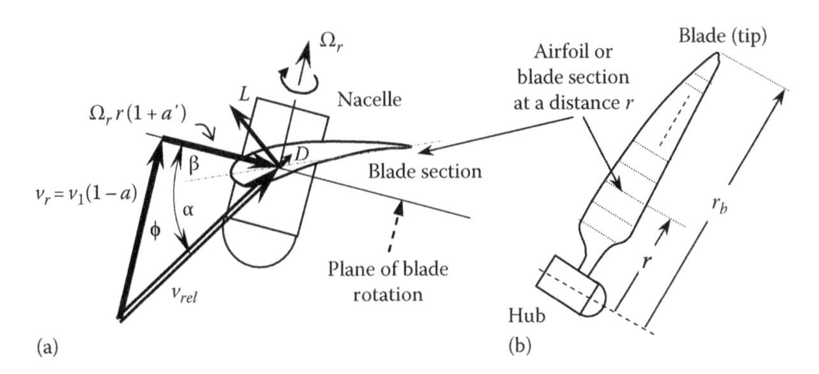

FIGURE 12.4
Wind velocity triangle. (a) View of an airfoil section from the top of the blade, and (b) view of the hub and the entire length of the blade.

As is seen in Equations 12.17 and 12.18, using the axial induction factor a, the axial velocity of the air through the rotor, v_r, is given by $v_r = (1-a)v_1$, and the axial downstream velocity of the air, v_2, is given by $v_2 = (1-2a)v_1$.

Similarly, there is a so-called rotational induction factor a'. At each radius r, the a' factor[317] can be calculated according to $a'(1+a')(\Omega_r r/v_1)^2 = a(1-a)$; the rotational velocity of the air through the rotor is given by $a'\Omega_r r$, and the rotational downstream velocity of the air is given by $2a'\Omega_r r$. For a nonrotating wake $a' = 0$. The velocity triangle is represented in Figure 12.4 and Equation 12.29.

The relative wind speed v_{rel}, as shown in Figure 12.4, is the effective wind velocity felt by each perpendicular section (airfoil) of the blade, at a distance r ($0 \le r \le r_b$) from the rotor axis and is given by

$$v_{rel} = \sqrt{v_r^2 + [\Omega_r r(1+a')]^2} = \sqrt{[v_1(1-a)]^2 + [\Omega_r r(1+a')]^2} \quad (12.29)$$

and the corrected aerodynamic coefficient C_p[317] is

$$C_P = \frac{8}{\lambda^2}\int_0^{\lambda} a'(1-a)x^3 dx, \quad \text{where } x = \frac{\Omega_r r}{v_1};\ a'(1+a')x^2 = a(1-a) \quad (12.30)$$

The rotating kinetic energy captured by the air flow behind the rotor results in less energy extraction by the WT. In general, turbines with higher torque (slower rotors for the same power) impart a higher kinetic energy to the wake behind the rotor, which implies a lower energy extraction by the WT. Figure 12.5 shows the maximum power coefficient C_{Pmax} versus the tip speed ratio λ, where $\lambda = \Omega_r r_b/v_1$, r_b is the blade radius, v_1 is the undisturbed far upstream wind speed, and Ω_r is the rotational speed of the rotor.

12.2.3 Wind Turbine Blades and Rotor Terminology

The cross sections of a WT blade from the root to the tip are called *airfoils* (see Figures 12.4b, 12.6, and 12.7) and usually have different shapes. The basic terms used to characterize an airfoil are shown in Figure 12.6, looking down from the blade tip to the root blade (as in Figure 12.4a).

The flow velocity increases over the upper convex surface of the airfoil, resulting in lower pressure (suction) compared with the higher pressure of the lower concave side of

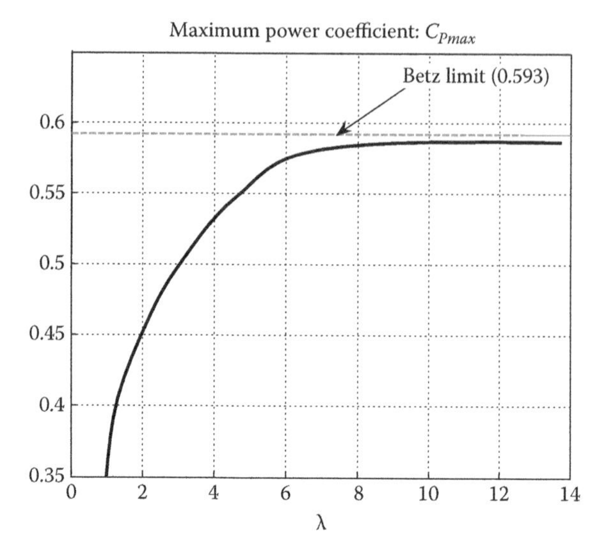

FIGURE 12.5
Maximum power coefficient C_{Pmax} versus the tip speed ratio λ.

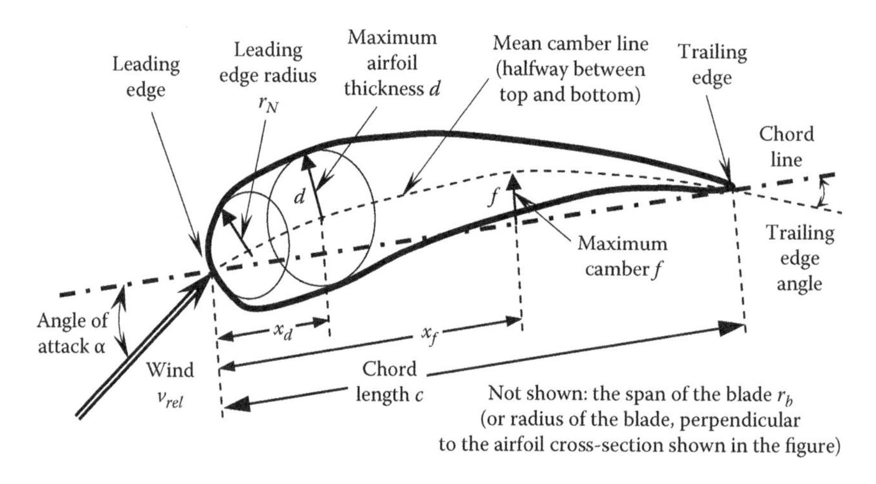

FIGURE 12.6
A basic description of an airfoil.

the airfoil. This phenomenon is usually resolved into two forces (lift L and drag D) and a pitch moment M that acts along the chord at a distance of $c/4$ from the leading edge (the quarter chord; see Figure 12.7).

The lift force L is defined as the force per unit span applied to the wind perpendicular to the direction of the incoming airflow. It is the consequence of the unequal pressure of the upper and lower airfoil surfaces.

The drag force D is defined as the force per unit span applied to the wind parallel to the direction of the incoming airflow. It is the consequence of the viscous friction forces and the unbalance of the incoming flow.

The pitch moment M is defined as the resulting torque per unit span applied to the wind at a distance of $c/4$ from the leading edge (nose-up positive). Note that M is always negative in practice (nose-down).

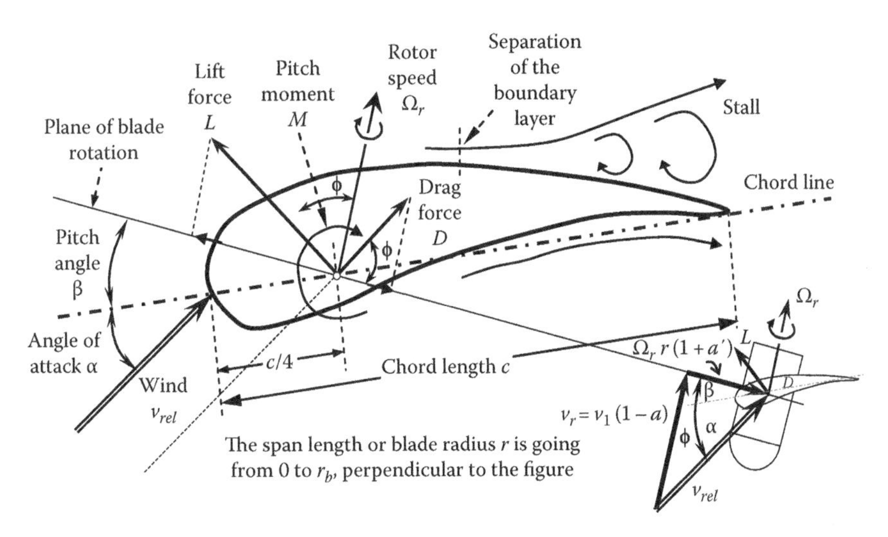

FIGURE 12.7
Airfoil cross-section, perpendicular to the blade radius, at a distance r from the rotor axis (see Figure 12.4 for r).

The lift coefficient C_L, drag coefficient C_D, and pitching moment coefficient C_M at each airfoil (each r) are defined, respectively, as

$$C_L = \frac{L}{0.5\rho v_{rel}^2 c} = \frac{\text{Lift force/unit span}}{\text{Dynamic force/unit span}} \tag{12.31a}$$

$$C_D = \frac{D}{0.5\rho v_{rel}^2 c} = \frac{\text{Drag force/unit span}}{\text{Dynamic force/unit span}} \tag{12.31b}$$

$$C_M = \frac{M}{0.5\rho v_{rel}^2 c} = \frac{\text{Pitching moment/unit span}}{\text{Dynamic force/unit span}} \tag{12.31c}$$

where
ρ is the density of the air
v_{rel} is the relative wind speed at that r
c is the airfoil chord length
r_b is the span or radius, with A_a the projected plane form area, $A_a = cr_b$, (see Figure 12.4b)

According to the areas and lengths of the rotor and blades, the following parameters define the geometrical rotor blade shape:

- *Rotor solidity* = total blade plane form area/rotor swept area (%)
- *Aspect ratio* = (rotor radius)^2/plane form area of a rotor blade
- *Taper* = chord length at blade tip/chord length at blade root
- The wind applies a force at every airfoil cross-section (at each radius r, $0 \le r \le r_b$, N blades) that,
  - turns the rotor: $dT_r = N[L \sin(\phi) - D \cos(\phi)]rdr$
  - bends the tower (y_t direction): $dF_T = N\,[L \cos(\phi) + D \sin(\phi)]dr$

TABLE 12.1

Properties of Some Airfoils

Airfoil	FFA-W3	64.6XX	NACA 63.4XX	FW84-W
λ_{opt} at C_{Pmax}	8	8.75	7	9
Solidity (%)	4.8	4.4	5.0	4.2
Aspect ratio	19	20.7	19	15.2
Taper	0.14	0.18	0.21	0.25
Thickness ratio d/c at 0.7 span (%)	16	18	20	17
Rotor radius	35	62	21	40
Number of blades N	3	3	3	2
Manufacturer/type	Vestas V66	LM 61.5	LM 19	Aeolus II

There are many types of airfoils with different nomenclature according to the geometry and aerodynamic properties: NACA, FX, LS, SERI, FFA series, etc. Table 12.1 lists the properties of some existing rotors of WTs.

12.2.3.1 Additional Symbols and Terminology

r_b	Maximum rotor radius (blade radius or span, $0 \leq r \leq r_b$)
α	Angle of attack: between chord line and relative wind v_{rel} (function of r)
β	Pitch angle: angle between chord line and plane of rotation; $\beta = \beta_{p0} + \beta_T$, where β_{p0} is the blade pitch angle at the root and β_T is the blade twist angle
ϕ	The angle between the plane of rotation and the relative wind v_{rel} (function of r) and is given by $\phi = \alpha + \beta = \text{atan}\{v_r/[\Omega_r r(1+a')]\} = \text{atan}\{[v_1(1-a)]/[\Omega_r r(1+a')]\}$
v_{rel}	The relative wind speed at r is given by $v_{rel} = \sqrt{[v_1(1-a)]^2 + [\Omega_r r(1+a')]^2}$
ρ	Air density. It varies with altitude, pressure, and temperature so that $\rho = (P_{at}/RT)\exp[-(gz)/(RT)]$, where P_{at} is the atmospheric pressure (101,325 Pa at sea level), R is the specific gas constant for air (287.05 J/kg/K), g is the gravity constant (9.81 m/s^2 at sea level), T is the temperature (K), and z is the altitude above sea level (m)
	At sea level and $T = 288.15$ K, the standard air density is $\rho = 1.225$ kg/m^3
C_P	Aerodynamic power coefficient of the entire rotor
C_T	Aerodynamic thrust coefficient of the entire rotor
a	Axial flow induction factor
a'	Rotational flow induction factor
Ω_r	Rotor angular speed
$\dot{y}_t$	Axial velocity of the nacelle caused by tower bending
F_T	Thrust force applied by the wind on the rotor area
T_r	Aerodynamic torque applied by the wind on the rotor
T_g	Antagonist electrical torque applied on the shaft
λ	Tip speed ratio, $\lambda = \Omega_r r_b/v_1$
r_p	$=(2/3)r_b$, which is the distance from the center of the rotor to the center of pressure, or the point where the equivalent lumped force F_T is applied. This lumped force summarizes the distributed force applied by the wind on the blade. Considering the rotor to be made of a series of concentric annuli of width dr, the root flap-wise bending moment M_{root} for a WT with N blades is $M_{root} = \frac{1}{N}\int_0^{r_b} r\,dF_T = \frac{1}{N}\int_0^{r_b} r(0.5\rho C_T v_1^2\, 2\pi r\,dr) = \frac{2}{3N}F_T r_b$
	Defining $M_{root} = F_T r_p/N$, then $r_p = (2/3)r_b$ (see Figures 12.11 and 12.12)

12.2.4 Effect of Drag and Number of Blades

In the previous analysis, it is assumed there are a sufficient number of blades on the rotor for every particle of air passing through the rotor disc to interact with a blade, and there is no drag. However, the maximum power coefficient C_{Pmax} is smaller in the real case, where the WT rotor presents a small number of blades N and where there are drag losses. In that case, the maximum power coefficient C_{Pmax} versus the tip speed ratio λ is given by[306]

$$C_{Pmax} = 0.593\lambda\left[\lambda + \frac{1.32 + ((\lambda - 8)/20)^2}{N^{2/3}}\right]^{-1} - \frac{0.57\lambda^2}{(C_L/C_D)(1 + 1/2N)} \tag{12.32}$$

where

- N is the number of blades
- C_L is the lift coefficient
- C_D is the drag coefficient

See Figure 12.8 for $C_L/C_D = 76$ and $N = 5, 4, 3$.

12.2.5 Actual Wind Turbines

Considering frictional drag, a small number of blades, tip blade losses, and rotating wakes in the out-coming airflow, the manufacturers usually give the actual aerodynamic power coefficient C_P as a function of the pitch angle β and the tip speed ratio λ (see Figure 12.9), which is defined as follows:

$$\lambda = \frac{\Omega_r r_b}{v_1} \tag{12.33}$$

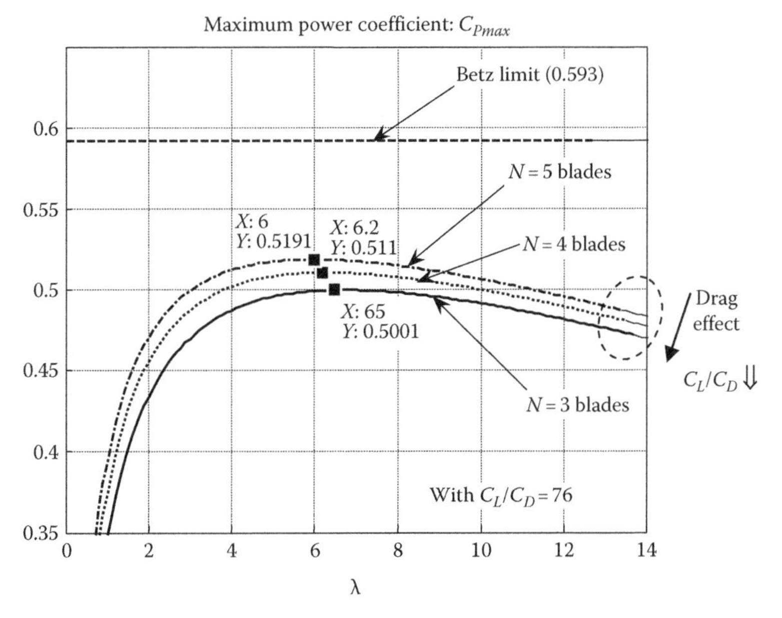

FIGURE 12.8
Maximum power coefficient C_{Pmax} versus the tip speed ratio λ.

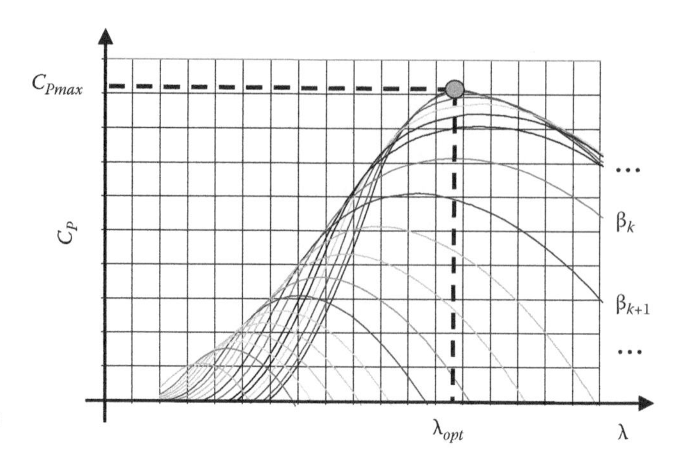

FIGURE 12.9
Aerodynamic power coefficient C_P as a function of λ and β.

A numerical approximation of the aerodynamic power coefficient C_P is given by the following equations:

$$C_P(\lambda,\beta) = c_1\left(\frac{c_2}{\lambda_i} - c_3\beta - c_4\right)\exp\left(\frac{-c_5}{\lambda_i}\right)$$
$$\lambda_i = \left(\frac{1}{\lambda + c_6\beta} - \frac{c_7}{\beta^3+1}\right)^{-1} \tag{12.34}$$

where

$c_1 = 0.39$
$c_2 = 116$
$c_3 = 0.4$
$c_4 = 5$
$c_5 = 16.5$
$c_6 = 0.089$
$c_7 = 0.035$

In this approach, it is considered that the variation of the axial displacement of the nacelle $\dot{y}_t$ and the variation of the axial displacement of the blades $\dot{y}r_p$ do not affect significantly the wind speed at the rotor.

The mechanical power extracted by the WT and the force (thrust) and torque that the air exerts on the WT rotor are given by the following:

For the entire rotor

$$P = \frac{1}{2}\rho\pi r_b^2 C_P(\lambda,\beta)\,v_1^3 = T_r\Omega_r \tag{12.35}$$

$$F_T = \frac{1}{2}\rho\pi r_b^2 C_T(\lambda,\beta)v_1^2 = N\int_{r_{root}}^{r_b}\left[L\cos(\phi) + D\sin(\phi)\right]dr \tag{12.36}$$

$$T_r = \frac{\rho\pi r_b^2 C_P(\lambda,\beta) v_1^3}{2\Omega_r} = N \int_{r_{root}}^{r_b} \left[L\sin(\phi) - D\cos(\phi)\right] r dr \tag{12.37}$$

For each airfoil (each r)

$$L = 0.5\rho A_a C_L v_{rel}^2; \quad D = 0.5\rho A_a C_D v_{rel}^2 \tag{12.38}$$

12.3 Mechanical Models

In Sections 12.3.2 and 12.3.3 the Euler–Lagrange method (energy-based approach)[320,331–334] is applied to obtain general mechanical State Space models of both (1) the DD WT and (2) the DFIG WT respectively.

The transfer matrix representations of the DD and the DFIG WT systems are obtained in Section 12.3.4, and the rotor speed transfer functions in Section 12.3.5.

12.3.1 Euler–Lagrange Energy-Based Description

12.3.1.1 Symbols and Terminology

According to Figures 12.11 and 12.12 the main symbols and terminology for the DD and DFIG WTs are listed in Tables 12.2 through 12.4.

Description of some of the variables in Tables 12.2 through 12.4 is as follows:

$m_1 = m_h + m_n + c_t m_t + (1 - c_b) N m_b$ is the equivalent mass for movement y_t

$m_2 = c_b N m_b$ is the equivalent mass for movement $\gamma r_b + y_t$

$c_t = 0.2357$, $0.0437 \le c_b \le 0.2357$

$K_t = 3EI_{At}/h^3$ is the tower stiffness coefficient, where E is the elastic module (N/m^2) and the inertia of the transversal section of the tower is $I_{At} = \int r^2 dA$ given by $I_{At} \approx (\pi/64)(D^4 - d^4)$

$K_b = 3EI_{Ab}/r_b^3$ is the blade stiffness coefficient. E is the elastic module (N/m^2) where $I_{Ab} = \int r^2 dA$ is the inertia of transversal section of the blade given by $I_{Ab} \approx (\pi/64)(D^4 - d^4)$

$K_s = G\pi(D^4 - d^4)/(32\,L)$ is the shaft stiffness coefficient, where G is the angular elastic module

Steel: $E = 210 \times 10^9$ N/m^2, $G = 80 \times 10^9$ N/m^2, glass fiber: $E = 45 \times 10^9$ N/m^2

TABLE 12.2

DD Machines (without Gearbox)

K_s	transmission stiffness coefficient (Nm/rad)	B_s	transmission damping coefficient (Nms/rad)

TABLE 12.3

Common Parameters

v_1	upstream wind speed (m/s)	N	number of blades (—)
m_t	mass of the tower (kg)	m_n	mass of the nacelle (kg)
m_h	mass of the hub (kg)	m_b	mass of each blade (kg)
r_b	blade radius (m)	h	tower height (m)
y_t	axial displacement nacelle (m)	γ	angular displacement blade (rad)
K_t	tower stiffness coefficient (N/m)	B_t	tower damping coefficient (Ns/m)
K_b	blade stiffness coefficient (N/m)	B_b	blade damping coefficient (Ns/m)
I_r	moment of inertia of elements at Ω_r (rotor, blades, hub, shaft, etc.) (kg m^2)	I_g	moment of inertia of elements at Ω_g (generator, shaft, etc.) (kg m^2)
θ_r	rotor angular position (rad)	θ_g	generator angular position (rad)
$\Omega_r = \dot{\theta}_r$	rotor angular speed (rad/s)	$\Omega_g = \dot{\theta}_g$	generator angular speed (rad/s)
T_r	aerodynamic torque applied by the wind on the rotor (see Equation 12.37) (Nm)	T_g	antagonic electrical torque applied on the shaft (Nm)
F_T	thrust force applied by the wind on the rotor (see Equation 12.36) (N)		
r_p	distance from the center of the rotor to the center of pressure, or point where the equivalent lumped force F_T is applied. $r_p = (2/3)\, r_b$ (m)		

TABLE 12.4

DFIG Machines (with Gearbox)

K_l	low-speed shaft torsional stiffness coefficient (Nm/rad)	B_l	low-speed shaft torsional damping coefficient (Nms/rad)
K_h	high-speed shaft torsional stiffness coefficient (Nm/rad)	B_h	high-speed shaft torsional damping coefficient (Nms/rad)
θ_l	angular position of the gearbox low-speed part (rad)	θ_h	angular position of the gearbox high-speed part (rad)
T_l	torque applied to the gearbox by the low-speed shaft (Nm)	T_h	torque applied to the gearbox by the high-speed shaft (Nm)
R_t	gear ratio (—)	I_w	equivalent moment of inertia of gearbox elements, at θ_l (kg m^2)

A description of the tower and blades, modeled as a beam, is shown in Figure 12.10 where v is the beam's mass per unit length; D and d are the outer and inner section diameters; L_{v0} is the value of y when $v = 0$; and L is the length of the tower, blade, or shaft, and where

$$E_{kinetic} = \int_0^L \dot{x}_L^2\, dm = \frac{1}{2}\int_0^L \frac{y^4(3L-y)^2}{4L^6}\dot{x}_L^2\left(a - \frac{a}{L_{v0}}y\right)dy = \frac{1}{2}c_i(aL)\dot{x}_L^2 = \frac{1}{2}m_{eq}\,\dot{x}_L^2$$

$$\text{if } L_{v0} = \infty, \Rightarrow v = \left(a - \frac{a}{L_{v0}}y\right) = a = \text{constant}, \Rightarrow c_b \text{ or } c_t = c_i = 0.2357$$

$$\text{if } L_{v0} = L, \Rightarrow v = \left(a - \frac{a}{L_{v0}}y\right) = \left(a - \frac{a}{L}y\right), \Rightarrow c_b \text{ or } c_t = c_i = 0.0437$$

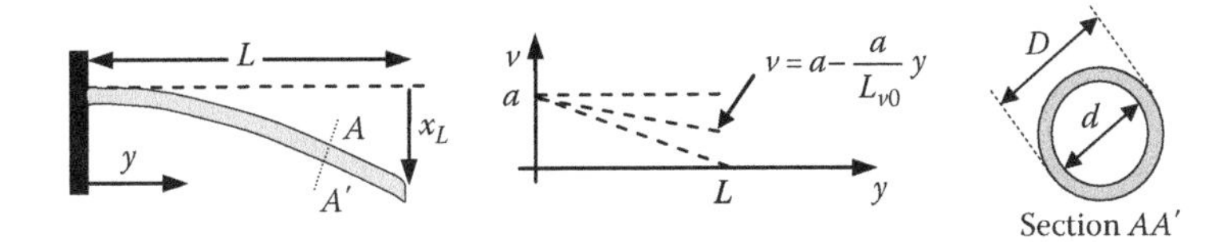

FIGURE 12.10
Blade or tower modeled as a beam.

12.3.1.2 Equations of Motion

The WT dynamic equations of motion that describe the behavior of the system, under the influence of external forces (wind) and as a function of time, are developed in this section as a set of mechanical differential equations with the following classic matrix form, which is derived later on:

$$\mathcal{M}\ddot{q} + \mathcal{C}\dot{q} + \mathcal{K}q = \mathcal{Q}(\dot{q}, q, u, t) \tag{12.39}$$

where
- q are the generalized coordinates
- $\dot{q}$ and $\ddot{q}$ are their time derivatives
- $\mathcal{M}$ is the inertia matrix (with the inertia coefficients of the system)
- $\mathcal{C}$ is the dissipative matrix (with the damping coefficients of the system)
- $\mathcal{K}$ is the stiffness matrix (with the stiffness coefficients of the system)
- Θ is the applied generalized forces
- u are the inputs (forces, torques)
- t is the time

The equations of motion in Lagrangian mechanics are the Lagrange equations of the second kind, also known as the Euler–Lagrange equations.[320,331–334] Note that E_k is used for kinetic energy and E_p for potential energy. D_n is the dissipation function to include nonconservative forces, Q_i are the conservative generalized forces, and q_i are the generalized coordinates. Defining L as the Lagrangian function $L = E_k - E_p$, the Euler–Lagrange equation is as follows:

$$\frac{d}{dt}\left(\frac{\partial L}{\partial \dot{q}_i}\right) - \frac{\partial L}{\partial q_i} + \frac{\partial D_n}{\partial \dot{q}_i} = Q_i, \quad i = 1, 2, \ldots \text{number}_{\text{degrees of freedom}} \tag{12.40}$$

12.3.2 Mechanical Dynamics of Direct-Drive Wind Turbines (Gearless)

This section presents a four-mass model for the DD WT. The approach considers 4 degrees of freedom: rotor angle, generator angle, axial displacement of the nacelle, and angular displacement of the blades.

Figure 12.11 shows a variable-speed direct-drive (gearless) WT with a full-power AC–DC–AC converter. A shaft connects a large rotor inertia at one end (blades) with a generator at the other end. The wind applies an aerodynamic torque T_r on the rotor. The power electronic converter applies an antagonic electrical torque T_g on the shaft.

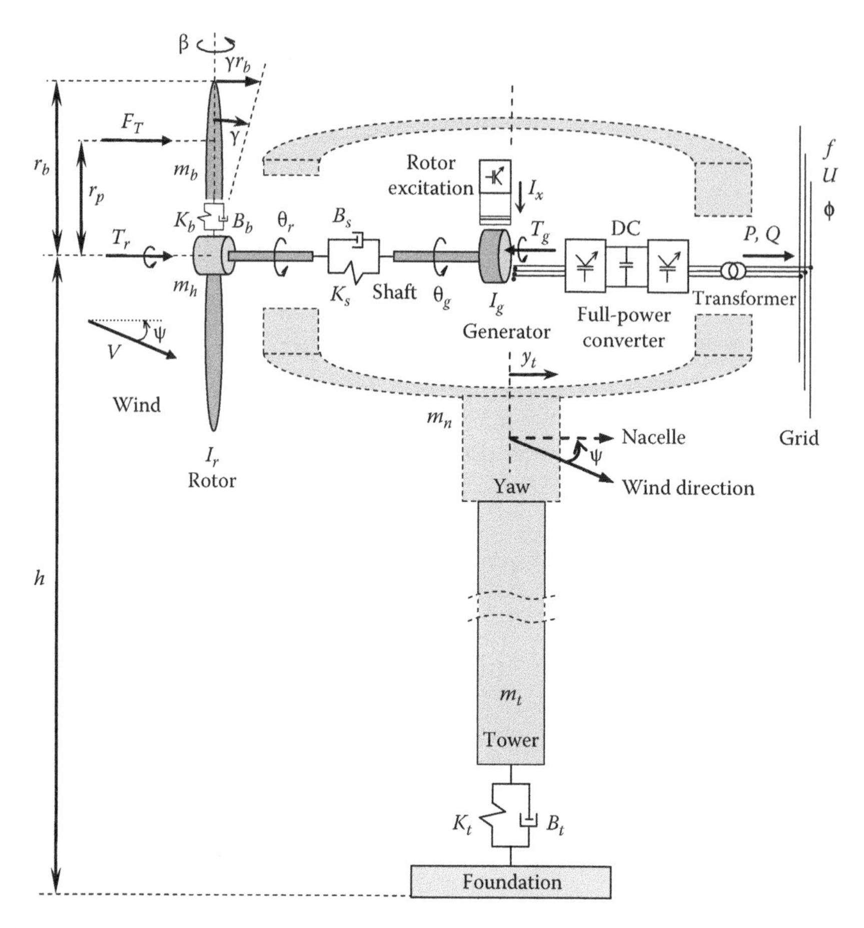

FIGURE 12.11
A variable-speed DD WT (gearless) with full-power AC–DC–AC converter.

The rotor presents a moment of inertia I_r. The shaft has a torsional stiffness coefficient K_s and a viscous damping coefficient B_s. The generator shows a moment of inertia I_g.

The rotor angle is θ_r, the rotor speed is $\Omega_r = \dot{\theta}_r$, and the generator angle is θ_g. Also, ψ is the yaw angle error (nacelle-wind angle) and β is the pitch angle (blades). The excitation current, I_x, is introduced in the rotor, and the active and reactive powers, P and Q, respectively, are supplied to the grid. f, U, and ϕ are the frequency, voltage, and power factor at the grid connection point, respectively. The equations of the DD are presented next.

1. *Generalized coordinates, q_i*

$$q = [q_i] = [y_t \quad \gamma \quad \theta_r \quad \theta_g]^T \tag{12.41}$$

where
y_t is the axial displacement of the nacelle
γ is the angular displacement of the blades out of the plane of rotation
θ_r is the rotor angular position
θ_g is the generator angular position

2. *Energy and dissipation function equations, E_k, E_p, D_n*
   The kinetic energy E_k, potential energy E_p, and dissipation function D_n of the DD WT shown in Figure 12.11 are derived as follows:

$$E_k = \frac{m_1}{2}\dot{y}_t^2 + \frac{m_2}{2}(r_b\dot{\gamma} + \dot{y}_t)^2 + \frac{I_r}{2}\dot{\theta}_r^2 + \frac{I_g}{2}\dot{\theta}_g^2 \tag{12.42}$$

$$E_p = \frac{K_t}{2}y_t^2 + \frac{NK_b}{2}(r_b\gamma)^2 + \frac{K_s}{2}(\theta_r - \theta_g)^2 \tag{12.43}$$

$$D_n = \frac{B_t}{2}\dot{y}_t^2 + \frac{NB_b}{2}(r_b\dot{\gamma})^2 + \frac{B_s}{2}(\dot{\theta}_r - \dot{\theta}_g)^2 \tag{12.44}$$

3. *Lagrange equations terms*
   Based on Equations 12.41 through 12.44, the terms of the Euler–Lagrange equation, Equation 12.40, where $\partial E_p/\partial \dot{q}_i = 0$ and $\partial E_k/\partial q_i = 0$, are as follows:

$$\frac{d}{dt}\left(\frac{\partial L}{\partial \dot{q}_i}\right) = \frac{d}{dt}\begin{bmatrix} \dfrac{\partial E_k}{\partial \dot{y}_t} \\ \dfrac{\partial E_k}{\partial \dot{\gamma}} \\ \dfrac{\partial E_k}{\partial \dot{\theta}_r} \\ \dfrac{\partial E_k}{\partial \dot{\theta}_g} \end{bmatrix} = \frac{d}{dt}\begin{bmatrix} (m_1 + m_2)\dot{y}_t + m_2 r_b\dot{\gamma} \\ m_2 r_b^2\dot{\gamma} + m_2 r_b\dot{y}_t \\ I_r\dot{\theta}_r \\ I_g\dot{\theta}_g \end{bmatrix} = \begin{bmatrix} m_1 + m_2 & m_2 r_b & 0 & 0 \\ m_2 r_b & m_2 r_b^2 & 0 & 0 \\ 0 & 0 & I_r & 0 \\ 0 & 0 & 0 & I_g \end{bmatrix}\begin{bmatrix} \ddot{y}_t \\ \ddot{\gamma} \\ \ddot{\theta}_r \\ \ddot{\theta}_g \end{bmatrix} = \mathcal{M}\ddot{q} \tag{12.45}$$

$$-\frac{\partial L}{\partial q_i} = \frac{\partial E_p}{\partial q_i} = \begin{bmatrix} \dfrac{\partial E_p}{\partial y_t} \\ \dfrac{\partial E_p}{\partial \gamma} \\ \dfrac{\partial E_p}{\partial \theta_r} \\ \dfrac{\partial E_p}{\partial \theta_g} \end{bmatrix} = \begin{bmatrix} K_t y_t \\ NK_b r_b^2\gamma \\ K_s\theta_r - K_s\theta_g \\ K_s\theta_g - K_s\theta_r \end{bmatrix} = \begin{bmatrix} K_t & 0 & 0 & 0 \\ 0 & NK_b r_b^2 & 0 & 0 \\ 0 & 0 & K_s & -K_s \\ 0 & 0 & -K_s & K_s \end{bmatrix}\begin{bmatrix} y_t \\ \gamma \\ \theta_r \\ \theta_g \end{bmatrix} = \mathcal{K}q \tag{12.46}$$

$$\frac{\partial D_n}{\partial \dot{q}_i} = \begin{bmatrix} \dfrac{\partial D_n}{\partial \dot{y}_t} \\ \dfrac{\partial D_n}{\partial \dot{\gamma}} \\ \dfrac{\partial D_n}{\partial \dot{\theta}_r} \\ \dfrac{\partial D_n}{\partial \dot{\theta}_g} \end{bmatrix} = \begin{bmatrix} B_t\dot{y}_t \\ NB_b r_b^2\dot{\gamma} \\ B_s\dot{\theta}_r - B_s\dot{\theta}_g \\ B_s\dot{\theta}_g - B_s\dot{\theta}_r \end{bmatrix} = \begin{bmatrix} B_t & 0 & 0 & 0 \\ 0 & NB_b r_b^2 & 0 & 0 \\ 0 & 0 & B_s & -B_s \\ 0 & 0 & -B_s & B_s \end{bmatrix}\begin{bmatrix} \dot{y}_t \\ \dot{\gamma} \\ \dot{\theta}_r \\ \dot{\theta}_g \end{bmatrix} = \mathcal{C}\dot{q} \tag{12.47}$$

$$\mathcal{Q} = \begin{bmatrix} F_T \\ r_p F_T \\ T_r \\ -T_g \end{bmatrix} = \begin{bmatrix} 1 & 0 & 0 \\ r_p & 0 & 0 \\ 0 & 1 & 0 \\ 0 & 0 & -1 \end{bmatrix} \begin{bmatrix} F_T \\ T_r \\ T_g \end{bmatrix} = \mathcal{R}u \tag{12.48}$$

Using Equations 12.45 through 12.48, Equation 12.40 becomes the original equation (12.39): $\mathcal{M}\ddot{q} + \mathcal{C}\dot{q} + \mathcal{K}q = \mathcal{Q}(\dot{q}, q, u, t)$

4. *Rearrangement of the equation of motion*
   Now, with very simple manipulation, Equation 12.39 adopts the form

$$\ddot{q} = -\mathcal{M}^{-1}\mathcal{C}\dot{q} - \mathcal{M}^{-1}\mathcal{K}q + \mathcal{M}^{-1}\mathcal{R}u \tag{12.49}$$

5. *Description of the system in State Space*
   The following three vectors are defined:

$$\text{State variables:} \quad x = [y_t \quad \gamma \quad \theta_r \quad \theta_g \quad \dot{y}_t \quad \dot{\gamma} \quad \dot{\theta}_r \quad \dot{\theta}_g]^T \tag{12.50}$$

$$\text{Inputs:} \quad u = [F_T \quad T_r \quad T_g]^T \tag{12.51}$$

$$\text{Outputs:} \quad y = [\dot{y}_t \quad \dot{\gamma} \quad \dot{\theta}_r \quad \dot{\theta}_g]^T \tag{12.52}$$

Equation 12.49 takes the form of a classical State Space description of the system ($\dot{x} = Ax + Bu$; $y = Cx$) so that the State equations are

$$\underset{8\times1}{\dot{x}} = \left[\begin{array}{c|c} \underset{4\times4}{\mathbf{0}} & \underset{4\times4}{\mathbf{I}} \\ \hline -\underset{4\times4}{\mathcal{M}^{-1}}\underset{4\times4}{\mathcal{K}} & -\underset{4\times4}{\mathcal{M}^{-1}}\underset{4\times4}{\mathcal{C}} \end{array}\right] \underset{8\times1}{x} + \left[\begin{array}{c} \underset{4\times3}{\mathbf{0}} \\ \hline \underset{4\times4}{\mathcal{M}^{-1}}\underset{4\times3}{\mathcal{R}} \end{array}\right] \underset{3\times1}{u}$$

$$\underset{4\times1}{y} = \left[\begin{array}{c|c} \underset{4\times4}{\mathbf{0}} & \underset{4\times4}{I} \end{array}\right] \underset{8\times1}{x} \tag{12.53}$$

Note that Equation 12.53 is a system with three inputs and four outputs (12 transfer functions). The calculation of the inverse of $\mathcal{M}$ yields

$$\mathcal{M}^{-1} = \begin{bmatrix} \dfrac{1}{m_1} & \dfrac{-1}{m_1 r_b} & 0 & 0 \\ \dfrac{-1}{m_1 r_b} & \dfrac{m_1 + m_2}{m_1 m_2 r_b^2} & 0 & 0 \\ 0 & 0 & \dfrac{1}{I_r} & 0 \\ 0 & 0 & 0 & \dfrac{1}{I_g} \end{bmatrix} \tag{12.54}$$

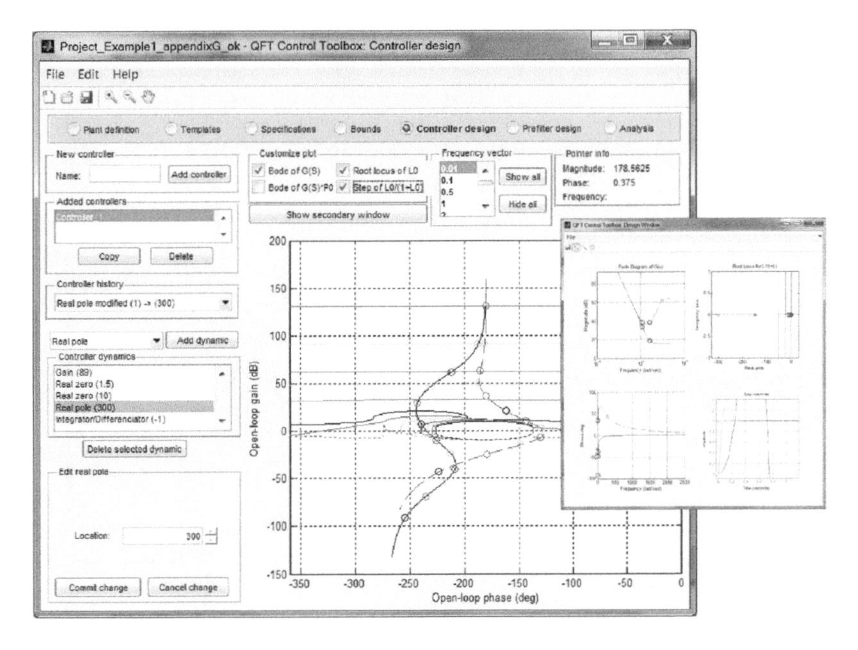

FIGURE 1.7
The QFT control toolbox for MATLAB®. An interactive object-oriented tool for quantitative robust controller design. Developed by Garcia-Sanz, Mauch, Philippe et al., at the Public University of Navarra, Case Western Reserve University and the European Space Agency. (From Garcia-Sanz, M. et al., QFT control toolbox: An interactive object-oriented Matlab CAD tool for quantitative feedback theory, in *Sixth IFAC Symposium on Robust Control Design, ROCOND'09*, Haifa, Israel; Garcia-Sanz, M. et al., QFT control toolbox: An interactive object-oriented Matlab CAD tool for controller design, Research project, European Space Agency ESA-ESTEC, Public University of Navarra, Pamplona, Spain.)

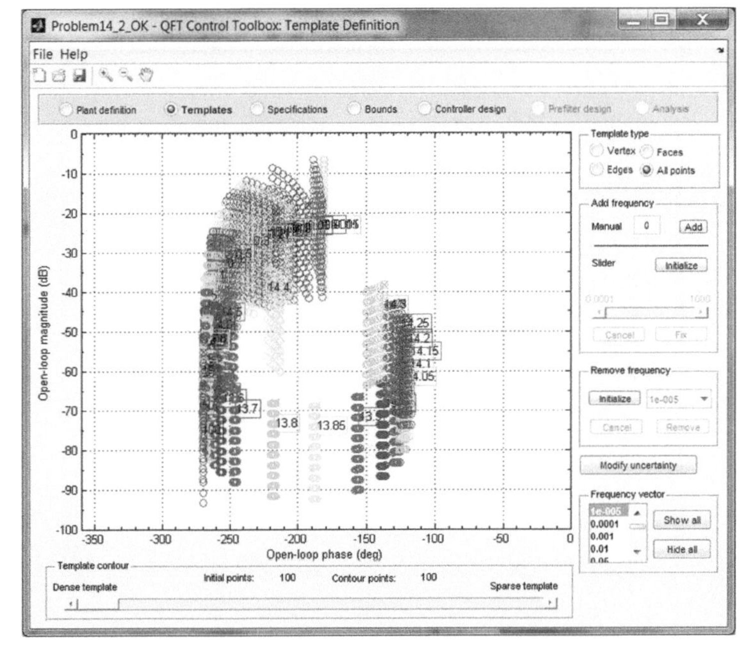

FIGURE 14.6
Templates calculation for $P(s)$.

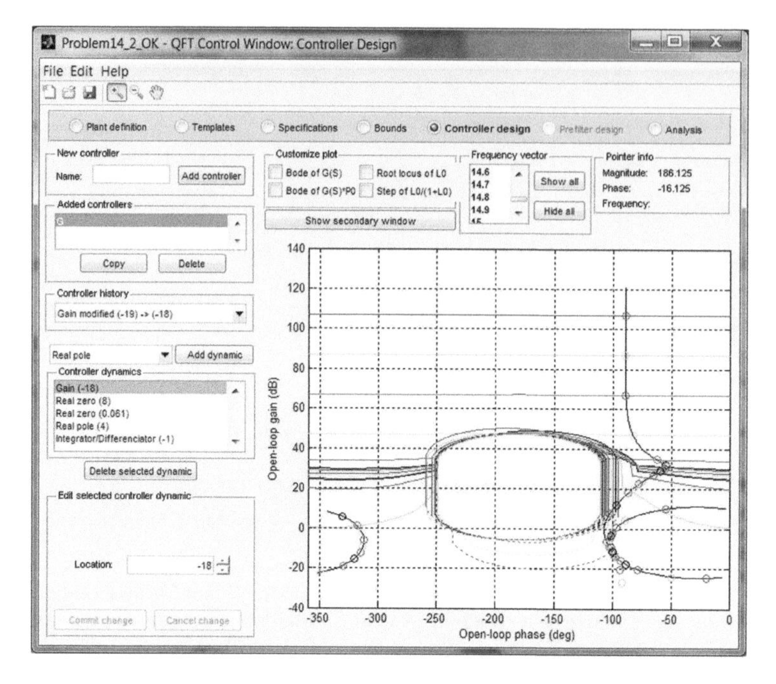

FIGURE 14.11
QFT bounds and loop shaping: $L_0(s) = P_0(s)\ G(s)$.

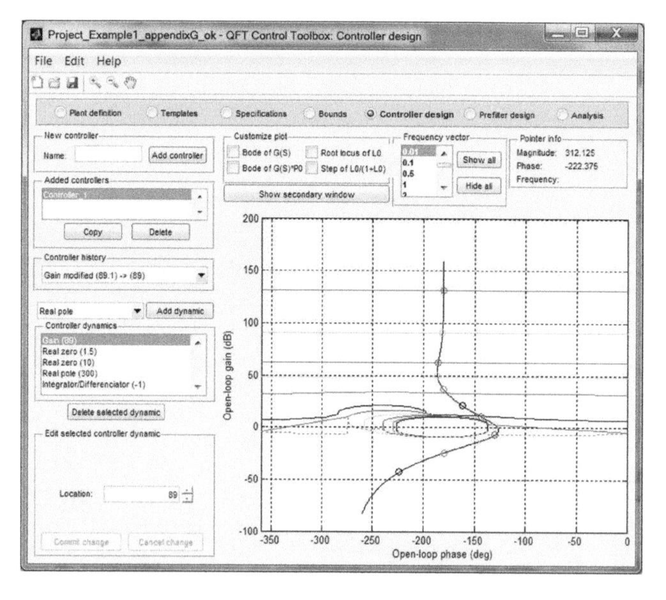

FIGURE F.27
Controller design window.

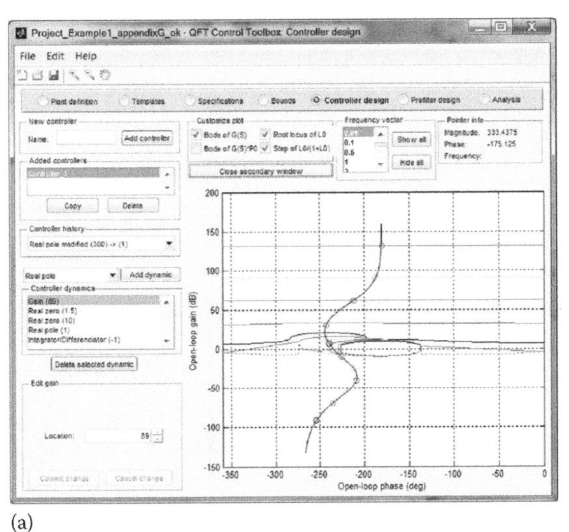

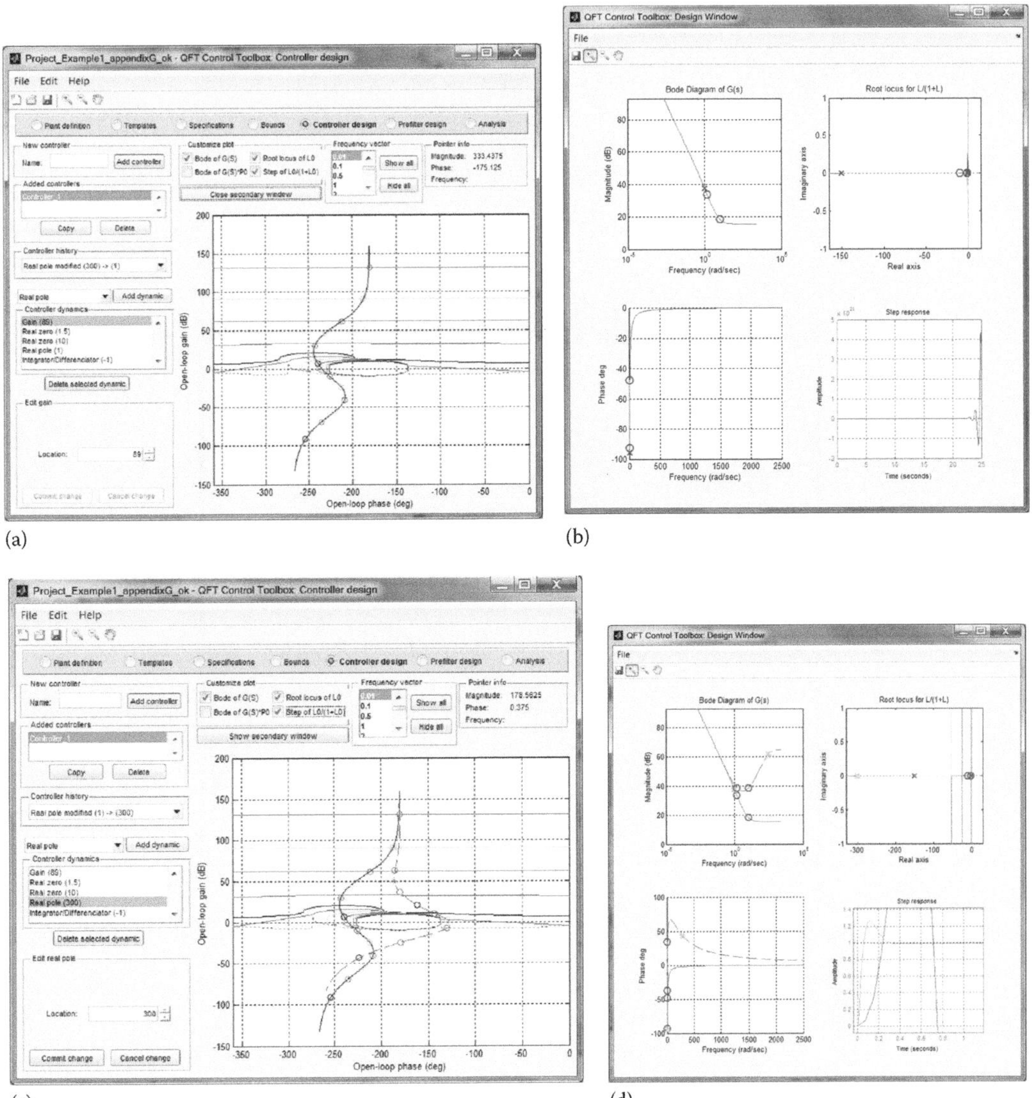

(a)

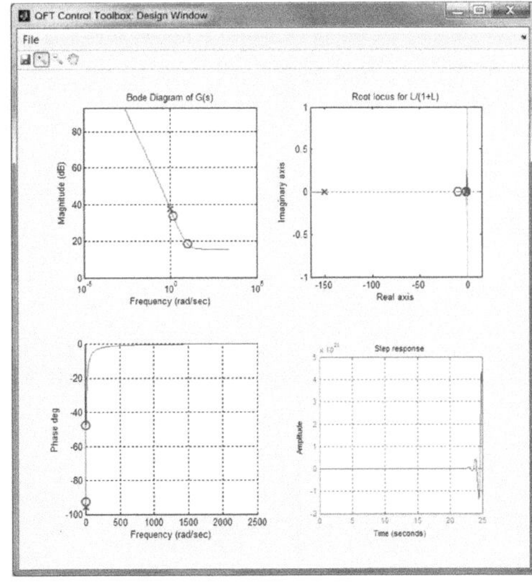

(b)

(c)

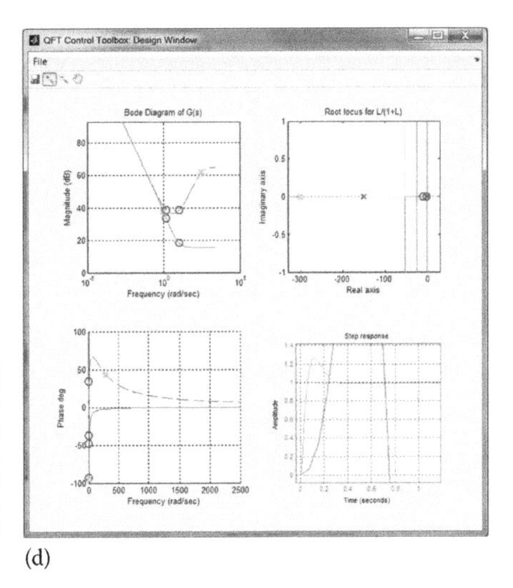

(d)

FIGURE F.32
An example of the graphic dynamic edition. (a) Controller design windows, Nichols chart, (b) secondary window, Bode diagram, root locus and step time response, (c) controller design window after moving a pole from $p = 1$ to $p = 300\{1/[(s/p)+1]\}$, (d) secondary window after moving that pole.

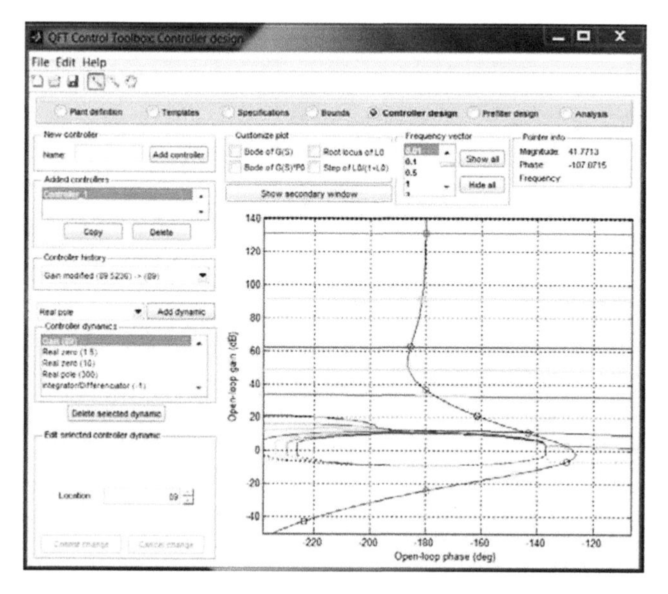

FIGURE G.7
Controller design (Nichols chart): Example 1.

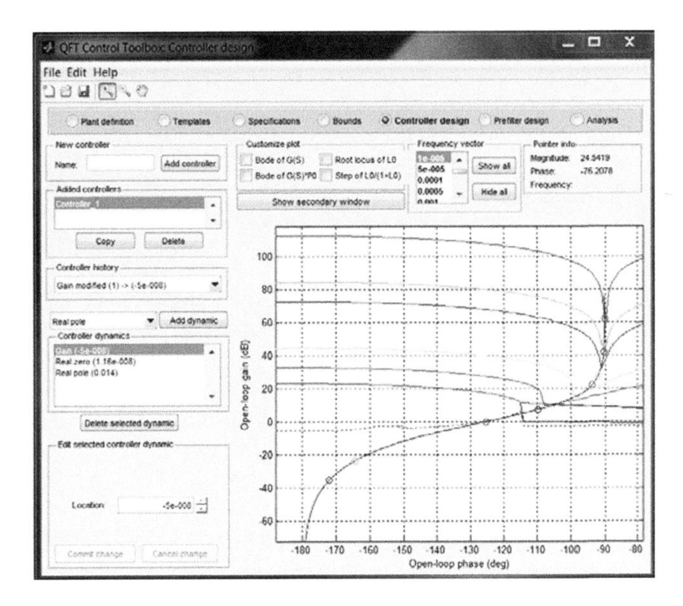

FIGURE G.13
Controller design (Nichols chart): Example 2.

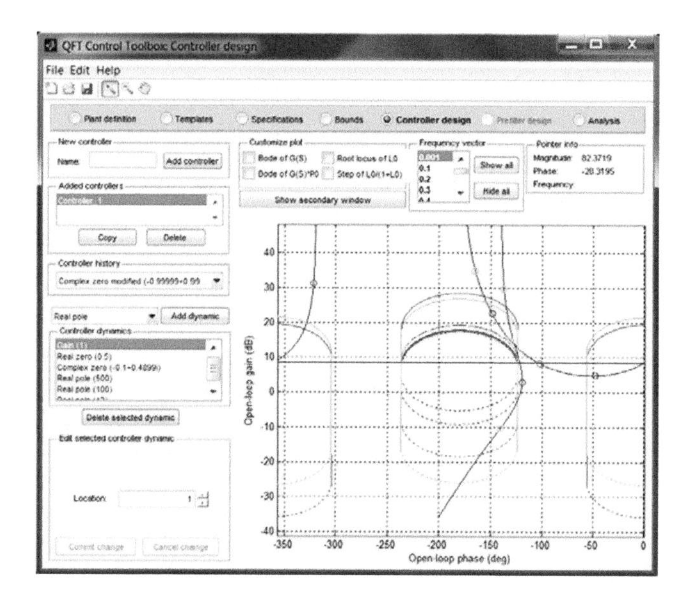

FIGURE G.18
Controller design (Nichols chart): Example 3.

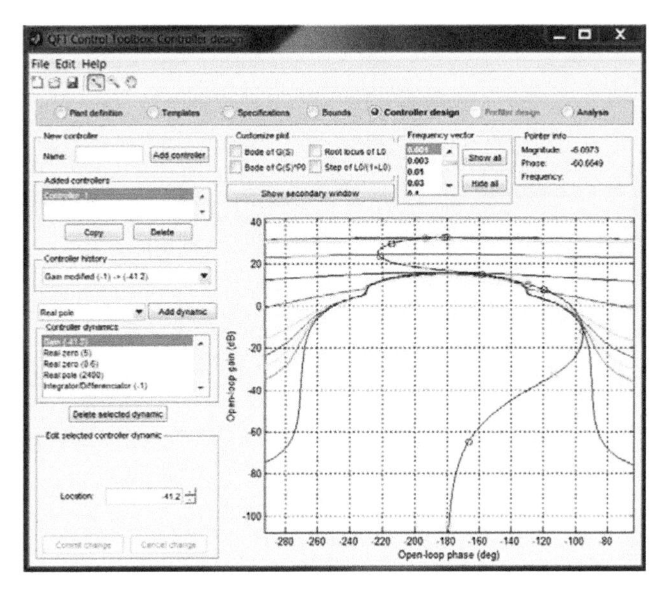

FIGURE G.21
Controller design (Nichols chart): Example 4.

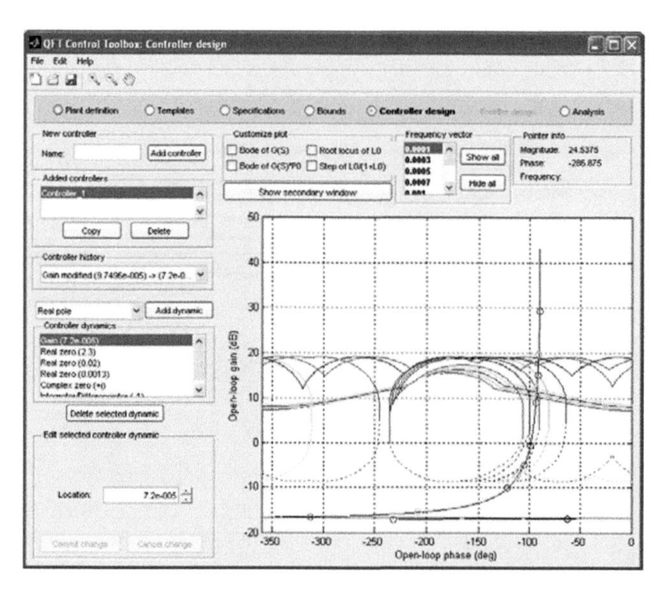

FIGURE G.25
Controller design (Nichols chart): Example 5.

Thus, the matrices of the State Space description are

$$A=\begin{bmatrix} 0 & 0 & 0 & 0 & 1 & 0 & 0 & 0 \\ 0 & 0 & 0 & 0 & 0 & 1 & 0 & 0 \\ 0 & 0 & 0 & 0 & 0 & 0 & 1 & 0 \\ 0 & 0 & 0 & 0 & 0 & 0 & 0 & 1 \\ -\dfrac{K_t}{m_1} & \dfrac{NK_b r_b}{m_1} & 0 & 0 & -\dfrac{B_t}{m_1} & \dfrac{NB_b r_b}{m_1} & 0 & 0 \\ \dfrac{K_t}{m_1 r_b} & -\dfrac{(m_1+m_2)NK_b}{m_1 m_2} & 0 & 0 & \dfrac{B_t}{m_1 r_b} & -\dfrac{(m_1+m_2)NB_b}{m_1 m_2} & 0 & 0 \\ 0 & 0 & -\dfrac{K_s}{I_r} & \dfrac{K_s}{I_r} & 0 & 0 & -\dfrac{B_s}{I_r} & \dfrac{B_s}{I_r} \\ 0 & 0 & \dfrac{K_s}{I_g} & -\dfrac{K_s}{I_g} & 0 & 0 & \dfrac{B_s}{I_g} & -\dfrac{B_s}{I_g} \end{bmatrix} \tag{12.55}$$

$$B=\begin{bmatrix} 0 & 0 & 0 \\ 0 & 0 & 0 \\ 0 & 0 & 0 \\ 0 & 0 & 0 \\ \dfrac{1}{3m_1} & 0 & 0 \\ \dfrac{2m_1-m_2}{3m_1 m_2 r_b} & 0 & 0 \\ 0 & \dfrac{1}{I_r} & 0 \\ 0 & 0 & -\dfrac{1}{I_g} \end{bmatrix}; \quad C=\begin{bmatrix} 0 & 0 & 0 & 0 & 1 & 0 & 0 & 0 \\ 0 & 0 & 0 & 0 & 0 & 1 & 0 & 0 \\ 0 & 0 & 0 & 0 & 0 & 0 & 1 & 0 \\ 0 & 0 & 0 & 0 & 0 & 0 & 0 & 1 \end{bmatrix} \tag{12.56}$$

12.3.3 Mechanical Dynamics of DFIG Wind Turbines (with Gearbox)

This section presents a five-mass model for the DFIG WT. The approach considers 5 degrees of freedom (rotor angle, generator angle, gearbox angle, axial displacement of the nacelle, and angular displacement of the blades) and five independent masses (one for the rotor, one for the generator, one for the gearbox, one for the blades, and one for the tower).

The method is more complete than the classical one-mass model, which only considers one lumped mass (inertia) for all the rotating components of the drive-train, or the classical two-mass model, which considers two masses (inertias) for the low-speed and high-speed shafts respectively.

Figure 12.12 shows the variable-speed DFIG WT. A two-part horizontal drive shaft connects, through a gearbox, a large rotor inertia at one end (blades and hub) with a smaller inertia (generator) at the other end. The wind applies an aerodynamic torque T_r on the low-speed shaft. The generator applies an antagonist electrical torque T_g on the high-speed shaft.

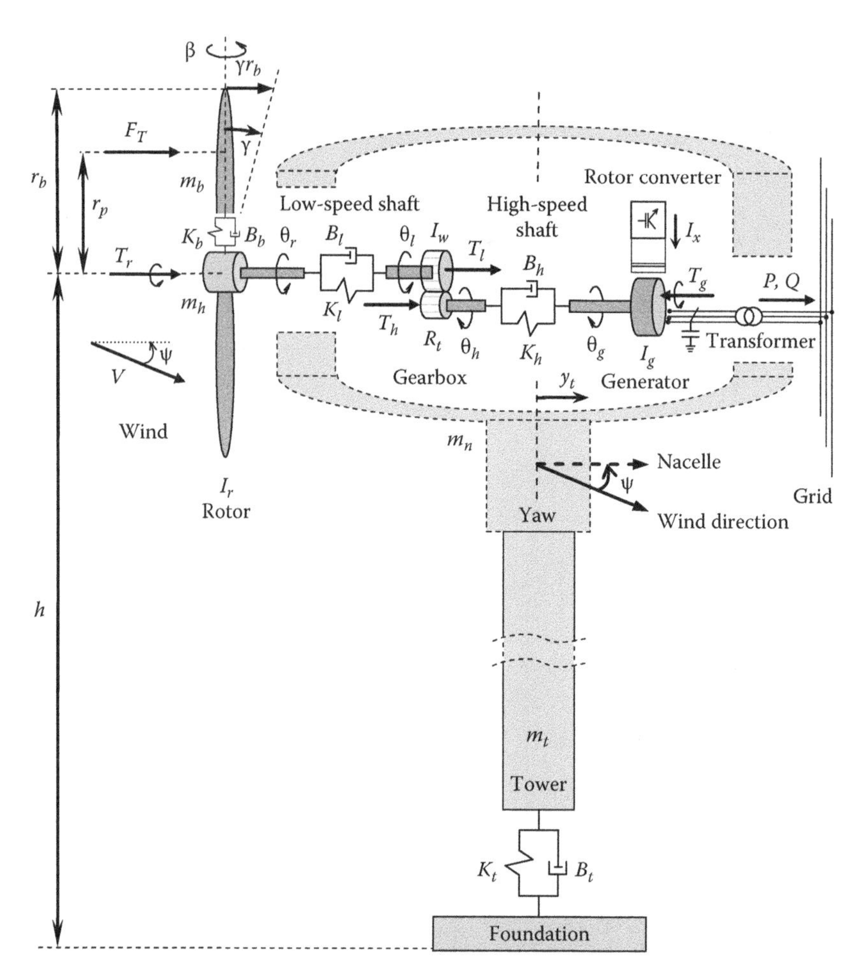

FIGURE 12.12
A classical variable-speed DFIG WT.

The rotor (blades and hub) presents a moment of inertia I_r. The gearbox shows an equivalent moment of inertia I_w at the low-speed part. The low-speed shaft has a torsional stiffness coefficient K_l and a torsional damping coefficient B_l. The increasing ratio between the low- and high-speed parts in the gearbox is R_t. The high-speed shaft has a torsional stiffness coefficient K_h and a torsional damping coefficient B_h. The generator shows a moment of inertia I_g. The rotor angle is θ_r, the rotor speed $\Omega_r = \dot{\theta}_r$, the generator angle θ_g, and the generator speed $\Omega_g = \dot{\theta}_g$. The angles at the gearbox are θ_l at the low-speed part and θ_h at the high-speed part.

Note that two torques are applied to the gearbox: T_l by the low-speed shaft and T_h by the high-speed shaft. ψ is the yaw angle error (nacelle-wind angle) and β the pitch angle (blades). I_x is the excitation current introduced in the rotor, P and Q the active and reactive power supplied to the grid, respectively, and f, U, and ϕ the frequency, voltage, and power factor at the grid connection point, respectively. The main mechanical equations of the DFIG are presented next.

1. *Generalized coordinates, q_i*

$$q = [q_i] = \begin{bmatrix} y_t & \gamma & \theta_r & \theta_g & \theta_l \end{bmatrix}^T \tag{12.57}$$

where
y_t is the axial displacement of the nacelle
γ is the angular displacement of the blades out of the plane of rotation
θ_r is the rotor angular position
θ_g is the generator angular position
θ_l is the gearbox low-speed shaft position

2. *Energy and dissipation function equations, E_k, E_p, D_n*
The kinetic energy E_k, potential energy E_p, and dissipation function D_n of the DFIG WT shown in Figure 12.12 are derived as follows:

$$E_k = \frac{m_1}{2}\dot{y}_t^2 + \frac{m_2}{2}(r_b\dot{\gamma} + \dot{y}_t)^2 + \frac{I_r}{2}\dot{\theta}_r^2 + \frac{I_g}{2}\dot{\theta}_g^2 + \frac{I_w}{2}\dot{\theta}_l^2 \tag{12.58}$$

$$E_p = \frac{K_t}{2}y_t^2 + \frac{N}{2}K_b(r_b\gamma)^2 + \frac{K_l}{2}(\theta_r - \theta_l)^2 + \frac{K_h}{2}(R_t\theta_l - \theta_g)^2 \tag{12.59}$$

$$D_n = \frac{B_t}{2}\dot{y}_t^2 + \frac{N}{2}B_b(r_b\dot{\gamma})^2 + \frac{B_l}{2}(\dot{\theta}_r - \dot{\theta}_l)^2 + \frac{B_h}{2}(R_t\dot{\theta}_l - \dot{\theta}_g)^2 \tag{12.60}$$

where

$$\theta_h = R_t\theta_l \tag{12.61}$$

$$\dot{\theta}_h = R_t\dot{\theta}_l \tag{12.62}$$

3. *Lagrange equations terms*
Based on Equations 12.58 through 12.60, the terms of the Euler–Lagrange equation (12.40) are as follows (note that $\partial E_p/\partial \dot{q}_i = 0$ and $\partial E_k/\partial q_i = 0$):

$$\frac{d}{dt}\left(\frac{\partial L}{\partial \dot{q}_i}\right) = \frac{d}{dt}\begin{bmatrix} \dfrac{\partial E_k}{\partial \dot{y}_t} \\ \dfrac{\partial E_k}{\partial \dot{\gamma}} \\ \dfrac{\partial E_k}{\partial \dot{\theta}_r} \\ \dfrac{\partial E_k}{\partial \dot{\theta}_g} \\ \dfrac{\partial E_k}{\partial \dot{\theta}_l} \end{bmatrix} = \begin{bmatrix} m_1 + m_2 & m_2 r_b & 0 & 0 & 0 \\ m_2 r_b & m_2 r_b^2 & 0 & 0 & 0 \\ 0 & 0 & I_r & 0 & 0 \\ 0 & 0 & 0 & I_g & 0 \\ 0 & 0 & 0 & 0 & I_w \end{bmatrix} \begin{bmatrix} \ddot{y}_t \\ \ddot{\gamma} \\ \ddot{\theta}_r \\ \ddot{\theta}_g \\ \ddot{\theta}_l \end{bmatrix} = \mathcal{M}\ddot{q} \tag{12.63}$$

$$-\frac{\partial L}{\partial q_i} = \frac{\partial E_p}{\partial q_i} = \begin{bmatrix} \frac{\partial E_p}{\partial y_t} \\ \frac{\partial E_p}{\partial \gamma} \\ \frac{\partial E_p}{\partial \theta_r} \\ \frac{\partial E_p}{\partial \theta_g} \\ \frac{\partial E_p}{\partial \theta_l} \end{bmatrix} = \begin{bmatrix} K_t & 0 & 0 & 0 & 0 \\ 0 & NK_b r_b^2 & 0 & 0 & 0 \\ 0 & 0 & K_l & 0 & -K_l \\ 0 & 0 & 0 & K_h & -K_h R_t \\ 0 & 0 & -K_l & -K_h R_t & K_l + K_h R_t^2 \end{bmatrix} \begin{bmatrix} y_t \\ \gamma \\ \theta_r \\ \theta_g \\ \theta_l \end{bmatrix} = \mathcal{K} q \quad (12.64)$$

$$\frac{\partial D_n}{\partial \dot{q}_i} = \begin{bmatrix} \frac{\partial D_n}{\partial \dot{y}_t} \\ \frac{\partial D_n}{\partial \dot{\gamma}} \\ \frac{\partial D_n}{\partial \dot{\theta}_r} \\ \frac{\partial D_n}{\partial \dot{\theta}_g} \\ \frac{\partial D_n}{\partial \dot{\theta}_l} \end{bmatrix} = \begin{bmatrix} B_t & 0 & 0 & 0 & 0 \\ 0 & NB_b r_b^2 & 0 & 0 & 0 \\ 0 & 0 & B_l & 0 & -B_l \\ 0 & 0 & 0 & B_h & -B_h R_t \\ 0 & 0 & -B_l & -B_h R_t & B_l + B_h R_t^2 \end{bmatrix} \begin{bmatrix} \dot{y}_t \\ \dot{\gamma} \\ \dot{\theta}_r \\ \dot{\theta}_g \\ \dot{\theta}_l \end{bmatrix} = \mathcal{C} \dot{q} \quad (12.65)$$

$$\mathcal{Q} = \begin{bmatrix} F_T \\ r_p F_T \\ T_r \\ -T_g \\ 0 \end{bmatrix} = \begin{bmatrix} 1 & 0 & 0 \\ r_p & 0 & 0 \\ 0 & 1 & 0 \\ 0 & 0 & -1 \\ 0 & 0 & 0 \end{bmatrix} \begin{bmatrix} F_T \\ T_r \\ T_g \end{bmatrix} = \mathcal{R} u \quad (12.66)$$

Using Equations 12.63 through 12.66, Equation 12.40 becomes Equation 12.39 or 12.49: $\mathcal{M}\ddot{q} + \mathcal{C}\dot{q} + \mathcal{K}q = \mathcal{Q}(\dot{q}, q, u, t)$, $\ddot{q} = -\mathcal{M}^{-1}\mathcal{C}\dot{q} - \mathcal{M}^{-1}\mathcal{K}q + \mathcal{M}^{-1}\mathcal{R}u$, respectively.

4. *Description of the system in the State Space*
   The following three vectors are defined:

$$\text{State variables: } x = [y_t \quad \gamma \quad \theta_r \quad \theta_g \quad \theta_l \quad \dot{y}_t \quad \dot{\gamma} \quad \dot{\theta}_r \quad \dot{\theta}_g \quad \dot{\theta}_l]^T \quad (12.67)$$

$$\text{Inputs: } u = [F_T \quad T_r \quad T_g]^T \quad (12.68)$$

$$\text{Outputs: } y = [\dot{y}_t \quad \dot{\gamma} \quad \dot{\theta}_r \quad \dot{\theta}_g \quad \dot{\theta}_l]^T \quad (12.69)$$

Equation 12.49 takes the form of a classical State Space description of the system ($\dot{x} = Ax + Bu$; $y = Cx$), so that

$$\underset{10\times1}{\dot{x}} = \left[\begin{array}{c|c} \underset{5\times5}{\mathbf{0}} & \underset{5\times5}{I} \\ \hline -\underset{5\times5}{\mathcal{M}^{-1}}\underset{5\times5}{\mathcal{K}} & -\underset{5\times5}{\mathcal{M}^{-1}}\underset{5\times5}{\mathcal{C}} \end{array}\right]\underset{10\times1}{x} + \left[\begin{array}{c} \underset{5\times3}{\mathbf{0}} \\ \hline \underset{5\times5}{\mathcal{M}^{-1}}\underset{5\times3}{\mathcal{R}} \end{array}\right]\underset{3\times1}{u}$$
$$\underset{5\times1}{y} = \left[\begin{array}{c|c} \underset{5\times5}{\mathbf{0}} & \underset{5\times5}{I} \end{array}\right]\underset{10\times1}{x} \tag{12.70}$$

Note that Equation 12.70 is a system with three inputs and five outputs (15 transfer functions). The calculation of the inverse of $\mathcal{M}$ yields

$$\mathcal{M}^{-1} = \begin{bmatrix} \frac{1}{m_1} & \frac{-1}{m_1 r_b} & 0 & 0 & 0 \\ \frac{-1}{m_1 r_b} & \frac{m_1+m_2}{m_1 m_2 r_b^2} & 0 & 0 & 0 \\ 0 & 0 & \frac{1}{I_r} & 0 & 0 \\ 0 & 0 & 0 & \frac{1}{I_g} & 0 \\ 0 & 0 & 0 & 0 & \frac{1}{I_w} \end{bmatrix} \tag{12.71}$$

Thus, the matrices of the State Space description are

$$A = \begin{bmatrix} 0 & 0 & 0 & 0 & 0 & 1 & 0 & 0 & 0 & 0 \\ 0 & 0 & 0 & 0 & 0 & 0 & 1 & 0 & 0 & 0 \\ 0 & 0 & 0 & 0 & 0 & 0 & 0 & 1 & 0 & 0 \\ 0 & 0 & 0 & 0 & 0 & 0 & 0 & 0 & 1 & 0 \\ 0 & 0 & 0 & 0 & 0 & 0 & 0 & 0 & 0 & 1 \\ -\frac{K_t}{m_1} & \frac{NK_b r_b}{m_1} & 0 & 0 & 0 & -\frac{B_t}{m_1} & \frac{NB_b r_b}{m_1} & 0 & 0 & 0 \\ \frac{K_t}{m_1 r_b} & -\frac{(m_1+m_2)NK_b}{m_1 m_2} & 0 & 0 & 0 & \frac{B_t}{m_1 r_b} & -\frac{(m_1+m_2)NB_b}{m_1 m_2} & 0 & 0 & 0 \\ 0 & 0 & -\frac{K_l}{I_r} & 0 & \frac{K_l}{I_r} & 0 & 0 & -\frac{B_l}{I_r} & 0 & \frac{B_l}{I_r} \\ 0 & 0 & 0 & -\frac{K_h}{I_g} & \frac{K_h R_t}{I_g} & 0 & 0 & 0 & -\frac{B_h}{I_g} & \frac{B_h R_t}{I_g} \\ 0 & 0 & \frac{K_l}{I_w} & \frac{K_h R_t}{I_w} & -\frac{K_l + K_h R_t^2}{I_w} & 0 & 0 & \frac{B_l}{I_w} & \frac{B_h R_t}{I_w} & -\frac{B_l + B_h R_t^2}{I_w} \end{bmatrix} \tag{12.72}$$

$$
\boldsymbol{B} = \begin{bmatrix} 0 & 0 & 0 \\ 0 & 0 & 0 \\ 0 & 0 & 0 \\ 0 & 0 & 0 \\ 0 & 0 & 0 \\ \frac{1}{m_1}\left(1 - \frac{r_p}{r_b}\right) & 0 & 0 \\ \frac{1}{m_1 r_b}\left(\frac{(m_1 + m_2) r_p}{m_2 r_b} - 1\right) & 0 & 0 \\ 0 & \frac{1}{I_r} & 0 \\ 0 & 0 & -\frac{1}{I_g} \\ 0 & 0 & 0 \end{bmatrix}; \quad \boldsymbol{C} = \begin{bmatrix} 0 & 0 & 0 & 0 & 0 & 1 & 0 & 0 & 0 & 0 \\ 0 & 0 & 0 & 0 & 0 & 0 & 1 & 0 & 0 & 0 \\ 0 & 0 & 0 & 0 & 0 & 0 & 0 & 1 & 0 & 0 \\ 0 & 0 & 0 & 0 & 0 & 0 & 0 & 0 & 1 & 0 \\ 0 & 0 & 0 & 0 & 0 & 0 & 0 & 0 & 0 & 1 \end{bmatrix} \tag{12.73}
$$

Note that by making $R_t = 1$; $K_l = K_s$; $B_l = B_s$; $I_w = I_g$; $K_h = 0$; $B_h = 0$; $\theta_l = \theta_g$ the mechanical model of the DFIG WT (see Equations 12.72 and 12.73) becomes the mechanical model of the DD WT (see Equations 12.55 and 12.56).

12.3.4 Wind Turbine Transfer Matrix

The transfer matrix description $\boldsymbol{G}(s)$ of the WT, valid for both the classical variable-speed DFIG and the multipole DD is calculated in this section. The general State Space model representation calculated in the last section and the well-known transformation $\boldsymbol{G}(s) = \boldsymbol{C}(s\boldsymbol{I} - \boldsymbol{A})^{-1}\boldsymbol{B}$ for $\boldsymbol{y}(s) = \boldsymbol{G}(s)\boldsymbol{u}(s)$ are used to derive the transfer function $\boldsymbol{G}(s)$:

$$
\boldsymbol{y}(s) = \begin{bmatrix} \dot{y}_t(s) \\ \dot{\gamma}(s) \\ \Omega_r(s) \\ \Omega_g(s) \\ \Omega_l(s) \end{bmatrix} = \begin{bmatrix} \mu_{11}(s) & 0 & 0 \\ \mu_{21}(s) & 0 & 0 \\ 0 & \mu_{32}(s) & \mu_{33}(s) \\ 0 & \mu_{42}(s) & \mu_{43}(s) \\ 0 & \mu_{52}(s) & \mu_{53}(s) \end{bmatrix} \begin{bmatrix} F_T(s) \\ T_r(s) \\ T_g(s) \end{bmatrix} = \boldsymbol{G}(s)\boldsymbol{u}(s) \tag{12.74}
$$

with

$$
\mu_{ij}(s) = \frac{n_{\mu ij}(s)}{d_{\mu ij}(s)}, \quad i = 1,\ldots,5;\ j = 1,\ldots,3 \tag{12.75}
$$

$$
n_{\mu 11}(s) = s[m_2(r_b - r_p)s^2 + NB_b r_b s + NK_b r_b] \tag{12.76}
$$

$$
\begin{aligned} d_{\mu 11}(s) = r_b[&m_1 m_2 s^4 + (NB_b m_2 + B_t m_2 + NB_b m_1)s^3 \cdots \\ &+ (K_t m_2 + B_t NB_b + NK_b m_1 + NK_b m_2)s^2 \cdots \\ &+ (B_t NK_b + NB_b K_t)s + K_t NK_b] \end{aligned} \tag{12.77}
$$

$$n_{\mu 21}(s) = s[(m_1 r_p + m_2(r_p - r_b))s^2 + B_t r_p s + K_t r_p] \tag{12.78}$$

$$d_{\mu 21}(s) = r_b d_{\mu 11}(s) \tag{12.79}$$

$$\begin{aligned} n_{\mu 32}(s) = {} & I_g I_w s^4 + (B_h I_w + I_g B_l + I_g B_h R_t^2)s^3 \cdots \\ & + (B_h B_l + I_g K_l + I_g K_h R_t^2 + K_h I_w)s^2 \cdots \\ & + (K_h B_l + B_h K_l)s + K_h K_l \end{aligned} \tag{12.80}$$

$$\begin{aligned} d_{\mu 32}(s) = {} & s[I_r I_w I_g s^4 + (I_r I_g B_l + B_h I_r I_w + I_r I_g B_h R_t^2 + B_l I_w I_g)s^3 \cdots \\ & + (K_l I_w I_g + B_h I_r B_l + B_l I_g B_h R_t^2 + K_h I_r I_w + I_r I_g K_l + B_l B_h I_w \cdots \\ & + I_r I_g K_h R_t^2)s^2 + (B_l K_h I_w + K_l I_g B_h R_t^2 + K_h I_r B_l + B_h I_r K_l \cdots \\ & + K_l B_h I_w + I_g B_l K_h R_t^2)s + (I_g K_l K_h R_t^2 + K_h I_r K_l + K_l K_h I_w)] \end{aligned} \tag{12.81}$$

$$n_{\mu 33}(s) = -B_h R_t B_l s^2 - R_t(K_h B_l + B_h K_l)s - K_h R_t K_l \tag{12.82}$$

$$d_{\mu 33}(s) = d_{\mu 32}(s) \tag{12.83}$$

$$n_{\mu 42}(s) = -n_{\mu 33}(s) \tag{12.84}$$

$$d_{\mu 42}(s) = d_{\mu 32}(s) \tag{12.85}$$

$$\begin{aligned} n_{\mu 43}(s) = {} & -I_r I_w s^4 - (I_r B_h R_t^2 + B_l I_w + I_r B_l)s^3 \cdots \\ & - (B_l B_h R_t^2 + I_r K_l + I_r K_h R_t^2 + K_l I_w)s^2 \cdots \\ & - (B_l K_h R_t^2 + K_l B_h R_t^2)s - K_l K_h R_t^2 \end{aligned} \tag{12.86}$$

$$d_{\mu 43}(s) = d_{\mu 32}(s) \tag{12.87}$$

$$n_{\mu 52}(s) = B_l I_g s^3 + (B_l B_h + I_g K_l)s^2 + (B_l K_h + K_l B_h)s + K_l K_h \tag{12.88}$$

$$d_{\mu 52}(s) = d_{\mu 32}(s) \tag{12.89}$$

$$n_{\mu 53}(s) = -I_r B_h R_t s^3 - R_t(K_h I_r + B_l B_h)s^2 - R_t(B_l K_h + K_l B_h)s - K_l K_h R_t \tag{12.90}$$

$$d_{\mu 53}(s) = d_{\mu 32}(s) \tag{12.91}$$

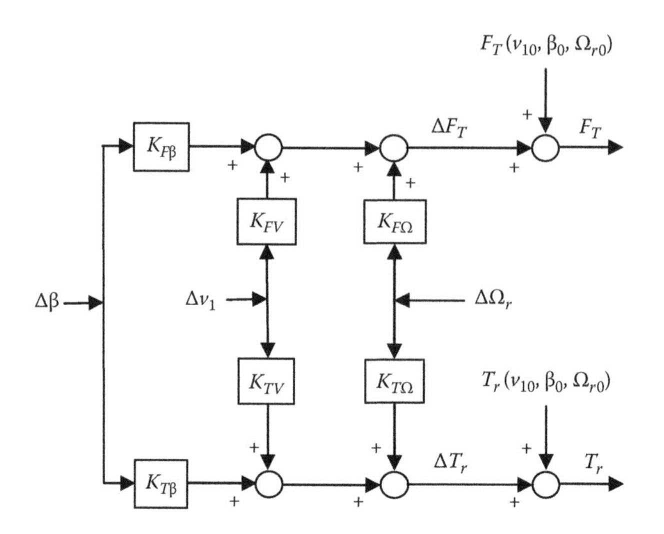

FIGURE 12.13
Linearization of F_T and T_r around the working point (v_{10}, β_0, Ω_{r0}).

According to the aerodynamic equations (12.36) and (12.37) and the characteristics of C_P and C_T, the inputs of Equation 12.74, F_T and T_r, depend on v_1, β, and Ω_r in a nonlinear way. If the aerodynamic part of these equations is linearized around a working point (v_{10}, β_0, Ω_{r0}) and the bias components are ignored (see Figure 12.13), then the inputs F_T and T_r can be described by a transfer matrix whose elements are just gains, so that

$$\begin{bmatrix} F_T(s) \\ T_r(s) \end{bmatrix} = \begin{bmatrix} K_{F\Omega} & K_{FV} & K_{F\beta} \\ K_{T\Omega} & K_{TV} & K_{T\beta} \end{bmatrix} \begin{bmatrix} \Omega_r(s) \\ v_1(s) \\ \beta(s) \end{bmatrix} \tag{12.92}$$

where the gains are calculated by using the C_T and C_P curves and Equations 12.36 and 12.37, so that

$$K_{F\Omega} = \left.\frac{\partial F_T(t)}{\partial \Omega_r(t)}\right|_0 = \frac{1}{2}\rho\pi r_b^2 \left.\frac{\partial C_T}{\partial \Omega_r}\right|_0 v_{10}^2 \tag{12.93}$$

$$K_{FV} = \left.\frac{\partial F_T(t)}{\partial v_1(t)}\right|_0 = \frac{1}{2}\rho\pi r_b^2 \left[\left.\frac{\partial C_T}{\partial v_1}\right| v_{10}^2 + 2v_{10}C_{r0}\right] \tag{12.94}$$

$$K_{F\beta} = \left.\frac{\partial F_T(t)}{\partial \beta(t)}\right|_0 = \frac{1}{2}\rho\pi r_b^2 \left.\frac{\partial C_T}{\partial \beta}\right|_0 v_{10}^2 \tag{12.95}$$

$$K_{T\Omega} = \left.\frac{\partial T_r(t)}{\partial \Omega_r(t)}\right|_0 = \frac{1}{2}\rho\pi r_b^2 \left(\left.\frac{\partial C_P}{\partial \Omega_r}\right|_0 \frac{1}{\Omega_{r0}} - C_{P0}\frac{1}{\Omega_{r0}^2}\right) v_{10}^3 \tag{12.96}$$

$$K_{TV} = \left.\frac{\partial T_r(t)}{\partial v_1(t)}\right|_0 = \frac{1}{2}\rho\pi r_b^2 \frac{1}{\Omega_{r0}}\left(\left.\frac{\partial C_p}{\partial v_1}\right|_0 v_{10}^3 + 3C_{P0}v_{10}^2\right) \tag{12.97}$$

$$K_{T\beta} = \left.\frac{\partial T_r(t)}{\partial \beta(t)}\right|_0 = \frac{1}{2}\rho\pi r_b^2 \frac{1}{\Omega_{r0}}\left.\frac{\partial C_P}{\partial \beta}\right|_0 v_{10}^3 \tag{12.98}$$

By combining this linearized description of the aerodynamic part, Equations 12.92 through 12.98, with the transfer matrix of the mechanical part, Equations 12.74 through 12.91, the transfer matrix for the overall WT system is obtained. Now, the manipulated inputs (actuators) are the blade pitch angle β and the electrical torque T_g and the external disturbance input is the wind speed v_1, so that

$$\begin{bmatrix} \dot{y}_t(s) \\ \dot{\gamma}(s) \\ \Omega_r(s) \\ \Omega_g(s) \\ \Omega_l(s) \end{bmatrix} = \begin{bmatrix} \mu_{11}(s)\dfrac{\mu_{32}(s)(K_{F\Omega}K_{T\beta} - K_{F\beta}K_{T\Omega}) + K_{F\beta}}{1-\mu_{32}(s)K_{T\Omega}} & \mu_{11}(s)\dfrac{\mu_{33}(s)K_{F\Omega}}{1-\mu_{32}(s)K_{T\Omega}} \\ \mu_{21}(s)\dfrac{\mu_{32}(s)(K_{F\Omega}K_{T\beta} - K_{F\beta}K_{T\Omega}) + K_{F\beta}}{1-\mu_{32}(s)K_{T\Omega}} & \mu_{21}(s)\dfrac{\mu_{33}(s)K_{F\Omega}}{1-\mu_{32}(s)K_{T\Omega}} \\ \mu_{32}(s)\dfrac{K_{T\beta}}{1-\mu_{32}(s)K_{T\Omega}} & \mu_{33}(s)\dfrac{1}{1-\mu_{32}(s)K_{T\Omega}} \\ \mu_{42}(s)\dfrac{K_{T\beta}}{1-\mu_{32}(s)K_{T\Omega}} & \dfrac{\mu_{42}(s)\mu_{33}(s)K_{T\Omega} + \mu_{43}(s) - \mu_{43}(s)\mu_{32}(s)K_{T\Omega}}{1-\mu_{32}(s)K_{T\Omega}} \\ \mu_{52}(s)\dfrac{K_{T\beta}}{1-\mu_{32}(s)K_{T\Omega}} & \dfrac{\mu_{52}(s)\mu_{33}(s)K_{T\Omega} + \mu_{53}(s) - \mu_{53}(s)\mu_{32}(s)K_{T\Omega}}{1-\mu_{32}(s)K_{T\Omega}} \end{bmatrix} \begin{bmatrix} \beta(s) \\ T_g(s) \end{bmatrix}$$

$$+ \begin{bmatrix} \mu_{11}(s)\dfrac{\mu_{32}(s)(K_{F\Omega}K_{TV} - K_{FV}K_{T\Omega}) + K_{FV}}{1-\mu_{32}(s)K_{T\Omega}} \\ \mu_{21}(s)\dfrac{\mu_{32}(s)(K_{F\Omega}K_{TV} - K_{FV}K_{T\Omega}) + K_{FV}}{1-\mu_{32}(s)K_{T\Omega}} \\ \mu_{32}(s)\dfrac{K_{TV}}{1-\mu_{32}(s)K_{T\Omega}} \\ \mu_{42}(s)\dfrac{K_{TV}}{1-\mu_{32}(s)K_{T\Omega}} \\ \mu_{52}(s)\dfrac{K_{TV}}{1-\mu_{32}(s)K_{T\Omega}} \end{bmatrix} v_1(s) \tag{12.99}$$

The transfer functions of the actuators are

$$\beta(s) = A_\beta(s)\beta_d(s) \tag{12.100}$$

$$T_g(s) = A_T(s)T_{gd}(s) \tag{12.101}$$

where

β_d is the demanded blade pitch angle
T_{gd} is the demanded electrical torque

both calculated by the control system and where $A_\beta(s)$ and $A_T(s)$ are the transfer functions from the control signals (β_d, T_{gd}) to the actual value of the actuators (β, T_g). Thus, by use of Equations 12.99 through 12.101, the following equation is obtained:

$$\begin{bmatrix} \dot{y}_t(s) \\ \dot{\gamma}(s) \\ \Omega_r(s) \\ \Omega_g(s) \\ \Omega_l(s) \end{bmatrix} = \boldsymbol{P}(s) \begin{bmatrix} \beta_d(s) \\ T_{gd}(s) \end{bmatrix} + \boldsymbol{D}(s) v_1(s) \tag{12.102}$$

where the plant matrix is

$$\boldsymbol{P}(s) = \begin{bmatrix} \mu_{11}(s)\dfrac{\mu_{32}(s)\left(K_{F\Omega}K_{T\beta} - K_{F\beta}K_{T\Omega}\right) + K_{F\beta}}{1-\mu_{32}(s)K_{T\Omega}} A_\beta(s) & \mu_{11}(s)\dfrac{\mu_{33}(s)K_{F\Omega}}{1-\mu_{32}(s)K_{T\Omega}} A_T(s) \\ \mu_{21}(s)\dfrac{\mu_{32}(s)\left(K_{F\Omega}K_{T\beta} - K_{F\beta}K_{T\Omega}\right) + K_{F\beta}}{1-\mu_{32}(s)K_{T\Omega}} A_\beta(s) & \mu_{21}(s)\dfrac{\mu_{33}(s)K_{F\Omega}}{1-\mu_{32}(s)K_{T\Omega}} A_T(s) \\ \mu_{32}(s)\dfrac{K_{T\beta}}{1-\mu_{32}(s)K_{T\Omega}} A_\beta(s) & \mu_{33}(s)\dfrac{1}{1-\mu_{32}(s)K_{T\Omega}} A_T(s) \\ \mu_{42}(s)\dfrac{K_{T\beta}}{1-\mu_{32}(s)K_{T\Omega}} A_\beta(s) & \dfrac{\mu_{42}(s)\mu_{33}(s)K_{T\Omega} + \mu_{43}(s) - \mu_{43}(s)\mu_{32}(s)K_{T\Omega}}{1-\mu_{32}(s)K_{T\Omega}} A_T(s) \\ \mu_{52}(s)\dfrac{K_{T\beta}}{1-\mu_{32}(s)K_{T\Omega}} A_\beta(s) & \dfrac{\mu_{52}(s)\mu_{33}(s)K_{T\Omega} + \mu_{53}(s) - \mu_{53}(s)\mu_{32}(s)K_{T\Omega}}{1-\mu_{32}(s)K_{T\Omega}} A_T(s) \end{bmatrix} \tag{12.103}$$

and the disturbance transfer matrix is

$$\boldsymbol{D}(s) = \begin{bmatrix} \mu_{11}(s)\dfrac{\mu_{32}(s)\left(K_{F\Omega}K_{TV} - K_{FV}K_{T\Omega}\right) + K_{FV}}{1-\mu_{32}(s)K_{T\Omega}} \\ \mu_{21}(s)\dfrac{\mu_{32}(s)\left(K_{F\Omega}K_{TV} - K_{FV}K_{T\Omega}\right) + K_{FV}}{1-\mu_{32}(s)K_{T\Omega}} \\ \mu_{32}(s)\dfrac{K_{TV}}{1-\mu_{32}(s)K_{T\Omega}} \\ \mu_{42}(s)\dfrac{K_{TV}}{1-\mu_{32}(s)K_{T\Omega}} \\ \mu_{52}(s)\dfrac{K_{TV}}{1-\mu_{32}(s)K_{T\Omega}} \end{bmatrix} \tag{12.104}$$

12.3.5 Rotor Speed Wind Turbine Transfer Functions

The rotational speed Ω_r of the WT rotor is continuously modified (1) by the actuators (demanded control signals for the blade pitch angles β_d and for the electrical torque T_{gd}), (2) by the external disturbance inputs (wind speed v_1), and (3) through the dynamics introduced in the last section (rotor speed Ω_r itself).

By using (1) the WT transfer matrix description (see Equations 12.102 through 12.104); (2) the expression of $\mu_{32}(s)$, given by Equations 12.80 and 12.81; (3) the expression of $\mu_{33}(s)$, given by Equations 12.82 and 12.83; (4) the static coefficients $K_{T\Omega}$, K_{TV}, and $K_{T\beta}$ given by Equations 12.96, 12.97, and 12.98, respectively; and (5) the actuator transfer functions $A_\beta(s)$ and $A_T(s)$, given by Equations 12.100 and 12.101, respectively, the transfer functions for the rotor speed $\Omega_r(s)$ from the demanded blade pitch angles $\beta_d(s)$, the demanded electrical torque $T_{gd}(s)$, and the wind speed $v_1(s)$, are all obtained, so that

$$\Omega_r(s) = \frac{1}{1-\mu_{32}(s)K_{T\Omega}}\left\{\mu_{32}(s)\left[K_{T\beta}A_\beta(s)\beta_d(s) + K_{TV}v_1(s)\right] + \mu_{33}(s)A_T(s)T_{gd}(s)\right\} \tag{12.105}$$

which can be expressed as

$$\Omega_r(s) = F_1(s)v_1(s) + F_2(s)\beta_d(s) + F_3(s)T_{gd}(s) \tag{12.106}$$

where

$$F_1(s) = \frac{K_{TV}n_{\mu 32}(s)}{d_{tf}(s)} \tag{12.107}$$

$$F_2(s) = \frac{K_{T\beta}n_{\mu 32}(s)A_\beta(s)}{d_{tf}(s)} \tag{12.108}$$

$$F_3(s) = \frac{n_{\mu 33}(s)A_T(s)}{d_{tf}(s)} \tag{12.109}$$

and

$$\begin{aligned} n_{\mu 32}(s) = {} & I_g I_w s^4 + (B_h I_w + I_g B_l + I_g B_h R_t^2)s^3 \cdots \\ & + (B_h B_l + I_g K_l + I_g K_h R_t^2 + K_h I_w)s^2 \cdots \\ & + (K_h B_l + B_h K_l)s + K_h K_l \end{aligned} \tag{12.110}$$

$$n_{\mu 33}(s) = -B_h R_t B_l s^2 - R_t(K_h B_l + B_h K_l)s - K_h R_t K_l \tag{12.111}$$

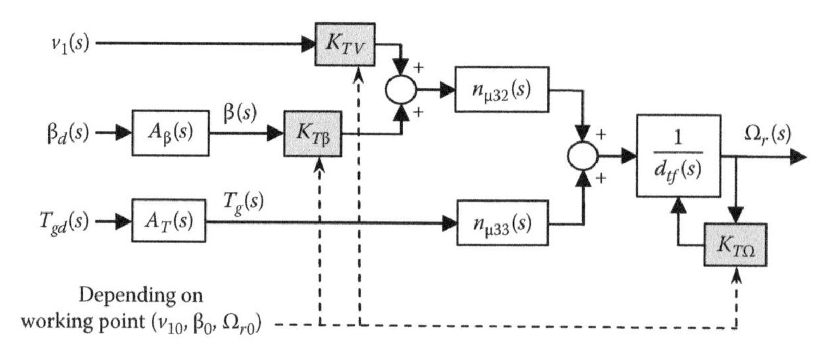

FIGURE 12.14
Rotor speed linear transfer functions from the wind speed, demanded pitch angle, and electrical torque (K_{TV}, $K_{T\beta}$, $K_{T\Omega}$ have uncertainty).

$$\begin{aligned}
d_{tf}(s) &= d_{\mu32}(s) - n_{\mu32}(s)K_{T\Omega} \\
&= I_r I_w I_g s^5 + [(I_r I_g B_l + B_h I_r I_w + I_r I_g B_h R_T^2 + B_l I_w I_g) - (I_g I_w)K_{T\Omega}]s^4 \ldots \\
&\quad + [(K_l I_w I_g + B_h I_r B_l + B_l I_g B_h r_t^2 + K_h I_r I_w + I_r I_g K_l + B_l B_h I_w + I_r I_g K_h R_t^2) \\
&\quad - (B_h I_w + I_g B_l + I_g B_h R_t^2)K_{T\Omega}]s^3 \cdots \\
&\quad + [(B_l K_h I_w + K_l I_g B_h R_t^2 + K_h I_r B_l + B_h I_r K_l + K_l B_h I_w + I_g B_l K_h R_t^2) \\
&\quad - (B_h B_l + I_g K_l + I_g K_h R_t^2 + K_h I_w)K_{T\Omega}]s^2 \cdots \\
&\quad + [(I_g K_l K_h R_t^2 + K_h I_r K_l + K_l K_h I_w) - (K_h B_l + B_h K_l)K_{T\Omega}]s \\
&\quad - (K_h K_l)K_{T\Omega}
\end{aligned} \tag{12.112}$$

Figure 12.14 represents graphically the rotor speed linear transfer functions given by Equations 12.105 through 12.112. In other words, it represents the dynamic model that explains how (1) the two control signals, demanded blade pitch angle $\beta_d(s)$ and demanded electrical torque $T_{gd}(s)$, and (2) the external disturbances, wind speed $v_1(s)$, affect the rotor speed $\Omega_r(s)$.

Due to the nonlinear characteristic of the aerodynamic model of the WT (see T_r, with Equation 12.37 and the curve of $C_P(\lambda,\beta)$ shown in Figure 12.9), the coefficients K_{TV}, $K_{T\beta}$, and $K_{T\Omega}$ that have been used to linearize the system present different constant values depending on the working point (v_{10}, β_0, Ω_{r0}) (see Equations 12.96 through 12.98).

In particular, as the wind speed changes from low to high values, the following parameters of the model change (see also Figures 12.15 through 12.17):

- The DC gain of the transfer function $F_1(s) = \Omega_r(s)/v_1(s)$ depends on the ratio $K_{TV}/K_{T\Omega}$. The gain decreases as the wind speed increases. Typically, it varies so that

$$\frac{gain\_F_1(s)\ (\text{at } v_1 = 25\,\text{m/s})}{gain\_F_1(s)\ (\text{at } v_1 = 12\,\text{m/s})} \approx 0.016 \tag{12.113}$$

- The DC gain of the transfer function $F_2(s) = \Omega_r(s)/\beta_d(s)$ depends on the ratio $K_{T\beta}/K_{T\Omega}$. The gain increases as the wind speed increases. Typically, it varies so that

$$\frac{gain\_F_2(s)\ (\text{at } v_1 = 25\,\text{m/s})}{gain\_F_2(s)\ (\text{at } v_1 = 12\,\text{m/s})} \approx 2.5 \tag{12.114}$$

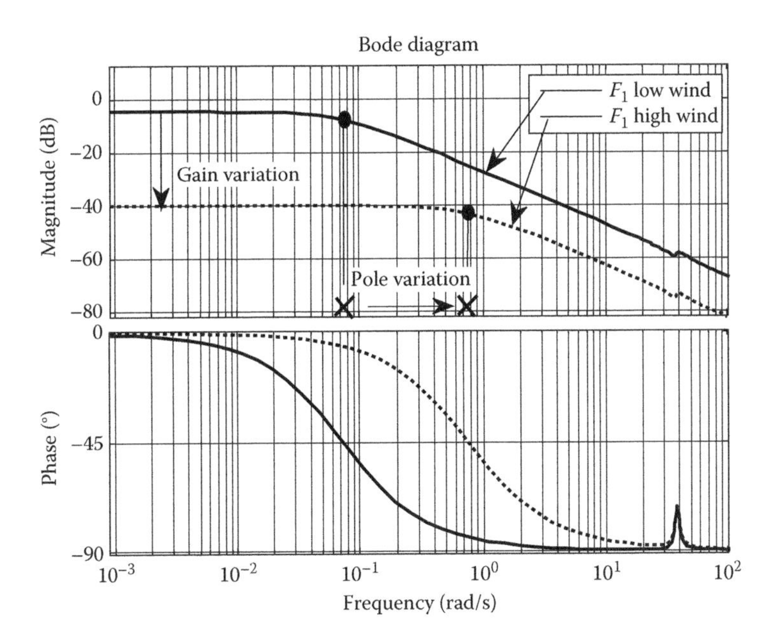

FIGURE 12.15
Bode diagram of $F_1(s) = \Omega_r(s)/v_1(s)$ for low- and high-wind speed.

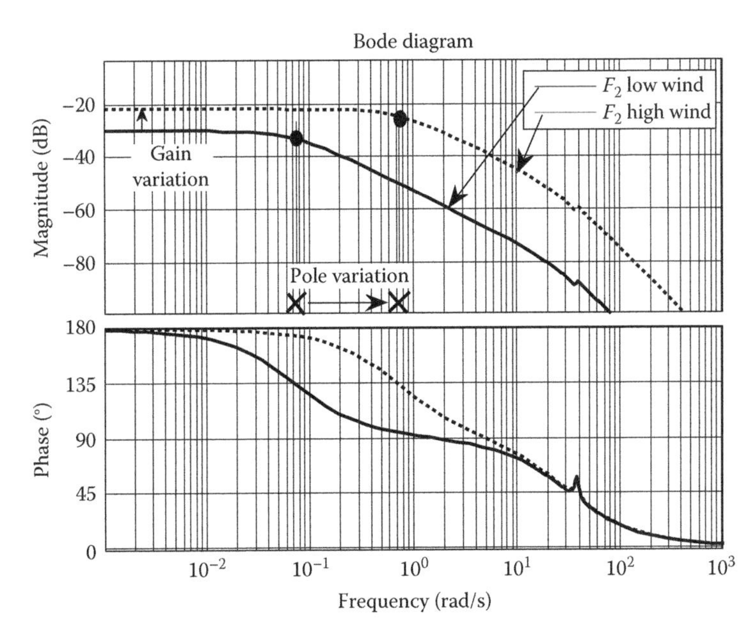

FIGURE 12.16
Bode diagram of $F_2(s) = \Omega_r(s)/\beta_d(s)$ for low- and high-wind speed.

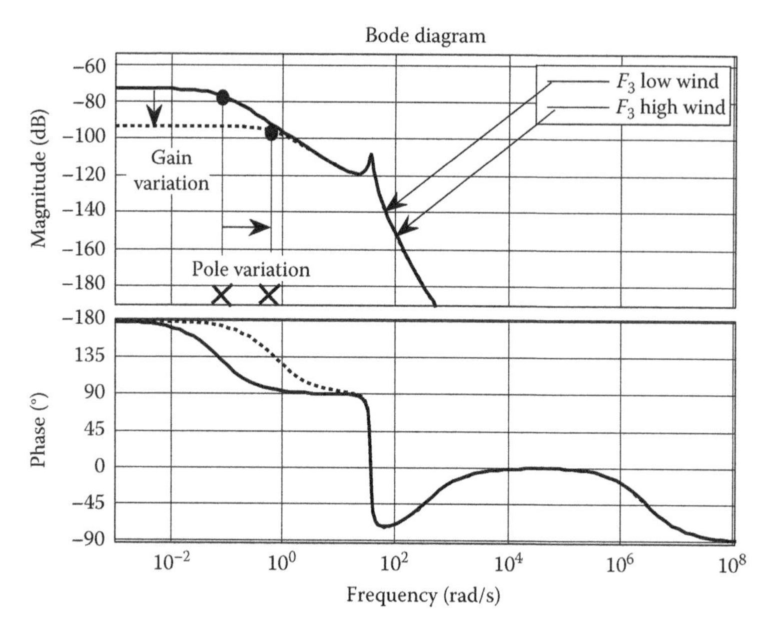

FIGURE 12.17
Bode diagram of $F_3(s) = \Omega_r(s)/T_{gd}(s)$ for low- and high-wind speed.

- The DC gain of the transfer function $F_3(s) = \Omega_r(s)/T_{gd}(s)$ depends on the ratio $1/K_{T\Omega}$. The gain decreases as the wind speed increases. Typically, it varies so that

$$\frac{gain\_F_3(s)\,(\text{at } v_1 = 25\text{ m/s})}{gain\_F_3(s)\,(\text{at } v_1 = 12\text{ m/s})} \approx 0.1 \tag{12.115}$$

- The three transfer functions $F_1(s)$, $F_2(s)$, and $F_3(s)$ have the same denominator. Its dominant (slower) dynamics is a Laplace first-order polynomial, where the absolute value of the location of the dominant pole increases as the wind speed increases. Typically, it varies so that

$$\frac{Pole\,(\text{at } v_1 = 25\text{ m/s})}{Pole\,(\text{at } v_1 = 12\text{ m/s})} \approx 10 \tag{12.116}$$

- The parameters $K_{T\beta}$ and $K_{T\Omega}$, according to Equations 12.96 and 12.98, are typically $K_{T\beta} < 0$ and $K_{T\Omega} < 0$. The pitch and torque control systems are usually designed with these premises. However, both parameters can be positive under some conditions (see the $C_P(\lambda,\beta)$ curves in Figure 12.9), and in those cases the system becomes unstable.

The linear model, described in Figure 12.14 and Equations 12.105 through 12.112, is a useful representation to (1) design the robust/nonlinear controller and (2) understand how the demanded blade pitch angle $\beta_d(s)$, the demanded electrical torque $T_{gd}(s)$, and the wind speed $v_1(s)$ affect the rotor speed $\Omega_r(s)$ (see Equations 12.113 through 12.116 and Figures 12.15 through 12.17). To be consistent with the linear approach and the Laplace Transform,

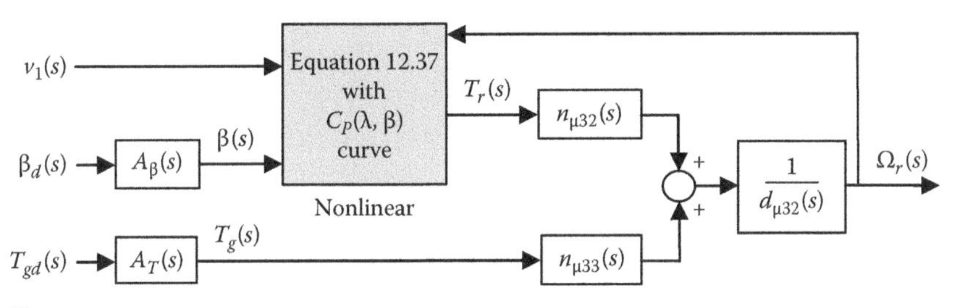

FIGURE 12.18
Nonlinear system diagram (*true model*) for simulation. Output: rotor speed; inputs: wind speed, demanded pitch angle, demanded electrical torque.

in this representation the parameters K_{TV}, $K_{T\beta}$, $K_{T\Omega}$ take constant values that depend on the working point (v_{10}, β_0, Ω_{r0}) and present uncertainty.

For the nonlinear (*true-model*) simulation of the rotor speed $\Omega_r(s)$, Equations 12.37, 12.74, 12.75, 12.80 through 12.83, 12.100, 12.101, and the complete curve $C_P(\lambda,\beta)$ must be used, according to the block diagram in Figure 12.18. In this case, the nonlinear part is represented by T_r, with Equation 12.37 and the curve $C_P(\lambda,\beta)$. The linear part is represented by Equations 12.74, 12.75, 12.80 through 12.83, 12.100, and 12.101.

12.4 Summary

The aerodynamics and the mechanical models, State Space and transfer matrix representations, that describe the dominant characteristics of a WT were introduced in this chapter. The models presented the main fundamentals to (1) understand the dynamical behavior of both the DD and the DFIG WTs and (2) design the appropriate control systems.

13

Electrical Modeling of Wind Turbines

13.1 Introduction

This chapter introduces the electrical models for the dominant dynamics of wind turbines (WTs) connected to the electrical power system, as a first step to understand its electrical dynamical behavior and to design the required controllers. Section 13.2 derives the dynamic models for squirrel cage induction generators (SCIG), doubly fed induction generators (DFIG), and direct-drive synchronous generators (DD). Section 13.3 summarizes the most common power electronic converters. Section 13.4 describes typical WT power quality characteristics, and Section 13.5 introduces some specifics for wind energy integration into the grid.

13.2 Electrical Models

13.2.1 Electrical Machine and Park's Transformation

The dynamic model of an electrical machine is described by a set composed of two differential equations.[308,327] The first one (Equation 13.1) describes the relationship between the voltage at the terminals of the windings and the current in the winding conductors through the self and mutual inductances:

$$\boldsymbol{U} = -\boldsymbol{R}\boldsymbol{I} - \frac{d\boldsymbol{L}(\theta)}{dt}\boldsymbol{I} - \boldsymbol{L}(\theta)\frac{d\boldsymbol{I}}{dt} \tag{13.1}$$

where

$\boldsymbol{U}$ and $\boldsymbol{I}$ are column matrices with the voltage and current of each winding, respectively
the matrix $\boldsymbol{L}$ represents the self and mutual inductances of the windings
the matrix $\boldsymbol{R}$ represents the resistances of the windings
θ is the shaft angle

The second differential equation (Equation 13.2) explains the relationship between the electrical torque T_g at the shaft and the variation of the magnetic fields in the electrical machine:

$$T_g = \frac{1}{2}\boldsymbol{I}^T \frac{d\boldsymbol{L}(\theta)}{d\theta}\boldsymbol{I} \tag{13.2}$$

Both equations explain the dynamic behavior of any electrical machine. However, they are difficult to analyze because the self and mutual inductances of the windings depend on the shaft angle θ, which results in θ varying in time.

To facilitate the analysis and to eliminate this dependency, it is possible to project the nonrotating (or stator) reference frame (*abc*) onto a new rotating (or rotor) reference frame (*dq*0). This can be done by applying the well-known *Park's transformation*.[308,327] It is based on the following expressions:

$$\begin{bmatrix} A_d \\ A_q \\ A_0 \end{bmatrix} = \sqrt{\frac{2}{3}} \begin{bmatrix} \cos(\theta) & \cos\left(\theta - \frac{2\pi}{3}\right) & \cos\left(\theta + \frac{2\pi}{3}\right) \\ \sin(\theta) & \sin\left(\theta - \frac{2\pi}{3}\right) & \sin\left(\theta + \frac{2\pi}{3}\right) \\ \frac{1}{\sqrt{2}} & \frac{1}{\sqrt{2}} & \frac{1}{\sqrt{2}} \end{bmatrix} \begin{bmatrix} A_a \\ A_b \\ A_c \end{bmatrix}$$

or

$$\boldsymbol{A}_{dq0} = \boldsymbol{P}_{dq0}(\theta)\boldsymbol{A}_{abc} \tag{13.3}$$

$$\begin{bmatrix} A_a \\ A_b \\ A_c \end{bmatrix} = \sqrt{\frac{2}{3}} \begin{bmatrix} \cos(\theta) & \sin(\theta) & \frac{1}{\sqrt{2}} \\ \cos\left(\theta - \frac{2\pi}{3}\right) & \sin\left(\theta - \frac{2\pi}{3}\right) & \frac{1}{\sqrt{2}} \\ \cos\left(\theta + \frac{2\pi}{3}\right) & \sin\left(\theta + \frac{2\pi}{3}\right) & \frac{1}{\sqrt{2}} \end{bmatrix} \begin{bmatrix} A_d \\ A_q \\ A_0 \end{bmatrix}$$

or

$$\boldsymbol{A}_{abc} = \boldsymbol{P}_{dq0}(\theta)^{-1}\boldsymbol{A}_{dq0} \tag{13.4}$$

where

- $\boldsymbol{A}$ can be the current i, the voltage u, or the flux linkage Ψ
- *abc* is the three-phase (AC) reference frame (fixed to the stator)
- *dq*0 is the direct-quadrature-zero (DC) reference frame (fixed to the rotor), according to Figure 13.1

Note that Equations 13.3 and 13.4 are also known as the *power invariant Park's transformation* for Figure 13.1, where the power calculated in the *dq*0 coordinate system is the same as that in the *abc* system. Other different and also quite common choice is to use the coefficient 2/3 instead of sqrt(2/3). The choice of 2/3 is made to maintain the same length of the voltage and current vectors for sinusoidal steady state.

The *Park's transformation* can be interpreted in geometrical terms as the projection of the three sinusoidal *abc* quantities onto two axes *dq* rotating with the same angular velocity as the rotor $\Omega_g = d\theta/dt$. Note that (1) d is the so-called direct axis, (2) q is the quadrature axis, (3) 0 is the homopolar component, (4) d and q are perpendicular, and (5) a, b, and c are 120° apart. The aim of the *Park's transformation* is to eliminate the dependence of the machine inductances $\boldsymbol{L}$ on the angular displacement θ.

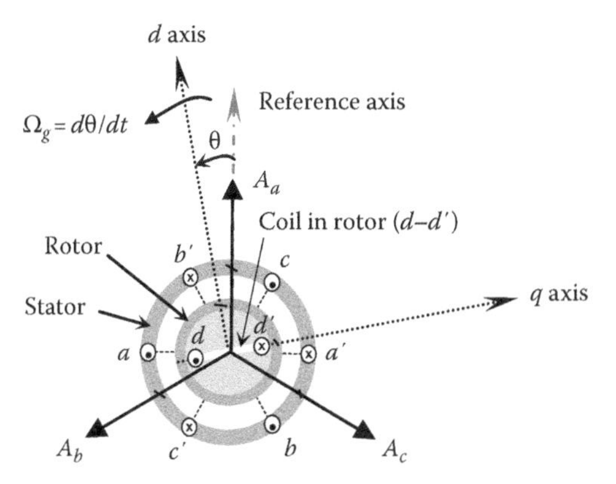

FIGURE 13.1
Electrical machine: *abc* reference frame fixed to the stator and *dq*0 reference frame fixed to the rotor. Stator coils: *a–a′*, *b–b′*, *c–c′*. Rotor coil: *d–d′*.

13.2.2 Squirrel Cage Induction Generator

The configuration of the SCIG is shown in Figure 13.2. Due to its simplicity, and use of appropriate blade airfoils, the machine operates as a constant-speed WT. In the *dq*0 reference frame the voltage equations of the SCIG are as follows (see also Figure 13.3)[308]:

$$u_{ds} = -R_s i_{ds} - \omega_s \Psi_{qs} + \frac{d\Psi_{ds}}{dt} \tag{13.5}$$

$$u_{qs} = -R_s i_{qs} + \omega_s \Psi_{ds} + \frac{d\Psi_{qs}}{dt} \tag{13.6}$$

$$u_{dr} = 0 = -R_r i_{dr} - \sigma\omega_s \Psi_{qr} + \frac{d\Psi_{dr}}{dt} \tag{13.7}$$

$$u_{qr} = 0 = -R_r i_{qr} + \sigma\omega_s \Psi_{dr} + \frac{d\Psi_{qr}}{dt} \tag{13.8}$$

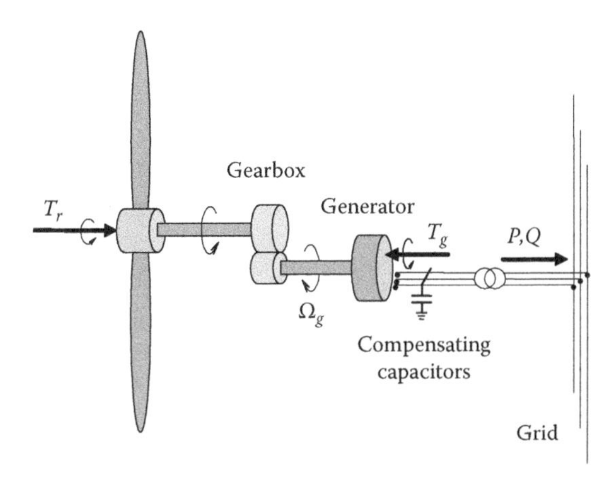

FIGURE 13.2
Squirrel cage induction generator.

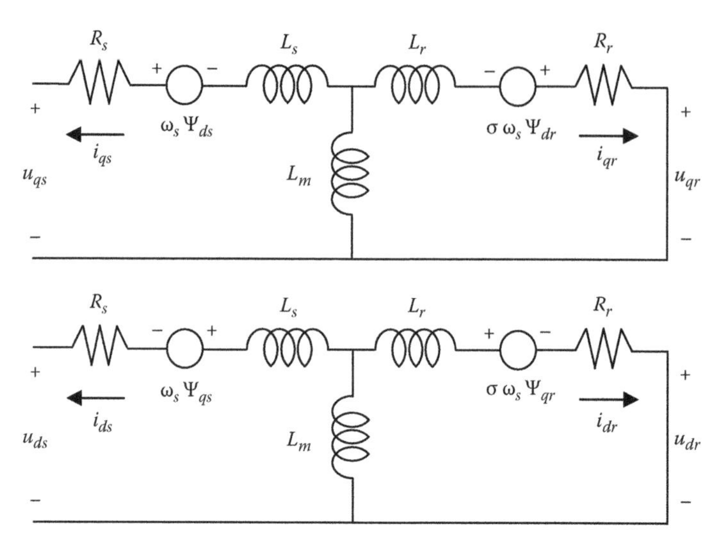

FIGURE 13.3
SCIG equivalent circuits in the *dq*0 frame.

where

- the subscripts "s" stands for stator, "r" for the rotor, "d" for the direct axis, "q" for the quadrature axis
- u_{ds}, u_{qs}, u_{dr}, and u_{qr} are the voltages at the direct and quadrature axes of the stator and the direct and quadrature axes of the rotor, respectively
- i_{ds}, i_{qs}, i_{dr}, and i_{qr} are the currents at the direct and quadrature axes of the stator and direct and quadrature axes of the rotor, respectively
- Ψ_{ds}, Ψ_{qs}, Ψ_{dr}, and Ψ_{qr} are the fluxes at the direct and quadrature axes of the stator and direct and quadrature axes of the rotor, respectively
- R_s is the stator resistance, R_r the rotor resistance, ω_s the synchronous speed, $\omega_s = 2\pi f$, p the number of poles, f the grid frequency, Ω_g the rotor speed, and σ the slip, which is defined as follows:

$$\sigma = 1 - \frac{p}{2}\frac{\Omega_g}{\omega_s} \tag{13.9}$$

Using generator convention, the stator currents are positive when flowing toward the network, and real and reactive powers are positive when fed into the grid (see Figure 13.3). Utilizing this convention the *flux linkages equations* are as follows:

$$\Psi_{ds} = -(L_s + L_m)i_{ds} - L_m i_{dr} \tag{13.10}$$

$$\Psi_{qs} = -(L_s + L_m)i_{qs} - L_m i_{qr} \tag{13.11}$$

$$\Psi_{dr} = -(L_r + L_m)i_{dr} - L_m i_{ds} \tag{13.12}$$

$$\Psi_{qr} = -(L_r + L_m)i_{qr} - L_m i_{qs} \tag{13.13}$$

The *electrical torque equation* is

$$T_g = (\Psi_{qr} i_{dr} - \Psi_{dr} i_{qr}) \tag{13.14}$$

Contrary to the DD and DFIG cases, the electrical torque T_g cannot be controlled in this case.

The equations for the active power generated and the reactive power consumed are as follows:

$$P = u_{ds} i_{ds} + u_{qs} i_{qs} \tag{13.15}$$

$$Q = u_{qs} i_{ds} - u_{ds} i_{qs} \tag{13.16}$$

In the SCIG system, the reactive power exchanged with the grid (power factor) cannot be controlled. This condition means that the node voltage (voltage at the grid connection point) cannot be controlled either.

Note that the system can only consume and not generate reactive power. The reactive power exchanged with the grid depends on the rotor speed, terminal voltage, and active power generation. This fact is an important disadvantage. In some cases, with large SCIG wind farms, the reactive power consumption may cause severe node voltage drops. Figure 13.2 shows the typical compensating capacitors connected to the terminals of the SCIG to reduce this voltage drop problem.

13.2.3 Doubly Fed Induction Generator

The configuration of the DFIG is shown in Figure 13.4. It is a variable-speed machine, where the frequency generated at the grid connection point is the addition of the rotor frequency f_r and the converter frequency f_c. The latter (f_c) is controlled by the converter to allow a variable rotor frequency f_r, so that $f_{grid} = f_r + f_c$. In the $dq0$ reference frame the equations of the DFIG are as follows (see also Figure 13.5)[308]:

Voltage equations

$$u_{ds} = -R_s i_{ds} - \omega_s \Psi_{qs} + \frac{d\Psi_{ds}}{dt} \tag{13.17}$$

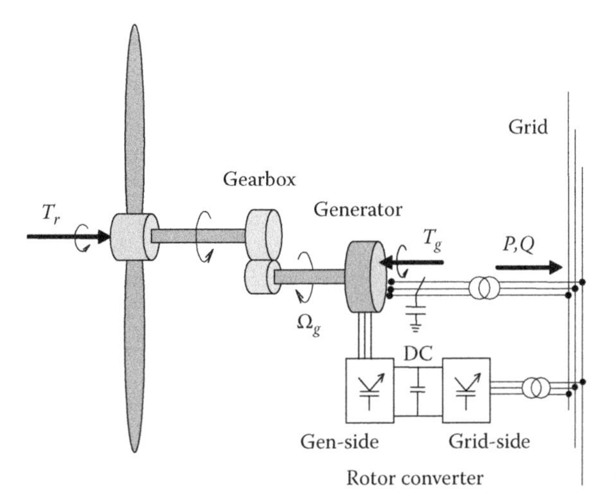

FIGURE 13.4
Doubly fed induction generator.

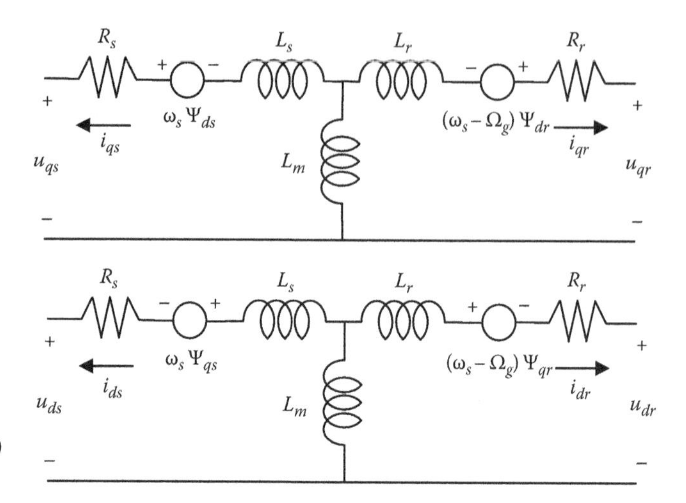

FIGURE 13.5
DFIG equivalent circuits in the $dq0$ frame.

$$u_{qs} = -R_s i_{qs} + \omega_s \Psi_{ds} + \frac{d\Psi_{qs}}{dt} \tag{13.18}$$

$$u_{dr} = -R_r i_{dr} - \sigma\omega_s \Psi_{qr} + \frac{d\Psi_{dr}}{dt} \tag{13.19}$$

$$u_{qr} = -R_r i_{qr} + \sigma\omega_s \Psi_{dr} + \frac{d\Psi_{qr}}{dt} \tag{13.20}$$

Flux linkages equations

$$\Psi_{ds} = -(L_s + L_m) i_{ds} - L_m i_{dr} \tag{13.21}$$

$$\Psi_{qs} = -(L_s + L_m) i_{qs} - L_m i_{qr} \tag{13.22}$$

$$\Psi_{dr} = -(L_r + L_m) i_{dr} - L_m i_{ds} \tag{13.23}$$

$$\Psi_{qr} = -(L_r + L_m) i_{qr} - L_m i_{qs} \tag{13.24}$$

Rotor slip

$$\sigma = \frac{\omega_s - \Omega_g}{\omega_s} \tag{13.25}$$

Electrical torque equation
Neglecting the stator resistance and assuming that the d-axis coincides with the maximum of the stator flux results in the following equation:

$$T_g = \left(\frac{p}{2}\right)(\Psi_{dr} i_{qr} - \Psi_{qr} i_{dr}) = \left(\frac{p}{2}\right) L_m (i_{qs} i_{dr} - i_{ds} i_{qr}) \tag{13.26}$$

The electrical torque T_g is one of the control variables, as shown in Equations 12.68 and 12.99. It can be modified properly with the rotor converter and the related control system (see also Equation 12.101 and Section 11.4.5.1).

Equation for active power injected into the grid is given by

$$P = u_{ds}i_{ds} + u_{qs}i_{qs} + (u_{dr}i_{dr} + u_{qr}i_{qr})\eta_{conv} \tag{13.27}$$

The reactive power fed into the grid depends on both the generator and the converter with the control strategy. The following notations are utilized in obtaining Equation 13.27:

- u_{qs}, u_{ds} represent the three-phase stator voltages in the $dq0$ reference frame
- i_{qs}, i_{ds} represent the three-phase stator currents in the $dq0$ reference frame
- Ψ_{qs}, Ψ_{ds} represent the three-phase stator flux linkages in the $dq0$ reference frame
- u_{qr}, u_{dr} represent the three-phase rotor voltages in the $dq0$ reference frame
- i_{qr}, i_{dr} represent the three-phase rotor voltages in the $dq0$ reference frame
- Ψ_{qr}, Ψ_{dr} represent the three-phase rotor flux linkages in the $dq0$ reference frame
- R_s, R_r are the stator and rotor resistances of machine per phase, respectively
- L_s, L_r are the leakage inductances of stator and rotor windings, respectively
- ω_s, Ω_g are the synchronous (grid) and rotor angular frequency, respectively
- T_g is the electromagnetic torque; T_r is the mechanical input torque; P, Q are the stator-side active and reactive power, respectively
- p is the number of poles
- η_{conv} is the efficiency of the converter

In the DFIG system, the reactive power exchanged with the grid (and the power factor) can be partially controlled by the manipulation of the rotor current with the rotor converter. However, this is not a full-range control, because the rotor current is primarily used to generate the electrical torque T_g to control the rotor speed Ω_g. This fact determines the capacity that is left to circulate the current to generate or to consume the reactive power. The node voltage (voltage at the grid connection point) can also be partially controlled in the same manner.

13.2.4 Direct-Drive Synchronous Generator

The configuration of the DD is shown in Figure 13.6. It is a variable-speed machine where the frequency at the grid is created by the inverter (grid-side converter) from the DC link.

Considering the effect of damper circuits in the electrical machine to be negligible, the equations of DD synchronous generator in the $dq0$ reference frame are as follows (see also Figure 13.7)[308,327]:

Voltage equations

$$u_{0s} = -R_s i_{0s} + \frac{d\Psi_{0s}}{dt} \tag{13.28}$$

$$u_{ds} = -R_s i_{ds} - \Omega_g \Psi_{qs} + \frac{d\Psi_{ds}}{dt} \tag{13.29}$$

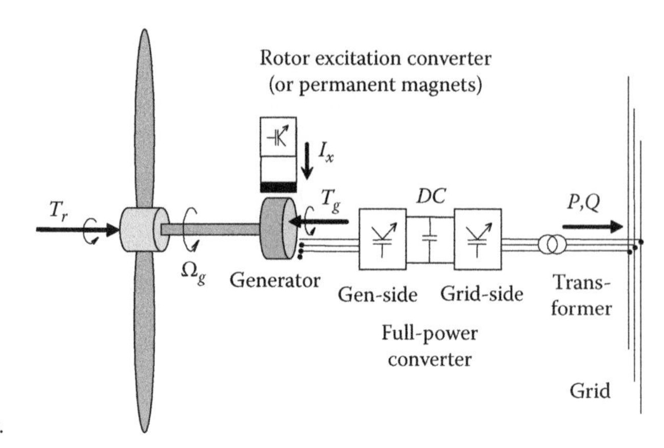

FIGURE 13.6
Direct-drive synchronous generator.

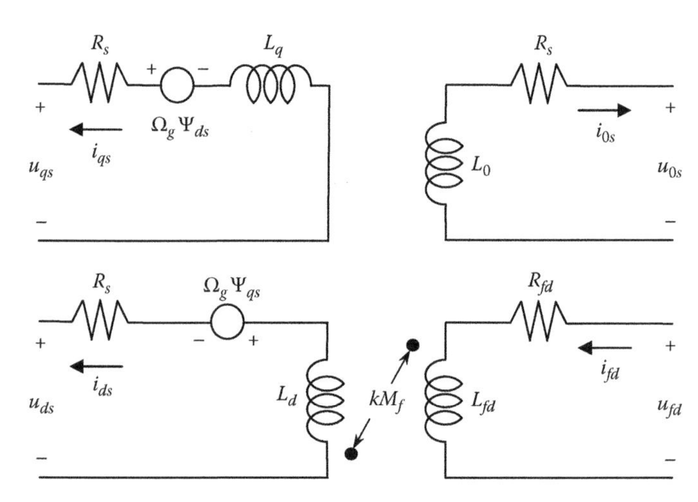

FIGURE 13.7
DD circuits in the $dq0$ frame.

$$u_{qs} = -R_s i_{qs} + \Omega_g \Psi_{ds} + \frac{d\Psi_{qs}}{dt} \tag{13.30}$$

$$u_{fd} = R_{fd} i_{fd} + \frac{d\Psi_{fd}}{dt} \tag{13.31}$$

where "*fd*" indicates field quantities.

In case of a balanced synchronous operation, the homopolar current and voltage are zero ($i_{0s} = 0$ and $u_{0s} = 0$), and then the zero-sequence equation (13.28) disappears.

In practice, quite often the following conditions exist: $d\Psi_{ds}/dt \ll \Omega_g \Psi_{qs}$ and $d\Psi_{qs}/dt \ll \Omega_g \Psi_{ds}$. This implies that the terms $d\Psi_{ds}/dt$ and $d\Psi_{qs}/dt$ can be neglected in Equations 13.29 and 13.30, respectively, thus resulting in the following equations.

Flux linkages equations

$$\Psi_{0s} = -L_0 i_{0s} \tag{13.32}$$

$$\Psi_{ds} = -L_d i_{ds} + kM_f i_{fd} \tag{13.33}$$

$$\Psi_{qs} = -L_q i_{qs} \tag{13.34}$$

$$\Psi_{fd} = L_{fd} i_{fd} - kM_f i_{ds} \tag{13.35}$$

Inductances equations

$$L_0 = L_s - 2M_s \tag{13.36}$$

$$L_d = L_s + 2M_s + \frac{3}{2} L_m \tag{13.37}$$

$$L_q = L_s + 2M_s - \frac{3}{2} L_m \tag{13.38}$$

For the case utilizing permanent magnets, there is no excitation current i_{fd}, voltage u_{fd}, or flux linkage Ψ_{fd}, and Equation 13.33 becomes

$$\Psi_{ds} = -L_d i_{ds} + \Psi_{pm} \tag{13.39}$$

where Ψ_{pm} is the flux of the permanent magnets on the rotor.

Electrical torque equation

$$T_g = \left(\frac{p}{2}\right)\left(\Psi_{ds} i_{qs} - \Psi_{qs} i_{ds}\right) \tag{13.40}$$

The electrical torque T_g is one of the control variables, as shown in Equations 12.51 and 12.99. It can be modified properly by the use of full-power converter and control system (see also Equation 12.101 and Section 11.4.5.1). This situation results in the following equations:

Equation for the active power generated by the synchronous generator

$$P = u_{0s} i_{0s} + u_{ds} i_{ds} + u_{qs} i_{qs} \tag{13.41}$$

Equation for the reactive power exchange between generator and converter

$$Q_{gc} = u_{qs} i_{ds} - u_{ds} i_{qs} \tag{13.42}$$

Equation for the net reactive power exchange between the converter and the grid

$$Q_{cgr} = u_{qc} i_{dc} - u_{dc} i_{qc} \tag{13.43}$$

where the subscript "*c*" stands for converter.

In the DD system the reactive power exchanged with the grid (power factor) is independent of the characteristics of the generator, which is fully decoupled from the grid.

In fact, the full-power converter is able to control independently and totally the power factor at both the generator and the grid sides. The node voltage (voltage at the grid connection point) can be partially controlled in the same manner.

13.3 Power Electronic Converters

The development of new high-power electronic components in the 1980s and 1990s (like high-power diodes, high-power thyristors, integrated gate commutated thyristors [IGCTs], insulated gate bipolar transistor [IGBTs], etc.) opened new avenues for WT designs, including the DFIG and DD machines. There are currently many converter topologies and several generator–converter combinations available, with different cost, efficiency, control possibilities, complexity, and power consumption features. The most common combinations are shown in Tables 13.1 and 13.2.

Cases DFIG-3 and DD-4, with IGBT-based (or IGCT-based) converters at both sides (generator and grid sides), also called *back-to-back* converters, give full control flexibility, which allows two independent control loops (d and q axes) at each side. Cases DFIG-4 and DD-5, with matrix converters, are not very common in commercial applications due to the circuit complexity and the corresponding higher maintenance requirements. Cases DFIG-1, DFIG-2, DD-1, and DD-2 are cheaper but give less control flexibility.

For additional information about power electronic converters, please see Refs. 324 and 325 for design and topology, and 323 and 328 for modeling and controller design.

TABLE 13.1

DFIG Converter Topologies (See Figure 13.4)

	Generator Side	Grid Side
DFIG-1	Diode bridge rectifier	Thyristor-based inverter
DFIG-2	Diode bridge rectifier	IGBT or IGCT inverter
DFIG-3	IGBT or IGCT rectifier	IGBT or IGCT inverter
DFIG-4	Matrix converter	

TABLE 13.2

DD Converter Topologies (See Figure 13.6)

	Generator Side	Grid Side
DD-1	Diode bridge rectifier	Thyristors inverter
DD-2	Diode bridge rectifier	IGBT or IGCT inverter
DD-3	Diode bridge rectifier, DC chopper	IGBT or IGCT inverter
DD-4	IGBT or IGCT rectifier	IGBT or IGCT inverter
DD-5	Matrix converter	
Rotor options	With excitation converter	
	With permanent magnets	

13.4 Power Quality Characteristics

The main standards and guidelines for power quality of WTs are the IEC 61400-21 (*Wind Turbine Generator Systems, Part 21: Measurement and Assessment of Power Quality Characteristics of Grid Connected Wind Turbines*)[305,337] and the MEASNET guideline (*Power Quality Measurement Procedure of Wind Turbines*).[340] This section describes some of the main concepts of power quality measurements for WTs.

13.4.1 Power Flicker

Power fluctuations may cause corresponding voltage fluctuations on the grid, which may cause annoying changes in the luminance from lamps. This effect is denoted as flicker. Flicker from WTs occurs in both continuous mode (where the power fluctuations are produced by the wind speed variation, tower shadow effects, and pitch control) and switching mode (where the power fluctuations are produced by the startup and shutdown operations of the WT). Flicker is properly defined in standards IEC 61000-4-15, IEC 60868, IEC 61400-21.[305,337]

For continuous operation, the *flicker coefficient* is a normalized measure of the maximum flicker emission, so that

$$\text{flicker coefficient} = c(\psi_k, v_a) = P_{st}\frac{S_k}{S_n} \quad (13.44)$$

For switching operation, the *flicker step factor* is a normalized measure of the flicker emission due to a single switching operation, so that

$$\text{flicker step factor} = k(\psi_k) = \frac{P_{st}T_p^{0.31}S_k}{300S_n} \quad (13.45)$$

where

- P_{st} is the short-term flicker emission from the WT, measured with a flickermeter over a period of 10 min
- S_n is the rated apparent power of the WT
- S_k is the short-circuit apparent power of the grid
- T_p is the duration of the voltage variation due to the switching operation

The flicker coefficient and the flicker step factor have to be calculated for a grid impedance phase angle of ψ_k = 30°, 50°, 70°, and 85° and the flicker coefficient also for annual average wind speed of v_a = 6.0, 7.5, 8.5, and 10 m/s. The flicker calculation requires the three instantaneous measurements of voltages and currents.

13.4.2 Harmonics and Interharmonics

Harmonics are voltage or current sinusoidal components with frequencies that are multiples (h) of the grid frequency (h50 Hz or h60 Hz). Interharmonics are components with frequencies located between the harmonics of the grid frequency. Harmonics and interharmonics are caused by nonlinear loads, power electronics, etc., and distort the voltage and current waveform. They can produce overheating and the failure of the grid equipment. Harmonics and interharmonics are properly defined in the standard IEC 61000-4-7 and Amendment 1.[337] According to that standard, the *harmonic current* produced at several sources connected to a common point is given by

$$\text{Harmonic current of order } h = I_h = \left(\sum_k I_{h,k}^{\alpha} \right)^{1/\alpha} \tag{13.46}$$

where

I_h is the harmonic current of order h

$I_{h,k}$ is the harmonic current of order h from source k

α is an exponent so that, for $h < 5$, $\alpha = 1$; for $5 \leq h \leq 10$, $\alpha = 1.4$; and for $h > 10$, $\alpha = 2$

The *total harmonic distortion* (THD) *of the voltage* is defined as

$$\text{Total harmonic distortion} = THD = \left(\sum_{h=2}^{40} \left(\frac{U_h}{U_n} \right)^2 \right)^{1/2} \tag{13.47}$$

where

U_h is the RMS voltage amplitude of the harmonic of order h

U_n is the nominal phase-to-phase voltage

Harmonic and interharmonic current measurements require great accuracy (better than 0.1%) and a high frequency value (up to 9 kHz in MEASNET standards).[340] Note that the IEC standards require measuring integer harmonic currents up to the 50th order. In addition, the typical power electronic converters are usually systems that are commutated at 2–3 kHz and produce mainly interharmonic currents.

13.4.3 Power Peaks

The maximum power output of a WT over a specific averaging time of continuous operation (without stopping and restarting the machine) is the so-called power peak. The IEC standards only require measuring the 10 min power peak.[337]

13.4.4 Reactive Power and Power Factor

The IEC standards also require measuring both 10 and 1 min average values of the reactive power over the whole active power range at the output of the WT. Instead of reactive power, other standards require measuring the power factor (i.e., the ratio of active power and apparent power, where apparent power vector is the active plus reactive power vectors). Usually, SCIG machines have a power factor of around 0.96. DFIG and DD machines can control the power factor and quite often have a power factor of 1.00.

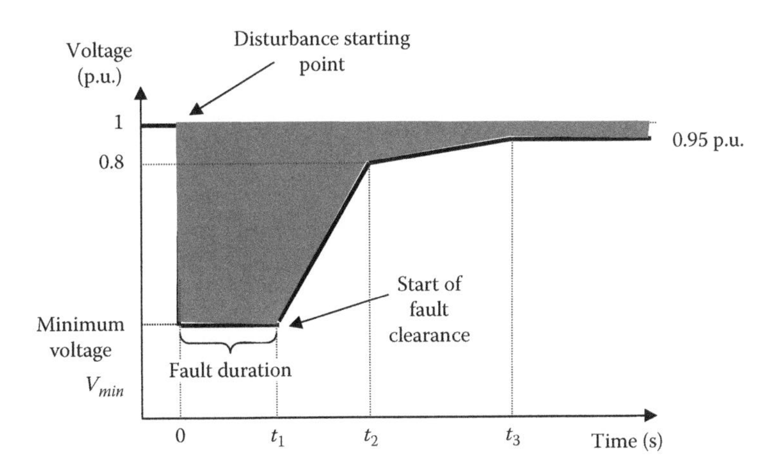

FIGURE 13.8
Voltage dip description.

13.4.5 Voltage Dips

A voltage dip is a sudden reduction of the voltage to a value between 1% and 90% of the nominal value, followed by a recovery after a short period of time (see Figure 13.8). Voltage dips are properly defined in standard EN 50160. They can affect all the three phases (symmetrical) or two phases or only one phase (unsymmetrical).

Voltage dips force classical WTs to stop working immediately. As wind energy penetration increases, the sudden disconnection of many wind farms (due to the voltage dips) may affect significantly the behavior of the grid, producing transients in the frequency and voltage that eventually could generate a global blackout.

In the last few years, new standards to regulate this important problem have been developed in Europe and in countries with large wind energy penetration percentage (see, e.g., REE standards and Canary Island standards in Spain, and Eo.N standards in Germany). The standards require the machines to keep working during and immediately after the voltage dips. As a consequence, new technology to allow the WTs working under voltage dips is being developed (see, for instance, Section 15.5).

13.5 Wind Farms Integration in the Power System

13.5.1 Capacity Factor of a Wind Farm

The capacity factor (CF) of a wind farm (or a WT) is the ratio of the energy produced in a year to the energy that could have been produced at continuous maximum power (rated power) operation during that year. It depends on the wind resources at the location and the type of WT. CF is usually in the range 0.25 (at locations with low average wind speed) through 0.4 (at locations with high average wind speed). The *annual equivalent full load hours* (at rated power) is 8760 CF and corresponds to 2190 equivalent hours for $CF = 0.25$ and to 3504 equivalent hours for $CF = 0.4$.

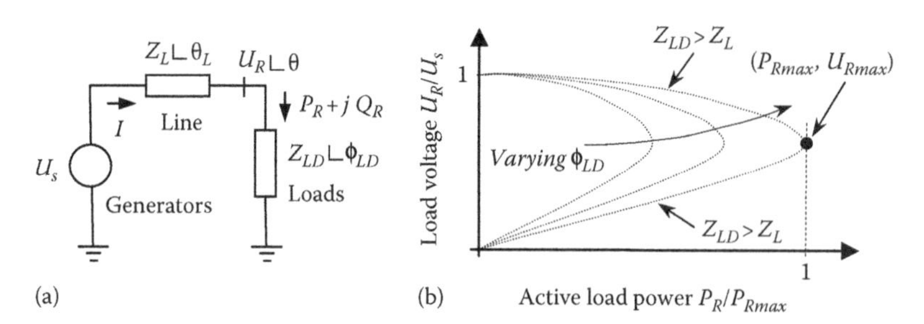

FIGURE 13.9
(a) Two-terminal power transmission system and (b) nose curve for power transmission evaluation, with $U_s = 1$ p.u.

13.5.2 Limited Transmission Capacity

Electrical lines and transmission systems are designed and installed to connect power generators with electrical loads.

For economic reasons, they are not usually overdimensioned and therefore present some limits in the transmission capacity, which can be significant in some particular cases.

Main factors that limit the transmission capacity are thermal limits of the conductors and voltage and frequency grid stability considerations. During some extreme cases at very high wind energy production (windy days), low local electric loads, grid stability degradation, etc., the capacity of the electrical lines could be limited.[323]

The maximum possible power transfer between the wind farms and the loads depends on the line and load impedances (see Figure 13.9a). Figure 13.9b shows the different regions of operation, depending on the line and load impedance relationship (upper curve for $Z_L < Z_{LD}$, and lower curve for $Z_L > Z_{LD}$). The curves can change with the load power factors ϕ_{LD} and with the line length. The power transmission limit, also called *maximum loadability point*, is at the nose point (P_{Rmax}, U_{Rmax}).[308]

If the wind farm and the other power generators (P_R) try to exceed the *maximum loadability point* (P_{Rmax}, U_{Rmax}) or if the load power factor (ϕ_{LD}) decreases too much, then the grid behavior (voltage stability) can be seriously deteriorated, and the grid becomes unstable, ending with a blackout.[308]

13.5.3 Grid Control

There are two main control loops in any electrical power system: the voltage and the frequency control loops. The former (voltage) is mainly a local variable, and the latter (frequency) a global variable. The voltage can be regulated by injecting reactive power from the local generators and the frequency by injecting active power from any generator.[323]

Classical WTs have been designed primarily to generate the maximum amount of possible energy, following the wind speed. Additional power factor considerations, such as compensating capacitors for the SCIG or a particular current in the rotor converter for the DFIG or a specific controller in the grid-side converter for the DD, are added to generate at levels close to the requested unitary power factor $(\cos(\phi) = 1)$.

New WTs in the near future, in a higher wind energy penetration factor scenario, will have to offer the grid operator new control features. These features, such as reactive power control and active power control, will allow the operator to use that capacity to help in the regulation of the voltage and frequency grid control loops. DD systems with full-power converters connected to the grid are in a very good position to meet these future requirements (see Chapter 15, Sections 15.4 and 15.5).

13.6 Summary

The electrical models for WTs connected to the grid, including SCIG, DFIG, DD, and a summary of the most common power electronic converters were discussed in this chapter. The typical WT power quality characteristics including flicker, harmonics and interharmonics, power peaks, reactive power and power factors, voltage dips, as well as wind farm grid integration issues such as capacity factor (CF), limited transmission capacity, and grid control were also discussed.

14

Advanced Pitch Control System Design

14.1 Introduction

The design of advanced pitch control systems for a variable-speed pitch-controlled DFIG (doubly fed induction generator) wind turbines (WTs) is presented in this chapter. The application of the robust QFT (quantitative feedback theory) control techniques and nonlinear switching methods previously introduced is developed in detail in three illustrative cases.

The first example, presented in Section 14.2, derives (1) the specific dynamical models of the pitch control system for the DFIG WT presented in Chapter 12 (equations, parameters, and uncertainty); (2) the required performance specifications, QFT templates, bounds and loop-shaping controller design, all according to Chapters 2 and 3; and (3) the frequency- and time-domain validation of the control system. The QFT Control Toolbox (*QFTCT*), provided with this book and introduced in Appendix F,[149,150] is used to perform the design of the control system.

The second example, presented in Section 14.3, applies the practical robust-switching control design methodology introduced in Chapter 8 to the design of a pitch control system for a DFIG WT. The method is capable of optimizing performance and stability simultaneously, going beyond the classical linear limitations while giving a solution for the well-known robustness–performance trade-off. The results show significant improvements in comparison with the classical linear control, combining (1) high reliability and robustness, (2) minimum mechanical fatigue, and (3) performance optimization, all under turbulent wind conditions.[140]

Finally, Section 14.4 introduces a switching robust control strategy for pitch control systems working in a large nonlinear (large parameter variation) environment. Overall, this chapter reviews and utilizes the main concepts and methodologies introduced in this book. In particular,

- Introduction to QFT (Chapter 2)
- MISO analog QFT techniques (Chapter 3)
- Nonlinear switching control techniques (Chapter 8)
- Wind turbine control objectives and strategies (Chapter 11)
- Aerodynamics and mechanical modeling of wind turbines (Chapter 12)
- Template generation (Appendix A)
- Inequality bound expressions (Appendix B)
- Essentials for loop shaping (Appendix D)
- *QFTCT* provided with this book (Appendix F)[149,150]

14.2 QFT Robust Control Design

WTs are complex, flexible, and multidisciplinary systems that work under very variable environmental conditions. This example introduces the design of a pitch control system to regulate the rotor speed Ω_r of a variable-speed pitch-controlled DFIG WT in *Region 3* (Figure 11.5 with $v_1 \geq v_r$), rejecting the wind disturbances (variation of v_1) by moving the pitch angle β_d of the turbine blades.

14.2.1 Model, Parameters, and Uncertainty Definition

The dynamic model that describes the dominant characteristics of a DFIG WT is introduced in Chapter 12. Figure 14.1 shows a classical DFIG WT diagram with a two-part horizontal drive shaft that connects, through a gearbox, a large rotor inertia at one end (blades) with a generator at the other end. The wind applies an aerodynamic torque T_r on the rotor. The grid and power electronics apply an electrical torque T_g on the generator's shaft.

The rotational speed Ω_r of the WT rotor is continuously modified (1) by the controllers and actuators, which modify the blade pitch angles β_d and the electrical torque T_{gd}; (2) by the external disturbance inputs, like the wind speed v_1; and (3) through the dynamics of the rotor speed Ω_r itself. The transfer functions of the rotor speed $\Omega_r(s)$ from the demanded blade pitch angles $\beta_d(s)$, the demanded electrical torque $T_{gd}(s)$, and the wind speed $v_1(s)$ are introduced in Chapter 12 (see Equations 12.106 through 12.112) and are repeated here for convenience (Equations 14.1 through 14.7).

$$\Omega_r(s) = F_1(s)v_1(s) + F_2(s)\beta_d(s) + F_3(s)T_{gd}(s) \tag{14.1}$$

where

$$F_1(s) = \frac{K_{TV} n_{\mu 32}(s)}{d_{tf}(s)} = D_1(s) \tag{14.2}$$

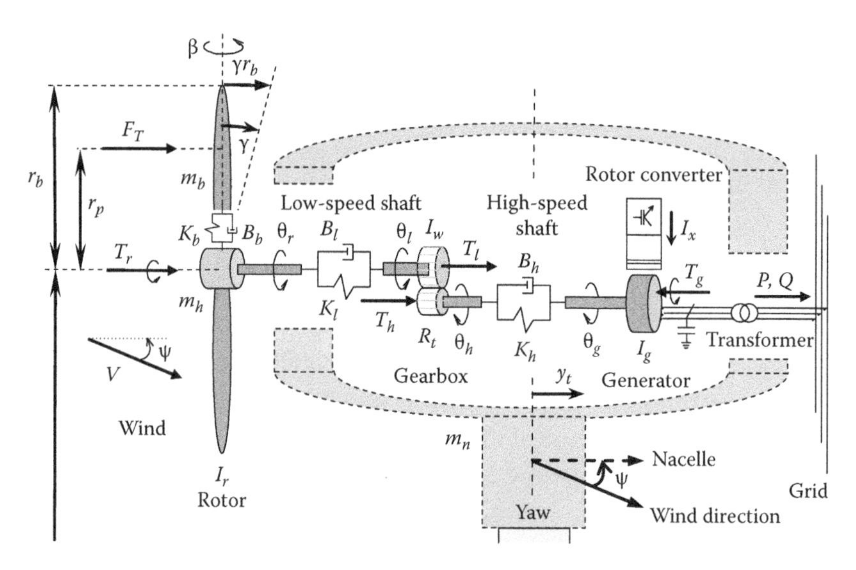

FIGURE 14.1
Variable-speed pitch-controlled DFIG WT.

$$F_2(s) = \frac{K_{T\beta} n_{\mu 32}(s) A_\beta(s)}{d_{tf}(s)} = P(s) \tag{14.3}$$

$$F_3(s) = \frac{n_{\mu 33}(s) A_T(s)}{d_{tf}(s)} = H(s) \tag{14.4}$$

and

$$n_{\mu 32}(s) = I_g I_w s^4 + (B_h I_w + I_g B_l + I_g B_h R_t^2)s^3 + (B_h B_l + I_g K_l + I_g K_h R_t^2 + K_h I_w)s^2 + (K_h B_l + B_h K_l)s + K_h K_l \tag{14.5}$$

$$n_{\mu 33}(s) = -B_h R_t B_l s^2 - R_t(K_h B_l + B_h K_l)s - K_h R_t K_l \tag{14.6}$$

$$\begin{aligned} d_{tf}(s) &= d_{\mu 32}(s) - n_{\mu 32}(s) K_{T\Omega} \\ &= I_r I_w I_g s^5 + [(I_r I_g B_l + B_h I_r I_w + I_r I_g B_h R_T^2 + B_l I_w I_g) - (I_g I_w) K_{T\Omega}] s^4 \cdots \\ &\quad + [(K_l I_w I_g + B_h I_r B_l + B_l I_g B_h r_t^2 + K_h I_r I_w + I_r I_g K_l + B_l B_h I_w + I_r I_g K_h R_t^2) \\ &\quad - (B_h I_w + I_g B_l + I_g B_h R_t^2) K_{T\Omega}] s^3 \cdots \\ &\quad + [(B_l K_h I_w + K_l I_g B_h R_t^2 + K_h I_r B_l + B_h I_r K_l + K_l B_h I_w + I_g B_l K_h R_t^2) \\ &\quad - (B_h B_l + I_g K_l + I_g K_h R_t^2 + K_h I_w) K_{T\Omega}] s^2 \cdots \\ &\quad + [(I_g K_l K_h R_t^2 + K_h I_r K_l + K_l K_h I_w) - (K_h B_l + B_h K_l) K_{T\Omega}] s \\ &\quad - (K_h K_l) K_{T\Omega} \end{aligned} \tag{14.7}$$

The block diagrams associated with the DFIG WT model (Equations 14.1 through 14.7) are presented in Figure 14.2a and b. Note that in this figure $K_{T\Omega}$, K_{TV}, and $K_{T\beta}$ are varying in the nonlinear real-world case. These diagrams are used to introduce the plant models, $P(s)$ and $D_1(s)$, in the *QFTCT*, as shown in Figures 14.4 and 14.5.

For this example, the parameters of the selected DFIG WT model (Equations 14.1 through 14.7), and the associated uncertainty, are presented in Figure 14.3 and Table 14.1.

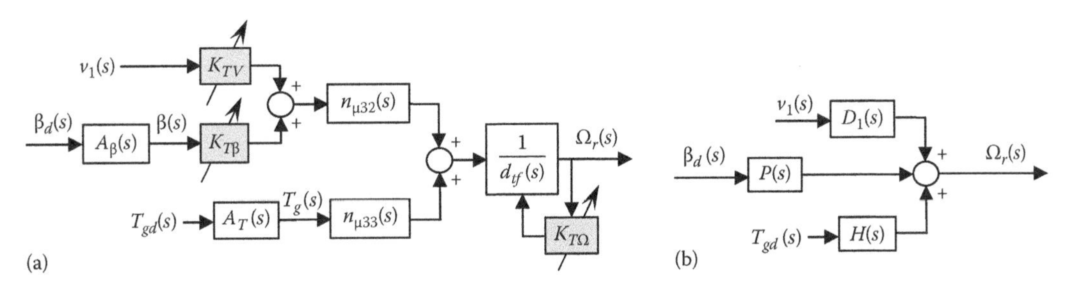

FIGURE 14.2
Open-loop pitch control system block diagram, DFIG turbine: (a) detailed block diagram and (b) reduced block diagram based on Equations 14.2, 14.3, and 14.4.

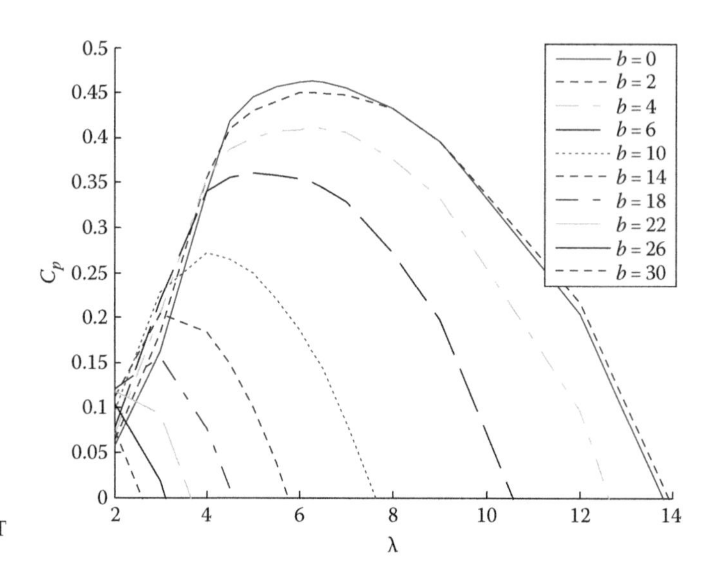

FIGURE 14.3
Cp/λ curve for the selected 40 m WT blade (beta in degrees).

TABLE 14.1
WT Parameters

Power to the grid (MW)	2.0	r_b (blade radius) (m)	40.0
ρ (air) (kg/m^3)	1.225	K_l (Nm), shaft low speed	160×10^6
N (number of blades)	3	B_l (Nms), shaft low speed	25,000
Ω_{r_nom} (rad/s)	1.8850	K_h (Nm), shaft high speed	2.3×10^4
T_{gd_max} (Nm), shaft high speed	14,000	B_h (Nms), shaft high speed	36
R_t (gearbox ratio)	83.33	I_g (kg m^2), shaft high speed	60
I_r (kg m^2), shaft low speed	5.5×10^6	I_w (kg m^2), shaft high speed	0.0
$K_{T\beta}$ min (Nm/deg)	-17.53×10^4	$K_{T\beta}$ max (Nm/deg)	-1.18×10^4
$K_{T\Omega}$ min (Nm/(rad/s))	-14×10^5	$K_{T\Omega}$ max (Nm/(rad/s))	-3.75×10^5
K_{TV} min (Nm/m/s)	2.12×10^5	K_{TV} max (Nm/m/s)	2.61×10^5
$A_\beta(s)$	1.0	$A_T(s)$	1.0

Note: The parameters $K_{T\Omega}$, K_{TV}, and $K_{T\beta}$ have been calculated from Figure 14.3 and according to Equations 12.96, 12.97, and 12.98, respectively, and for a scenario of wind velocity between 12 and 17 m/s.

This leads to the following expressions:

$$n_{\mu 32}(s) = 1.6499 \times 10^7 s^3 + 1.9183 \times 10^{10} s^2 + 6.3350 \times 10^9 s + 3.68 \times 10^{12} \tag{14.8}$$

$$n_{\mu 33}(s) = -7.4997 \times 10^7 s^2 - 5.2790 \times 10^{11} s - 3.0665 \times 10^{14} \tag{14.9}$$

$$\begin{aligned} d_{tf}(s) = {} & 9.0743 \times 10^{13} s^4 + (1.0551 \times 10^{17} - 1.6499 \times 10^7 K_{T\Omega}) s^3 \cdots \\ & + (3.7482 \times 10^{16} - 1.1983 \times 10^{10} K_{T\Omega}) s^2 \cdots \\ & + (2.1773 \times 10^{19} - 6.3350 \times 10^9 K_{T\Omega}) s - 3.68 \times 10^{12} K_{T\Omega} \end{aligned} \tag{14.10}$$

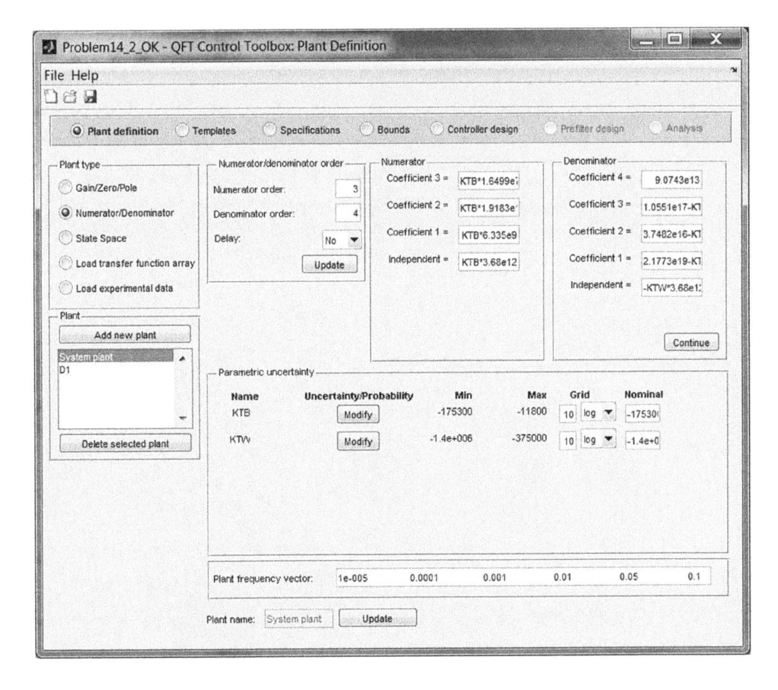

FIGURE 14.4
WT model: plant description $P(s)$ (Equation 14.3).

By looking at the Bode diagrams for the $F_1(s)$, $F_2(s)$, and $F_3(s)$ transfer functions the frequencies of interests (frequency vector) are the following:

$$w = [0.00001\ 0.0001\ 0.001\ 0.01\ 0.05\ 0.1\ 0.11\ 0.12\ 0.15\ 0.17\ 0.2\ 0.21\ 0.3\ 0.5\ 0.6\ 0.7\ 1\ 2\ 3\ 5\ 10\ 11\ 12\ 13\ 13.1\ 13.2\ 13.3\ 13.4\ 13.5\ 13.6\ 13.7\ 13.8\ 13.85\ 13.9\ 13.95\ 14\ 14.05\ 14.1\ 14.15\ 14.2\ 14.25\ 14.3\ 14.4\ 14.5\ 14.6\ 14.7\ 14.8\ 14.9\ 15\ 16\ 17\ 18\ 19\ 20\ 50\ 100]\ \text{rad/s}.$$

Note that the frequency vector is very much populated between 13 and 15 rad/s to capture the existence of resonances (flexible modes) (see, e.g., the loop in Figure 14.6 and the peak in Figure 14.12).

The plant definition windows of the *QFTCT*[149,150] are shown in Figures 14.4 and 14.5 for the plant model $P(s)$ (Equations 14.3, 14.5, and 14.7) and the disturbance model $D_1(s)$ (Equations 14.2, 14.5, and 14.7), respectively.

The minimum values of $K_{T\Omega}$, K_{TV}, and $K_{T\beta}$ are selected for the nominal plants $P_0(s)$ and $D_{10}(s)$. The templates calculated with the *QFTCT* are shown in Figure 14.6. Note that (1) they start at −180° at very low frequency because of the negative sign and type-0 characteristic of $P(s)$; (2) they present a loop in the range 13–15 rad/s because of a resonance (flexible mode); and (3) they end at −270° at high frequency because of the negative sign of $P(s)$ and the fact that the difference of order between denominator and numerator is $3 - 2 = 1$.

14.2.2 Performance Specifications

The main objectives of the pitch control system (see Figure 14.7) in *Region 3* (see Figure 11.5 with $v_1 \geq v_r$, $v_r = 12$ m/s typically) are (1) regulating the rotor speed at the rated (nominal) value $\Omega_{r\_ref} = \Omega_{r\_nom}$ (=1.8850 rad/s = 18 rpm in this case), (2) rejecting the wind disturbances

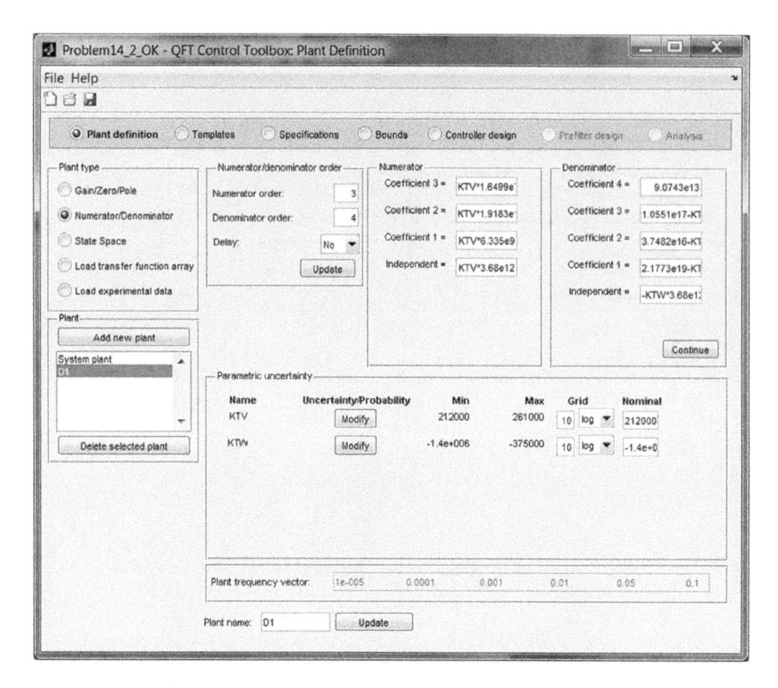

FIGURE 14.5
WT model: disturbance description $D_1(s)$ (Equation 14.2).

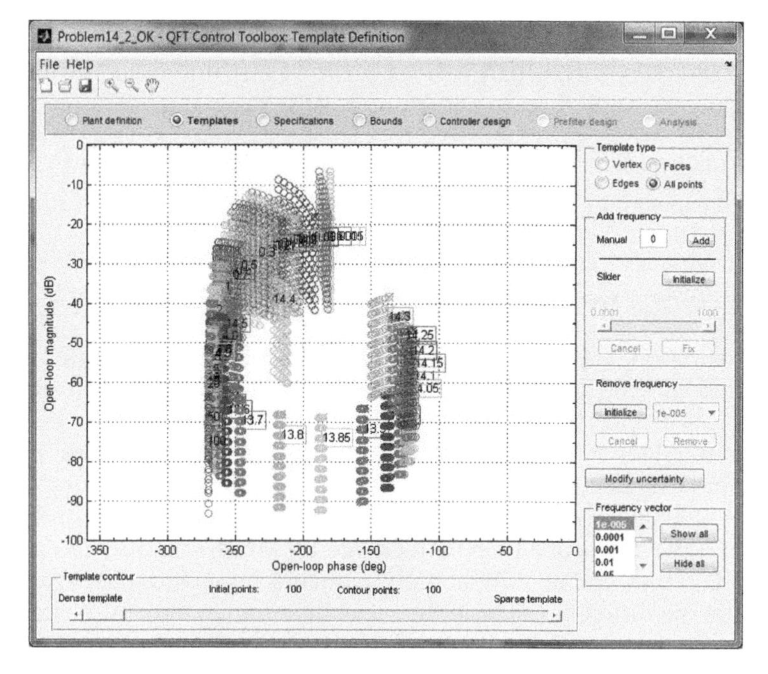

FIGURE 14.6
(See color insert.) Templates calculation for $P(s)$.

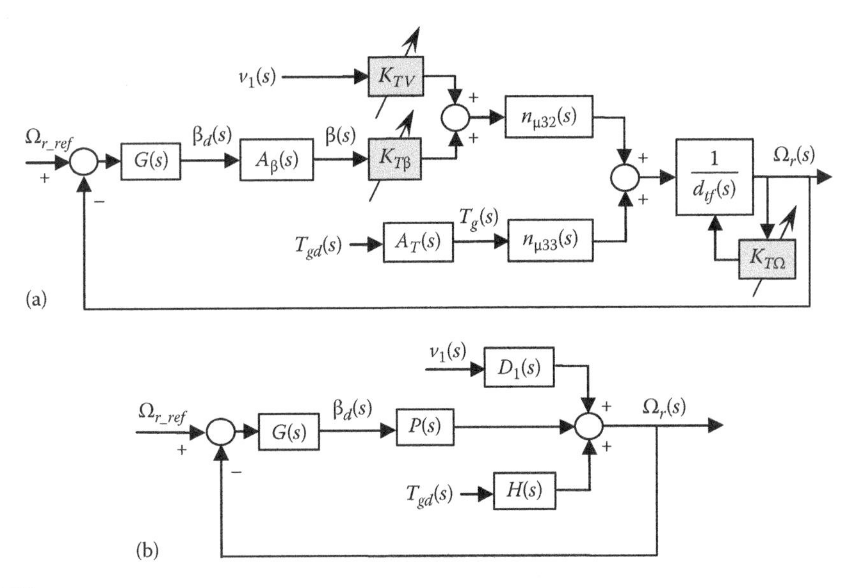

FIGURE 14.7
Pitch control system for the DFIG WT: (a) detailed block diagram and (b) reduced block diagram.

(variation of v_1), and (3) avoiding overspeed situations (with significant $\Delta\Omega_r$) that can be dangerous for the turbine.

In *Region 3*, the electrical torque T_{gd} is considered to be constant at T_{gd_max}, and the pitch angle β_d is the variable to regulate the rotor speed Ω_r.

The rotor speed model of the WT (as it is presented in Chapter 12, Figure 12.14, Equations 14.1 through 14.7, and Figure 14.2) and the structure of the pitch control system are shown in Figure 14.7a and b. Note that $K_{T\Omega}$, K_{TV}, and $K_{T\beta}$ are varying in the nonlinear real-world case.

The specifications considered here include stability and disturbance rejection and are defined in Equations 14.11 and 14.12.

Specification 1 (*stability*)

$$\left|\frac{\Omega_r(j\omega)}{\Omega_{r\_ref}(j\omega)}\right| = \left|\frac{P(j\omega)G(j\omega)}{1+P(j\omega)G(j\omega)}\right| \le \mu = 1.07, \quad \forall\omega \tag{14.11}$$

The stability specification, $\mu = 1.07$ in magnitude (Equation 14.11), is introduced in the *QFTCT*, as shown in Figure 14.8. It implies a gain margin of 5.73 dB and a phase margin of 55.7°.

Specification 2 (*disturbance rejection*)

$$\left|\frac{\Omega_r(j\omega)}{v_1(j\omega)}\right| = \left|\frac{D_1(j\omega)}{1+P(j\omega)G(j\omega)}\right| \le \left|\frac{j\omega}{15j\omega+1}\right|, \quad \forall\omega \tag{14.12}$$

The disturbance rejection specification is introduced in the *QFTCT*, as shown in Figure 14.9, by using the option "Defined by user," which has the expression $|[A(s) + B(s)G(s)]/[C(s) + D(s)G(s)]| \le \delta(\omega)$, $G(s)$ being the controller. The values are given as $A = D_1$, $B = 0$, $C = 1$, and $D = P$.

The bounds calculated by the *QFTCT* are shown in Figure 14.10, where we can see the intersection of the worst-case scenario bounds for the stability and disturbance rejection cases and for the frequencies of interest.

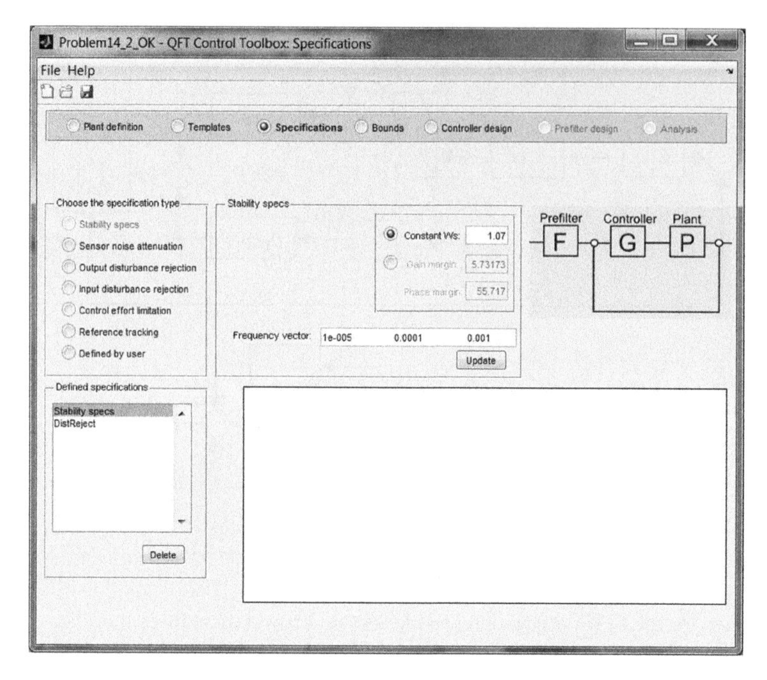

FIGURE 14.8
Specifications: stability.

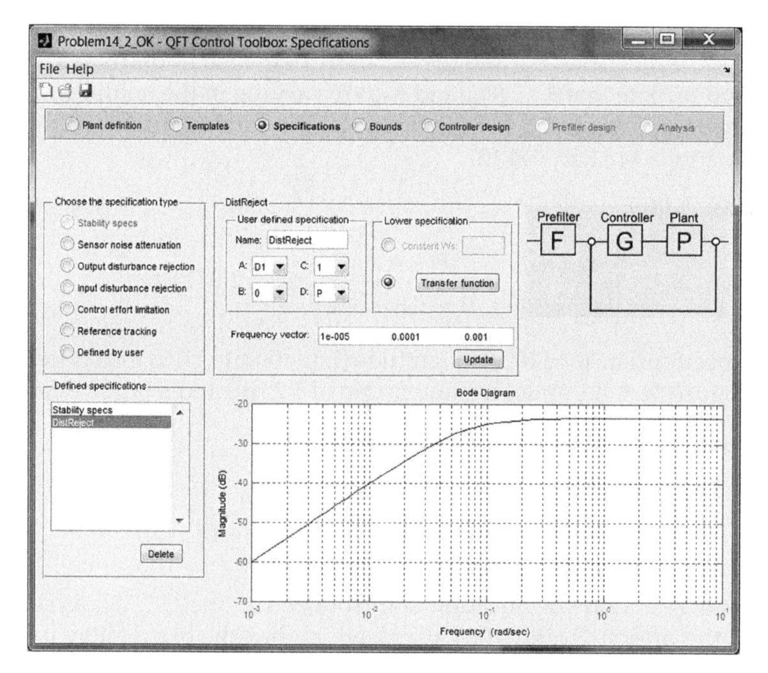

FIGURE 14.9
Specifications: disturbance rejection ("Defined by user" case).

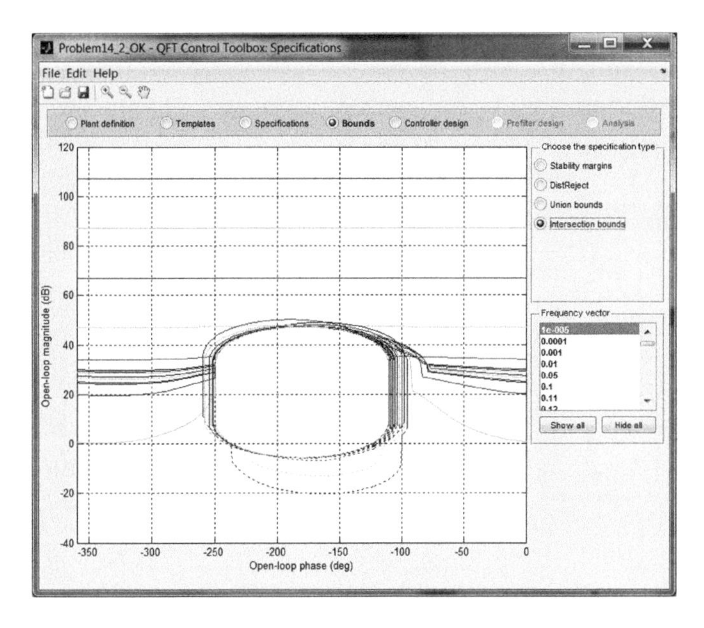

FIGURE 14.10
QFT bounds calculation: worst-case scenario and intersection.

14.2.3 Controller Loop Shaping

The controller $G(s)$, calculated by using the loop-shaping window of the *QFTCT*, is shown in Equation 14.13 and Figure 14.11. It is a proportional-integral-derivative (PID) structure with a first-order low-pass filter. The controller meets all the QFT bounds requirements at every frequency of interest, as is seen in Figure 14.11:

$$G(s) = \frac{-18((s/8)+1)((s/0.061)+1)}{s((s/4)+1)} \tag{14.13}$$

14.2.4 Performance Validation

Two parallel paths to validate the designed controller $G(s)$ (Equation 14.13) are determined by the *QFTCT*: a frequency-domain analysis and a time-domain analysis.

Figure 14.12 shows the frequency-domain analysis of the stability specification (Equation 14.11) (dashed line), and the worst cases within the uncertainty for every frequency (solid line). The solid line does not represent a single transfer function, but a representation of the worst-case result (the higher magnitude in this case) at each frequency, obtained from any possible combination of parameters within the uncertainty. At every frequency of interest the worst-case result (solid line) is below the specification (dashed line). This proves the controller meets the stability specification.

Similarly, Figure 14.13 shows the frequency-domain analysis of the disturbance rejection specification (Equation 14.12) (dashed line), and the worst cases within the uncertainty for every frequency (solid line). The controller also meets the disturbance rejection specification because at every frequency of interest the worst-case result (solid line) is below the specification (dashed line).

Figure 14.14 shows the time response $\Omega_r(t)$ of the nominal plant $P_0(s)$ with the designed controller $G(s)$ for the closed-loop system of Figure 14.7, when a step input is introduced in

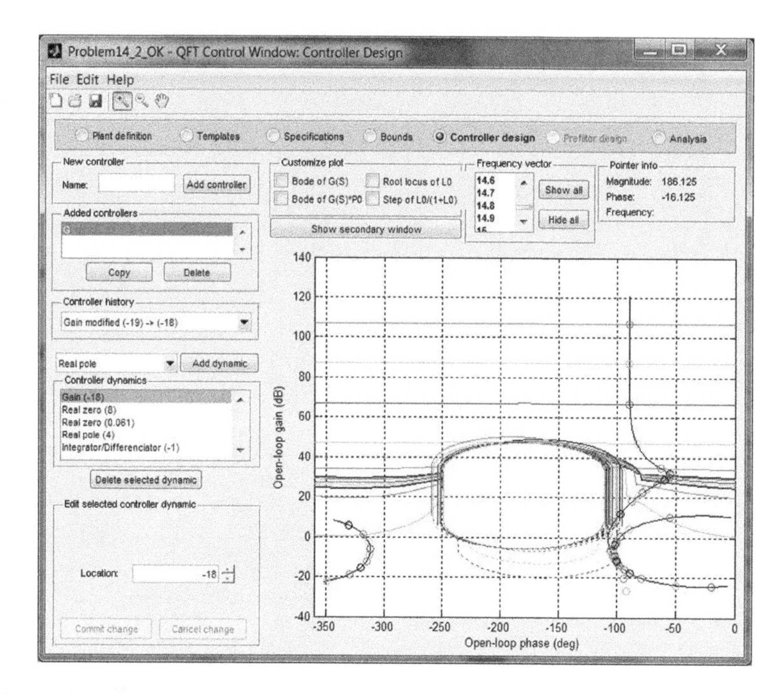

FIGURE 14.11
(See color insert.) QFT bounds and loop shaping: $L_0(s) = P_0(s)\ G(s)$.

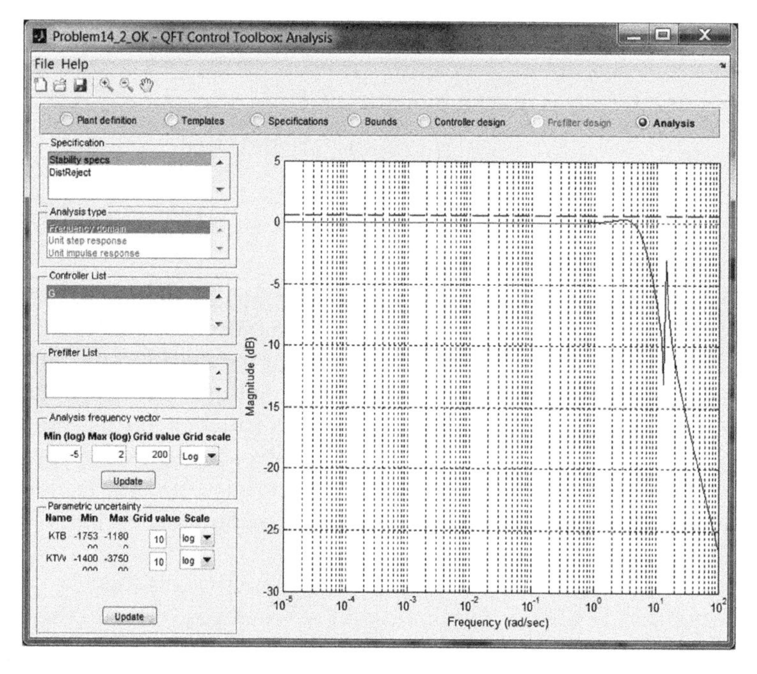

FIGURE 14.12
Validation in the frequency domain: stability.

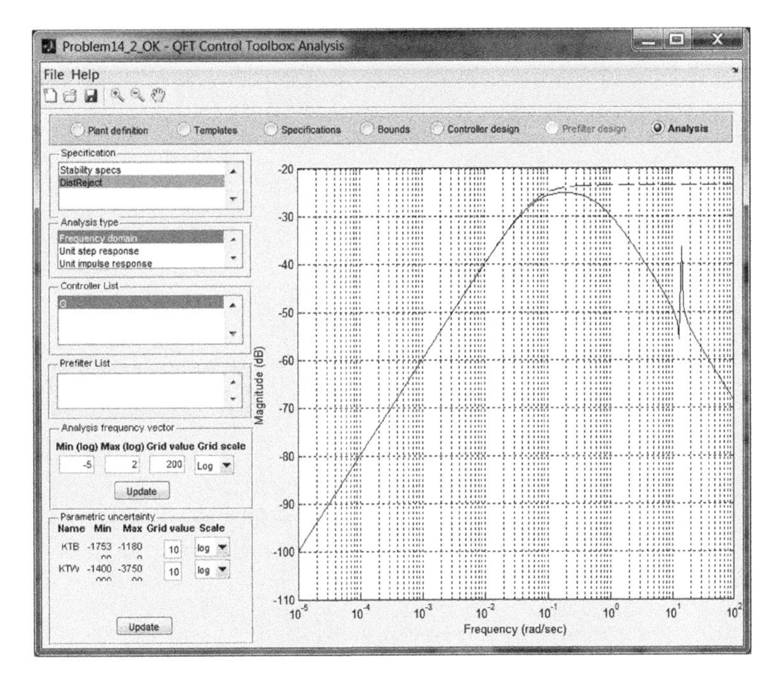

FIGURE 14.13
Validation in the frequency domain: disturbance rejection.

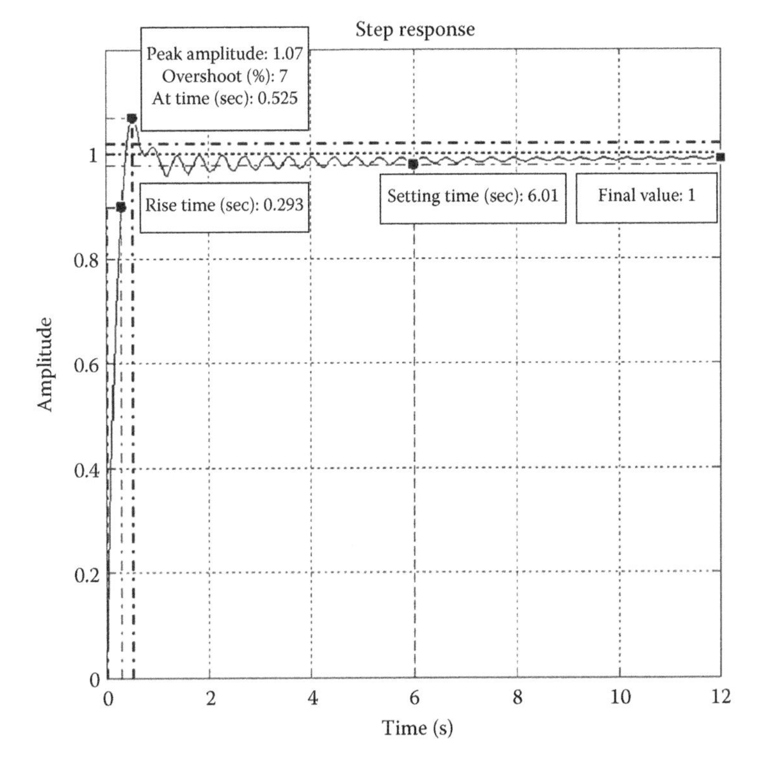

FIGURE 14.14
Validation in the time domain: response to step input reference.

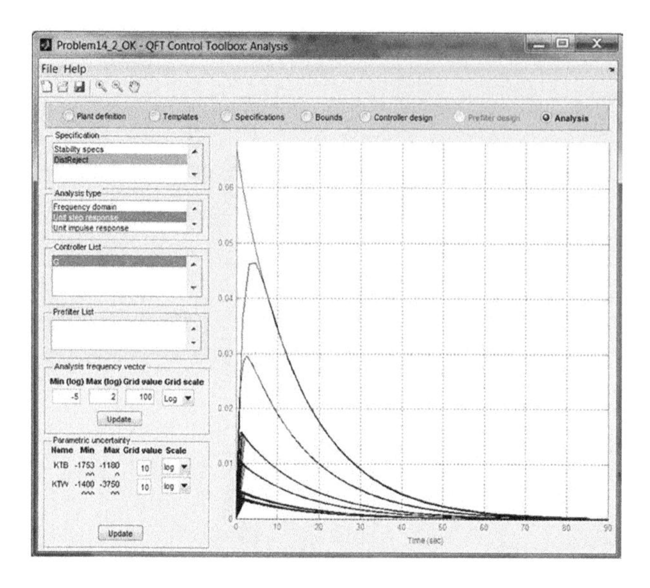

FIGURE 14.15
Validation in the time domain: response to step input disturbance.

the reference $\Omega_{r\text{-}ref}$ of the control system. The overshoot obtained is about 7%, the rise time nearly 0.3s and the settling time 6 s.

Analogously, Figure 14.15 shows the time response $\Omega_r(t)$ of the nominal plant $P_0(s)$ with the designed controller $G(s)$ shown in Figure 14.7, when a step input is introduced in the disturbance input $v_1(t)$ of the control system.

Finally, Figure 14.17 shows the time response $\Omega_r(t)$ of the plant $P(s)$ with the designed controller $G(s)$ in a Simulink® closed-loop simulation (Figure 14.7), when a gusty wind, with a 9.8% turbulence intensity (Equation 14.14 and Figure 14.16), is introduced in the disturbance input v_1 of the control system:

$$TI(\text{Turbulence intensity}) = \frac{\text{std}(v_1)}{\text{mean}(v_1)} = 9.48\% \tag{14.14}$$

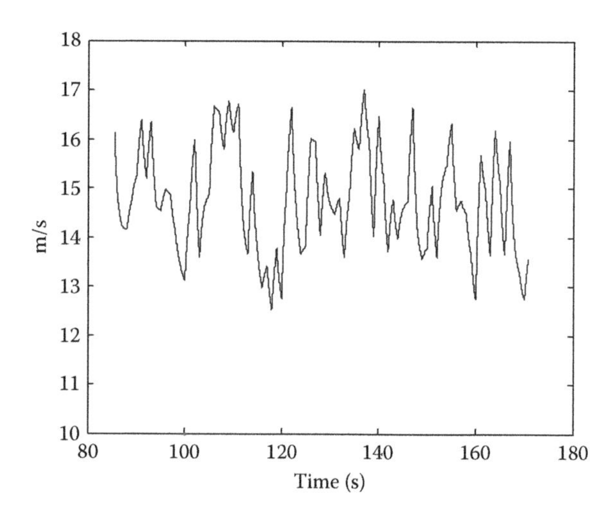

FIGURE 14.16
Wind speed $v_1(t)$.

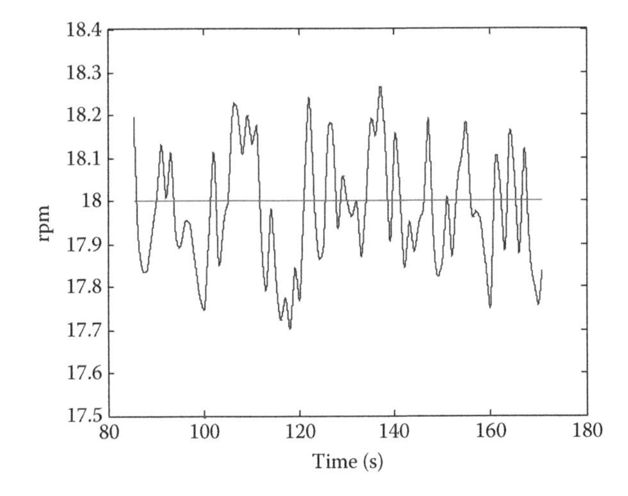

FIGURE 14.17
Validation in the time domain: rotor speed $\Omega_r(t)$ response to wind speed $v_1(t)$ of Figure 14.16.

14.2.5 A More General/Flexible Model Definition

As described in Section F.2.2, Appendix F, the plant models and the additional transfer functions (*SystemPlant* and D_1 defined in Section 14.2.1, Figures 14.4 and 14.5) can also be described by using the option "Load Transfer Function Array" in the plant definition window of the *QFTCT*.

In the example just described (Sections 14.2.1 through 14.2.4), we used the plant definition window of the QFT Toolbox and defined the parameter uncertainty as shown in Figures 14.4 and 14.5. Complementarily, we could apply the option "Load Transfer Function Array," by using the *m-file* defined in the following code [*Case* (a)]:

Case (a)

```
c=1;
KTWmin=-1.4e+006; KTWmax=-375000;
KTBmin=-175300; KTBmax=-11800;
KTVmin=212000; KTVmax=261000;
for KTW=real(logspace(log10(KTWmin),log10(KTWmax),10))
    for KTB=real(logspace(log10(KTBmin),log10(KTBmax),10))
        for KTV=real(logspace(log10(KTVmin),log10(KTVmax),5))
            denomP=[9.0743e13 1.0551e17-KTW*1.6499e7 ...
                3.7482e16-KTW*1.1983e10 2.1773e19-KTW*6.335e9 ...
                -KTW*3.68e12];
            P(1,1,c)=tf(KTB*[1.6499e7 1.9183e10 6.335e9 ...
                3.68e12],denomP);
            D1(1,1,c)=tf(KTV*[1.6499e7 1.9183e10 6.335e9 ...
                3.68e12],denomP);
            c=c+1;
        end
    end
end
```

where P(1,1,c) is the *SystemPlant* and D1(1,1,c) is the additional transfer function D_1 used to define the specifications ("Defined by user") in the specifications window (see Appendix F, Section F.2.2 for more details). The frequency vector is the same as in Section 14.2.1:

TABLE 14.2

Parametric Uncertainty without Cross-Dependence: *Case* (a)

$K_{T\beta}$ **(min)** (Nm/deg)	-17.53×10^4	$K_{T\beta}$ **(max)** (Nm/deg)	-1.18×10^4
$K_{T\Omega}$ **(min)** (Nm/(rad/s))	-14×10^5	$K_{T\Omega}$ **(max)** (Nm/(rad/s))	-3.75×10^5
K_{TV} **(min)** (Nm/m/s)	2.12×10^5	K_{TV} **(max)** (Nm/m/s)	2.61×10^5

ω = [0.00001 0.0001 0.001 0.01 0.05 0.1 0.11 0.12 0.15 0.17 0.2 0.21 0.3 0.5 0.6 0.7 1 2 3 5 10 11 12 13 13.1 13.2 13.3 13.4 13.5 13.6 13.7 13.8 13.85 13.9 13.95 14 14.05 14.1 14.15 14.2 14.25 14.3 14.4 14.5 14.6 14.7 14.8 14.9 15 16 17 18 19 20 50 100] rad/s.

In this way, by using the "Load Transfer Function Array" option, we generate the same results (templates, bounds, etc.) as the ones presented in Section 14.2.1, with the Numerator/Denominator option of the plant definition window.

However, this particular method is much more flexible and general than the one used in Section 14.2.1. It allows the user to deal with any plant, as for the systems described by different structures (uncertainty in the structure) and/or with interdependence among the uncertain parameters, etc.

As an example to illustrate this option, a model for the WT pitch system of this section is described in a more detailed manner. Instead of defining the uncertainty of the parameters $K_{T\Omega}$, K_{TV}, and $K_{T\beta}$ independently, as shown in Table 14.1 (repeated here for convenience as Table 14.2 [*Case* a]), they are defined by the uncertainty including a cross-dependence among the parameters, as shown in Table 14.3 (*Case* b). In other words, if $K_{T\Omega}$ has a particular interval of uncertainty, then the other two parameters, K_{TV} and $K_{T\beta}$, are related by a specific interval of uncertainty. For instance, for v_1 = 12 m/s, if the first parameter $K_{T\Omega} \in [-4.62 \times 10^5, -3.75 \times 10^5]$, then the other two parameters are $K_{TV} \in [2.18 \times 10^5, 2.26 \times 10^5]$ and $K_{T\beta} \in [-2.52 \times 10^4, -1.18 \times 10^4]$. In this case, with this definition, the final uncertainty of the system is smaller, and the design of the controller can reach more demanding specifications (according to the classical trade-off between uncertainty and specifications).

The new *m-file* that defines the system with such a cross-dependence among the uncertainty of the parameters (Table 14.3) is as follows:

Case (b)

```
c=1;

% Case: v1 = 12m/s
KTWmin=-4.62e5; KTWmax=-3.75e5; KTBmin=-2.52e4; KTBmax=-1.18e4;
KTVmin=2.18e5; KTVmax=2.26e5;
```

TABLE 14.3

Parametric Uncertainty with Cross-Dependence: *Case* (b)

v_1	β_0	$K_{T\Omega}$ **(min)**	$K_{T\Omega}$ **(max)**	K_{TV} **(min)**	K_{TV} **(max)**	$K_{T\beta}$ **(min)**	$K_{T\beta}$ **(max)**
12	0	-4.62×10^5	-3.75×10^5	2.18×10^5	2.26×10^5	-2.52×10^4	-1.18×10^4
13	5.72	-4.81×10^5	-4.5×10^5	2.12×10^5	2.14×10^5	-8.42×10^4	-6.21×10^4
14	8.2	-7.22×10^5	-6.07×10^5	2.32×10^5	2.35×10^5	-9.50×10^4	-9.51×10^4
15	10.47	-9.68×10^5	-7.38×10^5	2.26×10^5	2.41×10^5	-13.54×10^4	-11.65×10^4
16	12.22	-11.35×10^5	-9.25×10^5	2.43×10^5	2.61×10^5	-14.10×10^4	-14.11×10^4
17	13.97	-14×10^5	-10.72×10^5	2.54×10^5	2.61×10^5	-17.53×10^4	-14.67×10^4

```
% Case: v1 = 13m/s
KTWmin=-4.81e5; KTWmax=-4.5e5; KTBmin=-8.42e4; KTBmax=-6.21e4;
KTVmin=2.12e5; KTVmax=2.14e5;

% ... to Case: v1 = 17m/s ...

for KTW=real(logspace(log10(KTWmin),log10(KTWmax),10))
    for KTB=real(logspace(log10(KTBmin),log10(KTBmax),10))
        for KTV=real(logspace(log10(KTVmin),log10(KTVmax),5))
            denomP=[9.0743e13 1.0551e17-KTW*1.6499e7 ...
                3.7482e16-KTW*1.1983e10 2.1773e19-KTW*6.335e9 ...
                -KTW*3.68e12];
            P(1,1,c)=tf(KTB*[1.6499e7 1.9183e10 6.335e9 ...
                3.68e12],denomP);
            D1(1,1,c)=tf(KTV*[1.6499e7 1.9183e10 6.335e9 ...
                3.68e12],denomP);
            c=c+1;
        end
    end
end
```

Figure 14.18 compares the templates of *Cases* (a) and (b) for $\omega = 0.001$ rad/s. Figure 14.18a shows the template calculated with the *m-file* of *Case* (a) and Table 14.2, without a cross-dependence among the uncertainty of the parameters $K_{T\Omega}$, K_{TV}, and $K_{T\beta}$. This template is also shown in Figure 14.6.

Similarly, Figure 14.18b shows the template calculated with the *m-file* of *Case* (b) and Table 14.3, with a cross-dependence among the uncertainty of the parameters $K_{T\Omega}$, K_{TV}, and $K_{T\beta}$.

As is shown in Figure 14.18, the templates are smaller in *Case* (b), because the cross-dependence among the parameters reduces the uncertainty and thus the area of the templates.

As a result, and for the same specifications, the QFT bounds in *Case* (b) are less demanding than in *Case* (a). Then, the user is able to design a more active/aggressive controller, or to require more demanding specifications if necessary. Finally, for the same specifications, the loop shaping in *Case* (b) is easier to design than in *Case* (a).

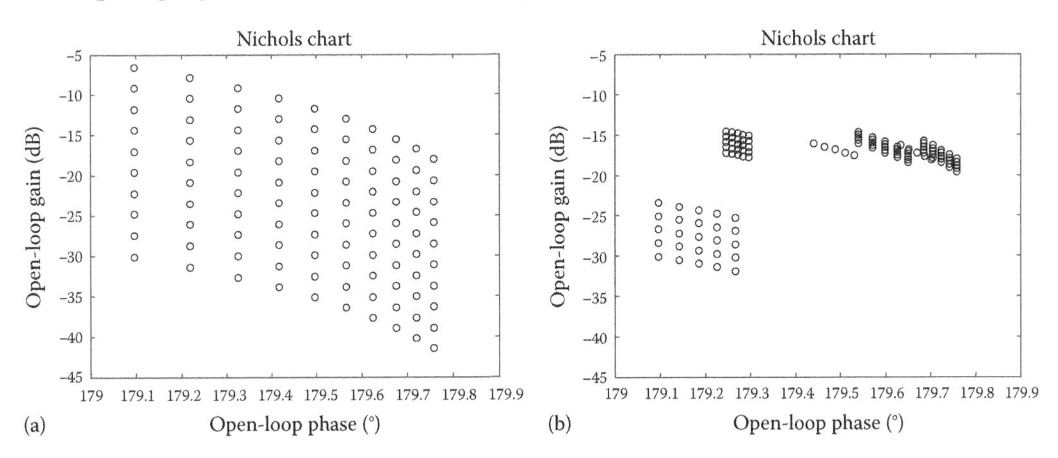

FIGURE 14.18
Templates for $\omega = 0.001$ rad/s: (a) without cross-dependence and (b) with cross-dependence among the uncertainty.

14.3 Nonlinear Switching Multi-Objective Design[140]

This section presents an additional example to design the pitch control system of a variable-speed DFIG WT, now by using the robust-switching methodology introduced in Chapter 8 to simultaneously achieve high reliability (robustness), minimum mechanical fatigue, and performance optimization, all under turbulent wind conditions.

14.3.1 Model, Parameters, and Uncertainty Definition

The dynamic model that describes the dominant characteristics of the variable-speed pitch-controlled DFIG WT is the same as described by Equations 14.1 through 14.7 and shown in Figures 14.1 and 14.2.

14.3.2 Performance Specifications

The main objective of the controller at the above-rated region (*Region 3*, Figure 11.5, $v_1 \geq v_r = 12\,\text{m/s}$) is to regulate the rotor speed Ω_r at $\Omega_{r\text{-}ref} = \Omega_{r\_nom}$, rejecting the wind disturbances Δv_1 and avoiding overspeed situations $\Delta\Omega_r$ that can be potentially dangerous for the turbine. In this region, the electrical torque T_{gd} is considered to be constant and the pitch angle β_d variable to regulate the rotor speed Ω_r. Mechanical fatigue must be also considered during the design process.

According to the methodology introduced in Chapters 2 and 3, a preliminary QFT robust controller is designed. The specifications considered here include stability, disturbance rejection, and control effort limitation.

Specification 1 (Stability)

$$\left|\frac{\Omega_r(j\omega)}{\Omega_{r\_ref}(j\omega)}\right| = \left|\frac{P(j\omega)G(j\omega)}{1+P(j\omega)G(j\omega)}\right| \leq \mu = 1.0828, \quad \forall\omega \tag{14.15a}$$

which implies a gain margin of 5.68 dB and a phase margin of 55°.

Specification 2 (Disturbance rejection)

$$\left|\frac{\Omega_r(j\omega)}{v_1(j\omega)}\right| = \left|\frac{D_1(j\omega)}{1+P(j\omega)G(j\omega)}\right| \leq \left|\frac{4j\omega}{12.5j\omega+1}\right|, \quad \forall\omega \tag{14.15b}$$

The disturbance rejection specification is introduced in the *QFTCT*, as shown in Figure 14.9, by using the option *"Defined by user,"* which has the expression $|[A(s) + B(s)G(s)]/[C(s) + D(s)G(s)]| \leq \delta(\omega)$, $G(s)$ being the controller to be designed. The values $A = D_1$, $B = 0$, $C = 1$, and $D = P$ are selected for this design.

Specification 3 (Control effort limitation)

$$\left|\frac{\beta_d(j\omega)}{v_1(j\omega)}\right| = \left|\frac{G(j\omega)D_1(j\omega)}{1+P(j\omega)G(j\omega)}\right| \leq \left|\frac{20((j\omega/2)+1)}{((j\omega/0.2)+1)((j\omega/0.5)+1)}\right| \tag{14.15c}$$

The control effort specification is also introduced in the *QFTCT* by using the option *"Defined by user"* (Figure 14.9), with the expression $|[A(s) + B(s)G(s)]/[C(s) + D(s)G(s)]| \leq \delta(\omega)$, $A = 0$, $B = D_1$, $C = 1$, $D = P$, and $G =$ the controller.

By looking at the Bode diagram of the system transfer function, the main frequencies of interests (frequency vector) are the following:

$$\omega = [0.05\ 0.1\ 0.2\ 0.5\ 0.6\ 1\ 10]\,\mathrm{rad/s}. \qquad (14.16)$$

14.3.3 Controller Design

Step 1 The first controller $G_0(s)$ is designed by using conventional QFT techniques, as in Section 14.2.3, according to the methodology introduced in Chapters 2 and 3:

$$G_0(s) = \frac{-18.6(s+0.1)(s+1.75)}{s(s+0.65)} \qquad (14.17)$$

The loop shaping in the Nichols chart, with the QFT bounds and the open-loop transfer function for the nominal plant [$L_0(s) = P_0(s)\ G_0(s)$], are shown in Figure 14.19.

Step 2 Now, starting from $G_0(s)$, the extreme controllers $G_1(s)$ and $G_2(s)$, for low and high errors, respectively, are designed observing the guidelines given in Chapter 8. The elements to be varied are the gain and a zero. The new controllers are

$$G_1(s) = \frac{-11(s+0.4)(s+1.75)}{s(s+0.65)} \qquad (14.18)$$

$$G_2(s) = \frac{-37(s+0.03)(s+1.75)}{s(s+0.65)} \qquad (14.19)$$

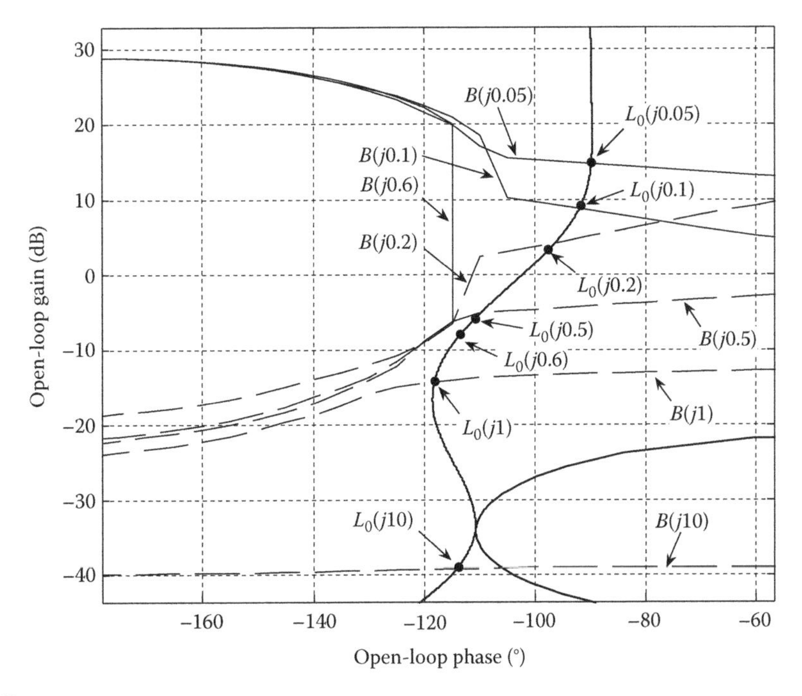

FIGURE 14.19
QFT bounds and loop shaping: $L_0(s) = P_0(s)\ G_0(s)$.

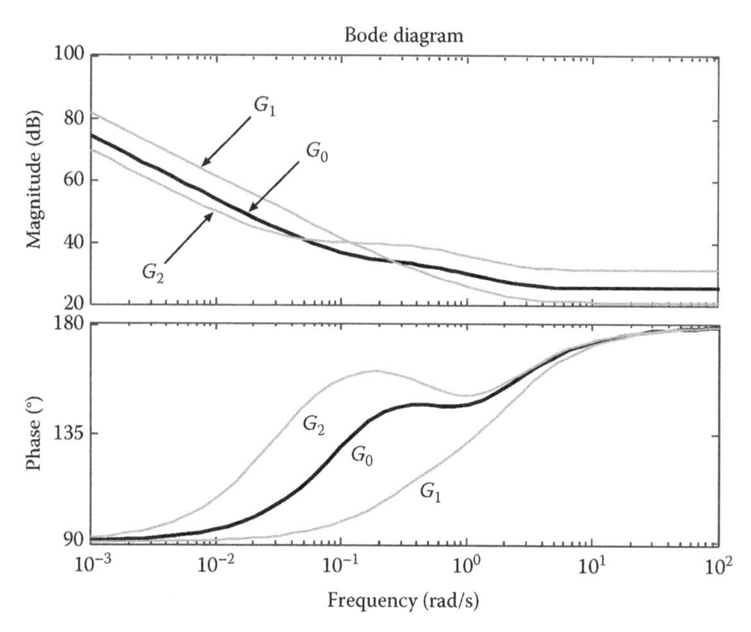

FIGURE 14.20
Bode plot of the three controllers: $G_0(s)$, $G_1(s)$, $G_2(s)$.

and their frequency-domain characteristics can be seen in the Bode plot of Figure 14.20. As it is appreciated, $G_1(s)$ presents higher low-frequency gain, and its bandwidth is smaller, while these characteristics are inverted in $G_2(s)$. Their robust stability margins are checked with QFT tools.

Step 3 Figure 14.21 shows the templates $T[1 + L_1(j\omega)]$ and $T[1 + L_2(j\omega)]$ at various representative frequencies (0.01, 0.1, and 3 rad/s) and the Nichols plot of each nominal function $[1 + L_{10}(j\omega)]$ and $[1 + L_{20}(j\omega)]$. As there is no variation in any pole of the controller, the angle α is zero for all frequencies, and the only aspect to be checked is that, in the path from each point of the first template to its corresponding point of the second template, the maximum horizontal distance between two points is not higher than 90° (see Equation 8.18 with $\alpha = 0$). For most of the templates, this is easy to confirm at a glance, because they are very close.

Step 4 The switching controller presents the following expression:

$$G_{swi}(s) = \frac{-(11+26k)(s+0.4-0.37k)(s+1.75)}{s(s+0.65)} \tag{14.20}$$

where the parameter k is given by a function $P \rightarrow [0, 1]$ of the error signal $e(t)$. In order to reduce possible impulse effects, a smooth function (see Equation 14.21) is selected instead of a relay-type or saturation-type function:

$$k = 1 - \exp\left(-\frac{e(t)}{0.001}\right) \tag{14.21}$$

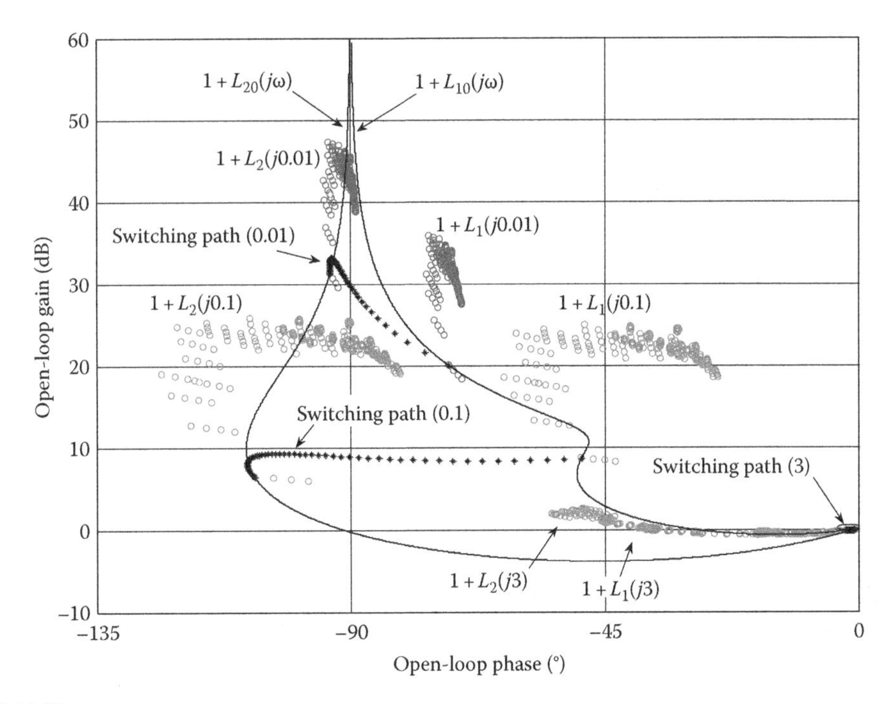

FIGURE 14.21
Stability study of the switching system on the Nichols chart.

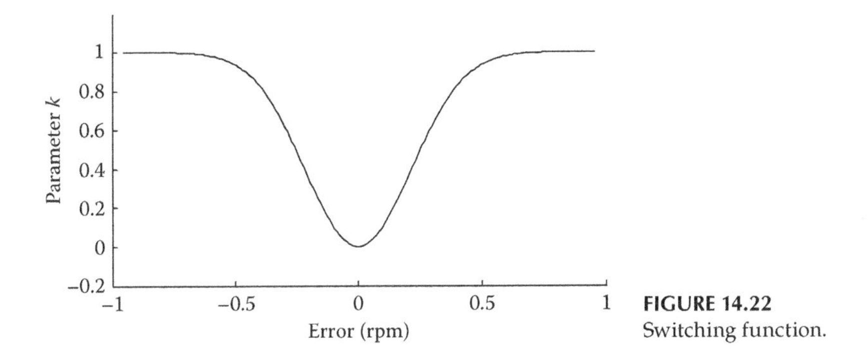

FIGURE 14.22
Switching function.

where the values of the parameters are adjusted for optimal performance by simulation. Some further research can be devoted to finding the optimum shape for the switching functions. The shape of the switching function is shown in Figure 14.22.

14.3.4 Performance Validation

The compensators are validated by using a realistic and nonlinear WT simulator based on a 1.7MW DFIG WT described according to the model introduced in Chapter 12. The design is verified in the time domain for the most representative plants picked from the set of uncertain plants.

Figure 14.23 shows the response to a step disturbance in the wind velocity $v_1(t)$ from 12 to 15m/s. The response of the controller designed with the new methodology (G_{swi}) combines

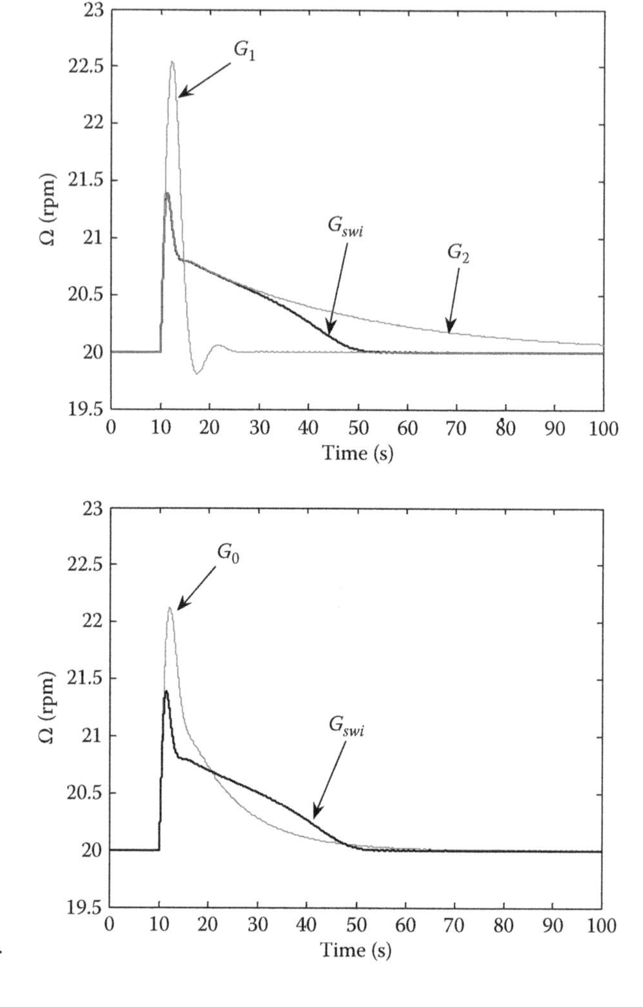

FIGURE 14.23
Responses to a step disturbance (wind gust).

the best characteristics of the extreme controllers (G_1 and G_2), producing a considerable improvement with respect to the response of the original QFT design (G_0).

The new design is also tested with a real wind speed input for 300 s. The signal, shown in Figure 14.24, is obtained from field measurements and presents a turbulence intensity of about 10%.

Once the response is obtained with the four controllers, some representative functions are calculated to compare the results (Table 14.4). The integral of the absolute error (IAE) and the integral of the squared error (ISE) inform about the disturbance rejection capacity, and the maximum absolute error is related with the most dangerous situation produced in terms of overspeed (first row, Table 14.4).

The mean of the control action derivative (second row, Table 14.4) gives an idea about the frequency and abruptness of the changes in the control action, which have great importance in the mechanical fatigue. The robust stability margins of each controller are also calculated (third row, Table 14.4).

All these coefficients are calculated for the preliminary QFT controller G_0, the extreme controllers for small error G_1 and for high error G_2, and the switching controller G_{swi}. Looking at the results, it is clear that the switching controller G_{swi} has as good results as G_2,

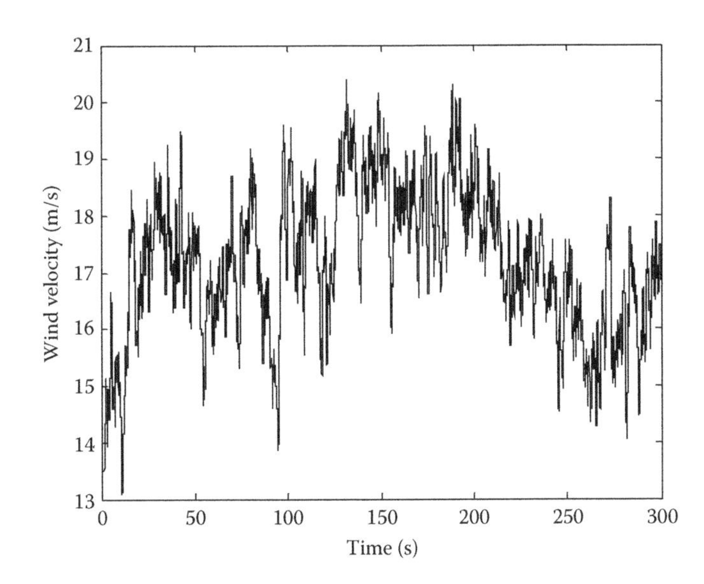

FIGURE 14.24
Experimental wind data $v_1(t)$.

TABLE 14.4
Results with the Four Controllers

		G_0	G_1	G_2	G_{swi}
Error	IAE	815.39	994.96	628.44	**616.52**
	ISE	36.43	58.23	20.69	**18.84**
	max(|e|)	1.323	1.959	0.869	**0.871**
Fatigue	mean(du/dt)	0.6154	0.5267	0.8975	**0.7988**
Robust stability	PM (deg)	55	36	56	**∈ [36, 56]**
	GM (dB)	5.68	4.17	5.75	**∈ [4.17, 5.75]**

or even better, in disturbance rejection, with less cost in terms of fatigue, and maintaining similar robust stability margins. To complete the information about mechanical fatigue, a rainflow algorithm is used to count the number of cycles in the control signal. The G_{swi} controller reduces 13.25% of the number of cycles of G_2 on the main amplitudes.

14.4 Nonlinear Robust Control Design for Large Parameter Variation

The mechanical power extracted by the WT and the force (thrust) and torque that the air exerts on the WT rotor are all given by nonlinear equations, as described in Chapter 12 (see Equations 12.35, 12.36, and 12.37 for P, F_T, and T_r, respectively, and curves of $C_p(\lambda, \beta)$ and $C_T(\lambda, \beta)$ shown in Figures 12.2, 12.9, and 14.3). For convenience, Equations 12.35, 12.36, 12.37, and 12.38 are repeated here as Equations 14.22, 14.23, 14.24, and 14.25, respectively.

For the entire rotor

$$P = \frac{1}{2}\rho\pi r_b^2 C_P(\lambda, \beta) v_1^3 = T_r \Omega_r \tag{14.22}$$

$$F_T = \frac{1}{2}\rho\pi r_b^2 C_T(\lambda,\beta)v_1^2 = N\int_{r_{root}}^{r_b}[L\cos(\phi)+D\sin(\phi)]\,dr \tag{14.23}$$

$$T_r = \frac{\rho\pi r_b^2 C_P(\lambda,\beta)v_1^3}{2\Omega_r} = N\int_{r_{root}}^{r_b}[L\sin(\phi)-D\cos(\phi)]\,r\,dr \tag{14.24}$$

For each airfoil (each r)

$$L = 0.5\rho A_a C_L v_{rel}^2;\quad D = 0.5\rho A_a C_D v_{rel}^2 \tag{14.25}$$

Note that the functions P, F_T, and T_r depend on v_1, β, and Ω_r in a nonlinear way. If the aerodynamic part of these equations is linearized around a working point (v_{10}, β_0, Ω_{r0}), and the bias components are ignored (as shown in Chapter 12 and Figure 12.13), the functions P, F_T, and T_r can be described by a transfer matrix whose nonlinear elements are just the gains $K_{T\Omega}$, K_{TV}, $K_{T\beta}$, $K_{F\Omega}$, K_{FV}, and $K_{F\beta}$, as shown in Equations 12.102 through 12.104.

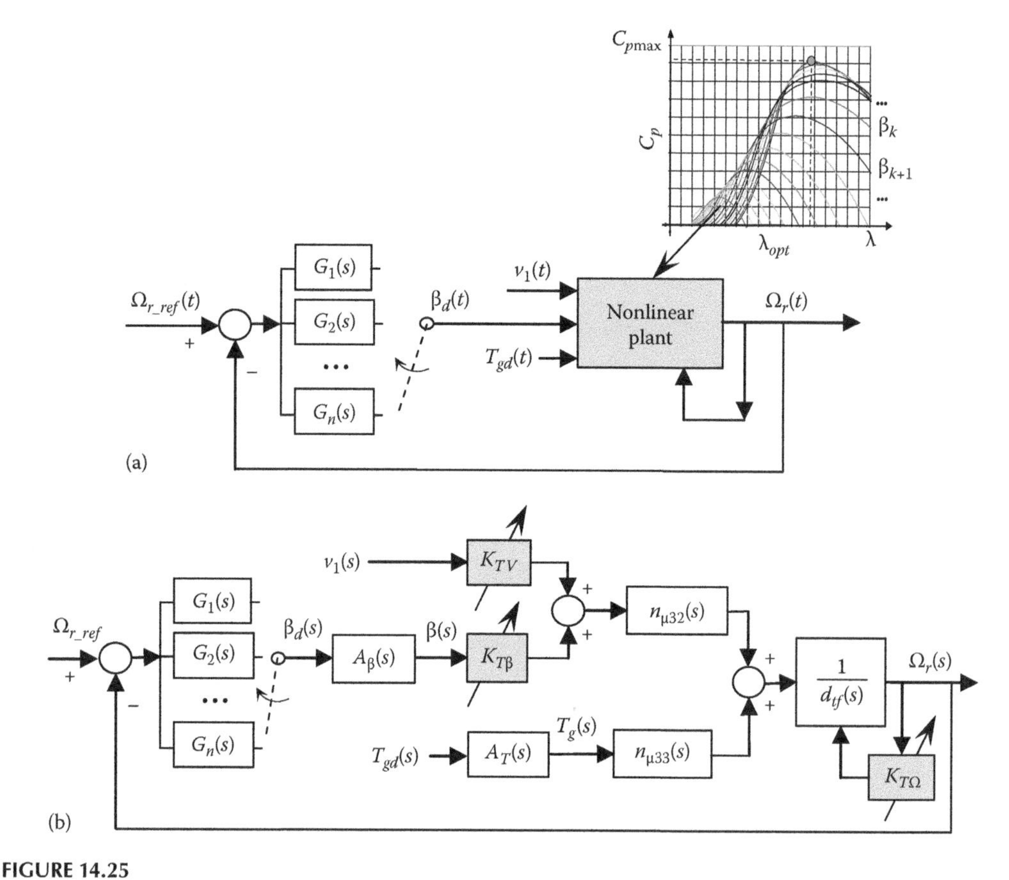

FIGURE 14.25
Nonlinear QFT robust pitch control system: (a) nonlinear description and (b) linear description with parametric uncertainty.

If the variation of such gains is not very large, as for instance in the problems of Sections 14.2 and 14.3, then the nonlinearities can be considered as parametric uncertainty, and it is then possible to design just one robust controller.

If, on the contrary, the variation of the gains is very large, as it usually occurs when the wind speed changes from typically 12 to 25 m/s (see Equations 12.113 through 12.116), then the nonlinearities (Figure 14.25a) can still be considered as parametric uncertainty (Figure 14.25b), but splitting that uncertainty into smaller regions. In this case a robust controller is designed for each new region. Then, a control structure (Figure 14.25) to switch among these robust controllers in real time, depending on the region where the system evolves, is applied. Chapter 8 and Section 14.3 present the methodology to analyze the switching robust control system.

14.5 Summary

Three advanced pitch control system examples for variable-speed DFIG WTs were presented in this chapter.

The first example applied the QFT robust control technique introduced in Chapters 2 and 3 and derived the specific dynamical models of the pitch control system (see Chapter 12), the required performance specifications, QFT templates, bounds, loop-shaping controller design, and the frequency- and time-domain validation. The example used the *QFTCT* for MATLAB® provided with this book and introduced in Appendix F.[149,150]

The second example applied the robust-switching methodology introduced in Chapter 8 to simultaneously achieve (1) high reliability and robustness, (2) minimum mechanical fatigue, and (3) performance optimization, all under turbulent wind conditions. It designed a pitch control system for a DFIG WT, going beyond the classical linear limitations while giving a solution for the well-known robustness–performance trade-off.[140]

Finally, the third example introduced a switching control structure of QFT robust controllers for pitch control systems working in a large nonlinear (large parameter variation) environment.

15

Experimental Results with the Direct-Drive Wind Turbine TWT-1.65[174,349,366,379]

15.1 Introduction

It is well known that variable-speed direct-drive (gearless) wind turbines (WTs) have in general more capabilities and better performance than conventional gearbox-drive machines. Some of the major wind turbine manufacturers in the world have identified this technology as the most appropriate for offshore applications. However, international companies that design multimegawatt variable-speed direct-drive WTs nowadays are but a few, due to the novelty and technological complexity of the system. Among others are Enercon (Germany), M.Torres[349] (Spain), General Electric (United States), and Siemens (Germany).

This chapter presents some experimental control system results of the multimegawatt variable-speed direct-drive multipole synchronous WTs, Torres wind turbine (TWT)-1.65/70 (Class Ia), TWT-1.65/77 (Class IIa), and TWT-1.65/82 (Class IIIa). The first project was started in 1998, and the first prototype was installed in Cabanillas (Navarra, Spain) and began its operation in April 2001. Since then many WTs of this family have been installed in different countries, and a large amount of experimental data have been collected. The design of the controllers was made by using advanced quantitative feedback theory (QFT)-robust and nonlinear-switching control strategies based on both mathematical modeling and analysis of the experimental data. This chapter introduces the main advantages of the direct-drive multipole system and shows some of the most representative experimental results with the direct-drive TWT with the QFT and switching controllers, under medium and extreme wind conditions and connected to different grid situations.[174,349,366,379]

Section 15.2 describes the variable-speed direct-drive multipole synchronous TWT machines (TWT-1.65/70-Class Ia, TWT-1.65/77-Class IIa, TWT-1.65/82-Class IIIa, TWT-2.5/90-Class Ia, TWT-2.5/100-Class IIa, and TWT-2.5/103-Class IIIa). Some significant experimental results are presented in Section 15.3 for the TWT pitch control and rotor speed control systems under typical and extreme wind conditions. This is followed by Section 15.4, which presents experimental results of the behavior of the TWT machines when dealing with different grid integration problems, with advanced reactive power control systems, voltage control systems, and active power control systems. Finally, Section 15.5 introduces new solutions to survive under voltage dip problems in the grid.

15.2 Variable-Speed Direct-Drive Torres Wind Turbine Family

The TWT machine is a multimegawatt direct-drive gearless variable-speed pitch-controlled multipole synchronous generator. The first design was started in 1998 and was developed by Manuel Torres, founder of M.Torres Group (www.mtorres.es),[349] with the collaboration of Professor Garcia-Sanz and a multidisciplinary group of engineers.

Taking advantage of the multidisciplinary capabilities of the M.Torres Group,[349] the design avoided the classical way of working, where every engineering team (mechanical, aerodynamic, electrical, electronics, and control engineers) works independently in a sequential manner to merge all the design proposals at a very late stage (see Figure 15.1). On the contrary, the design of the WTs was carried out according to an integrated design philosophy, where the engineering teams work together in a simultaneous and concurrent manner from the very beginning. This integrated view of the design (*concurrent engineering*) requires that *control engineers* play a central role from the very beginning of the project, to coordinate the different disciplines and achieve a better system's dynamics, controllability, and optimum design (see Figure 15.1).

As any new energy system, the design of the new multimegawatt direct-drive TWT machines involves many aspects—engineering concepts, economics, reliability issues, efficiency, certification, regulations, implementation, marketing, interaction with grid and environment, maintenance, etc.—that deeply affect each other in a continuous feedback loop, which is called the *project road map* (see Figure 15.2). In other words, any single aspect in this big picture can affect the rest of the areas, and eventually be the critical bottleneck of the entire project. For instance, a specific airfoil for the blades could affect the rotor speed, which could require modifying the electrical design of

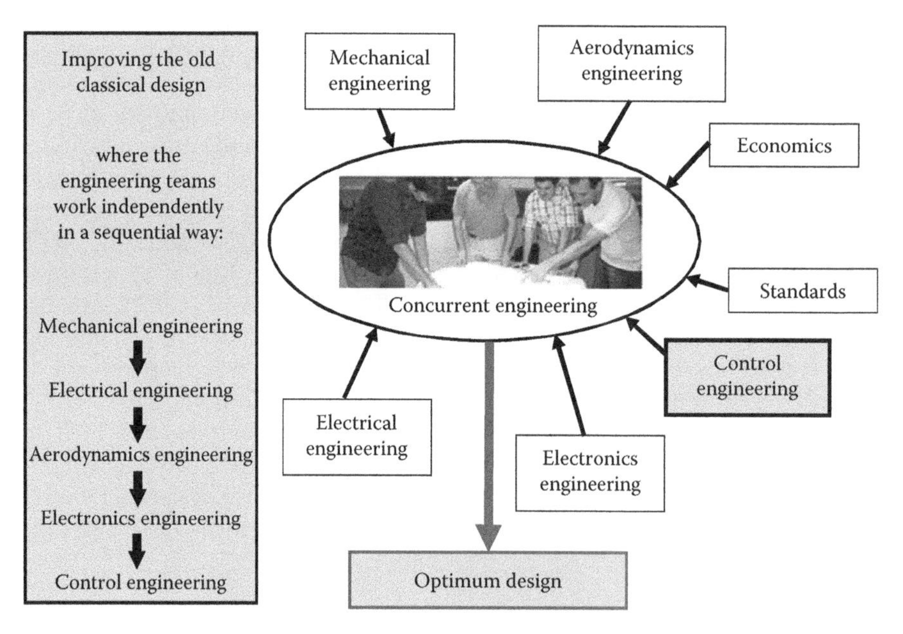

FIGURE 15.1
Concurrent engineering design.

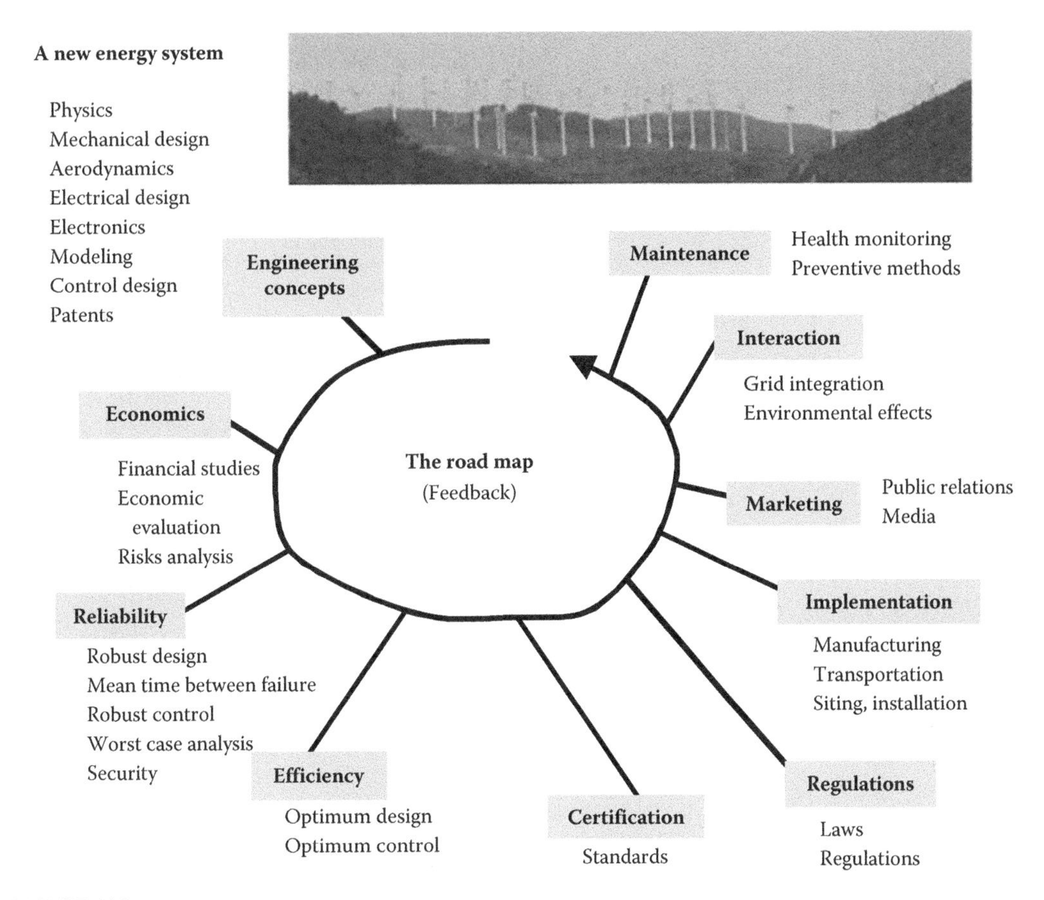

FIGURE 15.2
The big picture of the design of a new energy system.

the generator, which could change the economics, and/or the reliability, and/or the efficiency, and/or the maintenance of the project.

Analogously, a marketing or a regulation issue could require a noise level reduction, which eventually will need modifying the rotor speed or the blade design, which again could require modifying the electrical design of the generator, which can change the economics, and/or the reliability, and/or the efficiency, and/or the maintenance of the project. Also, a logistic transport limitation or a economic specification could require shorter blades, which in turn will affect the rotor speed, which will require modifying the electrical design of the generator, which will then require changing the mechanical loads, the structure of the tower, the foundation, the economics, and/or the reliability, and/or the efficiency, and/or the maintenance of the entire project.

This big picture definitely requires the concurrent engineering way of thinking and working, with the leadership of a multidisciplinary engineer, as the *control engineer*, to be able to achieve an optimum design that takes into account all the aspects of the road map.

In addition to the road map, the design of the TWTs was made keeping in mind four critical objectives: (a) high reliability at every working condition, (b) optimum energy efficiency at every wind speed, (c) maximum electrical power quality at the output, and (d) low cost maintenance.

TABLE 15.1
TWT-1.65/70, Class Ia, Characteristics

Rotor		**Operation Data**	
Diameter	70 m	Wind class	Ia, according to GL
Swept area	3870 m^2	Cut-in wind speed	3 m/s
Number of blades	3	Cut-out wind speed (1)	25 m/s more than 10 min
Position, rotational direction	Upwind, clockwise	Cut-out wind speed (2)	30 m/s more than 0.1 s
Range of rotor speed	Variable: 6–22 rpm	Survival speed	70 m/s
Pitch control	Independent electrical blade actuators and energy stored in each blade	**Generator**	
		Type	Direct-drive multipole synchronous generator
Rotor brake	Disk and callipers	Poles	Wound cupper poles
Tower		Excitation	Independent electrical
Type	Tubular conical steel	Voltage	690 V
Hub height	71 m	**Nacelle**	
Corrosion protection	Epoxy coating	Construction	Monocoque in steel
Access to nacelle	Elevator and ladder	Yaw	Electrical AC motors
Converter		Corrosion protection	Epoxy coating
Power to the grid	1.65 MW	**General**	
Power electronics	Two three-phase symmetric and reversible bridges, controlled with IGBTs	Control	Advanced QFT-robust and nonlinear-switching and predictive strategies
Power quality	According to IEC-61400-21	Lightning protection	in blades and structure
Voltage dips ride-through	REE, BOC, E.ON, etc.	Design life span	20 years
Reactive power	cos $\phi \geq 0.90$ or more, L/C	Hub	Cast iron
Frequency	50, 60 Hz	Pitch system	Electrical AC motors

Source: M.Torres Group, Navarra, Spain, http://www.mtorres.es/

Since 1998, the TWT concept has been applied to six different models, according to two power output possibilities (1.65 and 2.5 MW) and the main three Germanisher Lloyd's Standards (GL) wind speed class scenarios (Class Ia, IIa, and IIIa) (see Table 10.3).[349] The models use six different rotor diameters: 70, 77, 82, 90, 100, and 103 m. In other words, the machines are TWT-1.65/70-Class Ia, TWT-1.65/77-Class IIa, TWT-1.65/82-Class IIIa, TWT-2.5/90-Class Ia, TWT-2.5/100-Class IIa, and TWT-2.5/103-Class IIIa (see Tables 15.1 through 15.4). All of them include advanced control systems based on QFT-robust and nonlinear-switching control strategies.

Utilizing the integrated design philosophy (Figure 15.2), the TWTs' design was undertaken following the most demanding Wind Turbine International Standards, see Chapter 10, including all the recommendations for performance, safety, mechanical loads, electrical issues, voltage dips, power quality, etc. of

- European Wind Energy Association (EWEA)[339]
- American Wind Energy Association (AWEA)[338]
- International Electrotechnical Commission Standards, IEC 61400 (*Wind Turbine Generator Systems*)[337]

TABLE 15.2

TWT-1.65/77, Class IIa, Characteristics

Rotor	
Diameter	77 m
Swept area	4630 m^2
Number of blades	3
Position, rotational direction	Upwind, clockwise
Range of rotor speed	Variable: 6–20 rpm
Pitch control	Independent electrical blade actuators and energy stored in each blade
Rotor brake	Disk and callipers
Tower	
Type	Tubular conical steel
Hub height	71 m
Corrosion protection	Epoxy coating
Access to nacelle	Elevator and ladder
Converter	
Power to the grid	1.65 MW
Power electronics	Two three-phase symmetric and reversible bridges controlled with IGBTs
Power quality	According to IEC-61400-21
Voltage dips ride-through	REE, BOC, E.ON, etc.
Reactive power	$\cos \phi \geq 0.90$ or more, L/C
Frequency	50, 60 Hz
Operation Data	
Wind class	IIa, according to GL
Cut-in wind speed	3 m/s
Cut-out wind speed (1)	25 m/s more than 10 min
Cut-out wind speed (2)	30 m/s more than 0.1 s
Survival speed	59.5 m/s
Generator	
Type	Direct-drive multipole synchronous generator
Poles	Wound cupper poles
Excitation	Independent electrical
Voltage	690 V
Nacelle	
Construction	Monocoque in steel
Yaw	Electrical AC motors
Corrosion protection	Epoxy coating
General	
Control	Advanced QFT-robust and nonlinear-switching and predictive strategies
Lightning protection	In blades and structure
Design life span	20 years
Hub	Cast iron
Pitch system	Electrical AC motors

Source: M.Torres Group, Navarra, Spain, http://www.mtorres.es/

- GL's Standards (*Regulation for the Certification of Wind Energy Conversion Systems*)[347]
- MEASNET European standards for measurements procedures[340]
- Voltage dips standards of the Spanish Ministry, Canary Islands, IEC and E.ON Netz Grid Code[337]
- Electromagnetic compatibility standards[337]
- Manufacturing quality standards certified under ISO-9001, etc.

Figure 15.3 shows the installation of the first two TWT prototypes in Cabanillas and Unzue (Navarra, Spain). Since 1998 many TWT machines have been installed and a large amount of experimental data have been collected to improve the dynamic model and the control system of the wind turbines.[349,366]

The mechanical design of the TWT machines was optimized using finite element calculations and test bed experimentation. The study focused on the analysis of fatigue, mechanical stress, and structural resonances of each critical element of the system.

TABLE 15.3

TWT-1.65/82, Class IIIa, Characteristics

Rotor		**Operation Data**	
Diameter	82 m	Wind class	IIIa, according to GL
Swept area	5365 m^2	Cut-in wind speed	3 m/s
Number of blades	3	Cut-out wind speed (1)	25 m/s more than 10 min
Position, rotational direction	Upwind, clockwise	Cut-out wind speed (2)	30 m/s more than 0.1 s
Range of rotor speed	Variable: 6–18 rpm	Survival speed	52.5 m/s
Pitch control	Independent electrical blade actuators and energy stored in each blade	**Generator**	
		Type	Direct-drive multipole synchronous generator
Rotor brake	Disk and callipers	Poles	Wound cupper poles
Tower		Excitation	Independent electrical
Type	Tubular conical steel	Voltage	690 V
Hub height	71 m	**Nacelle**	
Corrosion protection	Epoxy coating	Construction	Monocoque in steel
Access to nacelle	Elevator and ladder	Yaw	Electrical AC motors
Converter		Corrosion protection	Epoxy coating
Power to the grid	1.65 MW	**General**	
Power electronics	Two three-phase symmetric and reversible bridges controlled with IGBTs	Control	Advanced QFT-robust and nonlinear-switching and predictive strategies
Power quality	According to IEC-61400-21	Lightning protection	In blades and structure
Voltage dips ride-through	REE, BOC, E.ON, etc.	Design life span	20 years
Reactive power	cos $\phi \geq 0.90$ or more, L/C	Hub	Cast iron
Frequency	50, 60 Hz	Pitch system	Electrical AC motors

Source: M.Torres Group, Navarra, Spain, http://www.mtorres.es/

Figure 15.4 shows a 3D drawing of the nacelle and Figure 15.5 shows the finite element gridding of the generator.

The electrical design and control diagram of the TWT is shown in Figure 15.6. It consists of two three-phase reversible IGBT (*insulated gate bipolar transistor*) converters that connect the grid with the multipole synchronous generator, also designed by M.Torres (see Figure 15.7).

The design was optimized using finite element calculations, advanced electrical simulators, and a special test bed developed for full-power experimentation (see Figure 15.8). Figure 15.9 shows the nacelle with the multipole synchronous generator and the hub with the pitch control systems.

The TWT machines are also monitored by a SCADA (*Supervisory Control and Data Acquisition*) system (see Figure 15.10 and Section 11.4.7). This is basically a computer system that monitors the WTs and wind farms. It runs in continuous and/or batch modes, collects information from all the sensors of the machines, grid, and environment, generates periodically reports based on the real-time information (power generation, wind and weather conditions, grid conditions, critical machine variables, etc.), and communicates remotely with the users.

TABLE 15.4

TWT-2.5/90, Class Ia; TWT-2.5/100, Class IIa; and TWT-2.5/103, Class IIIa Characteristics

Rotor		**Operation Data**	
Diameter	90, 100, 103 m	Wind class	Ia, IIa, IIIa, (GL)
Swept area	6362, 7854, 8332 m^2	Cut-in wind speed	3 m/s
Number of blades	3	Cut-out wind speed (Ia, IIa)	25 m/s more than 10 min
Position, rotational direction	Upwind, clockwise	Cut-out wind speed (IIIa)	20 m/s more than 10 min
Range of rotor speed	Variable: 6–17 rpm	Survival speed	70, 59.5, 52.5 m/s
Pitch control	Independent electrical blade actuators and energy stored in each blade	**Generator**	
Rotor brake	Disk and callipers	Type	Direct-drive multipole synchronous generator
Tower		Poles	Wound cupper poles
Type	Tubular conical steel	Excitation	Independent electrical
Hub height	100 m	Voltage	690 V
Corrosion protection	Epoxy coating	**Nacelle**	
Access to nacelle	Elevator and ladder	Construction	Monocoque in steel
Converter		Yaw	Electrical AC motors
Power to the grid	2.5 MW	Corrosion protection	Epoxy coating
Power electronics	Two three-phase symmetric and reversible bridges controlled with IGBTs	**General**	
		Control	Advanced QFT-robust and nonlinear-switching and predictive strategies
Power quality	According to IEC-61400-21		
Voltage dips ride-through	REE, BOC, E.ON, etc.	Lightning protection	In blades and structure
Reactive power	cos $\phi \geq 0.90$ or more, L/C	Design life span	20 years
		Hub	Cast iron
Frequency	50, 60 Hz	Pitch system	Electrical AC motors

Source: M.Torres Group, Navarra, Spain, http://www.mtorres.es/

FIGURE 15.3
TWT-1.65: first and second prototypes. (Courtesy of M.Torres Group, Navarra, Spain, http://www.mtorres.es/)

FIGURE 15.4
TWT-1.65 3D drawing. (Courtesy of M.Torres Group, Navarra, Spain, http://www.mtorres.es/)

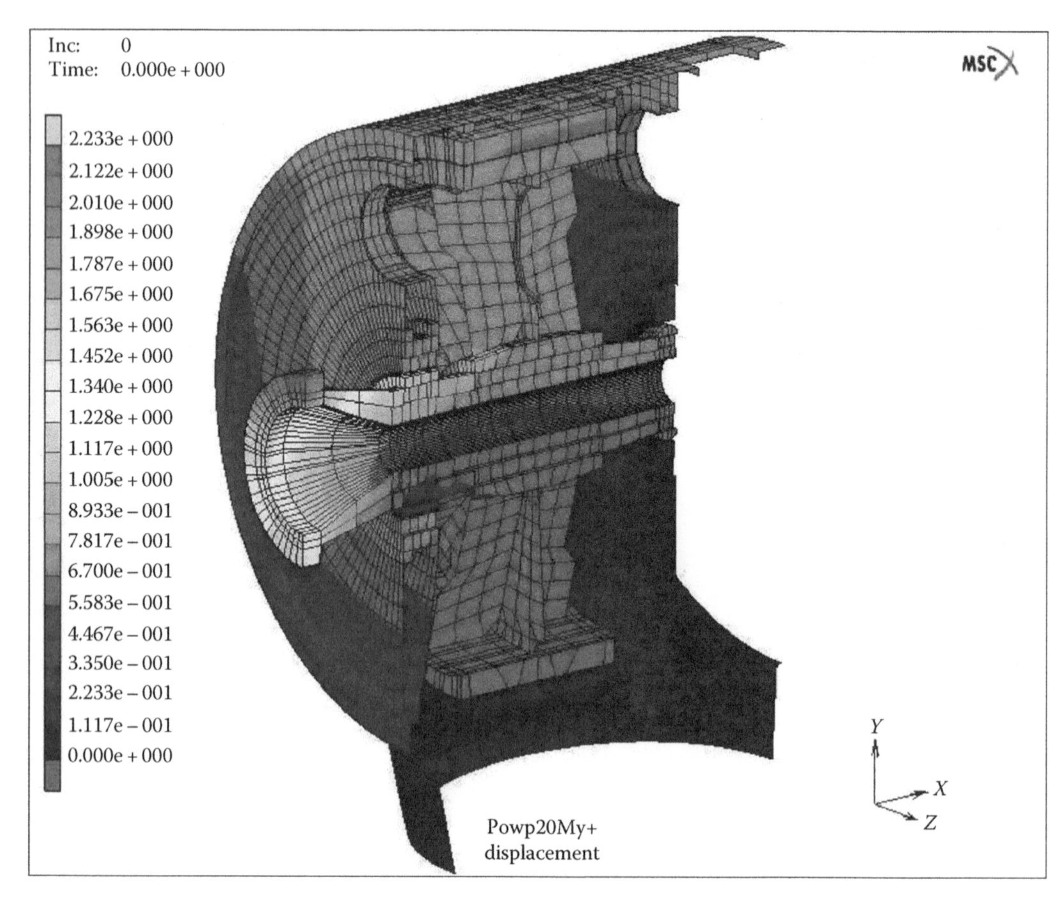

FIGURE 15.5
Finite element analysis of the electrical generator.

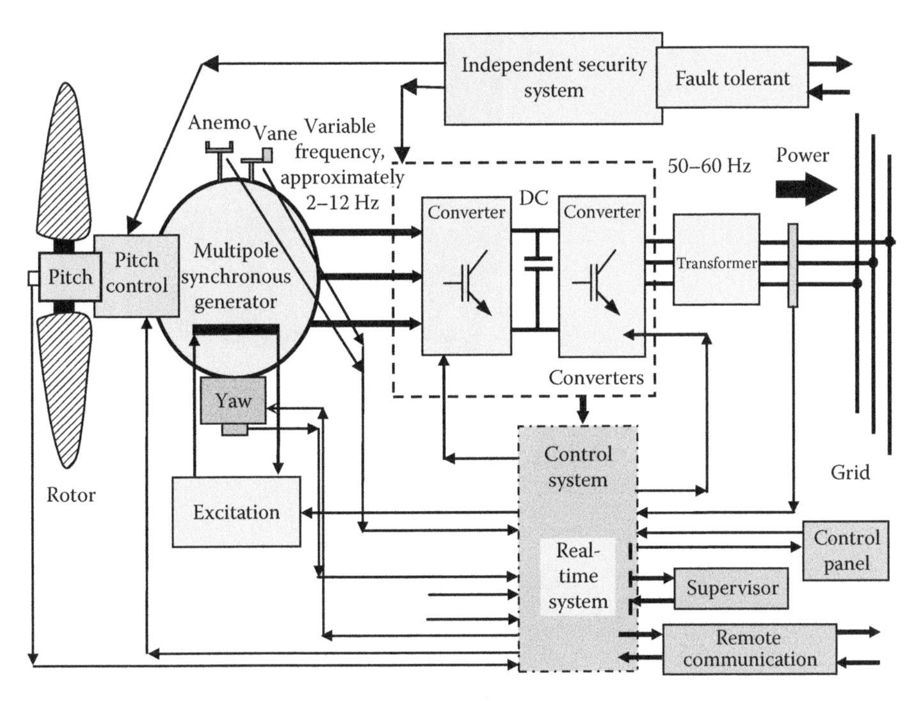

FIGURE 15.6
Electrical and control diagram of the TWT-1.65.

FIGURE 15.7
Rotor, TWT multipole synchronous generator. (Courtesy M.Torres Group, Navarra, Spain.)

FIGURE 15.8
Full-power test bed for the TWTs. (Courtesy of M.Torres Group, Navarra, Spain, http://www.mtorres.es/)

FIGURE 15.9
TWT-1.65: nacelle with the multipole synchronous generator and hub with the pitch control systems. (Courtesy of M.Torres Group, Navarra, Spain, http://www.mtorres.es/)

The main advantages of the variable-speed direct-drive multipole WT are as follows:

- *Power quality optimization*: The machine is able to control the power factor, following the grid operator demands. The system also reduces the harmonics and flicker level, fulfilling the electrical standards.
- *Reliable behavior under voltage dips*: The machine is able to keep the system connected under voltage dips, fulfilling the most demanding international standards. This fact improves grid stability and decreases the risk of blackouts, which allows for a greater penetration of wind energy into the grid.
- *High efficiency*: It produces more energy at every wind sector because the system is able to follow the maximum aerodynamic efficiency C_p at every wind speed (see Figure 11.5, Regions 1 and 2).
- *Longer machine life span*: The system minimizes the mechanical fatigue by increasing the damping of the electrical torque with the controller.
- *Continuous improvements*: It is easy to introduce new improvements and parameterize the system for every wind farm by just modifying the software and retuning the controllers.

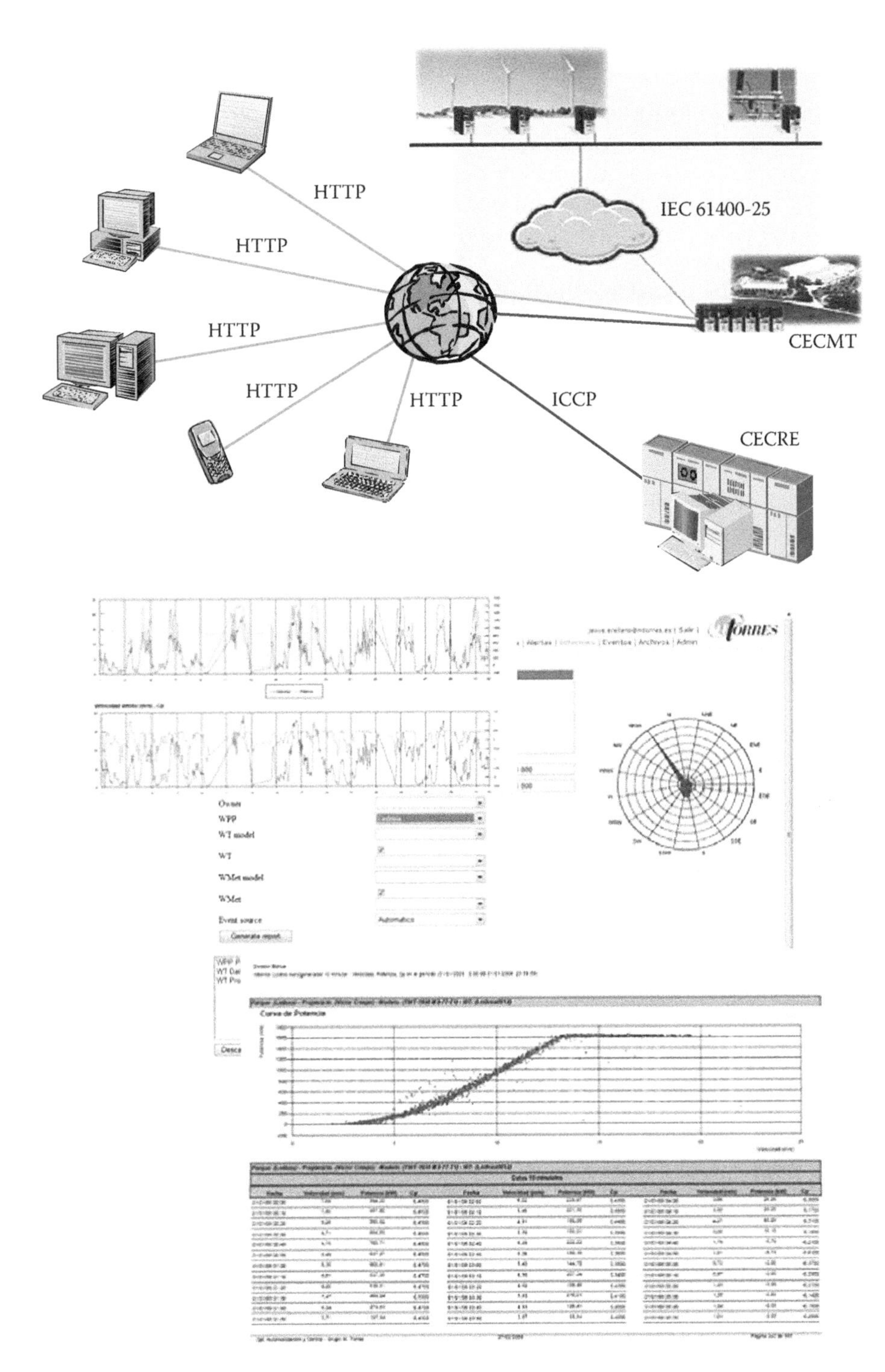

FIGURE 15.10
SCADA system of the TWTs' machines and wind farms. (Courtesy of M.Torres Group, Navarra, Spain, http://www.mtorres.es/)

15.3 Torres Wind Turbine Pitch and Rotor Speed Control Results

This section shows some significant experimental results obtained with the TWT pitch control and rotor speed control systems under typical and extreme wind conditions. Among the main control objectives of the TWT are (a) the optimization of the power efficiency at every wind speed, (b) the attenuation of the transient mechanical loads and fatigue stresses, (c) the reduction of the electrical harmonics and flicker, and (d) the robustness against parameter variation.

Some critical problems arise in the design of any WT control system. These problems are as follows: (a) the difficulty to work safely with random and extreme gusts, (b) the complexity introduced by the nonlinear, multivariable, and time-variable mathematical model (see Chapter 12), and (c) the impossibility to obtain a direct measurement of the wind speed experienced by the turbine, because of the high uncertainty in the anemometer measurement, the strong influence of the blade movement on it, the proximity of the nacelle, etc.[364]

These motivations require the control system designer to combine advanced control strategies such as QFT-robust control techniques, adaptive nonlinear-switching schemes, multivariable methodology, and predictive elements.

The controller design is made using advanced QFT-robust and nonlinear-switching control techniques, based on both mathematical modeling and analysis of experimental data. The QFT templates of the system are first calculated from the physical model of the WT, taking into account model structure uncertainty and parameter uncertainty. Once the first machines were installed, a large amount of operational data were collected (about 15000 data per second per WT) to model the machine dynamics under very different situations. This large set of models (different model structures with uncertainty) was used to validate and increase the size of the QFT templates, which was the basis for continuously improving the controllers' performance.[366]

Among the variables that need to be controlled, (see Figure 15.6), are the electrical power, electrical torque, rotational rotor speed, pitch angle and pitch rotational speed (blades), yaw angle, power factor, generator current, DC voltage, current excitation, temperatures, etc. This section contains some experimental results of the rotor speed control at rated (nominal) speed.

The control system of a pitch-regulated variable-speed WT aims to maintain the rotor speed within a permitted range in above-rated wind speed, Figure 11.5, *Region 3*, with $v_1 \geq v_r$. However, this requirement can be difficult to meet with conventional controllers

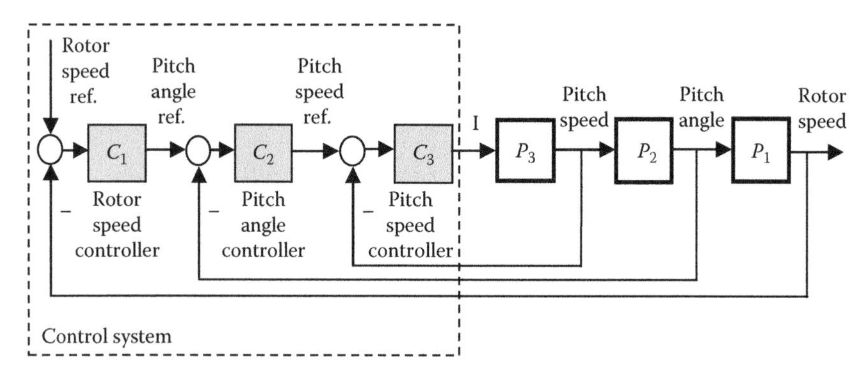

FIGURE 15.11
TWT rotor speed control system block diagram.

during extreme gusts, particularly for large-scale machines where the pitch actuation capability may be comparatively quite limited.

This section investigates the experimental results of the QFT-robust and nonlinear-switching controllers with the variable-speed direct-drive TWT-1.65/72. Figure 15.11 shows the basic block diagram of the three cascade controllers used to control the rotor speed: rotor speed controller $C_1(s)$, pitch angle controller $C_2(s)$, and pitch speed controller $C_3(s)$. Two experimental representative cases, at medium wind speed (Case 1) and at very high wind speed (Case 2), are presented:

Case 1: Medium wind speed. Cabanillas Wind Farm (Spain). February 04, 2003

Figure 15.12 shows large wind gusts over a medium average wind speed context (15 m/s), with excursions going from 13 to 19 m/s. The control system shows a good performance, following the rotor speed set point ($\Omega_{r\_ref} = 20$ rpm) correctly and rejecting the wind gust disturbances with a smooth movement of the blade pitch angles.

Case 2: Very high wind speed. Cabanillas Wind Farm (Spain). April 06, 2003

Figure 15.13 shows large wind gusts over a high average wind speed context (24 m/s), with excursions up to 30 m/s. The control system also shows a good performance, following

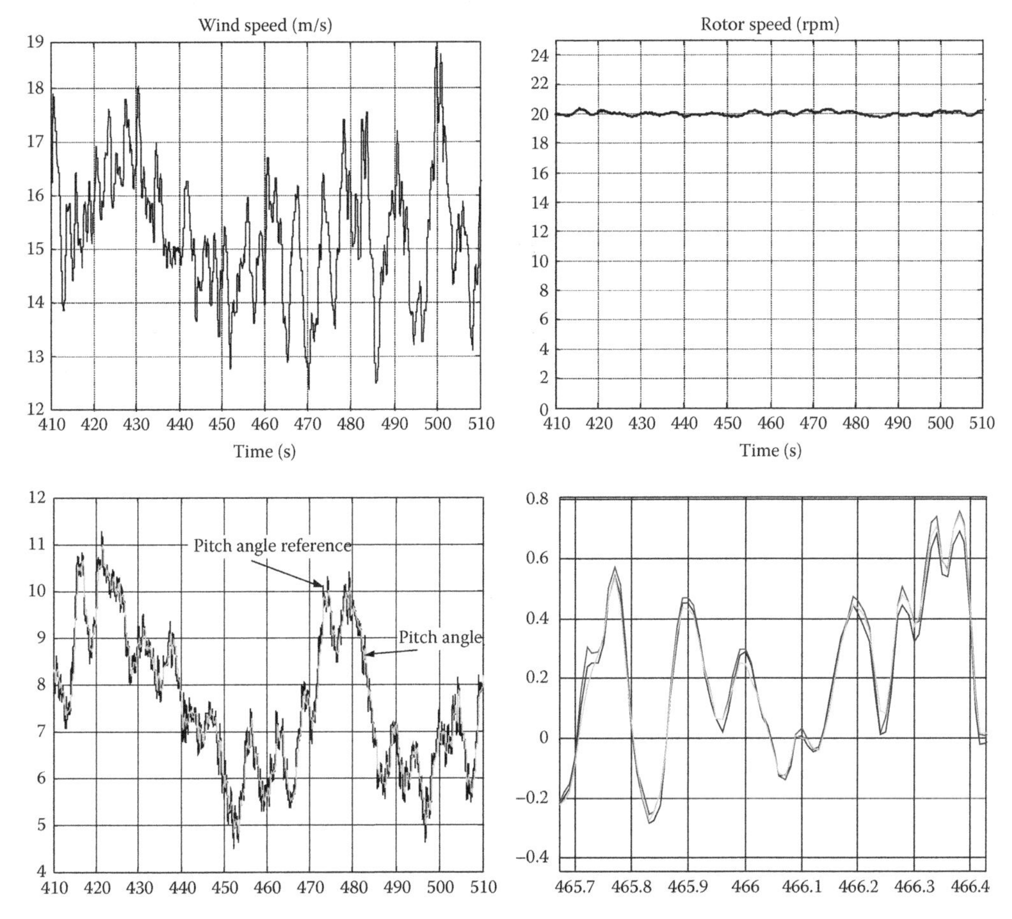

FIGURE 15.12
Experimental results. Control with medium wind speed: Case 1, TWT.

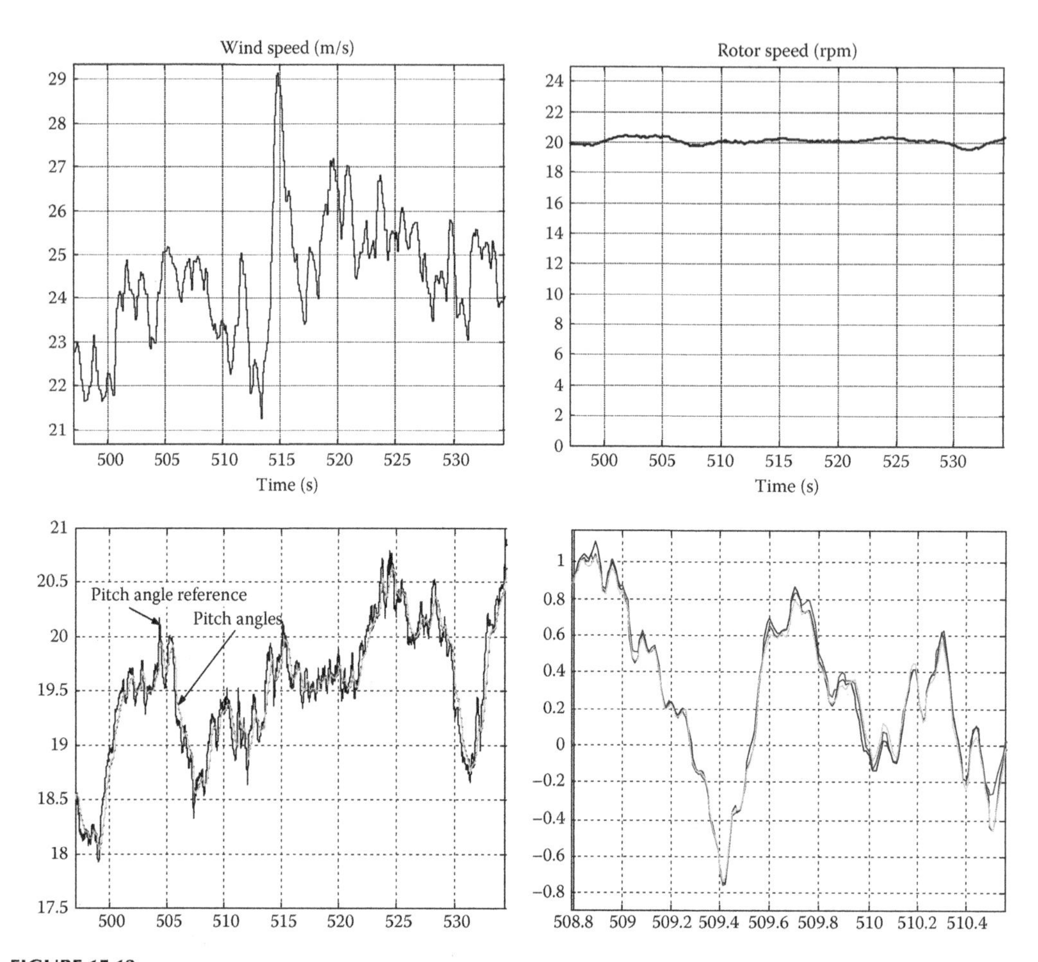

FIGURE 15.13
Experimental results. Control under high wind speed: Case 2, TWT.

correctly the rotor speed set point (Ω_{r_ref} = 20 rpm) and rejecting the wind gust disturbances with a smooth movement of the blade pitch angles.

Both figures (Figures 15.12 and 15.13) show how the QFT-robust and nonlinear-switching control strategies implemented in the TWT-1.65/70 (Class Ia) are able to deal with very different operating points (from low to very high wind speed conditions), avoiding overspeed situations and controlling the blade pitch angles with a smooth movement (low mechanical fatigue).

15.4 Wind Farm Grid Integration: Torres Wind Turbine Results

This section presents some experimental results of the TWT machines when dealing with different grid integration and advanced regulation problems.

As every WT, the main objective of the TWTs is to produce the maximum amount of energy at each wind speed (as explained in Sections 11.4.5.1 and 11.4.5.2). However, in addition to this primary objective, the direct-drive technology, with full-power converters

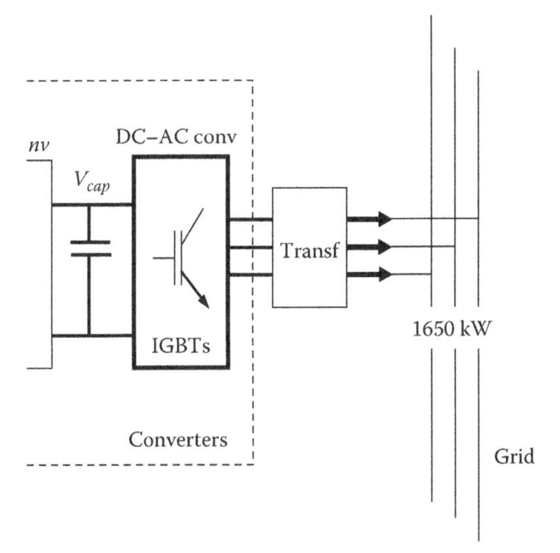

FIGURE 15.14
TWT: DC link and grid-side converter to control the grid.

connected to the grid, gives the possibility to help the grid operator to control the electrical grid. The DC link and the grid-side converter of the WT (see Figure 15.14), along with the appropriate control systems for the IGBT circuits, give the TWT machines some special and very unique functions to control actively (a) the reactive power introduced into the grid—even without wind, (b) the voltage at the grid node connection—also even without wind, (c) the active power introduced into the grid, (d) the grid frequency—provided that the number of WTs is high enough to give a significant power in comparison to the grid power, and (e) give solutions to survive under voltage dip grid problems.

These advanced functions are connected, at WT level or at wind farm level (see Figure 15.15) to the grid operator control system, to help the real-time control of the grid. In this case, the grid operator can request to the wind farm in real time some specific active power set point (P_{ref_farm}), or reactive power set point (Q_{ref_farm}), or voltage set point (V_{grid_ref}),

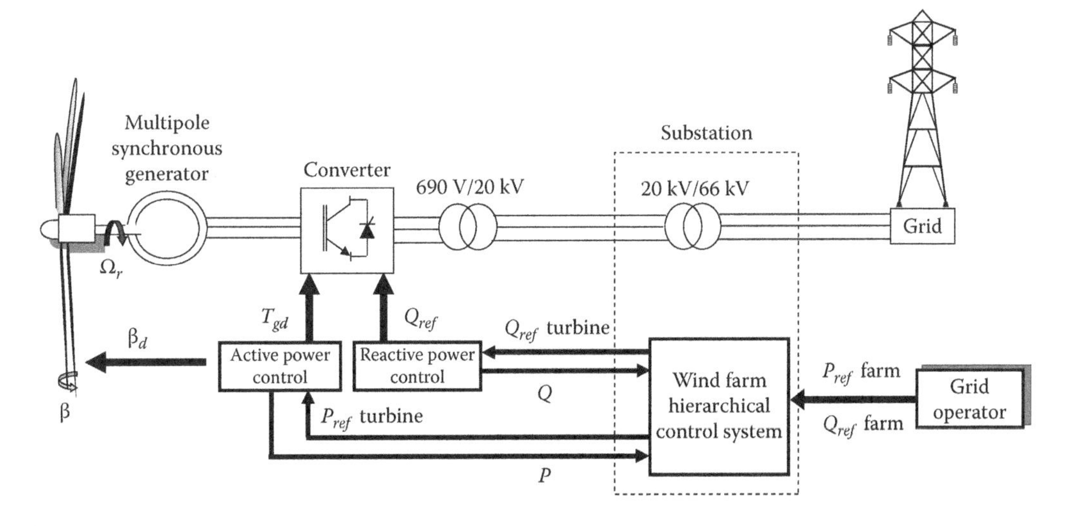

FIGURE 15.15
Active and reactive power TWT wind farm control following the grid operator request at the central grid control system.

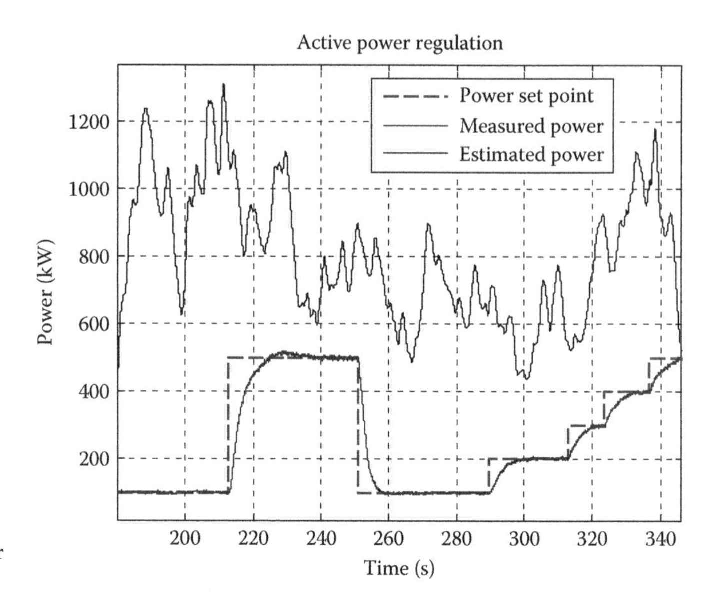

FIGURE 15.16
Experimental results. Active power control with the TWT.

or frequency set point (f_{grid_ref}), or a combination of these. The only limitation is the total current of the wind farm (or total current of the WTs), which is the addition of the active and reactive currents, and the power available at the wind (for active power control and frequency control). These TWT advanced functions are especially important when dealing with isolated and/or weak power systems.

Figure 15.16 shows some experimental results of the active power control system of the TWT-1.65, following an active power step-type set point ($P_{ref\_turbine}$) generated by the *wind farm hierarchical control system* (Figure 15.15), following the grid operator request (P_{ref_farm}). The actual active power (P) achieved by the pitch and torque control systems of the WT follows very well the set points ($P_{ref\_turbine}$). The figure also shows the active power in the current wind, called estimated power, which depends on the wind speed.

Figure 15.17 shows some experimental results of the reactive power control system of the TWT-1.65, following a reactive power step-type set point ($Q_{ref\_turbine}$) generated by the *wind farm hierarchical control system* (Figure 15.15), following the grid operator request (Q_{ref_farm}). The actual reactive power (Q) achieved by the IGBT's control system of the grid-side converter follows very well the set points ($Q_{ref\_turbine}$). This can be also done with the cos(ϕ) control—or power factor control. The reactive power control can be achieved even without wind, because reactive power does not require actual power.

Figure 15.19 shows some experimental results of the voltage control system of the TWT-1.65, following a voltage step-type set point (V_{grid_ref}) requested by the grid operator (Figure 15.18). This reference enters into a cascade structure, as the reference for the *wind farm hierarchical control system*, which calculates the set points to be demanded at each WT of the wind farm (Q_{ref} (i), $i = 1,2,\ldots,n$), at the reactive power control loop shown in Figure 15.15, which will be controlled by the IGBT's control systems of the grid-side converter of the WT. This is achieved even without wind, because reactive power does not require actual power. The first plot of Figure 15.19 shows the reactive power control loop (Q following Q_{ref}) and the second plot shows the voltage control loop (V_{grid} following V_{grid_ref}) needed to achieve the voltage control.

Figure 15.20 shows another possible *grid operator control system structure* to regulate the active P and reactive Q power of the TWT wind farms, while also integrating the voltage V_{grid} control loop. As is seen, the grid operator (1) sends to the *wind farm control system* the

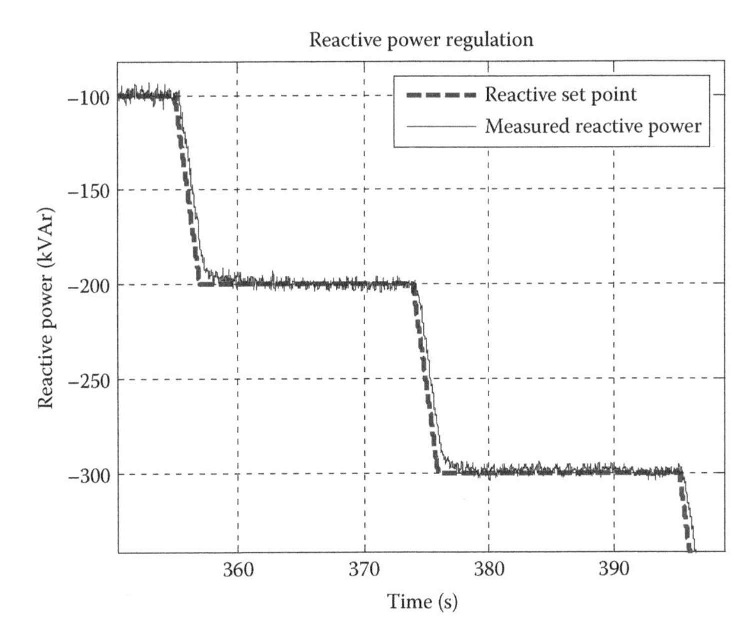

FIGURE 15.17
Experimental results. Reactive power control with the TWT.

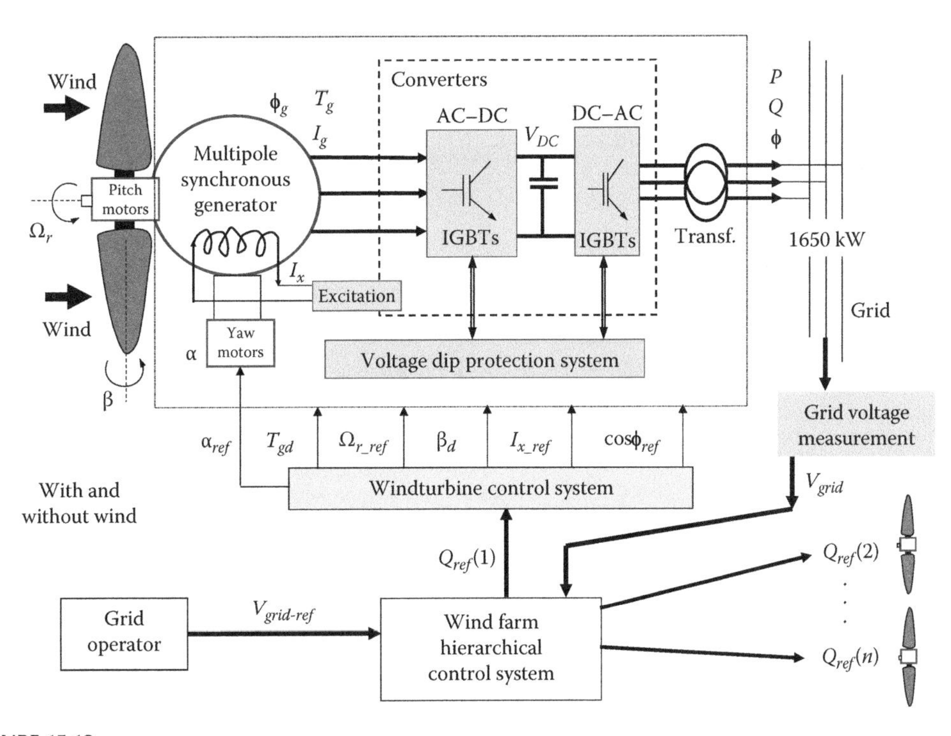

FIGURE 15.18
Voltage grid control with a TWT wind farm following the grid operator request at the central grid control system.

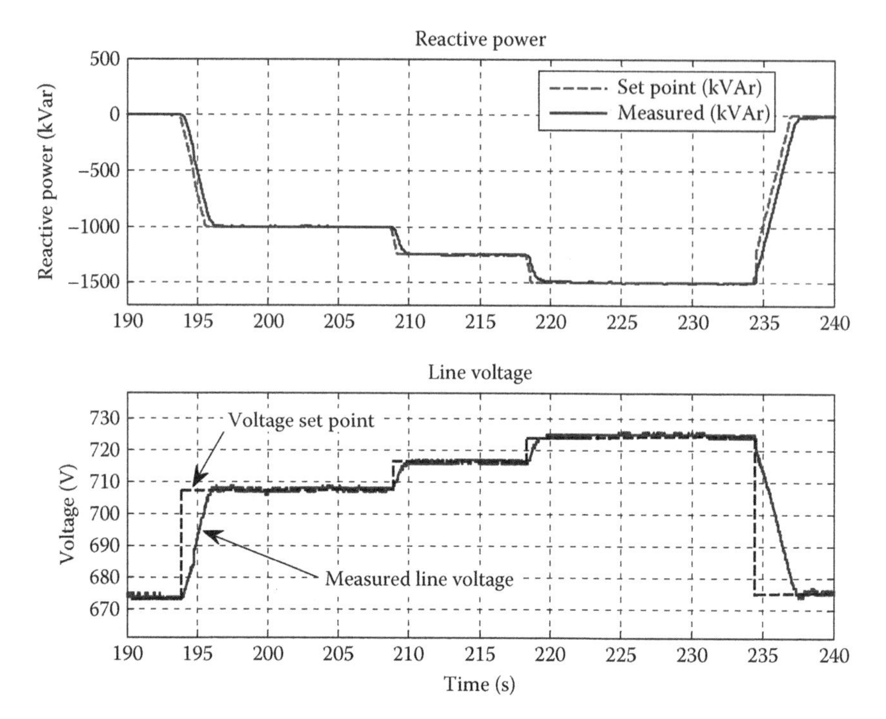

FIGURE 15.19
Experimental results TWT: reactive power and voltage control.

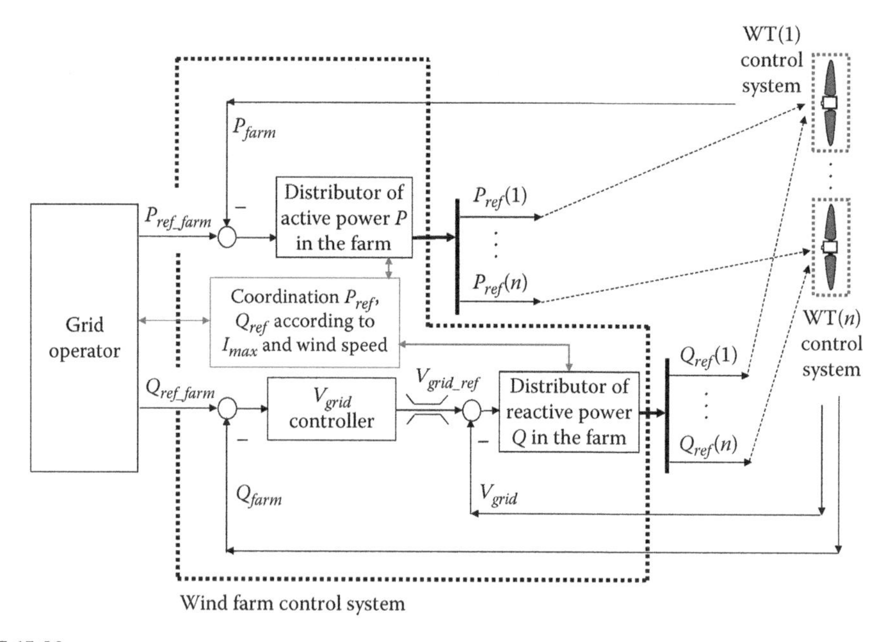

FIGURE 15.20
Grid operator control system of the TWT wind farm: active power, reactive power, and voltage grid control loops.

set points for active P_{ref_farm} and reactive Q_{ref_farm} power and (2) receives information about the actual active P and reactive Q power achieved, both total powers (of the wind farm) and specific powers (of each WT), as well as the possibilities of the wind farm according to the existing wind speed and the maximum total current available in the converters.

With this information the grid operator decides how much active power and how much reactive power need to be requested of the wind farm typically to achieve grid stability specifications and to meet transmission line evacuation limitations.

The *coordination block* of the *wind farm control system* helps with the distribution between active and reactive power of the wind farm, and the *active power distributor block* and the *reactive power distributor block* split the requests among the WTs: ($P_{ref}(i), i = 1,2,\ldots,n$; $Q_{ref}(i), i = 1,2,\ldots,n$).

Finally, the TWTs, due to their very flexible possibilities, can reconfigure their *WT control systems* to follow any active and reactive power set point requested (see Figures 15.16 and 15.17), always within the wind speed and maximum total current possibilities.

15.5 Voltage Dip Solutions: Torres Wind Turbine Results

Faced with the increase in installed wind power (higher levels of wind energy penetration) and the compulsory need to preserve the stability of the electricity grid, the grid operators have warned about the enormous importance of the behavior of the wind generators when voltage dips occur in the grids.

Classical WTs are not able to control the energy when a voltage dip occurs, and thus they immediately stop working, needing about 4 min to restart again. If the percentage of wind power generation in the grid is significant, this massive disconnection of generators creates a large transient in the grid frequency (and grid voltage) that eventually could result in a blackout.

This fact has given rise to the harshening of the connection requirements and standards for wind generators in the main European countries that utilize wind energy; see, for instance, the following standards: P.O. 12.3 from REE (Spain), standard E.ON Netz 2006-04-01 (Germany), standard IEC 61.400-21:2006 88/261/CD (International), and the *Canary Island Official Journal* No. 178, p. 16,334 dated September 14, 2004 (Spain). These requirements demand that a wind generator should not be disconnected under a brief voltage drop (voltage dip) of different depths (see Figure 15.21), guaranteeing the injection of reactive power and the nonconsumption of active power during the event.

The M.Torres multipole synchronous wind generator, TWT-1.65, has a voltage dip protection system (brake chopper + resistance bench), integrated into the full-power converter, which enables it to satisfy all the regulations mentioned earlier (see Figure 15.18). This system has been contrasted in the field with real experimental dips at full power (see Figure 15.22), and the tests have been certified by the Spanish official body LCOE in September 2006, fulfilling the following standards: Standard P.O. 12.3, Red Eléctrica Española, REE (Spain); Standard E.ON Netz 2006-04-01 (Germany); Standard IEC 61.400-21:2006 88/261/CD (International); *Canary Island Official Journal*, No. 178, p. 16,334 dated September 14, 2004 (Spain).

Figures 15.23 through 15.25 show some experimental data of the tests, up to 100% fault depth (V_{min} = 0 V) for t_1 = 0.5 s, with correct clearance and with intake of reactive current and nonconsumption of active current. In addition, Figure 15.26 shows how fast the TWT system is reacting to the fault, injecting reactive power (reactive current, I_q) into the grid in only 20 ms after the fault starts, to help the voltage grid recover as soon as possible.

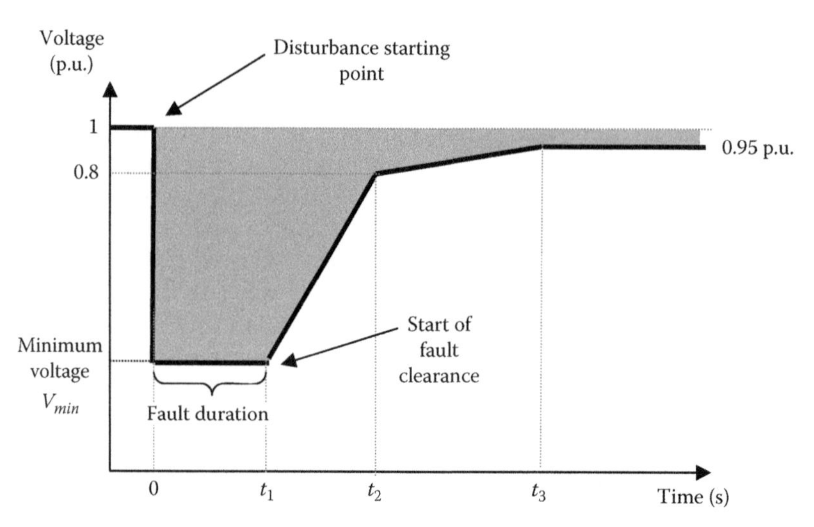

FIGURE 15.21
Voltage dip definition.

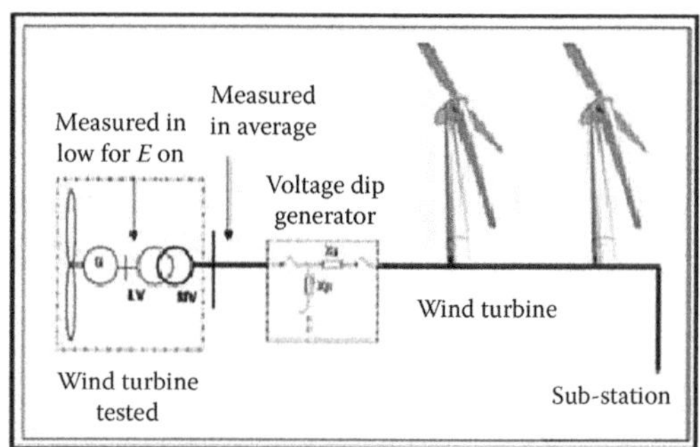

FIGURE 15.22
Experimental assays: voltage dip generator and TWT.

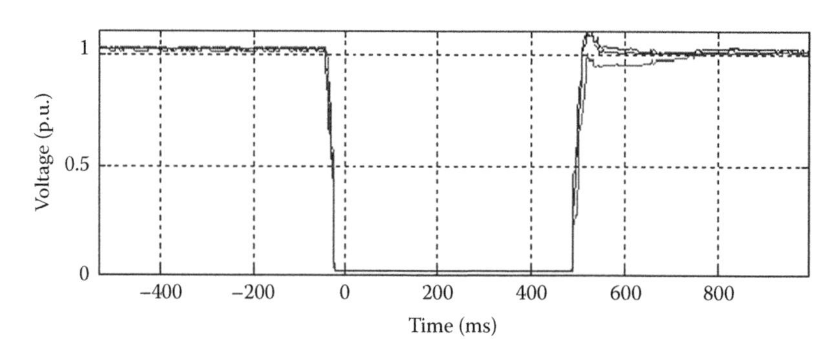

FIGURE 15.23
Experimental results: voltage dip measure. TWT-1.65.

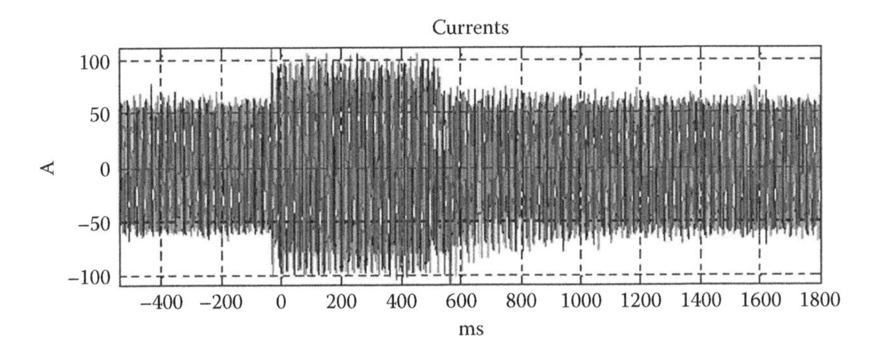

FIGURE 15.24
Experimental results. Currents under voltage dip. TWT-1.65.

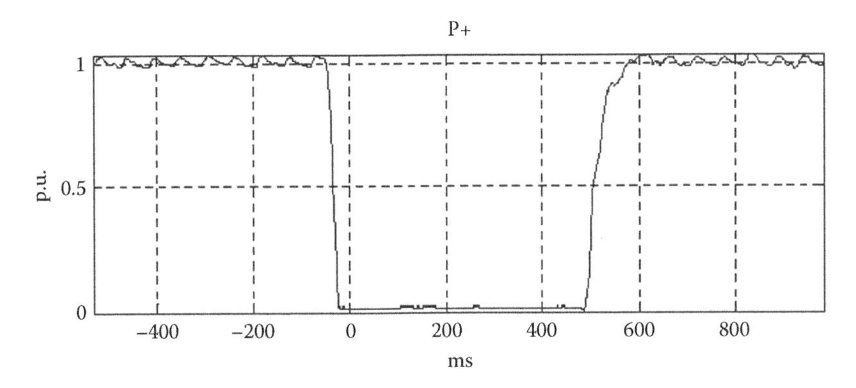

FIGURE 15.25
Experimental results. Active power under voltage dip. TWT-1.65.

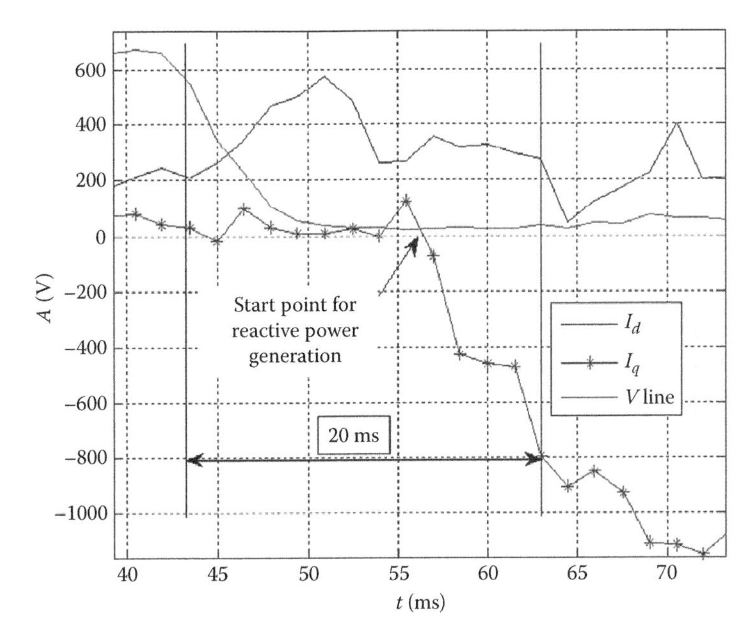

FIGURE 15.26
Experimental results. Active and reactive currents under voltage dip. TWT-1.65.

15.6 Summary

This chapter introduced experimental control system results of the multimegawatt variable-speed direct-drive (gearless) multipole synchronous WTs, TWTs, designed by M.Torres[349] with the collaboration of Professor Mario Garcia-Sanz[355] and a multidisciplinary group of engineers (see Figure 15.27).

The design of the controllers is made by using advanced QFT-robust and nonlinear-switching control strategies based on both mathematical modeling and analysis of the experimental data.

The main advantages of the direct-drive multipole system were discussed, and some of the most representative experimental results with the variable-speed direct-drive TWTs with the QFT and switching controllers, under medium and extreme wind conditions, and connected to different grid situations were presented.

Section 15.2 described the TWT machines (TWT-1.65/70-Class Ia, TWT-1.65/77-Class IIa, TWT-1.65/82-Class IIIa, TWT-2.5/90-Class Ia, TWT-2.5/100-Class IIa, and TWT-2.5/103-Class

FIGURE 15.27
The direct-drive (gearless) variable-speed multipole synchronous generator, with full-power converter, TWT-1.65. (Courtesy M.Torres Group, Navarra, Spain, http://www.mtorres.es/)

IIIa), and Section 15.3 showed significant experimental results with the TWT pitch control and rotor speed control systems under typical and extreme wind conditions. Section 15.4 presented experimental results of the behavior of the TWT machines when dealing with different grid integration problems, with advanced reactive power control systems, voltage control systems, and active power control systems. Section 15.5 introduced new solutions of the TWT to survive under voltage dip problems in the grid.

16

Blades Manufacturing: MIMO QFT Control for Industrial Furnaces[178,375,386]

16.1 Introduction

In the last two decades, the use of composite materials has grown considerably. The rapid upscaling of the fiber-reinforced composites in the industry, as well as their lightweight, design flexibility, strength, stiffness, low fatigue, and corrosion resistance, has made companies increase their interest in this field. Composite applications range from space exploration to alternative energies. This chapter addresses the temperature control of a multivariable industrial furnace destined to manufacture (polymerize) large composite pieces in general and wind turbine blades in particular (see Figure 16.1).

An accurate and homogeneous control of the temperatures involved in the blade polymerization process is essential to guarantee the cured product quality and to meet the industrial requirements. The first non-diagonal multiple-input-multiple-output quantitative feedback theory (MIMO QFT) control design methodology (*Method 3*), presented in Sections 6.2 through 6.8, is applied in this chapter to regulate the furnace temperature despite the disturbances (exothermic reactions) and loop interactions.

16.2 Composite Materials

Due to the characteristics of the polymers, composites are also making their way through the alternative energy market, which has adopted such lightweight materials in its blade designs. On the whole, obtaining the required mechanical properties for any kind of large composite materials is a critical and complex process that requires the application of temperatures, and sometimes pressures, for a finite time period.[321,326] Besides, the uniformity of temperature through the whole piece plays a key role in the overall quality of the cured product. But this homogeneity does not come easily. The coupling effects due to heat transmission between adjoining zones in the furnace must be taken into account. In addition, an important risk involved with using resins is the exothermic reaction that results during the polymerization process of the composite. Due to the thickness of these kinds of pieces and the low thermal conductivity of the resin, the heat generated by the exothermic reaction is not adequately removed. As a consequence, this exothermic cure can cause matrix microcracking and severe damage to the whole part if it is not properly controlled.

FIGURE 16.1
Wind turbine blade manufactured by M.Torres with the industrial furnace under study in this chapter. (Courtesy of M.Torres Group, Navarra, Spain, http://www.mtorres.es/)

All of this means that an accurate control of the temperature along the total length of the piece is essential to optimize the curing process of large composites. Within this framework, this chapter (a) models the dynamics of an industrial furnace utilized in composite-blade curing applications,[375] (b) develops a non-diagonal MIMO robust compensator for the temperature control, and (c) validates the controller experimentally in an existing 1 MW industrial furnace.[178,386] The furnace is an original Siflexa type (see Figure 16.2) and is located at the M.Torres Group Headquarters (Spain). A sequential non-diagonal MIMO QFT design methodology (*Method 3*), according to Sections 6.2 through 6.8, is proposed to improve reliability and control performance in terms of disturbance rejection (exothermic reactions) and tracking reference, despite the interactions among loops and the large parametric uncertainty of the process.

The chapter is structured as follows. Section 16.3 describes the industrial furnace, Section 16.4 develops the mathematical model of the furnace, and Section 16.5 shows the results of the parameter estimation for the model. In Section 16.6, the sequential non-diagonal MIMO QFT methodology (*Method 3*), introduced in Chapter 6, is applied to design the multivariable compensator for the furnace. The results of the experimental validation are presented in Section 16.7.

FIGURE 16.2
Industrial furnace for wind turbine blades. (Courtesy of M.Torres Group, Navarra, Spain, http://www.mtorres.es/)

16.3 Industrial Furnace Description

The system of interest is a large electrical furnace (4.03 m × 1.925 m × 37.6 m, extendible to a larger configuration if needed), of 1 MW power, employed to cure large fiber-reinforced epoxy-matrix composite pieces, destined for supporting aerodynamic structures and/or designed as wind turbine blades (see Figures 16.1 and 16.2). The furnace contains seven heating zones, each one having an independent power source composed of (a) electric resistive heaters, (b) a fan, and (c) a gate, all used to control the inner temperature.

The furnace is a long stainless steel room, which has seven *virtual* areas that are approximately 5 m long, except the first and the last one whose lengths are 4.8 and 7.8 m, respectively. These zones are called *virtual* because neither walls nor panels separate them.

Due to the piece length, the heating chamber must be big enough in order to process a whole load at a time. Furnace cross section and dimensions are shown in Figure 16.3. All the furnace areas are equipped with groups of heating. Every group is composed of a set of electrical resistive heaters, placed above the load, at the top surface of the furnace. An induced electric current flows through these resistors where it is dissipated as heat and transferred to the workpiece homogeneously by convection. A fan located over these heaters is used to produce a pressure jump, which creates the forced convection by raising the air velocity and therefore the heat transfer. In addition, there is a small gate located on the side that allows the entry of external air during the cooling process.

Several standard thermocouples attached to the walls, one for each zone, measure the temperature inside the furnace, and 20 more sensors evenly placed over the load give its surface temperature. A summary of the most important features of the furnace, dimensions and data, is shown in Table 16.1. The furnace is embedded in a control loop as shown in Figure 16.4. The power delivered by each heating element, q_i, $i = 1,\ldots,7$, is operated by a controller that regulates the heating of a zone of the furnace. The output is the temperature T_i, $i = 1,\ldots,7$, supplied by the sensing modules, consisting of seven thermocouples positioned strategically along the length of the furnace and connected to the data acquisition system, which records the measurements.

In many industrial furnace applications, the thermal load distribution of the pieces to be polymerized is homogeneous along the furnace volume. This fact very much simplifies the control system, allowing the temperatures to be controlled by independent and equal single-input-single-output proportional-integral-derivative (SISO PID) controllers, one for each zone.

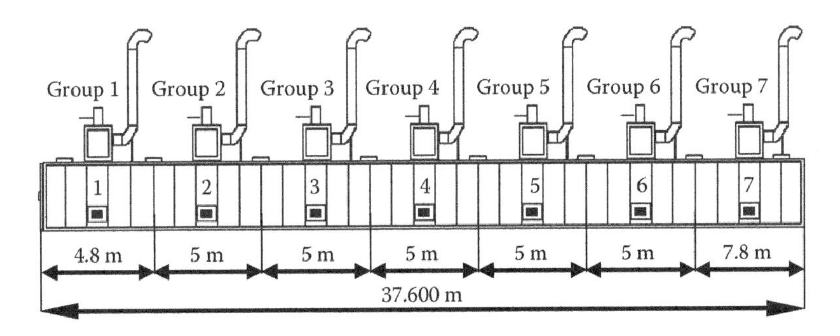

FIGURE 16.3
Furnace cross section (Siflexa industrial furnace).

TABLE 16.1

Furnace Characteristics

Heating chamber size ($W \times H \times L$)		4.03 m × 1.925 m × 37.6 m
Total power		900 kW
Voltage		380 V
Zones		7
Gates		7
Heating elements	Number	7
	Name	Resistor
	Power	85 kW
Fans	Number	7
	rpm	1500
Thermocouples		7 (on the walls)
		20 (on the load)

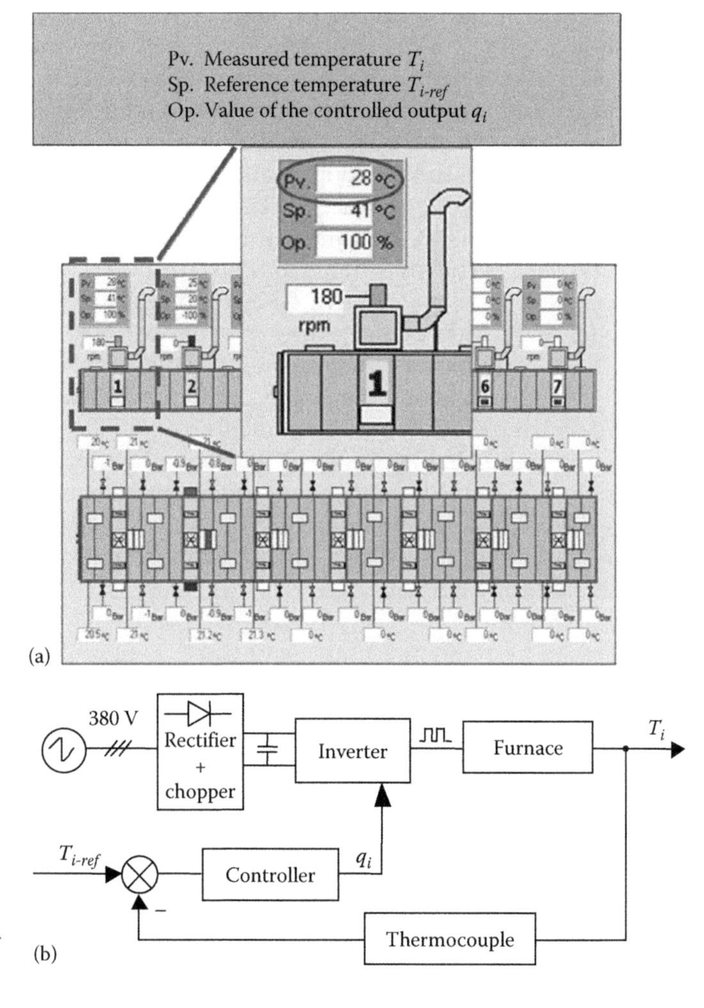

FIGURE 16.4
Furnace control system: (a) SCADA and (b) block diagram.

However, if the pieces to be polymerized have a heterogeneous thermal load distribution, as in the case of wind turbine blades (the mass and volume of the blade change significantly from the root to the tip), then under-loaded zones will transfer heat to those that have more load. This problem leads to couplings among the control loops, requiring advanced MIMO controllers to optimize the curing process while considering the load weight and its distribution inside the furnace.

16.4 Furnace Model[178,386]

This section develops the mathematical model of the industrial furnace. The multivariable furnace is considered as an electrical pi-model,[195] represented by the scheme shown in Figure 16.5.

Kirchoff's current and voltage laws are used to determine heat fluxes and temperatures, respectively. By applying the inspection method for nodal analysis of electrotechnical circuits,[335] the following system is obtained:

$$Q(s) = P^*(s)T(s) \tag{16.1}$$

where

$T(s)$ is a 7×1 vector that represents the air temperatures read by the thermocouples in each zone (analog to electrical voltages)

$Q(s)$ is a 7×1 vector that symbolizes the heat flux supplied by the resistors (analog to electrical currents)

$P^*(s)$ is a 7×7 tri-diagonal matrix that represents the admittances p^*_{ij} (reciprocal of the impedances z_{ij}), as seen in Equation 16.3, where the diagonal p^*_{ii} are the sum of the admittances ($1/z_{ii}$, $1/z_{ij}$) of each of the branches attached to node i and the off-diagonal p^*_{ij} are the sum of the admittances ($1/z_{ij}$) of all the branches joining nodes i and j, and multiplied by minus one[335]:

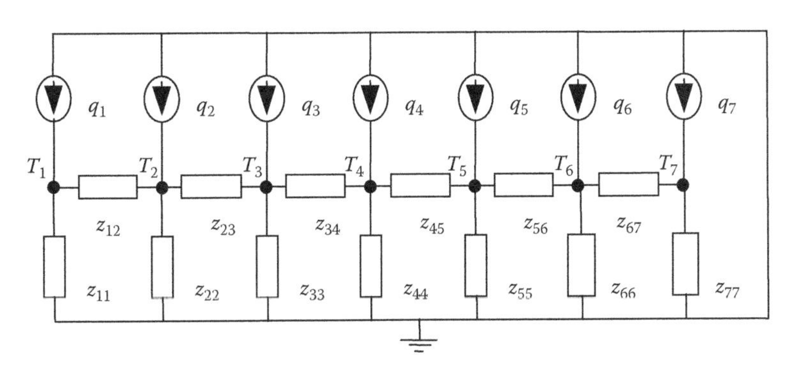

FIGURE 16.5
Electrical pi-model for the furnace.

$$T(s) = \begin{bmatrix} T_1(s) \\ T_2(s) \\ T_3(s) \\ T_4(s) \\ T_5(s) \\ T_6(s) \\ T_7(s) \end{bmatrix}; \quad Q(s) = \begin{bmatrix} q_1(s) \\ q_2(s) \\ q_3(s) \\ q_4(s) \\ q_5(s) \\ q_6(s) \\ q_7(s) \end{bmatrix} \tag{16.2}$$

$$P^*(s) = \begin{bmatrix} \frac{1}{z_{11}(s)} + \frac{1}{z_{12}(s)} & \frac{-1}{z_{12}(s)} & \cdots & 0 \\ \frac{-1}{z_{12}(s)} & \frac{1}{z_{12}(s)} + \frac{1}{z_{22}(s)} + \frac{1}{z_{23}(s)} & \cdots & 0 \\ \cdots & \cdots & \cdots & \cdots \\ 0 & 0 & \cdots & \frac{1}{z_{67}(s)} + \frac{1}{z_{77}(s)} \end{bmatrix} \tag{16.3}$$

The analytical model of each zone is based on a heat transfer study.[322] This allows us to obtain simple expressions, which are compared with the experimental results in Section 16.5. By a thorough study of the process, it can be noticed that forced convection is the dominant heat exchange mechanism inside the furnace. Although radiation contributes to the energy exchange, below 700°C it can be disregarded.[322,330] Conduction heat transfer is considered through the walls and load. Quite a few basic heat transfer equations are used to obtain the model that describes the furnace behavior. First of all, several hypotheses are set in order to reduce complexity and shorten the operations:

- It is assumed that the material properties are uniform in the whole part.
- As the records of the temperature measurements are long (9 h), the slight effects of the external environment are taken into consideration. In this model, this effect is assumed a constant external temperature because its variation is very small.

Apart from these assumptions, the uniformity of surface load temperature facilitates the calculations. Experimental temperatures measured at different points of the load surface are plotted in Figure 16.6. As it is seen, they are all very similar, not only in the initial time but also during all the cure cycles. This uniformity is really helpful to the analysis because it makes the analysis quicker and more straightforward.

Once all the energy balances that occur inside the furnace have been established,[195,322,330,375] an electrical circuit is used to model the heat transmission differential equations that describe every zone of the furnace (see Figure 16.7). The thermal characteristics of the

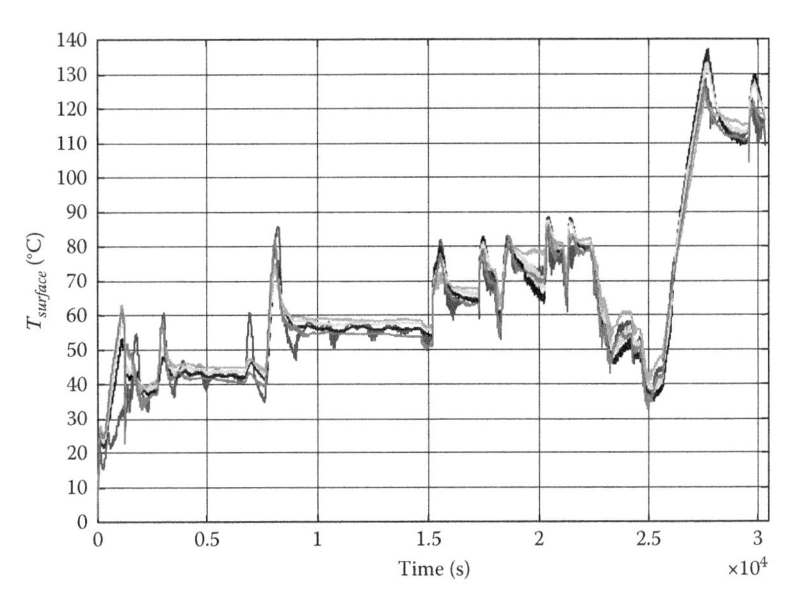

FIGURE 16.6
Surface temperatures.

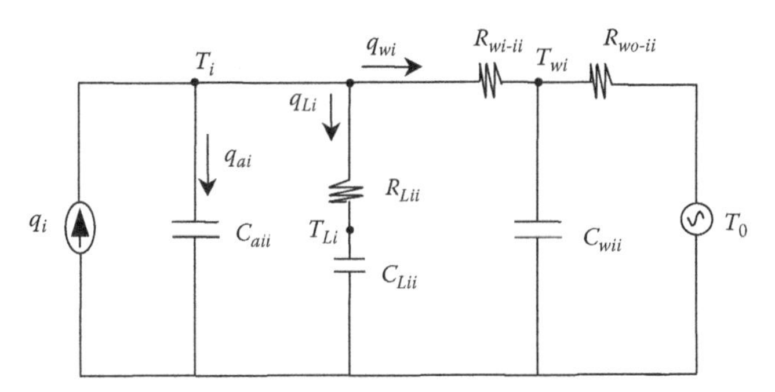

FIGURE 16.7
Thermal–electric analogy model for each zone *i*.

furnace vary continuously in time and space. As this furnace exhibits high thermal capacity, a lumped approach is used to build a thermal capacitance model.[368]

The thermal–electrical equivalencies shown in Figure 16.7 are as follows: voltages represent temperatures, currents are heat fluxes, electrical resistances correspond to heat transmission resistances, and finally electrical capacities symbolize thermal capacities.[195,322,375] The physical meaning of every element is shown in Table 16.2.

The heat balance equations that govern the process are formulated from the electric analogy point of view. Global heat balance or heat energy conservation is given by Equation 16.4. The input heat flux (q_i), provided in each zone, is equal to that absorbed by the workpiece (Equation 16.6) and by the air (Equation 16.5) and to the portion lost through the walls of the chamber (Equation 16.7):

$$q_i = q_{ai} + q_{Li} + q_{wi} \tag{16.4}$$

$$q_{ai} = C_{aii} \frac{dT_i}{dt} \tag{16.5}$$

TABLE 16.2

Nomenclature

Temperature (K)	Film coefficient (W/m² °C)
T_L = load internal temperature	h = furnace air convection coefficient
T_w = wall internal temperature	Density (kg/m³)
T_i = inside air furnace temperature	ρ_a = 1.225 = air density
T_0 = outside air temperature	ρ_L = 1.6 10³ = Load density
Heat flux per zone (kW)	ρ_w = wall material density
q_i = total heat flux from a resistance group	Specific heat (kJ/kg °C)
q_a = heat flux in the air inside the furnace	c_{pair} = 1.01 = air specific heat
q_c = load heat flux	c_{pc} = 712 10⁻³ = load specific heat
q_p = conduction losses through the walls	c_{pw} = wall specific heat
Resistance (m² °C/W)	Volume, area, and length
$R_L = (1/hA_L) + (\mathrm{Ln}(Re/Ri)/2\pi kL)$ = load thermal resistance	V_a = air volume (m³)
R_{wi} = input wall heat conduction resistance	V_L = load volume (m³)
R_{wo} = output wall heat conduction resistance	L = load length (m)
Capacity (kJ/°C)	A_L = load surface area (m²)
$C_w = V_w\, \rho_w\, c_{pw}$ = wall heat capacity	Re = external load radio (m)
$C_a = V_a\, \rho_a\, c_{pa}$ = internal air thermal capacity	Ri = internal load radio (m)
$C_L = V_L\, \rho_{aL} c_{pL}$ = load thermal capacity	Conductivity (W/m °C)
	k = load thermal conductivity

$$q_{Li} = C_{Lii}\frac{dT_{Li}}{dt} = \frac{(T_i - T_{Li})}{R_{Lii}} \tag{16.6}$$

$$q_{wi} = \frac{(T_i - T_{wi})}{R_{wi-ii}} = \frac{C_{wii}dT_{wi}}{dt} + \frac{(T_{wi} - T_0)}{R_{wo-ii}} \tag{16.7}$$

Combining Equations 16.4 through 16.7, and considering $R_{wi\text{-}ii} \approx 0$ and $R_{wo\text{-}ii} \approx \infty$, the differential equation that describes the furnace is written as

$$q_i = (C_{wii} + C_{aii})R_{Lii}C_{Lii}\frac{d^2T_i}{dt^2} + (C_{wii} + C_{aii} + C_{Lii})\frac{dT_i}{dt} \tag{16.8}$$

which when translated into the Laplace s-domain, with zero initial conditions, results in the following:

$$\frac{T_{ii}(s)}{q_{ii}(s)} = z_{ii}(s) = \frac{R_{Lii}C_{Lii}s + 1}{s\left[(C_{aii} + C_{wii})R_{Lii}C_{Lii}s + (C_{aii} + C_{wii} + C_{Lii})\right]} \tag{16.9}$$

Equation 16.9 represents the thermal impedance within one zone, also called $z_{ii}(s)$ in Figure 16.5 and Equation 16.3. A similar expression can be developed with the aim of describing the adjoining zone thermal impedance $z_{ij}(s)$, so that

$$\frac{T_{ij}(s)}{q_{ij}(s)} = z_{ij}(s) = \frac{R_{Lij}C_{Lij}s + 1}{s\left[(C_{aij} + C_{wij})R_{Lij}C_{Lij}s + (C_{aij} + C_{wij} + C_{Lij})\right]} \tag{16.10}$$

where the parameters in the equation are calculated as follows:

$$R_{ij} = \frac{R_{ii} + R_{ij}}{2}; \quad C_{ij} = \frac{C_{ii} + C_{ij}}{2} \tag{16.11}$$

Equations 16.10 and 16.11 describe the effect that the heat flux of one area has on the temperature measured in the contiguous area. The impedance $z_{ij}(s)$ between two zones is the same whether the heat is transferred from one zone or from the other. Consequently, $R_{ij} = R_{ji}$ and $C_{ij} = C_{ji}$.

All these results are substituted in the matrix $\boldsymbol{P}^*(s)$ (Equation 16.3). Note that the expression of the direct matrix $\boldsymbol{P}(s)$, which is obviously the inverse of $\boldsymbol{P}^*(s)$, is extremely complex, with 49 elements of a very high order (70/70). However, that expression is not needed here. The MIMO QFT methodology uses $\boldsymbol{P}^*(s)$, which has been already calculated in Equations 16.3 and 16.9 through 16.11.

16.5 Estimation of Furnace Parameters[375]

This section presents the results of the estimation of the $\boldsymbol{P}^*(s)$ model parameters, applying the estimation theory[199,217] to a large collection of experiments and real data, obtained from operating the furnace loaded with large wind turbine blades. Due to difficulties in the estimation (multivariable characteristics, loop coupling, low input excitation, and low sensor accuracy), the procedure was divided into two steps. Firstly, (a) the parameters of the thermal impedances of each zone $z_{ii}(s)$ are estimated by running a set of closed-loop (PID control) experiments that maintain equal temperature in all the zones (Figure 16.8a), avoiding heat transfer between zones. The sampling time is fixed at 120 s. The temperature resolution becomes 1°C.

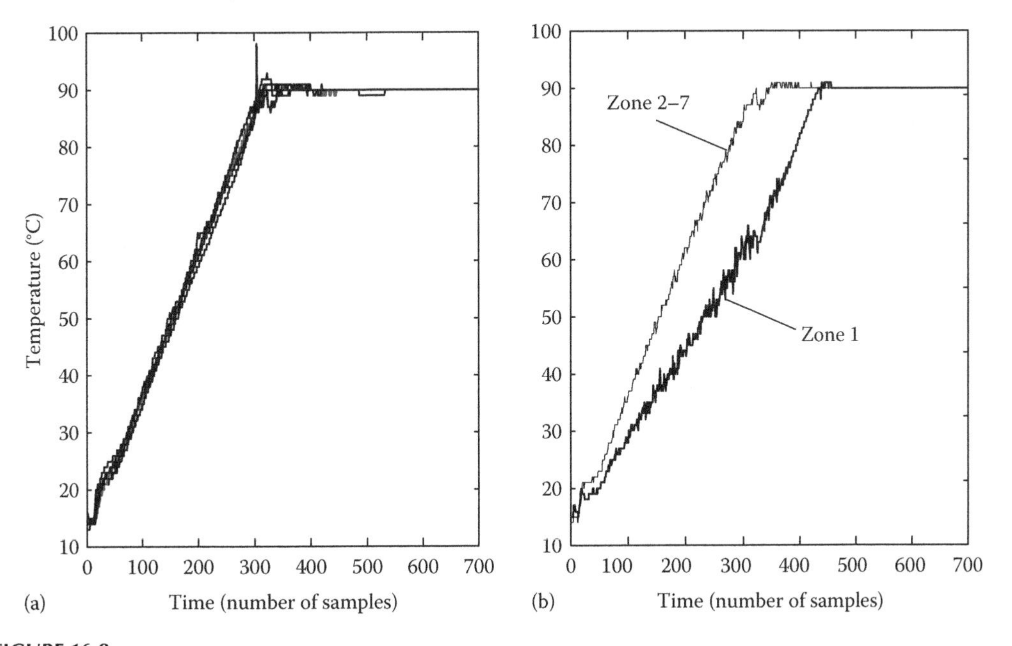

FIGURE 16.8
Real experiments for parameter estimation. (a) Keeping equal temperatures in all zones (b) varying temperature in zone 1.

TABLE 16.3

Parameters for the Zone Model (z_{ii})

	Zone *ii*						
	1	2	3	4	5	6	7
R_{Lii}	41.75	12.95	8.62	6.48	5.05	4.12	2.5
C_{Lii}	99.11	616.3	1380	2456.1	4227	5970.5	12003.7
C_{aii}	45.01	46.34	45.53	44.38	42.5	40.65	34.24
C_{wii}	4432.4	5046	5046	5046	5046	5046	5512.1

Units according to Table 16.2.

TABLE 16.4

Parameters for the Coupling Model (z_{ij})

	Between Zones *ij*					
	12	23	34	45	56	67
R_{Lij}	27.35	10.78	7.55	5.76	4.58	63.31
C_{Lij}	357.7	998.5	1918.4	3341.8	5100	8987
C_{aij}	45.67	45.93	45	43.44	41.57	37.44
C_{wij}	4740	5046	5046	5046	5046	5280

Units according to Table 16.2.

Table 16.3 shows the results for the parameters of $z_{ii}(s)$, $i = 1, 2, \ldots, 7$. Then, (b) the parameters of the adjoining zone thermal impedances $z_{ij}(s)$ are estimated by running another set of closed-loop (PID control) experiments that maintain equal temperature in all the zones but one (Figure 16.8b), creating a heat transfer between zone 1 and zone 2. Table 16.4 shows the results for the parameters of $z_{ij}(s)$, $i = 1, 2, \ldots, 7$.

16.6 MIMO QFT Controller Design[178,386]

This section develops the controllers for the first four zones of the industrial furnace, a 4×4 MIMO system, applying the sequential non-diagonal MIMO QFT methodology introduced in Sections 6.2 through 6.8, as *Method 3*.[93,97]

16.6.1 Model and Parametric Uncertainty

The multivariable model of the industrial furnace developed in Sections 16.4 and 16.5 is shown again by Equations 16.12 through 16.18 and Table 16.5:

$$\begin{bmatrix} q_1(s) \\ q_2(s) \\ q_3(s) \\ q_4(s) \end{bmatrix} = \begin{bmatrix} p_{11}^*(s) & p_{12}^*(s) & 0 & 0 \\ p_{12}^*(s) & p_{22}^*(s) & p_{23}^*(s) & 0 \\ 0 & p_{23}^*(s) & p_{33}^*(s) & p_{34}^*(s) \\ 0 & 0 & p_{34}^*(s) & p_{44}^*(s) \end{bmatrix} \begin{bmatrix} T_1(s) \\ T_2(s) \\ T_3(s) \\ T_4(s) \end{bmatrix} \tag{16.12}$$

TABLE 16.5

Parameters and Uncertainty

Parameter	Min	Max
a_{ij}	100	2000
b_{ij}	4×10^6	9×10^6
c_{ij}	4000	1×10^4

where

$$p_{11}^*(s) = [z_{11}(s)]^{-1} + [z_{12}(s)]^{-1}$$

$$p_{22}^*(s) = [z_{12}(s)]^{-1} + [z_{22}(s)]^{-1} + [z_{23}(s)]^{-1}$$

$$p_{33}^*(s) = [z_{23}(s)]^{-1} + [z_{33}(s)]^{-1} + [z_{34}(s)]^{-1}$$

$$p_{44}^*(s) = [z_{34}(s)]^{-1} + [z_{44}(s)]^{-1}$$

$$p_{12}^*(s) = p_{21}^*(s) = -[z_{12}(s)]^{-1}$$

$$p_{23}^*(s) = p_{32}^*(s) = -[z_{23}(s)]^{-1}$$

$$p_{34}^*(s) = p_{43}^*(s) = -[z_{34}(s)]^{-1}$$

As is seen in Equation 16.3, the diagonal elements p_{ii}^* s are the sum of the admittances ($1/z_{ii}$, $1/z_{ij}$) of each of the branches attached to node i (see Figure 16.5). The off-diagonal elements p_{ij}^* s are the sum of the admittances ($1/z_{ij}$) of all the branches joining nodes i and j, and multiplied by minus one, so that

$$\frac{1}{z_{ij}(s)} = \frac{d_{ij}(s)}{n_{ij}(s)}, \quad i, j = 1, 2, \ldots, 7 \tag{16.13}$$

where

$$d_{ij}(s) = s(b_{ij}s + c_{ij}), \quad i, j = 1, 2, \ldots, 7 \tag{16.14}$$

$$n_{ij}(s) = (a_{ij}s + 1), \quad i, j = 1, 2, \ldots, 7 \tag{16.15}$$

and the terms b_{ij}, c_{ij}, and a_{ij} are the design parameters that depend on the part to be manufactured (blades) and its spatial distribution along the furnace, so that

$$a_{ij} = R_{Lij}C_{Lij} \tag{16.16}$$

$$b_{ij} = (C_{aij} + C_{wij}) R_{Lij} C_{Lij} \tag{16.17}$$

$$c_{ij} = (C_{aij} + C_{wij} + C_{Lij}) \tag{16.18}$$

In other words

$$p_{11}^*(s) = [z_{11}(s)]^{-1} + [z_{12}(s)]^{-1} = \frac{s[(b_{11}s + c_{11})(a_{12}s + 1) + (b_{12}s + c_{12})(a_{11}s + 1)]}{(a_{11}s + 1)(a_{12}s + 1)} \tag{16.19}$$

$$\begin{aligned} p_{22}^*(s) &= [z_{12}(s)]^{-1} + [z_{22}(s)]^{-1} + [z_{23}(s)]^{-1} \\ &= \frac{s[(b_{22}s + c_{22})(a_{12}s + 1)(a_{23}s + 1) + (b_{12}s + c_{12})(a_{22}s + 1)(a_{23}s + 1) + (b_{23}s + c_{23})(a_{22}s + 1)(a_{12}s + 1)]}{(a_{12}s + 1)(a_{22}s + 1)(a_{23}s + 1)} \end{aligned} \tag{16.20}$$

$$\begin{aligned} p_{33}^*(s) &= [z_{23}(s)]^{-1} + [z_{33}(s)]^{-1} + [z_{34}(s)]^{-1} \\ &= \frac{s[(b_{23}s + c_{23})(a_{33}s + 1)(a_{34}s + 1) + (b_{33}s + c_{33})(a_{23}s + 1)(a_{34}s + 1) + (b_{34}s + c_{34})(a_{23}s + 1)(a_{33}s + 1)]}{(a_{23}s + 1)(a_{33}s + 1)(a_{34}s + 1)} \end{aligned} \tag{16.21}$$

$$p_{44}^*(s) = [z_{34}(s)]^{-1} + [z_{44}(s)]^{-1} = \frac{s[(b_{34}s + c_{34})(a_{44}s + 1) + (b_{44}s + c_{44})(a_{34}s + 1)]}{(a_{34}s + 1)(a_{44}s + 1)} \tag{16.22}$$

$$p_{12}^*(s) = p_{21}^*(s) = -[z_{12}(s)]^{-1} = \frac{-s(b_{12}s + c_{12})}{(a_{12}s + 1)} \tag{16.23}$$

$$p_{23}^*(s) = p_{32}^*(s) = -[z_{23}(s)]^{-1} = \frac{-s(b_{23}s + c_{23})}{(a_{23}s + 1)} \tag{16.24}$$

$$p_{34}^*(s) = p_{43}^*(s) = -[z_{34}(s)]^{-1} = \frac{-s(b_{34}s + c_{34})}{(a_{34}s + 1)} \tag{16.25}$$

As seen in Sections 16.4 and 16.5, the quantities C_a, C_L, and C_w are the internal air, load, and wall thermal capacities (kJ/°C), respectively, whereas R_L is the load thermal resistance (°C/kW). Due to the variety of parts to be manufactured (different blades or parts of the blades), there is significant uncertainty in the parameters. Table 16.5 shows the results for some typical furnace load situations for the 4 × 4 configuration (see Equations 16.12 and 16.19 through 16.25).

16.6.2 Control Specifications

The desired closed-loop performance specifications, described here, are determined for the following frequencies of interest:

$$\omega = [10^{-6}, 10^{-5}, 2\times10^{-5}, 5\times10^{-5}, 10^{-4}, 5\times10^{-4}, 10^{-3}, 10^{-2}, 0.1] \text{ rad/s}$$

16.6.2.1 Robust Stability

The robust stability specification for each channel is given by

$$|t_{ii}(s)| \le 1.2 \quad \text{for } i = 1,2,3,4, \ \forall\omega \tag{16.26}$$

which results in at least a 50° lower phase margin and a lower gain margin of at least 1.833 (5.26 dB).

16.6.2.2 Reference Tracking

The response of the output temperature $T_i(t)$, following the set point $T_{iref}(t)$, is required to lie between specified upper and lower bounds $B(t)_U$ and $B(t)_L$, respectively, or $B_U(\omega)$ and $B_L(\omega)$ in the frequency domain:

$$B_{Lii}(\omega) \le |t_{ii}^{y/r}(s)| \le B_{Uii}(\omega) \quad \text{for } i = 1,2,3,4 \quad \text{and} \quad \omega \le 0.0005 \text{ rad/s} \tag{16.27}$$

$$B_{Uii}(\omega) = \left|\frac{0.001^2}{s^2 + 0.018s + 10^{-6}}\right| \quad \text{for } i = 1,2,3,4 \quad \text{and} \quad \omega \le 0.0005 \text{ rad/s} \tag{16.28}$$

$$B_{Lii}(\omega) = \left|\frac{9\times10^{-8}}{0.2s^3 + s^2 + 0.0006s + 9\times10^{-8}}\right| \quad \text{for } i = 1,2,3,4 \quad \text{and} \quad \omega \le 0.0005 \text{ rad/s} \tag{16.29}$$

16.6.2.3 Disturbance Rejection at Plant Input

The robust disturbance rejection at the plant input must satisfy

$$|t_{ii}^{y/di}(s)| \le \left|\frac{0.25}{s^2 + 0.3s + 0.35}\right| \quad \text{for } i = 1,2,3,4 \quad \text{and} \quad \omega \le 0.0005 \text{ rad/s} \tag{16.30}$$

where $t_{ii}^{y/dt}$ is the transfer function between the inner temperature $T_i(t)$ and the disturbance $d_i(t)$.

16.6.3 MIMO Design Procedure

For the sake of clarity, in this section the sequential non-diagonal MIMO QFT control design methodology introduced as *Method 3* in Sections 6.2 through 6.8 is summarized.[93,97]

Given a $n \times n$ multivariable plant, $\boldsymbol{P} = [p_{ij}(s)]$, where $\boldsymbol{P} \in \mathcal{P}$ and $\mathcal{P}$ is the set of possible plants due to the uncertainty, and given the matrices $\boldsymbol{P}_{di}$ and $\boldsymbol{P}_{do}$, which represent the disturbances at the plant input and output, respectively, the *Method 3* is used to design the compensator matrix $\boldsymbol{G} = [g_{ij}(s)]$ and the prefilter $\boldsymbol{F} = [f_{ij}(s)]$. The reference vector $\boldsymbol{r}'$ and the external

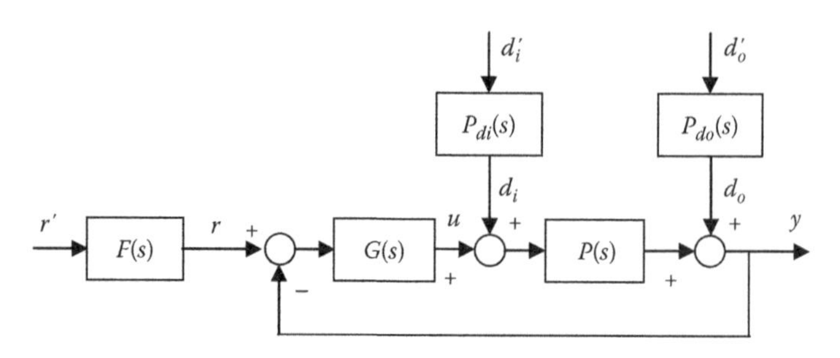

FIGURE 16.9
Structure of a 2DOF (two-degree-of-freedom) MIMO system.

disturbance vectors at the plant input $\boldsymbol{d}_i'$ and the plant output $\boldsymbol{d}_o'$ are the inputs of the system. The output vector $\boldsymbol{y}$ represents the variables to be controlled (see Figure 16.9). The plant inverse, denoted by $\boldsymbol{P}^*$, and the compensator $\boldsymbol{G}$ are presented as follows:

$$\boldsymbol{P}^{-1} = \boldsymbol{P}^* = \boldsymbol{\Lambda} + \boldsymbol{B} = \begin{bmatrix} p_{11}^* & 0 & 0 \\ 0 & \cdots & 0 \\ 0 & 0 & p_{nn}^* \end{bmatrix} + \begin{bmatrix} 0 & \cdots & p_{1n}^* \\ \cdots & 0 & \cdots \\ p_{n1}^* & \cdots & 0 \end{bmatrix} \tag{16.31}$$

$$\mathbf{G} = \mathbf{G}_d + \mathbf{G}_b = \begin{bmatrix} g_{11} & 0 & 0 \\ 0 & \cdots & 0 \\ 0 & 0 & g_{nn} \end{bmatrix} + \begin{bmatrix} 0 & \cdots & g_{1n} \\ \cdots & 0 & \cdots \\ g_{n1} & \cdots & 0 \end{bmatrix} \tag{16.32}$$

The compensator design method only needs to fulfill the hypothesis given by Equation 16.33 and is composed of the following four steps (see Figure 16.10):

$$\left|\left(p_{ij}^* + g_{ij}\right)t_{jj}\right| \gg \left|\left(p_{ik}^* + g_{ik}\right)t_{kj}\right| \quad \text{for } k \neq j \text{ and in the bandwidth of } t_{jj} \tag{16.33}$$

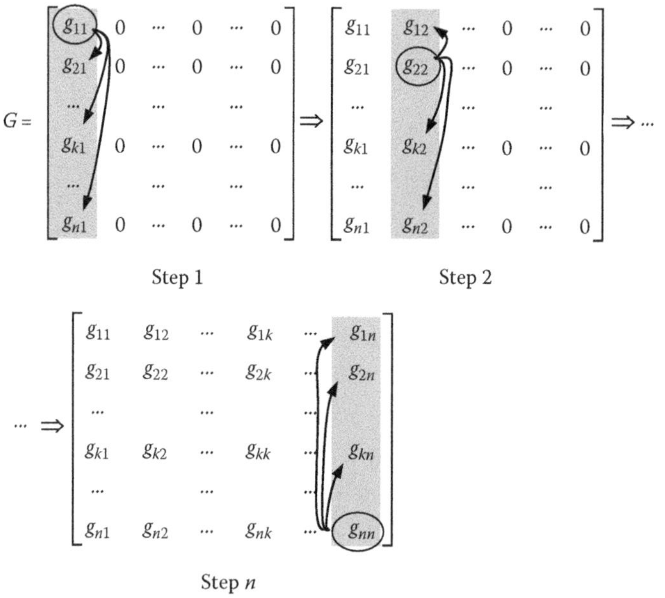

$$G = \begin{bmatrix} g_{11} & 0 & \cdots & 0 & \cdots & 0 \\ g_{21} & 0 & \cdots & 0 & \cdots & 0 \\ \cdots & & \cdots & & \cdots & \\ g_{k1} & 0 & \cdots & 0 & \cdots & 0 \\ \cdots & & \cdots & & \cdots & \\ g_{n1} & 0 & \cdots & 0 & \cdots & 0 \end{bmatrix} \Rightarrow \begin{bmatrix} g_{11} & g_{12} & \cdots & 0 & \cdots & 0 \\ g_{21} & g_{22} & \cdots & 0 & \cdots & 0 \\ \cdots & & \cdots & & \cdots & \\ g_{k1} & g_{k2} & \cdots & 0 & \cdots & 0 \\ \cdots & & \cdots & & \cdots & \\ g_{n1} & g_{n2} & \cdots & 0 & \cdots & 0 \end{bmatrix} \Rightarrow \cdots$$

Step 1 Step 2

$$\cdots \Rightarrow \begin{bmatrix} g_{11} & g_{12} & \cdots & g_{1k} & \cdots & g_{1n} \\ g_{21} & g_{22} & \cdots & g_{2k} & \cdots & g_{2n} \\ \cdots & & \cdots & & \cdots & \\ g_{k1} & g_{k2} & \cdots & g_{kk} & \cdots & g_{kn} \\ \cdots & & \cdots & & \cdots & \\ g_{n1} & g_{n2} & \cdots & g_{nk} & \cdots & g_{nn} \end{bmatrix}$$

Step n

FIGURE 16.10
n stages of the non-diagonal MIMO QFT design (*Method* 3).

Step A. Input/output pairing and loop ordering. First, the methodology begins paring the plant inputs and outputs with the relative gain analysis (RGA) technique, where $RGA(j\omega) = \mathbf{P}(j\omega) \otimes (\mathbf{P}^{-1}(j\omega))^T$, and where $\otimes$ denotes element-by-element multiplication (Schur product) (see Section 5.3 for RGA).

This is followed by arranging the matrix $\mathbf{P}^*$ so that $(p_{11}^*)^{-1}$ has the smallest phase margin frequency, $(p_{22}^*)^{-1}$ the next smallest phase margin frequency, and so on.

Step B. Design of the diagonal compensator elements g_{kk}. This design of the element g_{kk} is calculated using the standard QFT loop-shaping technique for the inverse of the equivalent plant $\left(p_{kk}^{*e}\right)^{-1}$ in order to achieve robust stability and robust performance specifications. The equivalent plant satisfies the recursive relationship given by

$$\left[p_{ii}^{*e}\right]_k = \left[p_{ii}^*\right]_{k-1} - \frac{\left(\left[p_{i(i-1)}^*\right]_{k-1} + \left[g_{i(i-1)}\right]_{k-1}\right)\left(\left[p_{(i-1)i}^*\right]_{k-1} + \left[g_{(i-1)i}\right]_{k-1}\right)}{\left[p_{(i-1)(i-1)}^*\right]_{k-1} + \left[g_{(i-1)(i-1)}\right]_{k-1}}, \quad i \geq k, \ [\mathbf{P}^*]_{k=1} = \mathbf{P}^* \tag{16.34}$$

At this point, the design has to fulfill two stability conditions: (a) $L_i(s) = g_{ii}(s)\left(p_{ii}^{*e}\right)^{-1}$ has to satisfy the Nyquist encirclement condition and (b) no RHP pole-zero cancellations have to occur between $g_{ii}(s)$ and $\left(p_{ii}^{*e}\right)^{-1}$.

Step C. Design of the non-diagonal compensator elements g_{ij}. The $(n-1)$ non-diagonal elements g_{ik} $(i \neq k, i = 1, 2, \ldots, n)$ of the *kth* compensator column are designed to minimize the cross-coupling terms c_{ik} given by Equations 6.31 through 6.33. The optimum compensator elements (see Table 16.6) are utilized in order to achieve this goal.

Once the design of $\mathbf{G}(s)$ has been completed, the design has to fulfill two more stability conditions: (c) no Smith–McMillan pole-zero cancellations have to occur between $\mathbf{P}(s)$ and $\mathbf{G}(s)$ and (d) no Smith–McMillan pole-zero cancellations have to occur in $|\mathbf{P}^*(s) + \mathbf{G}(s)|$.

Step D. Prefilter. The design of a prefilter $\mathbf{F}(s)$ is necessary in case of reference-tracking specifications. Once the full matrix compensator $\mathbf{G}(s)$ is designed, the prefilter does not present any difficulty because the final $\mathbf{T}_{y/r}$ function shows less loop interaction. Therefore, the prefilter $\mathbf{F}(s)$ is a matrix diagonal.

The method is now applied to design the controllers for the 4 × 4 industrial furnace case, according to the model (Section 16.6.1) and specifications (Section 16.6.2) previously introduced.

TABLE 16.6

Off-Diagonal Controller g_{ij} $(i \neq j)$ to Minimize the Interaction

Tracking: $g_{ij} = F_{pd}\left(g_{jj}\dfrac{p_{ij}^{*N}}{p_{jj}^{*N}}\right)$	(16.35)
Disturbance rejection at plant input: $g_{ij} = F_{pd}\left(-p_{ij}^{*N}\right)$	(16.36)
Disturbance rejection at plant output: $g_{ij} = F_{pd}\left(g_{jj}\dfrac{p_{ij}^{*N}}{p_{jj}^{*N}}\right)$	(16.37)

Step A. Input/output pairing

First the RGA matrix is calculated. The selected pairings are $[q_i - T_i]$ with $i = 1, 2, 3, 4$. With this arrangement, the hypothesis presented in Equation 16.33 is also fulfilled. Now, by canceling the off-diagonal controllers where the RGA matrix has zeros (or very low gains in the general case), the controller structure is given by

$$\mathbf{G}(s) = \begin{bmatrix} g_{11}(s) & g_{12}(s) & 0 & 0 \\ g_{21}(s) & g_{22}(s) & g_{23}(s) & 0 \\ 0 & g_{32}(s) & g_{33}(s) & g_{34}(s) \\ 0 & 0 & g_{43}(s) & g_{44}(s) \end{bmatrix} \tag{16.38}$$

Step B1. Diagonal compensator $g_{11}(s)$

The design of the compensator $g_{11}(s)$ (see Figure 16.11), for $\left(p_{11}^{*}\right)^{-1}$, is given by

$$g_{11}(s) = \frac{12(s+0.0006)(s+10^{-7})}{s(s+0.8)(s+0.05)(s+0.001)} \tag{16.39}$$

Step C1. Non-diagonal compensators $g_{21}(s)$ and $g_{12}(s)$

In this case the compensator matrix is symmetric, which means that $g_{ij}(s) = g_{ji}(s)$. Note that this property is only achieved due to the symmetric condition of the $\boldsymbol{P}^{*}(s)$ matrix, given by Equation 16.3, that describes the furnace. Because the principal specification is the reference tracking (see Equations 16.27 through 16.29), the non-diagonal compensator is calculated by using Equation 16.35, so that

$$g_{21}(s) = g_{12}(s) = \frac{-0.008s(s+0.000125)}{(s+0.002)(s+0.05)^2} \tag{16.40}$$

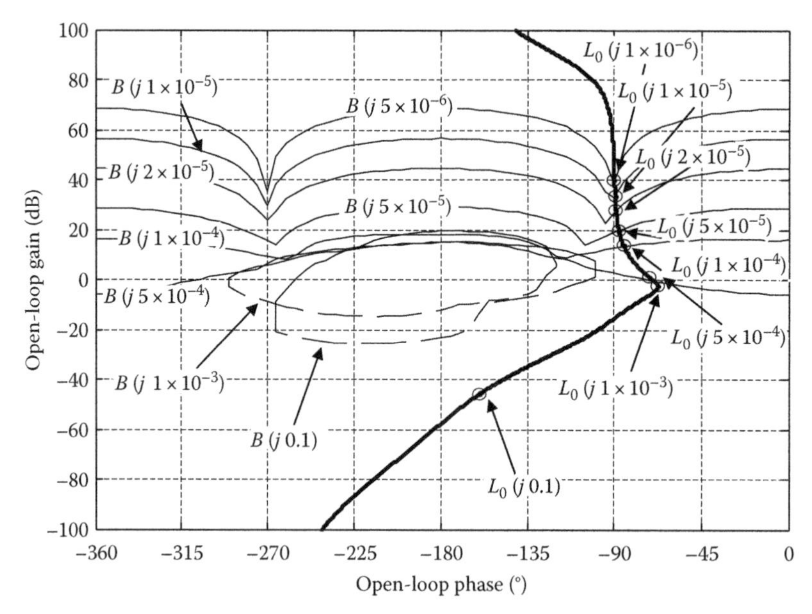

FIGURE 16.11
Designed nominal loop transfer function $L_0(s) = \left(p_{11}^{*}(s)\right)^{-1} g_{11}(s)$.

Step B2. Diagonal compensator $g_{22}(s)$
Once the first column is designed, the equivalent plant used now takes into account the compensators designed in the previous steps:

$$\left[p_{22}^{*e}\right]_2 = \left[p_{22}^{*}\right]_1 - \frac{\left(\left[p_{21}^{*}\right]_1 + \left[g_{21}\right]_1\right)\left(\left[p_{12}^{*}\right]_1\right)}{\left[p_{11}^{*}\right]_1 + \left[g_{11}\right]_1} \tag{16.41}$$

The diagonal compensator $g_{22}(s)$ (see Figure 16.12) for $\left[p_{22}^{*e}\right]_2^{-1}$ is

$$g_{22}(s) = \frac{3(s+10^{-7})}{(s+0.3)(s+0.04)} \tag{16.42}$$

Step C2. Non-diagonal compensators $g_{32}(s)$ and $g_{23}(s)$
Similar to *Step* C1, $g_{32}(s) = g_{23}(s)$, so that

$$g_{32}(s) = g_{23}(s) = \frac{-s(s+0.0009)}{(s+0.003)(s+0.5)^2} \tag{16.43}$$

Step B3. Diagonal compensator $g_{33}(s)$
The equivalent plant $\left[p_{33}^{*e}\right]_3$ takes into account the compensators already designed and is given by

$$\left[p_{33}^{*e}\right]_3 = \left[p_{33}^{*}\right]_2 - \frac{\left(\left[p_{32}^{*}\right]_2 + \left[g_{32}\right]_2\right)\left(\left[p_{23}^{*}\right]_2\right)}{\left[p_{22}^{*e}\right]_2 + \left[g_{22}\right]_2} \tag{16.44}$$

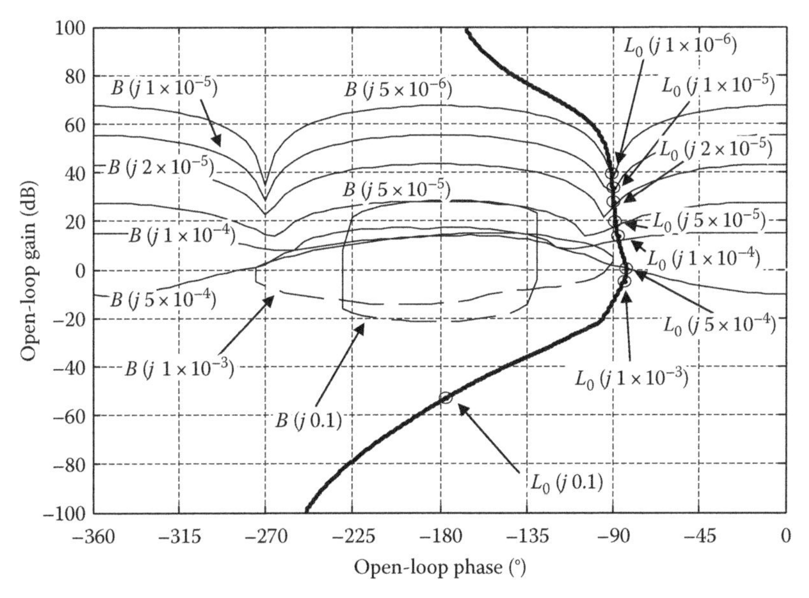

FIGURE 16.12
Designed nominal loop transfer function $L_0(s) = \left(p_{22}^{*}(s)\right)^{-1} g_{22}(s)$.

Note that now the expression of $\left[p_{22}^{*e}\right]_2$ is slightly different from that shown in Equation 16.41 due to the inclusion of the controllers of the second column, $g_{12}(s)$, in the equation, so that

$$\left[p_{22}^{*e}\right]_2 = \left[p_{22}^{*}\right]_1 - \frac{\left(\left[p_{21}^{*}\right]_1 + \left[g_{21}\right]_1\right)^2}{\left[p_{11}^{*}\right]_1 + \left[g_{11}\right]_1} \tag{16.45}$$

The diagonal compensator $g_{33}(s)$ (see Figure 16.13) for $\left[p_{33}^{*e}\right]_3^{-1}$ is given by

$$g_{33}(s) = \frac{9(s+0.002)(s+6\times 10^{-8})}{s(s+0.4)(s+0.18)(s+0.008)} \tag{16.46}$$

Step C3. Non-diagonal compensators $g_{43}(s)$ and $g_{34}(s)$
Similar to *Steps* C1 and C2,

$$g_{43}(s) = g_{34}(s) = \frac{-s(s+0.0008)}{(s+0.001)(s+0.5)^2} \tag{16.47}$$

Step B4. Diagonal compensator $g_{44}(s)$
The equivalent plant $\left[p_{44}^{*e}\right]_4$ for computing the diagonal compensator $g_{44}(s)$, which contains all the compensators designed in the previous steps, is given by

$$\left[p_{44}^{*e}\right]_4 = \left[p_{44}^{*}\right]_3 - \frac{\left(\left[p_{43}^{*}\right]_3 + \left[g_{43}\right]_3\right)\left(\left[p_{34}^{*}\right]_3\right)}{\left[p_{33}^{*e}\right]_3 + \left[g_{33}\right]_3} \tag{16.48}$$

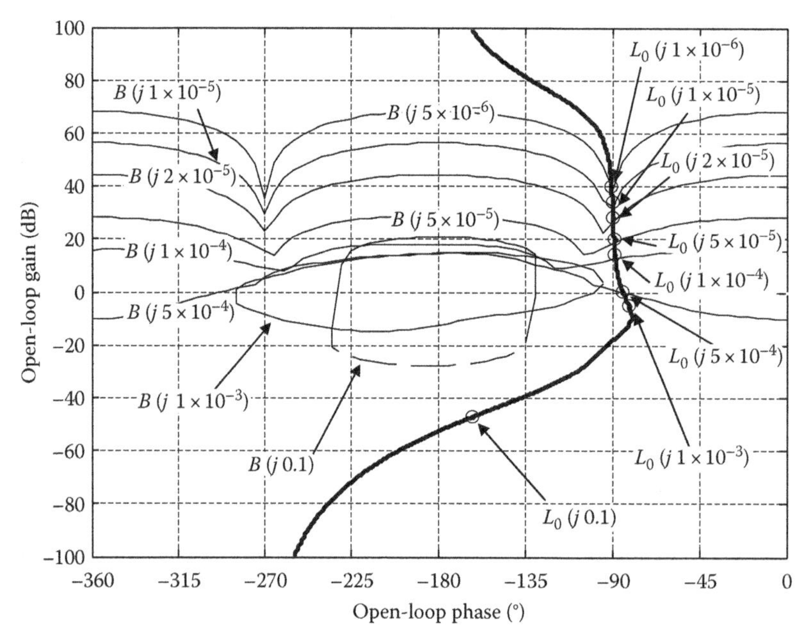

FIGURE 16.13
Designed nominal loop transfer function $L_0(s) = \left(p_{33}^{*}(s)\right)^{-1} g_{33}(s)$.

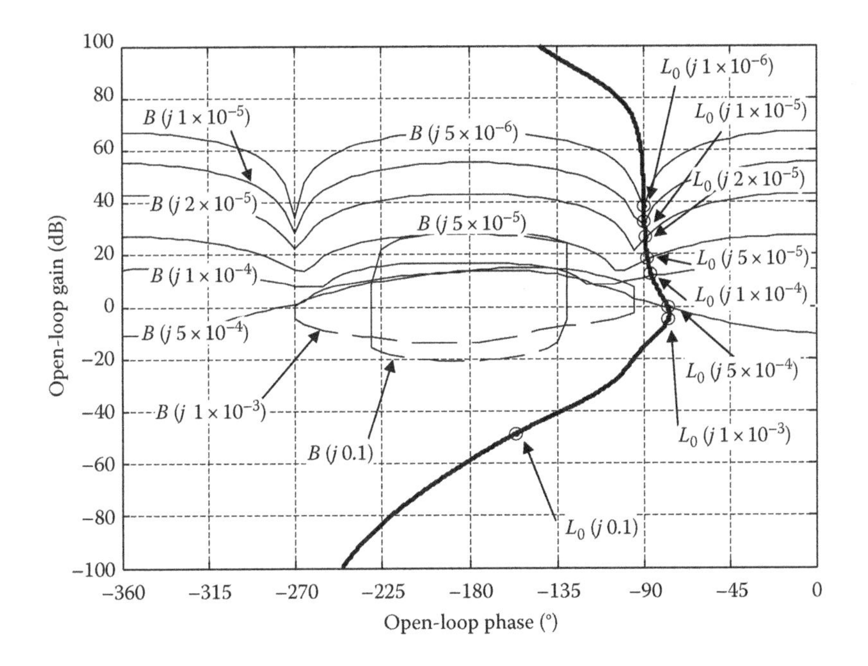

FIGURE 16.14
Designed nominal loop transfer function $L_0(s) = \left(p_{44}^*(s)\right)^{-1} g_{44}(s)$.

Note that now the expression of $\left[p_{33}^{*e}\right]_3$ is also slightly different from that shown in Equation 16.45, so that

$$\left[p_{33}^{*e}\right]_3 = \left[p_{33}^*\right]_2 - \frac{\left(\left[p_{43}^*\right]_2 + \left[g_{43}\right]_2\right)^2}{\left[p_{22}^*\right]_2 + \left[g_{22}\right]_2} \tag{16.49}$$

The diagonal compensator g_{44} (see Figure 16.14) for $\left[p_{44}^{*e}\right]_4^{-1}$ is given by

$$g_{44}(s) = \frac{2(s + 10^{-7})}{s(s + 0.5)(s + 0.007)} \tag{16.50}$$

16.7 Experimental Results[178,386,349]

The matrix compensator $\boldsymbol{G}(s)$ designed in Section 16.6 (Equations 16.38 through 16.40, 16.42, 16.43, 16.46, 16.47, and 16.50) is now validated by using the industrial furnace of 1 MW described in Sections 16.3 through 16.5.

The results obtained provide preliminary indications of the feasibility of the proposed theory to track the temperature reference while rejecting the disturbances (exothermic processes). The transient responses of the four closed-loop systems, with set point changes and external disturbances added at the plant input in the second loop, are shown in Figures 16.15 through 16.18. At t = 24,000 s a disturbance of 50 kW (like an exothermic processes) is added at plant input (second loop). As is seen in the plots, the furnace temperatures follow accurately the reference signals, the disturbance is rejected properly, and the coupling among loops is reduced successfully.

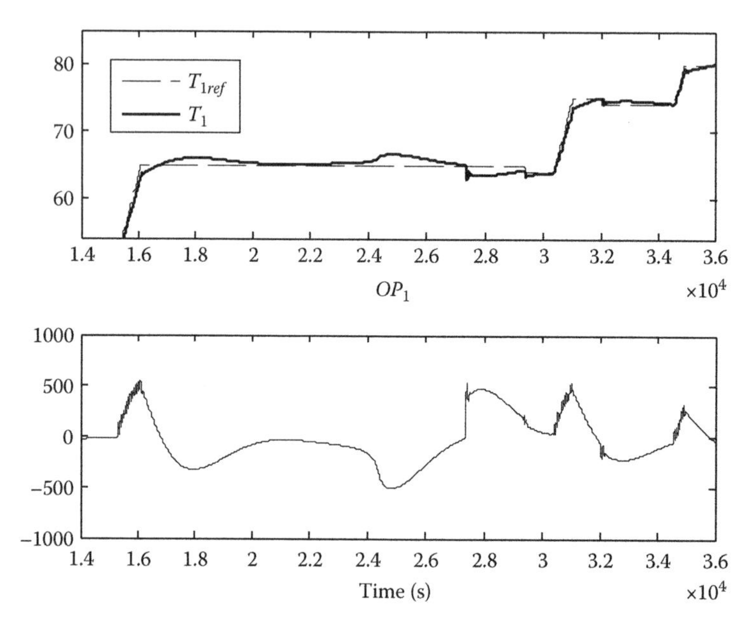

FIGURE 16.15
Experimental results for the first loop (T_1, T_{1ref}, q_1) of the furnace.

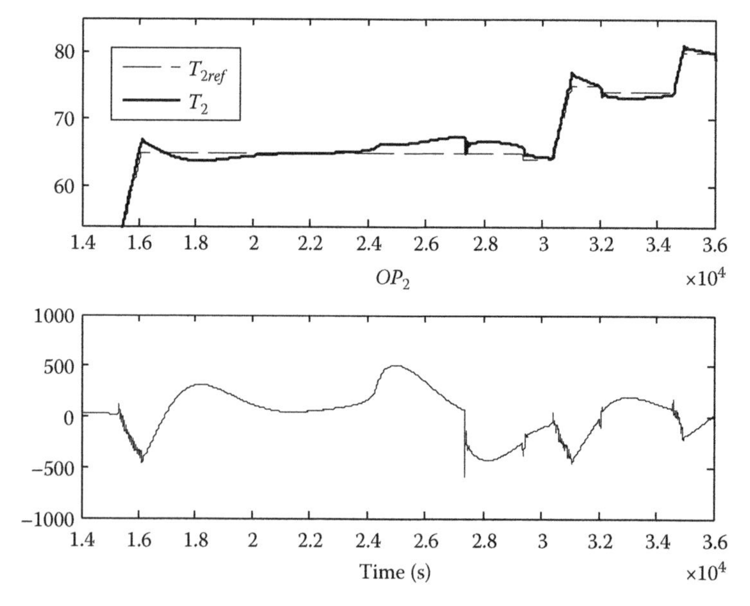

FIGURE 16.16
Experimental results for the second loop (T_2, T_{2ref}, q_2) of the furnace.

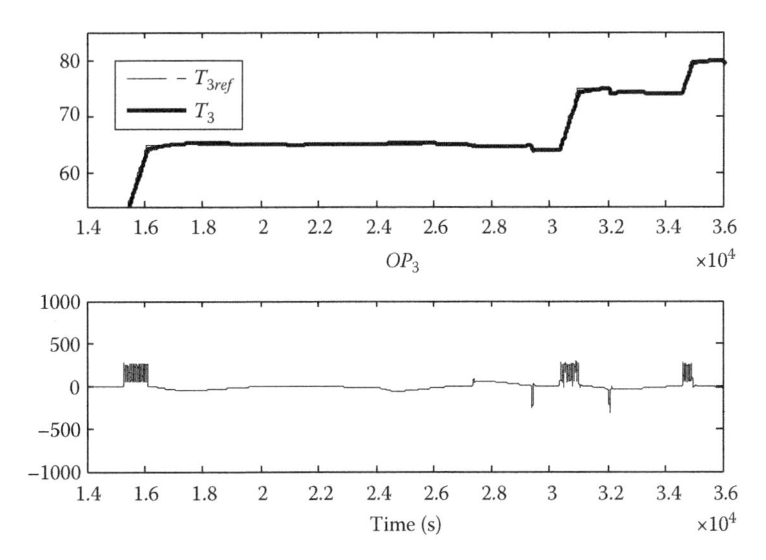

FIGURE 16.17
Experimental results for the third loop (T_3, T_{3ref}, q_3) of the furnace.

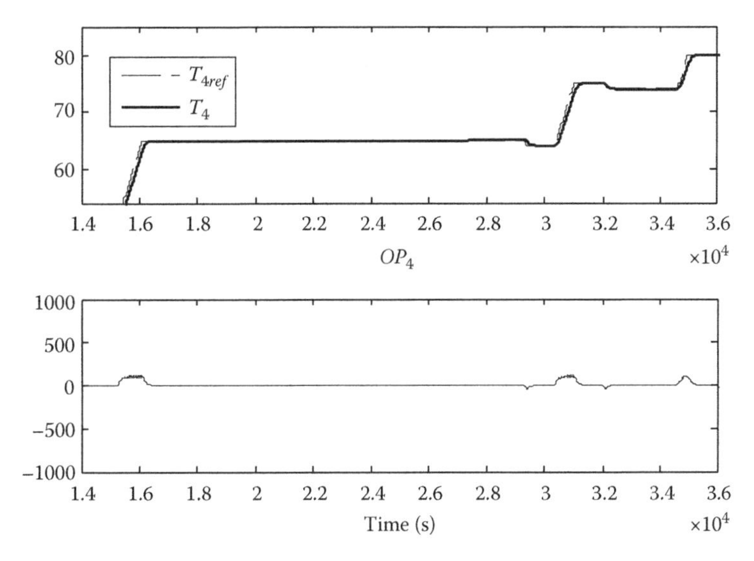

FIGURE 16.18
Experimental results for the fourth loop (T_4, T_{4ref}, q_4) of the furnace.

16.8 Summary

The sequential MIMO QFT methodology applied in this chapter is proven to work well for controlling the temperatures inside a multivariable industrial furnace for wind turbine blades manufacturing. The designed controller is found to be effective in achieving the demanded performance specifications. It not only copes with plant uncertainties but also enhances the rejection of the typical exothermic reactions of the polymerization process. Moreover, the coupling between the control loops due to the heterogeneous thermal loads in the furnace is successfully mitigated by the controller.

17

Smart Wind Turbine Blades[433–498]

17.1 Introduction

Attenuating mechanical loads on the wind turbine (WT) rotor offers great reduction to the total cost of WTs, as well as better maintenance and life span. This fact is particularly important when dealing with offshore WTs, which demand larger blades and as a consequence deal with bigger aerodynamic and mechanical loads. Some manufacturers are currently developing new super massive projects, such as the 15 MW Azimut project, just announced by Gamesa (Spain); the 10 MW project announced by Sway and Enova (Norway), with 145 m rotor diameter; the E-126 WT developed by Enercon (Germany), with 6–7 MW and 125 m rotor diameter; or the 5 MW WT developed by RePower (Germany), with 126 m rotor diameter, among others.

Every load reduction on the WT rotor can produce a multiple chain-effect benefit in the machine, simplifying the blades' structure and weight, and then the drive-train, tower design, foundation, and the required deployment vessel for the offshore installation, as well as increasing the life span and system reliability, and reducing the maintenance costs.

Smart blade systems, often named in popular terms "smart structures" or "smart rotor control," are very effective in reducing the loads on the WT by means of more sophisticated load control techniques. They consist of locally distributed aerodynamic control systems with built-in intelligence on the blades.

Section 17.2 introduces a general description of the smart blade concept for both passive and active control systems, including possible sensors and actuators. Section 17.3 presents a brief presentation of the history of the research made in the field in the last few years.

17.2 General Description

The aerodynamic and structural loads applied to a WT during operation are aeroelastically coupled and randomly variable due to several reasons: (1) the horizontal and vertical wind shear, (2) the wind turbulence and rotational variation, (3) the tower shadow, (4) the yaw and tilt misalignments, (5) the gravity forces, (6) the wakes of other turbines, etc.

Design loads on WTs are typically divided into extreme loads and fatigue loads. Fatigue loads are a key factor for the WT design. Reducing fatigue loads can result in a significant reduction in cost, increasing life span, simplifying required materials and maintenance costs, and improving system reliability.

TABLE 17.1

Sensors for Smart Blade Active Control Systems

Sensor	Where	To Measure
Encoder	Main shaft	Rotational rotor speed
Bending strains	Blade root	Local dynamic deformations
Strain gauges	Blade root	Local static deformations (static strain)
Fiber optics	Blade	Local static deformations
Piezoceramic patch (PZT)	Blade root	Local flap-wise deflections
Accelerometers	Blade tip	Change in acceleration of deflecting tip
LIDAR system	Nacelle or field	Upwind incoming flow wind
Angle of attack	Leading edge	Local blade flow
Relative wind velocity	Leading edge	Local blade flow

Smart blade systems are capable of considerably reducing extreme and fatigue loads on WTs by means of two main methods: passive and active control.

Passive load control counteracts changes in wind speed through passively adapting aeroelastic response of the rotor blades. Some advanced solutions like tension–torsion coupling, bend–twist coupling, and sweep–twist coupling are still under investigation. In general, all of them are inherently open loop, being effective only over limited operating conditions.[455,485,486]

With active load control, the blade is adapted by adjusting its aerodynamic properties (angle of attack or lift coefficient) based on specific real-time measurements. They are inherently closed-loop control techniques that utilize lumped or distributed sensors (see Table 17.1) that measure the current state of the rotor along with dynamic models, and real-time control strategies in microprocessors, to alter the blades' response through

TABLE 17.2

Actuators for Smart Blade Active Control Systems

Actuator Devices	Commanded by	Where	To Change
Fixed ailerons	—	Outer 30% of the blade span	Lift
Variable ailerons	Piezoelectric-based benders	Outer 30% of the blade span	Lift
Microtabs	Microelectrical mechanical	Near the trailing edge of the airfoil	The effective camber of the airfoil, providing changes in lift
Trailing edge flaps, with variable geometry—rigid, soft curved, and strongly curved	Piezoelectric-based benders	Trailing edge of the airfoil	Aerodynamic control surfaces and lift
Variable camber airfoils	Piezoelectric-based benders	Along the blade, with embedded active fiber composites	Aerodynamic control surfaces (shape morphing) and lift
Active twist control	Piezoelectric-based benders	Along the blade, with embedded active fiber composites	Moveable aerodynamic control surfaces and lift
Individual pitch control	Pitch motors	Blade root	Individual blade lift
Moving tips	Piezoelectric-based benders	Blade tip	Lift

innovative lumped or distributed actuators (see Table 17.2) that change the aerodynamic surfaces of the blades.[455,485,486]

Consequently, active control techniques offer significantly more flexibility, especially when dealing with unsteady changes in a flow state. Individual pitch control (IPC), with a pitch angle adjustment per blade instead of a collective one, is the most advanced smart blade active control technique applied nowadays.

17.3 Some History

Although some preliminary investigations of smart blade active control for WTs using distributed sensors and actuators on the rotor have been made during the 1990s, it is still a new, innovative, and ongoing part of research at various wind energy research institutes.

Initial investigations were performed by the National Renewable Energy Laboratory (NREL) during the 1990s, involving Migliore, Quandt, Miller, Huang, Stuart, Wright, and Butterfield, among others. They included (a) fixed ailerons distributed on WT blades for power regulation and aerodynamic braking purposes[438,439,443] and (b) variable ailerons installed on the outer 30% of the blade span, with PID closed-loop controllers and look-up tables for the aerodynamics of the ailerons, and the aeroelastic code FAST for simulation purposes.[441,442]

Contemporaneously, other smart blade active control investigations focused only on periodic loads, such as (a) cyclic pitch control (1P cyclic change in pitch), developed by Geyler, Kuik, and Wegerif and (b) higher harmonic control (pitch actions with multiples of nP), developed by Hammond, Chopra, and Gessow (1P means once per revolution, nP n-times per revolution).[434,446]

Nevertheless, in the early 2000, Bossanyi described how difficult it was to achieve a real load reduction by only using simple rotor encoder measurements and IPC, mainly because of the dominance of stochastic wind turbulences and variation of wind shear and up-flow.[369] In 2003 and 2005, again Bossanyi introduced further IPC techniques for load reduction, including additional load sensors on the blades (strain gauges and accelerometers) to superimpose an additional (individual) pitch demand to the collective pitch.[376,462]

In 2003, Marrant and van Holten developed a project entitled "Smart Dynamic Rotor Control for Large Offshore WTs," funded by the Dutch Technology Foundation STW.[455] It is a detailed inventory of rotor design options, possible load reductions, aerodynamic control devices, actuators, and smart materials. Similarly, in 2007, Barlas developed another study about the state of the art in smart rotor blades and rotor control—UPWIND work package.[485]

In 2005, van Engelen and van der Hooft presented a parameterization technique of feedback loops for IPC around the 1,2,3P frequencies for load reduction.[470] Also in 2005, Larsen, Madsen, and Thomsen introduced active load reduction techniques by using IPC, based on local blade flow measurements from the angle of attack and the relative wind velocity.[466]

In the period 2001–2007, van Dam, Yen D T, Smith, Nakafuji, Michel, Morrison, Baker, Standish, and Chow, among others, investigated microtabs devices.[449,453,460,488] They are small translational devices (microelectrical mechanical tabs [MEM tabs]) placed near the trailing edge of the airfoil and actuated/controlled by small integrated electronic circuits. The deployment of such tabs changes the effective camber of the airfoil, providing changes in lift.

Additional techniques were introduced in 2003 and 2007 by van der Hooft, van Engelen, Hand, Wright, Fingersh, and Harris based on (a) the estimation or (b) the direct measure of the upwind incoming flow wind with a LIDAR system and the reaction of the pitch actuators with a feed forward control system.[457,490]

In 2005–2007, Gaunaa, Troldborg, Basualdo, and Bak, among others, investigated trailing edge flap devices with variable geometry (rigid, soft curved, and strongly curved) and piezoelectric actuators at the Riso lab.[461,469,473,484]

In 2006, McCoy and Griffin at NREL published a detailed comparison between IPC and microtab control techniques.[476]

In 2005–2008, van Holten, Marrant, Joncas, Bergsma, Beukers, van Wingerden, Hulskamp, Barlas, van Kuik, Molenaar, Bersee, and Verhaegen published several papers about the scaled wind tunnel and the smart blade wind project, DUWIND, developed at the Delft University of Technology. They compared four different smart rotor blade concepts: (a) trailing edge flaps, (b) microtabs, (c) camber control, and (d) active twist, all of them commanded by piezoelectric actuators.[465,475,486,491,495]

Complementary, successful experiences in helicopter rotor control during the last decade,[440,447,448,452] with promising results in simulations, scale experiments,[450] and full-scale applications,[451,478] are technical solutions that could be very applicable to smart blade WT control systems and should be more closely investigated.

17.4 Summary

Mechanical load attenuation on the WT rotor can produce great reduction to the total cost of WTs during the design process, as well as better maintenance and life span. Smart blade systems are effective ways to reduce aerodynamic and structural loads applied to the WT during operation by means of passive and/or active load control systems built in the blades. This chapter presented a general description of the smart blade concept for both passive and active control systems, describing possible sensors and actuators and giving a brief description of the research made in the field in the last few years.

18

Offshore Wind Energy: Overview[410–432]

18.1 Introduction

Offshore wind energy is one of the most promising and largest potential energy sources for many countries. Strategically speaking, offshore wind can become one of the main sources of energy in some northern European countries, such as the United Kingdom, which has identified a potential offshore power capacity of 48 GW in its coasts. The European Wind Energy Association has identified a target for the EU members of 40 GW offshore wind power to be installed by 2020 and 150 GW by 2030.[339]

Additionally, the estimated potential offshore power capacity in the United States, in a band extending out to 50 nautical miles from the coastline (including the two coasts and the Great Lakes), has been distributed in three depth categories: about 1070 GW in shallow water (up to 30 m in depth), 620 GW in medium water (30–60 m in depth), and 2400 GW in deep water (deeper than 60 m). According to NREL, the target for offshore wind power installed in the United States by 2030 is between 54 and 89 GW.[318,342]

In comparison with onshore systems, some advantages for offshore wind energy include more reduced wind turbulence and higher mean wind speed; reduction of the turbine size constraints due to transportation limitations and visual concerns; less noise limitations, which brings the possibility to operate at higher blade tip speed, with better efficiency and lower torque-associated costs, etc.

However, offshore projects cost significantly more than land-based turbine systems (see Tables 9.2 and 18.2 for a quantitative comparison). Although it is difficult to give precise numbers due to global market and regulatory uncertainties, it can be estimated that in 2011 the ICC/P_r for onshore projects has been about 2.2 million dollars per MW of installed power and for offshore projects between 3.5 and 5.5 million dollars per MW (depending on water depth and distance to shoreline). At the same time, the *levelized cost of energy* (*LCOE*) for onshore has been about 10 cents of dollar per kWh of energy produced and for offshore projects about 22 cents of dollar per kWh. The objective of the U.S. Department of Energy (DOE) is to reduce the cost of offshore wind to values between 7 and 9 cents of dollar per kWh by 2030, with a medium-term objective of 13 cents of dollar per kWh by 2020. Such cost reductions to a competitive level will require a very significant research effort in the next few decades.

This chapter analyzes the main current projects and characteristics of offshore wind energy systems. Section 18.2 presents the history of offshore platforms, including the most successful systems installed by the oil and natural gas industry in the last century. Section 18.3 introduces the major current offshore wind energy projects with fixed platforms in the world. Section 18.4 analyzes the specific characteristics, challenges, and tools for offshore floating wind turbines (WTs). Finally, Section 18.5 summarizes the main conclusions of the chapter.

18.2 History of Offshore Platforms

Offshore platforms have been already demonstrated as successful and reliable systems by the oil and natural gas industry for more than a century. The first offshore oil platform was built in the freshwaters of the Grand Lake St. Marys, Ohio, around 1891. The first offshore oil platform in salt water was built in the Santa Barbara Channel, California, around 1896.

In the last 50 years, more than 5500 oil and gas offshore platforms have been installed in the Gulf of Mexico. During the same period, around 1700 of these structures were removed. The current number of offshore platforms in the Gulf is around 3800, most of them being fixed structures in water depths up to 530 m (1750 ft) and some of them floating platforms (FPs) in waters up to 2440 m (8050 ft). The Gulf of Mexico has the world's deepest oil platform, which is the floating system *Perdido,* a spar platform in a water depth of 2,438 m (7,996 ft), developed by Shell Oil Co.; the tallest free-standing structure in the world, which is the nonfloating compliant tower *Petronius Platform,* a 610 m structure in a water depth of 531 m (1,742 ft), developed by Chevron and Marathon Oil; and the deepest ever successful test well to date, the *Jack 2 platform,* which carried out a test being 2,133 m (7,000 ft) below sea level for a total depth of 8,572 m (28,116 ft), developed by Chevron, Devon Energy, and StatoilHydro.

Other very large offshore nonfloating compliant towers and fixed platforms are as follows: *Baldpate Platform,* in a water depth of 502 m (1647 ft), Gulf of Mexico, McDermott and HMC; *Bullwinkle Platform,* in a water depth of 413 m (1355 ft), Gulf of Mexico, Shell; *Pompano Platform,* in a water depth of 393 m (1289 ft), Gulf of Mexico, BP; *Benguela-Belize Lobito-Tomboco Platform,* in a water depth of 390 m (1279 ft), Belize, Chevron; *Tombua Landana Platform,* in a water depth of 366 m (1200 ft), Angola, Chevron; *Harmony Platform,* in a water depth of 366 m (1200 ft), California, ExxonMobil; *Troll A Platform,* in a water depth of 303 m (994 ft), North Sea, Norway, Statoil, which is the tallest construction that has ever been moved to another position; and *Gulfaks C Platform,* in a water depth of 217 m (712 ft), North Sea, Norway, Statoil.

A very well-known FP is unfortunately the *BP Deepwater Horizon,* a massive floating system able to operate in waters up to 2,400 m (7,872 ft) deep and drill down to 9,100 m (30,000 ft). On April 20, 2010, methane gas from the well expanded and exploded, killing 11 workers. After burning for 36 h, the platform sank on April 22, 2010, producing the largest oil spill ever in the Gulf of Mexico.

Different technical approaches have been proposed for the design of offshore platforms, mainly depending on water depth. The most common fixed and FP solutions are

1. *Jack-up systems,* which use legs that can be lowered (water depth up to 170 m, 560 ft)
2. *Fixed platforms* and *gravity-based substructures,* built on steel or concrete legs, anchored directly onto the seabed (water depth up to 530 m, 1740 ft)
3. *Compliant towers,* which are slender, flexible tower with a pile foundation (water depth from 460 to 910 m, 1500 to 3000 ft)

4. *Semisubmersible platforms*, with hulls having sufficient buoyancy to cause the structure to float and keep it upright (water depth from 60 to 3,000 m, 200 to 10,000 ft)
5. *Tension leg platforms* (TLP), which are FPs tethered to the seabed (water depth up to 2000 m, 6500 ft)
6. *Spar platforms*, moored to the seabed like the TLPs, but with additional counterweight at the bottom and mooring lines with adjusting tension (water depth up to 2440 m, 8000 ft)

18.3 Offshore Wind Farms

The first offshore wind system in history was the *Vindeby* wind farm, installed in Denmark in 1991, with 11 Bonus 450 kW WTs. The current deepest fixed-foundation offshore wind system is the *Beatrice* wind farm, installed in the United Kingdom in 2007, with 2 Repower 5 MW prototypes, in a 45 m water depth. The current largest offshore wind system in the world is the *Thanet* wind farm, installed in Kent, United Kingdom, in 2010, with 300 MW of power and 100 Vestas V90-3MW machines. The project covers an area of 35 km^2 (500 m between turbines, 800 m between rows). Water depth is 20–25 m and distance from shore 11 km.

Other projects currently operational are *Horns Rev II* (209 MW, with 91 Siemens 2.3–93 machines, water depth 9–17 m, distance from shore 30 km), Denmark, 2009; *Rodsand II* (207 MW, with 90 Siemens 2.3–93 machines, water depth 6–12 m, distance from shore 8.9 km), Denmark, 2010; *Lynn and Inner Dowsing* (194 MW, with 54 Siemens 3.6–107 machines, water depth up to 18 m, distance from shore 5.2 km), the United Kingdom, 2008; *Robin Rigg* (180 MW, with 60 Vestas V90-3MW machines, water depth 3–21 m, distance from shore 9 km), the United Kingdom, 2010.

At the end of 2010, 39 European wind farms with 1136 WTs were operational, with a power capacity of about 2.96 GW, in waters off of Belgium, Denmark, Finland, Germany, Ireland, the Netherlands, Norway, Sweden, and the United Kingdom.

Projects under construction are *London Array* (630 MW, with 175 Siemens 3.6–107 machines), the United Kingdom; *Greater Gabbard* (504 MW, with 140 Siemens 3.6–107 machines), the United Kingdom; *Bard 1* (400 MW, with 80 Bard 5.0 machines), Germany; *Yellow Sea* (1000 MW, with 200 machines of 5 MW), South Korea, etc. These projects will be dwarfed by subsequent wind farms that are in the pipeline in the U.K. Crown State Round 3, like the *Dogger Bank* wind farm (9000 MW), the *Norfolk Bank* wind farm (7200 MW), the *Irish Sea* wind farm (4200 MW), *Horn Sea* wind farm (4000 MW), *Firth of Forth* wind farm (3500 MW), etc. The province of Ontario in Canada is also pursuing several offshore wind farms in the Great Lakes with *Trillium power wind*, with over 400 MW.

The first offshore wind farm in the United States is going to be the *Cape Wind* project (Massachusetts), with 130 WTs, sited at 13 km offshore, over 77 km^2, in a water depth between 2 and 18 m, to produce 468 MW of power (estimated cost: 2.6 billion dollars; power purchase agreement, *PPA*: 20.7 cts$/kWh).

It is expected to reach a total offshore wind power capacity of 75 GW worldwide by 2020. The current average nameplate capacity of offshore WTs is 2.3–5 MW. It is projected to increase to 8–10 MW in the next few years, with rotor diameters more than 150 m. The typical foundations are monopile type (usually large steel tubes with a wall thickness of up to 60 mm and diameter of 6 m), for waters up to 30 m deep (100 ft); a tripod type or a steel jacket type, for waters from 20 to 80 m deep (65 to 260 ft); and FPs, for deeper water.[417]

18.4 Offshore Floating Wind Turbines

As seen in the previous section, FP structures have been already demonstrated as a successful and reliable system by the oil and natural gas industry over many decades. The first FP project dates back to 1977, and the world's deepest oil platform is the floating system, *Perdido*, a spar platform in a water depth of 2438 m (7996 ft). However, the economics for floating WT platforms have yet to be demonstrated.

The design of an FP system usually includes a combination of the following four passive restoring and damping mechanisms (Figure 18.1): [414,419]

1. *Buoyancy*, which is the upward acting force, exerted by the fluid, that opposes the object's weight and equals the weight of displaced fluid, according to the Archimedes' principle (like a barge-type platform)
2. *Ballast*, which is a device to improve stability, providing vertical separation of the center of gravity (lower) and the center of buoyancy (higher)
3. *Mooring*, which is a device used to hold secure an object by means of cables, anchors, or lines (like the tension leg platform)
4. *Viscous damping*, which is a device used to add drag and damping to the structure movement (like damping plates and drag elements)

Any FP contains those four passive mechanisms to some extent. The FP could be classified according to these mechanisms, as a point inside the tetrahedron shown in Figure 18.2.

In addition to the four passive restoring and damping mechanisms, the FP can also include active control systems to stabilize and damp the structure. These active control systems are varied and can include, among others, (a) systems with variable buoyancy, (b) systems with variable ballast, (c) mooring systems with variable tension, (d) systems with

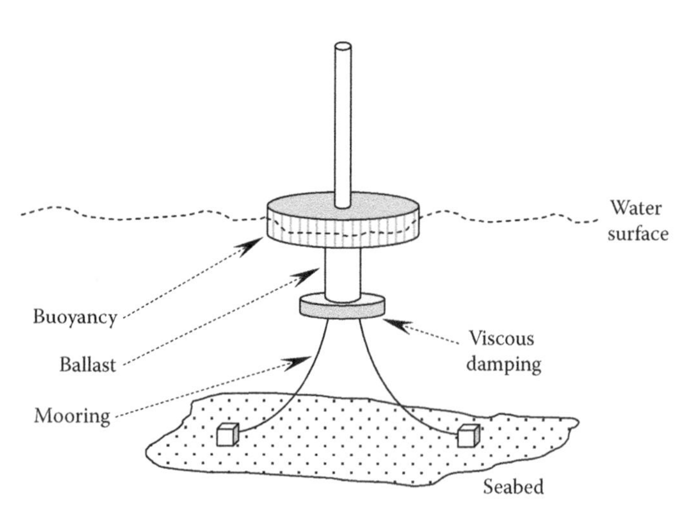

FIGURE 18.1
Passive restoring and damping mechanisms for an FP.

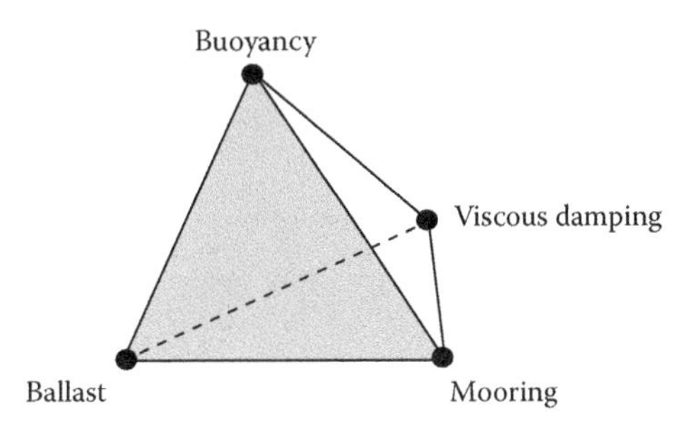

FIGURE 18.2
Tetrahedron for passive restoring and damping mechanisms. (From Butterfield, S. et al., Engineering challenges for floating offshore wind turbines, in *2005 Copenhagen Offshore Wind Conference*, Copenhagen, Denmark, 2005; Wayman, E.N. et al., Coupled dynamic modeling of floating wind turbine systems, in *2006 Offshore Technology Conference*, Houston, TX, 2006.)

variable damping, (e) extra motors and actuators, (f) smart blades systems, (g) individual pitch control systems, (h) active torque control systems, etc.

Apart from the elastic deformations and the WT degrees of freedom (DOF), the FP movement can be characterized as a 6DOF framework with the following coordinates (see also Figure 18.3): *FP position*: x (surge), y (sway), and z (heave); *FP angle*: *roll* (rotation around x axis or side-to-side), *pitch* (rotation around y axis or fore-aft or tilt rotation), and *yaw* (rotation around z axis).[416,419]

The first full-scale deepwater floating WT in history was installed by Siemens and StatoilHydro at the southwest coast of Norway, in 2009. It is a *HyWind* model, with 2.3 MW of power capacity, in 220 m deep water (722 ft) at 12 km from land. Other characteristics

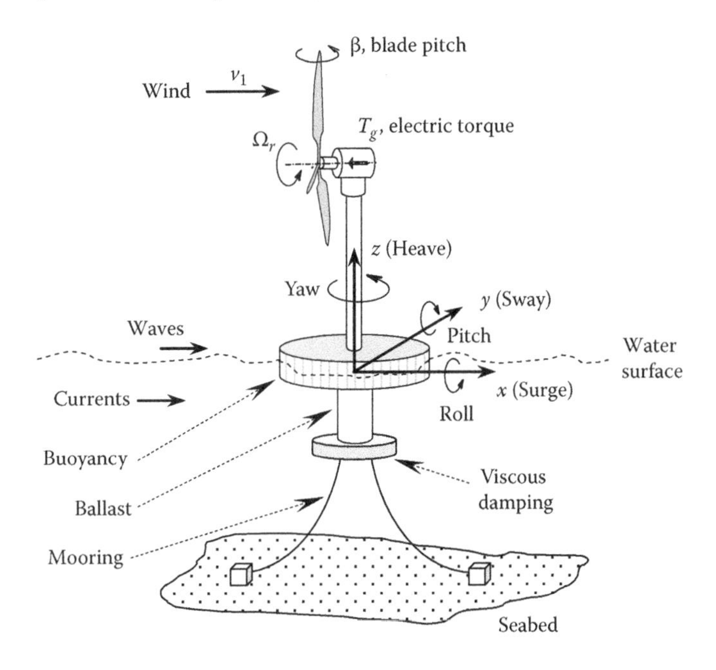

FIGURE 18.3
Floating WT movements and coordinates. (From Jonkman, J.M. and Sclavounos, P.D., Development of fully coupled aeroelastic and hydrodynamic models for offshore wind turbines, in *2006 ASME Wind Energy Symposium*, Reno, NV, 2006, 24 pp.; Wayman, E.N. et al., Coupled dynamic modeling of floating wind turbine systems, in *2006 Offshore Technology Conference*, Houston, TX, 2006.)

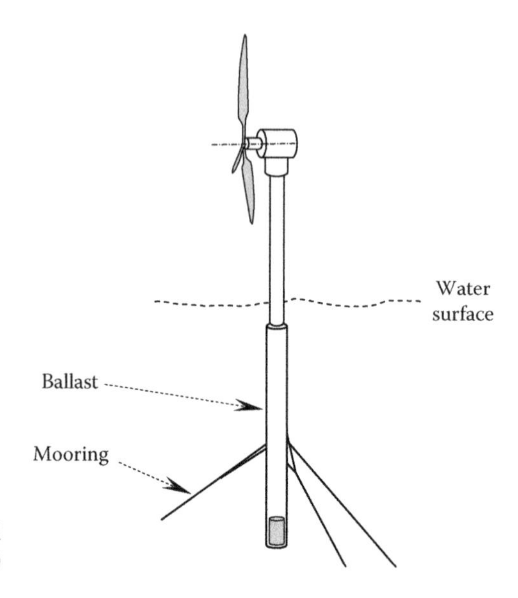

FIGURE 18.4
HyWind offshore floating WT. (From Larsen, T.J. and Hanson, T.D., *J. Phys., Conference Series*, 75, 012073, 1, 2007.)

of the machine are as follows: weight of turbine, 138 tons; turbine height, 65 m; rotor diameter, 82.4 m; floatation element below sea surface, 100 m; total water displacement, 5300 m^3; diameter at sea surface, 6 m; diameter of floatation element, 8.3 m; sea depth range, 120–700 m; and number of anchor moorings, 3 lines (see Figure 18.4).[424]

Another offshore floating WT currently in progress is the *WindFloat* project, which is a three-legged floating foundation structure for multimegawatt generators (see Figure 18.5). It has been designed to accommodate a 5 MW or larger WT (with a rotor diameter around 126 m) on one of the columns of the hull, with minimal modifications to the nacelle and rotor. The three legs provide buoyancy to support the turbine and stability from the water plane inertia. Horizontal plates at the bottom of each of the columns increase the added mass, shift the natural period away from the wave energy, and increase the viscous damping in roll, pitch, and heave. An active ballast control system moves water from column to column to compensate for the mean wind loading on the turbine, being able to transfer

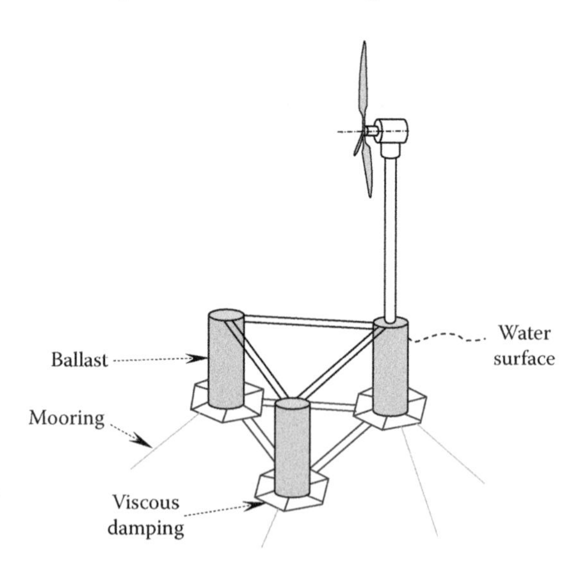

FIGURE 18.5
WindFloat offshore floating WT. (From Roddier, D. et al., *J. Renewable Sustainable Energy, Am. Inst. Phys.*, 2, 033104/1, 2010.)

TABLE 18.1

Sea States Definition

Sea State	H_s (m)	T_m (s)
1	0.09	2.0
2	0.67	4.8
3	2.44	8.1
4	5.49	11.3
5	10.00	13.6

Source: Wayman, E.N. et al., Coupled dynamic modeling of floating wind turbine systems, in *2006 Offshore Technology Conference*, Houston, TX, 2006.

up to 200 ton of ballast water in approximately 30 min. Six mooring lines anchor the platform to the site. The total water displacement is 7105 m^3.[431]

According to NREL, the sea waves can be characterized by two parameters, the wave height H_s and the mean period T_m, resulting in five different sea state levels, as shown in Table 18.1.[419]

The simulations of offshore floating WTs have to be performed including the worst-case scenarios by combining sea states (ocean waves), water currents, tides, ice, and wind velocity conditions.

The main components for the dynamics of the floating WT are shown in Figure 18.6, where block F includes the components related to the FP and mooring system, hydrodynamics and waves, currents, tides, and ice. The rest of the blocks are similar to ones of the onshore WT, as introduced in Chapters 12 and 13. They include the aerodynamics, mechanical, and electrical models. The control system interacts with the rotor dynamics, the power generation system, the nacelle dynamics, the platform dynamics, and the mooring dynamics.[422,423,428]

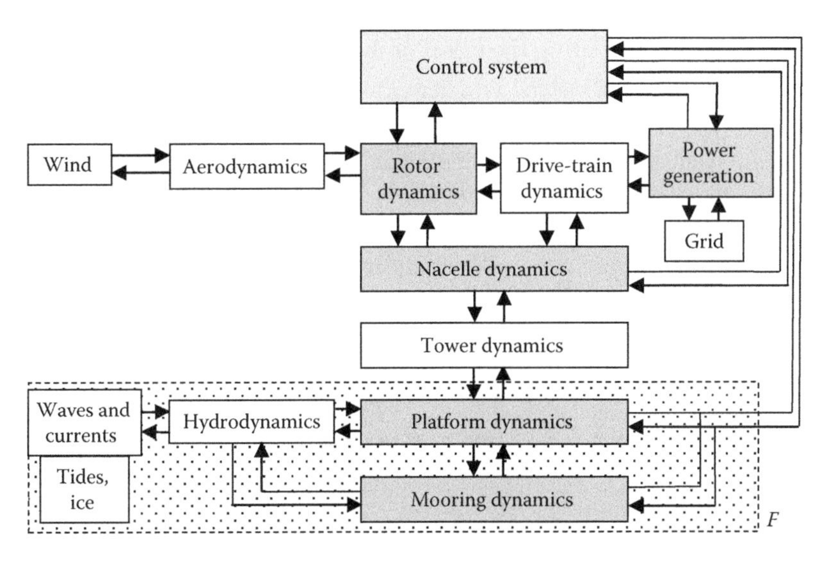

FIGURE 18.6
Control system and dynamics for offshore floating WTs.

The numerical modeling of offshore floating WTs (Figure 18.6) includes different approaches to simulate the mechanical structures, the aerodynamics, the hydrodynamics, and the mooring lines. In a general overview, the following is presented: [410]

(a) *Structural dynamics*: for calculating forces, loads, modes, and motion of the structure
  (a.1) *Modal analysis*, to calculate the fundamental mode shapes and frequencies of the structure
  (a.2) *Multibody system approach*, which uses rigid or flexible elements and the Newton–Euler or Lagrange's equations of motion
  (a.3) *Finite element modeling*, where the structure is discretized into a mesh of finite elements interconnected at nodes
(b) *Aerodynamics*: for calculating aerodynamic forces on the blades
  (b.1) *Blade element momentum* (BEM) *theory*, which uses the Bernoulli's theorem, can be improved by including tip and hub loss factors (Prandtl from propeller theory) and dynamic inflow theory
  (b.2) *Computational fluid dynamics codes*, which uses the Navier–Stokes equations
(c) *Hydrodynamics*: for calculating hydrodynamic loading on submerged structure
  (c.1) *Air wave theory*, to calculate the hydrodynamic loading on a submerged structure in the time domain. It works with linear sea states and calculates wave particle kinematics and dynamic pressure. The theory represents the wave elevation as a sinusoid propagating with a constant amplitude and period.
  (c.2) *Morison's equation*, to calculate the hydrodynamic loads acting on the support structure in terms of wave particle velocities and accelerations. This can include linear radiation, diffraction, and hydrostatics problem in the time domain.
  (c.3) *Linear hydrodynamic equations*, where the submerged body does not exert any influence on the surrounding fluid either in terms of diffraction or radiation.
  (c.4) *MacCamy–Fuchs approximation*, to calculate the effects of wave diffraction in linear irregular sea states in the time domain.
(d) *Mooring lines*: for calculating the effect of the mooring lines in the structure
  (d.1) *Force–displacement method*, which models the mooring lines for floating WTs by applying nonlinear spring stiffness for all 6DOF, including damping matrices
  (d.2) *Quasi-static approach*, where nonlinear mooring line restoring forces and tensions are solved, including elasticity
  (d.3) *Full dynamic modeling*, which gives an accurate representation of the drag and inertia of mooring lines and their effect on the FP

Several design tools to model offshore floating WTs in a fully coupled time-domain dynamic analysis are currently available. Among others, the following can be mentioned: [410]

FAST (fatigue, aerodynamics, structures, and turbulence)[342]

- Developed by NREL
- Structural dynamics: (a.1) and (a.2)
- Aerodynamics: (b.1). AeroDyn package

- Hydrodynamics: (c.1) and (c.2). HydroDyn package
- Mooring lines: (d.2)
- Additional modules: Charm3D, TimeFloat, etc.

GH Bladed[348]

- Developed by Garrad Hassan
- Structural dynamics: (a.3)
- Aerodynamics: (b.1)
- Hydrodynamics: (c.1), (c.2), and (c.4)
- Mooring lines: (d.1)

SIMO (simulation of marine operations)

- Developed by MARINTEK
- Can be combined with HAWC2, developed by Riso[345]
- Structural dynamics: (a.3). RIFLEX code
- Aerodynamics: (b.1)
- Hydrodynamics: (c.1) and (c.2)
- Mooring lines: (d.1)

3Dfloat

- Developed by Norwegian University of Life Sciences
- Structural dynamics: (a.2)
- Aerodynamics: (b.1)
- Hydrodynamics: (c.1) and (c.2)
- Mooring lines: (d.1)

ADAMS (automatic dynamic analysis of mechanical systems)

- Developed by MSC. Software Corporation
- Structural dynamics: (a.2)
- Aerodynamics: (b.1). AeroDyn package
- Hydrodynamics: (c.1) and (c.2). HydroDyn package
- Mooring lines: (d.1) or (d.2)

A quick analysis of the existing design tools shows that all of them use the BEM theory (b.1) to calculate aerodynamic forces on the WT rotor, the air wave theory (c.1) and the Morison's equation (c.2) to calculate the hydrodynamic loads acting on the support structure, and force–displacement (d.1) or quasi-static methods (d.2) to calculate the effect of the mooring lines in the structure.

Floating WTs are the technical solution to work in deep and maybe even shallow waters. However, the considerable movements of an FP (1) decrease significantly the aerodynamic rotor efficiency and then the energy production; (2) require a greater structural reinforcement, increasing the required floating WT capital cost; and (3) increase the levels of mechanical fatigue through the floating WT's life span, increasing the maintenance costs. Both efficiency and cost strongly depend on how the floating WT reacts to the external forces (wind, waves, currents, ice, etc.), making the control system essential.

Among the main control challenges for an FP system, the following can be mentioned:

1. The FP system has much lower natural frequencies, which could eventually interact with the blade control system. In onshore WTs, the lowest natural frequency is usually the first lateral tower bending mode, with values around 0.4 Hz in multimegawatt machines. In offshore floating WTs, the lowest natural frequencies are usually the first surge bending mode (around 0.01 Hz), the first platform pitch bending mode (around 0.035 Hz), and the first platform heave bending mode (around 0.037 Hz). The typical blade pitch control bandwidth is around 0.1 Hz or less.
2. Unfavorable coupling between tower motion (platform pitch) and blades' motion (blade pitch control), with potentially "negative" damping (blades pitch faster than tower motion and the derivative *dTrust/dWindSpeed* < 0), opposite to onshore (with derivative *dTrust/dWindSpeed* > 0), can cause instability.
3. Motions and natural frequencies vary with water depth, tide, and ice.
4. The offshore floating WT is intrinsically a nonlinear multiple-input-multiple-output (MIMO) system.

The robust quantitative feedback theory (QFT) control design methodology, presented in the first part of this book, is a good tool to design the control systems for the floating WT. It is a natural framework to deal with the mentioned control challenges, including a frequency-domain approach [needed to deal with points (1) and (2)], a robust control approach [needed to deal with point (3)], and a MIMO approach [needed to deal with point (4)].

Some new control ideas and mechanisms for floating WT have appeared in the last few years. Among others that can be emphasized are *individual pitch control* (IPC) versus collective pitch control (CPC),[430] *active torque control* versus fixed torque,[432] *active ballast control* versus fixed ballast,[431] systems with *variable buoyancy*, mooring systems with *variable tension*, systems with *variable damping*, and *extra motors and actuators*, etc.

Table 18.2 shows the economic breakdown of two 5.0 MW floating WTs: a *tension leg platform* (TLP), where the FP is stabilized and held upright by tensioned cables attached to the legs and the seabed, and a *tri-floater platform*, where the FP relies on a large volume for buoyancy and stability, being anchored to the seabed by using loose (catenary) mooring lines that prevent the platform from drifting but do not provide tension. Note that, in opposition to what has been stated in Table 9.2, here the costs of the foundation (now the FP and the mooring system) are included in the *turbine capital cost* (*TCC*) category instead of the *balance of plant* (*BoP*).

The first two columns of Table 18.2 present an estimation (in percentage) of the cost of the two floating WTs. The third and forth columns show how an advanced control system could reduce the cost of the floating WTs.

An *integrated control and system design solution* (concurrent engineering approach) is the way to attenuate the mechanical fatigue loads on the system and to reduce the movement of the platform. As a consequence, a multiple chain-effect benefit in the cost and efficiency

TABLE 18.2

Floating WTs Economics for *Tri-floater* and *TLP* (%)

		Current Floating WT		**New Floating WT**		**Advanced Control Systems**	
						Estimated Reduction	
		Tri-floater	TLP	Tri-floater	TLP	Estimated Reduction (per unit)	
Power (*Pr*)		5.0MW	5.0MW	5.0MW	5.0MW		
Rotor diameter (*rb*)		90m	90m	90m	90m		
Hub height (*h*)		80m	80m	80m	80m		
		%	%	%	%	(1-M)	M
A.1 Rotor		**4.1**	**4.5**	**3.5**	**3.8**	0.84	0.16
	Blades (3)	2.8	3.0	2.2	2.4	0.80	0.20
	Hub	0.6	0.7	0.5	0.6	0.90	0.10
	Pitch mechanism and bearings	0.8	0.8	0.7	0.8	0.95	0.05
A.2 Drive-train, nacelle		**12.4**	**13.4**	**12.0**	**13.1**	0.97	0.03
	Low-speed shaft	0.5	0.6	0.5	0.5	0.95	0.05
	Bearings	0.3	0.3	0.3	0.3	0.95	0.05
	Gearbox	3.5	3.8	3.4	3.7	0.95	0.05
	Mechanical brake	0.1	0.1	0.0	0.1	0.95	0.05
	Generator	1.8	2.0	1.8	2.0	1.00	0.00
	Power electronics	2.3	2.5	2.3	2.5	1.00	0.00
	Yaw drive and bearings	0.4	0.4	0.4	0.4	0.95	0.05
	Main frame	1.5	1.6	1.4	1.5	0.95	0.05
	Electrical connections	1.3	1.4	1.3	1.4	1.00	0.00
	Hydraulic, cooling system	0.4	0.4	0.4	0.4	1.00	0.00
	Nacelle cover	0.3	0.4	0.3	0.4	1.00	0.00
A.3 Control, safety system, condition monitoring		**0.5**	**0.6**	**0.5**	**0.6**	1.00	0.00
A.4 Tower		**3.6**	**3.9**	**2.5**	**2.7**	0.70	0.30
A.5 Floating platform		**14.4**	**5.3**	**10.8**	**4.0**	0.75	0.25
A.6 Mooring system		**10.9**	**14.5**	**8.2**	**10.9**	0.75	0.25
A.	*TCC*	**45.9**	**42.3**	**37.5**	**35.1**		

(*continued*)

TABLE 18.2 (continued)

Floating WTs Economics for *Tri-floater* and *TLP* (%)

B.1 Transportation		**0.5**	0.5	0.4	0.4	0.80	0.20
B.2 Assembly and installation		**6.0**	5.3	4.8	4.3	0.80	0.20
B.3 Electrical lines and connections		**13.0**	**14.1**	13.0	14.1	1.00	0.00
B.4 Engineering, permits, and others		5.2	5.7	5.2	5.7	1.00	0.00
B.5 Decommissioning, scour protect		2.4	2.6	1.9	2.1	0.80	0.20
B.	*BoP*	**27.0**	**28.2**	**25.2**	**26.5**		
C.1 Levelized replacement cost (20 years) × 0.6	(*LRC* × 20 years, *CF* = 0.3)	**5.9**	**6.4**	4.1	4.5	0.70	0.30
C.2 Regular maintenance cost (20 years) × 0.6	(*O&M* × 20 years, *CF* = 0.3)	8.2	8.9	5.7	6.2	0.70	0.30
C.3 Fees (20 years)	(*Fees* × 20 years)	**13.1**	**14.2**	13.1	14.2	1.00	0.00
C.	*Operating expenses* (20 years)	**17.1**	**29.5**	**22.9**	**24.9**	*CF*(1 − EL)*Av*	
A + *B* + *C* = *TCC* + *BoP* + *Operation expenses* (%) =		**100.0**	**100.0**	**85.7**	**86.5**	Improvement = 1.20 (0.2)	
CF(1 − EL)*Av* (considering capacity factor, losses & availability) =		0.28	0.28	0.336	0.336		
AEP_{net} = net annual energy production = 8760 *Pr* [*CF*(1 − EL)*Av*], (kWh/year)		12264000	12264000	14716800	14716800	DFR = 0.11	
LCOE − (*DRF ICC* + *LCR* + *O&M* + Fees)/AEP_{net} ($/kWh)		**0.2179**	**0.1973**	**0.1560**	**0.1429**		
LCOE in %				**71.56**	**72.45**	Tax reduction, C.1, C.2	
Cost reduction in *LCOE* in %				**28.44**	**27.55**	of 40% = 0.6 cost	

2011 baseline cost in Table Tri-floater: 1% = $285175, TLP: 1% = $262175 aprox. (U.S. dollar).
Current tri-Floater, aprox.: *ICC*/*Pr* = (*TCC* + *BoP*)/*Pr* = (285,175 * 72.9)/5.0 = 4.16 M$/MW.
Current TLP, aprox.: *ICC*/*Pr* = (*TCC* + *BoP*)/*Pr* = (262,175 * 70.5)/5.0 = 3.70 M$/MW.

of the machine can be achieved, as shown in detail in Table 18.2, third and fourth columns. It includes three main aspects: (1) a simplification of the structure and then costs of rotor, drive-train, tower, FP, mooring system, transportation, assembly and installation; (2) an increase of the life span and system reliability, reducing the replacement and maintenance costs; and (3) an increase in the aerodynamic efficiency and energy production.

Based on an analysis of the *Dutch tri-floater* design and the *NREL tension-leg-platform* design, the estimated overall reduction in *LCOE* for these offshore floating WTs that result from proposed advanced control systems is in the range of 28% and 27%, respectively, as shown in Table 18.2.

18.5 Summary

This chapter analyzed the main projects and characteristics of offshore wind energy systems, including (a) the history of offshore oil and natural gas platforms; (b) the major current offshore wind energy projects with fixed platforms in the world; and (c) the main specific characteristics, challenges, and tools for offshore floating WTs.

19

Airborne Wind Energy Systems[499–517]

19.1 Introduction

Airborne wind energy (AWE) systems are essentially tethered flying wind turbines (WTs) that combine a number of known and several innovative technologies into a unique method of collecting clean, renewable energy. AWE systems represent an exceptional engineering challenge. They are worth exploring as a source of power. Their design and flexibility allow them to be deployed in areas that are otherwise unsuitable for traditional WTs.[499–517]

The main objective of this chapter is to present a first approach to AWE systems. Section 19.2 gives an overview of the field, and Section 19.3 presents the design of a novel airborne WT: *the EAGLE system*.[508–511]

19.2 Overview of Airborne Wind Energy Systems

AWE applications are vast, ranging from providing power to villages in the developing world, where the massive cost of implementing an electrical grid and utility-grade power plants is prohibitive, to providing a means of auxiliary power to ships and offshore drilling platforms, where an ocean separates the platform or ship from a source of external power, to the remote industrial locations surrounded by rough terrain that makes transportation and installation of traditional electrical infrastructure impossible.

The common objective of AWE systems is to provide electrical power to locations where traditional means are impossible: mini-grid applications, off-grid combined wind and diesel solutions, rapid deployment systems to disaster areas, isolated farms, remote areas, cell towers, exploration equipment, backup power for water pumps in remote mines, etc. The social benefits from providing electrical power on a mini-grid basis gives an opportunity to supply a better quality of life to those that would otherwise be denied electricity, allow survival of crippled ships or drilling platforms, or enable industrial development to areas that would otherwise stay unrealized.

Some of the main advantages of AWE systems are (a) lower cost per kWh than other wind energy systems, since there is no need of a tower and they allow a much simpler foundation; (b) double capacity factor of conventional WTs, mainly if the AWE system flies at high altitude or moves in a large area; (c) greater mobility; and (d) deployment to more sites.

TABLE 19.1
Classification of Some Current AWE Systems

	Altitude		Generator		Weight		Aerodynamics			
Type	**(1.1)**	**(1.2)**	**(2.1)**	**(2.2)**	**(3.1)**	**(3.2)**	**(4.1)**	**(4.2)**	**(4.3)**	**(4.4)**
Joby Energy	x			x		x			x	
Makani Power	x			x		x			x	
Magenn Power	x			x	x					x
SkyWindpower		x		x		x	x			
Ampyx Power	x		x			x			x	
Kitenrg		x	x			x		x		
EAGLE system	x			x	x				x	x

Among the principal difficulties are (a) aviation concerns by regulatory authorities in Europe and the United States and (b) technical challenges, the design of the control system being one of the most difficult problems.

In the last few years, different designs for AWE systems have been proposed. They can be classified according to some of the following characteristics (see also Table 19.1):

- *Altitude*
  (1.1) Low and medium
  (1.2) High, more than 600 m above ground
- *Generator position*
  (2.1) On the ground
  (2.2) Aloft
- *Weight*
  (3.1) Lighter than air
  (3.2) Heavier than air
- *Aerodynamics*
  (4.1) Helicopter type
  (4.2) Kite type
  (4.3) Wing type
  (4.4) Aerostat type or airships

There are a dozen major R&D efforts worldwide, a number of inventors working on small-scale products and an international Airborne Wind Energy Consortium.[350] Among the main companies and current products under development are (see also Table 19.1)

- *Joby Energy* (www.jobyenergy.com), California.[512] The AWE system consists of a two twin airframes supporting an array of turbines. The turbines are connected to motor-generators, which produce thrust during takeoff and generate power during crosswind flight. A tether transmits electricity and moors the system to the ground. Characteristics of the first prototype are as follows: the generator power is 30 kW, the aircraft wingspan 2.5 m, and the operating height between 150 and 600 m (Classification: (1.1), (2.2), (3.2), and (4.3)).

- *Makani Power* (www.makanipower.com), California.[513] The AWE system consists of a wing-type fiberglass kite tethered to the ground. It operates in circular paths. The energy is extracted by small rotors that drive high-speed brushless DC generators aloft. Some power electronic converters on ground convert the energy to conventional 50 or 60 Hz. Characteristics of the first prototypes are as follows: the generator power is 10 kW, the weight 50 kg, and the operating height between 200 and 500 m (Classification: (1.1), (2.2), (3.2), and (4.3)).
- *Magenn Power* (www.magenn.com), California.[514] The AWE system, so-called MARS (Magenn Air Rotor System), consists of a lighter-than-air tethered WT, with a blimp full of helium that rotates about a horizontal axis in response to wind, generating electrical energy aloft. A tether transmits the electrical energy to ground. The horizontal rotation of the blimp also creates a "Magnus effect," which provides additional lift and improves the controllability. Characteristics of the first prototypes are as follows: the operating height is between 150 and 300 m, the cut-in wind speed 3 m/s, the cutoff wind speed 24 m/s, and the rated wind speed 12 m/s. There are some options for generator power: 4, 10, 25, and 100 kW. The 100 kW model is the only commercially available. It has a diameter of 13.7 m, a length of 30.5 m, and blades of 6.5 m. The volume of helium in the airship is 5660 m^3 (Classification: (1.1), (2.2), (3.1), and (4.4)).
- *SkyWindpower* (www.skywindpower.com), California.[515] The AWE system consists of a tethered elementary helicopter with no cabin and four contra-rotating rotors in a H-shape frame. The rotors are powered by electricity from the ground to have the craft reach altitude. Then, the craft is tilted by command, and the wind turns the rotors, thus both holding up the craft and generating power that is transmitted back to the ground by means of a tether. The target for the operating height is between 2000 and 9000 m. A 12 kW prototype is undergoing testing (Classification: (1.2), (2.2), (3.2), and (4.1)).
- *Ampyx Power* (www.ampyxpower.com), the Netherlands.[516] The AWE system consists of an airplane and a ground station containing an electric motor, which can also work as a generator. During the power generating phase the plane flies patterns downwind of the ground station, generating energy while pulling on a cable that connects the ground station and the plane. When the maximum cable length is reached, the plane dives toward the ground station and the ground station relies in the cable. Characteristics of the first prototype are as follows: the generator power is 10 kW, the aircraft wingspan 5.5 m, the aircraft weight 25 kg, the operating height 300 m, the cable diameter 4 mm, and the aircraft loading 4 kN (Classification: (1.1), (2.1), (3.2), and (4.3)).
- *Kitenrg* (www.kitenergy.net), Italy.[517] The AWE system consists of a lighter airfoil (kite) connected to the ground by two cables. Electricity is generated at ground level by converting the traction forces acting on the airfoil lines into electrical power, using suitable rotating mechanisms and electric generators placed on the ground. Characteristics of the first prototype are as follows: the kites have an area of 10–30 m^2, with cables up to 1000 m long. The permanent magnet synchronous motors/generators in ground are of 40 kW of peak power, and they are connected to a set of batteries at 340 V to accumulate the energy (Classification: (1.2), (2.1), (3.2), and (4.2)).

19.3 Eagle System

This section presents the novel, low-altitude, AWE system developed by Garcia-Sanz, White, and Tierno at Case Western Reserve University, Cleveland, Ohio: the so-called *EAGLE system* (Electrical Airborne Generator with a Lighter-than-air Eolian system).[508–511] It consists of a tethered and controllable airship supporting an independent counter-rotating wind power generator payload (see Figure 19.1).

The power generation system and lift system are separate and do not rely on each other for operation. The lift system is a lighter-than-air flyer, full of helium, with longitudinal and lateral control capabilities.

The flying WT is a variable-speed counter-rotating rotor system with a counter-rotating synchronous generator aloft. The tether serves to both anchor the device and transmit electricity to ground-based power electronics and grid collection (Classification: (1.1), (2.2), (3.1), (4.3), and (4.4)).

The *EAGLE system* utilizes the ailerons position, elevator position, rudder position, and unscratched length of the tether to control the longitudinal and lateral dynamics. The lifting system is designed with control and stability considerations as guiding principles. A unique approach is used to define the states and nonlinear model of the system, as traditional blimp and aircraft models are inadequate and not suited for this situation.

The energy collection system uses counter-rotating turbines connected to a generator, coupled solely through the magnetism of the stator and the rotor. The interior turbine rotor(s) is connected to the generator stator, and the exterior blade rotors are connected to the generator rotor as shown in Figures 19.1 and 19.2. Power electronic converters on ground convert the energy to 50 or 60 Hz.

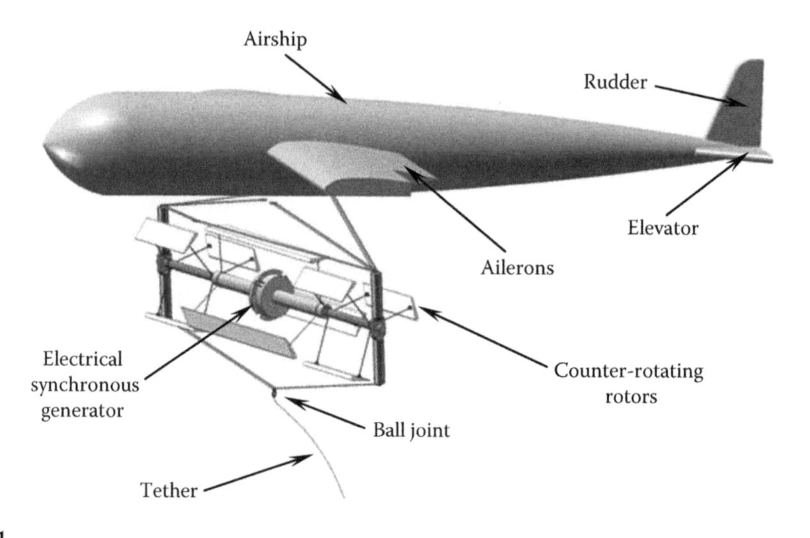

FIGURE 19.1
The EAGLE system. Here, the airship and the rotors are not in the same scale. (From Garcia-Sanz, White, N. and Tierno, N., Airborne wind turbine. US Provisional Patent No. 61/387,432, 2010.; White, N., Tierno, N. and Garcia-Sanz, M., A novel approach to airborne wind energy design and modeling. *IEEE EnergyTech Conference*, Cleveland, OH, 2011; Garcia-Sanz, M., White, N. and Tierno, N., A novel approach to airborne wind energy: Overview of the EAGLE system project. *63rd National Aerospace and Electronics Conference, NAECON*, Dayton, OH, 2011; Tierno, N., White, N. and Garcia-Sanz, M., Longitudinal flight control for a novel airborne wind energy system: Nonlinear and robust MIMO control design techniques. *ASME International Mechanical Engineering Congress & Exposition, IMECE*, Denver, CO, 2011.)

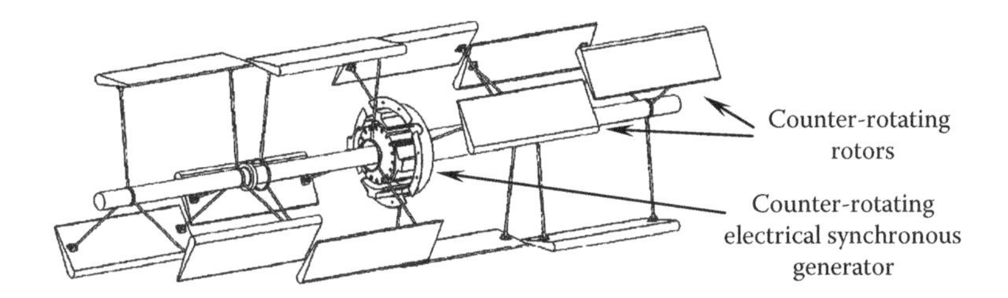

FIGURE 19.2
Energy collection rotors of the *EAGLE system*. The left generator housing is removed to demonstrate features of the system. (From Garcia-Sanz, White, N. and Tierno, N., Airborne wind turbine. US Provisional Patent No. 61/387,432, 2010.; White, N., Tierno, N. and Garcia-Sanz, M., A novel approach to airborne wind energy design and modeling. *IEEE EnergyTech Conference*, Cleveland, OH, 2011; Garcia-Sanz, M., White, N. and Tierno, N., A novel approach to airborne wind energy: Overview of the EAGLE system project. *63rd National Aerospace and Electronics Conference, NAECON*, Dayton, OH, 2011; Tierno, N., White, N. and Garcia-Sanz, M., Longitudinal flight control for a novel airborne wind energy system: Nonlinear and robust MIMO control design techniques. *ASME International Mechanical Engineering Congress & Exposition, IMECE*, Denver, CO, 2011.)

The internal and external turbines rotate in opposite directions, effectively doubling the generator relative speed and minimizing net angular momentum and gyroscopic effects. The magnetic, as opposed to mechanical link, between the stator and rotor removes mechanical constraints present with a geared option. Finally, symmetry of the energy collection systems results in a close to zero net torque about the yaw axis due to drag on the turbines.

Although not demonstrated in Figures 19.1 and 19.2, using helically curved blades results in a more constant net force in the x direction (wind direction) and z direction (vertical direction), as opposed to straight blades, because similar blade area is exposed over an entire revolution. This novel energy collection system exhibits beneficial properties for an AWE system where generating reaction forces are limited.

The airship presented here combines both a low-drag airship with aerodynamic wings to help provide stability. Stability when designing the *EAGLE system* is achieved through neutral buoyancy; static longitudinal stability; and collocating axially the center of gravity, center of buoyancy, and tether point. Neutral buoyancy occurs when the buoyancy force is greater than the weight force and is achieved by creating a large lighter-than-air structure.

The flyer is statically stable when the angle of attack changes and a moment is generated in the opposite direction to restore the angle of attack. This is achieved by ensuring that the overall moment coefficient is negative. The sign of the moment coefficient is a function of the neutral point, which in itself is a function of aerodynamic surfaces and the position of the center of gravity. The center of gravity is axially coincident with the center of buoyancy and the tether point.

Combining the wing with a traditional low-drag blimp shape allows for increased operational conditions and a more straightforward control system approach than if either shape was used independently.

Exclusively utilizing a lighter-than-air wing requires an enormous wing span to achieve neutral buoyancy; exclusively utilizing a traditional low-drag elliptical blimp shape causes large variations in tether angle. The hybrid design provides an ample amount of buoyant volume while still maintaining the tether angle.

Although using a sphere or ellipsoid rather than an elliptical blimp results in less surface area, this causes a much greater coefficient of drag. Simulation indicated that under zero wind conditions the *EAGLE system* is vertical and under the heaviest wind conditions the tether angle reaches a maximum of 3°.

To assist in the goal of neutral buoyancy and coaxial location of the center of gravity, center of buoyancy, and tether point, a thick cambered airfoil section, NACA 2418, is selected. The cambered shape allows lift to be generated at 0° angle of attack.

The material used for the tether is Kevlar K-49, which allows a diameter of 0.005 m to support forces of 30,000 N under 50 m/s wind conditions. A stiffness constant of 21,990 N/m for that tether is calculated by using basic stress and strain equations. The tether has also internal wires for power transmission.

The tether is secured to the ground and attached to the AWE system through a ball joint at the bottom of the energy collection system (see Figure 19.1). The ball joint helps the stability of the system, so that only forces and not torques are transmitted. Therefore, the net force transmitted to the tether from the airship is simply the sum of aerodynamic and gravitational forces.

Advanced longitudinal and lateral flight control systems are needed to increase the longevity of the system, specifically with regard to tether durability through increasing damping and making the system stable during periods of rapid wind transition. They are based on the nonlinear switching and robust sequential MIMO QFT (multiple-input-multiple-output quantitative feedback theory) control techniques presented in the first part of the book (see also similar successful applications developed by the authors, in Ref. 183).

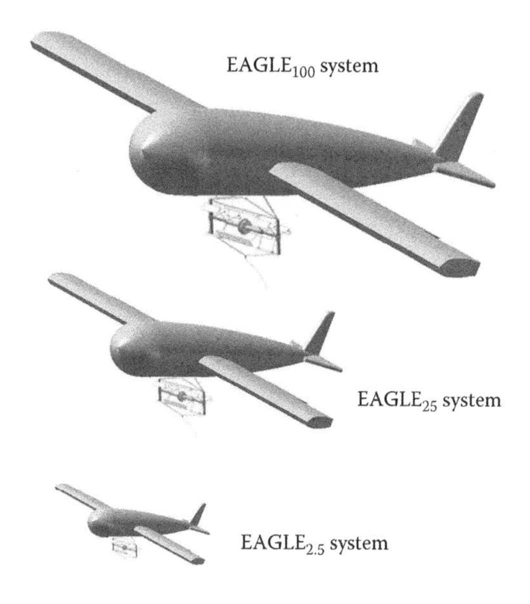

FIGURE 19.3
The EAGLE system: 2.5, 25, and 100 kW units. (From Garcia-Sanz, White, N. and Tierno, N., Airborne wind turbine. US Provisional Patent No. 61/387,432, 2010.; White, N., Tierno, N. and Garcia-Sanz, M., A novel approach to airborne wind energy design and modeling. *IEEE EnergyTech Conference*, Cleveland, OH, 2011; Garcia-Sanz, M., White, N. and Tierno, N., A novel approach to airborne wind energy: Overview of the EAGLE system project. *63rd National Aerospace and Electronics Conference, NAECON*, Dayton, OH, 2011; Tierno, N., White, N. and Garcia-Sanz, M., Longitudinal flight control for a novel airborne wind energy system: Nonlinear and robust MIMO control design techniques. *ASME International Mechanical Engineering Congress & Exposition, IMECE*, Denver, CO, 2011.)

TABLE 19.2

Main Characteristics of the *EAGLE System*

Rater Power Output (kW)	2.5	25	100
Wing area (m^2)	56	203	504
Wing span (m)	16	31.3	48
Net lift at zero wind (N)	92	431	2756
Length (m)	18.5	35	55.5
Buoyant volume (m^3)	78.5	538	2120
Blimp surface area (m^2)	210	760	1900
Energy collection mass (kg)	40	400	1600
Blimp mass (kg)[a]	30	108	270
Tether mass (kg)	2.8	11	71
Total mass (kg)	73	520	1940
kg/kW	29	21	19

Operating height (m): between 150 and 300; operational wind speed (m/s): cutin, 3 m/s; cutoff, 24 m/s; and rated, 12 m/s; number of counter-rotating rotors: three (Figure 19.1) or four (Figure. 19.2) depending on configuration.

[a] In this table, blimp mass solely accounts for blimp skin and not additional structural requirements.

The *EAGLE system* has been design so far for three different environments and sizes: 2.5, 25, and 100 kW of power (see Figure 19.3). The main characteristics of them are shown in Table 19.2.

19.4 Summary

This chapter analyzed the main projects and characteristics of AWE systems, including (a) an overview of the field and (b) the design of a novel airborne WT: *the EAGLE system.*

Appendix A: Templates Generation[38]

A.1 Analysis of the Template Contour

One of the main tasks when designing QFT robust controllers is the definition of the plant frequency response, taking into account the complete n-dimensional space of model plant uncertainty. Consider an uncertain plant P, where $P \in \mathcal{P}$ and $\mathcal{P}$ is the set of possible plants due to uncertainty, such that

$$\mathcal{P} = \left\{ \begin{matrix} P(j\omega_i, a_0, \ldots, a_l, b_0, \ldots, b_m),\ a_0 \in [a_{0\_min}, a_{0\_max}], \ldots b_m \in [b_{m\_min}, b_{m\_max}]; \\ a_i, b_i \in \Re,\ \omega_i \in \Re^+, \quad i, l, m \in N \end{matrix} \right\} \tag{A.1}$$

$\mathcal{T}\ P(j\omega_i)$ is the associated *template* for $\omega = \omega_i$, that is to say, the set of complex numbers representing the frequency response of the uncertain plant at a fixed frequency ω_i in the Nichols chart. In other words, it is the projection of the n-dimensional parameter space, through the transfer function $P(j\omega_i)$ at a given frequency ω_i, onto $\Re^2$, the Nichols chart (see Figure A.1 and Refs. 33 through 41).

In general, it is not true that the projection of the contour of the uncertain parameter space is the boundary of the template. In fact, it could happen that inner points of the parameter space, after the projection, are points of the boundary of the template.

In this context, assuming that the rank of the Jacobian matrix $\boldsymbol{M}$ (see Equation A.2) of the projection of the uncertain parameter space onto the complex map $\Re^2$, according to the plant transfer function P, is 2, the inner points of the initial space are also inner points of the projected space as illustrated in Figure A.1.[38] To be more precise, research of the minimum dimension k that represents the template is reduced to the study of the rank of several submatrices of $\boldsymbol{M}$, so that

$$\boldsymbol{M} = \begin{bmatrix} \dfrac{\partial Re[P(j\omega_i, a_0, \ldots, a_l, b_0, \ldots, b_m)]}{\partial a_0} & \cdots & \dfrac{\partial Re[P(j\omega_i, a_0, \ldots, a_l, b_0, \ldots, b_m)]}{\partial b_m} \\ \dfrac{\partial Im[P(j\omega_i, a_0, \ldots, a_l, b_0, \ldots, b_m)]}{\partial a_0} & \cdots & \dfrac{\partial Im[P(j\omega_i, a_0, \ldots, a_l, b_0, \ldots, b_m)]}{\partial b_m} \end{bmatrix}_{2xn} \tag{A.2}$$

where $n = l + m + 2$.

To find such a minimum dimension k, an iterative procedure from $k = n - 1$ to $k = 1$ is introduced.[38] It involves symbolic calculus. For each step, it is checked whether the rank

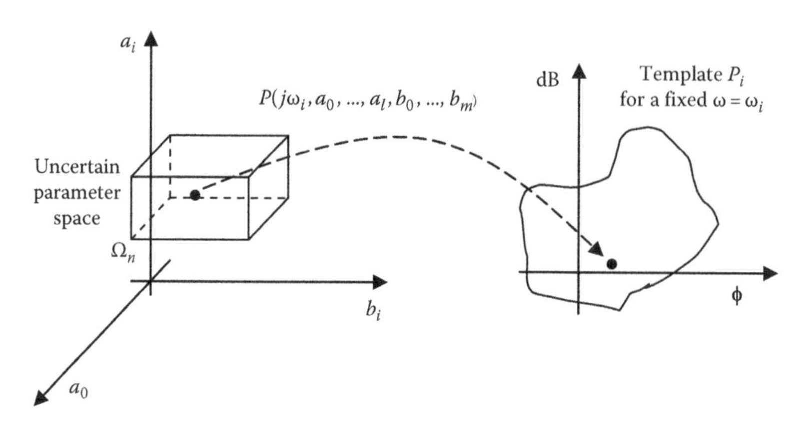

FIGURE A.1
Projection of $\mathbf{\Omega}_n$ onto $\Re^2$.

of the $\binom{n}{k+1}$ possible submatrices of $k + 1$ columns of $\boldsymbol{M}$ is 2 for every combination of the $k + 1$ uncertain parameters that correspond with those $k + 1$ columns of each submatrix. Such a checking must be satisfied for the whole 2^{n-k-1} possible combinations of both minimum and maximum of the $n - k - 1$ non-considered parameters in each submatrix.

When the aforementioned condition is satisfied, the projection of the inner points of the $\binom{n}{k+1}2^{n-k-1}$ sets $\mathbf{\Omega}_{k+1}$ onto $\Re^2$ will be within the template contour. Hence the $\binom{n}{k}2^{n-k}$ sets $\mathbf{\Omega}_k$ will be enough to define the template contour. The number of points that could be removed from the template is $\sum_{i=k+1}^{n}\left\{\binom{n}{i}2^{n-i}(r-2)^i\right\}$, where i corresponds to an iteration that fulfills the rank condition and where k goes from $n - 1$ to 1.

A.2 Example

In order to clarify the aforementioned ideas, let us consider an example that does not fulfill the rank condition (rank($\boldsymbol{M}$) = 2). This case corresponds to the following second-order system with time delay:

$$P(s) = \frac{1}{s^2 + 2\zeta\omega_n s + \omega_n^2}\exp(-\tau s) \tag{A.3}$$

where

$\zeta = 0.02$
$\omega_n = [0.7, 1.2]$
$\tau = [0, 2]$

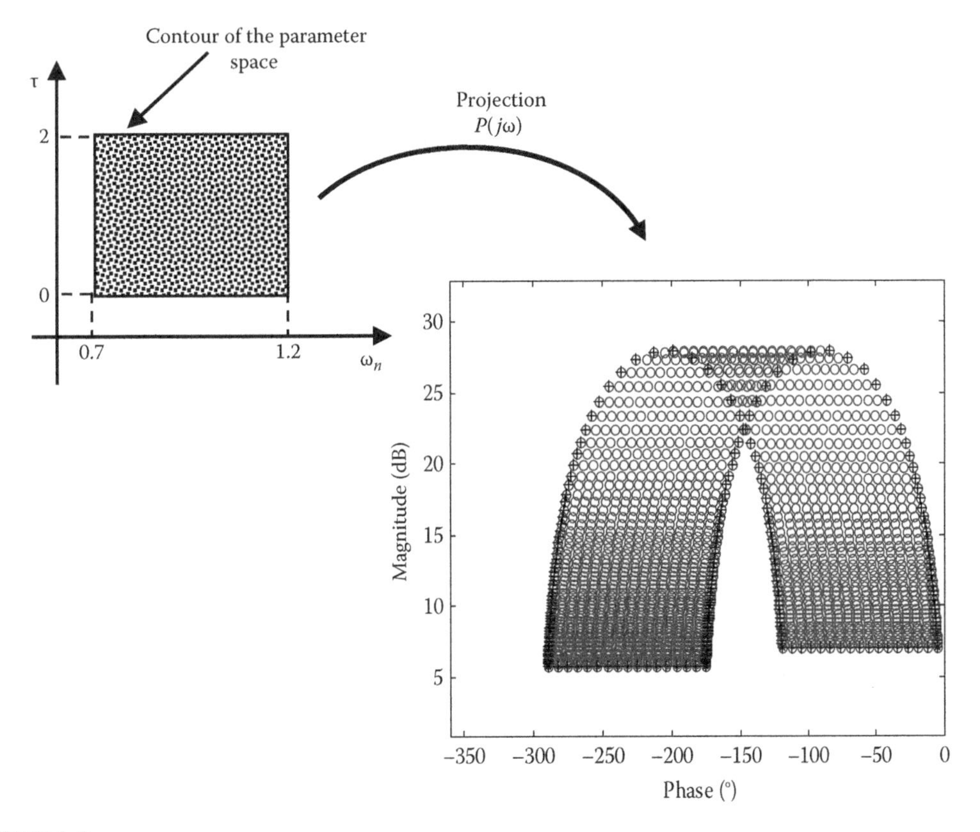

FIGURE A.2
Projection of the whole two-dimensional parameter space (o) and of only the parameter space contour (+) onto $\Re^2$ (Nichols chart).

Figure A.2 shows the template that corresponds to the frequency of 1 rad/s. Cross points (+) in the figure are the projections of the parameter space contour and circle points (o) are the projections of the whole parameter space. It is observed that the projection of the contour (+) of the parameter space does not match the boundary at the top of the template. Using the previous method[38] it is possible to detect the problem. The top points of the template make rank(M) < 2.

Appendix B: Inequality Bound Expressions[48,76]

B.1 Robust Control Performance Specifications

Consider the two-degree-of-freedom (2DOF) feedback system shown in Figure B.1. In a general case, the transfer function $P(s)$ represents an uncertain plant, where $P \in \mathcal{P}$, and $\mathcal{P}$ is the set of possible plants due to uncertainty. The compensator $G(s)$ and the prefilter $F(s)$ are designed to meet robust stability and robust performance specifications, following the desired reference $R(s)$, rejecting the disturbances $D_{1,2}(s)$ and the signal noise $N(s)$, using a limited control signal $U(s)$, and minimizing the "cost of feedback" (excessive bandwidth).

A fundamental step in the QFT methodology is the representation of the control objectives, modified with the model plant uncertainty, by some lines at every frequency on the Nichols chart. These lines are called *bounds* or *Horowitz–Sidi bounds* (see Refs. 42 through 63). They synthesize the performance specifications and the model uncertainty and allow the designer to use only one plant, the nominal plant $P_0(s)$, to design the compensator (controller).

Initially, designers used to calculate the bounds by using graphical manipulation of the specifications and the *templates* on the Nichols chart, as described in Chapter 3. In 1993, Chait and Yaniv[48] developed an iterative algorithm to compute the bounds through quadratic inequalities. Afterwards the software CAD packages implemented that algorithm or similar approaches for solving the bounds representation (see also Appendix C).[144,145,147,149,150]

Closed-loop specifications are usually described in terms of frequency functions $\delta_k(\omega)$ that are imposed on the magnitude of the system transfer functions $|T_k(j\omega)|$.

Table B.1 shows the stability and performance specifications in terms of transfer functions: $|T_k(j\omega)| \leq \delta_k(\omega)$, $k = 1,\ldots,5$. $\delta_1(\omega)$ restricts the transfer function $|L/(1 + L)|$, $L = PG$, implying conditions on the robust stability, the control effort in the input disturbance rejection ($|U/D_1|$), and the sensor noise attenuation ($|Y/N|$) (Equation B.1). $\delta_2(\omega)$ and $\delta_3(\omega)$ constrain the transfer functions $|1/(1 + L)|$ and $|P/(1 + L)|$, respectively, for output disturbance rejection ($|Y/D_2|$) (Equation B.2) and input disturbance rejection ($|Y/D_1|$) (Equation B.3). $\delta_4(\omega)$ restricts the control signal $|G/(1 + L)|$ for the system output disturbance rejection ($|U/D_2|$), the noise attenuation ($|U/N|$), and the tracking of reference signals ($|U/RF|$) (Equation B.4). The upper $\delta_{5sup}(\omega)$ and lower $\delta_{5inf}(\omega)$ models constrain the signal tracking ($|Y/R|$) (Equation B.5).

B.2 Quadratic Inequalities

Every plant in the ω_i-template can be expressed in its polar form as $P(j\omega_i) = pe^{j\theta} = p\angle\theta$. Likewise, the compensator polar form is $G(j\omega_i) = ge^{j\phi} = g\angle\phi$. By substituting these in Equations B.1 through B.5 and rearranging the expressions, the quadratic inequalities of Equations B.6 through B.10 in Table B.2 are calculated.

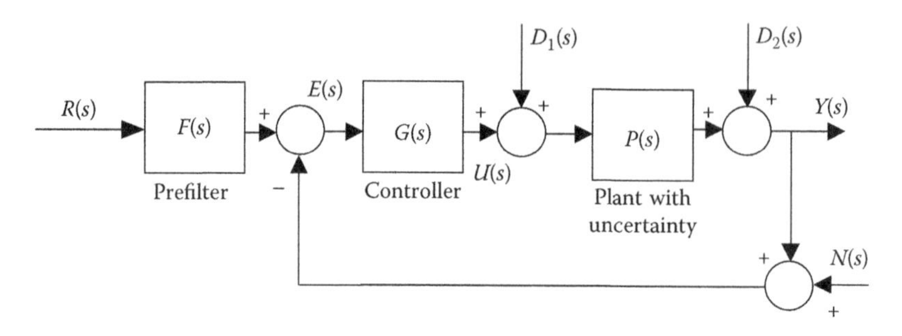

FIGURE B.1
2DOF control system structure.

TABLE B.1
Feedback Control Specifications

Transfer Functions and Specification Models	Equation
$\lvert T_1(j\omega)\rvert = \left\lvert\frac{L(j\omega)}{1+L(j\omega)}\right\rvert = \left\lvert\frac{U(j\omega)}{D_1(j\omega)}\right\rvert = \left\lvert\frac{Y(j\omega)}{N(j\omega)}\right\rvert \le \delta_1(\omega),\quad \omega\in\Omega_1$	(B.1)
$\lvert T_2(j\omega)\rvert = \left\lvert\frac{1}{1+L(j\omega)}\right\rvert = \left\lvert\frac{Y(j\omega)}{D_2(j\omega)}\right\rvert \le \delta_2(\omega),\quad \omega\in\Omega_2$	(B.2)
$\lvert T_3(j\omega)\rvert = \left\lvert\frac{P(j\omega)}{1+L(j\omega)}\right\rvert = \left\lvert\frac{Y(j\omega)}{D_1(j\omega)}\right\rvert \le \delta_3(\omega),\quad \omega\in\Omega_3$	(B.3)
$\lvert T_4(j\omega)\rvert = \left\lvert\frac{G(j\omega)}{1+L(j\omega)}\right\rvert = \left\lvert\frac{U(j\omega)}{D_2(j\omega)}\right\rvert = \left\lvert\frac{U(j\omega)}{N(j\omega)}\right\rvert = \left\lvert\frac{U(j\omega)}{R(j\omega)F(j\omega)}\right\rvert \le \delta_4(\omega),\quad \omega\in\Omega_4$	(B.4)
$\delta_{5inf}(\omega) \le \lvert T_5(j\omega)\rvert = \left\lvert F(j\omega)\frac{L(j\omega)}{1+L(j\omega)}\right\rvert = \left\lvert\frac{Y(j\omega)}{R(j\omega)}\right\rvert \le \delta_{5sup}(\omega),\quad \omega\in\Omega_5$	(B.5a)
$\frac{\max_P \lvert T(j\omega)\rvert}{\min_P \lvert T(j\omega)\rvert} = \frac{\left\lvert\frac{P_d(j\omega)G(j\omega)}{1+P_d(j\omega)G(j\omega)}\right\rvert}{\left\lvert\frac{P_e(j\omega)G(j\omega)}{1+P_e(j\omega)G(j\omega)}\right\rvert} \le \delta_5(\omega) = \frac{\delta_{5sup}(\omega)}{\delta_{5inf}(\omega)},\quad \omega\in\Omega_5$	(B.5b)

TABLE B.2
Bound Quadratic Inequalities

k	Bound Quadratic Inequality	Equation
1	$p^2\left(1-\frac{1}{\delta_1^2}\right)g^2 + 2p\cos(\phi+\theta)g + 1 \ge 0$	(B.6)
2	$p^2g^2 + 2p\cos(\phi+\theta)g + \left(1-\frac{1}{\delta_2^2}\right) \ge 0$	(B.7)
3	$p^2g^2 + 2p\cos(\phi+\theta)g + \left(1-\frac{p^2}{\delta_3^2}\right) \ge 0$	(B.8)
4	$\left(p^2-\frac{1}{\delta_4^2}\right)g^2 + 2p\cos(\phi+\theta)g + 1 \ge 0$	(B.9)
5	$p_e^2p_d^2\left(1-\frac{1}{\delta_5^2}\right)g^2 + 2p_ep_d\left(p_e\cos(\phi+\theta_d) - \frac{p_d}{\delta_5^2}\cos(\phi+\theta_e)\right)g + \left(p_e^2 - \frac{p_d^2}{\delta_5^2}\right) \ge 0$	(B.10)

TABLE B.3
g-Bound Formulation

k	$G_{1,2}$ Bound Expressions	Equation
1	$g_{1,2}=\frac{1}{p(1-(1/\delta_1^2))}\left(-\cos(\phi+\theta)\mp\sqrt{\cos^2(\phi+\theta)-\left(1-\frac{1}{\delta_1^2}\right)}\right)$	(B.11)
2	$g_{1,2}=\frac{1}{p}\left(-\cos(\phi+\theta)\mp\sqrt{\cos^2(\phi+\theta)-\left(1-\frac{1}{\delta_2^2}\right)}\right)$	(B.12)
3	$g_{1,2}=\frac{1}{p}\left(-\cos(\phi+\theta)\mp\sqrt{\cos^2(\phi+\theta)-\left(1-\frac{p^2}{\delta_3^2}\right)}\right)$	(B.13)
4	$g_{1,2}=\frac{1}{p\left(1-\left(1/p^2\delta_4^2\right)\right)}\left(-\cos(\phi+\theta)\mp\sqrt{\cos^2(\phi+\theta)-\left(1-\frac{1}{p^2\delta_4^2}\right)}\right)$	(B.14)
5	$g_{1,2}=\frac{-1}{p_e p_d\left(1-\left(1/\delta_5^2\right)\right)}\left(p_e\cos(\phi+\theta_d)-\frac{p_d}{\delta_5^2}\cos(\phi+\theta_e)\right)$ $\mp\sqrt{\left(p_e\cos(\phi+\theta_d)-\frac{p_d}{\delta_5^2}\cos(\phi+\theta_e)\right)-\left(1-\frac{1}{\delta_5^2}\right)\left(p_e^2-\frac{p_d^2}{\delta_5^2}\right)}$	(B.15)

For every frequency ω_i there is a constant $\delta_k = \delta_k(\omega_i)$, and for a fixed plant $p\angle\theta$ in the ω_i-template and a fixed controller phase ϕ in $[-360°, 0°]$, the unknown parameter of the inequalities in Table B.2 is the controller magnitude g. Then, solving the equality $ag^2 + bg + c = 0$ the set of ω_i-bounds for $\{\delta_{k=1,\ldots,5}\}$ is computed. The two possible solutions, g_1 and g_2, of the quadratic inequalities of Table B.2, for every feedback problem in Table B.1, are shown in Table B.3. Table B.4 shows the algorithm to compute the bounds.

B.3 Bound Typologies

Choosing the real and positive solutions of g_1 and g_2 in Table B.3 as effective compensator restrictions, the bounds are classified into four cases: o, n, ō, and u *typology bounds*[76] (see Figure B.2 and Table B.5).

A bound plotted with a *solid line* implies that $G(j\omega)$, or $L_0(j\omega)$, must lie above or on it to meet the specification, while a bound plotted with a *dashed line* means that $G(j\omega)$, or $L_0(j\omega)$, must lie below or on it.

The simultaneous meeting of general robust feedback control requirements (robust reference tracking, disturbance rejection, robust stability, and robust control effort minimization) can be analyzed with these bound typologies.[76]

TABLE B.4

Algorithm to Compute the Bounds

1. Discretize the frequency domain ω into a finite set $\Omega_k = \{\omega_i, i = 1, \ldots, n\}_k$.
2. Establish the uncertain plant models $\{P(j\omega)\}$ and map its boundary for each frequency $\omega_i \in \Omega_k$ on the Nichols chart: They are the n templates $P(j\omega_i)\}$, $i = 1, \ldots, n$. Each one contains m points or plants: $P(j\omega_i) = \{P_r(j\omega_i) = p\angle\theta, r = 0, \ldots, m-1\}$.
3. Select one plant as the nominal plant $P_0(j\omega_i) = p_0\angle\theta_0$.
4. The compensator is $G(j\omega_i) = g\angle\phi$. Define a range, Φ, for the compensator's phase ϕ, and discretize it, like $\phi \in \Phi = [-360°: 5°: 0°]$.
5. First *For-loop*: (for the frequencies $\Omega_k = \{\omega_i, i = 1, \ldots, n\}_k$)
   Choose a single frequency $\omega_i \in \Omega_k$.
6. Second *For-loop*: (for the phases $\phi \in \Phi = [-360°: 5°: 0°]$)
   Choose a single controller's phase $\phi \in \Phi$.
7. Third *For-loop*: (for the m plants $P_r(j\omega_i)$, $r = 0, \ldots, m-1$)
   Choose a single plant: $P_r(j\omega_i) = p\angle\theta$.
8. Compute the maximum $g_{max} = g_{max}(P_r)$ and the minimum $g_{min} = g_{min}(P_r)$ of the two roots g_1 and g_2 that solve the k quadratic inequality.
9. Choose the most restrictive of the m $g_{max}(P_r)$ and the m $g_{min}(P_r)$. Thus, $g_{max}(P)$ and $g_{min}(P)$ are obtained. They are the maximum and minimum bound points for the controller magnitude g at a phase ϕ.
10. The union of $g_{max}(P)$ and $g_{min}(P)$ along $\phi \in \Phi$ gives $g_{max}\angle\phi$ and $g_{min}\angle\phi$ respectively, for each frequency ω_i.
11. The open loop transmission is $L_0(j\omega_i) = l_0\angle\psi_0$.
    Now set $l_{0max}\angle\psi_0 = p_0 g_{max}\angle\phi$ and $l_{0min}\angle\psi_0 = p_0 g_{min}\angle\phi$, being $\psi_0 = \phi + \theta_0$, $\phi = [-360° : 5° : 0]$.
    These are the bounds: $B_k(j\omega_i)$, $\omega_i \in \Omega_k\}$.

Source: Chait, Y. and Yaniv, O., *Int. J. Robust Nonlinear Control*, 3(3), 47, 1993.

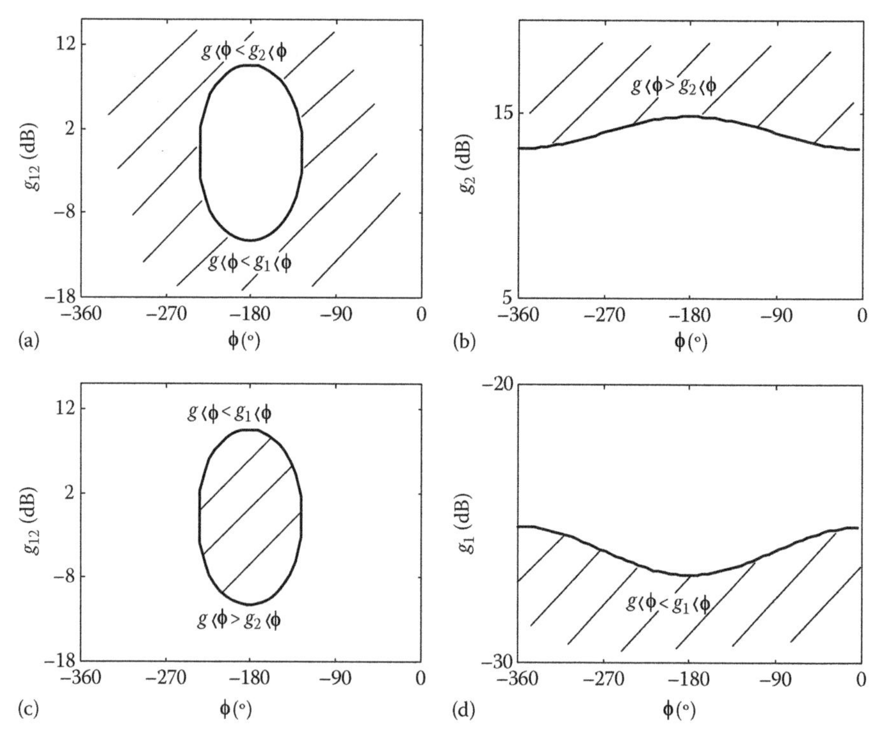

FIGURE B.2
Bound typologies: (a) typology o, (b) typology n, (c) typology $\bar{o}$, and (d) typlogy u.

TABLE B.5

Solutions to Quadratic Inequalities and Bound Typologies

sign $a = a(p,\delta)$	$g_{min} = \min(g_1, g_2)$	$g_{max} = \max(g_1, g_2)$	g	Bound	Typology
Any	Complex	Complex	$g \geq 0$	No	—
Any	Real, < 0	Real, ≤ 0	$g \geq 0$	No	—
≥0	Real, < 0	Real, ≥ 0	$g \geq g_{max}$	Upper	n
	Real, > 0	Real, ≥ 0	$g \geq g_{max}$ and $g \leq g_{min}$	Outer	o
< 0	Real, < 0	Real, ≥ 0	$g \leq g_{max}$	Lower	u
	Real, > 0	Real, ≥ 0	$g \leq g_{max}$ and $g \geq g_{min}$	Inner	$\bar{o}$

For every control objective, two bound types are possible with dependence on the specification model ratio and the uncertainty size. All in all, the following three different bound typologies are found:

1. Type n bound (single- or multi-valued upper bounds). It tries to obtain the feedback benefits by raising the open-loop gain at low and medium frequencies.
2. Type o bound (single- or multi-valued outer bounds). It encloses a forbidden area around the critical point (0 dB, −180°) to ensure a robust stability. It allows reducing the open-loop high-frequency gain to minimize the cost of feedback.
3. Type u bound (single- or multi-valued lower bounds). It explicitly imposes the open-loop gain reduction needed to reduce the control effort due to different inputs at any frequency, and in particular the cost of feedback at high frequencies.

On this basis, some general hints can be established to quantify the trade-off of the control requirements simultaneously achievable.[76] For a certain specification k, the least favorable intersection among the set of bounds for $\Im P(j\omega_i) = \{p\angle\theta\}$ is established in Table B.6 as $\max\{g_2\}$ for type n bounds, $\min\{g_1\}$ for type u bounds, or both for type o bounds, where g_1 and/or g_2 are the real and positive values calculated in Table B.3.

TABLE B.6

Bound Typologies for General Feedback Specifications

k	$\delta_k(\omega_i)$	Type	g-Bound; g_1 and g_2
1	$0 < \delta_1 < 1$	u	$(g\angle\phi) \leq (\min\{g_1\}\angle\phi)$; $\phi \in \Phi = [-360°, 0°]$
	$\delta_1 > 1$	o	$(g\angle\phi) \geq (\max\{g_2\}\angle\phi)$ $(g\angle\phi) \leq (\min\{g_1\}\angle\phi)$ $\phi \in \Phi_{12} = [-180° - \{\theta\} \mp \varepsilon]$
2	$0 < \delta_2 < 1$	n	$(g\angle\phi) \geq (\max\{g_2\}\angle\phi)$; $\phi \in \Phi = [-360°, 0°]$
	$\delta_2 > 1$	o	$(g\angle\phi) \geq (\max\{g_2\}\angle\phi)$; $(g\angle\phi) \leq (\min\{g_1\}\angle\phi)$ $\phi \in \Phi_{12} = [-180° - \{\theta\} \mp \varepsilon]$
3	$p > \delta_3$	n	$(g\angle\phi) \geq (\max\{g_2\}\angle\phi)$; $\phi \in \Phi = [-360°, 0°]$
	$p < \delta_3$	o	$(g\angle\phi) \geq (\max\{g_2\}\angle\phi)$; $(g\angle\phi) \leq (\min\{g_1\}\angle\phi)$ $\phi \in \Phi_{12} = [-180° - \{\theta\} \mp \varepsilon]$
4	$p < 1/\delta_4$	u	$(g\angle\phi) \leq (\min\{g_1\}\angle\phi)$; $\phi \in \Phi = [-360\}°, 0°]$
	$p > 1/\delta_4$	o	$(g\angle\phi) \geq (\max\{g_2\}\angle\phi)$; $(g\angle\phi) \leq (g_1\angle\phi)$ $\phi \in \Phi_{12} = [-180° - \{\theta\} \mp \varepsilon]$
5	$\frac{p_{max}}{p_{min}} > \delta_5$	n	$(g\angle\phi) \geq (\max\{g_2\}\angle\phi)$ $\phi \in \Phi = [-360°, 0°]$
	$\frac{p_{max}}{p_{min}} < \delta_5$	o	$(g\angle\phi) \geq (\max\{g_2\}\angle\phi)$; $(g\angle\phi) \leq (\min\{g_1\}\angle\phi)$ $\phi \in \Phi_{12} = [-180° - \{\theta\} \mp \varepsilon]$

Appendix C: Analytical QFT Bounds[61]

C.1 Introduction

This appendix describes an analytical formulation to compute quantitative feedback theory (QFT) bounds in one-degree-of-freedom (1DOF) feedback control problems. The approach is based on envelope curves and shows that a control performance specification can be expressed as a family of circumferences. Then, the controller bound is defined by the envelope curve of this family and can be obtained as an analytical function. Such an approach offers the possibility of studying the QFT bounds in an analytical way with useful properties. Gridding methods are avoided, resulting in a lower computational effort procedure. The formulation obtains analytical QFT bounds for 1DOF control cases and improves the accuracy of previous methods, allowing the designer to calculate multi-valued bounds as well.

QFT bounds are the limits (maximum and/or minimum) of the open-loop nominal transfer function $L_0(j\omega) = P_0(j\omega)G(j\omega)$ at each frequency that guarantee to meet the required closed-loop control performance specifications for all the plants in the uncertainty set (see Chapter 3 and Appendix B). QFT specifications for minimum-phase systems are formulated as inequalities of certain magnitude transfer functions for the desired closed-loop frequency response. Using these specifications and taking the templates into account (model plus uncertainty), the QFT bounds are computed, one for each robust control specification and frequency. An ω_i-template is the set of all possible values of the uncertain plant at frequency ω_i (see Chapter 3 and Appendix A). Only the contour of this ω_i-template is necessary to design the controller. This contour of the template at ω_i is labeled $\mathcal{T}_{\omega_i}$ (see Refs 33 through 41).

Originally, QFT bounds were calculated by using manual graphical manipulations of plant templates on the Nichols chart (see Chapter 3). Then, several researches developed some algorithms for automatic bound computation. Early bound generation algorithms used geometrical and/or search-based CAD techniques.[42,46,65,141] Afterward, more efficient numerical algorithms based on quadratic inequalities were proposed (see Appendix B).[48,105,107] These algorithms established a formal process that mapped plant uncertainty and feedback specifications into the bound expressions through quadratic inequalities. They form the basis of the bound computation in the Borghesani et al. QFT Toolbox[144] and the Garcia-Sanz et al. QFT Toolbox[149,150] (included in Appendix F. Visit http://cesc.case.edu or http://www.crcpress.com/product/isbn/9781439821794.). The Qsyn[145] is another QFT Toolbox solution that improves bound calculation by using algorithms based on the cross-section concept.[43,59,60] Algorithms based on quadratic inequalities are only able to find two values of the QFT bound (maximum and minimum values) at one phase $\phi \in [-2\pi, 0]$, whereas those based on cross section calculate all possible values (multi-valued bounds). Some examples of multi-valued bounds are shown for the robust tracking specification in Refs 59 and 60. Other techniques that improve the efficiency and accuracy of the algorithms are shown in Refs 49, 52, and 55. However, all of them present trade-offs between accuracy and computation time: They are based on computational nested loops for discrete sets according to uncertainty, frequency, and the bound phase.

This appendix extends the work presented in Ref. 61 and describes an analytical formulation that uses envelope curves to compute the QFT bounds for 1DOF control systems. It avoids former gridding methods, improving the accuracy and reducing the computational effort.

Section C.2 expounds the envelope method, whereas Section C.3 particularizes the solution for robust noise rejection and robust stability problems. Section C.4 solves the bound computation for control performance specifications based on the sensitivity function (i.e., the robust disturbance rejection at the plant output). The problem of robust disturbance rejection at the plant input is considered in Section C.5. Section C.6 makes the same with the control effort bound and Section C.7 presents a list of advantages in the use of the analytical bound formulation technique. Two examples for different robust control specifications illustrate the methodology in Section C.8. Finally, a summary is presented in Section C.9.

C.2 Envelope Method

Consider the classical 1DOF control diagram shown in Figure C.1, where the plant $P(s)$ has uncertainty, the reference input $r(t)$ is constant, and the remaining blocks are known.

A general SISO QFT control specification can be expressed as

$$\left|\frac{A(j\omega,\phi)+B(j\omega,\phi)G_{\omega_i}}{1+C(j\omega,\phi)G_{\omega_i}}\right| \leq \gamma(\omega_i) \quad \forall\omega_i \in \left|\omega_{min},\omega_{max}\right| \tag{C.1}$$

Note that $\gamma(\omega_i)$ is the upper limit of the control performance specification magnitude at the frequency ω_i, and $A = ae^{j\vartheta a} = A_r + jA_i$, $B = be^{j\vartheta b} = B_r + jB_i$, and $C = ce^{j\vartheta c} = C_r + jC_i$ are complex functions. They depend on the plant template values, which are given by $\mathcal{T}_{\omega_i}(\phi)$. In general, $\mathcal{T}_{\omega_i}(\phi)$ is supposed to be an analytical continuous and differentiable function of $\phi \in [0, 2\pi]$ that defines the template contour at ω_i.[41] Nevertheless, the methodology is also valid for discrete ϕ arrays (discrete templates). The expression $G_{\omega_i} = ge^{j\phi} = g_{re} + jg_{im}$ represents the controller bound for the $\gamma_k(\omega_i)$ control specifications, where $k = 1, 2, \ldots, \alpha$, where α represents the number of specified specifications. As long as each performance specification provides a $G_{\omega_i}^{(\alpha)}$, where

$G_{\omega_i}^{(1)}$ is the bound controller for the specification $\left|\dfrac{PG}{1+PG}\right| \leq \gamma_1$,

$G_{\omega_i}^{(2)}$ for $\left|\dfrac{1}{1+PG}\right| \leq \gamma_2$, $G_{\omega_i}^{(3)}$ for $\left|\dfrac{P}{1+PG}\right| \leq \gamma_3$, and $G_{\omega_i}^{(4)}$ for $\left|\dfrac{G}{1+PG}\right| \leq \gamma_4$

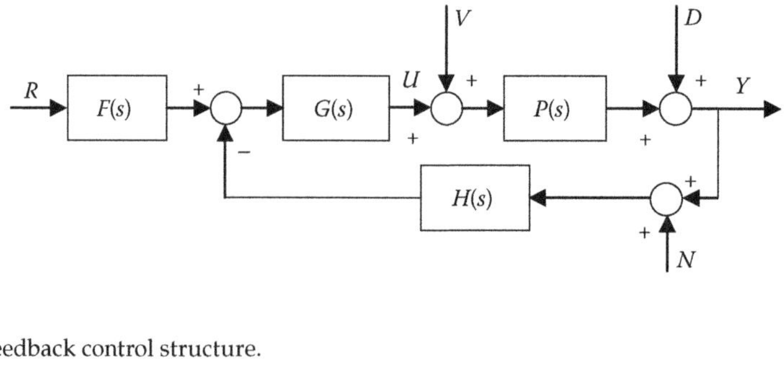

FIGURE C.1
Classical 2DOF feedback control structure.

then, without losing generality, it is assumed that $H(s) = 1$. γ_1, γ_2, γ_3, and γ_4 are magnitude frequency models of the desired closed-loop stability and performance specifications in a classical feedback control structure, where G is the controller and P is the plant. Equation C.1 is rewritten using imaginary and real parts as

$$\left| \frac{A_r + jA_i + (B_r + jB_i)(g_{re} + jg_{im})}{1 + (C_r + jC_i)(g_{re} + jg_{im})} \right| \leq \gamma \tag{C.2}$$

in order to obtain the quadratic inequality

$$\begin{aligned} &[(C_r^2 + C_i^2)\gamma^2 - (B_r^2 + B_i^2)](g_{re}^2 + g_{im}^2) + 2(\gamma^2 C_r - A_r B_r - A_i B_i) g_{re} \\ &-2(\gamma^2 C_i + A_i B_r - A_r B_i) g_{im} + \gamma^2 - A_r^2 - A_i^2 \geq 0 \end{aligned} \tag{C.3}$$

where A_r, A_i, B_r, B_i, C_r, and C_i can be a function of $\mathcal{T}_{\omega_i}$ and depend on a parameter ϕ used to track the plant template contour. Taking Equation C.3 as a mathematical expression that equals zero, a circumference family (one circumference for each ϕ value) is obtained. The structure of this family is

$$(g_{re} - g_{rc})^2 + (g_{im} - g_{ic})^2 = r^2 \tag{C.4}$$

where the center (real and imaginary parts) is defined by

$$g_{rc}(\phi) = \frac{-\gamma^2 C_r + A_r B_r + A_i B_i}{(C_r^2 + C_i^2)\gamma^2 - B_r^2 - B_i^2}; \quad g_{ic}(\phi) = \frac{\gamma^2 C_i + A_i B_r - A_r B_i}{(C_r^2 + C_i^2)\gamma^2 - B_r^2 - B_i^2} \tag{C.5}$$

and the radius is given by

$$r(\phi) = \frac{\gamma\sqrt{(z_1 - z_2)}}{(C_r^2 + C_i^2)\gamma^2 - B_r^2 - B_i^2} \tag{C.6}$$

where

$$z_1 = (A_r^2 + A_i^2)(C_r^2 + C_i^2) + B_r^2 + B_i^2$$

$$z_2 = 2(A_r B_r + A_i B_i) C_r - 2(A_i B_r - A_r B_i) C_i$$

If the plant does not have uncertainty, Equation C.4 is a circumference that delimitates the area where the controller can be defined in the Re–Im plane, that is, the controller bound G_{ω_i}. On the other hand, if there is uncertainty, the controller bound G_{ω_i} is the contour of the circumference family (the envelope curve). Therefore, the bound problem is reduced to computing the envelope curve of the family represented by Equation C.4. The circumference family can be written in a parametric form as

$$\begin{cases} g_{rc}(\phi) + r(\phi)\cos\theta \\ g_{ic}(\phi) + r(\phi)\sin\theta \end{cases} \tag{C.7}$$

where $\theta \in [0, 2\pi]$. Each point of the envelope curve is obtained by computing the intersection of two infinitesimally close circumferences. Then, an infinitesimally close circumference to Equation C.7 is

$$\begin{cases} g_{rc}(\phi+\Delta\phi)+r(\phi+\Delta\phi)\cos\theta \\ g_{ic}(\phi+\Delta\phi)+r(\phi+\Delta\phi)\sin\theta \end{cases} \tag{C.8}$$

$\Delta\phi$ being an infinitesimal ϕ. The intersection between Equations C.7 and C.8 yields

$$\begin{cases} g_{rc}(\phi)+r(\phi)\cos\theta_1 = g_{rc}(\phi+\Delta\phi)+r(\phi+\Delta\phi)\cos\theta_2 \\ g_{ic}(\phi)+r(\phi)\sin\theta_1 = g_{ic}(\phi+\Delta\phi)+r(\phi+\Delta\phi)\sin\theta_2 \end{cases} \tag{C.9}$$

As long as ϕ and $\phi + \Delta\phi$ are infinitesimally close, θ_1 and θ_2 are also infinitesimally close. Therefore, making $\theta_1 = \theta$ and $\theta_2 = \theta + \Delta\theta$, it is found that

$$\begin{cases} g_{rc}(\phi)+r(\phi)\cos\theta = g_{rc}(\phi+\Delta\phi)+r(\phi+\Delta\phi)\cos(\theta+\Delta\theta) \\ g_{ic}(\phi)+r(\phi)\sin\theta = g_{ic}(\phi+\Delta\phi)+r(\phi+\Delta\phi)\sin(\theta+\Delta\theta) \end{cases} \tag{C.10}$$

If it is supposed that $f(x + \Delta x) = f(x) + \Delta f(x) = f(x) + f'(x)\Delta x$, Equation C.10 can be reduced to

$$\begin{cases} g'_{rc}(\phi)\Delta\phi+r'(\phi)\cos\theta\Delta\phi = r(\phi)\sin\theta\Delta\theta+r'(\phi)\sin\theta\Delta\phi\Delta\theta \\ g'_{ic}(\phi)\Delta\phi+r'\sin\theta\Delta\phi = -r(\phi)\cos\theta\Delta\theta+r'(\phi)\cos\theta\Delta\phi\Delta\theta \end{cases} \tag{C.11}$$

where

$$g'_{rc}(\phi)=\frac{dg_{rc}(\phi)}{d\phi},\quad g'_{ic}(\phi)=\frac{dg_{ic}(\phi)}{d\phi},\quad \text{and}\quad r'(\phi)=\frac{dr(\phi)}{d\phi}$$

Obtaining $\Delta\theta$ from each equation and setting both of them equal to each other yields

$$\frac{g'_{rc}(\phi)+r'(\phi)\cos(\theta)}{r(\phi)\sin\theta+r'(\phi)\sin\theta\Delta\phi} = \frac{g'_{ic}(\phi)+r'(\phi)\sin(\theta)}{r(\phi)\cos\theta+r'(\phi)\cos\theta\Delta\phi} \tag{C.12}$$

The limit of Equation C.12 when $\Delta\phi \to 0$ provides

$$g'_{rc}(\phi)\cos\theta+g'_{ic}(\phi)\sin\theta+r'(\phi)=0 \tag{C.13}$$

By solving Equation C.13, the following is obtained:

$$\theta_{1,2}(\phi)=\arctan\left(\frac{-g'_{ic}(\phi)r'(\phi)\mp g'_{rc}(\phi)\sqrt{g'_{ic}(\phi)^2+g'_{rc}(\phi)^2-r'(\phi)^2}}{-g'_{rc}(\phi)r'(\phi)\pm g'_{ic}(\phi)\sqrt{g'_{ic}(\phi)^2+g'_{rc}(\phi)^2-r'(\phi)^2}}\right) \tag{C.14}$$

where

$$g'_{rc}(\phi)=\frac{dg_{rc}(\phi)}{d\phi},\quad g'_{ic}(\phi)=\frac{dg_{ic}(\phi)}{d\phi},\quad \text{and}\quad r'(\phi)=\frac{dr(\phi)}{d\phi}$$

If the template contour $\mathcal{T}_{\omega_i}$ is closed, the curve of the centers (g_{rc}, g_{ic}) is also closed. The two solutions in Equation C.14 provide the outer envelope curve and the inner one. The controller bound is defined by the inner envelope curve if the radius $r(\phi)$ of some circumference of the family is negative and it is defined by the outer one in the other case. Then, the controller bound G_{ω_i} can be expressed as a function of ϕ:

$$G_{\omega_i}(\phi) = \begin{cases} g_{rc}(\phi) + r(\phi)\cos[\theta(\phi)] \\ g_{ic}(\phi) + r(\phi)\sin[\theta(\phi)] \end{cases} \tag{C.15}$$

If $\mathcal{T}_{\omega_i}$ is an open curve, the two solutions and a portion of the extreme circumferences are necessary to calculate the envelope curve. Figure C.2 shows a family of circumferences (dashed lines) obtained from an open template. Solid black lines define the extreme circumferences (the two solutions of Equation C.14). The intersection of the extreme circumferences defines Equation C.15.

The previous calculation of the controller bound assumes that the intersection of two infinitesimally close circumferences exists. However, if the square root of Equation C.14 results in a complex number, some circumferences of the family are inner (or outer) to their infinitesimally close ones and there is no intersection. This situation arises when

$$g'_{ic}(\phi)^2 + g'_{rc}(\phi)^2 - r'(\phi)^2 < 0 \tag{C.16}$$

If the $\mathcal{T}_{\omega_i}(\phi)$ points that keep Equation C.16 do not generate bound points, then they should be eliminated. Thus, a new template contour, $\mathcal{T}^{\epsilon}_{\omega_i}$-labeled *equivalent template,* is obtained. The equivalent template is the set of all the points of the template contour $\mathcal{T}_{\omega_i}$ that generates all the points of the bound. Section C.4 explains how to calculate the controller bound.

Another particular case arises when the template has a deep concavity. Then, the bound obtained by using Equation C.15 could not be a Jordan curve (a simple and closed curve)

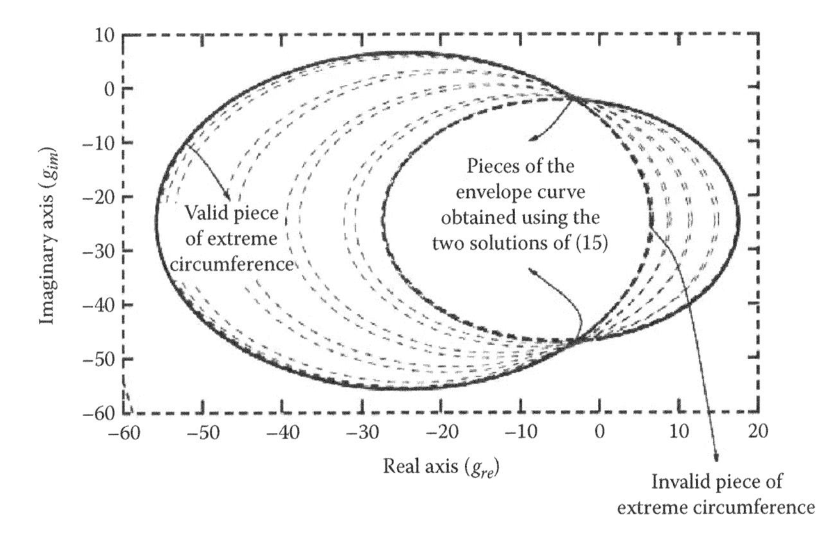

FIGURE C.2
Controller bound from an open template. Note that the curves shown are circumferences but the picture is on a scale that makes them seem ellipses.

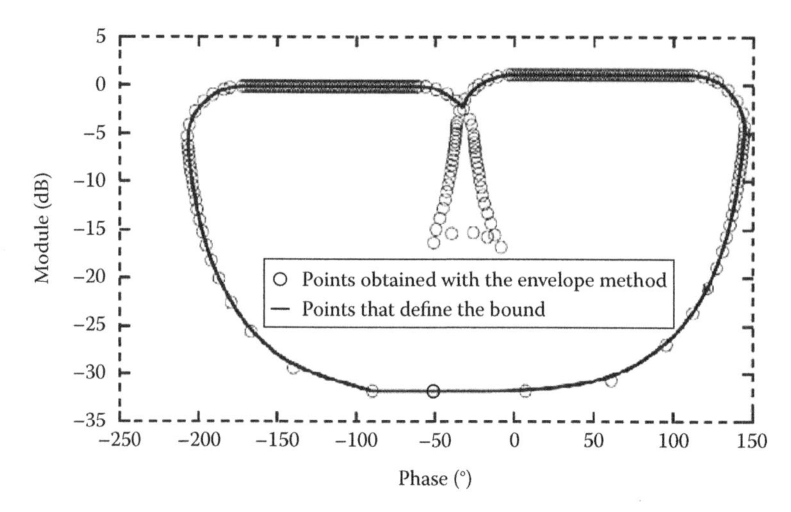

FIGURE C.3
Robust stability bound $B_{\omega_i}^{(1)} = P_0 G_{\omega_i}^{(1)}$ obtained by the envelope method using a template with a strong concavity.

and some pieces of this would not be relevant for the controller design. Figure C.3 illustrates this problem at $s = j\omega = j1$ for the uncertain plant

$$P(s) = \frac{e^{-Ts}}{s^2 + 0.04s + \omega_n^2}$$

which is based on the expression given in Section A.2,[38] where $T \in [0, 2]$, $\omega_n \in [0.7, 1.2]$, and a control performance specification given by

$$\left|\frac{PG}{1+PG}\right| < 1.82$$

The envelope method provides the curve marked with circles; the solid black line shows the effective QFT bound. A complementary analysis of this problem is found in Ref. 40.

Finally, if the template is nonconnected, a controller bound must be found for every piece of it. After that, the intersection of valid surfaces must be computed to find the final controller bound. An example is shown in Section C.4.

C.3 Robust Stability and Noise Rejection Bounds

Robust stability and robust noise attenuation specifications are defined as

$$\left|\frac{P(j\omega_i)G_{\omega_i}}{1+P(j\omega_i)G_{\omega_i}}\right| \le \gamma_1(\omega_i) \qquad \text{(C.17)}$$

where $\gamma_1(\omega_i) = \gamma_1$ is a real number for a fixed frequency ω_i. This number indicates the desired gain (in magnitude) and phase (in degrees) margins, so that $GM \ge 1 + (1/\gamma_1)$

and $PM \geq 180° - 2\,\mathrm{acos}\,[(0.5/\gamma_1^2) - 1]$, and the desired noise reduction at ω_i. Comparing with Equation C.1: $B = C = P(j\omega_i) = \mathcal{T}_{\omega_i} = p_{re} + jp_{im}$ and $A = 0$. Then, Equation C.17 can be written as

$$\left|\frac{(g_{re} jg_{im})[p_{re}(\phi) + jp_{im}(\phi)]}{1 + (g_{re} + jg_{im})[p_{re}(\phi) + jp_{im}(\phi)]}\right| \leq \gamma_1 \qquad \text{(C.18)}$$

The circumference family is

$$\left\{\frac{\left(\left(-p_{re}(\phi)\left(\gamma_1^2/\left(\gamma_1^2-1\right)\right)\right)/\left(p_{re}(\phi)^2+p_{im}(\phi)^2\right)\right)+\left(\left(\gamma_1/\left(\gamma_1^2-1\right)\right)/\left(\sqrt{p_{re}(\phi)^2+p_{im}(\phi)^2}\right)\cos(\theta)\right)}{\left(\left(p_{im}(\phi)\left(\gamma_1^2/\gamma_1^2-1\right)\right)/\left(p_{re}(\phi)^2+p_{im}(\phi)^2\right)\right)+\left(\left(\gamma_1/\left(\gamma_1^2-1\right)\right)/\left(\sqrt{p_{re}(\phi)^2+p_{im}(\phi)^2}\,\sin(\theta)\right)\right)}\right.$$

(C.19)

The solutions in Equation C.14 are valid if Equation C.16 is not satisfied. The worst case appears when the template comes from an uncertain gain because

$$r(\phi) = \frac{\left(\gamma_1/(\gamma_1^2-1)\right)}{\sqrt{p_{re}(\phi)^2 + p_{im}(\phi)^2}}$$

and its maximum variation $r'(\phi)$ appears with a maximum change of the module $\sqrt{p_{re}(\phi)^2 + p_{im}(\phi)^2}$. Thus, the most unfavorable template is

$$\mathcal{T}_{\omega_i}(\phi) = \begin{cases} m_1\phi \\ m_2\phi \end{cases} \qquad \text{(C.20)}$$

where m_1 and m_2 are constants. On the other hand, the most favorable case arises if the template comes from an uncertain delay in the dynamic plant. Then, the radius variation $r'(\phi)$ is zero because the module is a real number. Obtaining $g_{rc}(\phi)$, $g_{ic}(\phi)$, and $r(\phi)$ and substituting them into Equation C.16, the condition of intersection existence between two infinitesimally close circumferences is

$$\frac{\gamma_1^2}{(\gamma_1^2 - 1)(m_1^2 + m_2^2)\phi^4} > 0 \qquad \text{(C.21)}$$

If $\gamma_1 > 1$, the earlier expression is always true and the controller bound is obtained substituting Equation C.14 into Equation C.19. Figure C.4 shows the controller bound on the Re–Im plane for the plant

$$P(s) = \frac{e^{-Ts}}{s^2 + 0.04s + \omega_n^2}$$

and for the case described in Section C.2 (see Figure C.3), where the outer envelope curve has been selected. The circumference family and the $[g_{rc}(\phi), g_{ic}(\phi)]$ curve are also shown.

For feedback control performance specifications with $\gamma_1 < 1$, a procedure such as the one described in Section C.4 must be used.

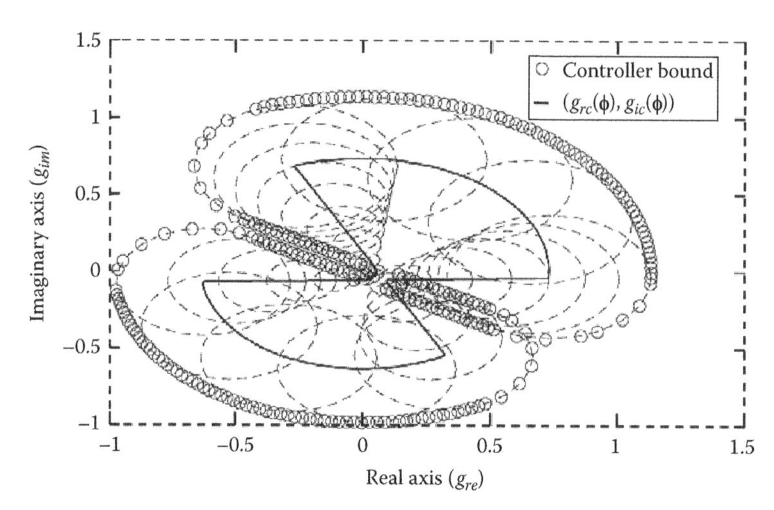

FIGURE C.4
Controller bound $[g_{re}(\phi), g_{im}(\phi)]$ and its circumference family.

C.4 Robust Sensitivity Bounds

The robust sensitivity specification imposes an upper limit in the magnitude so that

$$\left|\frac{1}{1+P(j\omega_i)G_{\omega_i}}\right| < \gamma_2 \tag{C.22}$$

According to Equation C.3, with $A = 1$, $B = 0$, and $C = P(j\omega_i) = \mathcal{T}_{\omega_i} = p_{re} + jp_{im}$, Equation C.22 becomes

$$(p_{im}(\phi)^2 + p_{re}(\phi)^2)(g_{im}^2 + g_{re}^2) + 2(p_{re}(\phi)g_{re} - p_{im}(\phi)g_{im}) + 1 - \frac{1}{\gamma_2^2} \geq 0 \tag{C.23}$$

The family of circumferences when Equation C.23 equals 0 results in the following equation:

$$\left\{\begin{array}{l} -\left(p_{re}(\phi)/\left(p_{re}(\phi)^2 + p_{im}(\phi)^2\right)\right) + \left(\gamma_2^{-1}/\left(\sqrt{p_{re}(\phi)^2 + p_{im}(\phi)^2}\right)\cos(\theta)\right) \\ \hline \left(p_{im}(\phi)/\left(p_{re}(\phi)^2 + p_{im}(\phi)^2\right)\right) + \left(\gamma_2^{-1}/\left(\sqrt{p_{re}(\phi)^2 + p_{im}(\phi)^2}\right)\sin(\theta)\right) \end{array}\right. \tag{C.24}$$

The intersection between two infinitesimally close circumferences must exist to accomplish the substitution of the θ parameter into Equation C.14. The less favorable case occurs when the radius variation reaches the maximum, that is, if the template comes from an uncertain gain

$$\mathcal{T}_{\omega_i}(\phi) = [p_{re}(\phi),\ p_{im}(\phi)] = (m_1\phi,\ m_2\phi)$$

where
- m_1 and m_2 are constants
- ϕ is the uncertain parameter

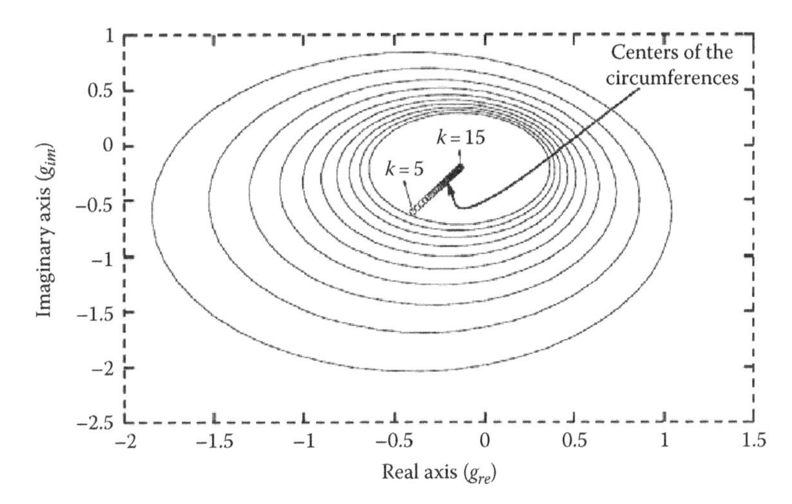

FIGURE C.5
Circumference family without intersections.

Under these conditions, Equation C.16 is evaluated obtaining

$$\frac{(\gamma_2^2 - 1)/\gamma_2^2}{\phi^4(m_1^2 + m_2^2)} < 0 \tag{C.25}$$

The inequality given by Equation C.25 is not achieved if γ_2 is greater than 1. Then, the envelope method can be applied by substituting Equation C.14 into Equation C.19. According to Equation C.25, if $\gamma_2 < 1$ then the square roots of Equation C.14 are complex.

Consider a plant given by $P(s) = k/(s + 2)$ with $k \in [5, 15]$, $\omega = 3$ rad/s, and $\gamma_2 = 0.5$, then the circumference family would be the one shown in Figure C.5. A similar behavior is obtained for the control performance specification in Equation C.17 with $\gamma_1 < 1$.

This proves that in each set of template points, with the same phase, only the lowest gain point of this set generates a point of the controller bound; the other points of this set can be eliminated. Figure C.6 illustrates this with an example where the circled points of the

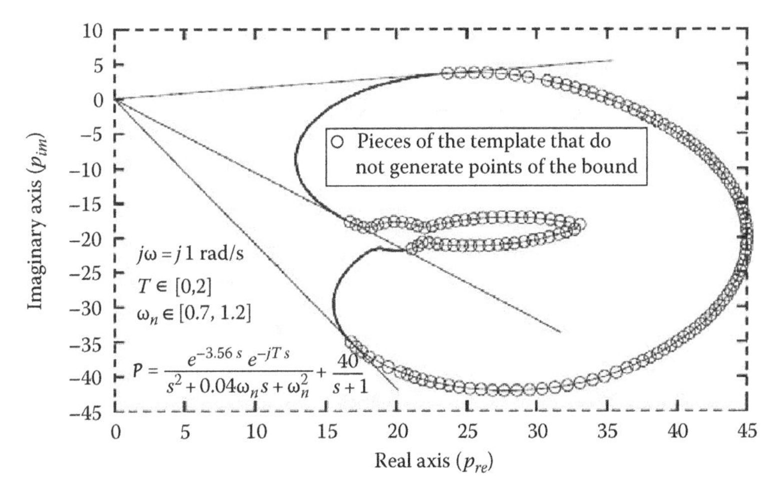

FIGURE C.6
Example of valid pieces of the template and nonvalid ones to calculate $G_{\omega_i}^2$.

template have no influence on the controller bound. Also, some of the rest of the points cannot generate bound points if some circumferences are within others. It depends on the value of γ_2 and is simple to determine by computation. Those template points that generate bound points define the equivalent template $\mathcal{T}_{\omega_i}^{\epsilon}$. Figure C.7 shows the equivalent template for two different γ_2 values.

If the equivalent template is open and differentiable in its extremes, the envelope curve is obtained without using pieces of the extreme circumferences. In Section C.2, it is explained how to calculate a controller bound from an open template that is nondifferentiable in its extremes. Defining $\mathcal{T}_{\omega_i}^{\epsilon}$ as

$$\mathcal{T}_{\omega_i}^{\epsilon} = \mathcal{T}_{\omega_i}(\phi) \quad \text{for } \phi \in [\phi_1, \phi_2] \tag{C.26}$$

where

$\mathcal{T}_{\omega_i}(\phi)$ is continuous and differentiable in $\phi \in [\phi_1 - \varepsilon_\phi, \phi_2 + \varepsilon_\phi]$

ε_ϕ is infinitesimal, for the angles

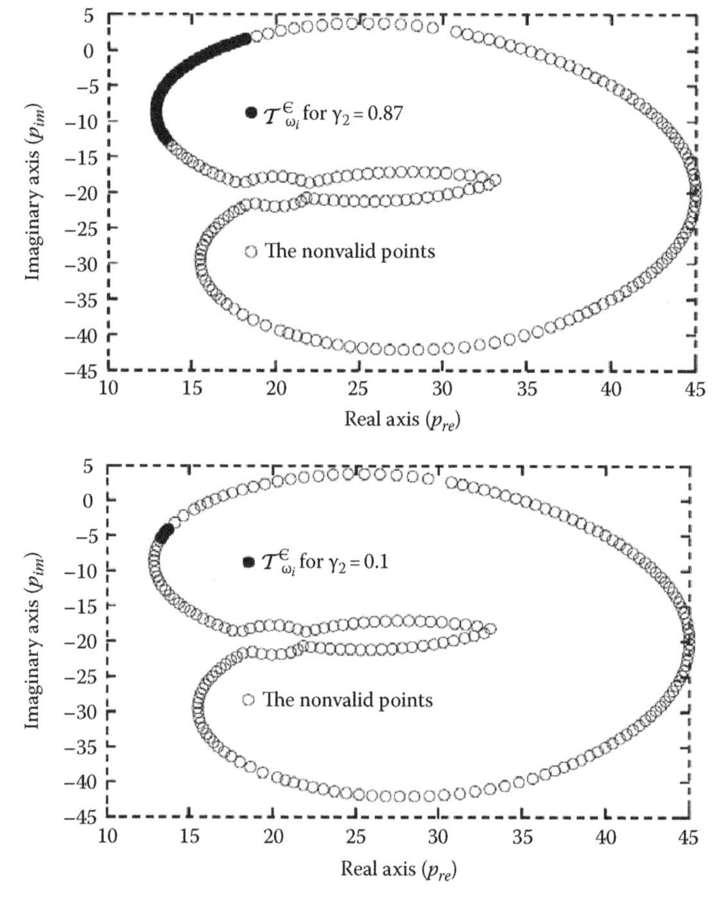

FIGURE C.7
Equivalent template $\mathcal{T}_{\omega_i}^{\epsilon}$ for two γ_2 values.

$$\theta_1 = \arctan\left(\frac{-g'_{ic}(\phi_1)r'(\phi_1) - g'_{rc}(\phi_1)\sqrt{g'_{ic}(\phi_1)^2 + g'_{rc}(\phi_1)^2 - r'(\phi_1)^2}}{-g'_{rc}(\phi_1)r'(\phi_1) + g'_{ic}(\phi_1)\sqrt{g'_{ic}(\phi_1)^2 + g'_{rc}(\phi_1)^2 - r'(\phi_1)^2}}\right)$$

$$\theta_2 = \arctan\left(\frac{-g'_{ic}(\phi_2)r'(\phi_2) + g'_{rc}(\phi_2)\sqrt{g'_{ic}(\phi_2)^2 + g'_{rc}(\phi_2)^2 - r'(\phi_2)^2}}{-g'_{rc}(\phi_2)r'(\phi_2) - g'_{ic}(\phi_2)\sqrt{g'_{ic}(\phi_2)^2 + g'_{rc}(\phi_2)^2 - r'(\phi_2)^2}}\right) \quad \text{(C.27)}$$

the following is obtained:

$$g_{rc}(\phi_1) + r(\phi_1)\cos(\theta_1) = g_{re}(\phi_2) + r(\phi_2)\cos(\theta_2)$$

$$g_{ic}(\phi_1) + r(\phi_1)\sin(\theta_1) = g_{ic}(\phi_2) + r(\phi_2)\sin(\theta_2) \quad \text{(C.28)}$$

This means that the controller bound can be completely computed by evaluating Equation C.15 with the points of $\mathcal{T}^{\epsilon}_{\omega_i}$. However, there are templates that are not continuous and differentiable in $[\phi_1 - \varepsilon_\phi, \phi_2 + \varepsilon_\phi]$. In these cases, it is necessary to take some points of the extreme circumferences as explained in Section C.2. Extreme circumferences are defined for $\phi = \phi_1$ and $\phi = \phi_2$. Figure C.8 shows an example where $\mathcal{T}^{\epsilon}_{\omega_i}$ is defined by two pieces that are not differentiable in their extremes.

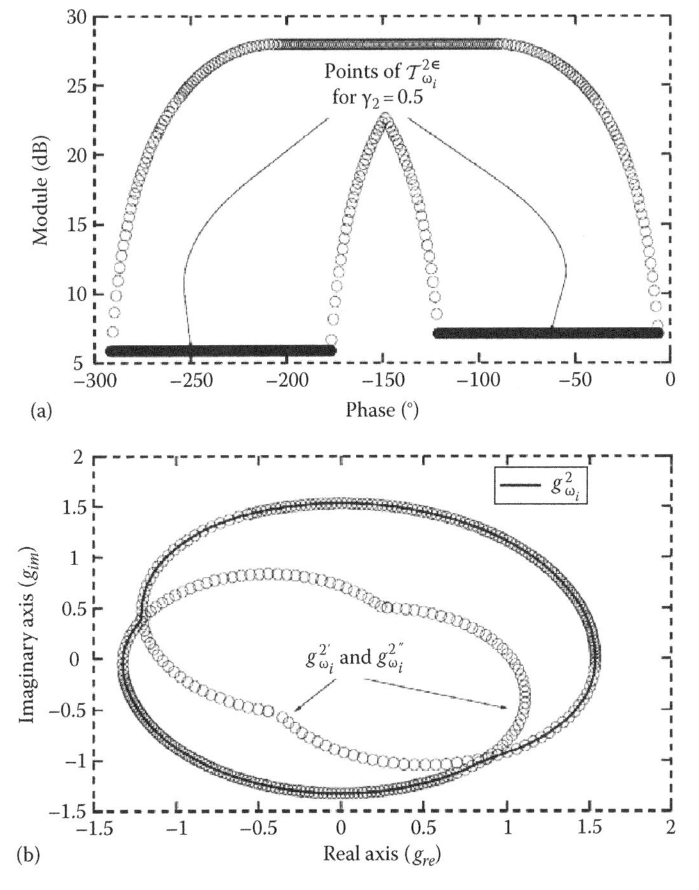

FIGURE C.8
$\mathcal{T}^{2\epsilon}_{\omega_i}$ compound of two pieces (a) and the final controller bound (b).

The two pieces of $\mathcal{T}_{\omega_i}^{\epsilon}$ (here $\mathcal{T}_{\omega_i}^{2\epsilon}$) define two "sub-bounds" of the controller, $G_{\omega_i}^{2'}$ and $G_{\omega_i}^{2''}$. The controller bound $G_{\omega_i}^{2}$ is the intersection. A controller bound defines an area of the Re–Im plane where the controller can be located. The intersection of the areas of different controller bounds gives the most unfavorable case.

C.5 Bounds for Robust Disturbance Rejection at Plant Input

The control specification for a robust disturbance rejection at the plant input specification is

$$\left|\frac{P}{1+PG}\right| \le \gamma_3$$

Then, $B(j\omega_i)=0$ and $A=C=\mathcal{T}_{\omega_i}(\phi)=p_{re}(\phi)+jp_{im}(\phi)$ in Equation C.1. By using Equations C.5 and C.6 and substituting them into Equation C.7, the circumference family becomes

$$\begin{cases} \dfrac{p_{re}(\phi)}{p_{re}(\phi)^2+p_{im}(\phi)^2}+\dfrac{1}{\gamma_3}\cos(\theta) \\ \dfrac{p_{im}(\phi)}{p_{re}(\phi)^2+p_{im}(\phi)^2}+\dfrac{1}{\gamma_3}\sin(\theta) \end{cases} \tag{C.29}$$

In this case, the radius is constant, $r(\phi) = (1/\gamma_3)$. Therefore, the inequality given by Equation C.16 is not fulfilled and Equation C.14 can be applied. See Ref. 61 for an example of the envelope method in this situation.

C.6 Robust Control Effort Bound

The control specification for the control effort is $|G/(1 + PG|$. For this case, substitute into Equation C.1 $A(j\omega) = 0$, $B(j\omega) = 1$, and $C=\mathcal{T}_{\omega_i}(\phi)=P_{re}(\phi)+jP_{im}(\phi)$. The inequality given by Equation C.3 becomes

$$(g_{re}^2+g_{im}^2)(\gamma_4^2(p_{re}^2+p_{im}^2)-1)+2\gamma_4^2 g_{re}p_{re}-2\gamma_4^2 g_{im}p_{im}+\gamma_4^2 \ge 0 \tag{C.30}$$

When this expression equals zero, the family of circumferences is

$$\begin{cases} \dfrac{-\gamma_4^2 p_{re}(\phi)}{\gamma_4^2[p_{re}(\phi)^2+p_{im}(\phi)^2]-1}+\dfrac{\gamma_4}{\gamma_4^2[p_{re}(\phi)^2+p_{im}(\phi)^2]-1}\cos(\theta) \\ \dfrac{\gamma_4^2 p_{im}(\phi)}{\gamma_4^2[p_{re}(\phi)^2+p_{im}(\phi)^2]-1}+\dfrac{\gamma_4}{\gamma_4^2[p_{re}(\phi)^2 p_{im}(\phi)^2]-1}\sin(\theta) \end{cases} \tag{C.31}$$

The radius is only a function of the module because $r = \gamma_4/(\gamma_4^2 m^2 - 1)$ where $m = \sqrt{p_{re}(\phi)^2 + p_{im}(\phi)^2}$. Then, its largest variation appears when $\mathcal{T}_{\omega_i}(\phi)$ is built from an uncertain gain; see Equation C.20. Thus, Equation C.16 becomes

$$\frac{(m_1^2 + m_2^2)\gamma_4^4}{[(m_1^2 + m_2^2)\phi^2\gamma_4^2 - 1]^2} < 0 \tag{C.32}$$

which is impossible to achieve. Therefore, each pair of two infinitesimally close circumferences always has at least one common point.

Despite Equation C.16 or C.32 not being met, it cannot be concluded that every point in the template contour $\mathcal{T}_{\omega_i}(\phi)$ generates a controller bound point; thus, further analysis is necessary. Some template contour points can achieve both

$$\gamma_4^2[p_{re}(\phi)^2 + p_{im}(\phi)^2] - 1 > 0 \quad \text{and} \quad \gamma_4^2[p_{re}(\phi)^2 + p_{im}(\phi)^2] - 1 < 0$$

The sign variation in this expression produces a high variation of the circumference center $[g_{rc}(\phi), g_{ic}(\phi)]$ and of the radius $r(\phi)$ sign. A positive radius implies that the outer envelope curve maintains the inequality of Equation C.30. A negative radius implies that Equation C.30 is satisfied by the inner envelope curve. If $\gamma_4^2(p_{re}(\phi)^2 + p_{im}(\phi)^2) - 1 = 0$, then the circumference becomes a line. Taking this into account, the template contour points are such that

$$\gamma_4^2[p_{re}(\phi)^2 + p_{im}(\phi)^2] - 1 < 0$$

which gives the less favorable bound. Figure C.9 shows the evolution of the circumference family when the sign of the earlier expression changes from positive to negative. Circumferences are bicolored (black and gray) to identify each side of the plane they divide, and the arrows indicate the part of the plane that maintains the control specification.

The following can be concluded: for those templates where every point keeps $\gamma_4^2[p_{re}(\phi)^2 + p_{im}(\phi)^2] - 1 > 0$, the controller bound is determined by the outer envelope curve of the family. For templates where every point keeps $\gamma_4^2[p_{re}(\phi)^2 + p_{im}(\phi)^2] - 1 < 0$, the controller bound is determined by the inner envelope curve of the family. For the templates where some points keep

$$\gamma_4^2[p_{re}(\phi)^2 + p_{im}(\phi)^2] - 1 > 0$$

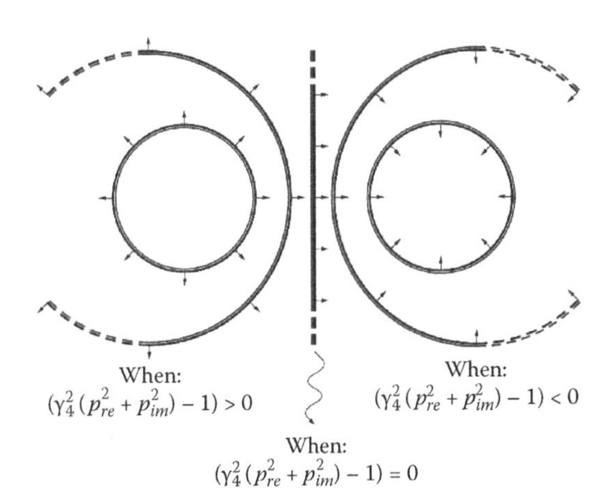

FIGURE C.9
Circumference evolution from positive to maintain negative $\gamma_4^2[p_{re}(\phi)^2 + p_{im}(\phi)^2] - 1$ values.

and others keep

$$\gamma_4^2[p_{re}(\phi)^2 + p_{im}(\phi)^2] - 1 < 0,$$

the controller bound is only determined by the template points that satisfy the last inequality (those that produce negative radius).

C.7 Remarks

The envelope method allows the designer to calculate analytical QFT bounds for robust performance and robust stability specifications with the following advantages:

- The new algorithm avoids the computational nested loops for the discrete value set of the controller's phase used in Borghesani QFT Toolbox[144] or the discrete value set of the nominal open loop in the cross-section method,[43,59,60,145] as well as the nested loops for the discrete value set of the plant required in all those methods. Hence, accuracy is improved and computational effort is significantly reduced.
- There is no restriction to compute multi-valued bounds. Thus, it is possible to calculate all the points that define the QFT bound at each phase.
- The solution bound is expressed as an analytical function on the independent variable, ϕ_i, used to track the plant template contour. Thus, the mathematical results are based on the use of $\mathcal{T}_{\omega_i}(\phi)$. This function of the variable ϕ can be obtained by approximating the discrete contour of the ω_i-template. In Ref. 41 it is shown how this discrete contour is approximated using Fourier series.
- The use of analytical functions for the bound expressions, apart from improving accuracy and reducing computational cost, gives the definition of the algebraic and geometrical properties and finds a relationship between the template points (equivalent template $\mathcal{T}_{\omega_i}^{\epsilon}$) and the bound points.
- To use the envelope method, the derivative of the template contour $\mathcal{T}_{\omega_i}$ must exist. Then, it is necessary to make the corners of $\mathcal{T}_{\omega_i}$ round. This is a simple task and produces a negligible error. A solution for making a discrete template contour defined by the array p or consecutive points is as follows:

```
k=2; angmin=5.5*pi/6;
while k<length(p),
  % Angle between two consecutive lines
  angulo=abs(angle(p(k-1)-p(k))-angle(p(k+1)-p(k)));
  if angulo>pi, angulo=2*pi-angulo; end;
  if angulo<angmin,
     p(k)=(p(k-1)+p(k)+p(k+1))/3;
     k=k-2;
  end;
  k=k+1;
end;
```

- The reference tracking specification is formulated by using the form $\gamma_{5inf} < |FPG/(1 + PG)| < \gamma_{5sup}$, which needs a 2DOF (prefilter and controller) control structure. When this double inequality is reduced to one of the form of Equation C.1, two uncertain transfer functions and two template contours arise, instead of just one as the envelope method formulates. That case is out of the scope of this appendix.

C.8 Examples

This section provides two examples of QFT bounds computation using the envelope method. In Section C.8.1 a case of robust sensitive reduction is studied. The example in Section C.8.2 shows how the method works calculating control effort bounds. The envelope method is compared with the Borghesani QFT Toolbox method[144] in both cases. Additional examples of robust stability and robust disturbance rejection at the plant input bounds can be found in Ref. 61.

C.8.1 Example 1: Robust Disturbance Rejection at the Plant Output

Consider the plant

$$P(s) = \frac{ke^{-0.3s}}{(s+a)(s+20)} \tag{C.33}$$

The aim is to obtain the bounds for robust disturbance rejection at the plant output at $\omega = 5\,\text{rad/s}: |1/(1 + PG)| < \gamma_2$ for the specifications $\gamma_2 = 0.1, 0.5,$ and 0.9.

First, the discrete template is computed with an adequate number of points (in this case a 400-element array). Then, g_{rc}, g_{ic}, and r are obtained using Equation C.24. After that, their derivatives are calculated. The examples presented in this section use the forward difference method to compute the derivatives, that is, $f'(x_1) = [f(x_1+e) - f(x_1)]/\varepsilon$. Note that the truncation error is of εth order, although its exact computation is unfeasible because the function $\mathcal{T}_{\omega_i}$ is unknown. The derivative is obtained from the 400-element array that defines the discrete contour template, performing the difference between the $n + 1$ and the n elements of this array. If the points defined in $\mathcal{T}_{\omega_i}$ are close enough, this approximation works fine, as it is shown in the comparative results.

Previously, the equivalent template for each specification value had to be calculated. Finally, the bounds can be obtained by substituting the θ parameter of Equation C.24 into Equation C.14. Figure C.10 shows the bounds.

The results obtained with the Borghesani QFT Toolbox[144] are bounds defined by an equidistant phases set (73 phases by default). The number of points obtained using the envelope method is equal to the number of points of the discrete template contour. Although $\mathcal{T}_{\omega_i}(\phi)$ is an analytical function, it must be discretized to be used with a computational program. Therefore, the number of points obtained with the envelope method is usually quite higher than the one obtained with the Borghesani QFT Toolbox[144] (because the number of template points can be several hundred).

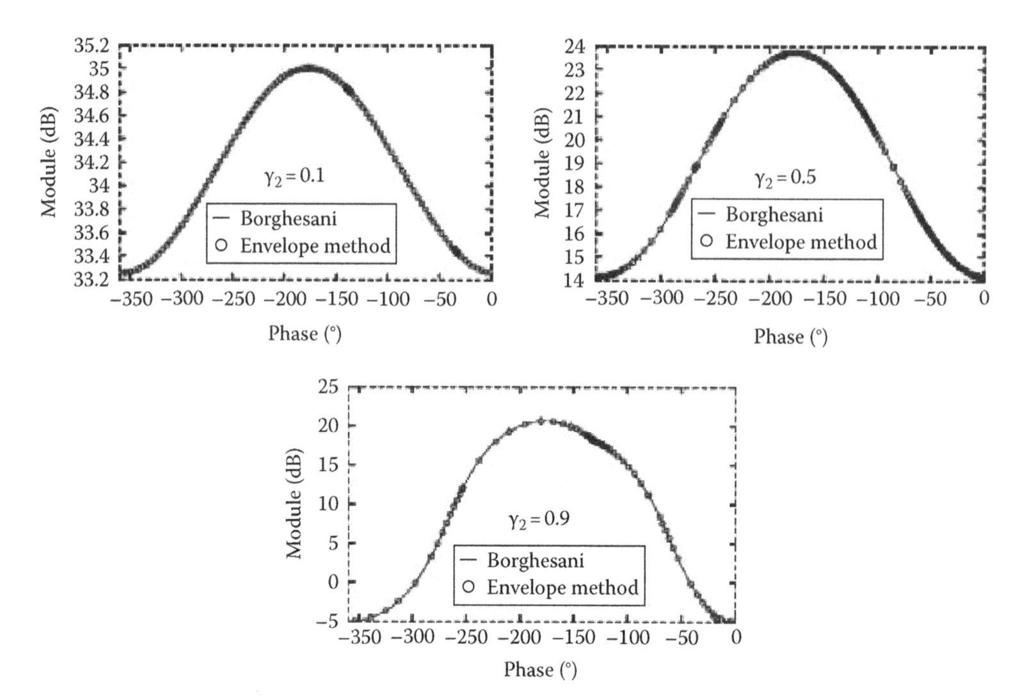

FIGURE C.10
Bounds for the specifications of Example 1.

The computation time required by the Borghesani QFT Toolbox[144] was 265 ms for the three bounds. The envelope method took 11 ms. The commands *tic-toc* are used, running Windows XP and MATLAB® on a PC with a 1.4 GHz Pentium IV processor and a 512 Mb RAM.

C.8.2 Example 2: Robust Control Effort Bound

It is desired to compute the robust control effort bound at $\omega = 5$ rad/s for the plant

$$P(s) = \frac{500e^{-0.3s}}{(s+a)(s+6b)} \tag{C.34}$$

where

$a \in [1,5]$
$b \in [1,10]$

The control effort specification at $\omega = 5$ rad/s is $|G/(1 + PG)| < 2$. In a similar way as was indicated in the previous example, the values of g_{rc}, g_{ic}, r, and their derivatives are computed. Using Equation C.31 the bound is obtained. Figure C.11 shows the different methodologies.

The computation time required with the envelope method was 20 times lower than the one required with the Borghesani QFT Toolbox.[144] Besides, this example shows that the envelope method is able to compute multi-valued bounds. Note that accuracy is also greater with the envelope method because there is no loss of bound extreme points.

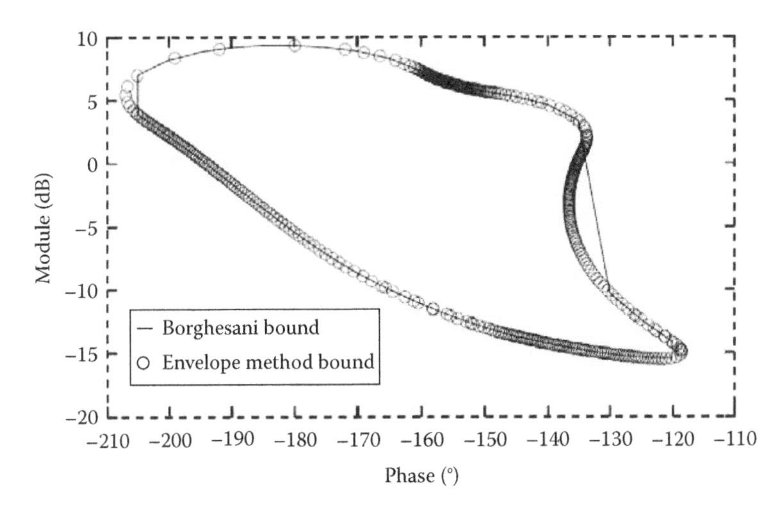

FIGURE C.11
Control effort bound computed using the Borghesani toolbox and the envelope method. (From Borghesani, C. et al., *Quantitative Feedback Theory Toolbox—For Use with MATLAB*, Terasoft, San Diego, CA, 1994, 2002.)

C.9 Summary

This appendix described an analytical formulation to compute QFT bounds based on envelope curves. The controller bound is defined by the envelope curve of a family of circumferences as an analytical function with useful properties. In comparison with other methods, computational effort does not increase geometrically with the number of discrete template points and discrete bound phase points, since (a) the bound phase is not discretized and (b) computational nested loops are not used to track both discrete template and phase sets (gridding methods). The significant reduction of computational effort allows the user to deal with a large number of template points and therefore improves accuracy. The method also allows the designer to calculate multi-valued bounds. The advantages are illustrated with two examples for robust disturbance rejection at the plant output and robust control effort specifications.

Appendix D: Essentials for Loop Shaping

D.1 Introduction

After calculating the stability and performance bounds in the QFT methodology, the following step consists of designing (loop shaping) the compensator $G(s)$ that allows the nominal open loop $L_0(s) = P_0(s)\,G(s)$ to satisfy all the specifications for all the plants within the uncertainty (intersection of bounds). This is probably one of the most difficult steps of the methodology for the beginner. For this reason, this appendix introduces a fundamental study of the most common elements that can be added to the compensator $G(s)$. Sections D.2 through D.11 show how these elements modify the position of $L_0(s)$ in the Nichols Chart (NC); Section D.12 analyses the stability, reformulating the Nyquist criterion for the NC; and Section D.13 reviews practical theorems on the existence of controllers.

D.2 Simple Gain: K

The effect of a simple gain K is to shift $L(j\omega)$ up or down if $K > 1$ or $K < 1$, respectively. The distance that $L(j\omega)$ is vertically shifted in the NC is $[20\log_{10}(K)]$ in dB. To make the calculations in dB much easier, bear in mind that multiplying $L(j\omega)$ by a gain $K = 2$ (or 1/2) corresponds to shifting it by a distance of $20\log_{10}(K) = 6$ dB (or -6 dB) in the NC (see Table D.1). Figure D.1 shows the effect of a simple gain $K = 2$, where $L_1 = KL_0$.

D.3 Real Pole: $1/[(s/p) + 1]$

The effect of a real pole, located at $s = -p$, is to shift $L(j\omega)$ down by a factor of $-10\log_{10}(1 + \omega^2/p^2)$ dB and left by $-\tan^{-1}(\omega/p)$ degrees. That is to say, at the frequency $\omega = p$ rad/s the pole shifts $L(j\omega)$ down by -3 dB and left by $-45°$. Figure D.2 shows the effect of a real pole $p = 1$, where $L_1(s) = \{1/[(s/p) + 1]\}\,L_0(s)$.

D.4 Real Zero: $[(s/z) + 1]$

The effect of a real zero, located at $s = -z$, is to shift $L(j\omega)$ up by a factor of $10\log_{10}(1 + \omega^2/z^2)$ dB and right by $\tan^{-1}(\omega/z)$ degrees. That is, at the frequency $\omega = z$ rad/s the zero shifts $L(j\omega)$ up by $+3$ dB and right by $+45°$. Figure D.3 shows the effect of a real zero at $z = 1$, where $L_1(s) = [(s/z) + 1]\,L_0(s)$.

TABLE D.1

Magnitude: Module versus dB

Module	**0.125**	**0.25**	**0.5**	**1**	**2**	**4**	**8**	**16**	**32**	**Multiply by 2**
dB	−18.1	−12.0	−6.0	0.0	6.0	12.0	18.1	24.1	30.1	=add + 6 dB
Module	**0.001**	**0.01**	**0.1**	**1**	**10**	**100**	**1,000**	**10,000**	**Multiply by 10**	
dB	−60.0	−40.0	−20.0	0.0	20.0	40.0	60.0	80.0	=add + 20 dB	

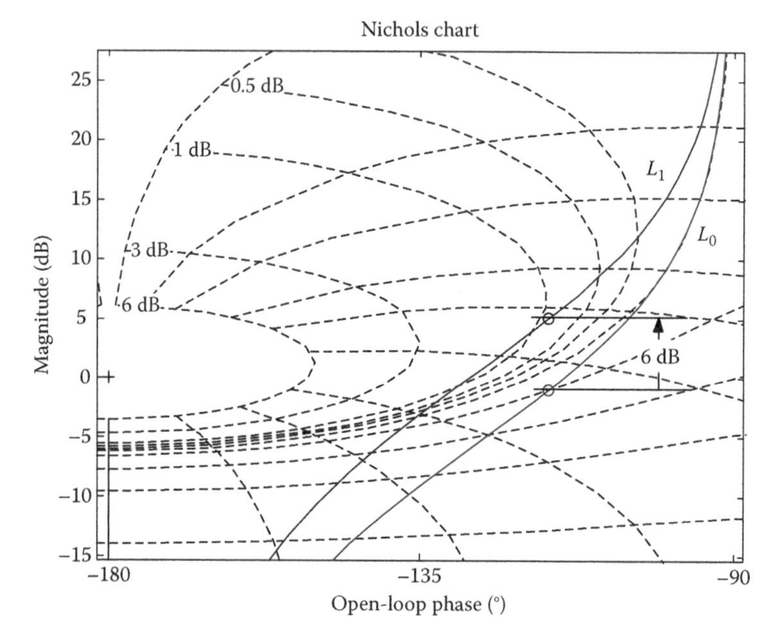

FIGURE D.1
Effect of a simple gain $K = 2$, where $L_1 = KL_0$. At every frequency, the gain shifts $L(j\omega)$ up by +6 dB.

D.5 Integrator: $1/s^n$; $n = 1, 2, \ldots$

The effect of n integrators is to shift $L(j\omega)$ left by $-90n°$ at every frequency. Notice that $L(j\omega)$ is not vertically shifted.

D.6 Differentiator: s^n; $n = 1, 2, \ldots$

The effect of n differentiators is to shift $L(j\omega)$ right by $+90n°$ at every frequency. Notice that $L(j\omega)$ is not vertically shifted.

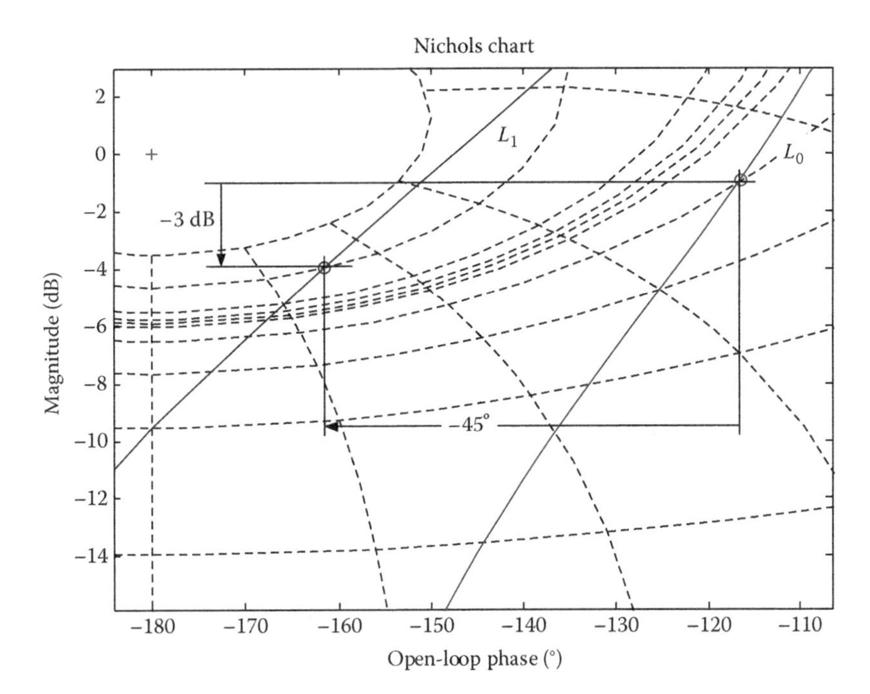

FIGURE D.2
Effect of a real pole $p = 1$, where $L_1(s) = \{1/[(s/p) + 1]\}\ L_0(s)$.

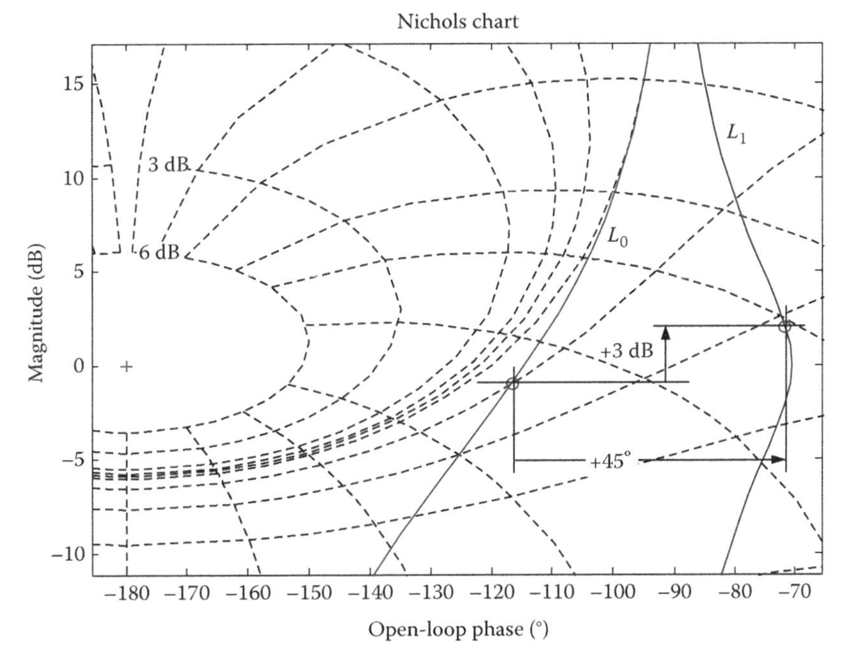

FIGURE D.3
Effect of a real zero $z = 1$, where $L_1 = [(s/z) + 1]\ L_0$.

D.7 Complex Pole: $\left[\omega_n^2/(s^2+2\zeta\omega_n s+\omega_n^2)\right]$ ($\zeta < 0.707$)

The effect of a complex pole, with a natural frequency ω_n and a damping factor ζ, is to shift $L(j\omega)$ up by a maximum magnitude (resonance) of $-20\log_{10}\left[2\zeta\sqrt{1-\zeta}\right]$dB at the frequency $\omega=\omega_n\sqrt{1-2\zeta^2}$ rad/s ($\zeta < 0.707$).

Furthermore, at that frequency, $L(j\omega)$ is shifted left by $-90^\circ+\sin^{-1}\left(\zeta/\sqrt{1-\zeta^2}\right)$ degrees. Figure D.4 shows the effect of a complex pole with $\omega_n = 1$ rad/s and $\zeta = 0.4$, where

$$L_1(s)=\left[\frac{\omega_n^2}{(s^2+2\zeta\omega_n s+\omega_n^2)}\right]L_0(s) \tag{D.1}$$

At the resonance frequency $\omega = 0.82$ rad/s, the complex pole shifts $L(j\omega)$ up by 2.7 dB and left by −89.5° (see Table D.2).

D.8 Complex Zero: $(s^2+2\zeta\omega_n s+\omega_n^2)/(\omega_n^2)$

The effect of a complex zero, with a natural frequency ω_n and a damping factor ζ, is to shift $L(j\omega)$ down by a maximum magnitude (resonance) of $20\log_{10}\left[2\zeta\sqrt{1-\zeta^2}\right]$dB at the

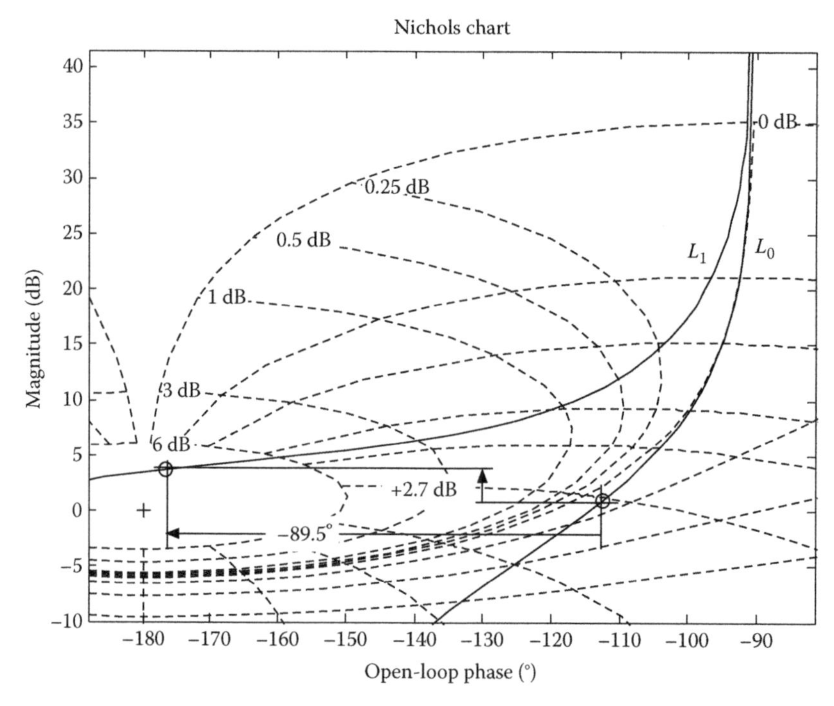

FIGURE D.4
Effect of a complex pole with $\omega_n = 1$ rad/s and $\zeta = 0.4$.

TABLE D.2
Magnitude and Phase at the Resonance Frequency $\omega = \omega_n\sqrt{1-2\zeta^2}$ rad/s ($\zeta < 0.707$)

ζ	**0.001**	**0.01**	**0.1**	**0.2**	**0.3**	**0.4**	**0.5**	**0.6**	**0.7**
M dB (resonance)	53.98	33.98	14.02	8.14	4.85	2.70	1.25	0.35	0.00
ϕ degrees (resonance)	−90.00	−89.99	−89.90	−89.79	−89.68	−89.55	−89.38	−89.15	−88.63

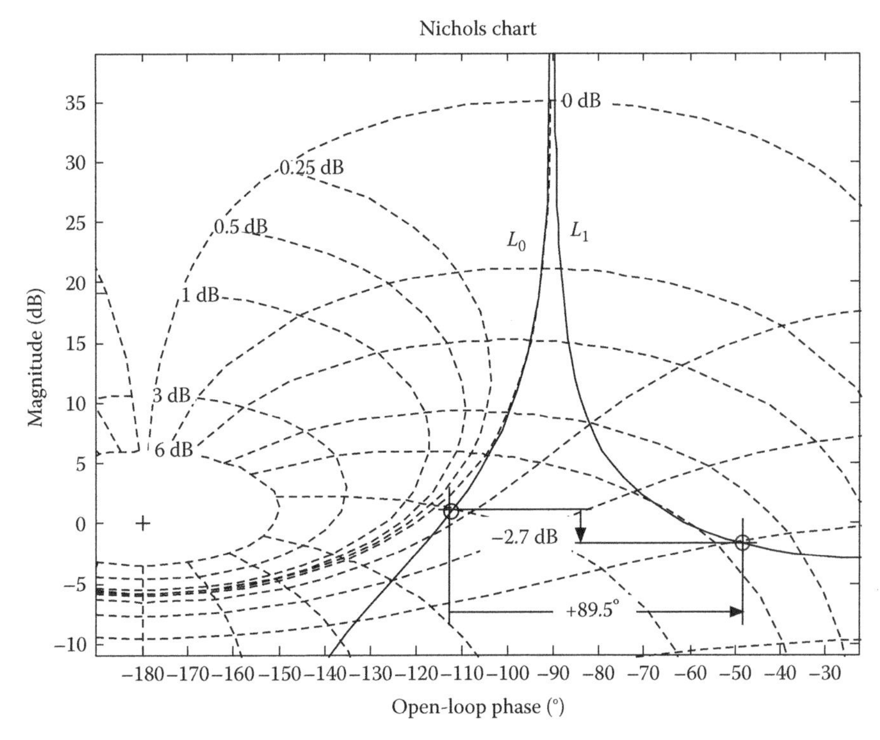

FIGURE D.5
Effect of a complex zero with $\omega_n = 1$ rad/s and $\zeta = 0.4$.

frequency $\omega = \omega_n\sqrt{1-2\zeta^2}$ rad/s ($\zeta < 0.707$). Furthermore, at that frequency, $L(j\omega)$ is shifted right by $90^\circ - \sin^{-1}\left[\zeta/\sqrt{1-\zeta^2}\right]$ degrees. Figure D.5 shows the effect of a complex zero with $\omega_n = 1$ rad/s and $\zeta = 0.4$, where

$$L_1(s) = \left[\frac{(s^2 + 2\zeta\omega_n s + \omega_n^2)}{\omega_n^2}\right] L_0(s) \tag{D.2}$$

At the resonance frequency $\omega = 0.82$ rad/s the complex zero shifts $L(j\omega)$ down by −2.7 dB and right by +89.5° (see Table D.3).

TABLE D.3

Magnitude and Phase at the Resonance Frequency $\omega = \omega_n\sqrt{1-2\zeta^2}$ rad/s ($\zeta < 0.707$)

ζ	0.001	0.01	0.1	0.2	0.3	0.4	0.5	0.6	0.7
M dB (resonance)	−53.98	−33.98	−14.02	−8.14	−4.85	−2.70	−1.25	−0.35	−0.00
ϕ degrees (resonance)	90.00	89.99	89.90	89.79	89.68	89.55	89.38	89.15	88.63

D.9 Lead Network: $[(s/z) + 1]/[(s/p) + 1]$, $(z < p)$

The effect of a lead network, with a zero located at $s = -z$ and a pole located at $s = -p$, $(z < p)$, is to shift $L(j\omega)$ right by a maximum phase of $\phi_{max} = 90 - 2\tan^{-1}\sqrt{z/p}$ degrees at the frequency $\omega = \sqrt{zp}$ rad/s. Furthermore, at that frequency, $L(j\omega)$ is shifted up by $10\log_{10}(p/z)$.

Figure D.6 shows the effect of a simple lead network with $p = 10$ and $z = 1$, where $L_1 = \{[(s/z) + 1]/[(s/p) + 1]\}L_0$. At the frequency $\omega = 3.16$ rad/s the lead network shifts $L(j\omega)$ up by +10 dB and right by +54.9° (see Table D.4).

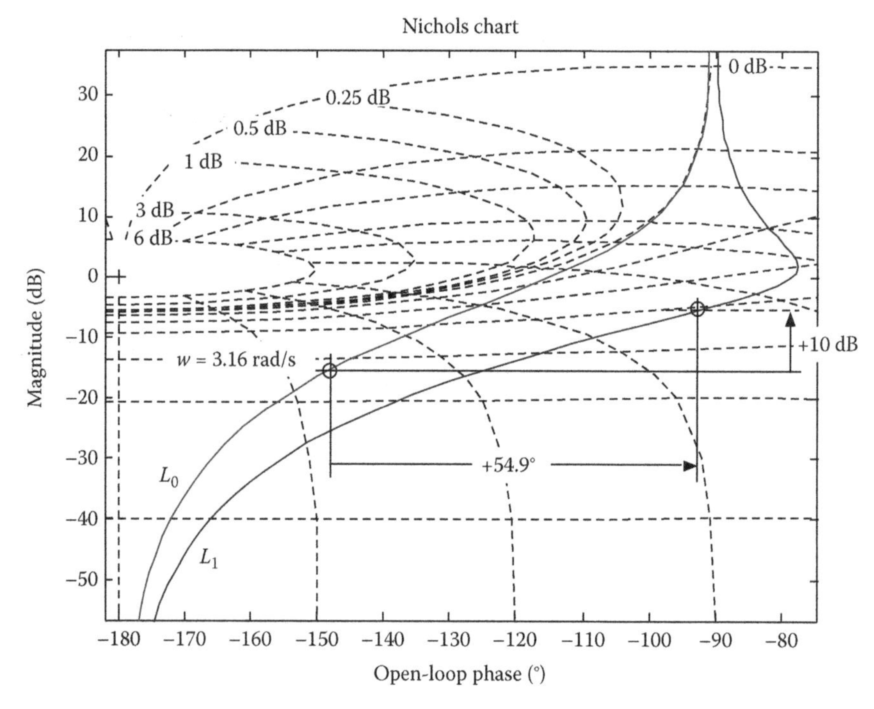

FIGURE D.6
Effect of a lead network with $p = 10$ and $z = 1$, where $L_1(s) = \{[(s/z) + 1]/[(s/p) + 1]\}\, L_0(s)$.

TABLE D.4

Magnitude and Phase at $\omega = \sqrt{zp}$ rad/s

z/p	0.001	0.01	0.1	0.2	0.3	0.4	0.5	0.6	0.7	0.8	0.9	1
ϕ_{max} (degrees)	86.4	78.6	54.9	41.8	32.6	25.4	19.5	14.5	10.2	6.4	3.0	0.0
M dB	30.00	20.00	10.00	6.99	5.23	3.98	3.01	2.22	1.55	0.97	0.46	0.00

D.10 Lag Network: [(s/z) + 1]/[(s/p) + 1] (z > p)

The effect of a lag network, with a zero located at $s = -z$ and a pole located at $s = -p$, $(z > p)$, is to shift $L(j\omega)$ left by a maximum phase of $\phi_{max} = -90 + 2\tan^{-1}\sqrt{z/p}$ degrees at the frequency $\omega = \sqrt{zp}$ rad/s (see Table D.5). Furthermore, at that frequency, $L(j\omega)$ is shifted down by $-10\log_{10}(p/z)$.

Figure D.7 shows the effect of a simple lag network with $p = 0.2$ and $z = 2$, where $L_1(s) = \{[(s/z) + 1]/[(s/p) + 1]\}L_0(s)$. At the frequency $\omega = 0.632$ rad/s the lead network shifts $L(j\omega)$ down by −10 dB and left by −54.9°.

TABLE D.5

Magnitude and Phase at $\omega = \sqrt{zp}$ rad/s

p/z	0.001	0.01	0.1	0.2	0.3	0.4	0.5	0.6	0.7	0.8	0.9	1
ϕ_{max} (degrees)	−86.4	−78.6	−54.9	−41.8	−32.6	−25.4	−19.5	−14.5	−10.2	−6.4	−3.0	0.0
M dB	−30.00	−20.00	−10.00	−6.99	−5.23	−3.98	−3.01	−2.22	−1.55	−0.97	−0.46	0.00

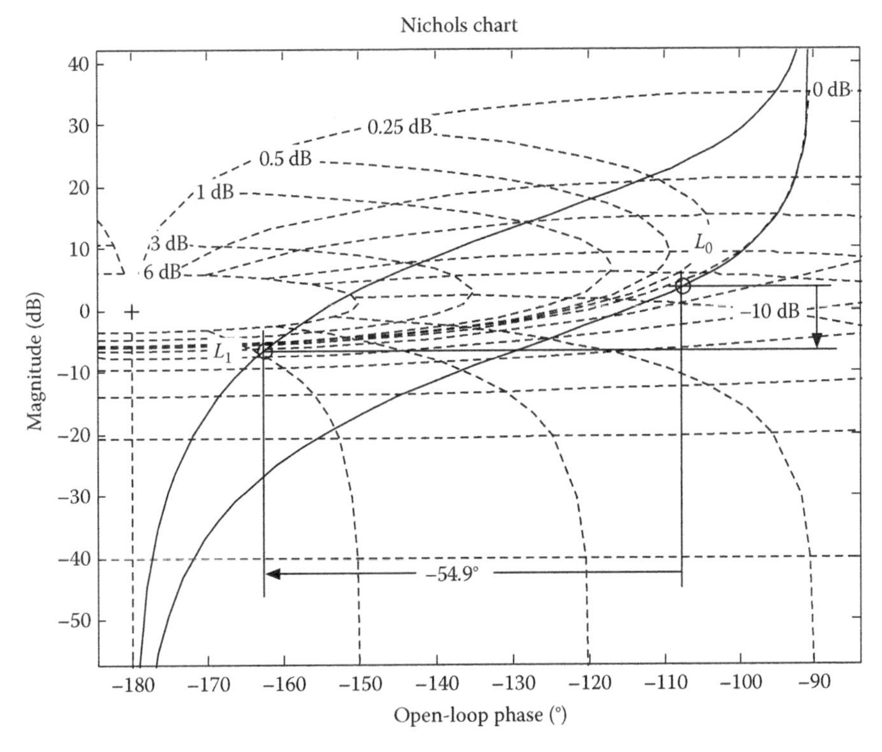

FIGURE D.7
Effect of a lag network with $p = 0.2$ and $z = 2$, where $L_1(s) = \{[(s/z) + 1]/[(s/p) + 1]\}L_0(s)$.

D.11 Notch Filter: $[s^2 + 2\zeta_1\omega_n s + \omega_n^2]/[s^2 + 2\zeta_2\omega_n s + \omega_n^2]$, $\zeta_1 < \zeta_2$

The effect of a notch filter, with a natural frequency ω_n and two damping factors ζ_1 and ζ_2, is to shift $L(j\omega)$ down by a maximum magnitude of $20\log_{10}(\zeta_1/\zeta_2)$ dB at the frequency $\omega = \omega_n$ rad/s (see Table D.6). At that frequency, $L(j\omega)$ is not horizontally shifted. Figure D.8 shows the effect of a notch filter with $\zeta_1 = 0.1$, $\zeta_2 = 1$, and $\omega_n = 2$ rad/s, where the notch filter shifts $L(j\omega)$ down by −20 dB and

$$L_1(s) = \left[\frac{(s^2+2\zeta_1\omega_n s+\omega_n^2)}{(s^2+2\zeta_2\omega_n s+\omega_n^2)}\right]L_0(s) \tag{D.3}$$

The transfer function $L_0(s)$ used in all the aforementioned examples is

$$L_0(s) = \frac{1}{\left[s(0.5s+1)\right]} \tag{D.4}$$

TABLE D.6

Magnitude and Phase at $\omega = \omega_n$ rad/s

ζ_1/ζ_2	0.001	0.01	0.1	0.2	0.3	0.4	0.5	0.6	0.7	0.8	0.9	1
M dB	−60.0	−40.0	−20.0	−13.98	−10.46	−7.96	−6.02	−4.44	−3.1	−1.94	−0.92	0.0
ϕ degrees	0	0	0	0	0	0	0	0	0	0	0	0

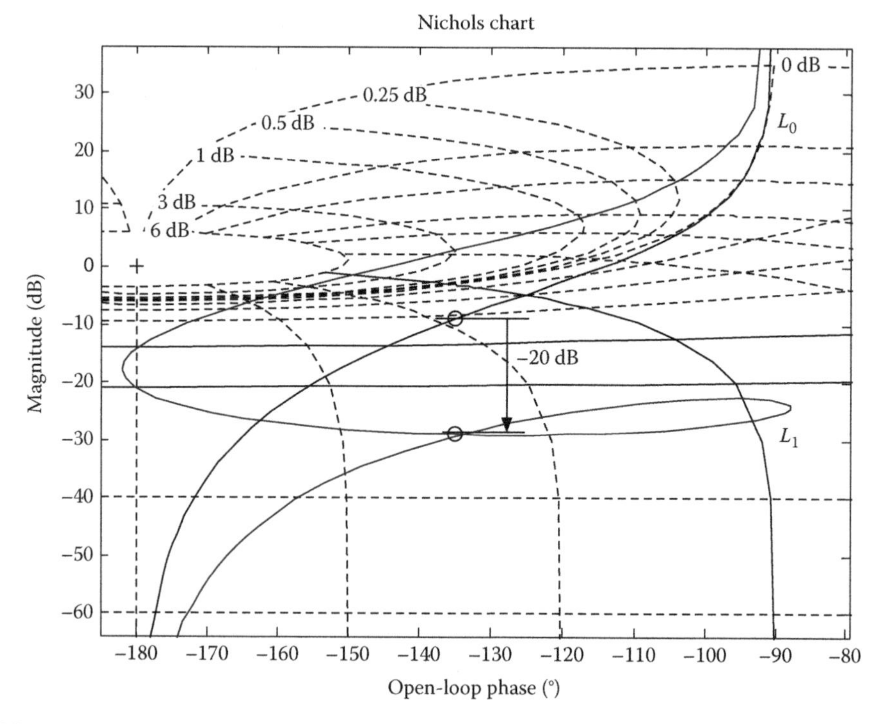

FIGURE D.8
Effect of a notch filter with $\zeta_1 = 0.1$, $\zeta_2 = 1$, and $\omega_n = 2$ rad/s.

D.12 Stability Analysis Using the Nichols Chart

The Nyquist stability criterion is a general methodology for determining absolute stability.[217] It is based on the principle of argument. The Nyquist criterion is especially useful for determining the stability of the closed-loop system $[Y(s)/R(s)]$ when the open loop $L(s)$ is given. It works well in any case, even in the so-called conditionally stable systems, where the simple gain and phase margins analysis does not work properly. Consider the closed-loop feedback system in Figure D.9. The closed-loop system transfer function is

$$\frac{Y(s)}{R(s)} = \frac{P(s)G(s)}{1+P(s)G(s)} = \frac{L(s)}{1+L(s)} \tag{D.5}$$

with the characteristic equation

$$1+L(s)=0 \tag{D.6}$$

Figure D.10 shows the Nyquist stability criterion, where Γ_c is the path encircling the area of the right-half s-plane (RHP). The stability of the closed-loop system can be determined by investigating how the path Γ_c is mapped by $L(s)$. The principle of argument states that the number of encirclements N in the positive direction (clockwise) around (–1,0) by the map of Γ_c, for the frequency range $(-\infty < \omega < +\infty)$, equals $N = z - p$, where p is the number of poles of $L(s)$ in the RHP and z the number of zeros (roots) of the characteristic equation—Equation D.6—in the RHP. Obviously, the closed-loop system is stable if and only if there

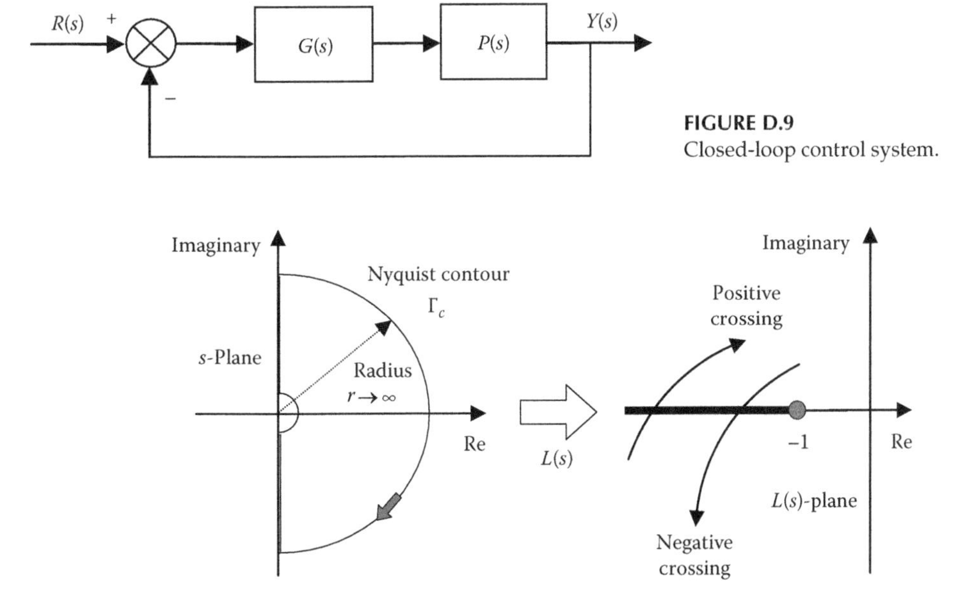

FIGURE D.9
Closed-loop control system.

FIGURE D.10
Nyquist stability criterion.

are no roots of the characteristic equation—Equation D.6—in the RHP, that is to say, $z = N + p = 0$. This general condition can be reformulated with the following two rules:

- *Rule 1*: A feedback control system is stable if and only if the Nyquist contour Γ_c when mapped in the $L(s)$-plane does not encircle the (–1,0) point ($N = 0$) and the number of poles of $L(s)$ in the RHP is zero ($p = 0$).
- *Rule 2*: A feedback control system is stable if and only if, for the Nyquist contour Γ_c when mapped in the $L(s)$-plane, the number of anticlockwise encirclements of the (–1,0) critical point (N) equals the number of poles of $L(s)$ with positive real parts (p).

In both cases $z = N + p = 0$. To determine N, the following four points in the $L(s)$-plane are utilized:

1. Count the net number of real-axis crossings to the left of the point (–1,0), with clockwise (decreasing phase) crossings being positive (see Figure D.10).
2. A real-axis ($r < -1$) crossing, with ω_0, $0 < \omega_0 < +\infty$, corresponds to two encirclements of the critical point, so $N = \pm 2$ (r is the real-axis crossing).
3. A real-axis ($r < -1$) crossing at $\omega_0 = 0$ corresponds to one encirclement of the critical point, so $N = \pm 1$.
4. An infinite positive imaginary value ($+\infty$) at $\omega_0 = +0$, (with an infinite negative imaginary value ($-\infty$) at $\omega_0 = -0$), with a projection in the real axis on the interval $r < -1$, corresponds to $N = +2$.

These conditions can be also reformulated for the Nichols diagram, the one that is used in QFT, with the following rules (see Figure D.11):

- *Rule 3*: A feedback control system is stable if and only if the single-sheeted Nichols plot of $L(j\omega)$ does not cross the ray $R_0 = \{(\phi, r)\colon \phi = -180°, r > 0\,\text{dB}\}$ ($N = 0$) when the number of poles of $L(j\omega)$ in the RHP is zero ($p = 0$).
- *Rule 4*: A feedback control system is stable if and only if the single-sheeted Nichols plot of $L(j\omega)$, for the frequency range ($0 < \omega < +\infty$), does not intersect the point $p_0 = \{\phi = -180°, r = 0\,\text{dB}\}$, and the net sum of its crossings of the ray $R_0 = \{(\phi, r)\colon \phi = -180°, r > 0\,\text{dB}\}$ equals p. The crossings and their corresponding sings in the Nichols diagram are illustrated in Figure D.11.

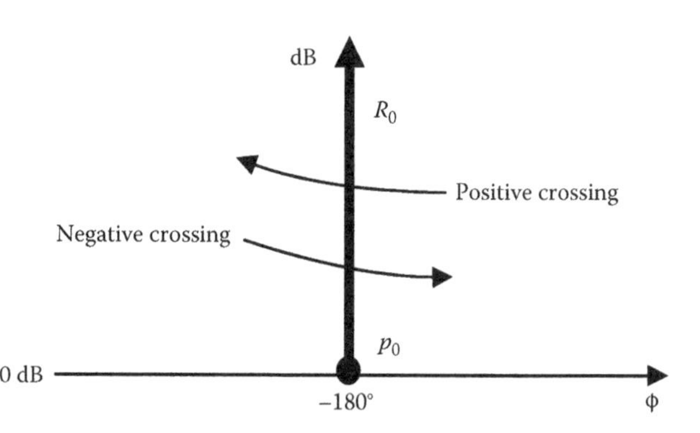

FIGURE D.11
Nyquist stability criterion on the Nichols diagram.

- *Rule 5*: A feedback control system is stable if and only if the multiple-sheeted Nichols plot of $L(j\omega)$, for the frequency range $(0 < \omega < +\infty)$, does not intersect the points $(2k + 1)\ p_0$, with $p_0 = \{\phi = -180°,\ r = 0\,\text{dB},\ k = 0, \pm1, \pm2\ldots\}$, and the net sum of its crossings of the rays $R_0 = \{(\phi, r)\text{: } \phi = -180(2k + 1)°,\ r > 0\,\text{dB},\ k = 0, \pm1, \pm2, \ldots\}$, equals p.

Rules 3, 4, and 5 are equivalent to Rules 1 and 2. In all the cases the system is stable if $z = N + p = 0$.

Now, to obtain N in the Nichols diagram, the following five points are utilized:

1. Count the net number of crossings N of $R_0 = \{(\phi, r)\text{: } \phi = -180(2k + 1)°,\ r > 0\,\text{dB},\ k = 0, \pm1, \pm2, \ldots\}$, with the left direction (decreasing phase) being positive and right direction (increasing phase) being negative (see Figure D.11).
2. A $R_0 = \{(\phi, r)\text{: } \phi = -180°,\ r > 0\,\text{dB}\}$ crossing of $L(j\omega)$ at the positive-frequency portion for ω_0, $0 < \omega_0 < +\infty$, corresponds to $N = \pm2$ [sign according to point (1)]. For the multiple-sheeted Nichols plot, every crossing of the rays $R_0 = \{(\phi, r)\text{: } \phi = -180(2k + 1)°,\ r > 0\,\text{dB},\ k = 0, \pm1, \pm2, \ldots\}$ adds ±2 to N [sign also according to Point (1)] (see Figure D.12a).
3. If $|L(j\omega_0)|$ is finite at $\omega_0 = 0\,\text{rad/s}$ and lies on $R_0 = \{(\phi, r)\text{: } \phi = -180(2k + 1)°,\ r > 0\,\text{dB},\ k = 0, \pm1, \pm2, \ldots\}$, then it adds ±1 to N (sign also according to first paragraph) (see Figure D.12b).
4. If $|L(j\omega_0)|$ is infinite at $\omega_0 = 0\,\text{rad/s}$ and lies in the range $\{(\phi, r)\text{: } -180(2k + 3)° < \phi < -180(2k + 1)°,\ r > 0\,\text{dB},\ k = 0, \pm1, \pm2, \ldots\}$, then it adds $+2(k + 1)$ to N (see Figure D.12c).
5. If $|L(j\omega_0)|$ is infinite at $\omega_0 = 0\,\text{rad/s}$ and approaches one of the rays $R_0 = \{(\phi, r)\text{: } \phi = -180(2k + 1)°,\ r > 0\,\text{dB},\ k = 0, \pm1, \pm2, \ldots\}$ when ω approaches 0, then it adds $+2k$ to N if $L(j\omega)$ goes to the right as ω increases and adds $+2(k + 1)$ to N if $L(j\omega_0)$ goes to the left as ω increases (see Figure D.12d).

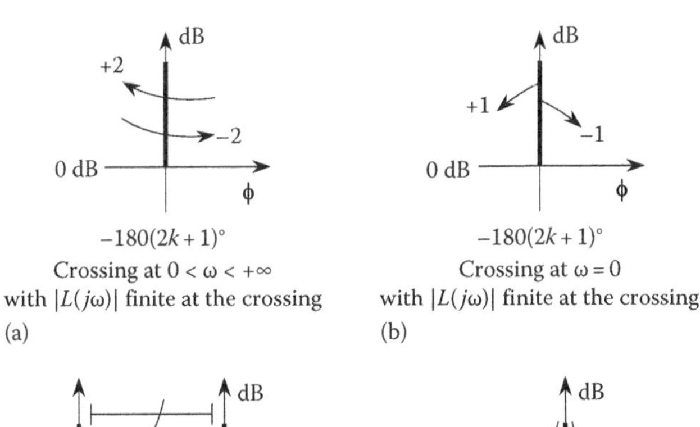

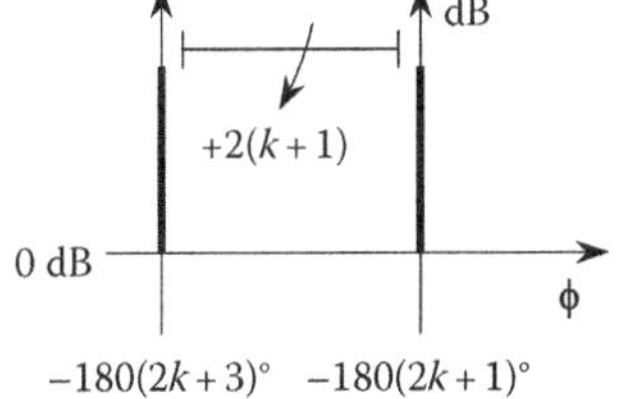

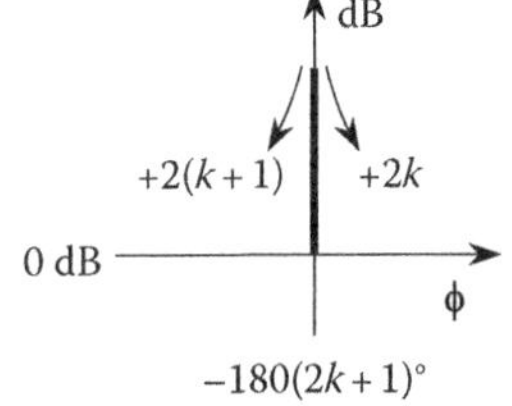

FIGURE D.12
How to obtain N in the Nichols diagram.

Examples

The following examples illustrate how to apply the Nyquist stability criterion with the aforementioned rules in the Nichols diagram directly.

First consider an open-loop transfer function according to Equation D.7. It is a conditionally stable plant. The number of poles p of $L(s)$ in the RHP is $p = 1$ (the pole at $s = +1$) and the number of encirclements $N = -1$ (see Figure D.12b, and Figures D.13 and D.14). This means that the closed-loop control system is stable since, $z = N + p = -1 + 1 = 0$:

$$L_0(s) = \frac{k}{(s-1)(s+2)(s+5)}, \quad k = 11 \tag{D.7}$$

Similarly, now consider an open-loop transfer function according to Equation D.8. It is also a conditionally stable plant. The number of poles p of $L(s)$ in the RHP is again $p = 1$ (due to the pole at $s = +1$) and the number of encirclements $N = +2-1 = +1$ (see Figures D.12a and b, D.15, and D.16). This means that the closed-loop control system is unstable since, $z = N + p = +1 + 1 = 2$:

$$L_0(s) = \frac{k}{(s-1)(s+2)(s+5)}, \quad k = 110 \tag{D.8}$$

Also consider an open-loop transfer function according to Equation D.9. The number of poles p of $L(s)$ in the RHP is $p = 0$ and the number of encirclements $N = +4-2-2 = 0$ (see Figures D.12a and c and D.17). This means that the closed-loop control system is stable since, $z = N + p = 0 + 0 = 0$:

$$L_0(s) = \frac{0.000002(10s+1)^6}{s^7(s+1)} \tag{D.9}$$

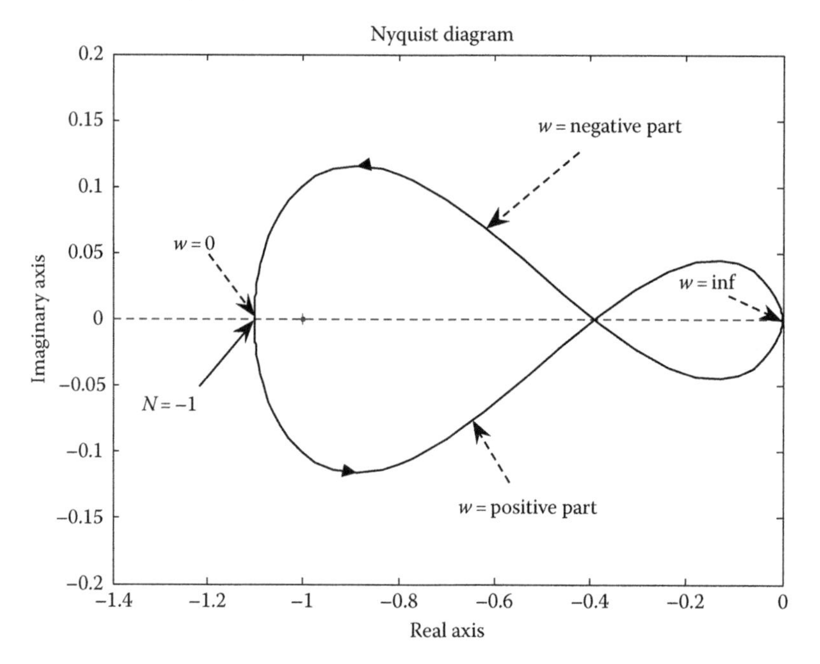

FIGURE D.13
Nyquist diagram for Equation D.7; $N = -1$.

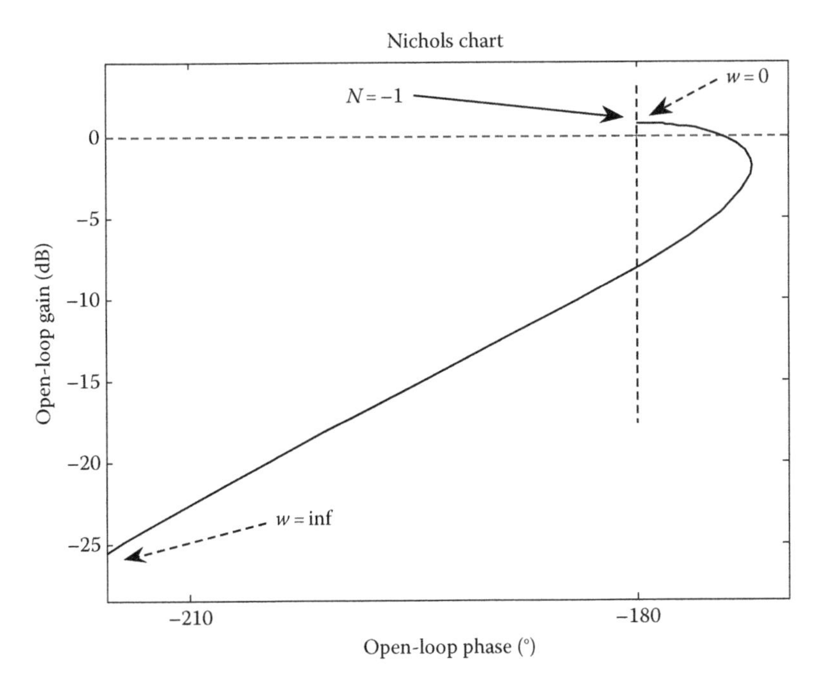

FIGURE D.14
Nichols diagram for Equation D.7; $N = -1$.

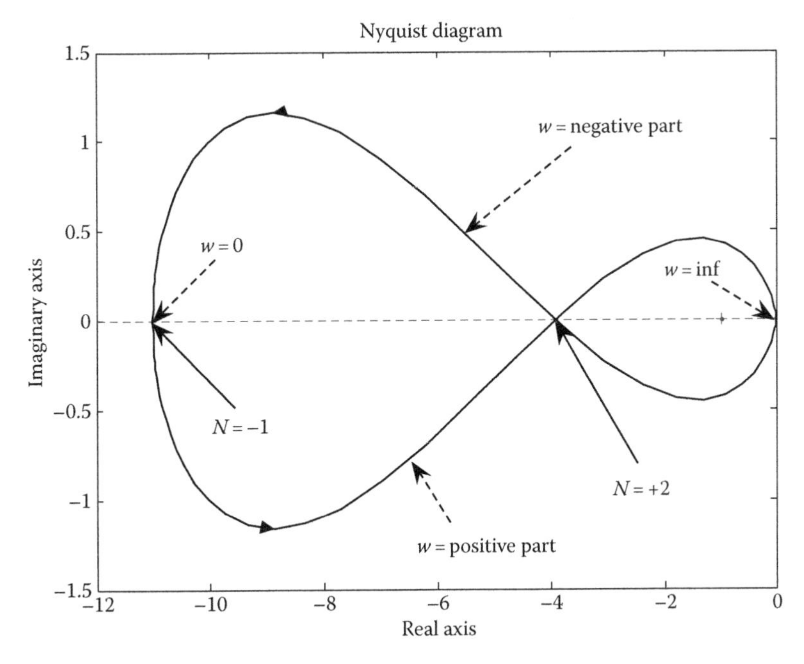

FIGURE D.15
Nyquist diagram for Equation D.8; $N = +2 - 1 = +1$.

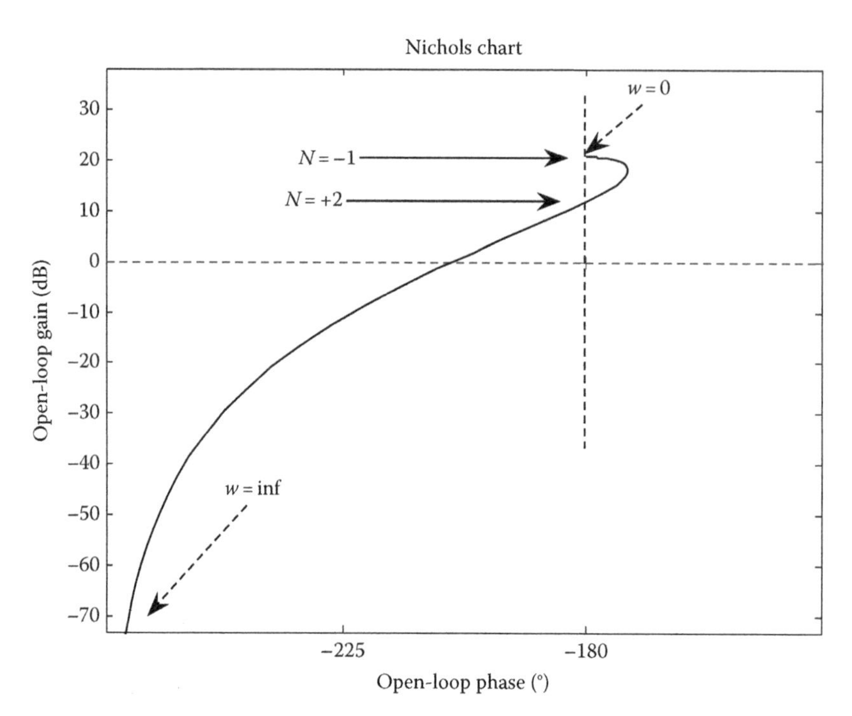

FIGURE D.16
Nichols diagram for Equation D.8; $N = +2 - 1 = +1$.

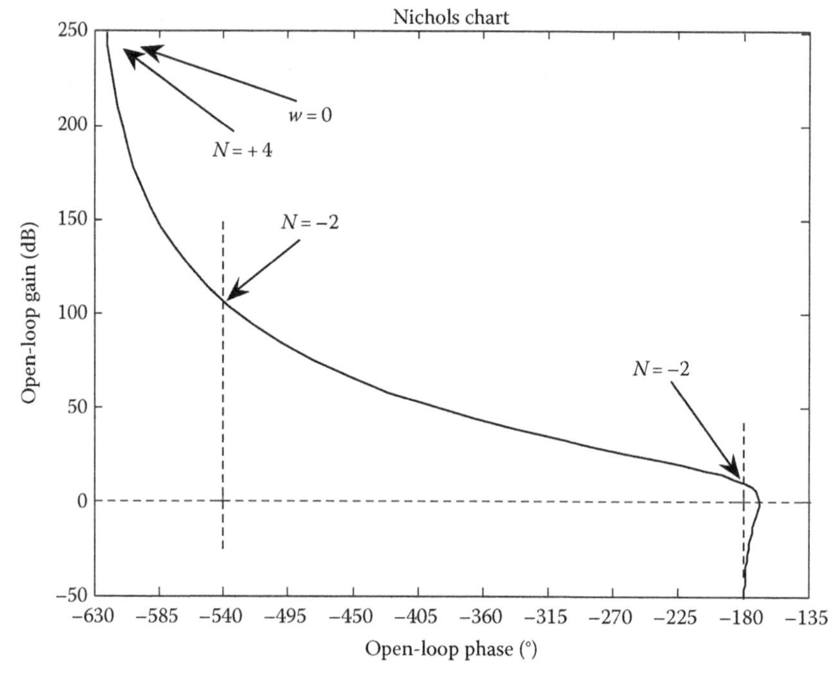

FIGURE D.17
Multiple-sheeted Nichols diagram for Equation D.9; $N = +4 - 2 - 2 = 0$.

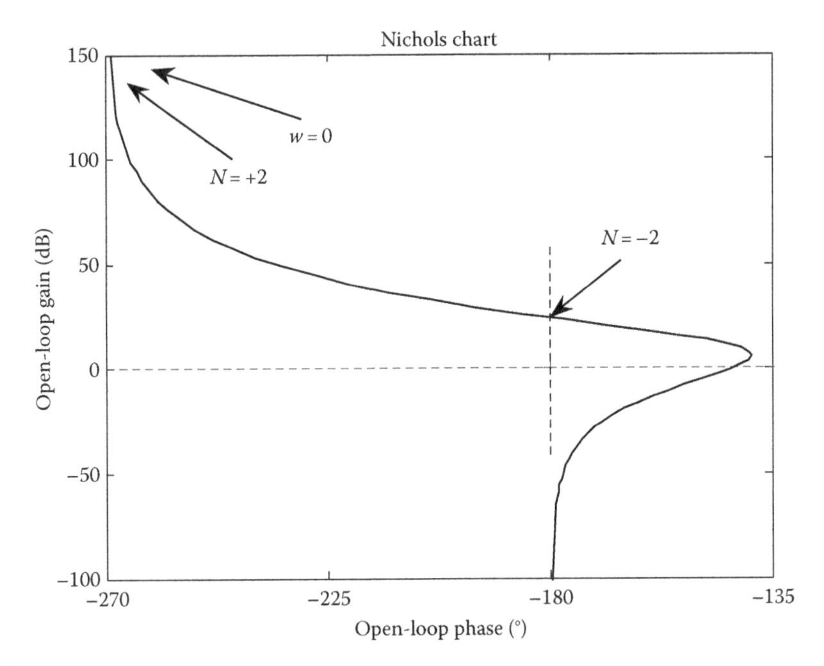

FIGURE D.18
Nichols diagram for Equation D.10; $N = +2 - 2 = 0$.

Finally, consider an open-loop transfer function according to Equation D.10. The number of poles p of $L(s)$ in the RHP is $p = 0$ and the number of encirclements $N = +2 - 2 = 0$ (see Figures D.12a and c and D.18). This means that the closed-loop control system is stable since, $z = N + p = 0 + 0 = 0$:

$$L_0(s) = \frac{0.01(10s+1)^2}{s^3(s+1)} \tag{D.10}$$

D.13 On the Existence of Controllers

D.13.1 Fundamental Theorem of Feedback Control[213]

"Every plant $P(s)$ of order n (order of the denominator) can be stabilized with a feedback compensator $G(s)$ of order no greater than $n - 1$." This is true if we consider the possibility of using not only stable compensators, but also unstable compensators if required.

D.13.2 Parity Interlacing Property[213]

A rational function $P(s)$ has the *parity interlacing property* (*p.i.p*) if $P(s)$ has an even number of poles between each pair of zeros of on the positive real axis (including infinity). The *p.i.p* condition is a necessary and sufficient condition for the existence of a stable stabilizing compensator $G(s)$. This is a very useful condition for unstable plants $P(s)$ with

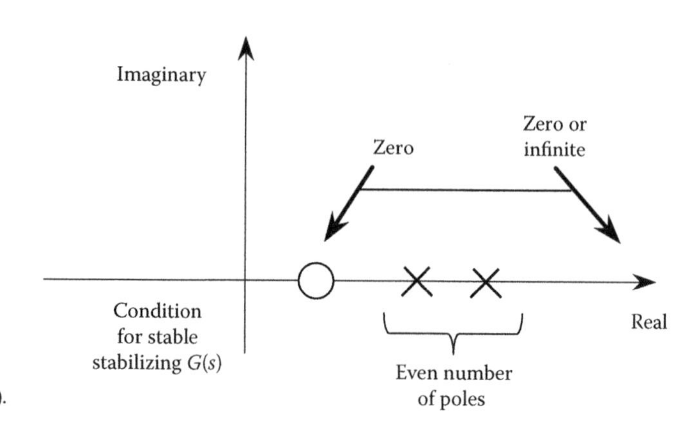

FIGURE D.19
Parity interlacing property for $P(s)$.

non-minimum-phase zeros. Of course, if the plant is stable, a stable compensator $G(s)$ can always be found (Figure D.19).

D.13.3 Passive Systems[213]

A plant with a transfer function $P(s)$ that satisfies the phase condition

$$-180^\circ \leq \text{Phase}\, P(s) \leq +180^\circ \tag{D.11}$$

is said to be a passive system (or dissipative system).

A passive system can always be robustly stabilized with a feedback strictly positive real compensator $G(s)$. Passive system also means that the physical plant has no internal sources of energy.

Appendix E: Fragility Analysis with QFT[70]

E.1 Introduction

In the last few years a significant number of papers related to controller fragility have appeared in the literature (see Refs. 70, 274–284). The fragility problem arises when the controller synthesis techniques and/or the digital implementation tend to produce control laws with a high sensitivity of the closed-loop stability to small changes in compensator/controller coefficients. As it is claimed in the literature, the controller fragility is mainly produced by using popular robust and optimal control synthesis methods, standard model-based identification techniques, certain parameterizations, and different levels of accuracy in compensator/controller implementation.

Thus, this appendix deals with the controller fragility problem. It introduces a practical method to analyze the controller resiliency/fragility to small changes in its coefficients.[70] The method simultaneously takes into account the plant model with parametric and nonparametric uncertainties, the robust stability/performance specifications, and the uncertainty in the compensator/controller coefficients. Based on the QFT (quantitative feedback theory) technique, the method studies the controller resiliency/fragility at every frequency and for both stability and any other control specification. The method starts with the standard QFT design steps, templates, bounds, loop shaping, etc. (or any other available technique), to design the robust controller. Once the nominal controller is designed, a set of new templates are then defined (the *extended templates*) that combine the model plant uncertainty with the controller uncertainty at every frequency. From these templates, and taking into account the previous robust stability specifications, a set of new bounds are defined (the *extended bounds*). Based on them, the loop-shaping procedure on the Nichols chart, with the same previous nominal plant and controller, is applied. Finally, a comparative analysis at every frequency of the bounds, extended bounds, and the nominal open-loop transfer functions is made to discover the controller resiliency/fragility that can be achieved.

This appendix is organized into five sections. Section E.2 presents a brief study of the fragility state of the art. Section E.3 introduces the practical method described above to analyze the controller fragility. In Section E.4 two examples are presented to illustrate the advantages of the proposed method. Finally, Section E.5 summarizes the conclusions.

E.2 State of the Art

The fragility problem directly relates to the issue of accuracy of controller implementation and typically requires a trade-off between the implementation accuracy and the performance deterioration.[275]

Some earlier researchers have expressed the existence of fragility in some controller realizations, originally using classical system sensitivity theory.[2,193,216]

E.2.1 Keel and Bhattacharyya's Work

In 1997, Keel and Bhattacharyya wrote an impressive paper entitled "Robust, Fragile or Optimal?,"[274] which shocked the international community and initiated a large set of comments and replies.[275–281,284] They demonstrated the inherent fragility of some well-known H_∞, H_2, L_2, and μ controllers[275] when small changes are introduced in the controller coefficients. Their research was based on the following premises.

On the one hand, the accuracy of any controller may be subjected to round-off errors in numerical computations and uncertainty due to analog/digital and digital/analog conversions (for digital controllers) or due to adjustment and tuning (in the case of analog controllers). Consequently, uncertainty of controller parameters is always needed to be taken into consideration.

On the other hand, some design techniques of robust controllers based on H_∞, H_2, L_2, and μ formulations[275] assume that the implementation is carried out exactly. At first sight, this assumption is legitimate taking into account the insignificant order of magnitude of controller uncertainty in comparison with that of the plant. However, as is shown in this appendix, the system can be destabilized or its behavior kept away from the design specifications due to small changes in the controller coefficients.

Because of that, Keel and Bhattacharyya proposed the parametric stability margin (ρ) as a fragility measure. It indicates the maximum permitted variation of the controller coefficients without destabilizing the closed-loop system. This measure is evaluated as the ratio of the biggest ball or hypersphere of the stability region in the parameter space. Its value is negligible in all the analyzed cases in their paper.[274] To back up these results, they calculated the phase and gain margins (classical measures of relative stability), obtaining the same conclusion. The methodologies of design are to be held responsible for fragility, since they do not consider the proximity of instability regions while determining the optimum solutions.

Consequently, Keel and Bhattacharyya suggested including the parametric stability margin (ρ) in the optimization algorithm of the design. Thus, there will exist a certain tolerance in controller coefficients. They had already put into practice this method in 1995 to analyze the resilience of a P and a PI controller,[209] and of a PID controller.[212]

It was in a subsequent publication[280] when they examined the sensitivity associated with the digital implementation of the same systems. Then they concluded that the stability margin increases with the sample time. They also stated that the hybrid configuration (closer to reality with a continuous plant controlled by a digital system through analog/digital and digital/analog converters) showed inaccuracy in the control signal value. It is caused by quantization errors (finite word length) of analog/digital conversions. Besides, instability may arise between samples, even with a wide stability margin of the system. The solution would therefore be to define a stability margin that captures the unstable behavior between samples of the hybrid system. However, this question is still unresolved by the parametric theory.

E.2.2 Jadbabaie, Chaouki, Famularo, and Dorato's Work

In 1998, A. Jadbabaie, T. Chaouki, D. Famularo, and P. Dorato[278] presented a robust and resilient state feedback controller, which is also optimal regarding a quadratic performance measure. The controller is premised on the fact that resilience is unachievable if only robustness is demanded. This new synthesis method consists in solving a system of linear matrix inequalities (LMIs) in which uncertainty is considered for both the plant and the controller. They also work on robust non-fragile linear-quadratic (LQ) controllers: the static-state feedback case.[276]

E.2.3 Haddad and Corrado's Work

Alternatively, two researchers, W.M. Haddad and J.R. Corrado,[277,282] presented a procedure to synthesize robust controllers of fixed structure that guaranteed no fragile designs given additive and/or multiplicative uncertainty of the controller gain. This rigorous method is based on the quadratic limits of a Lyapunov function and obtains a set of Riccati algebraic equations and a sufficient condition for stability, resilience, and robustness.

E.2.4 Yang, Wang, and Lin's Work

It is in 1999 that this team of researchers, G.H. Yang, J.L. Wang, and C. Lin,[283] showed the necessary and sufficient conditions for the design of robust and resilient H_∞ controllers considering additive uncertainty of the controller itself. This new approach searched for the optimum solution of a set of Riccati inequalities expressed in terms of the parameters with uncertainty that belong to both the plant and the controller.

E.2.5 Mäkilä's Work

These prior research groups solved the problem of redesigning their techniques. However, they neither analyzed the causes of fragility nor defined a measure of it. It is P.M. Mäkilä who, in both publications,[281,284] reconsidered other alternative reasons and provided new solutions to the problem. Basically, the origin of fragility is not found in some abstract theories such as instability regions but in two more practical approaches:

- *The application of an inadequate optimization criterion.* The McFarlane Glover method and the coprime factorization could be the origin of fragility, since they optimize the closed-loop response in terms of the four main basic operators of the system (i.e., the four transfer functions that relate the external and internal closed-loop variables).
- *The use of an incorrect controller implementation.* This was shown by carrying out some different realizations (companion, expanded, balanced, and standard) and examining how each response deviation gave different precision (measured in terms of the number of significant digits of the coefficients for each implementation).

As noted, one of the main measures employed in the literature to characterize the fragility is the parametric stability margin (ρ). It is an elegant mathematical device in the space of parameters. However, it has a high computational cost. Accordingly, the objective of this appendix is the identification of the controller fragility by means of a method that is transparent, quick, simple to interpret and applicable to any design. The QFT philosophy is thus introduced here in order to discern graphically and quantitatively the fragility or the resilience of a controller in the frequency domain.

With this background and with the aim of solving industrial problems, QFT is also propounded as a tool to analyze the effect of the controller implementation in an anticipated way.

E.3 Proposed QFT Methodology for Fragility Analysis

The proposed methodology[70] analyzes the controller resiliency or fragility to small changes in its coefficients. It simultaneously takes into account the plant model with parametric and nonparametric uncertainty, the robust stability and performance specifications, and the uncertainty in the coefficients of the controller. The method is based on QFT.

Consider an uncertain plant $\{P(s)\}$ and its respective robust controller $G_0(s)$, designed by means of any robust technique. It is verified that, under ideal conditions, the closed loop is robustly stable and the robust control specifications $\delta(\omega)$ are fulfilled for all the plants.

Now, consider that the uncertainty also affects the controller, which is represented as $\{G(s)\}$. The analysis of the fragility introduced by the controller is carried out by using the *extended plant approach*, that is,

$$\{P_e(s)\} = \{P(s)\}\{G(s)\} \tag{E.1}$$

Stability and robust performance are analyzed after choosing $P_e^o(s) = P_o(s)G_0(s)$ as the nominal extended plant. At this stage, a unitary controller, transparent with regard to the analyzed system, is introduced to control $P_e^o(s)$.

It should be remembered that, for a nominal plant, the QFT methodology converts system specifications and model plant uncertainty into a set of bounds for every frequency of interest that are to be fulfilled by the nominal open-loop transfer function.

In this context, the proposed methodology defines a set of new templates (the *extended templates*) that combine the model plant uncertainty $\{P(s)\}$ with the controller uncertainty $\{G(s)\}$ at every frequency of interest. By using these extended templates, and taking into account the previous robust stability specifications $\delta(\omega)$, a set of new bounds (the *extended bounds*) are defined and are now utilized in the loop-shaping procedure on the Nichols chart with the same nominal plant and controller $P_e^o(s)$ (see Figure E.1). Finally, a comparative analysis of the extended bounds and original bounds illustrates the controller resiliency/fragility at every frequency.

E.4 Design Procedure

The proposed resiliency/fragility analysis methodology is composed of the following 11 steps:

Step 1: Define the plant model with uncertainty $\{P(s)\}$.

Step 2: Select the nominal plant $P_0(s)$ within the set of plants defined by the uncertainty.

Step 3: Define the stability and performance specifications $\delta(\omega)$ required for the plant.

Step 4: Design a robust controller $G_0(s)$ using any available technique: QFT, H_∞, H_2, μ, etc.

Step 5: Select the interval of variation of the controller coefficients $\{G(s)\}$.

Step 6: Define the extended plant as $\{P_e(s)\} = \{P(s)\}\{G(s)\}$.

Step 7: Choose $P_e^o(s) = P_0(s)G_0(s)$ as the nominal extended plant.

Step 8: Generate the extended templates from the extended plant $\{P_e(s)\}$ and for the frequencies of interest (Figure E.1).

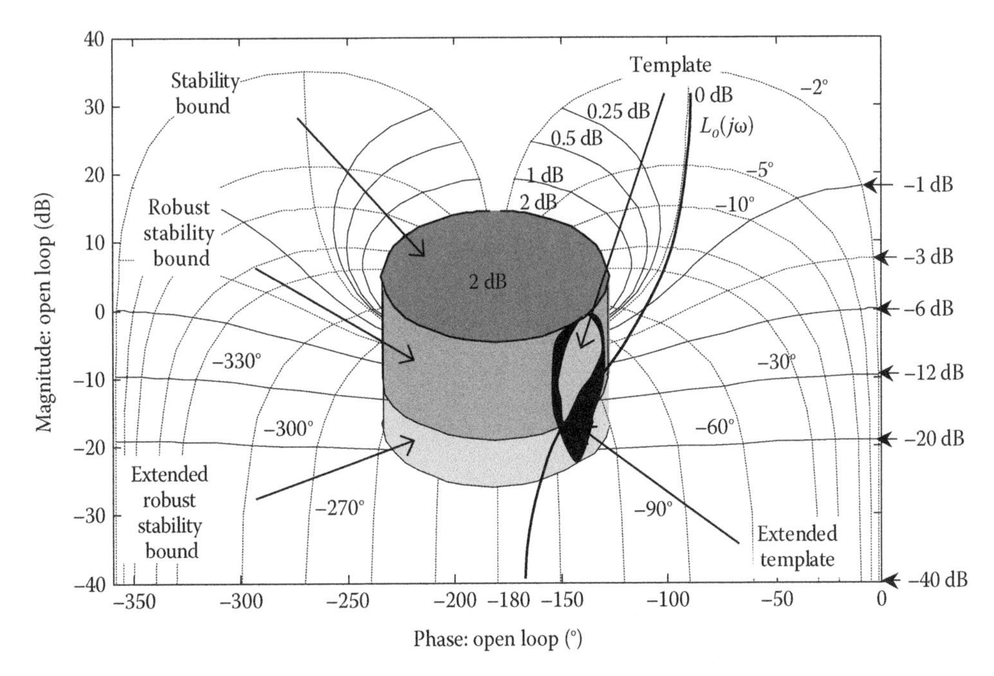

FIGURE E.1
Extended templates and extended stability bounds concepts on the Nichols chart.

Step 9: Calculate the extended bounds from the extended templates and the previous stability and performance specifications $\delta(\omega)$ (see Figure E.1).

Step 10: Apply a unitary controller, $G(s) = 1$, to shape $P_e^o(s)$ over the extended bounds.

Step 11: Being $L_e^o(s) = P_e^o(s)\ G(s)$, with $G(s) = 1$, if $L_0(s)$ penetrates the forbidden region delimited by the extended robust stability bounds, then the controller will be fragile at these frequencies. Otherwise it will be resilient (see Figure E.1).

E.5 Examples

In this section, two examples are presented to illustrate the proposed methodology. The first example consists of a simple and clarifying theoretical system. The second example applies the new method to the analysis of fragility of a system discussed by Keel and Bhattacharyya in Ref. 274.

E.5.1 Example 1 (A Simple System)

An elementary control system with parametric uncertainty is considered firstly. The controller is designed by means of QFT. The plant to be controlled is

$$\{P(s)\} = \frac{ka}{s(s+a)} \quad \begin{cases} k \in [1,10] \\ a \in [1,10] \end{cases} \tag{E.2}$$

The nominal plant is

$$P_0(s) = \frac{1}{s(s+1)} \tag{E.3}$$

The robust controller is designed by using QFT to give the robust stability specification $\delta_1(\omega) = 1.2$, which is equivalent to $GM = 1.8333\,\text{dB}$ and $PM = 50°$. The controller is

$$G_0(s) = \frac{k_G((1/z_1)s+1)((1/z_2)s+1)}{((1/p_1)s+1)((1/p_2)s+1)((1/p_3)s+1)} \tag{E.4}$$

where

$$\begin{array}{lll} k_G = 1.1 & z_2 = 2.7 & p_2 = 20 \\ z_1 = 23 & p_1 = 220 & p_3 = 18 \end{array}$$

The design is made by the inclusion of poles (p_1, p_2, p_3), zeros (z_1, z_2), and the gain k_G in the controller structure, so that the nominal open-loop transfer function $L_0(s)$ is adjusted to the stability bound at each frequency value while trying to minimize the controller gain, as in Figure E.2. The nominal open-loop transfer function becomes $L_0(s) = P_0(s)G_0(s)$.

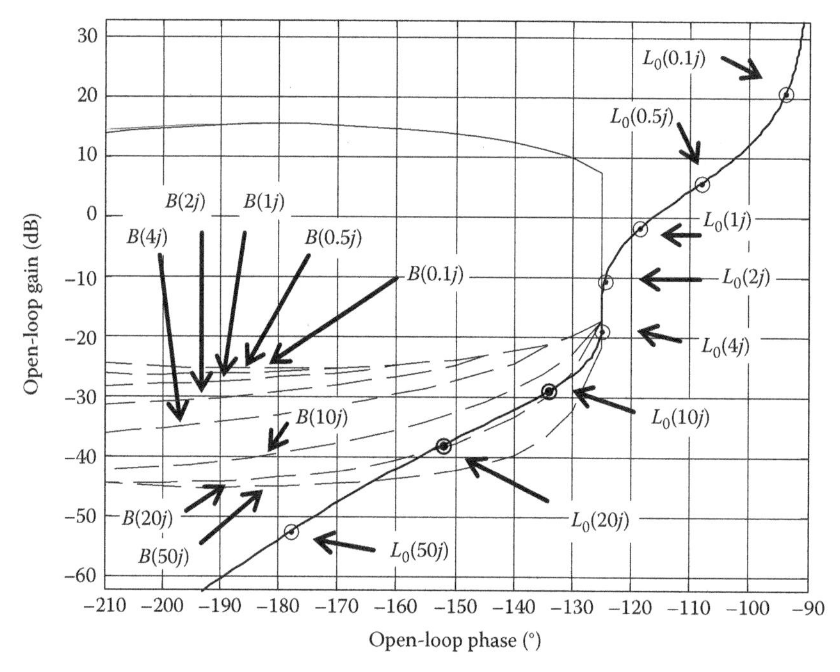

FIGURE E.2
Nichols chart with the stability bounds of the system $\{P(s)\}$ and the nominal open-loop function $L_0(s)$ with the designed controller $G_0(s)$.

The fragility analysis is now undertaken assuming the uncertainty of the controller parameters are

$$\begin{array}{lll} k_G \in [1.1,2] & z_2 \in [2.7,3] & p_2 \in [20,21] \\ z_1 \in [23,24] & p_1 \in [220,221] & p_3 \in [18,19] \end{array}$$

These intervals are arbitrarily taken in order to make clearer the effect of the fragility associated with the controller uncertainty. The model of the extended system results in the expression $\{P_e(s)\} = \{P(s)\}\ \{G(s)\}$, so that

$$P_e(s) = \frac{ka}{s(s+a)} \frac{k_G((1/z_1)s+1)((1/z_2)s+1)}{((1/p_1)s+1)((1/p_2)s+1)((1/p_3)s+1)} \tag{E.5}$$

Now, eight uncertain parameters exist, whose variation intervals are as follows:

$$\begin{array}{lll} k \in [1,10] & z_1 \in [23,24] & p_2 \in [20,21] \\ a \in [1,10] & z_2 \in [2.7,3] & p_3 \in [18,19] \\ k_G \in [1.1,2] & p_1 \in [220,221] & \end{array}$$

Next, the templates of this extended system are established; that is, the *extended templates*. From these, the new bounds (*extended bounds*) are drawn up for the same robust stability specification ($\delta_1(\omega) = 1.2$).

The nominal plant of the extended system is $P_e^o(s) = P_0(s)G_0(s)$. Note that the chosen extended nominal plant coincides with the previous nominal plant multiplied by the controller obtained utilizing the QFT design.

Subsequently, a unitary controller to shape $P_e^o(s)$ over the extended bounds is chosen as shown in Equation E.6. In this way, the open-loop transfer function remains the same as before and thus it enables the designer to make the comparison and determine the controller resilience. The performance of this new expanded nominal plant is checked out with regard to the extended bounds (stability bounds of the extended system), as shown in Figure E.3:

$$L_e^o(s) = P_e^o(s)G(s) = P_0(s)G_0(s) = L_0(s), \quad \text{with } G(s) = 1 \tag{E.6}$$

When comparing both Figures E.2 and E.3, one can notice that, for the controller given by Equation E.4, the extended system violates the stability requirements at the frequencies 10 and 20 rad/s ($L_0(j10)$ and $L_0(j20)$ are above $B_e(j10)$ and $B_e(j20)$, respectively).

In other words, the open-loop transfer function $L_0(s)$ penetrates the forbidden region delimited by the extended robust stability bounds, although the original open-loop transfer function (see Figure E.2) remained under the limits. In conclusion, it is evident that this is a fragile controller.

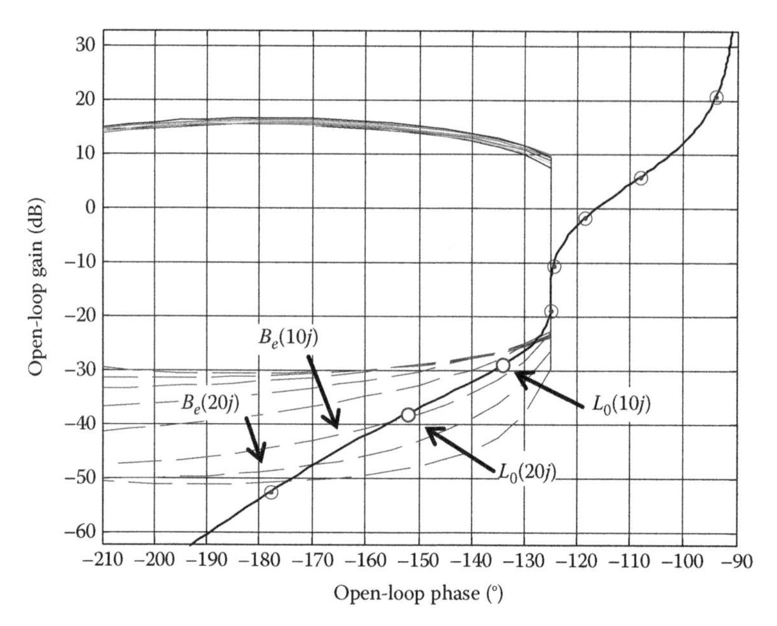

FIGURE E.3
Nichols chart of the extended stability bounds of the system $\{P_e(s)\}$ and the nominal open-loop function $L_0(s)$.

E.5.2 Example 2 (Example 2 in Ref. 274)

The fragility of the arbitrary controller, presented by Keel and Bhattacharyya in Ref. 274, is now analyzed for the uncertain unstable and non-minimum-phase plant:

$$\{P(s)\} = \frac{k(s-1)}{(s+1)(s-p)} \quad \begin{cases} p \in [1,2] \\ k \in [1,4] \end{cases} \tag{E.7}$$

The nominal plant is taken as

$$P_0(s) = \frac{(s-1)}{(s+1)(s-2)} = \frac{s-1}{s^2-s-2} \tag{E.8}$$

To design the proposed controller, the closed poles have been placed on a circle of radius $\sqrt{2}$ spaced equidistantly in the left-half plane. This controller is determined to be

$$G_0(s) = \frac{q_1^0 s + q_0^0}{s + p_0^0} \tag{E.9}$$

where the nominal controller coefficient values are

$$q_1^0 = 11.44974739$$

$$q_0^0 = 11.24264066$$

$$p_0^0 = -7.03553383$$

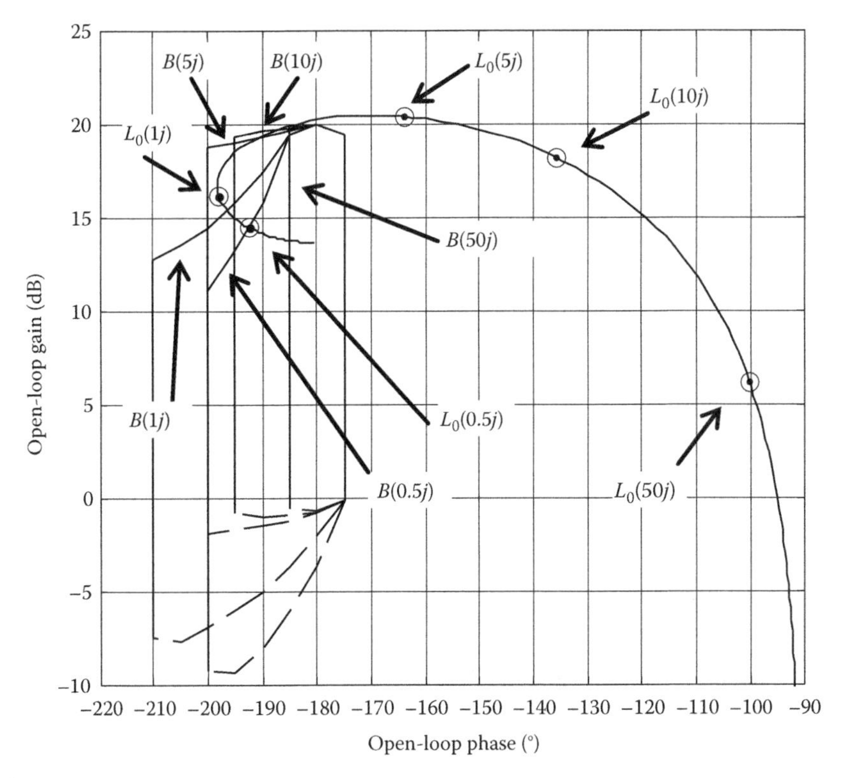

FIGURE E.4
Nichols chart with the stability bounds of the system $\{P(s)\}$ and the nominal open-loop function $L_0(s)$ with the designed controller $G_0(s)$.

By introducing the proposed fragility QFT method, a study of the robustness and fragility of this controller is to be determined. Examining iteratively the nominal open-loop transfer function $L_0(s) = P_0(s)\ G_0(s)$, it is clear that the system verifies the stability specification of $\delta_1(\omega) = 11$, equivalent to $GM = 1.0909$ dB and $PM = 6°$ (an example too close to instability, anyway). This stability specification is shown on the Nichols chart of Figure E.4. Note that the function $L_0(s)$ remains outside the bounds for all frequencies.

Keel and Bhattacharyya (Ref. 274) assumed the same minimum uncertainty in the controller coefficients will be able to destabilize the control system. In other words, the extended system that is now to be considered includes the uncertainty of both the plant and the controller $\{P_e(s)\} = \{P(s)\}\{G(s)\}$:

$$\{P_e(s)\} = \left(\frac{k(s-1)}{(s+1)(s-p)}\right)\left(\frac{q_1 s + q_0}{s + p_0}\right) \tag{E.10}$$

where the parameters vary within the intervals

$$\begin{array}{ll} q_1 \in [10.293425504, 11.44974739] & p \in [1,2] \\ q_0 \in [9.37759508, 11.24264066] & k \in [1,4] \\ p_0 \in [-7.5435313, -7.03553383] & \end{array}$$

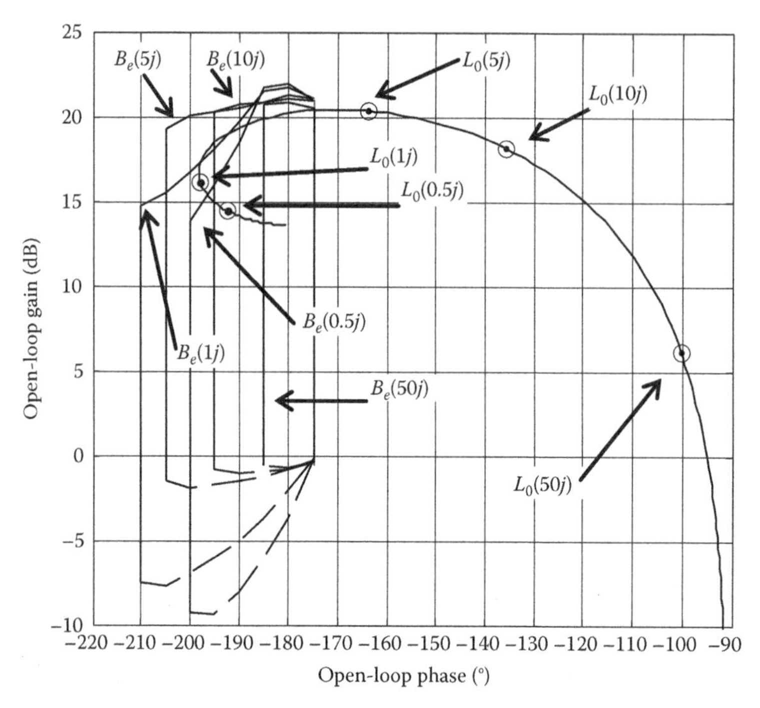

FIGURE E.5
Nichols chart of the extended stability bounds of the system $\{P_e(s)\}$ and the nominal open-loop function $L_0(s)$.

Next, an extended nominal open-loop transfer function is considered, which coincides with the previous nominal one $L_0(s)$. This results in the nominal extended plant:

$$P_e^o(s) = \left(\frac{s-1}{(s+1)(s-2)}\right)\left(\frac{11.44974739s + 11.2426066}{s - 7.03553383}\right) \tag{E.11}$$

On the basis of this nominal plant, the new extended templates are determined. As regards the extended bounds, they are calculated with the precedent stability specification, $\delta_1(\omega) = 11$, and the extended templates. Finally, Figure E.5 shows how the open-loop transfer function $L_0(s)$ does not respect the exclusion zone, as it is noticed at the frequencies 1 and 0.5 rad/s. Consequently, it can be stated that this is a fragile controller, since the design specifications are not fulfilled when uncertainty in the controller parameters is considered.

E.6 Summary

In this appendix, the controller fragility problem was studied. A new method was introduced to analyze the controller resiliency/fragility to small changes in its coefficients. The method simultaneously took into account the plant model with parametric and nonparametric uncertainties, the robust stability/performance specifications, and the uncertainty of the controller coefficients. Based on the QFT technique, the method studied the controller resiliency/fragility at every frequency and for both stability and any other control specification. The two examples showed the advantages of the proposed methodology. The analysis illustrated the practical use of the new technique.

Appendix F: QFT Control Toolbox: User's Guide[149,150]

F.1 Introduction

This appendix presents the interactive object-oriented CAD tool for quantitative feedback theory (QFT) controller design included with the book: the QFT Control Toolbox (*QFTCT*), version 1.0. It includes the latest technical quantitative robust control achievements included in the book within a user-friendly and interactive environment.

The toolbox has been developed by Professor Mario Garcia-Sanz and Augusto Mauch at the Public University of Navarra (UPNA) and Case Western Reserve University (CWRU), and by Dr. Christian Philippe, at the European Space Agency (ESA-ESTEC). It was also supported by MCyt (Ministerio de Ciencia y Tecnología, Spain).[149,150]

The *QFTCT* runs under MATLAB® and shows a special architecture based on seven principal windows (W1, *Plant definition*; W2, *Templates*; W3, *Control specifications*; W4, *Bounds*; W5, *Controller design*; W6, *Prefilter design*; W7, *Analysis*) and a common central memory.

It also includes a library of basic and advanced functions to be selected in the corresponding windows, allows a multitasking/threading operating system, offers a user-friendly and interactive environment by using an object-oriented programming, permits easily to rescale the problem from single-input-single-output (SISO) to multiple-input-multiple-output (MIMO), and uses reusable code.

The *QFTCT* CAD package includes the latest quantitative robust control system design achievements within a user-friendly and interactive environment. The main objective is to design and implement a 2DOF (two-degree-of-freedom) robust control system, see Figure F.1, for a plant with uncertainty, which satisfies the desired performance specifications, while achieving reasonably low loop gains (avoiding or minimizing bandwidth, sensor noise amplification, and control signal saturation).

The compensator $G(s)$ and a prefilter $F(s)$ are to be designed to meet robust stability and robust performance specifications and to deal with reference tracking $R(s)$, disturbance rejection $D(s)$, signal noise attenuation $N(s)$, and control effort minimization $U(s)$, while reducing the cost of the feedback (excessive bandwidth).

Figure F.2 lists the steps involved in the QFT design procedure. Figure F.3 represents the CAD flowchart of the MISO analog QFT design procedure, and Figure F.4 presents an overview of the QFT design process. The *QFTCT* files can be downloaded from the websites http://cesc.case.edu and http://www.crcpress.com/product/isbn/9781439821794.

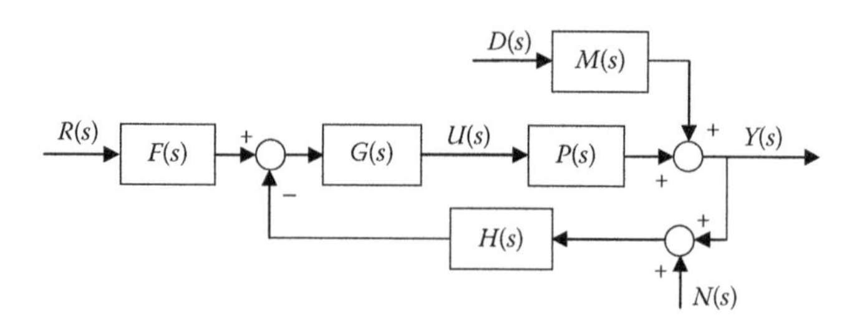

FIGURE F.1
MISO feedback control system.

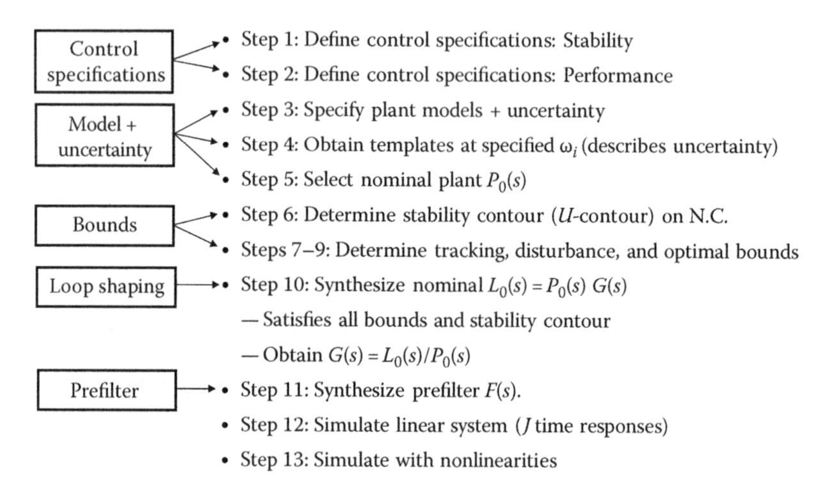

FIGURE F.2
QFT design procedure.

F.2 QFT Control Toolbox Windows

This section presents an overview of the QFT design procedure for this *QFTCT* CAD package.

Required products: MATLAB 7.9.0 (R2009b) or later version, Control System Toolbox 8.4 or later version.

Platform: PC or Mac.

Installation: To install the *QFTCT*, copy the folder containing the toolbox to a safe place (it can be any folder) and then add that folder to MATLAB paths (file → set path → add folder → save → close).

Start: To start a new session with the QFT Control Toolbox, type the following command in the MATLAB Command Window: *QFTCT*.

F.2.1 General Description

Windows: The toolbox contains windows of the form shown in Figure F.5. They fall into seven categories: W1, *Plant definition*; W2, *Templates*; W3, *Control specifications*; W4, *Bounds*; W5, *Controller design*; W6, *Prefilter design*; W7, *Analysis*. The windows have a "bar" at the top, listing these seven categories as shown in Figure F.6.

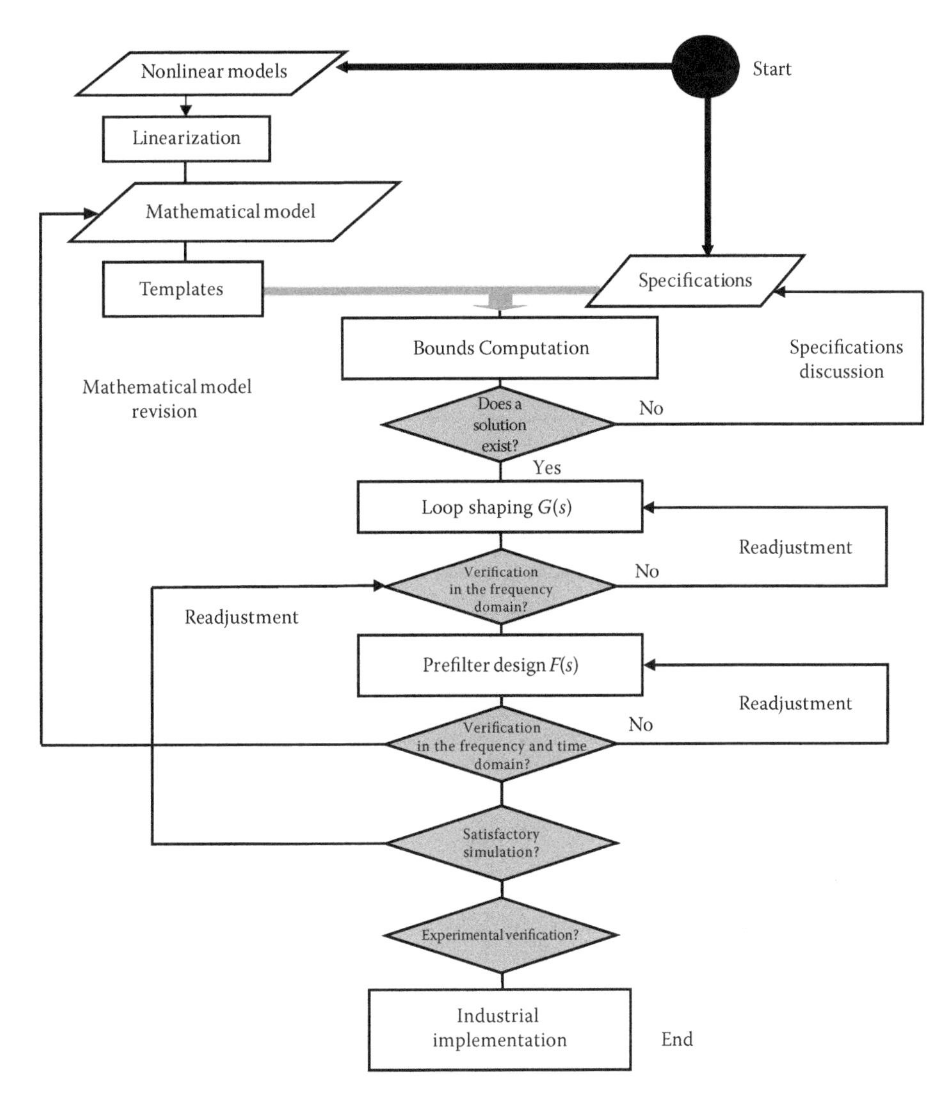

FIGURE F.3
QFT flowchart for a MISO control system.

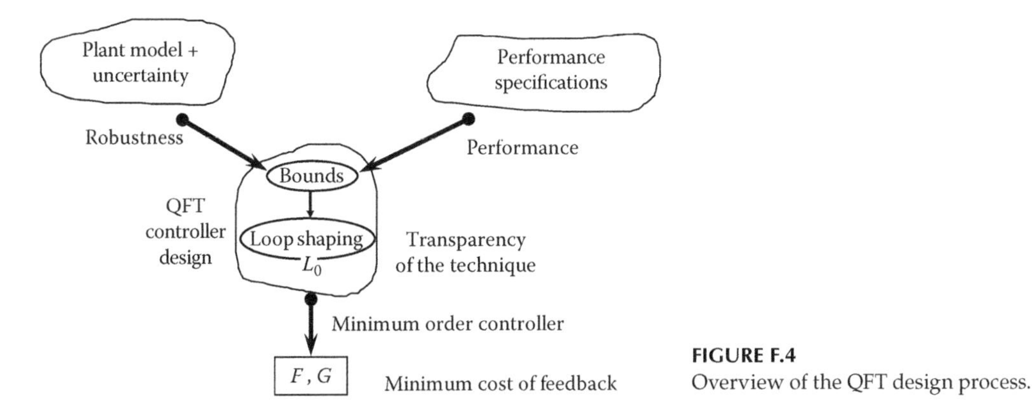

FIGURE F.4
Overview of the QFT design process.

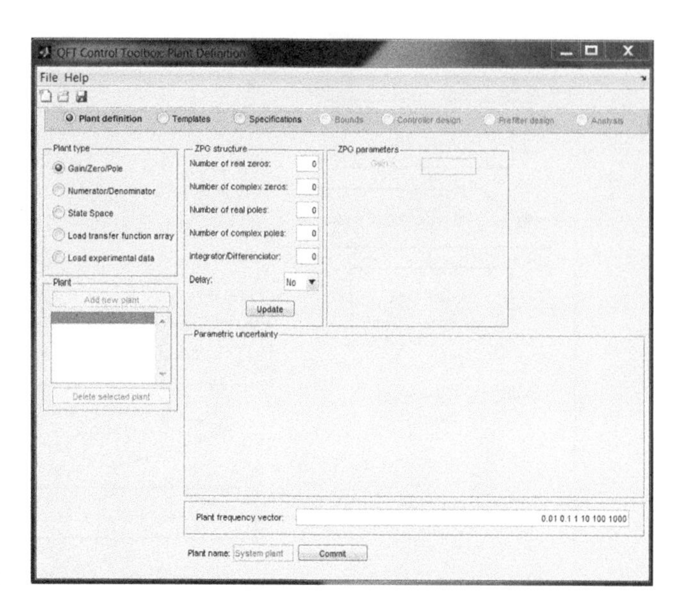

FIGURE F.5
Plant definition window.

FIGURE F.6
Window toolbar listing the seven window categories.

All the windows have a navigator panel (see Figure F.5) at the top that allows the user to change the active window. Not all the windows are available from the beginning, that is, some windows have prerequisites that have to be fulfilled in order to activate the windows. The seven categories are listed next, following the order that they have to be executed in the design process:

1. *Plant definition*: No prerequisites.
2. *Templates*: Needs Plant definition.
3. *Specifications*: Needs Plant definition.
4. *Bounds*: Needs Plant definition, Templates, and Specifications.
5. *Controller design*: Needs Plant definition, Templates, Specifications, and Bounds.
6. *Prefilter design*: Needs Plant definition, Templates, Specifications, Bounds, and Controller design. It is used only for reference-tracking problems.
7. *Analysis*: Needs Plant definition, Templates, Specifications, Bounds, and Controller design.

Toolbar: All the windows of the toolbox have a toolbar (see Figure F.7a), which allows the user to create a new project, open an existing one, and save the current project. The toolbar of the windows that have plots (Templates, Bounds, Controller design, Prefilter design, and Analysis) has controls to zoom in and out on the plots (see Figure F.7b).

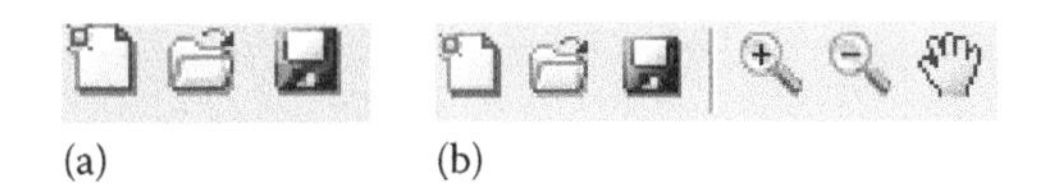

FIGURE F.7
Toolbars for Windows: (a) W1 and W3, (b) W2, W4, W5, W6, and W7.

Menu: All the windows have the following two menus:

1. *File*: Allows the user to create new projects, open existing projects, save the current project, and save the current project in a new file.
2. *Help*: Provides access to the User's Guide and to the information about the toolbox (authors, version, etc.).

In addition,

1. The windows that have plots have a menu item in the File menu that allows saving the plot in a .emf, .bmp, or .fig file.
2. Some window categories have specific menu items, for example,
   a. Templates. In the *File menu*, there is a menu item that allows the user to export the nominal plant as a transfer function.
   b. Controller design. In the *File menu* there are menu items that allow the user to (see Figure F.8)
      i. *Load a controller* from the hard drive (*.contr).
      ii. *Load a list of controllers* from the hard drive (*.contrList).
      iii. *Save selected controller* to the hard drive (*.contr).
      iv. *Save controller list* to the hard drive (*.contrList).
      v. *Export* the selected controller as a transfer function to workspace.
      vi. *Save figure* to the hard drive (*.emf or *.bmp or *.fig).
      vii. *Check the stability*. The MATLAB *isstable* function is used to verify the stability of the closed-loop system with the selected controller.

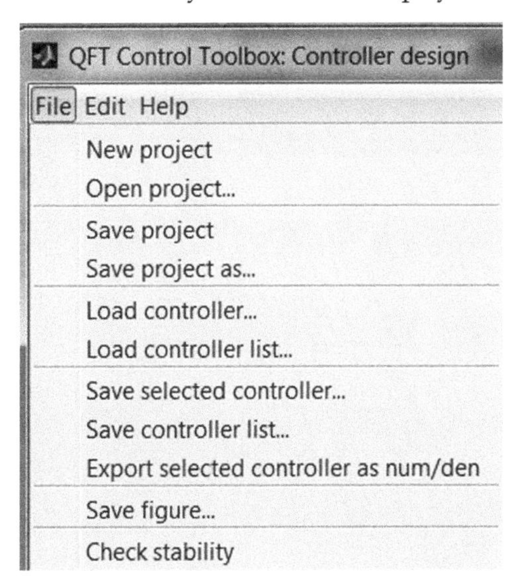

FIGURE F.8
File menu, Controller design window.

The controller design window also has an Edit menu that allows the user to

i. Undo changes
ii. Redo changes
iii. Set the frequency vector for L(s)

c. Prefilter design. Similarly, this window category has the following menu items in the *File menu*:

i. *Load prefilter* from the hard drive (*.prefltr).
ii. *Load prefilter list* from the hard drive (*.prefltrList).
iii. *Save selected prefilter* to the hard drive as (*.prefltr).
iv. *Save prefilter list* to the hard drive (*.prefltrList).
v. *Export* the selected prefilter as a transfer function to the workspace.
vi. *Save figure* to the hard drive (*.emf or *.bmp or *.fig).

F.2.2 Plant Definition Window

In the Plant definition window (W1), see Figure F.9, the user can define the plant model structure, the parameters, the uncertainty, and the frequencies of interest. There are several panels in this window. In the first one, Plant type, the user can select the way to describe the plant, which can be (1) gain/zero/pole transfer functions, (2) numerator/denominator transfer functions, (3) State Space representation, (4) transfer function multi-structure arrays, and (5) experimental data.

1. **Plant Type**
   a. Gain/zero/pole transfer function
   b. Numerator/denominator transfer function
   c. State Space
   d. Load transfer function array
   e. Experimental data

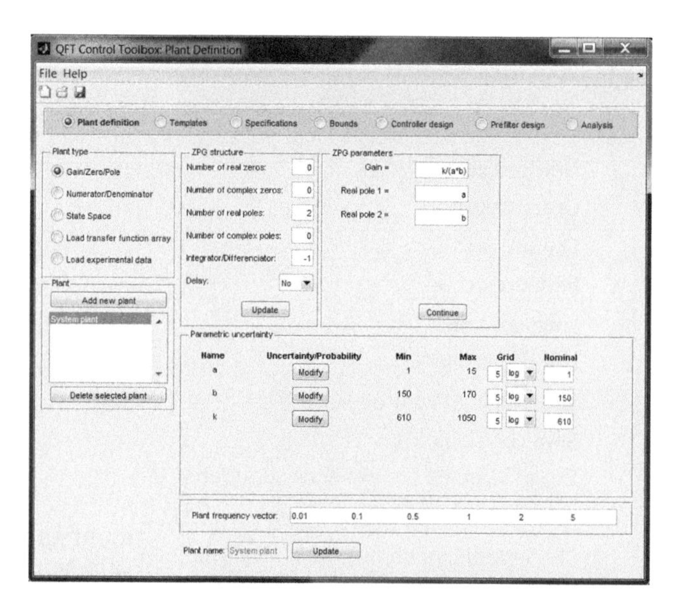

FIGURE F.9
Plant definition window.

TABLE F.1
Gain/Zero/Pole/Delay Element Syntax

Gain	k	Complex Zero	$(s/\omega_n)^2+(2\zeta/\omega_n)s+1,\ (\zeta<1)$
Real pole	$\dfrac{1}{(s/p)+1}$	Integrator	$\dfrac{1}{s^n}$
Real zero	$\left(\dfrac{s}{z}\right)+1$	Differentiator	s^n
Complex pole	$\dfrac{1}{(s/\omega_n)^2+(2\zeta s/\omega_n)s+1}\ (\zeta<1)$	Time delay	e^{-Ts}

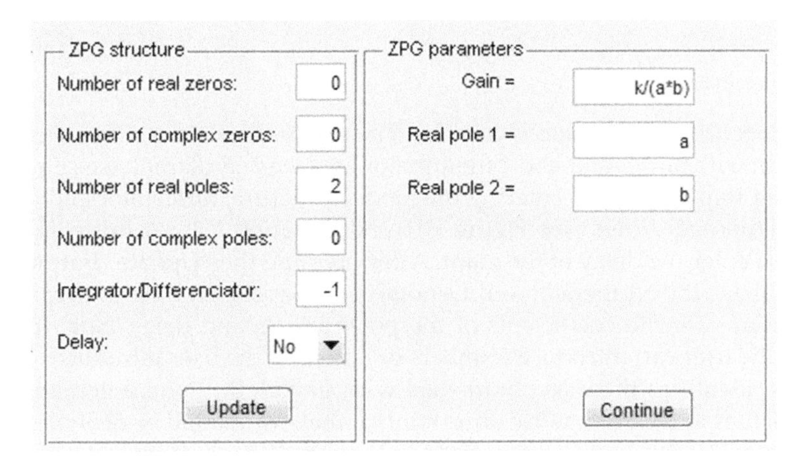

FIGURE F.10
ZPG structure and parameter panels.

a. *Gain/zero/pole transfer function*: The model structure and its elements are defined using the syntax listed in Table F.1 and Equation F.1. In the first step the user has to enter the structure of the plant: number of zeros, complex zeros, poles, and complex poles (see Figures F.9 and F.10). The user also has to enter the value of the integrator/differentiator element (0 if it is not used, a positive integer for differentiators, and a negative one for integrators) and specify if the plant has time delay. After this is accomplished, the user has to press the "Update" button. Then the toolbox updates the second ZPG panel, and the user can enter the expressions for the elements of the plant.

   Note that, at this point, the user can introduce numbers or letters. If the user introduces letters, the toolbox identifies them as parameters with uncertainty and automatically adds their names to the Parametric uncertainty panel, which will be defined afterward. In Figures F.9 and F.10 a gain is introduced as "*k*/(*ab*)," a real pole as "*a*," and another real pole as "*b*" (see Equation F.2 as well):

$$P(s)=\frac{k\,zeros(s)}{poles(s)}=\frac{k[(s/z_1)+1][(s/z_2)+1]\cdots[(s/\omega_{ni})^2+(2\zeta_i/\omega_{ni})s+1]\cdots}{s^r[(s/p_1)+1][(s/p_2)+1]\cdots[(s/\omega_{nj})^2+(2\zeta_j/\omega_{nj})s+1]\cdots}e^{-sT} \quad (F.1)$$

$$P(s)=\frac{k/(ab)}{s[(s/a)+1]\,[(s/b)+1]}=\frac{k}{s(s+a)(s+b)} \quad (F.2)$$

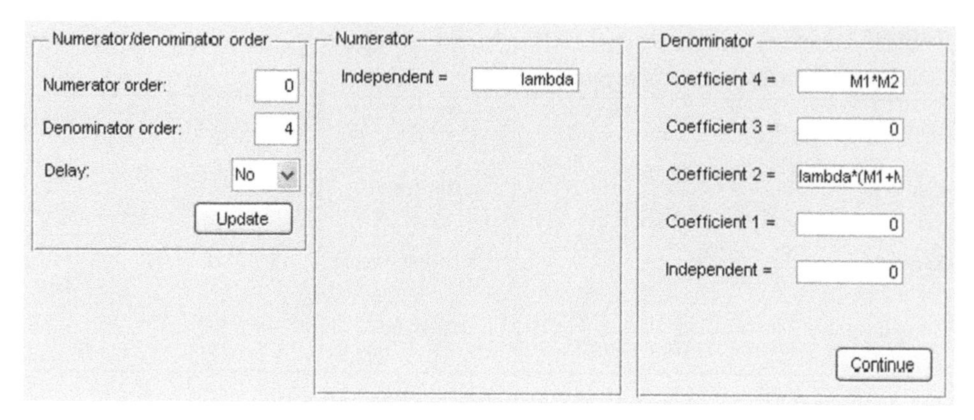

FIGURE F.11
Numerator/denominator panels.

b. *Numerator/denominator transfer function*: The plants can also be defined as a transfer function with numerator and denominator Laplace polynomials (see Equation F.3). The first step consists in entering the model structure: numerator and denominator polynomial orders (see Figure F.11 and Equation F.3). At this point the user can also enter the delay of the plant. After pressing the "Update" button, the toolbox updates the Numerator and Denominator panels, and the user can enter the expressions for the coefficients of the polynomials and delay. Note that, at this point, the user can introduce numbers or letters. If the user introduces letters, the toolbox identifies them as parameters with uncertainty and automatically adds their names to the Parametric uncertainty panel, which will be defined afterward. In Figure F.11, the expression defined in Equation F.4 is introduced:

$$P(s) = \frac{n(s)}{d(s)} = \frac{Ncoef_n s^n + Ncoef_{n-1} s^{n-1} + \cdots + Ncoef_1 s + Ncoef_{indep}}{Dcoef_m s^m + Dcoef_{m-1} s^{m-1} + \cdots + Dcoef_1 s + Dcoef_{indep}} e^{-sT} \quad \text{(F.3)}$$

$$P(s) = \frac{Lambda}{(M_1 M_2)s^4 + 0s^3 + Lambda(M_1 + M_2)s^2 + 0s + 0} \quad \text{(F.4)}$$

c. *State Space*: The plants can also be described by using a State Space representation. In the first step the user has to enter the structure or dimensions of the matrixes $\boldsymbol{A}$ $(n \times n)$, $\boldsymbol{B}$ $(n \times m)$, $\boldsymbol{C}$ $(l \times n)$, and $\boldsymbol{D}$ $(l \times m)$ (see Equation F.5 and Figure F.12). After pressing the "Update" button, the toolbox updates the State

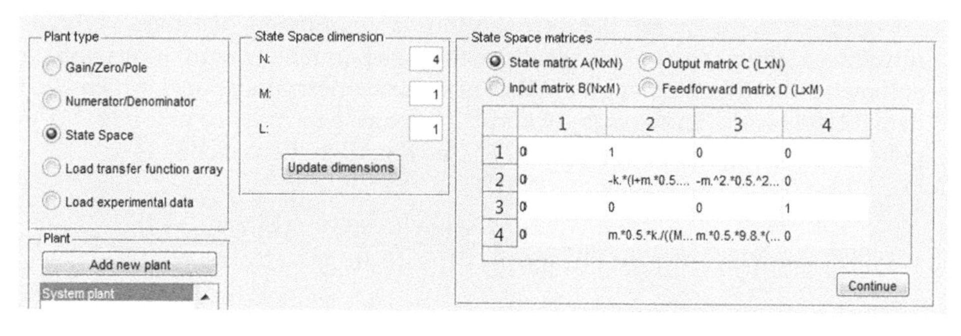

FIGURE F.12
State Space panels.

Space matrices panel, and the user can enter the expressions for the elements of the four matrices. Note that, at this point, the user can introduce numbers or letters. If the user introduces letters, the toolbox identifies them as parameters with uncertainty and automatically adds their names to the Parametric uncertainty panel, which is defined later:

$$\underset{n\times 1}{\dot{x}} = \underset{n\times n}{A}\, x + \underset{n\times m}{B}\, u$$
$$\underset{l\times 1}{y} = \underset{l\times n}{C}\, x + \underset{l\times m}{D}\, u \tag{F.5}$$

d. *Load transfer function array*: If the plant cannot be defined using the three previous structures, the user can also load an array of transfer functions from the hard drive. After uploading the array, the user has to enter the row of the array that represents the nominal plant. This is a very powerful tool that can define any kind of plant with different structures and parametric and nonparametric uncertainties. For the *System plant* the technique is as follows: (1) Run first in MATLAB an m.file like the one described next. (2) Then go to the workspace. (3) Click on "*P*" with right bottom and save as PP.mat (or any other name) in the hard drive. (4) Then go to the Plant definition window. (5) Click "Load transfer function array." (6) Select Nominal plant (usually num. 1). (7) Click "Import.mat." (8) Select in the hard drive PP.mat and click "Open." (9) Click "Commit." (10) The plant will appear in the list of plants as *System Plant*.

   For an additional plant, steps (1)–(6) are the same—now save as D.mat, for instance. Then (7) click "Add new plant." (8) Put a name in the "Plant name" cell at the bottom (e.g., DD). (9) Click "Import.mat." (10) Select in the hard drive D.mat and click "Open." (11) Click "Commit." (12) The plant will appear in the list of plants as *DD*.

   Example

```
c = 1;
for k=linspace(610,1050,3)
    for a=linspace(1,15,15)
        for b=linspace(150,170,2)
            P(1,1,c)=tf(k,[1 (a+b) a*b 0]);
            c=c+1;
        end
    end
end
```

Note that we have three nested "for" loops in the aforementioned example because there are three parameters with uncertainty: $k \in [610, 1050]$, $a \in [1, 15]$, $b \in [150, 170]$. We can also put another kind of grid, different from *linespace*, like *logspace* or others, or a mix. The definition of the plant in this example is as follows:

```
"P(1,1,c)=tf(k,[1 (a+b) a*b 0])",
```

which is $P(s) = k/[s^3 + (a+b)s^2 + ab\, s]$.

Note also that this is a very powerful option that allows the user to include in this line of the algorithm any kind of structure or expression, even many different structures (uncertainty in the structure) with parametric and non-parametric uncertainties, interdependence in the uncertain parameters, and many other special requirements.

e. *Load experimental data*: The user can also define the system by uploading experimental data in a frequency-response-data vector. The technique is as follows: (1) Prepare a frd (frequency-response-data model) system in MATLAB: *freq* = vector of frequencies in rad/s, *resp* = vector of complex numbers with the response of the system at each frequency ($a + jb$), and *experimentalData* = frd(*resp*,*freq*). Note that the name of the frd structure can be *experimentalData* or any other name.

For example, in MATLAB, *freq* = logspace(1,2), *resp* = 0.05*(*freq*). *exp(*i**2**freq*), experimentalData = *frd*(*resp*,*freq*) or a collection of real data "resp". (2) Then go to the workspace. (3) Click on "experimentalData" with right bottom and save as sys.mat (or any other name) in the hard drive. (4) Then go to the Plant definition window. (5) Click "Load experimental data". (6) Select Nominal plant (usually num.1). (7) Click "Import experimental…." (8) Select in the hard drive sys.mat and click "Open." (9) Click "Commit." (10) The plant will appear in the list of plants as *System Plant*.

2. **Expressions and parametric uncertainty**: As seen, it is possible to introduce alphanumeric expressions. The toolbox automatically recognizes letters as parameters with uncertainty. After entering the expressions of the plant (the parameters in the zpk model description, the coefficients in the numerator/denominator model description, etc.) and pressing the "Continue" button, a panel appears where the user is able to define the parametric uncertainty of the plant (see Figure F.13).

   This panel displays the following information of each parameter:

   a. The name of the parameter.
   b. A button that will open a window to modify the probability distribution of the parameter.
   c. The minimum and the maximum values of the parameter (given by its probability distribution).
   d. Number of points in the grid for each parameter (two is the minimum number, which means that there are only two points in the grid, the minimum and the maximum) and its distribution (logarithmic or lineal).
   e. The nominal value of the parameter (must be included in the minimum–maximum range of the parameter).

3. **Probability distribution of the parameters**: In order to define the probability distribution for the uncertainty of the parameters, the user needs to press the "Modify" button in the Parametric uncertainty panel and an emerging window will appear (see Figure F.14). There are three probability distributions available:

Parametric uncertainty

Name	Uncertainty/Probability	Min	Max	Grid		Nominal
a	Modify	1	15	5	log	1
b	Modify	150	170	5	log	150
k	Modify	610	1050	5	log	610

FIGURE F.13
Parametric uncertainty panel.

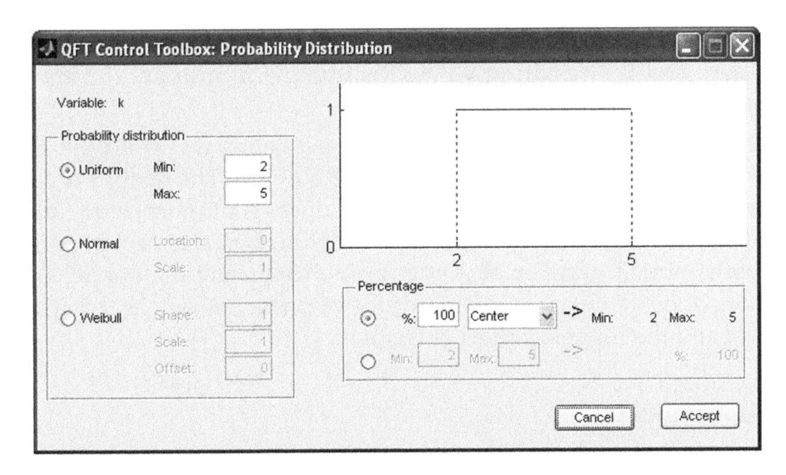

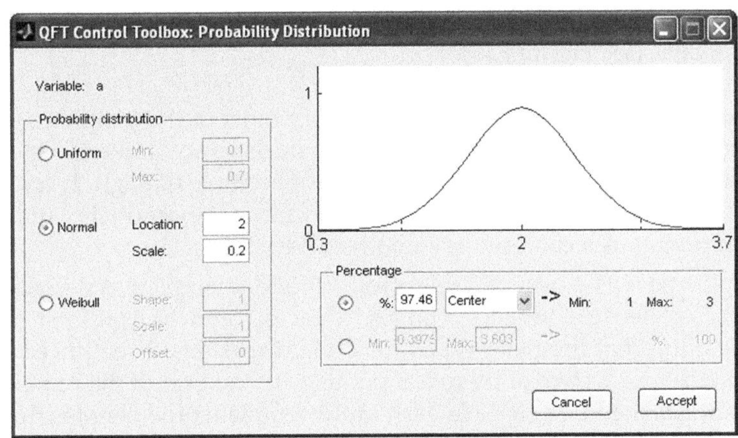

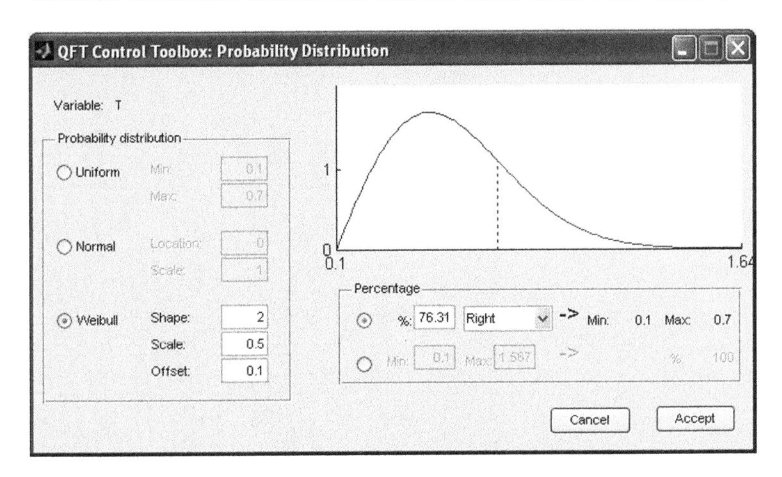

FIGURE F.14
Probability distribution: uniform, normal, and Weibull cases.

uniform, normal, and Weibull. After selecting the distribution and entering its parameters, a graphical representation of the distribution is plotted. After that, the user can define the percentage to be reached. This percentage is applied to the distribution, and the resulting values will be the minimum and the maximum values of the parameter.

4. **System plant frequency vector**: The user has to enter a vector with the frequencies of interest of the plant (see, e.g., Figure F.5 or F.9). The textbox accepts a number array or a MATLAB command that produces a number array (e.g., logspace (–2,3,100)). Note that it is very important to define well this vector, populating the vector with enough number of points in the frequency regions where there are resonances or quick changes in magnitude or phase, as well as selecting properly the lowest and the highest values of frequency.

F.2.3 Templates Window

A template is the representation of the frequency response of the plants, including the uncertainty, in the Nichols chart at a particular frequency. There is a specific window to define the templates (see Figure F.15).

1. **Number of template points**: The number of points of the templates depends on three parameters: template type, parametric uncertainty, and template contour. If the template is too sparse, then it may be not accurate enough. If too few points are used, the computed bounds are not relevant to the original plant description whose boundary is a continuous smooth curve.

   a. *Template type*: The user can select four different template types, ordered from sparse to dense: "Vertex," "Edges," "Faces," and "All points." Figure F.16 shows different template types for a plant that has three uncertain parameters (the grid of two of them is 5 points, and the grid of the remaining one is 10 points). The filled circle in each of the template type denotes the template's nominal point.

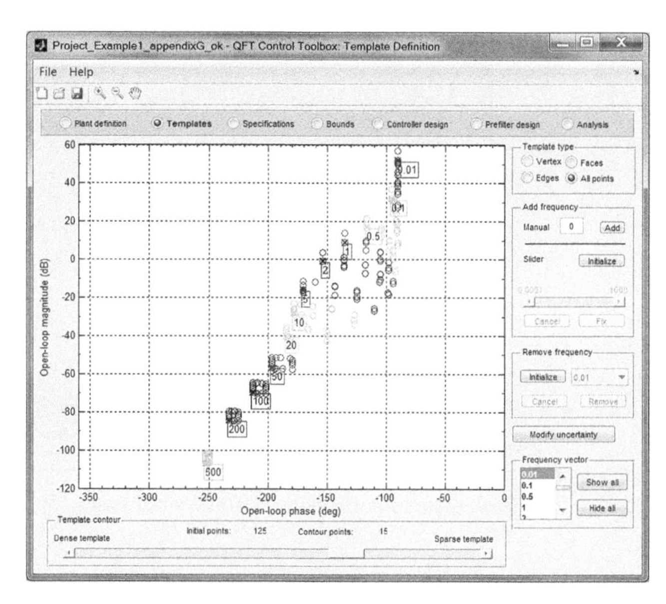

FIGURE F.15
Template definition window.

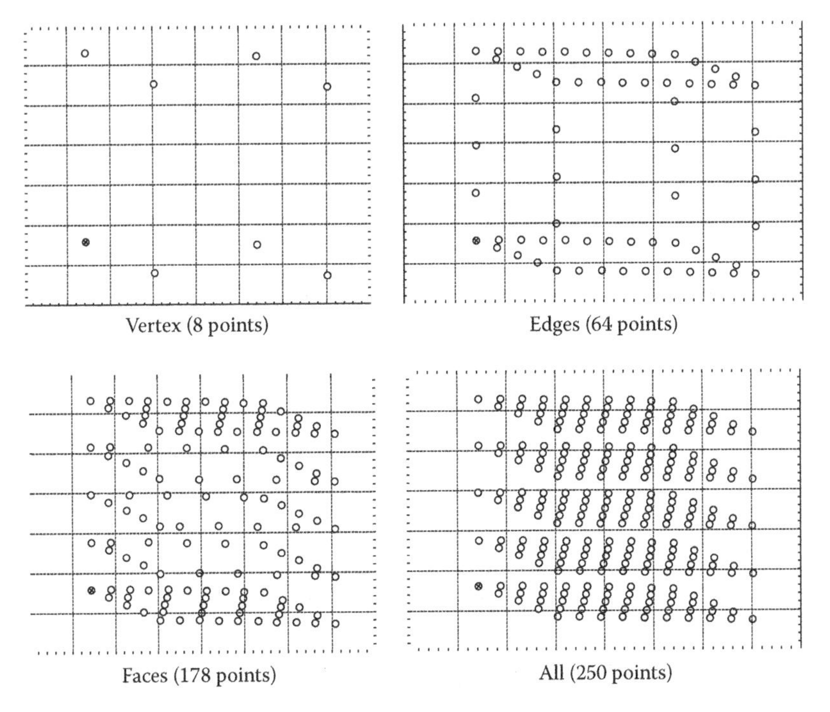

FIGURE F.16
Templates types.

b. *Grid of the parametric uncertainty variables*: The more points are sampled (Figure F.13) the denser the templates are (Note that if the template-type is "Vertex," this grid does not affect the template's figure).

c. *Template contour*: In the lower part of the window there is a slider, which can be used to adjust the contour of the templates. If the slider is moved to the right, the inner points of the template disappear.

   But if the slider is moved to the right too much then the *points of the contour* become sparser too, and the contour is not representative. Figure F.17 shows the template of a plant, which has initially 1000 points (a), and the template of the same plant after its contour has been adjusted (b).

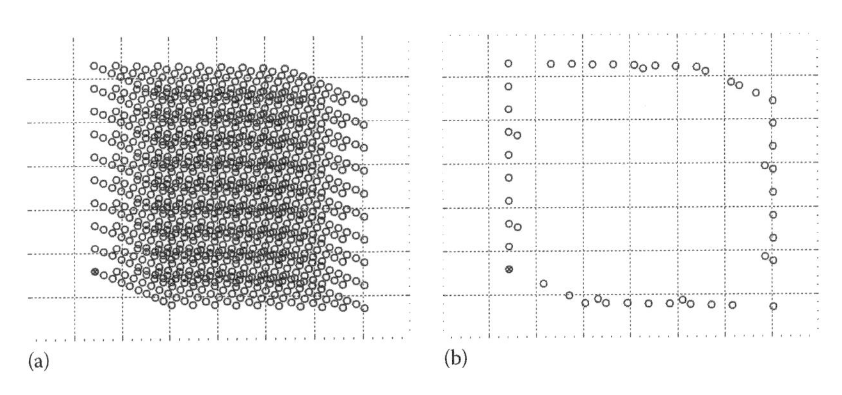

FIGURE F.17
Template without contour adjustment (1000 points) (a) and with contour adjustment (51 points) (b).

2. **Plant frequencies**: There is a template for each plant frequency. The user can add and remove frequencies using the Templates window.
   a. *Add frequency*: Frequencies can be added using two different methods:
      i. By entering the value of the new frequency in the textbox beside "Manual" and pressing the "Add" button (see Figure F.15).
      ii. By using the slider (see the frequency panel in Figure F.15). When the user clicks the "Initialize" button in the Add frequency panel, the slider is enabled and the user can previsualize the position and shape of the new template. The user can then use the slider to enter the value of the new frequency. Then, by clicking the "Fix" button, the new frequency is added to the list.
   b. *Remove frequency*: To remove a frequency, the user has to click the "Initialize" button in the Remove frequency panel. A list box appears and enables the user to select the frequency to be removed. The point markers of the template associated with the frequency are indicated by "x" instead of "o." To remove the selected frequency, click "Remove."
   c. *Frequency vector*: The user can use the Frequency vector panel (see Figure F.15) to change the visibility of the templates associated with the frequencies. If the user double clicks on a number in the frequency list that represents a template, the visibility of the template is switched (on/off). The user can use the "Show all" and "Hide all" buttons to show all the templates or to hide them all, respectively.
3. **Modify parametric uncertainty**: The user can modify the parametric uncertainty by pressing the "Modify uncertainty" button (Figure F.15). Then, an emerging window appears (see Figure F.18) where the user can modify the parametric uncertainty in the same way as in the Plant definition window. The user can see how the changes in the parametric uncertainty are applied in real time to the Templates window.
4. **Export nominal plant**: The Templates window has a menu item that allows the user to export the nominal plant as a transfer function. The nominal plant is saved in the current directory in a *.mat file. Afterward, the user can load that file to the MATLAB 's workspace (load *.mat) to use it.

F.2.4 Specifications Window

This window allows the user to introduce robust stability and performance control specifications (see Figure F.19). The panel Choose the specification type is used to select seven

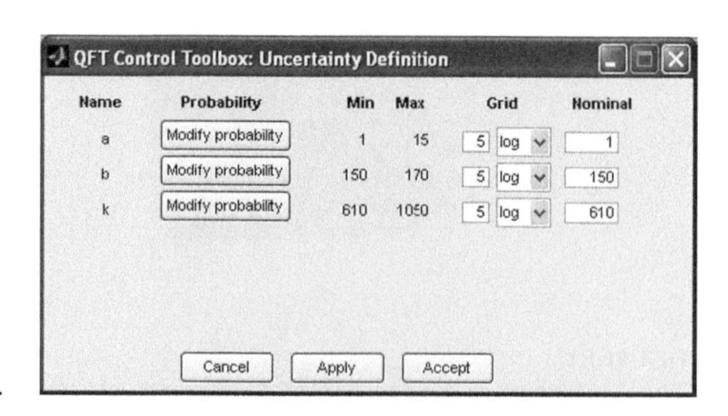

FIGURE F.18
Parameter uncertainty window.

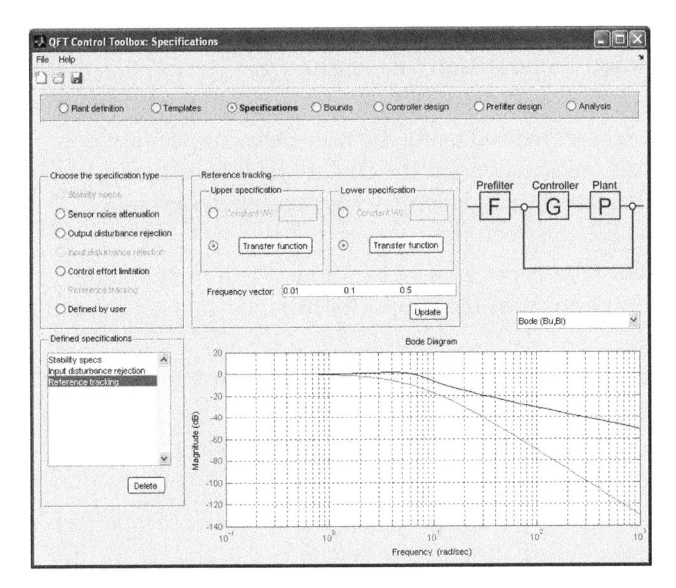

FIGURE F.19
Performance specifications window.

groups of specifications. The first six options are the classical specifications (corresponding to Equations F.6 through F.10), and the last one is a very general one (corresponding to Equation F.11), able to generate the first five specifications and many other possibilities, including additional transfer functions like $M(s)$ in Figure F.1 or others. See Table F.2 for more details.

TABLE F.2
Control System Specifications

$$|T_1(j\omega)| = \left|\frac{Y(j\omega)}{R(j\omega)F(j\omega)}\right| = \left|\frac{U(j\omega)}{D_1(j\omega)}\right| = \left|\frac{Y(j\omega)}{N(j\omega)}\right| = \left|\frac{Y(j\omega)}{R(j\omega)F(j\omega)}\right| =$$

$$= \left|\frac{P(j\omega)G(j\omega)}{1+P(j\omega)G(j\omega)}\right| \le \delta_1(\omega), \quad \omega \in \{\omega_1\} \tag{F.6}$$

$$|T_2(j\omega)| = \left|\frac{Y(j\omega)}{D_2(j\omega)}\right| = \left|\frac{1}{1+P(j\omega)G(j\omega)}\right| \le \delta_2(\omega), \quad \omega \in \{\omega_2\} \tag{F.7}$$

$$|T_3(j\omega)| = \left|\frac{Y(j\omega)}{D_1(j\omega)}\right| = \left|\frac{P(j\omega)}{1+P(j\omega)G(j\omega)}\right| \le \delta_3(\omega), \quad \omega \in \{\omega_3\} \tag{F.8}$$

$$|T_4(j\omega)| = \left|\frac{U(j\omega)}{D_2(j\omega)}\right| = \left|\frac{U(j\omega)}{N(j\omega)}\right| = \left|\frac{U(j\omega)}{R(j\omega)F(j\omega)}\right|$$

$$= \left|\frac{G(j\omega)}{1+P(j\omega)G(j\omega)}\right| \le \delta_4(\omega), \quad \omega \in \{\omega_4\} \tag{F.9}$$

$$\delta_{5inf}(\omega) \le |T_5(j\omega)| = \left|\frac{Y(j\omega)}{R(j\omega)}\right| = \left|F(j\omega)\frac{P(j\omega)G(j\omega)}{1+P(j\omega)G(j\omega)}\right| \le \delta_{5sup}(\omega), \quad \omega \in \{\omega_5\} \tag{F.10}$$

$$\left|\frac{A(j\omega)+B(j\omega)G(j\omega)}{C(j\omega)+D(j\omega)G(j\omega)}\right| \le \delta_6(\omega), \quad \omega \in \{\omega_6\} \tag{F.11}$$

1. **Predefined specifications:** (see Equations F.6 through F.10). The user can add specifications on any single-loop closed-loop relation (see Figure F.1).
2. **User-defined specifications:** (see Equation F.11). All the predefined specifications, except the reference tracking, can be expressed using the user-defined specifications. Moreover, the user can use these specifications in the design of cascaded loop and multiple loops and in systems that involve single-loop design at each design step. Looking at Figure F.1 and Table F.2, it is seen that
   a. Equation F.6 $[T_1(j\omega)]$ defines three types of specifications: (1) robust stability, (2) robust control effort limitation from the input disturbance, and (3) robust sensor noise attenuation.
   b. Equation F.7 $[T_2(j\omega)]$ defines two type of specifications: (1) robust rejection of disturbances at the output of the plant and (2) the sensitivity.
   c. Equation F.8 $[T_3(j\omega)]$ defines one type of specification: robust rejection of disturbances at the input of the plant.
   d. Equation F.9 $[T_4(j\omega)]$ defines three types of specifications: robust control effort limitation from (1) the output disturbance, (2) the sensor noise, and (3) the filtered reference signal.
   e. Equation F.10 $[T_5(j\omega)]$ defines one type of specification: robust reference tracking.
   f. Equation F.11 $[T_6(j\omega)]$ defines any specification from type $T_1(j\omega)$ to $T_4(j\omega)$, and many other options in general, where $A(j\omega)$, $B(j\omega)$, $C(j\omega)$, and $D(j\omega)$ can be defined by the user, from several options like 0, 1, $P(j\omega)$, or any other plant introduced in the Plant definition window (e.g., $M(j\omega)$ to define the specification $|M(j\omega)/[1 + P(j\omega)G(j\omega)]|$).

      The value of $\delta_i(\omega)$ denotes the magnitude of the objective (the specification) at every frequency of interest. Each specification can be defined for a different set of frequencies ω_i, always a subset of the original set of frequencies of interest.
3. **Defining a specification:** The user has to enter the value of the performance specification $\delta_i(\omega)$ ($\delta_{5inf}(\omega)$ and $\delta_{5sup}(\omega)$ in the reference-tracking case). The method to enter the value of $\delta_i(\omega)$ depends on the type of specification being defined:
   a. *Robust stability specs* (see Table F.2, Equation F.6, and Figure F.20): $\delta_1(\omega)$ is a constant (*Ws*). The user can enter directly either the value of $\delta_1(\omega) = \mu$, in magnitude, which represents the *M*-circle in the Nichols diagram (in this case the gain and phase margin are calculated automatically), or the gain *GM* or phase *PM* margins (in that case μ is calculated automatically). The following equations are used in the calculations:

      Gain margin: $GM \geq 1 + (1/\mu)$ (magnitude)

      Phase margin: $PM \geq 180° - \theta$ (deg)

      where μ is the M circle specification in magnitude, $M_{dB} = 20 \log_{10}(\mu)$, and $\theta = 2\cos^{-1}(0.5/\mu) \in [0, 180°]$.
   b. *Robust performance specs* (see Table F.2, Equations F.6 through F.9, and Figures F.21 and F.22): $\delta_i(\omega)$, $i = 1,2,3,4$ can be defined as a constant (see Figure F.21), a vector of constants of the same length as the specification frequency vector, or a transfer function (zero/pole/gain or num/den, see Figure F.22). In this case, if $\delta_i(\omega)$ is a transfer function, its Bode diagram is also displayed in the Specification window (similar to Figure F.19).

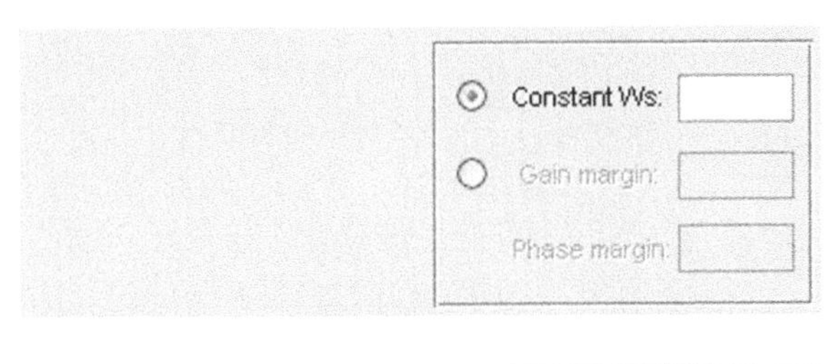

FIGURE F.20
Defining stability specifications.

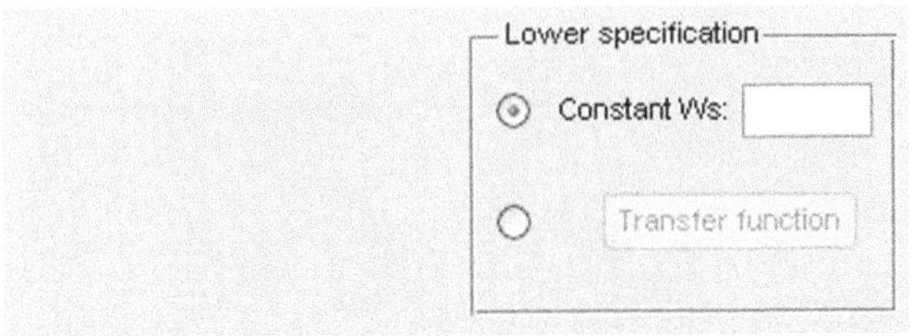

FIGURE F.21
Defining $\delta_i(\omega)$, $i = 1,2,3,4$, as a constant.

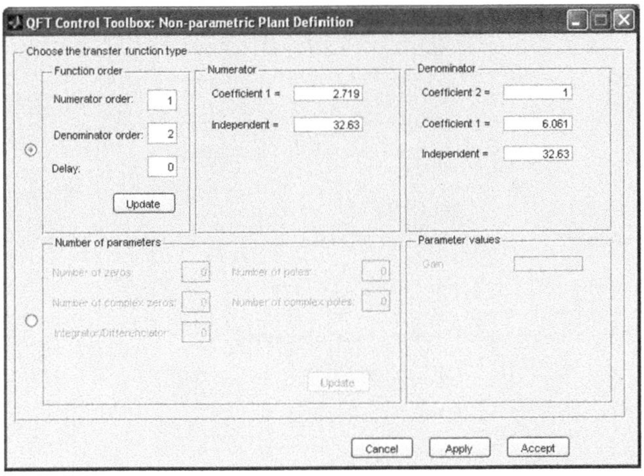

FIGURE F.22
Defining $\delta_i(\omega)$, $i = 1,2,3,4$, as a transfer function.

c. *Robust reference-tracking specs* (see Table F.2, Equation F.10, and Figures F.23 and F.24): In this case the user has to define $\delta_{5inf}(\omega) = \delta_{lower}(\omega)$ and $\delta_{5sup}(\omega) = \delta_{upper}(\omega)$ (see Figure F.23). Again, both values can be defined as constants, as a vector of constants, or as transfer functions. Besides, if this type of specification is selected, there are more plots available: the step response in the time domain of $\delta_{lower}(\omega)$ and $\delta_{upper}(\omega)$, the Bode diagram of $\delta_{upper}(\omega)-\delta_{lower}(\omega)$, see Figure F.23, and the Bode diagram of $\delta_{lower}(\omega)$ and $\delta_{upper}(\omega)$, see Figure F.19.

d. *Defined by user specs* (see Table F.2, Equation F.11): In this case the user not only has to enter $\delta_i(\omega)$ but also $A(j\omega)$, $B(j\omega)$, $C(j\omega)$, and $D(j\omega)$, for a specification like $|[A(j\omega) + B(j\omega)G(j\omega)]/[C(j\omega) + D(j\omega)G(j\omega)]| < \delta_i(\omega)$. They can be constants (0 or 1), the system plant $P(j\omega)$, or other auxiliary plants defined in the Plant definition window ($G(j\omega)$ is the controller).

4. **Frequency vector**: The user has to define a frequency vector for each specification. This vector must be a subset of the frequency vector of the system plant.
5. **Specification addition**: The user has to select in the Choose specification type panel, see Figure F.19 or F.25, the type of specification to be defined. Then the user can enter the parameters of the specification. When the data are entered the user presses the "Commit" button to add the specification.

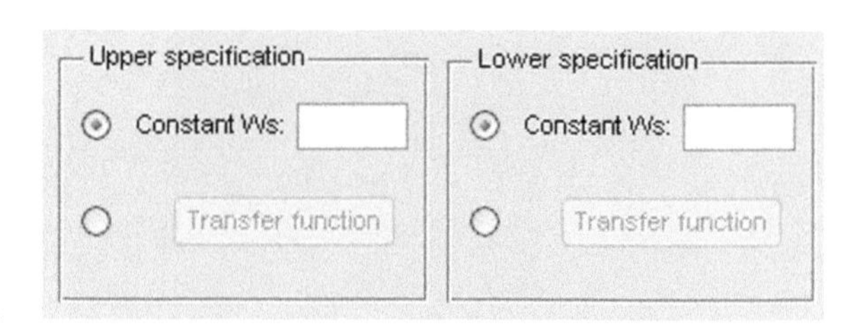

FIGURE F.23
Defining $\delta_{upper}(\omega)$ and $\delta_{lower}(\omega)$ specifications.

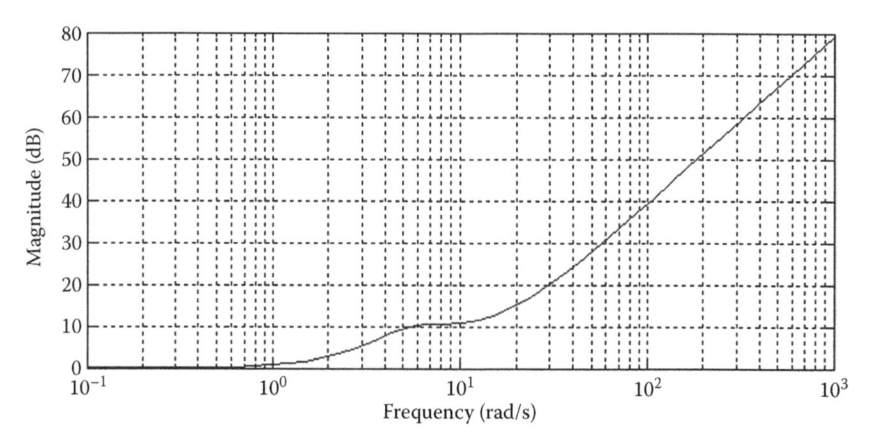

FIGURE F.24
Bode diagram of $\delta_{upper}(\omega)-\delta_{lower}(\omega)$.

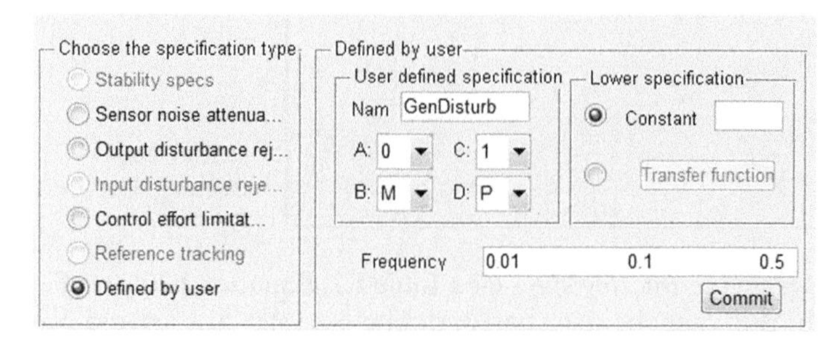

FIGURE F.25
Defining "Defined by user" specs, Equation F.11, for example, having $M(j\omega)$ from Plant definition window and doing $|M(j\omega)/[1 + P(j\omega)G(j\omega)]|$.

6. **Specification edition**: To edit a specification, the user has to select it in the Defined specifications panel. Then all the data of the specification will be displayed in the center panel of the window. Once the specification has been edited, the user has to press the "Update" button to apply the changes.
7. **Specification removal**: To remove a specification the user has to select it in the Defined specifications panel and press the "Delete" button.

F.2.5 Bounds Window

Given the plant templates and the control specifications, QFT converts closed-loop magnitude specifications ($T_1(j\omega)$ to $T_6(j\omega)$) into magnitude and phase constraints for a nominal open-loop function $L_0(j\omega)$. These constraints are called QFT bounds (see Figure F.26). After the design of the controller (next section), the nominal open-loop function $L_0(j\omega)$

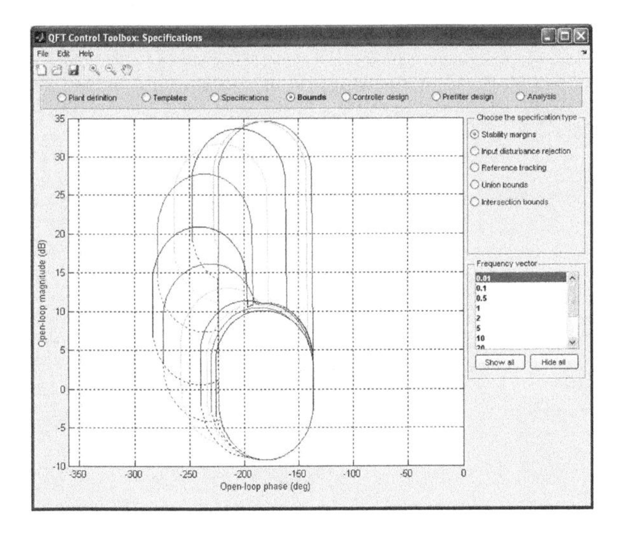

FIGURE F.26
Bounds window.

must remain above the solid-line bounds and below the dashed-line bounds at each specific frequency to fulfill the desired specifications.

The Bounds window (see Figure F.26) shows the QFT bounds of each specification defined in the Specification window, as well as the union of all the bounds and the intersection (worst-case scenario) of all the bounds.

By clicking "Show all" or "Hide all," the toolbox plots all the bounds or hides all the bounds. By double clicking on its corresponding frequency, the bound of that frequency is shown or hidden.

The user can also define the minimum, maximum, and step value of the phase vector by using the "Edit phase vector" submenu in the "Edit menu."

F.2.6 Controller Design Window

Once the user has introduced the information about the plant and the control specifications, and once the templates and bounds have been calculated, the next step involves the design (loop shaping) of the controller $G(s)$ so that the nominal open-loop transfer function $L_0(s) = P_0(s)\, G(s)$ meets the bounds (see Figure F.27). Generally speaking, the loop shaping, or $G(s)$ design, involves changing the gain and adding poles and zeros, either real or complex, until the nominal loop $L_0(s)$ lies near its bounds, more specifically above the solid-line bounds and below the dashed-line bounds at each frequency of interest.

Controller management

1. *Controller addition*: There are two ways of adding a new controller:
   a. To create the new controller from scratch, the user has to enter the name of the controller and press the "Add controller" button. If there is no other controller defined with that name, the new controller appears in the list of added controllers.
   b. If the user wants to add a new controller based on the dynamics of an existing controller, the user has to select it from the "Added controllers" list and press the "Copy" button. The user has to enter the name of the new controller in the emerging window.
2. *Controller removal*: To remove a controller, the user has to select it from the list of "Added controllers" and press the "Delete" button.

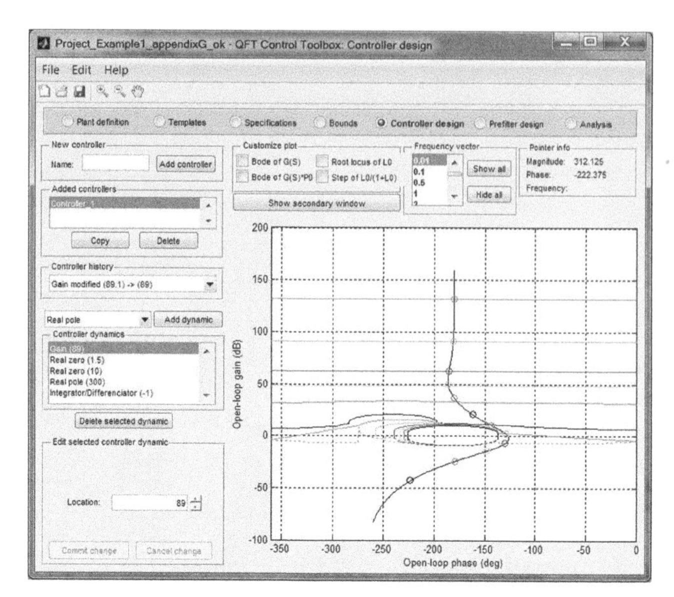

FIGURE F.27
(See color insert.) Controller design window.

Controller dynamics: When defining a controller, the user can work with the dynamics listed in Table F.3, according to Equation F.12:

$$L_0(s) = P_0(s)\prod_{i=0}^{n}[G_i(s)] \tag{F.12}$$

To add a new dynamic element $G_i(s)$ in the controller $G(s) = \Pi\, G_i(s)$, the user has to press the "Add dynamic" button. The dynamic is added with its predefined values. In the lower left corner of the window there is a panel that shows information about the selected dynamic. The information displayed depends on the type of the element, as shown in Figures F.28 through F.31.

TABLE F.3
Controller Elements, $G_i(s)$

Gain	k	2° order/2° order	$\frac{a_1s^2 + a_2s + 1}{b_1s^2 + b_2s + 1}$
Real pole	$\frac{1}{(s/p)+1}$	Integrator	$\frac{1}{s^n}$
Real zero	$\left(\frac{s}{z}\right)+1$	Differentiator	s^n
Complex pole	$\frac{1}{(s/\omega_n)^2 + (2\zeta/\omega_n)s + 1}$, $(\zeta < 1)$	Lead/lag network	$\frac{(s/z)+1}{(s/p)+1}$
Complex zero	$(s/\omega_n)^2 + (2\zeta/\omega_n)s + 1$, $(\zeta < 1)$	Notch filter	$\frac{(s/\omega_n)^2 + (2\zeta_1 s/\omega_n)s + 1}{(s/\omega_n)^2 + (2\zeta_2 s/\omega_n)s + 1}$

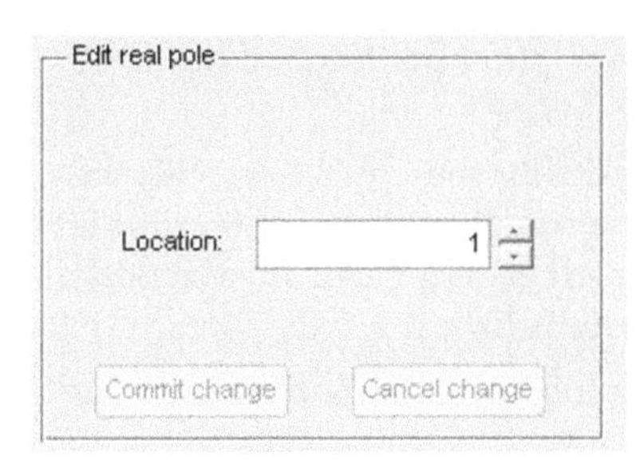

FIGURE F.28
Window for gain, zero, pole, and integrator/differentiator.

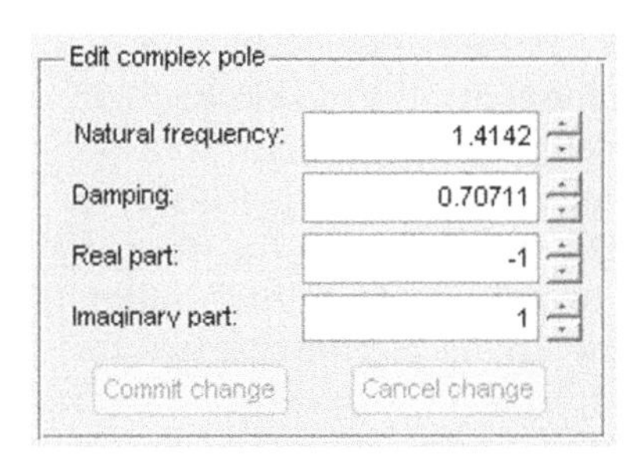

FIGURE F.29
Window for complex zero and complex pole.

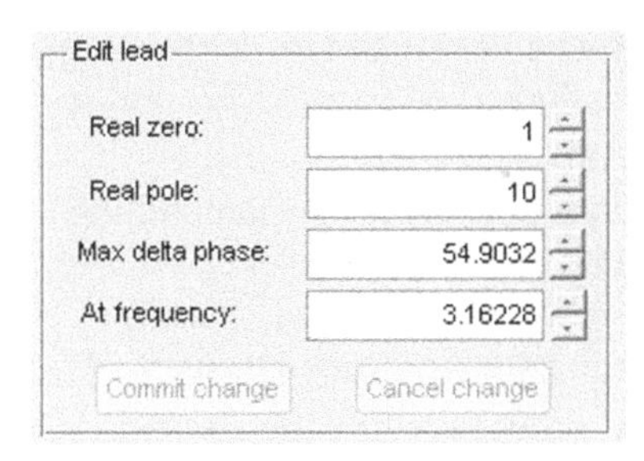

FIGURE F.30
Window for lead/lag element.

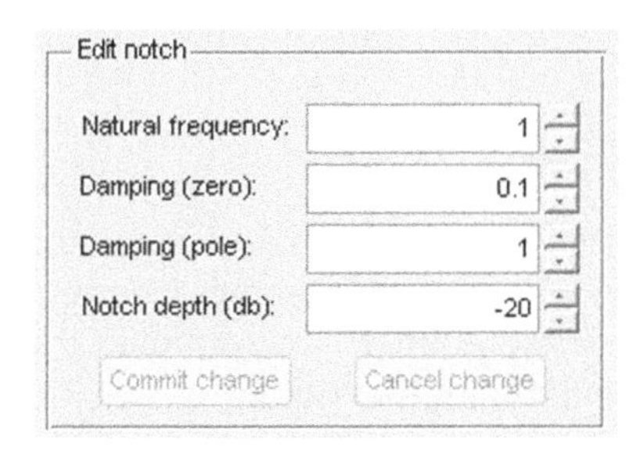

FIGURE F.31
Window for notch filter.

1. *Dynamic edition*:
    a. With textual tools. All the dynamics can be edited by entering their new values in the textboxes that appear in the Edit selected controller dynamic panel (see Figure F.27), as shown in Figures F.28 through F.31.
    b. With graphical tools. The dynamic elements $G_i(s)$ of the controller can be edited by interacting with the graphs with the mouse. This is a very powerful tool

that helps the user in the controller design. To use it, first the element has to be selected in the Controller dynamics panel. For example,

i. the *gain* can be modified by (1) dragging vertically the $L_0(j\omega)$ line with the mouse in the Nichols chart, (2) by dragging vertically the Bode magnitude plot of $G(s)$ in the secondary window, or (3) by dragging vertically the Bode magnitude plot of $G(s)$ $P_0(s)$ in the secondary window.

ii. a *zero* or a *pole* can be modified by (1) dragging the $L_0(j\omega)$ line to the right or to the left with the mouse in the Nichols chart, (2) by dragging to the right or to the left the selected zero or pole in the Bode magnitude plot of $G(s)$ in the secondary window, or (3) by dragging to the right or to the left the selected zero or pole in the Bode magnitude plot of $G(s)$ $P_0(s)$ in the secondary window.

2. *Dynamic removal*: The user can remove an element of the controller by selecting it in the Controller dynamics panel and pressing the "Delete selected dynamic" button.

   When the user edits some dynamics of the controller, the updated $L_0(j\omega)$ is shown in the Nichols plot as a red dashed line (Figure F.32c), meaning that the changes are provisional. To commit the changes the user has to press the "Commit" button. Then the red dashed line is drawn as a solid black line. To discard the changes, the user has to press the "Cancel" button.

   Figure F.32a through d illustrate an example of the graphic dynamic edition. Figure F.32a and b shows the Controller design window and the secondary window. To meet the bounds, the user changes the position of the selected real pole from $p = 1$ to $p = 300$. The user can do it by entering the new value in the textbox or by dragging horizontally the green "x" marked in the Bode plot. Figure F.32c and d shows how all the plots have been updated accordingly.

3. *Change history*: Each time the user adds, edits, or removes a dynamic element, the change is added to the Controller history panel (see Figure F.33). The user can undo and redo changes by selecting different entries in the Controller history panel, by pressing Ctrl + x and Ctrl + y respectively, and by using the Edit menu as well.

4. *Secondary window*: The user can open a secondary window with the following plots (see Figures F.32 and F.34): Bode diagram of $G(s)$, Bode diagram of $L_0(s)$, root locus of $L_0(s)$, and unit step response of $L_0(s)/[1 + L_0(s)]$. The secondary window can be resized, and the changes made to the selected controller are applied in real time to all the diagrams.

Pointer information

In the upper right corner of the window, there is a panel that shows information about the pointer (see Figure F.35). If the pointer is over the Nichols plot, the panel shows information about where the pointer is (magnitude and phase). If the pointer is placed over the $L_0(j\omega)$ line, the panel also includes the frequency ω associated with that point.

F.2.7 Prefilter Design Window

If the control problem requires reference-tracking specifications, then the Prefilter design window is active after the design of the $G(s)$ controller.

Figure F.36 shows the Prefilter design window with the following plots: the upper and lower reference-tracking specifications [$\delta_{upper}(\omega)$, $\delta_{lower}(\omega)$, dashed lines] (see Equation F.10 and Figure F.23) and the maximum and minimum cases of $L_0(s)F(s)/[1 + L_0(s)]$ over the plant uncertainty (lines between δ_{upper} and δ_{lower}).

The design of the prefilter involves obtaining the worst upper and lower closed-loop response cases of $L_0(s)F(s)/[1 + L_0(s)]$ over the plant uncertainty, which should be between the upper and lower reference-tracking functions to meet the specifications. The Prefilter design window is very similar to the Controller design window. The way in which the prefilters are added, edited, and removed is the same.

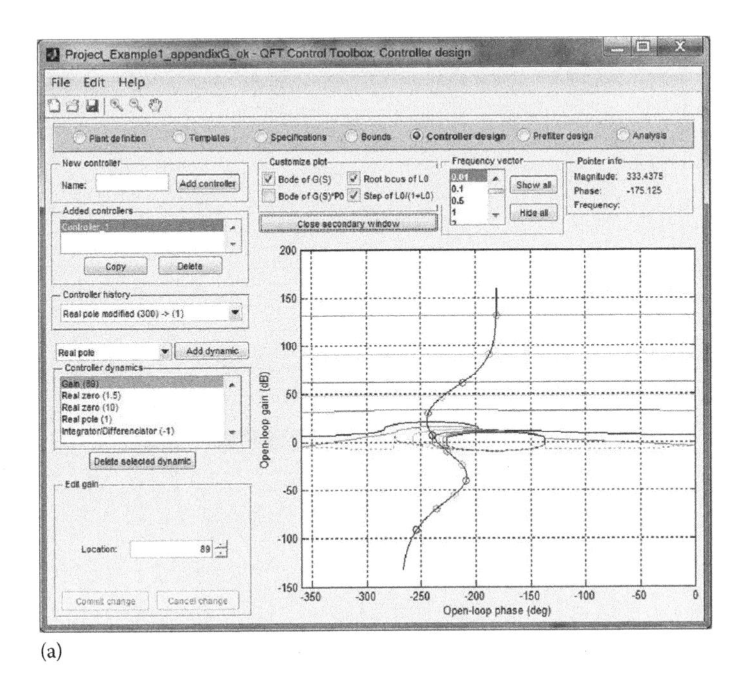

(a)

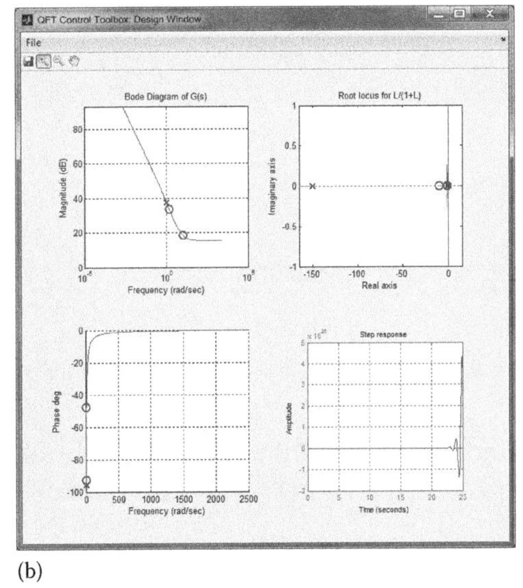

(b)

FIGURE F.32
(See color insert.) An example of the graphic dynamic edition. (a) Controller design window, Nichols chart, (b) secondary window, Bode diagram, root locus and step time response.

(continued)

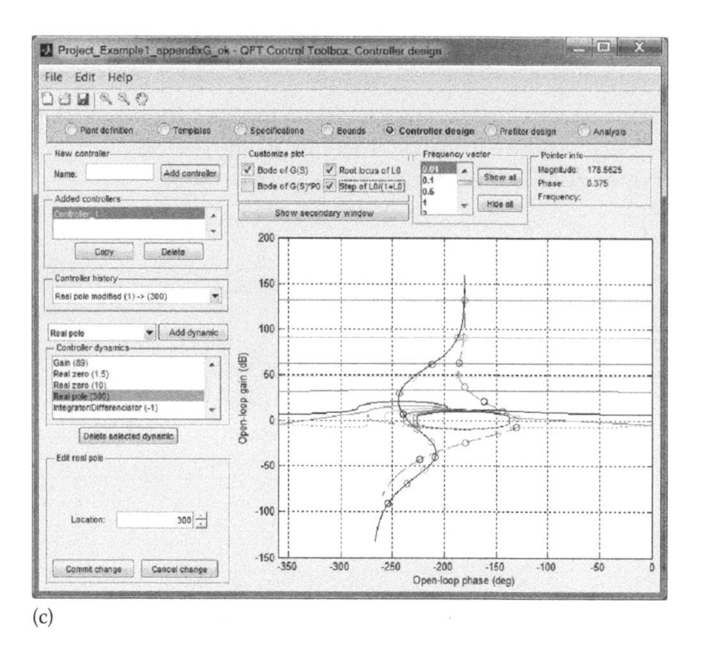

(c)

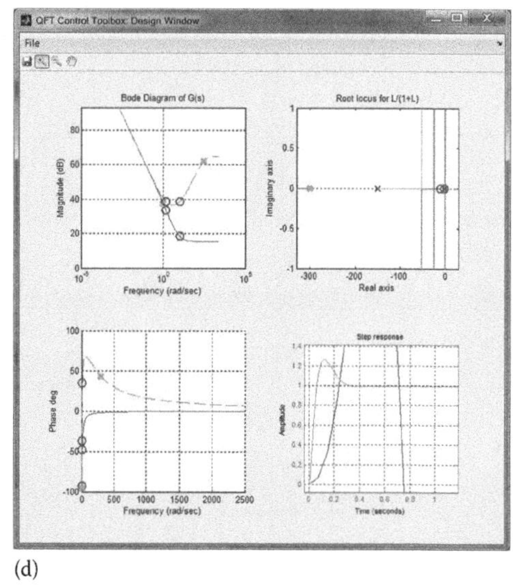

(d)

FIGURE F.32 (continued)
(See color insert.) (c) controller design window after moving a pole from $p = 1$ to $p = 300\{1/[(s/p)+1]\}$, (d) secondary window after moving that pole.

The Prefilter design window has an additional List of controllers panel that allows the user to select among the controllers designed in the previous window.

F.2.8 Analysis Window

Once the user finishes the controller (and prefilter) design, the Analysis window is active. The analysis is performed in both the frequency domain and the time domain. The window analyses the controller $G(s)$ and prefilter $F(s)$ in the worst-case scenario over the plant uncertainty.

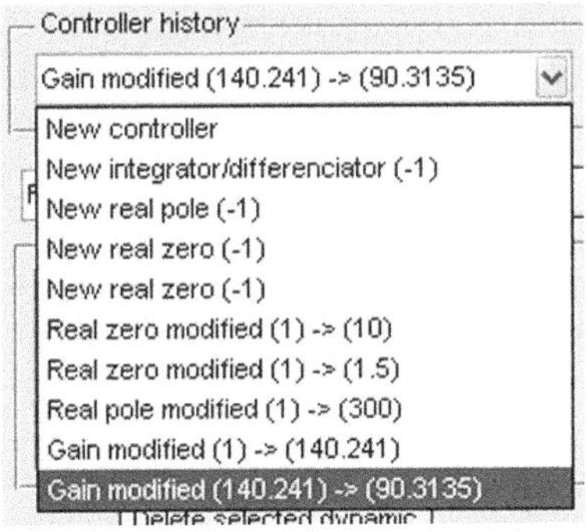

FIGURE F.33
Controller history panel.

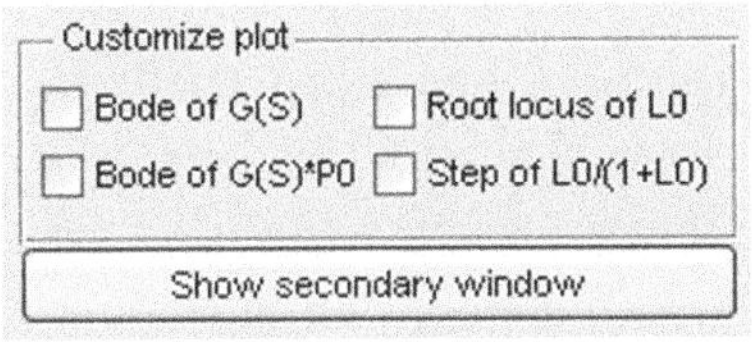

FIGURE F.34
Panel to open the secondary window.

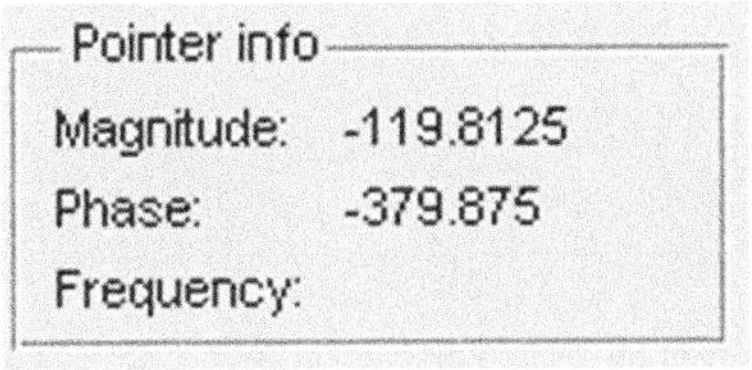

FIGURE F.35
Pointer info panel.

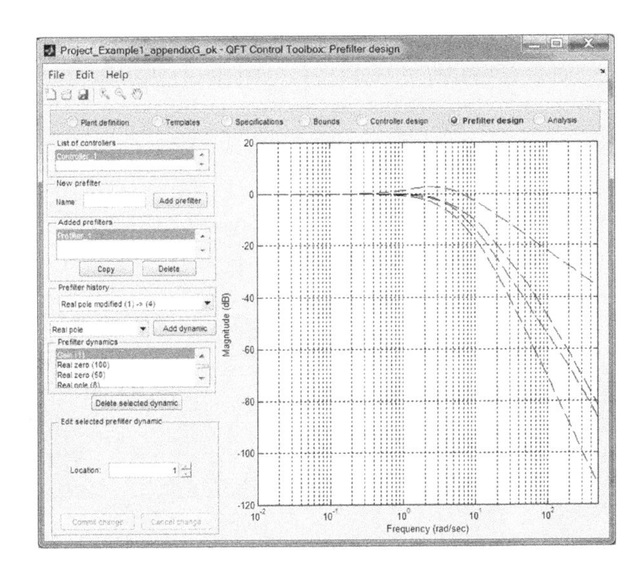

FIGURE F.36
Prefilter design window.

Types of analysis: The window allows the user to perform two types of analysis:

1. **Frequency-domain analysis panel**: This panel permits an analysis of the closed-loop response of the control system with respect to a specification defined in the specification window.

   The dashed line is the desired specification $\delta_i(\omega)$ and the solid line the worst case of the control system over the plant uncertainty at each frequency (see Figure F.37 for the analysis of the stability specification in the frequency domain). Note that the solid line is not a transfer function, but the worst case among all the transfer functions at every frequency over the plant uncertainty.
2. **Time-domain analysis panel**: This panel analyses the time response of the control system, with many plants over the plant uncertainty (see Figure F.38 for the analysis of the unit step response in the time domain). The window can apply

   a. A unit step at the reference signal, studying the performance of $L_0(s)F(s)/[1 + L_0(s)]$ or $L_0(s)/[1 + L_0(s)]$

   b. A unit impulse at the output of the plant, studying the performance of $1/[1 + L_0(s)]$

   The number of plants analyzed (number of lines plotted) depends on the values introduced in the Parametric uncertainty panel (see Figure F.38).

Controller/prefilter combinations: The user can select any combination of controllers (from the Controller List panel) and prefilters (from the Prefilter List panel) previously defined (if any).

Frequency vector panel for the analysis: The Analysis frequency vector panel allows the user to enter the frequency vector to be used in the frequency-domain analysis. If there are not enough points in the frequency vector, the resulting analysis may not be accurate enough (see Figures F.39 and F.40).

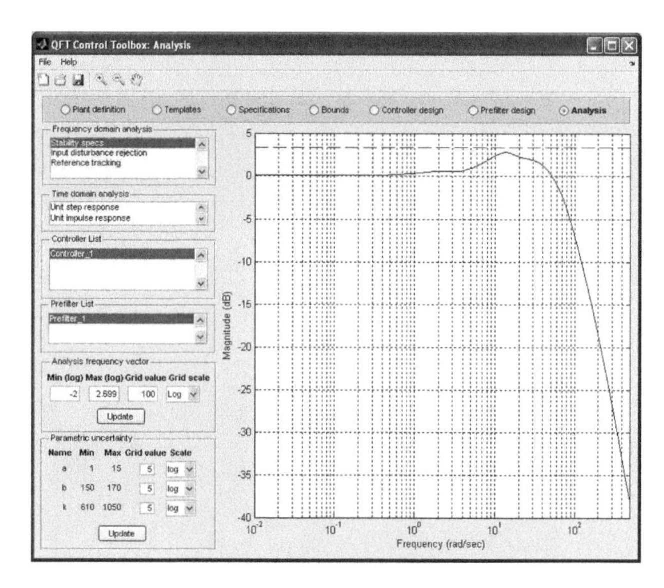

FIGURE F.37
Analysis window.

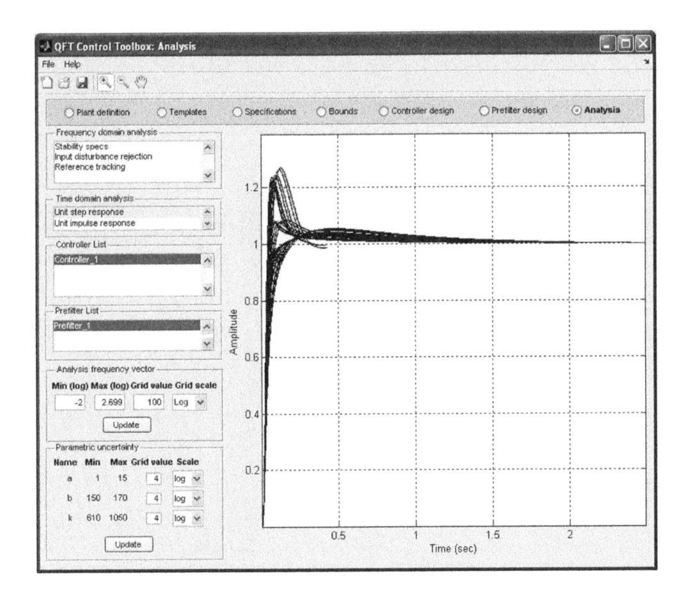

FIGURE F.38
Time-domain analysis.

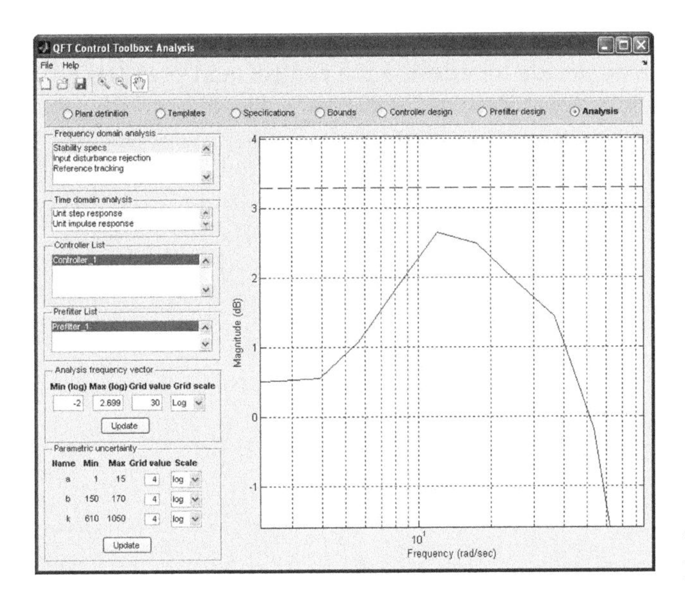

FIGURE F.39
Stability analysis with not enough points in the frequency vector.

Parametric uncertainty panel: As it was explained, this panel allows the user to modify the grid of the parametric uncertainty variables. Again, if too few points are selected, the analysis may not be accurate enough. On the other hand, if too many points are selected, the analysis may be slow. This panel allows the user to analyze the system both (1) in the points previously defined in the Plant definition window (with the parametric uncertainty panel) and (2) in new points of uncertainty, defined now in the Analysis window with the parametric uncertainty panel.

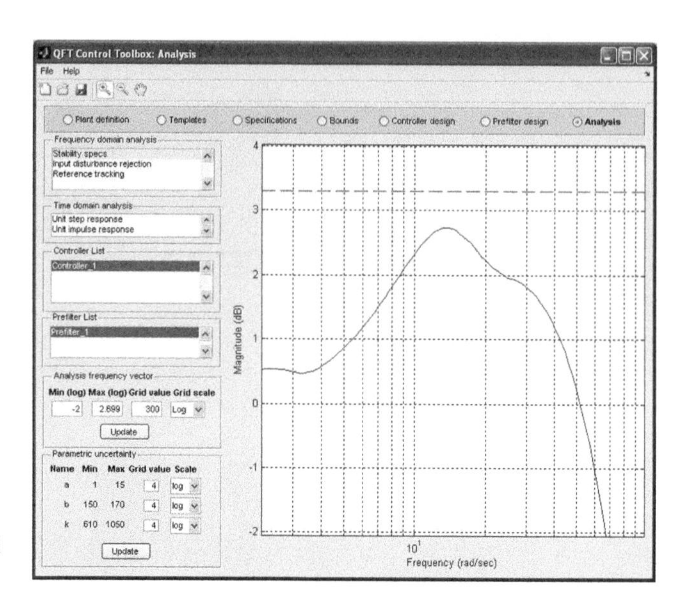

FIGURE F.40
Stability analysis with a more populated grid.

If the responses for all the *selected plants* satisfy the desired control performance specifications at both frequency and time domains, then the design is completed. If the design fails at any frequency or time, you may decide to redesign the control (and prefilter) or to redefine the plant model, uncertainty, or specifications definition.

Appendix G: Controller Design Examples

G.1 Introduction

This appendix presents a set of selected examples to help the reader understand the main concepts and procedures of the QFT Control Toolbox (*QFTCT*).[149,150]

The related control theory can be found along the book, in Chapters 2 through 8 and Appendices A through E. The *QFTCT* user's guide is in Appendix F. The selected examples are plants with model uncertainty and stability and performance robust specifications. They cover a wide range of model and control aspects. A classification of the selected examples is shown in Table G.1. Each of the examples is broken up into two parts: P1, setting up the problem; P2, solving the problem. The solutions of the examples with the *QFTCT* are included in the websites http://cesc.case.edu and http://www.crcpress.com/product/isbn/9781439821794.

G.2 DC Motor (Example 1, P1)

DC motors are direct current machines employed extensively in industrial applications, including wind turbines. Medium- and high-power DC motors are often employed in wind turbines for pitch control systems (to move the blades and change the aerodynamic coefficient of the rotor) and for yaw control systems (to move the nacelle following the wind direction).

In this example, a yaw control system for a wind turbine is considered (Figure G.1). It consists of DC motors (the actuators), the nacelle to be oriented (inertia), the bearings (friction and inertia), a vane (sensor), and the controller. The transfer function between the angle $y(s)$ to be controlled (output) and the voltage $u(s)$ applied to the DC motors/actuators is

$$\frac{y(s)}{u(s)} = P(s) = \frac{K_m}{s(Js+D)(L_a s+R_a)} = \frac{k}{s(s+a)(s+b)} \tag{G.1}$$

where

J is the inertia of the rotating elements
D is the viscous friction
K_m is the motor-torque constant
R_a is the resistance
L_a is the inductance of the motor armature

Combining these parameters, it is possible to lump the parameters in $k = K_m/(L_a J)$ as the gain, $1/a = J/D$ as the mechanical time constant, and $1/b = L_a/R_a$ as the electrical time

TABLE G.1

Examples, Controller Design

Example and File	Model	Characteristics	Specifications
1. DC motor Example1_appendixG.mat	$P(s)=\dfrac{k}{s(s+a)(s+b)}$	Stable, three parameters with uncertainty	Robust stability, reference tracking, disturbance rejection
2. Satellite Example2_appendixG.mat	$P(s)=\dfrac{k}{s^2}$	Two integrators, one parameter with uncertainty	Robust stability, reference tracking
3. Two carts coupled by spring Example3_appendixG.mat	$P(s)=\dfrac{1}{s^2(as^2+b)}$	Two imaginary poles, two integrators, three parameters with coupled uncertainty	Robust stability
4. Inverted pendulum Example4_appendixG.mat	$\dot{x}=\begin{bmatrix}0 & 1 & 0 & 0\\ 0 & -\dfrac{k(I+mL^2)}{(M+m)I+mL^2M} & -\dfrac{m^2L^2g}{(M+m)I+mL^2M} & 0\\ 0 & 0 & 0 & 1\\ 0 & \dfrac{mLk}{(M+m)I+mL^2M} & \dfrac{mLg(M+m)}{(M+m)I+mL^2M} & 0\end{bmatrix}x+\begin{bmatrix}0\\ \dfrac{I+mL^2}{(M+m)I+mL^2M}\\ 0\\ -\dfrac{mL}{(M+m)I+mL^2M}\end{bmatrix}u$ $y=[0 \quad 0 \quad 1 \quad 0]x$	Unstable, four parameters with uncertainty	Robust stability, disturbance rejection
5. Central heating system Example5_appendixG.mat	$P(s)=\dfrac{k}{\tau s+1}e^{-Ls}$	System with delay, three parameters with uncertainty	Robust stability, disturbance rejection

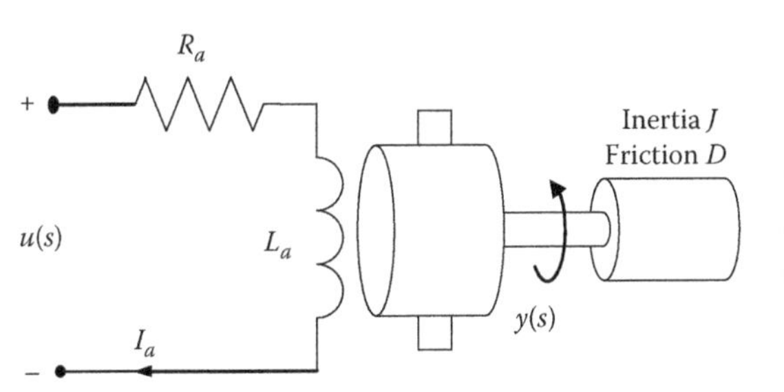

FIGURE G.1
DC motor and a yaw wind turbine system.

constant of the system. For a very small wind turbine, the parameters and the associated uncertainty for the system are $610 \le k \le 1050$, $1 \le a \le 15$, and $150 \le b \le 170$.

The problem is to synthesize a compensator $G(s)$ and a prefilter $F(s)$ for the DC motor defined earlier to achieve the following specifications:

1. *Robust stability*: a minimum phase margin angle of 40°
2. *Reference tracking*: given by tracking models $\delta_{upper}(s)$ and $\delta_{lower}(s)$ to satisfy $M_p = 1.30$ and $t_s = 1.92\,\text{s}$ (for the upper bound) and $t_s = 1.70\,\text{s}$ (for the lower bound)
3. *Disturbance rejection at plant input*: to satisfy $\alpha_p < |y_D(t_p)| = 0.1$ for $t_p = 65\,\text{ms}$ and for a unit disturbance step input $y_D(\infty) = 0$, $\omega < 10\,\text{rad/s}$

G.3 Satellite (Example 2, P1)

This example deals with the design of controllers to regulate the relative distance (dx, dy, dz) between two spacecrafts flying in formation in deep space and with no ground intervention. The second spacecraft is fixed, while the first spacecraft moves the thrusters (u_{x1}, u_{y1}, u_{z1}) to control the relative distance (dx, dy, dz) between both units, which is measured by radio frequency (RF) and laser-based metrology (Figure G.2). The system is described as

$$dx = x_2 - x_1, \quad dy = y_2 - y_1, \quad dz = z_2 - z_1 \tag{G.2}$$

$$\begin{bmatrix} dx \\ dy \\ dz \end{bmatrix} = \begin{bmatrix} p_{1x}(s) & 0 & 0 \\ 0 & p_{1y}(s) & 0 \\ 0 & 0 & p_{1z}(s) \end{bmatrix} \begin{bmatrix} u_{x1} \\ u_{y1} \\ u_{z1} \end{bmatrix} \tag{G.3}$$

which presents three independent single-input-single-output (SISO) systems, one for each axis (x, y, z). The plant models have uncertainty due to fuel consumption, so that

$$P_{1x}(s) = P_{1y}(s) = P_{1z}(s) = P_1(s) = \frac{-1}{m_1 s^2}, \quad m_1 \in [360, 460]\,\text{kg} \tag{G.4}$$

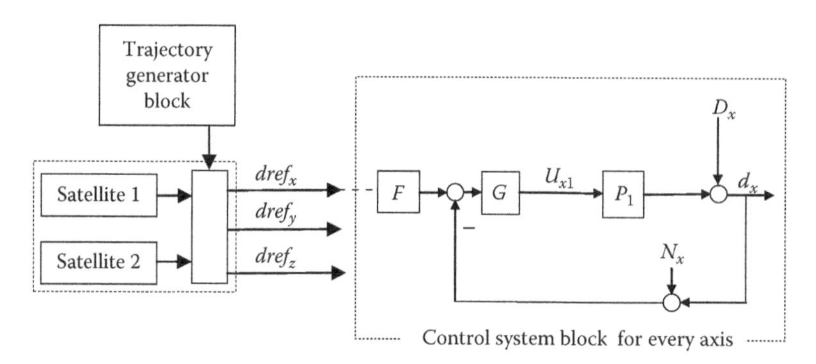

FIGURE G.2
Spacecraft control system.

Synthesize the compensator $G(s)$ and the prefilter $F(s)$ for the uncertain plant of the x-axis of the first spacecraft and for the following specifications:

1. *Robust stability*: $|(P_1(j\omega)G(j\omega))/(1 + P_1(j\omega)G(j\omega))| \leq 1.1 \forall\omega$, which involves a phase margin of at least 55° and a gain margin of at least 1.99 (5.9 dB)
2. *Reference tracking*: $T_{R_L}(\omega) \leq \left|\left(P_1(j\omega)G(j\omega)/\left(1+P_1(j\omega)G(j\omega)\right)\right)\right| \leq T_{R_U}(\omega)$ where

$$T_{R_U}(\omega) = \left|\frac{k\left((j\omega/a)+1\right)}{\left((j\omega/\omega_{n1})\right)^2 + (2\zeta/\omega_{n1})(j\omega)+1}\right|; \quad T_{R_L}(\omega) = \left|\frac{1}{\left((j\omega/\sigma_1)+1\right)\left((j\omega/\sigma_2)+1\right)\left((j\omega/\sigma_3)+1\right)}\right|$$

and where $k = 1$, $\omega_{n1} = 0.025$, $\zeta = 0.8$, $a = 0.035$, $\omega_{n2} = 0.03$, $\sigma_1 = 2\,\omega_{n2}$, $\sigma_2 = 0.3\,\omega_{n2}$, $\sigma_3 = \omega_{n2}$, for ω = [0.00001 0.00005 0.0001 0.0005 0.001 0.005 0.01 0.05 0.1 0.5 1.0 2.0 3.0] rad/s

G.4 Two Carts (Example 3, P1)

Consider the classical ACC benchmark problem shown in Figure G.3. It is a mechanical frictionless system composed of two carts of mass m_1 and m_2, coupled by a link (spring) of stiffness γ. The problem is to control the position $x_2(t)$ of the second cart by applying a force $u(t)$ to the first cart.

The transfer function between the position $x_2(t)$ of the second cart and the force $u(t)$ in the first cart is

$$\frac{x_2(s)}{u(s)} = P(s) = \frac{1}{s^2\left[(m_1 m_2)s^2 + \gamma(m_1 + m_2)\right]} \quad \text{(G.5)}$$

with parametric uncertainty $m_1 \in [0.9, 1.1]$, $m_2 \in [0.9, 1.1]$, and $\gamma \in [0.4, 0.6]$.

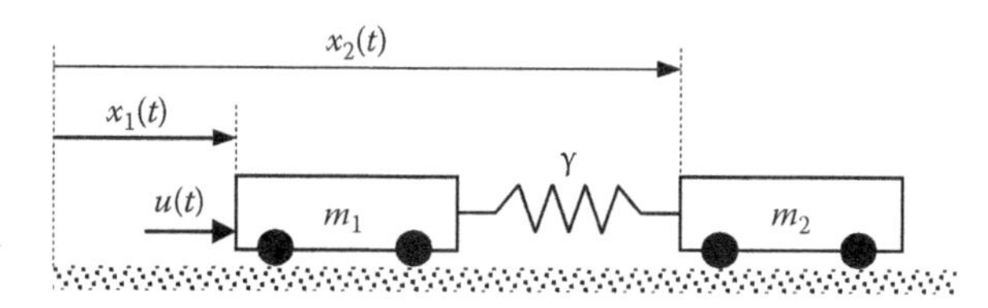

FIGURE G.3
ACC benchmark problem: two carts coupled by a spring.

Synthesize the compensator $G(s)$ to control the position $x_2(t)$, fulfilling the robust stability specification:

$$\left|\frac{P(j\omega)G(j\omega)}{1+P(j\omega)G(j\omega)}\right| \leq 1.2, \quad \forall\omega \in [0,\infty]$$

G.5 Inverted Pendulum (Example 4, P1)

Consider an inverted pendulum mounted on a cart, as shown in Figure G.4. The cart is controlled so that the mass m is always in the upright position. The equations that describe the system are

$$\begin{aligned} (M+m)\ddot{x} &= u - mL\ddot{\theta}\cos\theta + mL\dot{\theta}^2\sin\theta - k\dot{x} \\ I\ddot{\theta} &= mgL\sin\theta - mL^2\ddot{\theta} - mL\ddot{x}\cos\theta \end{aligned} \tag{G.6}$$

where
- θ is the pendulum angle to be controlled
- u is the force to be applied to the cart
- x is the linear position of the cart
- L is the stick length
- M is the cart mass
- m is the end mass
- I is the stick inertia
- k is the friction coefficient between the cart and the rail

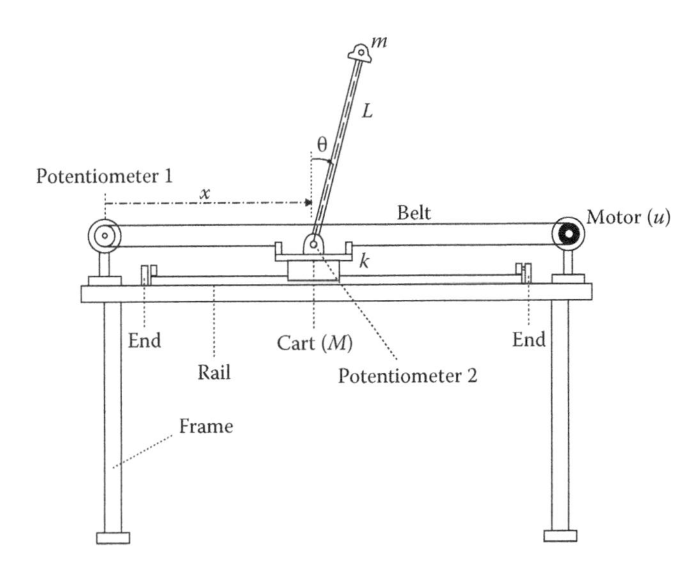

FIGURE G.4
Inverted pendulum.

The parameters are

$$m \in [0.1, 0.2]\,\text{kg}, \quad M \in [0.9, 1.1]\,\text{kg}, \quad I \in [0.005, 0.01]\,\text{kg m}^2$$
$$k \in [0.1, 0.2]\,\text{Ns/m}, \quad L = 0.5\,\text{m}, \quad g = 9.8\,\text{m/s}^2$$

By linearizing the system around the equilibrium point, $\theta(t) = 0$ and $d\theta(t)/dt = 0$, the equations are

$$\begin{aligned} &(M+m)\ddot{x} + mL\ddot{\theta} + k\dot{x} - u = 0 \\ &(I+mL^2)\ddot{\theta} + mL\ddot{x} - mgL\theta = 0 \end{aligned} \tag{G.7}$$

Given the state variables

$$\boldsymbol{x} = [x_1 \quad x_2 \quad x_3 \quad x_4]^T = [x \quad \dot{x} \quad \theta \quad \dot{\theta}]^T$$

the State Space description is given by

$$\begin{aligned} \dot{\boldsymbol{x}} &= \begin{bmatrix} 0 & 1 & 0 & 0 \\ 0 & -\dfrac{k(I+mL^2)}{(M+m)I+mL^2M} & -\dfrac{m^2L^2g}{(M+m)I+mL^2M} & 0 \\ 0 & 0 & 0 & 1 \\ 0 & \dfrac{mLk}{(M+m)I+mL^2M} & \dfrac{mLg(M+m)}{(M+m)I+mL^2M} & 0 \end{bmatrix}\boldsymbol{x} \\ &\quad + \begin{bmatrix} 0 \\ \dfrac{I+mL^2}{(M+m)I+mL^2M} \\ 0 \\ -\dfrac{mL}{(M+m)I+mL^2M} \end{bmatrix} u \\ y &= [0 \quad 0 \quad 1 \quad 0]\boldsymbol{x} \end{aligned} \tag{G.8}$$

Design a compensator $G(s)$ to keep the pendulum in the upright position for the following stability and performance specifications:

1. *Robust stability*: a minimum phase margin of 45°
2. *Disturbance rejection at plant output*:

$$\left|\frac{1}{1+P(j\omega)G(j\omega)}\right| < \left|\frac{0.05\left((j\omega/0.05)+1\right)}{(j\omega+1)}\right|$$

G.6 Central Heating System (Example 5, P1)

Consider a central heating system of a three-floor building (Figure G.5). The transfer function that describes the inner room temperature $T_r(t)$ in terms of the desired mixed water temperature $T_{md}(t)$, that comes from the mixing valve, is linearized around the equilibrium point and is

$$\frac{T_r(s)}{T_{md}(s)} = P(s) = \frac{k}{\tau s+1}e^{-Ls} \tag{G.9}$$

Using system identification techniques with experimental data, the obtained parameters are $k \in [40, 60]$, $\tau \in [800, 1200]$ s, and $L \in [100, 200]$ s.

Design a loop compensator $G(s)$ so that the system fulfills the following specifications:

1. *Robust stability*: $|P(j\omega)G(j\omega)/(1 + P(j\omega)G(j\omega))| \leq 1.2$ for ω = [0.0001 0.0003 0.0005 0.0007 0.001 0.003 0.005 0.007 0.01 0.03 0.05 0.07 0.1] rad/s
2. *Disturbance rejection at plant input*:

$$\left|\frac{P(j\omega)}{1+P(j\omega)G(j\omega)}\right| \leq 0.3$$

for ω = [0.0001 0.0003 0.0005 0.0007] rad/s

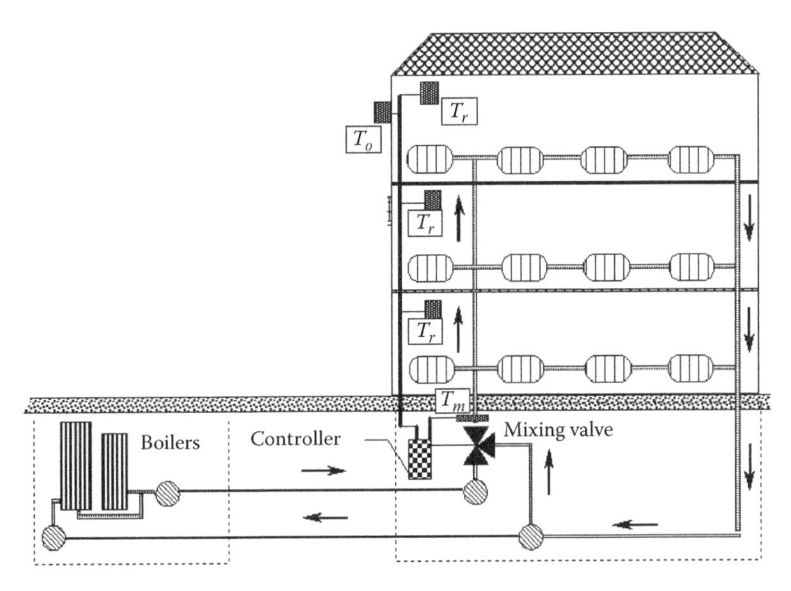

FIGURE G.5
Central heating system.

TABLE G.2
Controller Design Solutions

	$G(s)$	$F(s)$
1. DC motor	$G(s)=\frac{89(s/10+1)(s/1.5+1)}{s(s/300+1)}$	$F(s)=\frac{(s/50+1)(s/100+1)}{(s/4+1)(s/8+1)}$
2. Satellite	$G(s)=\frac{-5\times10^{-8}(s/(1.16\times10^{-8})+1)}{(s/0.014+1)}$	$F(s)=\frac{6.408e5s^3+1.632e4s^2+209.6s+1}{1.143e5s^3+9371s^2+202.9s+1}$
3. Two carts coupled by spring	$G(s)=\frac{(s/0.5+1)(4s^2+0.8s+1)}{(s/12+1)(s/100+1)(s/500+1)}$	None
4. Inverted pendulum	$G(s)=\frac{-41.2(s/0.6+1)(s/5+1)}{s(s/2400+1)}$	None
5. Central heating system	$G(s)=\frac{7.2\times10^{-5}(s/0.0013+1)(s/0.02+1)(s/2.3+1)}{s(s/1.1+1)(s/1.1+1)}$	None

G.7 Controller Design Solutions

Some possible solutions for the aforementioned examples, calculated by using the *QFTCT*,[149,150] are shown in Table G.2.

G.8 DC Motor, *QFTCT* Windows (Example 1, P2)

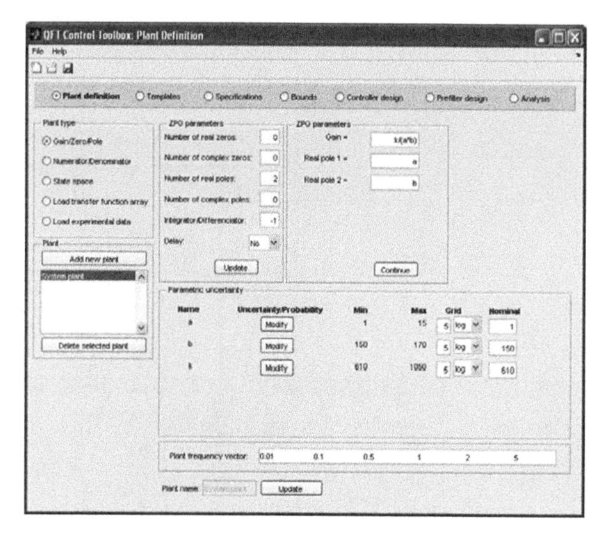

FIGURE G.6
Plant definition. A zero/pole/gain model: Example 1.

```
Stability specification:
M = 1.4619
Disturbance rejection at plant input specification:
nDR=[3.575 0];
dDR=[1 21.4 3.18];
DR=tf(nDR,dDR);
Reference tracking specification:
nTU=[8 10];
dTU=[1 6.4 10];
TU=tf(nTU,dTU);
nTL=25;
dTL=[0.08333 2 14.08 25];
TL=tf(nTL,dTL);
```

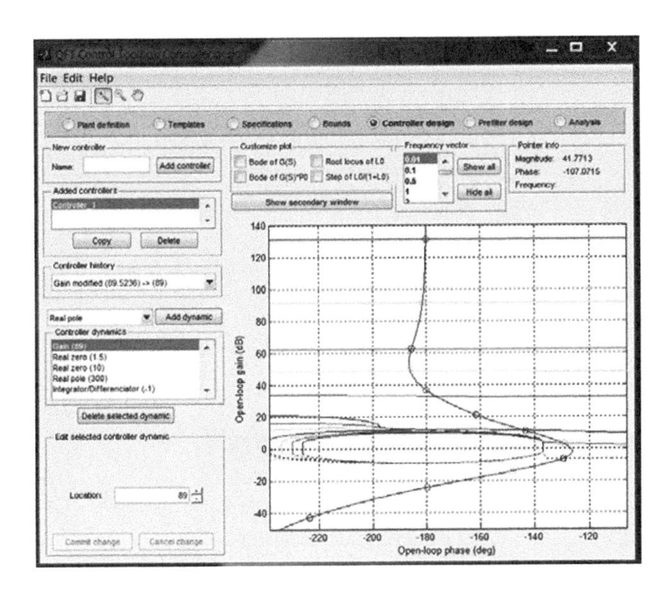

FIGURE G.7
(See color insert.) Controller design (Nichols chart): Example 1.

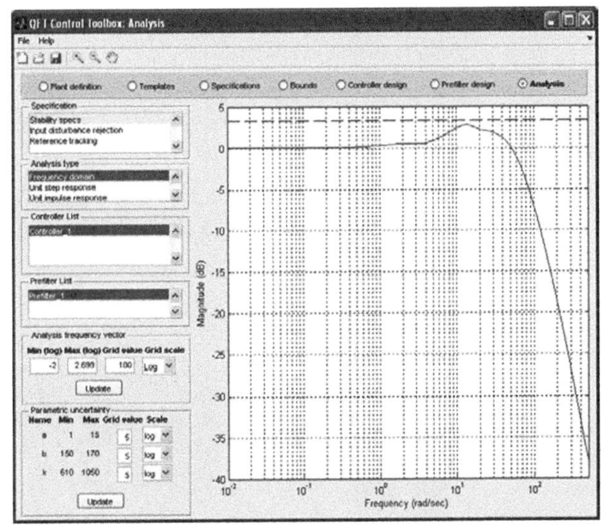

FIGURE G.8
Stability analysis in the frequency domain: Example 1.

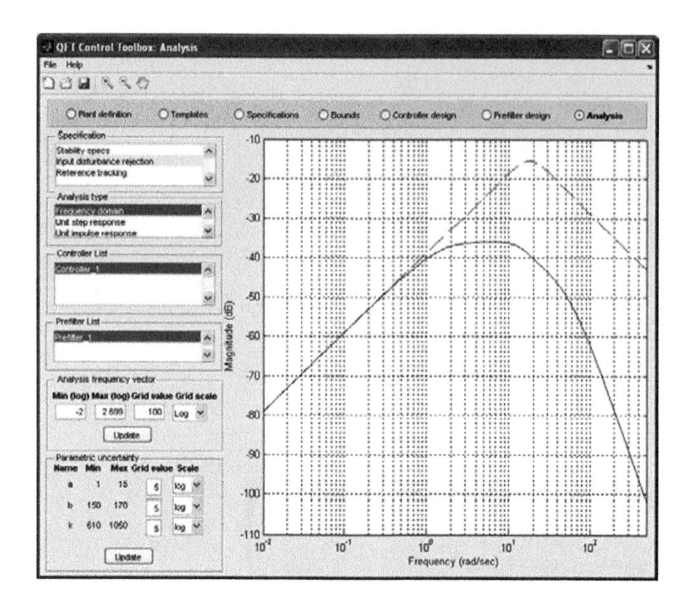

FIGURE G.9
Disturbance rejection plant input in the frequency domain: Example 1.

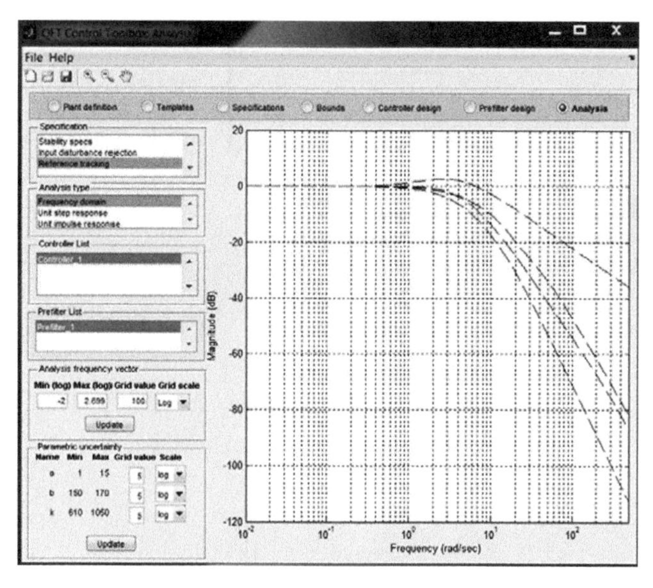

FIGURE G.10
Reference-tracking analysis in the frequency domain: Example 1.

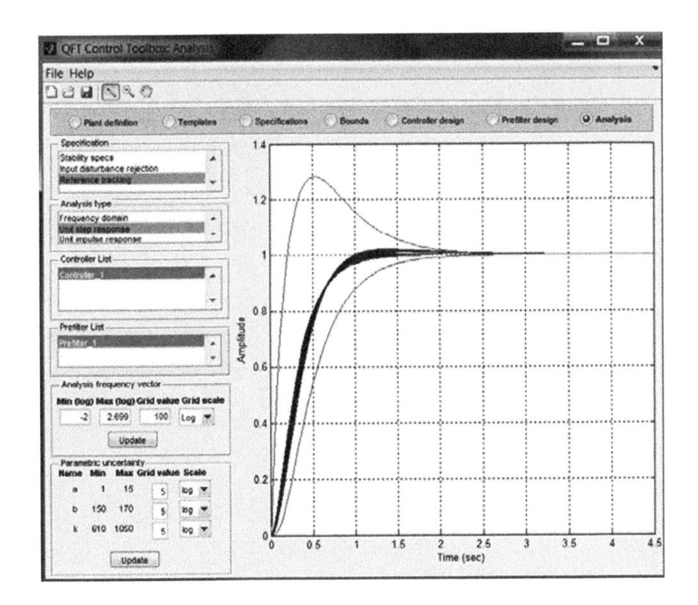

FIGURE G.11
Reference-tracking analysis in the time domain: Example 1.

G.9 Satellite, *QFTCT* Windows (Example 2, P2)

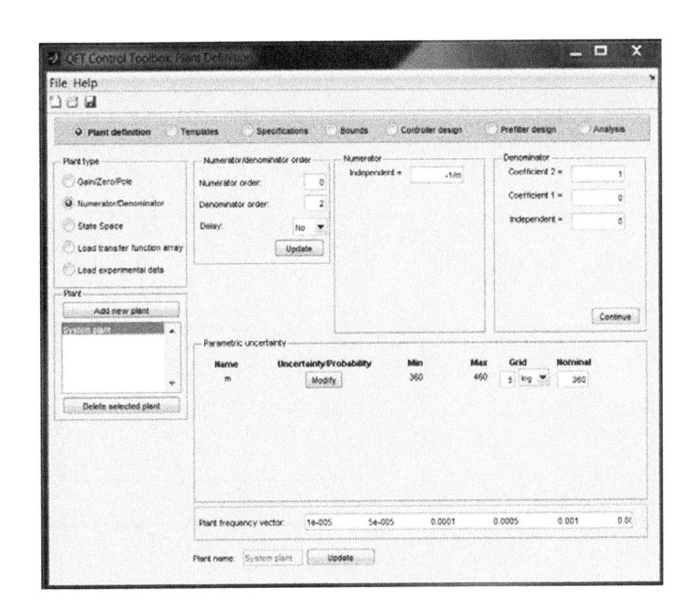

FIGURE G.12
Plant definition (numerator/denominator model): Example 2.

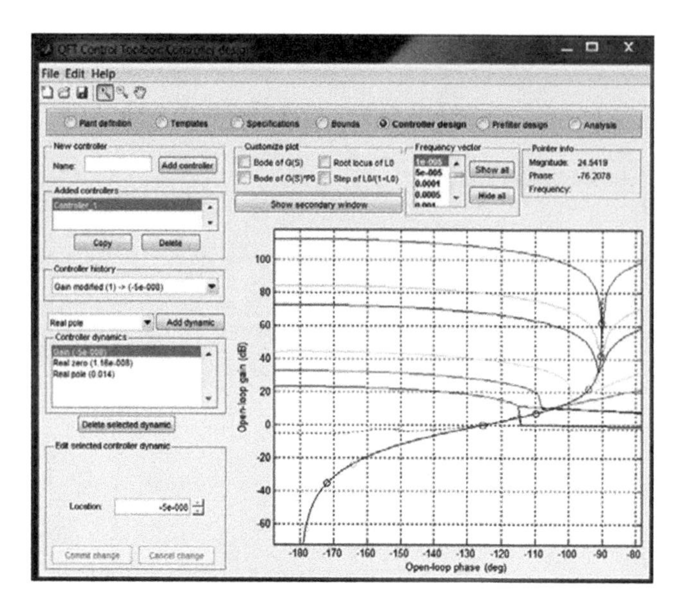

FIGURE G.13
(See color insert.) Controller design (Nichols chart): Example 2.

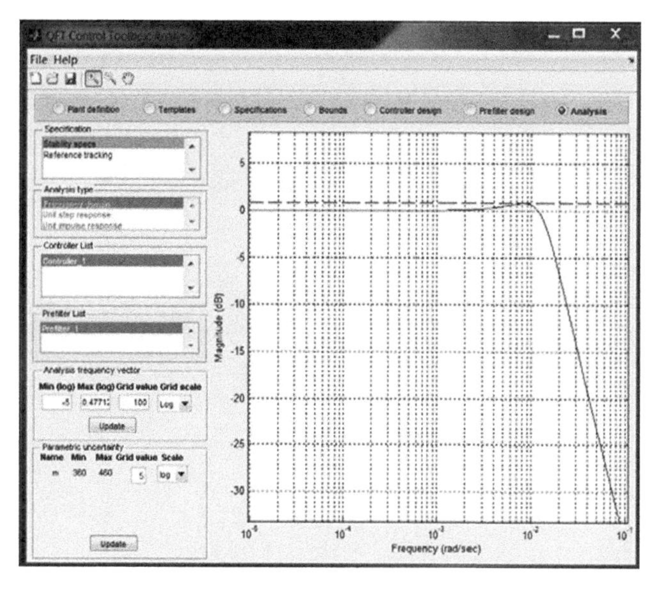

FIGURE G.14
Stability analysis in the frequency domain: Example 2.

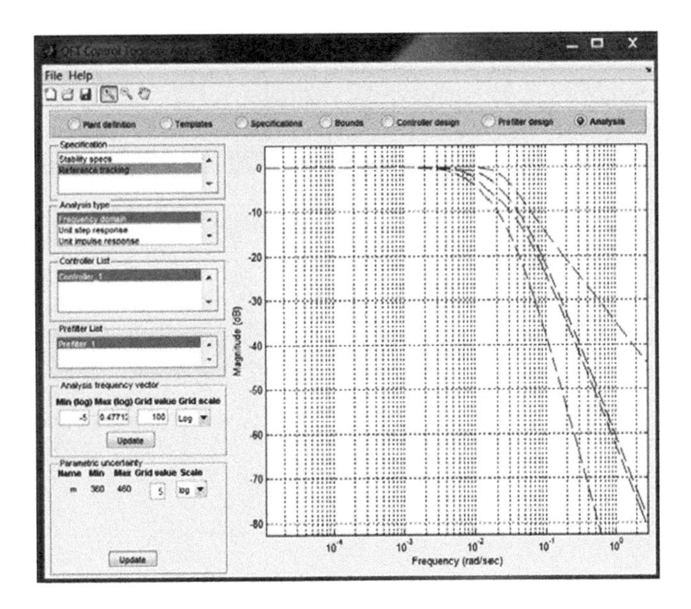

FIGURE G.15
Reference-tracking analysis in the frequency domain: Example 2.

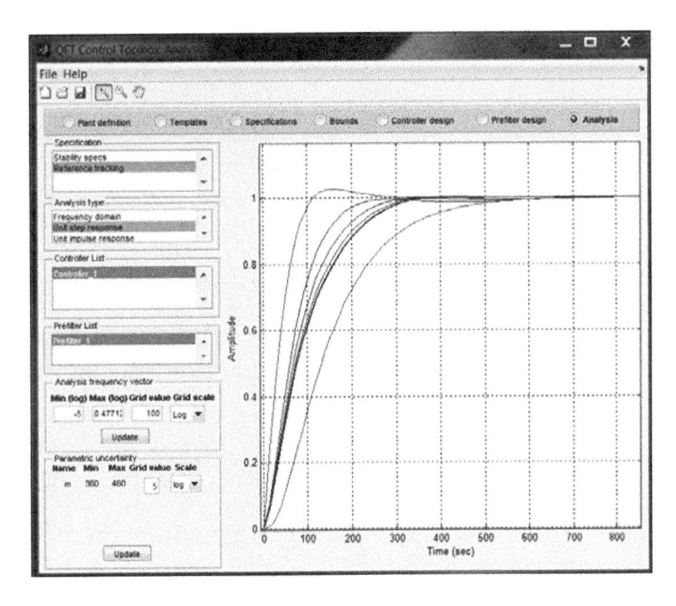

FIGURE G.16
Reference-tracking analysis in the time domain: Example 2.

G.10 Two Carts, *QFTCT* Windows (Example 3, P2)

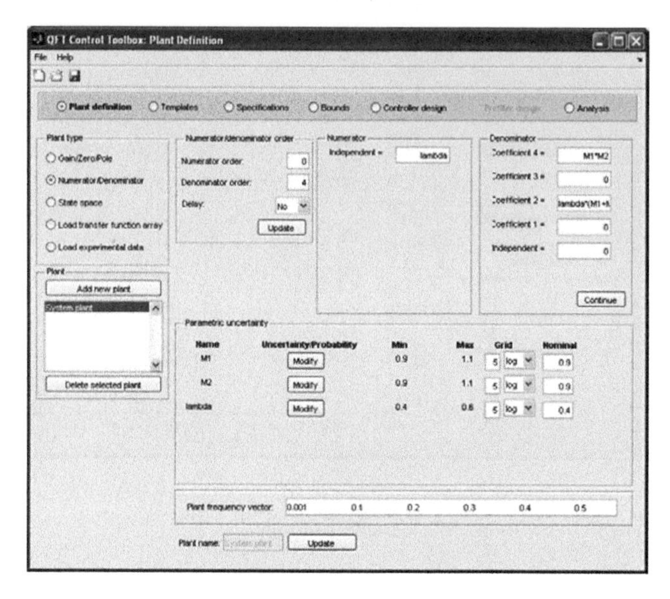

FIGURE G.17
Plant definition. A numerator/denominator model: Example 3.

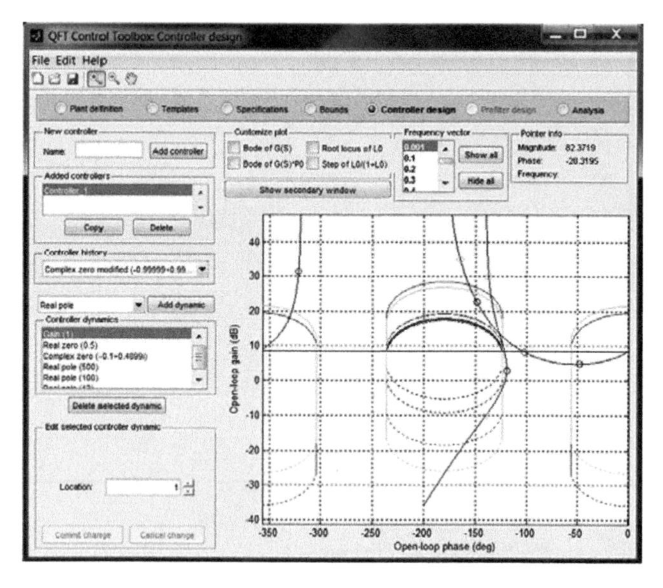

FIGURE G.18
(See color insert.) Controller design (Nichols chart): Example 3.

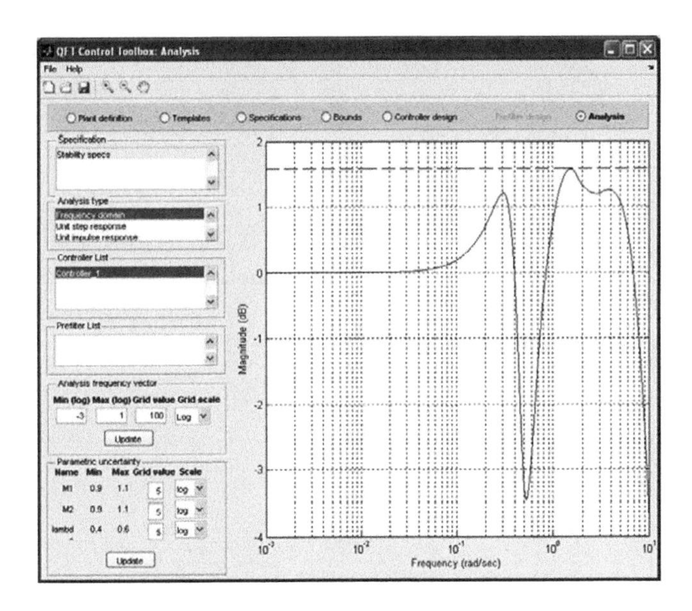

FIGURE G.19
Stability analysis: Example 3.

G.11 Inverted Pendulum, *QFTCT* Windows (Example 4, P2)

According to Section D.12, the number of poles p of $L(s)$ in the RHP is $p = 1$ (due to the pole at $s = +4.22$), and the number of encirclements (-2 at $\omega = 1.5$ rad/s, $+1$ at $\omega = 0$) is $N = -2 + 1 = -1$ (see Figure G.21). This means that the closed-loop control system is stable, $z = N + p = -1 + 1 = 0$.

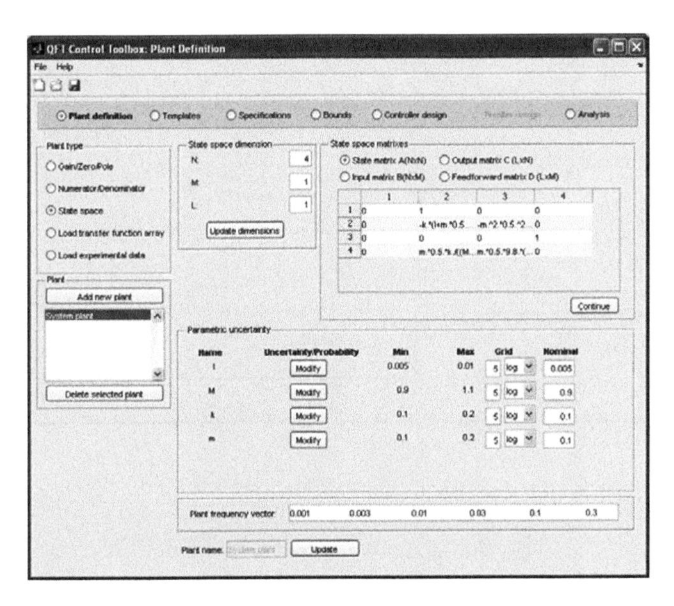

FIGURE G.20
Plant definition (a State Space model): Example 4.

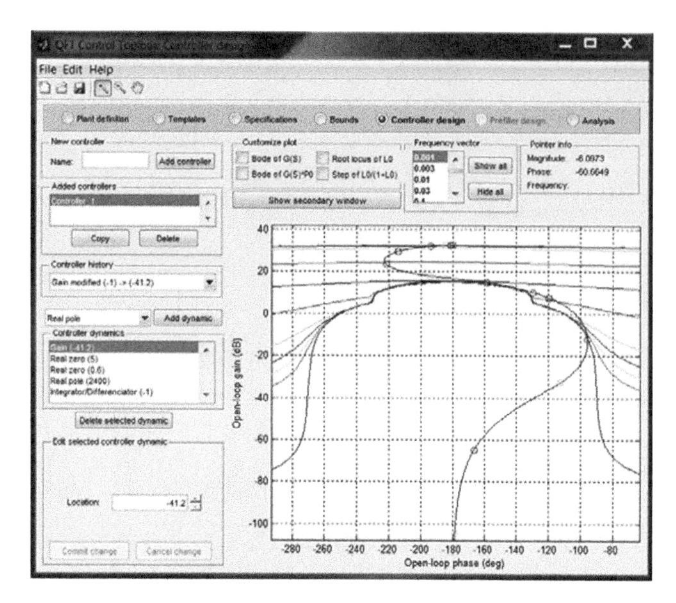

FIGURE G.21
(See color insert.) Controller design (Nichols chart): Example 4.

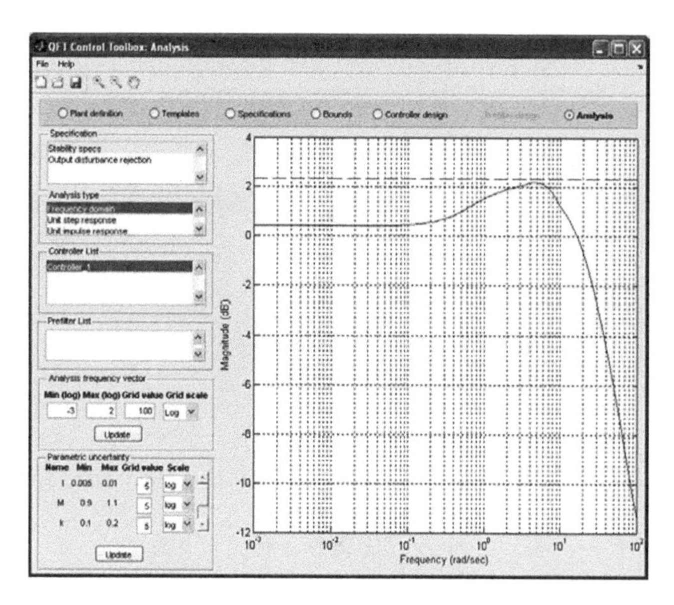

FIGURE G.22
Stability analysis: Example 4.

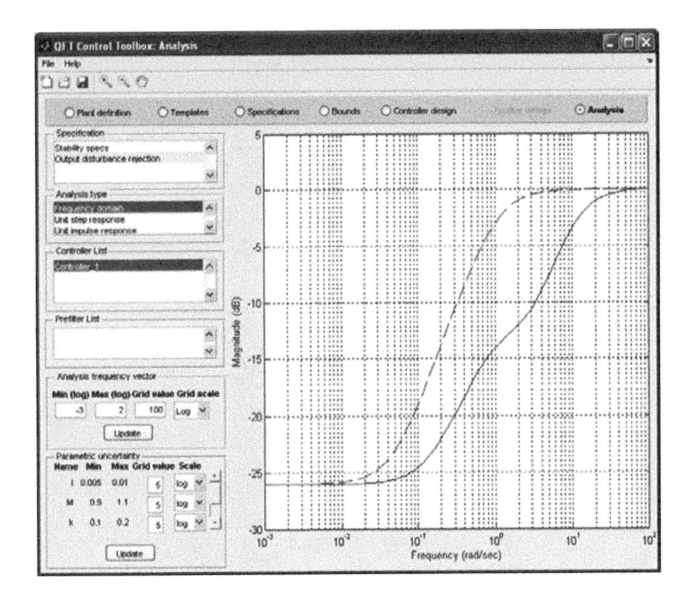

FIGURE G.23
Output disturbance rejection analysis: Example 4.

G.12 Central Heating System: *QFTCT* Windows (Example 5, P2)

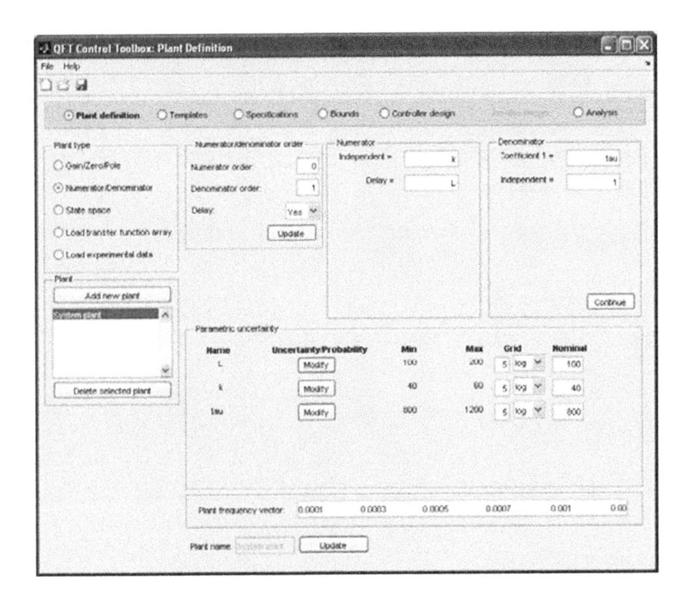

FIGURE G.24
Plant definition (a numerator/denominator model): Example 5.

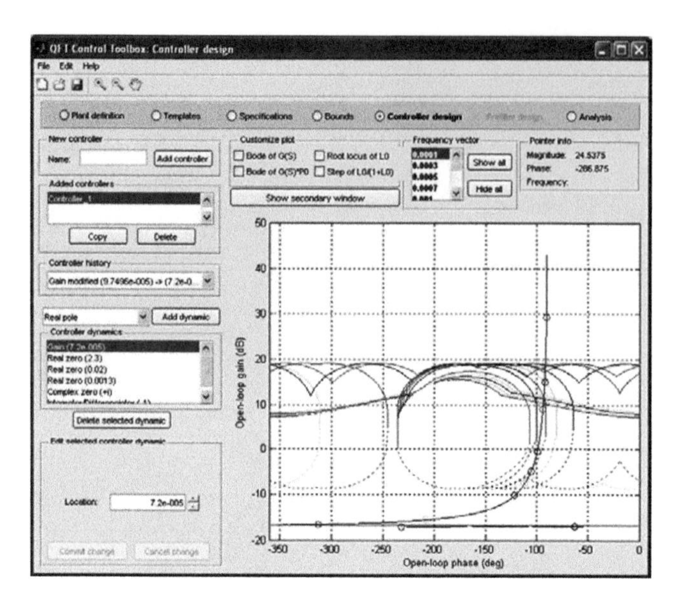

FIGURE G.25
(See color insert.) Controller design (Nichols chart): Example 5.

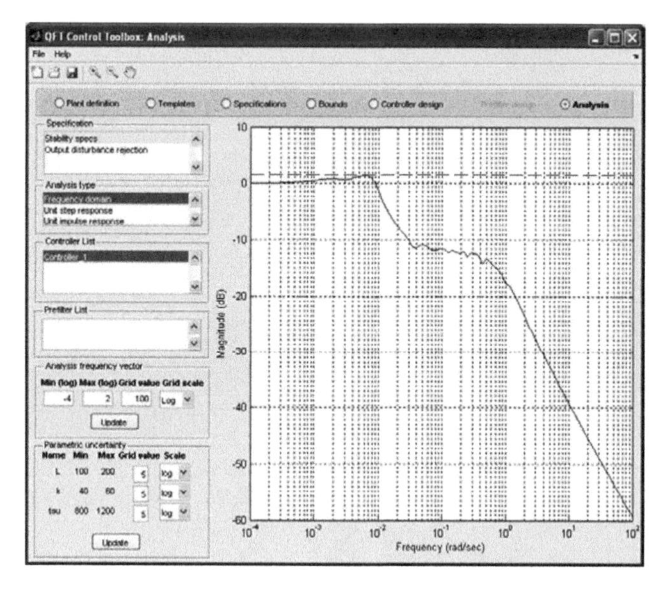

FIGURE G.26
Stability analysis: Example 5.

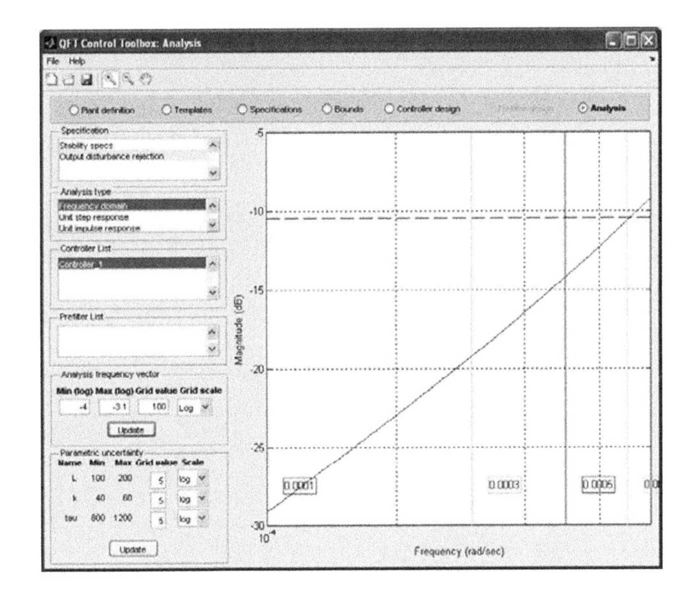

FIGURE G.27
Output disturbance rejection analysis: Example 5.

Appendix H: Conversion of Units

	Unit	Symbol	Conversion
International system of units (SI)	Meter	m	1 m = 3.2808 ft
	Meter	m	1 m = 39.37 in
	Meter	m	1 m = 1.0936 yd
	Square meter	m^2	1 m^2 = 10.7639 ft^2
	Square meter	m^2	1 m^2 = 2.4710×10^{-4} ac
	Square kilometer	km^2	1 km^2 = 0.3861 $mile^2$
	Cubic meter	m^3	1 m^3 = 35.3145 ft^3
	Cubic meter	m^3	1 m^3 = 264.17 gal
	Meter/second	m/s	1 m/s = 3.2808 ft/s
	Kilometer/hour	km/h	1 km/h = 0.6214 mile/h
	Kilogram	kg	1 kg = 2.2046 lb
	Celsius degree	°C	T(°C) = 5/9[T(°F) − 32]
	Watt	W	1 W = 1.3405×10^{-3} HP
American system of units (US)	Foot	ft	1 ft = 0.3048 m
	Inch	in	1 in = 2.54×10^{-2} m
	Yard	yd	1 yd = 0.9144 m
	Square foot	sq. ft (ft^2)	1 ft^2 = 9.2903×10^{-2} m^2
	Acre	ac	1 ac = 4.0469×10^3 m^2
	Square mile	sq. mi (mi^2)	1 $mile^2$ = 2.59 km^2
	Cubic foot	cu. ft (ft^3)	1 ft^3 = 2.8317×10^{-2} m^3
	Gallon	gal	1 gal = 3.7854×10^{-3} m^3
	Foot/second	ft/s	1 ft/s = 0.3048 m/s
	Mile/hour	mph (mile/h)	1 mile/h = 1.6093 km/h
	Pound	lb	1 lb = 0.4536 kg
	Fahrenheit degree	°F	T(°F) = (9/5)T(°C) + 32
	Horsepower	HP	1 HP = 746 W
Energy	Joule	J	1 J = 2.7778×10^{-7} kWh
	Kilowatt-hour	kWh	1 kWh = 3.6×10^6 J
Angle	Radian	rad	1 rad = (180/π)°
	Degree	°	1° = (π/180) rad
Frequency	Hertz	Hz	1 Hz = 60 rpm
	Revolutions per minute	rpm	1 rpm = (1/60) Hz
	Radians/second	rad/s	1 rad/s = 9.5493 rpm
	Revolutions per minute	rpm	1 rpm = 0.1047 rad/s

Problems

Chapter 2

2.1 Consider a plant transfer function given by

$$P(s) = \frac{k}{s(s+a)(s+b)}$$

with the parametric uncertainty

$$610 \le k \le 1050; \quad 1 \le a \le 15; \quad 150 \le b \le 170$$

(a) Plot $20 \log_{10}|P(j\omega)|$ and $\angle P(j\omega)$ versus ω on semilog graph paper for several plants for the frequency range of $0.05 < \omega < 500$ rad/s.

(b) Plot the plant templates on the Nichols chart for the following frequencies: 0.05, 1, 10, 100, 250, and 500 rad/s.

2.2 Consider a plant transfer function given by

$$P(s) = \frac{k(s+b)}{s(s+a)}$$

with the parametric uncertainty

$$0.5 \le k \le 4; \quad 0.5 \le a \le 1; \quad 1.1 \le b \le 1.6$$

(a) Plot $20 \log_{10}|P(j\omega)|$ and $\angle P(j\omega)$ versus ω on semilog graph paper for several plants for the frequency range of $0.05 < \omega < 500$ rad/s.

(b) Plot the plant templates on the Nichols chart for the following frequencies: 0.05, 1, 10, 100, 250, and 500 rad/s.

2.3 Consider a plant transfer function given by

$$P(s) = \frac{k(s+b)}{s^2(s+a)}$$

with the parametric uncertainty

$$1 \le k \le 5; \quad 0.5 \le a \le 1; \quad 2 \le b \le 5$$

(a) Plot $20 \log_{10}|P(j\omega)|$ and $\angle P(j\omega)$ versus ω on semilog graph paper for several plants for the frequency range of $0.05 < \omega < 100$ rad/s.

(b) Plot the plant templates on the Nichols chart for the following frequencies: 0.05, 0.3, 1, 5, 10, and 100 rad/s.

Chapter 3

3.1 Obtain the tracking models $T_{RU}(s)$ and $T_{RL}(s)$ to satisfy the following specifications:

(a) For the upper bound $M_p = 1.11$ and $t_s = 1.21$ s
For the lower bound $t_s = 1.71$ s

(b) For the upper bound $M_p = 1.13$ and $t_s = 1.17$ s
For the lower bound $t_s = 2$ s

(c) For the upper bound $M_p = 1.2$ and $t_s = 1.20$ s
For the lower bound $t_s = 1.48$ s

(d) For the upper bound $M_p = 1.15$ and $t_s = 1.34$ s
For the lower bound $t_s = 1.69$ s

Hint: M_p is overshoot, t_s is settling time.

3.2 Determine the disturbance-bound model $T_D(s)$, disturbance at the plant input, that satisfies the following specifications:

(a) $|y_D(t_p)| = 0.1$ for $t_p = 65$ ms

(b) $|y_D(t_p)| = 0.08$ for $t_p = 50$ ms

where the peak time t_p is

$$t_p = \frac{\cos^{-1}(\zeta)}{\omega_n\sqrt{1-\zeta^2}}$$

Hint: It is recommend the following transfer function structure:

$$T_D(s) = \frac{K_x s}{s^2 + 2\zeta\omega_n s + \omega_n^2} = \frac{K_x s}{(s+a)^2 + b^2}$$

with a parameter ζ selected in the range of $0.5 \le \zeta \le 0.7$.

3.3 Consider a plant transfer function given by

$$P(s) = \frac{k}{s(s+a)(s+b)}$$

with the parametric uncertainty

$$610 \le k \le 1050; \quad 1 \le a \le 15; \quad 150 \le b \le 170$$

(1) The tracking specifications are
- For the upper bound $M_p = 1.15$ and $t_s = 1.34$ s
- For the lower bound $t_s = 1.69$ s

(2) The disturbance-rejection specification (at plant input) is

$$\alpha_p < |y_D(t_p)| = 0.1 \quad \text{for } t_p = 65 \text{ ms}, \quad y_D(\infty) = 0, \quad \omega < 10 \text{ rad/s}.$$

(3) The minimum desired phase margin angle (γ) is 40°.

(a) Synthesize the compensator $G(s)$ and the prefilter $F(s)$.

(b) Obtain $y_R(t)$ and $y_{D1}(t)$ and compare their respective specification.

3.4 Consider a plant transfer function given by

$$P(s) = \frac{k}{s(s+a)(s+b)}$$

with the parametric uncertainty

$$300 \le k \le 480; \quad 1 \le a \le 4; \quad 120 \le b \le 150$$

(1) The disturbance-rejection specification (at plant input) is

$$\alpha_p < |y_D(t_p)| = 0.1 \quad \text{for } t_p = 65 \text{ ms}, \quad y_D(\infty) = 0, \quad \omega < 10 \text{ rad/s}.$$

(2) The minimum desired phase margin angle (γ) is 45°.

(a) Synthesize the compensator $G(s)$.

(b) Obtain $y_{D1}(t)$ and compare its respective specification.

3.5 Consider a plant transfer function given by

$$P(s) = \frac{k}{s(s+a)(s+b)}$$

with the parametric uncertainty

$$900 \le k \le 1200; \quad 2 \le a \le 10; \quad 120 \le b \le 150$$

(1) The tracking specifications are

- For the upper bound $M_p = 1.20$ and $t_s = 1.2\,\text{s}$
- For the lower bound $t_s = 1.48\,\text{s}$

(2) The disturbance-rejection specification (at plant input) is

$$\alpha_p < |y_D(t_p)| = 0.1 \quad \text{for } t_p = 65 \text{ ms}, \quad y_D(\infty) = 0, \quad \omega < 10 \text{ rad/s}.$$

(3) The minimum desired gain margin is 4 dB.

(a) Synthesize the compensator $G(s)$ and the prefilter $F(s)$.

(b) Obtain $y_R(t)$ and $y_{D1}(t)$ and compare their respective specification.

3.6 Consider a plant transfer function given by

$$P(s) = \frac{k(s+b)}{s(s+a)}$$

with the parametric uncertainty

$$0.5 \le k \le 4; \quad 0.5 \le a \le 1; \quad 1.1 \le b \le 1.6$$

(1) The tracking specifications are
- For the upper bound $M_p = 1.15$ and $t_s = 1.34$ s
- For the lower bound $t_s = 1.69$ s

(2) The disturbance-rejection specification (at plant input) is

$$\alpha_p < |y_D(t_p)| = 0.1 \quad \text{for } t_p = 65 \text{ ms}, \quad y_D(\infty) = 0, \quad \omega < 10 \text{ rad/s}.$$

(3) The minimum desired phase margin angle (γ) is 40°.

(a) Synthesize the compensator $G(s)$ and the prefilter $F(s)$.

(b) Obtain $y_R(t)$ and $y_{D1}(t)$ and compare their respective specification.

3.7 Consider a plant transfer function given by

$$P(s) = \frac{k}{s(s+a)(s+b)}$$

with the parametric uncertainty

$$3060 \le k \le 3360; \quad 3 \le a \le 9; \quad 140 \le b \le 170$$

(1) The tracking specifications are
- For the upper bound $M_p = 1.13$ and $t_s = 1.17$ s
- For the lower bound $t_s = 2.0$ s

(2) The disturbance-rejection specification (at plant output) is

$$\alpha_p < |y_D(t_p)| = 0.1, \quad y_D(\infty) = 0, \quad \omega < 20 \text{ rad/s}.$$

(3) The minimum desired gain margin is 5 dB.

(a) Synthesize the compensator $G(s)$ and the prefilter $F(s)$.

(b) Obtain $y_R(t)$ and $y_{D2}(t)$ and compare their respective specification.

3.8 Consider a plant transfer function given by

$$P(s) = \frac{k}{s(s+a)(s+b)}$$

with the parametric uncertainty

$$3060 \le k \le 3360; \quad 3 \le a \le 9; \quad 140 \le b \le 170$$

(1) The tracking specifications are
- For the upper bound $M_p = 1.11$ and $t_s = 1.21$ s
- For the lower bound $t_s = 1.71$ s

(2) The disturbance-rejection specification (at plant output) is

$$\alpha_p < |y_D(t_p)| = 0.1, \quad y_D(\infty) = 0, \quad \omega < 20 \text{ rad/s}.$$

(3) The minimum desired margin angle (γ) is 45°.

(a) Synthesize the compensator $G(s)$ and the prefilter $F(s)$.

(b) Obtain $y_R(t)$ and $y_{D2}(t)$ and compare their respective specification.

3.9 Consider a plant transfer function given by

$$P(s) = \frac{k}{s(s+a)(s+b)}$$

with the parametric uncertainty

$$3060 \le k \le 3360; \quad 3 \le a \le 9; \quad 140 \le b \le 170$$

(1) The tracking specifications are

- For the upper bound $M_p = 1.11$ and $t_s = 1.21$ s
- For the lower bound $t_s = 1.71$ s

(2) The disturbance-rejection specification (at plant output) is

$$T_D(s) = (s^2 + 3.0540 \text{ s})/(s^2 + 6.1090 \text{ s} + 9.5790), \quad \omega < 20 \text{ rad/s}.$$

(3) The minimum desired margin angle (γ) is 45°.

(a) Synthesize the compensator $G(s)$ and the prefilter $F(s)$.

(b) Obtain $y_R(t)$ and $y_{D2}(t)$ and compare their respective specification.

3.10 Consider the plant transfer function $P(s)$ given by

$$P(s) = \frac{ck(as^2 + ds + b)}{fs\left[cas^2 + d(c+a)s + b(c+a)\right]}$$

where the parametric uncertainty is

$$k \in [100,\ 800]; \quad a \in [1400,\ 11000];$$

$$b \in [58000,\ 115000]; \quad c = 65.6; \quad d = 377; \quad f = 3.07$$

(a) Obtain the complete template for the frequency $\omega = 10$ rad/s.

(b) Compare the result with the template generated by using only the contour of the parameter uncertainty space.

3.11 Repeat Problem 3.10 for the following transfer function:

$$P(s) = \frac{k}{s^2 + 2\zeta\omega_n s + \omega_n^2} e^{-\tau s}$$

where the parametric uncertainty is

$$k = 6; \quad \zeta = 0.1; \quad \omega_n \in [0.6,\ 1.4]; \quad \tau \in [0.1,\ 2].$$

and for the frequency $\omega = 0.9$ rad/s.

3.12 Repeat Problem 3.10 for the following transfer function:

$$P(s) = \frac{b_1 s^3 + b_2^2 s^2 + 12s + (12 + 5b_3)}{12s^4 + (b_4 + 5)s^3 + (b_5 b_1^2)s^2 + (b_3 + 8b_6)s + (3b_7)} e^{-\tau s}$$

where the parametric uncertainty is

$$b_1 \in [1,\ 2]; \quad b_2 \in [2,\ 3]; \quad b_3 \in [21,\ 25]; \quad b_4 \in [1,\ 9];$$

$$b_5 \in [1,\ 2]; \quad b_6 \in [1,\ 2]; \quad b_7 \in [1,\ 2]; \quad \tau \in [\pi/6,\ \pi/3]$$

and for the frequency $\omega = 10$ rad/s.

3.13 Repeat Problem 3.10 for the following transfer function:

$$P(s) = \frac{k}{as + b} e^{-\tau s}$$

where the parametric uncertainty is

$$k \in [1,\ 2]; \quad a \in [1,\ 5]; \quad b \in [10,\ 12]; \quad \tau \in [0.1,\ 0.2]$$

and for the frequency $\omega = 3$ rad/s.

3.14 Determine if the transfer function $P_1(s)$ with three uncertain parameters can be reduced to the transfer function $P_2(s)$ with only two related uncertain parameters in order to reduce the number of operations needed to compute the template:

$$P_1 = \left\{ \begin{vmatrix} P_1(s) = \dfrac{c}{as+b}, & a \in [1,5] \\ b \in [10,12], & c \in [1,2] \end{vmatrix} \right\}$$

$$P_2 = \left\{ \begin{vmatrix} P_2(s) = \dfrac{c/a}{s + b/a}, & (c/a) \in \left[\dfrac{\min c}{\max a} = 0.2, \dfrac{\max c}{\min a} = 2\right] \\ & (b/a) \in \left[\dfrac{\min b}{\max a} = 2, \dfrac{\max b}{\min a} = 12\right] \end{vmatrix} \right\}$$

Obtain, on the Nichols chart, the templates at $\omega = 5$ rad/s for both transfer functions. Explain the effect of the reduction (selection of P_2 instead of P_1), if any, in the controller design.

3.15 Consider the bench-top helicopter shown in Figure P3.15. It is a laboratory scale plant with 3 degrees of freedom (3DOF), roll angle ϕ, pitch angle θ, and yaw angle ψ, each one measured by an absolute encoder. Two electrical DC motors are attached to the helicopter body, making the two propellers turn. The total force F caused by aerodynamics makes the total system turn around an angle measured by an encoder. A counterweight of mass M helps the propellers lift the body weight due to mass m. The dynamics of the pitch angle is obtained by applying Lagrange's equations to the mechanical scheme, so that

$$Fl_1 - mg\,[(h+d)\sin\theta + l_1\cos\theta] + Mg(l_2 + l_3\cos\alpha)\cos\theta + Mg(l_3\sin\theta - h)\sin\theta - b_e\frac{d\theta}{dt} = J_e\frac{d^2\theta}{dt^2}$$

where
- h, d, l_1, l_2, and l_3 are lengths
- m is the sum of both motors' mass
- M is the counterweight mass
- b_e is the dynamic friction coefficient
- g is the gravity acceleration
- J_e is the inertial moment of the whole system around the pitch angle θ
- α is a fixed construction angle

The total nonlinear model obtained from the previous equation can be simplified by linearizing around the operational point $\theta_0 = 0$. It yields a second-order transfer function between the pitch angle $\theta(s)$ and the motor control signal $U(s)$ given by

$$P(s) = \frac{\theta(s)}{U(s)} = \frac{k\omega_n^2}{s^2 + 2\zeta\omega_n s + \omega_n^2}e^{-sT}$$

Using system identification techniques from experimental data, the parametric uncertainty is as follows: $k \in [0.0765, 0.132]$ rad/V; $\zeta \in [0.025, 0.05]$; $\omega_n \in [0.96, 1.58]$ rad/s; $T \in [0.09, 0.11]$ s. The motor control signal U presents a saturation limit of ± 10 V.

Synthesize the compensator $G(s)$ and the prefilter $F(s)$ to fulfill the following control objectives: (a) to minimize the reference tracking error, (b) to increase the damping and increase the system stability, (c) to reduce the overshoot, (d) to reject the high-frequency noise at the feedback sensors, and (e) to deal with the actuator constraints.

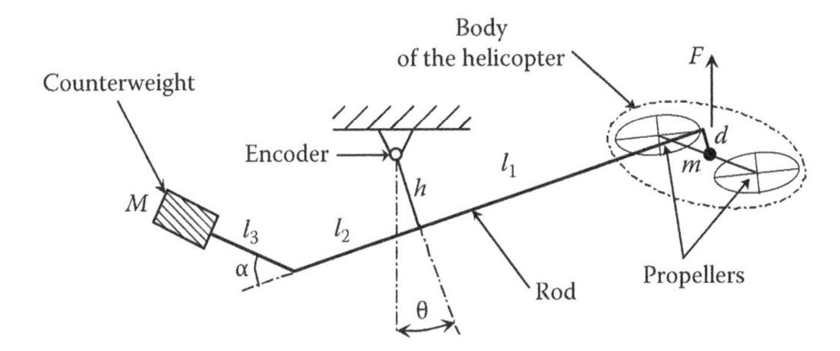

FIGURE P3.15
3DOF Lab helicopter: diagram for pitch angle.

3.16 Consider the second angle $\delta_2(s)$ of a 5DOF SCARA robot (Figure P3.16). The transfer function that describes $\delta_2(s)$ in terms of the motor signal $u_2(s)$ is

$$\frac{\delta_2(s)}{u_2(s)} = p_{22}(s) = \frac{(\alpha_1 + 2\alpha_3 h)s + \nu_1}{s\Delta(s)}\frac{1}{k}$$

where

$$\Delta = \chi_2 s^2 + \chi_1 s + \chi_0$$

and

$$\chi_2 = \alpha_2(\alpha_1 + 2\alpha_3 h) - (\alpha_2 + \alpha_3 h)^2$$

$$\chi_1 = \alpha_2 \nu_1 + \nu_2(\alpha_1 + 2\alpha_3 h)$$

$$\chi_0 = \nu_1 \nu_2$$

The parametric uncertainty is

$$k = 75 \text{ ct/N}\cdot\text{m}; \quad h \in [-1, 1]; \quad a_1 k \in [719, 813] \text{ ct s}^2/\text{rad}$$

$$\alpha_2 k \in [186, 200] \text{ ct s}^2/\text{rad}; \quad \alpha_3 k \in [134, 230] \text{ ct s}^2/\text{rad}$$

$$\nu_1 k \in [67, 381] \text{ ct s/rad}; \quad \nu_2 k \in [11.6, 91.9] \text{ ct s/rad}$$

Fixing the other 4DOF, design and synthesize a compensator $G(s)$ and a prefilter $F(s)$ to control the second angle $\delta_2(s)$, which fulfills the following performance specifications:

- *Robust stability*: $|P(j\omega)G(j\omega)/[1 + P(j\omega)G(j\omega)]| \leq 1.2, \forall\omega \in [0, \infty]$.
- *Control effort constraint*: The control signals have to be lower than 32,767 [ct] for a disturbance rejection at the plant output of 20°.
- *Disturbance rejection at plant input*: The maximum allowed error has to be 30° for torque disturbances of 1000 [ct].
- *Tracking specifications*: $|T(j\omega)| = |\{P(j\omega)G(j\omega)/[1 + P(j\omega)G(j\omega)]\}F(j\omega)|$ has to achieve tracking tolerances defined by $a(\omega) \leq |T(j\omega)| \leq b(\omega)$, where,

$$a(\omega) = \left|\frac{2.25}{[(j\omega)^2 + 4.5(j\omega) + 2.25][(j\omega)/10 + 1]}\right|, \quad b(\omega) = \left|\frac{12.25\{[(j\omega)/30] + 1\}}{[(j\omega)^2 + 5.25(j\omega) + 12.25]}\right|$$

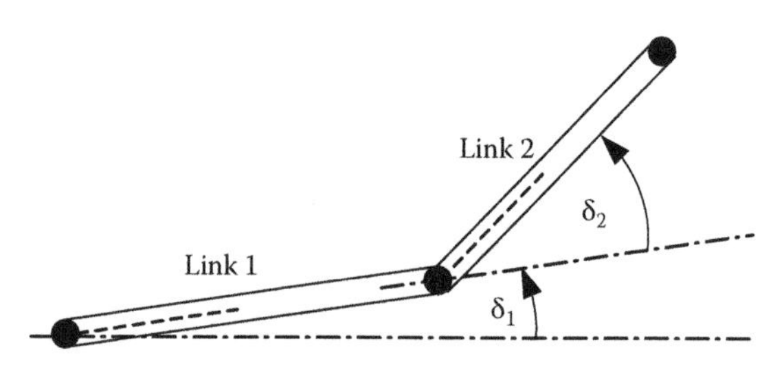

FIGURE P3.16
5DOF SCARA robot: diagram for links 1 and 2.

3.17 Consider a DC field-controlled motor whose nominal characteristics are showed in the Table P3.17 (Figure P3.17). Assuming that the angular velocity Ω is the signal (plant output) to be controlled, and the field voltage V_f, applied to the DC motor, is the actuator (plant input), this results in the following plant transfer function:

$$P(s) = \frac{\Omega}{V_f} = \frac{K_m/R_f}{Js+b}$$

where the electrical time constant L_f/R_f is neglected compared to the field time constant J/b.

Usually the motor parameters differ from the nominal ones due to temperature, aging, and some other special effects. Because of that, it is presumed a 20% variation in b and a 5% variation in K_m/R_f from their nominal values.

In addition, the larger divergence observed in the proposed real application takes place at the rotor inertia J. Its nominal value of $J = 0.01$ refers to unload driving. However, considering the load, J is expected to be in the range [0.01–0.1], between unload and full load.

Synthesize the compensator $G(s)$ and the prefilter $F(s)$ to fulfill the following specifications:

(1) *Robust stability:* $|P(j\omega)G(j\omega)/1 + P(j\omega)G(j\omega)| \leq 1.3, \forall\omega \in [0,\infty]$

(2) *Robust reference tracking*: $|T(j\omega)| = |(P(j\omega)G(j\omega)/1 + P(j\omega)G(j\omega))F(j\omega)|$ has to achieve tracking tolerances defined by $a(\omega) \leq |T(j\omega)| \leq b(\omega)$

where

$$b(\omega) = \left|\frac{0.66[(j\omega)+30]}{(j\omega)^2 + 4(j\omega) + 19.75}\right|$$

TABLE P3.17

Nominal Characteristics of a DC Field-Controlled Motor

Parameter	Nominal Values
Unload rotor inertia, J	0.01 kg m^2
Friction, b	0.1 N·m s
Motor constant, K_m	0.05 N·m/A
Field resistance, R_f	1 Ω
Field inductance, L_f	<<0.1 H

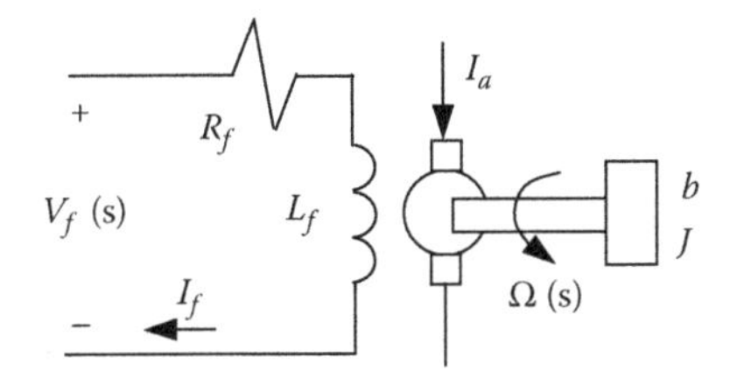

FIGURE P3.17
DC field-controlled motor: electric circuit diagram.

$$a(\omega) = \left| \frac{8400}{[(j\omega)+3][(j\omega)+4][(j\omega)+10]} \right|$$

(3) *Gain at high frequency*: Reduce the "cost of feedback":

$$|L(j\omega_{hf})| < -20 \text{ dB}, \quad \omega_{hf} \geq 100.$$

Chapter 4

4.1 Consider the classical benchmark problem shown in Figure P4.1. It is a mechanical frictionless system composed of two carts of mass M_1 and M_2, coupled by a link of stiffness γ. The problem is to control the position $x_2(t)$ of the second cart by applying a force $u(t)$ to the first cart.

The parametric uncertainty is given by

$$M_1 \in [0.9,\ 1.1]; \quad M_2 \in [0.9,\ 1.1]; \quad \gamma \in [0.4,\ 0.6]$$

Synthesize a controller $G(z)$ for a digital control system where the sampling time is $T = 0.01$ s, to control the position $x_2(t)$ fulfilling the following robust stability specification:

$$\left| \frac{P(j\omega)G(j\omega)}{1+P(j\omega)G(j\omega)} \right| \leq 1.2, \quad \forall \omega \in [0, \infty]$$

4.2 For the plant described in Problem 3.3, synthesize a controller $G(z)$ and a prefilter $F(z)$ for a digital control system with a sampling time: $T = 0.002$ s.

4.3 For the plant described in Problem 3.4, synthesize a controller $G(z)$ for a digital control system with a sampling time: $T = 0.001$ s.

4.4 For the plant described in Problem 3.5, synthesize a controller $G(z)$ and a prefilter $F(z)$ for a digital control system with a sampling time: $T = 0.002$ s.

4.5 For the plant described in Problem 3.6, synthesize a controller $G(z)$ and a prefilter $F(z)$ for a digital control system with a sampling time: $T = 0.002$ s.

4.6 For the plant described in Problem 3.8, synthesize a controller $G(z)$ and a prefilter $F(z)$ for a digital control system with a sampling time: $T = 0.002$ s.

4.7 For the plant described in Problem 3.9, synthesize a controller $G(z)$ and a prefilter $F(z)$ for a digital control system with a sampling time: $T = 0.01$ s.

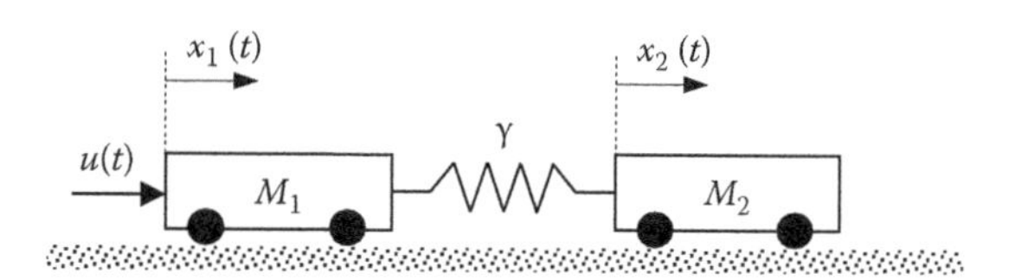

FIGURE P4.1
Two-carts mechanical system.

Chapter 5

5.1 Consider a 3×3 multiple-input-multiple-output (MIMO) system whose transfer function matrix is

$$\begin{bmatrix} y_1(s) \\ y_2(s) \\ y_3(s) \end{bmatrix} = \begin{bmatrix} \dfrac{0.1e^{-0.4s}}{0.92s+1} & \dfrac{2(3s+1)}{4s+1} & \dfrac{-1}{2s+1} \\ \dfrac{1e^{-0.1s}}{7s+1} & \dfrac{1}{3s+1} & \dfrac{-0.1e^{-0.2s}}{0.87s+1} \\ \dfrac{-2(s+1)}{0.92s+1} & \dfrac{-3e^{-0.4s}}{0.54s+1} & \dfrac{1e^{-0.3s}}{6s+1} \end{bmatrix} \begin{bmatrix} m_1(s) \\ m_2(s) \\ m_3(s) \end{bmatrix}$$

(a) Calculate the Bristol's relative gain analysis array.

(b) Find the best input/output pairing.

(c) Analyze the existing coupling between loops.

(d) Suppose that the three control loops are designed and working. In that case, study the effect of an instantaneous opening of the loop $y_2(s)$.

(e) Using *Method 1*, design a diagonal MIMO QFT compensator so that the system reaches a good level of stability, disturbance rejection, and decoupling (no uncertainty).

(f) Using *Method 2*, design a diagonal MIMO QFT compensator so that the system reaches a good level of stability, disturbance rejection, and decoupling (no uncertainty).

5.2 Consider two gaseous elements A and B. In Figure P5.2, the tank 1 has 80% of A (X_1 = 80%) and 20% of B, whereas tank 2 has 20% of A (X_2 = 20%) and 80% of B. The hydraulic system represented in the figure mixes two fluxes, F_1 and F_2, which come from the tanks 1 and 2, respectively. The product has a composition of X% of A and a flux F, so that

$$F = F_1 + F_2$$

$$FX = F_1X_1 + F_2X_2$$

The inputs are the fluxes F_1 and F_2, and the controlled outputs are the flux F and the concentration X of element A.

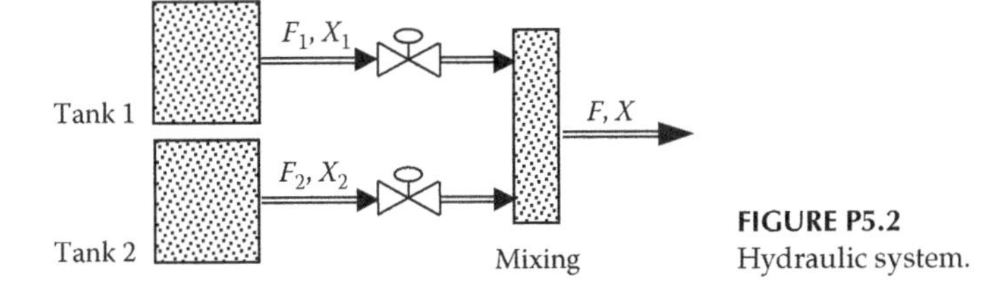

FIGURE P5.2
Hydraulic system.

In order to control around $X_{ref} = 60\%$ and $F_{ref} = 200\,\text{mol/h}$

(a) Calculate the Bristol's relative gain analysis array.

(b) Find the best input/output pairing.

(c) Analyze the existing coupling between loops.

(d) Find the coupling worst case according to the set point X_{ref}.

5.3 Consider a 2 × 2 linear multivariable system with uncertainty having the following transfer function matrix:

$$\mathbf{P}(s) = \begin{bmatrix} p_{11}(s) & p_{12}(s) \\ p_{21}(s) & p_{22}(s) \end{bmatrix} = \begin{bmatrix} \dfrac{10\alpha\beta}{2s+\beta} & \dfrac{-4\beta}{s+\beta} \\ \dfrac{2\alpha\gamma}{s+\gamma} & \dfrac{3\alpha\gamma}{0.5s+\gamma} \end{bmatrix}$$

$$\alpha \in [0.8, 1.2]; \quad \beta \in [8, 10]; \quad \gamma \in [5, 6]$$

(a) Calculate the Bristol's relative gain analysis array.

(b) Find the best input/output pairing.

(c) Analyze the existing coupling between loops.

(d) Using *Method 1*, design a diagonal MIMO QFT compensator so that the system fulfills the required specifications (i, ii, iii).

(e) Using *Method 2*, design a diagonal MIMO QFT compensator so that the system fulfills the required specifications (i, ii, iii).

(i) *Robust stability*

$$\left| \frac{\left(p_{ii}^{*e}\right)^{-1} g_{ii}}{1 + \left(p_{ii}^{*e}\right)^{-1} g_{ii}} \right| \le 1.4, \quad i = 1, 2$$

(ii) *Robust reference tracking*

$$\left|B_{ii}^{L}(j\omega)\right| \le \left|t_{ii}^{Y/R}(j\omega)\right| \le \left|B_{ii}^{U}(j\omega)\right|, \quad i = 1, 2$$

where

$$B_{11}^{L}(s) = \frac{16}{(s^2 + 7.6s + 16)(s/10 + 1)}$$

$$B_{11}^{U}(s) = \frac{16(s/15 + 1)}{s^2 + 3.6s + 16}$$

$$B_{22}^{L}(s) = \frac{100}{(s^2 + 16s + 100)(s/20 + 1)}$$

$$B_{22}^{U}(s) = \frac{100(s/25 + 1)}{s^2 + 14s + 100}$$

(iii) *Minimizing the coupling effects*

Reduce the interaction, $t_{12}^{Y/R}$ and $t_{21}^{Y/R}$, as much as possible.

5.4 Consider the popular highly interacting 2 × 2 distillation column (Figure P5.4), described by Skogestad and Postlethwaite as a benchmark problem, where the transfer function matrix is

$$\begin{bmatrix} y_1(s) \\ y_2(s) \end{bmatrix} = \frac{1}{1+75s}\begin{bmatrix} 0.878 & -0.864 \\ 1.082 & -1.096 \end{bmatrix}\begin{bmatrix} k_1e^{-sT_1} & 0 \\ 0 & k_2e^{-sT_2} \end{bmatrix}\begin{bmatrix} u_1(s) \\ u_2(s) \end{bmatrix}$$

and where the parameters are as follows: $k_1, k_2 \in [0.8, 1.2]$; $T_1, T_2 \in [0, 60]$ s.

The inputs are the reflux ($u_1 = L$) and the boil up ($u_2 = V$), and the controlled outputs are the top and bottom product compositions ($y_1 = y_D$ and $y_2 = x_B$).

(a) Calculate the Bristol's relative gain analysis array.

(b) Find the best input/output pairing.

(c) Analyze the existing coupling between loops.

(d) Using *Method 1*, design a diagonal MIMO QFT compensator so that the system fulfills the required specifications (i, ii).

(e) Using *Method 2*, design a diagonal MIMO QFT compensator so that the system fulfills the required specifications (i, ii).

(i) Closed-loop stability

(ii) For a unit step demand in channel 1 at $t = 0$, the plant output y_1 (tracking) and y_2 (interaction) should satisfy

- *Tracking*: $y_1(t) \geq 0.9$ for all $t \geq 30$ min
  - $y_1(t) \leq 1.1$ for all t
  - $0.99 \leq y_1(\infty) \leq 1.01$
- *Interaction*: $y_2(t) \leq 0.5$ for all t
  - $-0.01 \leq y_2(\infty) \leq 0.01$

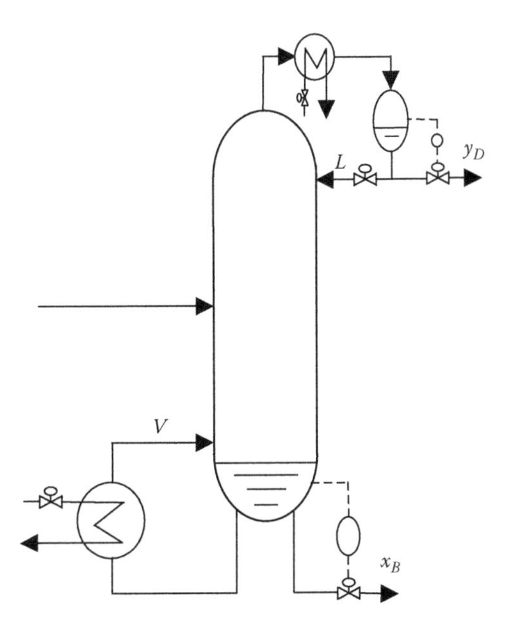

FIGURE P5.4
2 × 2 distillation column.

5.5 Consider a 2 × 2 linear multivariable system with uncertainty whose transfer function matrix is

$$P(s) = \begin{bmatrix} \frac{k_{11}}{s/p_{11}+1} & \frac{k_{12}}{s/p_{12}+1} \\ \frac{k_{21}}{s/p_{21}} & \frac{k_{22}}{s/p_{22}+1} \end{bmatrix}$$

where

$$\begin{aligned} &k_{11} \in [5,8], && p_{11} \in [100,150] \\ &k_{12} \in [1,2], && p_{12} \in [5,6] \\ &k_{21} \in [-2,-1], && p_{21} \in [5,6] \\ &k_{22} \in [5,8], && p_{22} \in [10,15] \end{aligned}$$

(a) Calculate the Bristol's relative gain analysis array.
(b) Find the best input/output pairing.
(c) Analyze the existing coupling between loops.
(d) Using *Method 1*, design a diagonal MIMO QFT compensator so that the system fulfills the required specifications (i, ii, iii).
(e) Using *Method 2*, design a diagonal MIMO QFT compensator so that the system fulfills the required specifications (i, ii, iii).
  (i) *Robust stability*: Stability of at least 50° lower phase margin and at least 1.8333 (5.26 dB) lower gain margin.
  (ii) *Rejection of disturbances at plant output*: The error should be lower than 0.5 with disturbances of 1.0 at plant output for frequencies lower than 120 rad/s for loop 1, and error lower than 0.5 with disturbances of 1.0 at plant output for frequencies lower than 8 rad/s for loop 2.
  (iii) *Control effort restriction*: The control signal should be lower than 0.2 when rejecting disturbances at plant output of 0.2 considering a bandwidth of 120 rad/s for loop 1 and of 8 rad/s for loop 2.

5.6 Consider a 2 × 2 linear multivariable system with uncertainty whose transfer function matrix is

$$P(s) = \begin{bmatrix} \frac{k_{11}}{\tau_{11}s+1} & \frac{k_{12}}{\tau_{12}s+1} \\ \frac{k_{21}}{\tau_{21}s+1} & \frac{k_{22}}{\tau_{22}s+1} \end{bmatrix}$$

where

$$\begin{aligned} &k_{11} \in [0.5,3], && \tau_{11} \in [0.5,3] \\ &k_{12} \in [-2.2,-1.8], && \tau_{12} \in [8,12] \\ &k_{21} \in [11,15], && \tau_{21} \in [3,8] \\ &k_{22} \in [2,7], && \tau_{22} \in [5,10] \end{aligned}$$

(a) Calculate the Bristol's relative gain analysis array.

(b) Find the best input/output pairing.

(c) Analyze the existing coupling between loops.

(d) Using *Method 1,* design a diagonal MIMO QFT compensator so that the system fulfills the required specifications (i, ii, iii).

(e) Using *Method 2,* design a diagonal MIMO QFT compensator so that the system fulfills the required specifications (i, ii, iii).

   (i) *Robust stability:* $|t_{ii}(j\omega)| \le 1.2$ for $i = 1, 2$, $\forall\omega$ where the terms $t_{ii}(j\omega)$ are the diagonal elements of the matrix $T_{y/r}$. This condition implies at least 50° lower phase margin and at least 1.833 (5.26 dB) lower gain margin.

   (ii) *Rejection of disturbances at plant output*

$$\left|t_{ii}^{y/do}(j\omega)\right| = \left|\frac{(j\omega)}{(j\omega)+10}\right|, \quad \omega < 50\,\text{rad/s}, \quad \text{for } i = 1,2$$

   (iii) *Minimizing the coupling effects:* c_{ij} as much as possible and for a range of frequencies: $\omega < 10^{-2}$ rad/s.

Chapter 6

6.1 The model of a spacecraft (*s/c*) flying in formation (Figure P6.1), relative to the center of mass of the formation (*c.m.f.*), is

$$\begin{bmatrix} x \\ y \\ z \end{bmatrix} = \frac{1}{m}\begin{bmatrix} \dfrac{1}{s^2+\omega_0^2} & \dfrac{2\omega_0}{s(s^2+\omega_0^2)} & 0 \\ \dfrac{-2\omega_0}{s(s^2+\omega_0^2)} & \dfrac{s^2-3\omega_0^2}{s^2(s^2+\omega_0^2)} & 0 \\ 0 & 0 & \dfrac{1}{s^2+\omega_0^2} \end{bmatrix}\begin{bmatrix} Q_x \\ Q_y \\ Q_z \end{bmatrix}$$

where

$\rho = [x, y, z]^T$ are the distances of the *s/c* to the *c.m.f* (Figure P6.1)

$[Q_x, Q_y, Q_z]^T$ are the actuator forces (thrusters)

Note that this is a 2×2 MIMO system (axes x and y) plus a SISO (single-input-single-output) system (axis z). Also note that although the p_{22} element of the system is non-minimum phase, the MIMO plant does not present RHP transmission zeros. Its Smith–McMillan decomposition yields a double integrator and a conjugate pole at $s = \pm j\omega_0$, but no transmission zeros.

For an *s/c* flying in formation in a circular geostationary Earth orbit, the parameters are as follows: mean orbit rate $\omega_0 \in [7.2637 \times 10^{-5}, 7.3208 \times 10^{-5}]$ rad/s and mass of the satellite $m \in [1600, 1650]$ kg.

Design a MIMO QFT controller according to *Method 4* to achieve the following specifications:

- *Matrix specifications on* ***S****(s) and* ***T****(s)*

$$\| \boldsymbol{T}(j\omega) \|_\infty < 2 \quad \text{and} \quad \| \boldsymbol{S}(j\omega) \|_\infty < 2$$

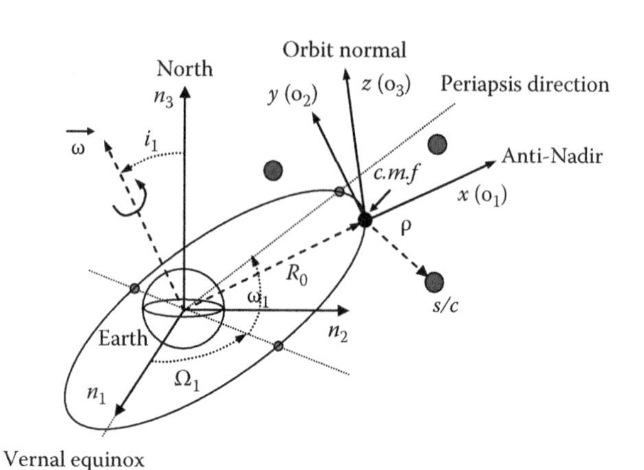

FIGURE P6.1
Multiple spacecraft flying in formation: Low Earth Orbit.

- *Classical loop-by-loop robust stability and performance specifications: Robust stability (Type 1)*

$$\left| \frac{g_{ii}^{\beta}(s)\left[\left(p_{ii}^{x}(s)\right)^{*_e}\right]^{-1}}{1+g_{ii}^{\beta}(s)\left[\left(p_{ii}^{x}(s)\right)^{*_e}\right]^{-1}} \right| \leq 1.1 \quad \forall \omega, i=1,2,3$$

- *Disturbance rejection at plant output (Type 2)*

$$\left| \frac{1}{1+g_{33}(s)p_{33}(s)} \right| \leq 2 \quad \forall \omega$$

- *Disturbance rejection at plant input (Type 3), $i = 1,2,3$*

$$\left| \frac{\left[\left(p_{ii}^{x}(s)\right)^{*_e}\right]^{-1}}{1+g_{ii}^{\beta}(s)\left[\left(p_{ii}^{x}(s)\right)^{*_e}\right]^{-1}} \right| \leq \left| \frac{20s}{\left(\frac{s}{0.008}+1\right)\left(\frac{s}{0.06}+1\right)\left(\frac{s}{0.6}+1\right)} \right| \quad \forall \omega < 1\ \text{rad/s}$$

6.2 Calculate a non-diagonal MIMO QFT compensator for Problem 5.1 by using *Method 3* in question (e) and *Method 4* in question (f).

6.3 Calculate a non-diagonal MIMO QFT compensator for Problem 5.3 by using *Method 3* in question (e) and *Method 4* in question (f).

6.4 Calculate a non-diagonal MIMO QFT compensator for Problem 5.4 by using *Method 3* in question (e) and *Method 4* in question (f).

6.5 Calculate a non-diagonal MIMO QFT compensator for Problem 5.5 by using *Method 3* in question (e) and *Method 4* in question (f).

6.6 Calculate a non-diagonal MIMO QFT compensator for Problem 5.6 by using *Method 3* in question (e) and *Method 4* in question (f).

Chapter 11

11.1 DC motors are direct current machines employed extensively in a variety of industrial applications, including wind turbines (WTs). Medium- and high-power DC motors are often employed in WTs for pitch control systems (to move the blades and change the aerodynamic coefficient of the rotor) and for yaw control systems (to move the nacelle following the wind direction). In this design problem, a yaw control system is utilized for a WT. It consists of DC motors (the actuators), the nacelle to be oriented (inertia), the bearings (friction and inertia), a vane (sensor), and the controller. The transfer function between the angle $y(s)$ to be controlled (output) and the voltage applied (input) to the DC motors' actuators $u(s)$ is

$$\frac{y(s)}{u(s)} = P(s) = \frac{K_m}{s(Js+D)(L_a s+R_a)} = \frac{K_0}{s(\tau_1 s+1)(\tau_2 s+1)}$$

where
- J is the inertia of the rotating elements
- D is the viscous friction
- K_m is the motor-torque constant
- R_a is the resistance
- L_a is the inductance of the motor armature

Combining these parameters, the lumped parameters are $K_0 = K_m/(R_a D)$ as the gain, $\tau_1 = J/D$ as the mechanical time constant, and $\tau_2 = L_a/R_a$ as the electrical time constant of the system. The parameters for the selected WT yaw system are as follows: $K_0 = 100\% \pm 5\%$, $\tau_1 = 1\% \pm 5\%$ s, and $\tau_2 = 0.1 \pm 5\%$ s.

Using quantitative feedback theory (QFT), design two controllers with the following structures (Figure P11.1a and b).

Case 1: [2-zeros 2-poles controller]

$$\frac{u(s)}{e(s)} = G_1(s) = K_c\frac{(a_1 s+1)(a_2 s+1)}{(a_3 s+1)(a_4 s+1)}$$

Case 2: (Practical 2DOF proportional-integral-derivative [PID] controller)

$$u(s) = \frac{k_P s+k_I}{s}[r(s)-y(s)] - \frac{k_D s}{(k_D/N)s+1}y(s)$$

where $u(s)$ is calculated from a PI that multiplies the error, and a derivative part with a filter that multiplies the output. In other words, $u(s) = [q_0(s)]^{-1}[q_1(s)r(s) - q_2(s)y(s)]$ with

$$q_0(s) = s\left(s+\frac{N}{k_D}\right);\quad q_1(s) = k_P s^2 + \left(k_I + \frac{Nk_P}{k_D}\right)s + \frac{Nk_I}{k_D}$$

$$q_2(s) = (N+k_P)s^2 + \left(k_I + \frac{Nk_P}{k_D}\right)s + \frac{Nk_I}{k_D}$$

to achieve the following specifications:

(1) *Steady-state specifications*: $E_{ss}(t) = 0$ for step changes in $r(t)$

(2) *Transient specifications*: Overshoot $\leq 55\%$ and a 2% settling time ≤ 3.5 s

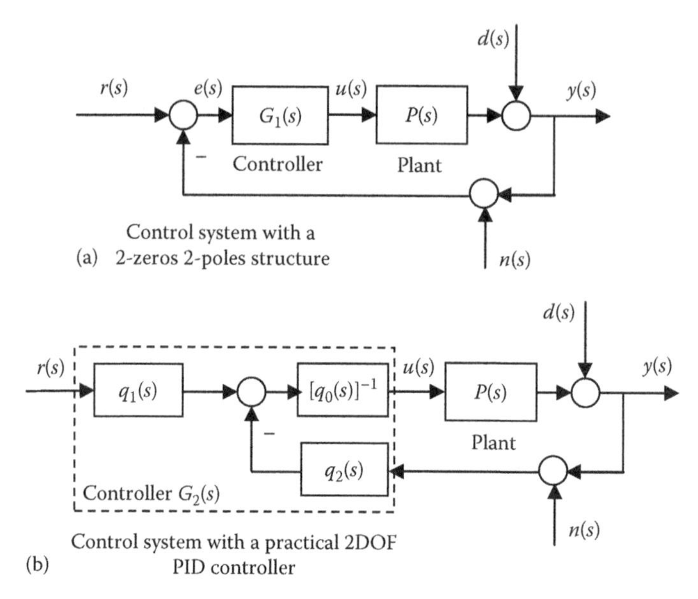

FIGURE P11.1
Control structures: (a) 2-zeros 2-poles, (b) practical 2DOF PID controller.

For both controller cases,

(a) Plot the Root locus of the controlled system and identify the final roots of the system.
(b) Plot the Bode diagram of the open-loop controlled system.
(c) Calculate the stability gain and phase margins.
(d) Calculate Bode diagram and the bandwidth of the closed-loop controlled system.
(e) For the closed-loop controlled system, plot the time response of the output $y(s)$ and the control signal $u(t)$ under a unit step input at the reference $r(t)$, [with $d(t) = 0$, $n(t) = 0$].
(f) For the closed-loop controlled system, plot the time response of the output $y(s)$ and the control signal $u(t)$ under a unit ramp input at the reference $r(t)$, [with $d(t) = 0$, $n(t) = 0$].
(g) For the closed-loop controlled system, plot the time response of the output $y(s)$ and the control signal $u(t)$ under a unit step disturbance input at $d(t)$, [with $r(t) = 0$, $n(t) = 0$].
(h) For the closed-loop controlled system, plot the time response of the output $y(s)$ and the control signal $u(t)$ under noise input at $n(t)$, [with $r(t) = 0$, $d(t) = 0$].
(i) Compare and discuss the results of both controllers.

11.2 A generic qualitative power curve for a variable-speed pitch-controlled WT is shown in Figure 9.3. Four regions and two areas are indicated in this figure. The power P of the WT, that is, the actual power supplied to the grid, which is the incoming wind power minus losses, separates the graph in two main areas: below and above rated power P_r. Below rated power (below rated wind speed, $v_1 < v_r$) the WT produces only a fraction of its total design power (see following equation), and, therefore, an optimization strategy to capture the maximum amount of energy at every wind speed needs to be performed. On the other hand, above rated power (above rated wind speed, $v_r < v_1$), where the wind speed has more power than the rated power P_r, a limitation strategy to generate only the rated power is required.

$$P = \frac{1}{2}\rho A_r C_p(t) v_1(t)^3 \eta = T_r(t)\Omega_r(t)\eta$$

where
- T_r is the mechanical torque at the shaft due to the wind
- Ω_r is the rotor speed
- ρ is the air density
- A_r is the rotor effective surface
- v_1 is the undisturbed upstream wind speed
- C_p is the aerodynamic power coefficient
- t is the time
- η is the mechanical and electrical efficiency

In Region 3 (see Figure 9.3), the rotor speed Ω_r is controlled by varying the pitch angle β of the three blades at the same time (collective pitch control). A simplified transfer function model of the WT, from the pitch command β_d (radians) to the rotor speed Ω_r (rpm), is presented by the following expression (Figure P11.2):

$$\frac{\Omega_r(s)}{\beta_d(s)} = P(s) = \frac{K_1\omega_g^2}{(\tau_g s + 1)(s^2 + 2\zeta_g\omega_g s + \omega_g^2)}$$

where
- $K_1 = -6800$
- $\tau_g = 4.8\,\text{s}$
- $\zeta_g = 0.006$
- $\omega_g = 19\,\text{rad/s}$

Commercial WTs often use PID controllers with low-pass filters for the rotor-speed/pitch control system. The transfer function representing the PID controller, with a filter in the derivative part, is given by

$$\frac{\beta_d(s)}{e(s)} = G(s) = k_P + \frac{k_I}{s} + \frac{k_D s}{(\gamma s + 1)}$$

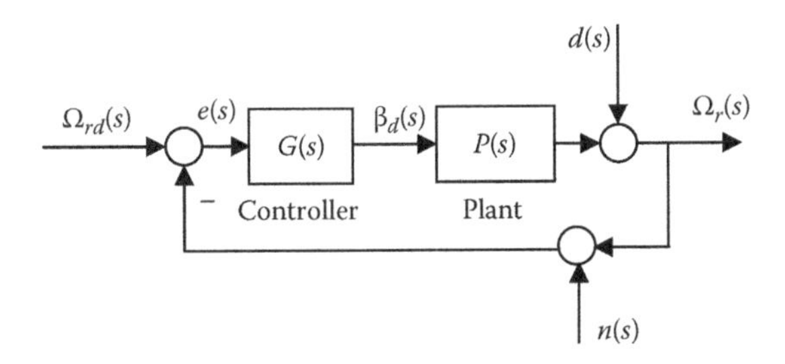

FIGURE P11.2
Control system diagram.

(a) Designing the PID controller requires selecting the coefficients k_P, k_I, k_D, and γ. Using QFT, select the coefficients for fast accuracy and control so that the following specifications are simultaneously achieved:

Stability: gain margin $GM \geq 6\text{dB}$; phase margin $30° \leq PM \leq 60°$

Transient response: rise time $t_r < 5\,\text{s}$; peak time $t_p < 11\,\text{s}$

Steady-state error: $E_{ss} = 0$ for step inputs.

(b) Plot the root locus of the controlled system. Find the final roots of the controlled system.

(c) Plot the Bode diagram of $L(s)$. Find the GM and PM achieved.

(d) Plot the time response of the control system under a step reference input and no wind disturbances.

(e) Plot the time response of the control system under a constant reference input and an impulse wind disturbance.

(f) Study and compare the direct effect of noise $N(t)$ of 60 Hz in the output $\beta_d(t)$ of the controller $G(s)$, with the filter of the derivative part and without that filter ($\gamma = 0$).

11.3 The rotational speed Ω_r of a WT rotor is continuously modified (1) by the actuators (demanded control signals for the blade pitch angles β_d and for the electrical torque *Tgd*); (2) by the external disturbance inputs (wind speed v_1); and (3) through the dynamics introduced in the last section (rotor speed Ω_r itself). The transfer function that describes the rotor speed $\Omega_r(s)$ of a doubly fed induction generator (DFIG) WT is calculated in Equations 12.105 through 12.112 and is repeated here for convenience (Table P11.3):

$$\Omega_r(s) = \frac{1}{1-\mu_{32}(s)K_{T\Omega}}\left\{\mu_{32}(s)\left[K_{T\beta}A_\beta(s)\beta_d(s) + K_{TV}v_1(s)\right] + \mu_{33}(s)A_T(s)T_{gd}(s)\right\}$$

which can be expressed as

$$\Omega_r(s) = F_1(s)v_1(s) + F_2(s)\beta_d(s) + F_3(s)T_{gd}(s)$$

TABLE P11.3

2.0 MW DFIG WT Characteristics

ρ (kg/m^3)	1.225	K_l (N·m), shaft low speed	160×10^6
N	3	B_l (N·m s), shaft low speed	25,000
Ω_{r_nom} (rad/s)	1.8850	K_h (N·m), shaft high speed	2.3×10^4
T_{gd_max} (N·m), shaft high speed	14,000	B_h (N·m s), shaft high speed	36
R_t	83.33	I_g (kg m^2), shaft high speed	60
I_r (kg m^2), shaft low speed	5.5×10^6	I_w (kg m^2), shaft high speed	0.0
$K_{T\beta}$ min	-17.53×10^4	$K_{T\beta}$ max	-1.18×10^4
$K_{T\Omega}$ min	-14×10^5	$K_{T\Omega}$ max	-3.75×10^5
K_{TV} min	2.12×10^5	K_{TV} max	2.61×10^5
$A_\beta(s)$	1.0	$A_T(s)$	1.0

Wind velocity varies between 12 and 17 m/s.

where

$$F_1(s) = \frac{K_{TV} n_{\mu 32}(s)}{d_{tf}(s)}$$

$$F_2(s) = \frac{K_{T\beta} n_{\mu 32}(s) A_\beta(s)}{d_{tf}(s)}$$

$$F_3(s) = \frac{n_{\mu 33}(s) A_T(s)}{d_{tf}(s)}$$

and

$$\begin{aligned} n_{\mu 32}(s) = {} & I_g I_w s^4 + (B_h I_w + I_g B_l + I_g B_h R_t^2)s^3 \cdots \\ & + (B_h B_l + I_g K_l + I_g K_h R_t^2 + K_h I_w)s^2 \cdots \\ & + (K_h B_l + B_h K_l)s + K_h K_l \end{aligned}$$

$$n_{\mu 33}(s) = -B_h R_t B_l s^2 - R_t(K_h B_l + B_h K_l)s - K_h R_t K_l$$

$$\begin{aligned} d_{tf}(s) &= d_{\mu 32}(s) - n_{\mu 32}(s) K_{T\Omega} \\ &= I_r I_w I_g s^5 + [(I_r I_g B_l + B_h I_r I_w + I_r I_g B_h R_T^2 + B_l I_w I_g) - (I_g I_w) K_{T\Omega}]s^4 \cdots \\ &\quad + [(K_l I_w I_g + B_h I_r B_l + B_l I_g B_h r_t^2 + K_h I_r I_w + I_r I_g K_l + B_l B_h I_w + I_r I_g K_h R_t^2) \\ &\quad - (B_h I_w + I_g B_l + I_g B_h R_t^2) K_{T\Omega}]s^3 \cdots \\ &\quad + [(B_l K_h I_w + K_l I_g B_h R_t^2 + K_h I_r B_l + B_h I_r K_l + K_l B_h I_w + I_g B_l K_h R_t^2) \\ &\quad - (B_h B_l + I_g K_l + I_g K_h R_t^2 + K_h I_w) K_{T\Omega}]s^2 \cdots \\ &\quad + [(I_g K_l K_h R_t^2 + K_h I_r K_l + K_l K_h I_w) - (K_h B_l + B_h K_l) K_{T\Omega}]s \\ &\quad - (K_h B_l + B_h K_l) K_{T\Omega} \end{aligned}$$

(a) Design a QFT compensator to control the rotor speed Ω_r by varying the pitch angle β_d in *Region* 3) and achieve the following specifications:
Stability: gain margin $GM \geq 6\,\text{dB}$; phase margin $PM \leq 60°$
Transient response: rise time $t_r < 5\,\text{s}$; peak time $t_p < 11\,\text{s}$
Steady-state error: $E_{ss} = 0$ for step inputs

(b) Plot the root locus of the controlled system. Find the final roots of the controlled system.

(c) Plot the Bode diagram of $L(s)$. Find the GM and PM achieved.

(d) Plot the time response of the control system under a step reference input and no wind disturbances.

(e) Plot the time response of the control system under a constant reference input and an impulse wind disturbance.

(f) Study and compare the direct effect of noise $N(t)$ of 60 Hz in the output $\beta_d(t)$ of the controller $G(s)$, with the filter of the derivative part and without this filter ($\gamma = 0$).

Answers to Selected Problems

Chapter 2

2.1 Figure A.P2.1 shows the plots [$Lm\ P(j\omega)$ and $\angle P(j\omega)$ versus ω] for nine different plants selected according to several sets of parameters within the uncertainty (see Table A.P2.1) and for the frequency range of $0.05 < \omega < 500$ rad/s.

In order to introduce the meaning of the quantitative feedback theory (QFT) templates, as it is explained in Chapter 3, the figures superimpose the variation of the plants due to the uncertainty at every frequency.

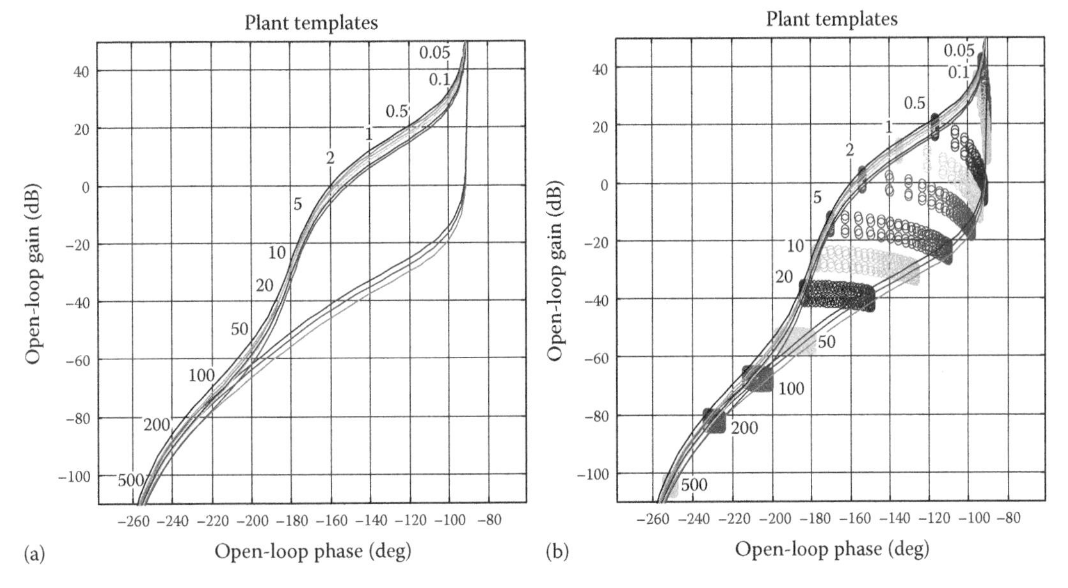

FIGURE A.P2.1
Nichols plot of $p(j\omega)$: (a) nine plants according to Table A.P2.1, (b) adding more plants, templates of $p(j\omega)$, at some frequencies of interest.

TABLE A.P2.1
Plant Definition according to Parametric Uncertainty

Plant	P_1	P_2	P_3	P_4	P_5	P_6	P_7	P_8	P_9
k	610	610	610	820	820	820	1050	1050	1050
a	1	15	1	1	15	1	1	15	1
b	150	150	170	150	150	170	150	150	170

Chapter 3

3.1

(a) $B_U(s)=\dfrac{2.222s+19.75}{s^2+5.333s+19.75}$; $B_L(s)=\dfrac{84.31}{s^3+13.16s^2+57.69s+84.31}$

(d) $B_U(s)=\dfrac{2.193s+30.07}{s^2+5.818s+30.07}$; $B_L(s)=\dfrac{88.66}{s^3+13.38s^2+59.66s+88.66}$

3.2

(a) $T_D(s)=\dfrac{3.423s}{s^2+18.7s+349.7}$

(b) $T_D(s)=\dfrac{3.559s}{s^2+24.3s+590.5}$

3.3

$$G(s)=2708.36\,\frac{(s+10.32)(s+1.61)}{s(s+300)};\quad F(s)=\frac{16}{(s+4)^2}$$

3.4

$$G(s)=5913.95\,\frac{(s+5.15)(s+1.97)}{s(s+400)}$$

3.6

$$G(s)=3138\,\frac{(s+6)}{s(s+104.6)};\quad F(s)=\frac{21}{(s+7)(s+3)}$$

3.7

$$G(s)=561.14\,\frac{(s+7.05)(s+1.82)}{s(s+240)};\quad F(s)=\frac{10.5}{(s+3)(s+3.5)}$$

3.8

$$G(s)=473.24\,\frac{(s+5.65)(s+1.87)}{s(s+200)};\quad F(s)=\frac{15}{(s+3)(s+5)}$$

3.10 (a) The complete template and (b) the simplified template using only the contour of the parameter uncertainty space are shown in respectively in Figures A.P3.10(a) and A.P3.10(b).

3.17 According to the defined stability and performance specifications and taking into account the parameter uncertainty, a solution for the compensator and the prefilter is

$$G(s)=\frac{37.71((s/0.1)+1)}{s(s/175.5+1)};\quad (s)=\frac{28}{s^2+11s+28}$$

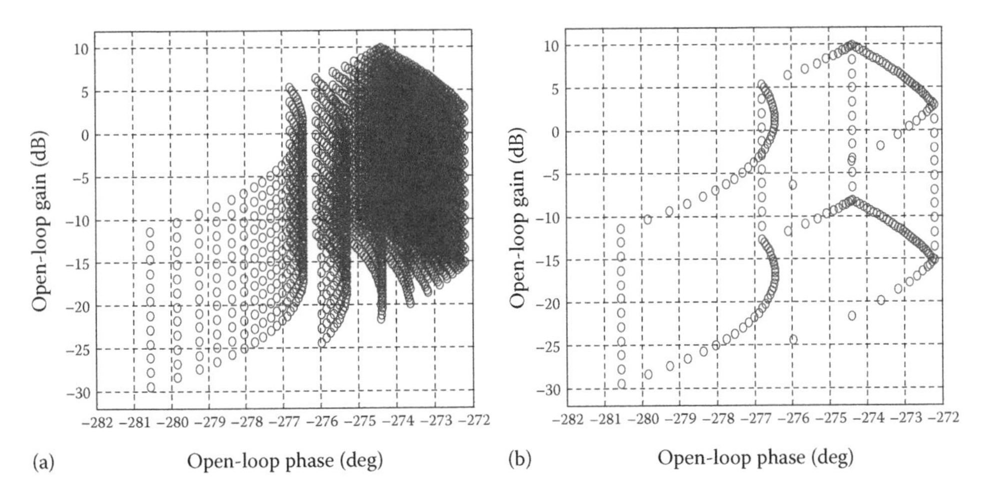

FIGURE A.P3.10
Nichols plot of $p(j\omega)$: (a) complete template, (b) simplified template using only the contour of the parameter uncertainty space.

Chapter 4

4.1 According to the defined stability specification and taking into account the parameter uncertainty, a solution for the compensator $G(s)$ and the corresponding controller $G(z)$ is

$$G(s) = \frac{0.8(s+1)((s/0.5)+1)^2((s/0.1)+1)}{s((s/150)+1)((s/112)+1)((s/90)+1)}$$

$$G(z) = \frac{(1.2350z^4 - 4.9150z^3 + 7.3340z^2 - 4.8630z + 1.2090) \times 10^7}{z^4 - 1.8040z^3 + 1.0060z^2 - 0.2167z + 0.01528}$$

Sampling time: $T = 0.01$ s for the Tustin discretization approach.

Chapter 6

6.1 The notation adopted here for transfer function expressions denotes the steady-state gain as a constant without parenthesis; simple poles and zeros as (ω), which corresponds to $[(s/\omega + 1)]$; poles and zeros at the origin as (0); conjugate poles and zeros as $[\zeta; \omega_n]$, with $[(s/\omega_n)^2 + (2\zeta/\omega_n)s + 1]$; n-multiplicity of poles and zeros as an exponent $()^n$. The solution is

Controller $\boldsymbol{G}_\alpha$

$$G_\alpha = \begin{bmatrix} \dfrac{3\left(\sqrt{3}\alpha\right)^2}{[\zeta;\alpha]} & \dfrac{-6\alpha\left(\sqrt{3}\alpha\right)^2}{(0)[\zeta;\alpha]} & 0 \\ \dfrac{2\alpha^{-1}(0)}{[\zeta;\alpha]} & \dfrac{3\left(\sqrt{3}\alpha\right)^2}{[\zeta;\alpha]} & 0 \\ 0 & 0 & 1 \end{bmatrix}$$

where

$\alpha = 7.2827 \times 10^{-5}$ rad/s

$\zeta = 0.8$

Controller $\mathbf{G}_\beta$

Loop 1: Design of g_{11}^{β}: $g_{11}^{\beta} = \dfrac{0.06(0.018)(0.0077)}{(0)(1.25)(1.2)}$

Loop 2: Design of g_{22}^{β}: $g_{22}^{\beta} = \dfrac{0.07(0.022)(0.0069)}{(0)(1.7)^2}$

Controller g_{33}: $g_{33} = \dfrac{0.06(0.014)(0.01)}{(0)(1.2)(1.1)}$

6.3 The RGA matrix of the plant is

$$\Lambda = \begin{bmatrix} \mu & 1-\mu \\ 1-\mu & \mu \end{bmatrix}, \quad \text{where } \mu \in [0.75;0.818] \text{ due to the uncertainty}$$

Thus, the input 1 controls the output 1, and the input 2 controls the output 2. The p_{11} is chosen to be the first loop to be closed because it has the lowest bandwidth compared to p_{22}. According to the defined stability and performance specifications and taking into account the parameter uncertainty, a solution for the compensator with *Method 3* is

$$G(s) = \begin{bmatrix} \dfrac{0.6496}{(s/0.8556+1)} & \dfrac{0.48}{(s/47+1)} \\ \dfrac{-0.43}{(s/0.8556+1)} & \dfrac{1.197}{(s/47.66+1)} \end{bmatrix}$$

6.6 According to the defined stability and performance specifications and taking into account the parameter uncertainty, a solution for the non-diagonal MIMO compensator with *Method 3* is

$$g_{11}(s) = \frac{7{,}550s + 2{,}718}{s^2 + 302s}$$

$$g_{21}(s) = \frac{-18{,}875s^2 - 10{,}570s - 1{,}359}{3s^3 + 907s^2 + 302s}$$

$$g_{22}(s) = \frac{26{,}745s^2 + 1{,}207 \times 10^3 s + 168{,}498}{s^3 + 277s^2 + 17{,}020s}$$

$$g_{12}(s) = \frac{7{,}430s^3 + 3.4 \times 10^5 s^2 + 1.8 \times 10^5 s + 18{,}725}{s^4 + 277.1s^3 + 17{,}047.7s^2 + 1{,}418s}$$

Chapter 11

11.1 A solution for the Case 1 [2-zeros 2-poles controller] is

$$\frac{u(s)}{e(s)} = G_1(s) = K_c \frac{(a_1 s+1)(a_2 s+1)}{(a_3 s+1)(a_4 s+1)}$$

$K_c = 0.042185$; $a_1 = 0.15$; $a_2 = 6.8$; $a_3 = 0.013$; $a_4 = 0.097$

A solution for the Case 2 [Practical 2 DoF PID controller] is

$$u(s) = \frac{k_P s + k_I}{s}[r(s) - y(s)] - \frac{k_D s}{(k_D/N)s+1} y(s) = \frac{[q_1(s)r(s) - q_2(s)y(s)]}{q_0(s)}$$

Prefilter

$$F_2(s) = \frac{q_1(s)}{q_2(s)} = \frac{(1+s+0.5041s^2)}{(1+1.2s+0.4624s^2)}$$

Controller

$$G_2(s) = \frac{q_2(s)}{q_0(s)} = 0.16266\frac{(1+0.96s+0.2809s^2)}{s(1+0.025s)}$$

References

This section compiles the main references used along the book in both, Part I (*Advanced Robust Control Techniques: QFT and Nonlinear Switching*) and Part II (*Wind Turbine Control*). It is arranged according to subject and chronologically within each subject.

Part I: Advanced Robust Control Techniques: QFT and Nonlinear-Switching

Books Related to QFT and Frequency Domain

1. Bode, H.W. (1945). *Network Analysis and Feedback Amplifier Design*. Van Nostrand Company, Princeton, NJ.
2. Horowitz, I. (1963). *Synthesis of Feedback Systems*. Academic Press, New York.
3. Horowitz, I. (1993). *Quantitative Feedback Design Theory (QFT)*. QFT Publication, 660 South Monaco Parkway, Denver, CO.
4. Yaniv, O. (1999). *Quantitative Feedback Design of Linear and Non-Linear Control Systems*. Kluwer Academic Publishers, Norwell, MA.
5. Sidi, M. (2002). *Design of Robust Control Systems: From Classical to Modern Practical Approaches*. Krieger Publishing, Malabar, FL.
6. Houpis, C.H., Rasmussen, S.J., and Garcia-Sanz, M. (2006). *Quantitative Feedback Theory: Fundamentals and Applications*, 2nd edn. A CRC Press book, Taylor & Francis, Boca Raton, FL.

Special Issues about QFT

7. Nwokah, O.D.I. (Guest Editor). (1994). Horowitz and QFT design methods special issue. *Int. J. Robust Nonlinear Control*, 4(1).
8. Houpis, C.H. (Guest Editor). (1997). Quantitative feedback theory special issue. *Int. J. Robust Nonlinear Control*, 7(6).
9. Eitelberg, E. (Guest Editor). (2001–2002). Isaac Horowitz special issue. *Int. J. Robust Nonlinear Control*, Part 1, 11(10); Part 2, 12(4).
10. Garcia-Sanz, M. (Guest Editor). (2003). Robust frequency domain special issue. *Int. J. Robust Nonlinear Control*, 13(7).
11. Garcia-Sanz, M. and Houpis, C.H. (Guest Editors). (2007). Quantitative feedback theory: In Memoriam of Isaac Horowitz, Special Issue. *Int. J. Robust Nonlinear Control*, 17(2–3).

International QFT Symposia

12. Houpis, C.H. and Chander, P. (Eds). (1992). *Proceedings of the 1st International Symposium on QFT and Robust Frequency Domain Methods*, Wright Patterson Airforce Base, Dayton, OH.
13. Nwokah, O.D.I. and Chander, P. (Eds). (1995). *Proceedings of the 2nd International Symposium on QFT and Robust Frequency Domain Methods*, Purdue University, West Lafayette, IN.

14. Petropoulakis, L. and Leithead, W.E. (Eds). (1997). *Proceedings of the 3rd International Symposium on QFT and Robust Frequency Domain Methods*, University of Strathclyde, Glasgow, Scotland, U.K.
15. Boje, E. and Eitelberg, E. (Eds). (1999). *Proceedings of the 4th International Symposium on QFT and Robust Frequency Domain Methods*, University of Natal, Durban, South Africa.
16. Garcia-Sanz, M. (Ed.). (2001). *Proceedings of the 5th International Symposium on QFT and Robust Frequency Domain Methods*, Public University of Navarra, Pamplona, Spain.
17. Boje, E. and Eitelberg, E. (Eds). (2003). *Proceedings of the 6th International Symposium on QFT and Robust Frequency Domain Methods*, University of Cape Town, Cape Town, South Africa.
18. Colgren, R. (Ed.). (2005), *Proceedings of the 7th International Symposium on Quantitative Feedback Theory and Robust Frequency Domain Methods*, University Kansas, Lawrence, KS.
19. Gutman, P.-O. (Ed.). (2007). *Proceedings of the 8th International Symposium on Quantitative Feedback Theory and Robust Frequency Domain Methods*, Weizmann Institute of Science, Rehovot, Israel.
20. From 2009 on, the *International Symposium on Quantitative Feedback Theory and Robust Frequency Domain Methods* merged into the *IFAC Symposium on Robust Control Design*. So, the *9th QFT International Symposium* merged into the *6th ROCOND*, June 2009, Haifa, Israel.

Tutorials about QFT

21. Horowitz, I. (1982). Quantitative feedback theory. *IEE Control Theory Appl.*, 129, 215–226.
22. Horowitz, I. (1991). Survey of quantitative feedback theory. *Int. J. Control*, 53(2), 255–291.
23. Houpis, C.H. (1996). Quantitative feedback theory (QFT) technique. In *The Control Handbook* (Ed. W.S. Levine). CRC Press, Boca Raton, FL, Chapter 44, pp. 701–717.
24. Garcia-Sanz, M. (2005). Control Robusto Cuantitativo: Historia de una Idea (in Spanish). *Revista Iberoamericana de Automatica e Informatica Industrial*, 2(3), 25–38.

About the QFT History

25. Horowitz, I.M. (1992). QFT—Past, present and future. Plenary address. In *Proceedings of the 1st International Symposium on QFT & Robust Frequency Domain Methods*, Dayton, OH, pp. 9–14.
26. Horowitz, I.M. (1999). Frequency response in control. Plenary. In *Proceedings of the 4th International Symposium on QFT and Robust Frequency Domain Methods*, Durban, South Africa, pp. 233–239.
27. Garcia-Sanz, M. (2001). QFT international symposia: Past, present and future. In Editorial of the *5th International Symposium on QFT and Robust Frequency Domain Methods*, Pamplona, Spain.
28. Horowitz, I.M. (2002). It was not easy: A personal view. *Int. J. Robust Nonlinear Control*, 12(4), 289–293.
29. Houpis, C.H. (2002). Horowitz: Bridging the gap. *Int. J. Robust Nonlinear Control*, 12(4), 293–302.

First QFT Papers

30. Horowitz, I.M. (1959). Fundamental theory of automatic linear feedback control systems. *I.R.E. Trans. Autom. Control*, 4, 5–19.
31. Horowitz, I.M. and Sidi, M. (1972). Synthesis of feedback systems with large plant ignorance for prescribed time-domain tolerances. *Int. J. Control*, 16(2), 287–309.
32. Horowitz, I.M. (1973). Optimum loop transfer function in single-loop minimum-phase feedback systems. *Int. J. Control*, 18(1), 97–113.

QFT Templates

33. Bartlett, A.C. (1993). Computation of the frequency response of systems with uncertain parameters: A simplification. *Int. J. Control*, 57(6), 1293–1309.
34. Bartlett, A.C., Tesi, A., and Vicino, A. (1993). Frequency response of uncertain systems with interval plants. *IEEE Trans. Autom. Control*, 38(6), 929–933.

35. Gutman, P.O., Baril, C., and Neuman, L. (1994). An algorithm for computing value sets of uncertain transfer functions in factored real form. *IEEE Trans. Autom. Control*, 39(6), 1268–1273.
36. Ballance, D.J. and Hughes, G. (1996). A survey of template generation methods for quantitative feedback theory. In *Proceedings of the UKACC International Conference on Control '96*, September 2–5, 1996, Stevenage, U.K., pp. 172–174.
37. Ballance, D.J. and Chen, W. (1998). Symbolic computation in value sets of plants with uncertain parameters. In *Proceedings of the UKACC International Conference on Control '98*, Swansea, U.K., pp. 1322–1327.
38. Garcia-Sanz, M. and Vital, P. (1999). Efficient computation of the frequency representation of uncertain systems. In *Proceedings of the 4th International Symposium on QFT and Robust Frequency Domain Methods*, Durban, South Africa, pp. 117–126.
39. Nataraj, P.S.V. and Sardar, G. (2000). Template generation for continuous transfer functions using interval analysis. *Automatica*, 36, 111–119.
40. Boje, E. (2000). Finding non-convex hulls of QFT templates. *Trans. ASME*, 122, 230–232.
41. Martin, J.J., Gil-Martinez, M., and Garcia-Sanz, M. (2007). Analytical formulation to compute QFT templates for plants with a high number of uncertain parameters. In *Proceedings of the 15th Mediterranean Conference on Control and Automation, MED'07*, Athens, Greece. http://med.ee.nd.edu

QFT Bounds

42. Longdon, L. and East, D.J. (1978). A simple geometrical technique for determining loop frequency bounds which achieve prescribed sensitivity specifications. *Int. J. Control*, 30(1), 153–158.
43. Bailey, F.N., Panzer, D., and Gu, G. (1988). Two algorithms for frequency domain design of robust control systems. *Int. J. Control*, 48(5), 1787–1806.
44. Brown, M. and Petersen, I.R. (1991). Exact computation of the Horowitz bound for interval plants. In *Proceedings of the 30th IEEE Conference on Decision and Control*, Brighton, U.K., pp. 2268–2273.
45. Nwokah, O.D.I., Jayasuriya, S., and Chait, Y. (1991). Parametric robust control by quantitative feedback theory. In *Proceedings of the American Control Conference*, Boston, MA, pp. 1975–1980.
46. Wang, G.G., Chen, C.W., and Wang, S.H. (1991). Equations for loop bound in quantitative feedback theory. In *Proceedings of the 30th Conference on Decision and Control*, Brighton, U.K., pp. 2968–2969.
47. Fialho, I.J., Pande, V., and Nataraj, P.S.V. (1992). Design of feedback systems using Kharitonov's segments in quantitative feedback theory. In *Proceedings of the 1st QFT Symposium*, Dayton, OH, pp. 457–470.
48. Chait, Y. and Yaniv, O. (1993). Multi-input/single-output computer-aided control design using the quantitative feedback theory. *Int. J. Robust Nonlinear Control*, 3(3), 47–54.
49. Zhao, Y. and Jayasuriya, S. (1994). On the generation of QFT bounds for general interval plants. *Trans. ASME*, 116, 618–627.
50. Chait, Y., Borghesani, C., and Zheng, Y. (1995). Single-loop QFT design for robust performance in the presence of non-parametric uncertainties. *Trans. ASME J. Dyn. Syst. Meas. Control*, 117(3), 420–425.
51. Moreno, J.C., Baños, A., and Montoya, J.F. (1997). An algorithm for computing QFT múltiple-valued performance bounds. In *Proceedings of the International Symposium on QFT and Robust Frequency Domain Methods*, Scotland, U.K., pp. 29–34.
52. Rodrigues, J.M., Chait, Y., and Hollot, C.V. (1997). An efficient algorithm for computing QFT bounds. *Trans. ASME*, 119, 548–552.
53. Eitelberg, E. (2000). Quantitative feedback design for tracking error tolerance. *Automatica*, 36, 319–326.

54. Nataraj, P. and Sardar, G. (2000). Template generation for continuous transfer functions using interval analysis. *Automatica*, 36(1), 111–119.
55. Nataraj, P. and Sardar, G. (2000). Computation of QFT bounds for robust sensitivity and gain-phase margin specifications. *Trans. ASME*, 122, 528–534.
56. Nataraj, P. (2002). Computation of QFT bounds for robust tracking specifications. *Automatica*, 38(2), 327–334.
57. Nataraj, P. (2002). Interval QFT: A mathematical and computational enhancement of QFT. *Int. J. Robust Nonlinear Control*, 12(4), 385–402.
58. Boje, E. (2003). Pre-filter design for tracking error specifications in QFT. *Int. J. Robust Nonlinear Control*, 13, 637–642.
59. Moreno, J.C., Baños, A., and Berenguel, M. (2006). Improvements on the computation of boundaries in QFT. *Int. J. Robust Nonlinear Control*, 16, 575–597.
60. Gutman, P.-O., Nordin, M., and Cohen, B. (2007). Recursive grid methods to compute value sets and Horowitz-Sidi bounds. *Int. J. Robust Nonlinear Control*, 17, 155–171.
61. Martin-Romero, J.J., Gil-Martinez, M., and Garcia-Sanz, M. (2009). Analytical formulation to compute QFT bounds: The envelope method. *Int. J. Robust Nonlinear Control*, 19(17), 1959–1971.
62. Yang, S.F. (2011). Generation of QFT bounds for robust tracking specifications for plants with affinely dependent uncertainties. *Int. J. Robust Nonlinear Control*, 21(3), 237–247.
63. Elso, J., Gil-Martinez, M., and Garcia-Sanz, M. (2011). Non-conservative QFT bounds for tracking error specifications. *Int. J. Robust Nonlinear Control*. In press, Doi: 10.1002/rnc.1804.

QFT Loop Shaping: Controller Synthesis

64. Gera, A. and Horowitz, I.M. (1980). Optimisation of the loop transfer function. *Int. J. Control*, 31(2), 389–398.
65. Thompson, D.F. and Nwokah, O.D.I. (1989). Stability and optimal design in quantitative feedback theory. In *Proceedings of the ASME Winter Annual Meeting Conference*, San Francisco, CA, ASME Paper No. 89-WA/DSC-39.
66. Thompson, D.F. and Nwokah, O.D.I. (1994). Analytic loop shaping methods in quantitative feedback theory. *Trans. ASME J. Dyn. Syst. Meas. Control*, 116(2), 169–177.
67. Chait, Y., Chen, Q., and Hollot, C.V. (1997). Automatic loop-shaping of QFT controllers via linear programming. In *Proceedings of the 3rd International Symposium on QFT and other Robust Frequency Domain Methods*, Glasgow, U.K., pp. 13–28.
68. Garcia-Sanz, M. and Guillen, J.C. (2000). Automatic loop-shaping of QFT robust controllers via genetic algorithms. In *Proceedings of the 3rd IFAC Symposium Robust Control Design*, Prague, Czech Republic.
69. Sidi, M.J. (2002). A combined QFT/H-infinity design technique for TDOF uncertain feedback systems. *Int. J. Control*, 75, 475–489.
70. Garcia-Sanz, M., Brugarolas, M.J., and Eguinoa, I. (2004). Quantitative analysis of controller fragility in the frequency domain. In *Proceedings of the 23rd IASTED International Symposium on Modelling, Identification and Control*, Grindelwald, Switzerland. http://www.iasted.org
71. Garcia-Sanz, M. and Oses, J.A. (2004). Evolutionary algorithms for automatic tuning of QFT controllers. In *Proceedings of the 23rd IASTED International Symposium Modelling, Identification and Control*, Grindelwald, Switzerland. http://www.iasted.org
72. Molins, C. and Garcia-Sanz, M. (2009). Automatic loop-shaping of QFT robust controllers. In *Proceedings of the 61st National Aerospace & Electronics Conference, NAECON*, July 2009, Dayton, OH. http://www.naecon.org
73. Garcia-Sanz, M. and Molins, C. (2010). Automatic loop-shaping of QFT robust controllers with multi-objective specifications via nonlinear quadratic inequalities. In *Proceedings of the 62nd National Aerospace & Electronics Conference, NAECON*, July 2010, Dayton, OH. http://www.naecon.org

Existence Conditions for QFT Controllers

74. Nwokah, O.D.I., Thompson, D.F., and Perez, R.A. (1990). On some existence conditions for QFT controllers, *DSC*, 24, 1–10.
75. Jayasuriya, S. and Zhao, Y. (1994). Stability of quantitative feedback designs and the existence of robust QFT controllers. *Int. J. Robust Nonlinear Control*, 4(1), 21–46.
76. Gil-Martinez, M. and Garcia-Sanz, M. (2003). Simultaneous meeting of robust control specifications in QFT. *Int. J. Robust Nonlinear Control*, 13(7), 643–656.

MIMO QFT

77. Horowitz, I.M. (1979). Quantitative synthesis of uncertain multiple input–output feedback systems. *Int. J. Control*, 30(1), 81–106.
78. Horowitz, I.M. and Sidi, M. (1980). Practical design of feedback systems with uncertain multivariable plants. *Int. J. Control*, 11(7), 851–875.
79. Horowitz, I.M. and Loecher, C. (1981). Design 3x3 multivariable feedback system with large plant uncertainty. *Int. J. Control*, 33, 677–699.
80. Horowitz, I.M., Neumann, L., and Yaniv, O. (1981). A synthesis technique for highly uncertain interacting multivariable flight control system (TYF16CCV). In *Proceedings of the Naecon Conference*, Dayton, OH, pp. 1276–1283.
81. Horowitz, I.M. (1982). Improved design technique for uncertain multiple input–output feedback systems. *Int. J. Control*, 36, 977–988.
82. Nwokah, O.D.I. (1984). Synthesis of controllers for uncertain multivariable plants for described time domain tolerances. *Int. J. Control*, 40, 1189–1206.
83. Yaniv, O. and Horowitz, I.M. (1986). A quantitative design method for MIMO linear feedback systems having uncertain plants. *Int. J. Control*, 43(2), 401–421.
84. Nwokah, O.D.I. (1988). Strong robustness in uncertain multivariable systems. In *Proceedings of the IEEE Conference on Decision and Control*, Austin, TX, pp. 2157–2164.
85. Park, M.S., Chait, Y., and Steinbuch, M. (1994). A new approach to multivariable quantitative feedback theory: Theoretical and experimental results. In *Proceedings of the 1994 American Control Conference–ACC '94*, Baltimore, MD, pp. 340–344.
86. Franchek, M.A. and Nwokah, O.D.I. (1995). Robust multivariable control of distillation columns using non-diagonal controller matrix. *DSC*, 57-1, 257–264, *IMECE, ASME Dynamics Systems and Control Division*.
87. Yaniv, O. (1995). MIMO QFT using non-diagonal controllers. *Int. J. Control*, 61(1), 245–253.
88. Franchek, M.A., Herman, P., and Nwokah, O.D.I. (1997). Robust nondiagonal controller design for uncertain multivariable regulating systems. *Trans. ASME J. Dyn. Syst. Meas. Control*, 119, 80–85.
89. Boje, E. and Nwokah, O.D.I. (2001). Quantitative feedback design using forward path decoupling. *Trans. ASME. J. Dyn. Syst. Meas. Control*, 123(1), 129–132.
90. Boje, E. (2002). Non-diagonal controllers in MIMO quantitative feedback design. *Int. J. Robust Nonlinear Control*, 12(4), 303–320.
91. Boje, E. (2002). Multivariable quantitative feedback design for tracking error specifications. *Automatica*, 38, 131–138.
92. De Bedout, J.M. and Franchek, M.A. (2002). Stability conditions for the sequential design of non-diagonal multivariable feedback controllers. *Int. J. Control*, 75(12), 910–922.
93. Garcia-Sanz, M. and Egaña I. (2002). Quantitative non-diagonal controller design for multivariable systems with uncertainty. *Int. J. Robust Nonlinear Control*, 12(4), 321–333.
94. Eitelberg, E. (2003). On multivariable tracking. In *Proceedings of the 6th International Symposium on Quantitative Feedback Theory*, December 3–5, 2003, Cape Town, South Africa, Vol. 2, pp. 514–519.

95. Kerr, M. and Jayasuriya, S. (2003). Sufficient conditions for robust stability in non-sequential MIMO QFT. In *Proceedings of the 42nd IEEE International Conference on Decision and Control*, Maui, HI.
96. Lan, C.Y., Kerr, M.L., and Jayasuriya, S. (2004). Synthesis of controllers for non-minimum phase and unstable systems using non-sequential MIMO quantitative feedback theory. In *Proceedings of the 2004 American Control Conference (AAC)*, Boston, MA, pp. 4139–4144.
97. Garcia-Sanz, M., Egaña, I., and Barreras, M. (2005). Design of quantitative feedback theory non-diagonal controllers for use in uncertain MIMO systems. *IEE Control Theory Appl.*, 152(2), 177–187.
98. Garcia-Sanz, M. and Eguinoa, I. (2005). Improved non-diagonal MIMO QFT design technique considering non-minimum phase aspects. In *Proceedings of the 7th International Symposium on QFT and Robust Frequency Domain Methods*, Lawrence, KS. http://www.ku.edu
99. Kerr, M.L., Jayasuriya S., and Asokanthan, S.F. (2005). On stability in non-sequential MIMO QFT designs. *ASME J. Dyn. Syst. Meas. Control*, 127(1), 98–104.
100. Kerr, M.L. and Jayasuriya, S. (2006). An improved non-sequential multi-input multi-output quantitative feedback theory design methodology. *Int. J. Robust Nonlinear Control*, 16(8), 379–395.
101. Mahdi Alavi, S.M., Khaki-Sedigh, A., Labibi, B., and Hayes, M.J. (2007). Improved multivariable feedback design for tracking error specifications. *IET Control Theory Appl.*, 1, 1046–1053.
102. Garcia-Sanz, M., Eguinoa, I., and Bennani, S. (2009). Non-diagonal MIMO QFT controller design reformulation. *Int. J. Robust Nonlinear Control*, 19(9), 1036–1064.

Time-Delay Systems: QFT Controller Design

103. Garcia-Sanz, M. and Guillen, J.C. (1999). Smith predictor for uncertain systems in the QFT framework. In *Progress in System and Robot Analysis and Control Design*. Lecture Notes in Control and Information Sciences, Ed. Vol. 243. Springer Verlag, New York, Chapter 20, pp. 243–250.

Digital QFT

104. Horowitz, I.M. and Liao, Y. (1986). Quantitative feedback design for sampled-data systems. *Int. J. Control*, 44, 665–675.
105. Yaniv, O. and Chait, Y. (1991). Direct robust control of uncertain sampled-data systems using the quantitative feedback theory. In *Proceedings of the ACC Conference*, June 26–28, 1991, Boston, MA, pp. 1987–1988.
106. Houpis, C.H. and Lamont, B.G. (1992). Discrete quantitative feedback technique. *Digital Control Systems: Theory, Hardware, Software*, 2nd edn. McGraw Hill, New York, Chapter 16.
107. Yaniv, O. and Chait, Y. (1993). Direct control design in sampled-data uncertain systems. *Automatica*, 29(2), 365–372.

Distributed Parameter Systems: QFT Controller Design

108. Horowitz, I.M. and Azor, R. (1983). Quantitative synthesis of feedback systems with distributed uncertain plants. *Int. J. Control*, 38(2), 381–400.
109. Horowitz, I.M. and Azor, R. (1984). Uncertain partially non-casual distributed feedback systems. *Int. J. Control*, 40(5), 989–1002.
110. Chait, Y., Maccluer C.R., and Radcliffe, C.J. (1989). A Nyquist stability criterion for distributed parameter systems. *IEEE Trans. Autom. Control*, 34(1), 90–92.
111. Horowitz, I.M., Kannai, Y., and Kelemen, M. (1989). QFT approach to distributed systems control and applications. In *Proceedings of ICCON '89. IEEE International Conference*, April 3–6, 1989, pp. 516–519.
112. Kelemen, M., Kanai, Y., and Horowitz, I.M. (1989). One-point feedback approach to distributed linear systems. *Int. J. Control*, 49(3), 969–980.
113. Kelemen, M., Kanai, Y., and Horowitz, I.M. (1990). Improved method for designing linear distributed feedback systems. *Int. J. Adapt. Control Signal Process.*, 4, 249–257.

114. Hedge, M.D. and Nataraj, P.S.V. (1995). The two-point feedback approach to linear distributed systems. In *Proceedings of the International Conference on Automatic Control*, Indore, India, pp. 281–284.
115. Garcia-Sanz, M., Huarte, A., and Asenjo, A. (2007). A quantitative robust control approach for distributed parameter systems. *Int. J. Robust Nonlinear Control*, 17(2–3), 135–153.

Non-Minimum Phase Systems: QFT Controller Design

116. Horowitz, I.M. and Sidi, M. (1978). Optimum synthesis of non-minimum phase systems with plant uncertainty. *Int. J. Control*, 27(3), 361–386.
117. Horowitz, I.M. (1979). Design of feedback systems with non-minimum phase unstable plants. *Int. J. Syst. Sci.*, 10, 1025–1040.
118. Horowitz, I.M. and Liao, Y. (1984). Limitations on non-minimum phase feedback systems. *Int. J. Control*, 40(5), 1003–1015.
119. Horowitz, I.M. (1986). The singular-G method for unstable non-minimum phase plants. *Int. J. Control*, 44(2), 533–541.
120. Horowitz, I.M., Oldak, S., and Yaniv, O. (1986). An important property of non-minimum phase multi-inputs multi-outputs feedback systems. *Int. J. Control*, 44(3), 677–688.
121. Chen, W. and Ballance, D. (1998). QFT design for uncertain non-minimum phase and unstable plants. In *Proceedings of the American Control Conference*, Philadelphia, PA, pp. 2486–2490.

Multi-Loop Systems: QFT Controller Design

122. Horowitz, I.M., Neumann, L., and Yaniv, O. (1985). Quantitative synthesis of uncertain cascade multi-input multi-output feedback systems. *Int. J. Control*, 42(2), 273–303.
123. Horowitz, I.M. and Yaniv, O. (1985). Quantitative cascade MIMO synthesis by an improved method. *Int. J. Control*, 42(2), 305–331.
124. Eitelberg, E. (1999). *Load Sharing Control*. NOYB Press, Durban, South Africa.
125. Baños, A. and Horowitz, I.M. (2000). QFT design of multi-loop nonlinear control systems. *Int. J. Robust Nonlinear Control*, 10(15), 1263–1277.

Nonlinear Systems: QFT Controller Design

126. Horowitz, I.M. (1976). Synthesis of feedback systems with non-linear time-varying uncertain plants to satisfy quantitative performance specifications. *IEEE Proc.*, 64, 123–130.
127. Horowitz, I.M. (1981). Quantitative synthesis of uncertain non-linear feedback systems with non-minimum phase inputs. *Int. J. Syst. Sci.*, 1(12), 55–76.
128. Horowitz, I.M. (1981). Improvements in quantitative non-linear feedback design by cancellation. *Int. J. Control*, 34(3), 547–560.
129. Breiner, M. and Horowitz, I.M. (1981). Quantitative synthesis of feedback systems with uncertain nonlinear multivariable plants. *Int. J. Syst. Sci.*, 12, 539–563.
130. Horowitz, I.M. (1982). Feedback systems with non-linear uncertain plants. *Int. J. Control*, 36, 155–171.
131. Horowitz, I.M. (1983). A synthesis theory for a class of saturating systems. *Int. J. Control*, 38(1), 169–187.
132. Horowitz, I.M. and Liao, Y. (1986). Quantitative non-linear compensation design for saturating unstable uncertain plants. *Int. J. Control*, 44, 1137–1146.
133. Oldak, S., Baril, C., and Gutman, P.O. (1994). Quantitative design of a class of nonlinear systems with parameter uncertainty. *Int. J. Robust Nonlinear Control*, 4(1), 101–117.
134. Baños, A. and Bailey, F.N. (1998). Design and validation of linear robust controllers for nonlinear plants. *Int. J. Robust Nonlinear Control*, 8(9), 803–816.

135. Baños, A. and Barreiro, A. (2000). Stability of non-linear QFT designs based on robust absolute stability criteria. *Int. J. Control*, 73(1), 74–88.
136. Baños, A., Barreiro, A., Gordillo, F., and Aracil, J. (2002). A QFT framework for nonlinear robust stability. *Int. J. Robust Nonlinear Control*, 12(4), 357–372.

Linear-Time-Variant Systems: QFT Controller Design

137. Horowitz, I.M. (1975). A synthesis theory for linear time-varying feedback systems with plant uncertainty. *IEEE Trans. Autom. Control*, AC-20, 454–463.
138. Yaniv, O. and Boneh, R. (1997). Robust LTV feedback synthesis for SISO nonlinear plants. *Int. J. Robust Nonlinear Control*, 7, 11–28.
139. Yaniv, O. (1999). Robust LTV feedback synthesis for nonlinear MIMO plants. *Trans. ASME*, 121, 226–232.
140. Garcia-Sanz, M. and Elso, J. (2009). Beyond the linear limitations by combining switching and QFT. Application to wind turbines pitch control systems. *Int. J. Robust Nonlinear Control*, 19(1), 40–58.

CAD Tools for QFT Controller Design

141. Bailey, F.N. and Hul, C.H. (1989). CACSD tools for loop gain-phase shaping design of SISO robust controllers. In *Proceedings of the IEEE Control System Society Workshop on Computer Aided Control System*, Berkeley, CA, pp. 151–157.
142. Sating, R.R. (1992). Development of an analog MIMO quantitative feedback theory (QFT) CAD package. MS thesis, AFIT/GE/ENG/92J-04, Air Force Institute of Technology, Wright Patterson AFB, OH.
143. Houpis, C.H. and Sating, R.R. (1997). MIMO QFT CAD package (Ver.3). *Int. J. Control*, 7(6), 533–549.
144. Borghesani, C., Chait, Y., and Yaniv, O. (1994, 2002). *Quantitative Feedback Theory Toolbox—For Use with MATLAB*. Terasoft, San Diego, CA.
145. Gutman, P.-O. (1996, 2001). *Qsyn—The Toolbox for Robust Control Systems Design for Use with Matlab*, User's manual, NovoSyn AB, Jonstorp, Sweden.
146. Houpis, C.H., Rasmussen, S.J., and Garcia-Sanz, M. (2001, 2006). CAD tool for controller design. In *Quantitative Feedback Theory: Fundamentals and Applications*, 2nd edn. Taylor & Francis, Boca Raton, FL.
147. Garcia-Sanz, M., Vital, P., Barreras, M., and Huarte, A. (2001). *InterQFT*. Public University of Navarra. Also as interactive tool for easy robust control design. In *IFAC International Workshop, Internet Based Control Education*, Madrid, Spain, pp. 83–88.
148. Diaz, J.M., Dormido, S., and Aranda, J. (2004). SISO-QFTIT, una herramienta software interactiva para diseño de controladores robustos usando QFT. *UNED*, Madrid, Spain.
149. Garcia-Sanz, M., Mauch, A., and Philippe, C. (2009). QFT control toolbox: An interactive object-oriented Matlab CAD tool for quantitative feedback theory. In *Proceedings of the 6th IFAC Symposium on Robust Control Design, ROCOND'09*, Haifa, Israel. http://www.technion.ac.il/~rocond09
150. Garcia-Sanz, M., Mauch, A., Philippe, C., Elso, J., Molins, C. (2008–2011). QFT Control Toolbox: An interactive object-oriented Matlab CAD tool for controller design. Research project, European Space Agency ESA-ESTEC, Public University of Navarra, Pamplona, Spain and Case Western Reserve University, Cleveland, OH.

Real-World Applications with QFT

151. Horowitz, I., Neumann, L., and Yaniv, O. (1981). A synthesis technique for highly uncertain interacting multivariable flight control system (TYF16CCV). In *Proceedings of the IEEE Naecon Conference*, Dayton, OH, pp. 1276–1283.
152. Horowitz, I.M. et al. (1982). *Multivariable Flight Control Design with Uncertain Parameters (YF16CCV)*. AFWAL-TR-83-3036, Air Force Wright Aeronautical Laboratories, Wright-Patterson AFB, OH.

153. Walke, J., Horowitz, I., and Houpis, C. (1984). Quantitative synthesis of highly uncertain MIMO flight control system for the forward swept wing X-29 aircraft. In *Proceedings of the IEEE Naecon Conference*, Dayton, OH, pp. 576–583.
154. Bossert, D.E. (1989). *Design of Pseudo-Continuous-Time Quantitative Feedback Theory Robot Controllers.* AFIT/GE/ENG/89D-2, Air Force Institute of Technology, Wright-Patterson AFB, OH.
155. Trosen, D.W. (1993). *Development of an Prototype Refueling Automatic Flight Control System Using Quantitative Feedback Theory.* AFIT/GE/ENG/93-J-03, Air Force Institute of Technology, Wright-Patterson AFB, OH.
156. Kelemen, M. and Bagchi, A. (1993). Modeling and feedback control of a flexible arm of a robot for prescribed frequency domain tolerances. *Automatica*, 29, 899–909.
157. Reynolds, O.R., Pachter, M., and Houpis, C.H. (1994). Design of a subsonic flight control system for the Vista F-16 using quantitative feedback theory. In *Proceedings of the American Control Conference*, Baltimore, MD, pp. 350–354.
158. Rasmussen, S.J. and Houpis, C.H. (1994). Development implementation and flight of a MIMO digital flight control system for an unmanned research vehicle using quantitative feedback theory. In *Proceedings of the ASME Dynamic Systems and Control, Winter Annual Meeting of ASME*, Chicago, IL. http://www.dsc-conference.org
159. Bentley, A.E. (1994). Quantitative feedback theory with applications in welding. *Int. J. Robust Nonlinear Control*, 4(1), 119–160.
160. Miller, R.B., Horowitz, I.M., Houpis, C.H., and Barfield, A.F. (1994). Multi-input, multi-output flight control system design for the YF-16 using nonlinear QFT and pilot compensation. *Int. J. Robust Nonlinear Control*, 4(1), 211–230.
161. Osmon, C., Pachter, M., and Houpis, C.H. (1996). Active flexible wing control using QFT. In *Proceedings of the IFAC 13th World Congress*, San Francisco, CA, Vol. H, pp. 315–320.
162. Franchek, M. and Hamilton, G.K. (1997). Robust controller design and experimental verification of I.C. engine speed control. *Int. J. Robust Nonlinear Control*, 7, 609–628.
163. Pachter, M., Houpis, C.H., and Kang, K. (1997). Modelling and control of an electro-hydrostatic actuator. *Int. J. Robust Nonlinear Control*, 7, 591–608.
164. Boje, E. and Nwokah, O.D.I. (1999). Quantitative multivariable feedback design for a turbofan engine with forward path decoupling. *Int. J. Robust Nonlinear Control*, 9(12), 857–882.
165. Garcia-Sanz, M. and Ostolaza, J.X. (2000). QFT-control of a biological reactor for simultaneous ammonia and nitrates removal. *Int. J. Syst. Anal. Modell. Simul., SAMS*, 36, 353–370.
166. Egaña, I., Villanueva, J., and Garcia-Sanz, M. (2001). Quantitative multivariable feedback design for a SCARA robot arm. In *Proceedings of the 5th International Symposium on QFT and Robust Frequency Domain Methods*, Pamplona, Spain, pp. 67–72.
167. Rueda, T.M. and Velasco, F.J. (2001). Robust QFT controller for marine course-changing control. In *Proceedings of the 5th International Symposium on QFT and Robust Frequency Domain Methods*, Pamplona, Spain, pp. 79–84.
168. Kelemen, M. and Akhrif, O. (2001). Linear QFT control of a highly nonlinear multi-machine power system. *Int. J. Robust Nonlinear Control*, 11(10), 961–976.
169. Bentley, A.E. (2001). Pointing control design for high precision flight telescope using quantitative feedback theory. *Int. J. Robust Nonlinear Control*, 11(10), 923–960.
170. Liberzon, A., Rubinstein, D., and Gutman, P.O. (2001). Active suspension for single wheel satin of on-road track vehicle. *Int. J. Robust Nonlinear Control*, 11(10), 977–999.
171. Garcia-Sanz, M., Guillen, J.C., and Ibarrola, J.J. (2001). Robust controller design for time delay systems with application to a pasteurisation process. *Control Eng. Pract.*, 9, 961–972.
172. Yaniv O., Fried O., and Furst-Yust, M. (2002). QFT application for headphone's active noise cancellation. *Int. J. Robust Nonlinear Control*, 12(4), 373–383.
173. Gutman, P.O., Horesh, E., Guetta, R., and Borshchevsky, M. (2003). Control of the aero-electric power station—An exciting QFT application for the 21st century. *Int. J. Robust Nonlinear Control*, 13(7), 619–636.

174. Torres, E. and Garcia-Sanz, M. (2004). Experimental results of the variable speed, direct drive multipole synchronous wind turbine: TWT1650, *Wind Energy*, 7(2), 109–118.
175. Garcia-Sanz, M. and Hadaegh, F.Y. (2004). *Coordinated Load Sharing QFT Control of Formation Flying Spacecrafts. 3D Deep Space and Low Earth Keplerian Orbit Problems with Model Uncertainty*, NASA-JPL, JPL Document, D-30052, Pasadena, CA.
176. Kerr, M. (2004). Robust control of an articulating flexible structure using MIMO QFT. PhD dissertation, University of Queensland, Brisbane, Queensland, Australia.
177. Barreras, M. and Garcia-Sanz, M. (2004). Multivariable QFT controllers design for heat exchangers of solar systems. In *Proceedings of the International Conference on Renewable Energy and Power Quality*, Barcelona, Spain. http://www.icrepq.com
178. Garcia-Sanz, M. and Barreras, M. (2006). Non-diagonal QFT controller design for a 3-input 3-output industrial Furnace. *Int. J. Dyn. Syst. Meas. Control ASME*, 128(2), 319–329.
179. Barreras, M., Villegas, C., Garcia-Sanz, M., and Kalkkuhl, J. (2006). Robust QFT tracking controller design for a car equipped with 4-wheel steer-by-wire. In *Proceedings of the IEEE International Conference on Control Applications, CCA*, Munich, Germany, pp. 1312–1317.
180. Garcia-Sanz, M., Eguinoa, I., Ayesa, E., and Martin, C. (2006). Non-diagonal multivariable robust QFT control of a wastewater treatment plant for simultaneous nitrogen and phosphorus removal. In *Proceedings of the Robust Control Design Conference, ROCOND'06, IFAC*, Toulouse, France. http://www.laas.fr/rocond06
181. Garcia-Sanz, M. and Hadaegh, F.Y. (2007). Load-sharing robust control of spacecraft formations: Deep space and low Earth elliptic orbits. *IET Control Theory Appl. (former IEE)*, 1(2), 475–484.
182. Kerr, M.L., Lan, C.Y., and Jayasuriya, S. (2007). Non-sequential MIMO QFT control of the X-29 aircraft using a generalized formulation. *Int. J. Robust Nonlinear Control*, 17(2–3), 107–134.
183. Garcia-Sanz, M., Eguinoa, I., Barreras, M., and Bennani, S. (2008). Non-diagonal MIMO QFT controller design for Darwin-type spacecraft with large flimsy appendages. *J. Dyn. Syst. Meas. Control ASME*. 130, 011006-1:011006-15.
184. Garcia-Sanz, M., Eguinoa, I., Gil-Martinez, M., Irizar, I., and Ayesa, E. (2008). MIMO quantitative robust control of a wastewater treatment plant for biological removal of nitrogen and phosphorus. In *Proceedings of the 16th Mediterranean Conference on Control and Automation, MED'08*, Ajaccio, France, pp. 541–546.
185. Garcia-Sanz, M. and Molins, C. (2008). QFT robust control of a Vega-type space launcher. In *Proceedings of the 16th Mediterranean Conference on Control and Automation*, Ajaccio, France, pp. 35–40.
186. Garcia-Sanz, M. (2009). QFT: New developments and advanced real-world applications. Plenary Session. In *Proceedings of the 6th IFAC Symposium on Robust Control Design*, Haifa, Israel, pp. 19–24.
187. Garcia-Sanz, M., Eguinoa, I., and Barreras, M. (2011). Advanced attitude and position MIMO robust control strategies for telescope-type spacecraft with large flexible appendages. In *Advances in Spacecraft Technologies*. INTECH, Rijeka, Croatia. ISBN: 978-953-7619-X-X.

Books Related to Control Engineering

188. Rosenbrock, H.H. (1970). *State-Space and Multivariable Theory*. Thomas Nelson, London, U.K.
189. Takahaschi, Y., Rabims, M., and Auslander, D. (1970). *Control and Dynamic Systems*. Addison Wesley, Boston, MA.
190. Rosenbrock, H.H. (1974). *Computer-Aided Control System Design*. Academic Press, New York.
191. Wolovich, W.A. (1974). *Linear Multivariable Systems*, Vol. 11. Springer-Verlag, New York.
192. Desoer, C.A. and Vidyasagar, M. (1975). *Feedback Systems: Input–Output Properties*. Academic Press, New York.
193. Frank, P.M. (1978). *Introduction to System Sensitivity Theory*. Academic Press, New York.

194. Postlethwaite, I. and MacFarlane, A.G.J. (1979). *A Complex Variable Approach to the Analysis of Linear Multivariable Feedback Systems*, Vol. 12. Springer-Verlag, New York.
195. Wellstead, P.E. (1979). *Introduction to Physical System Modelling*. Academic Press, New York.
196. Kailath, T. (1980). *Linear Systems*. Prentice-Hall, Englewood Cliffs, NJ.
197. Hung, Y.S. and MacFarlane, A.G.J. (1982). *Multivariable Feedback: A Quasi-Classical Approach*, Vol. 40. Springer-Verlag, Berlin.
198. McAvoy, T.J. (1983). *Interaction Analysis—Principles and Applications*. Instrument Society of America, Research Triangle Park, NC.
199. Ljung, L. (1987). *System Identification: Theory for the User*, Information and System Science Series. Prentice Hall, Upper Saddle River, NJ; Ljung, L. (1997). *System Identification Toolbox User's Guide*. The Mathworks, Inc., Natick, MA.
200. O'Reilly, J. (1987). *Multivariable Control for Industrial Applications*. Peter Peregrinus Ltd, London, U.K.
201. Franklin, G.F. and Powell, J.D. (1988). *Digital Control of Dynamic Systems*, 2nd edn. Addison-Wesley, Reading, MA.
202. Freudenberg, J.S. and Looze, D.P. (1988). *Frequency Domain Properties of Scalar and Multivariable Feedback Systems*. Springer-Verlag, Berlin, Germany.
203. Deshpande, P.B. (1989). *Multivariable Process Control*. Instrument Society of America, Research Triangle Park, NC.
204. Maciejowski, J.M. (1989). *Multivariable Feedback Design*. Addison-Wesley, Reading, MA.
205. Morari, M. and Zafiriou, E. (1989). *Robust Process Control*. Prentice-Hall, Englewood Cliffs, NJ.
206. Houpis, C.H. and Lamont, G. (1992). *Digital Control Systems: Theory, Hardware, Software*, 2nd edn. McGraw-Hill, New York.
207. Franklin, G.F., Powell, J.D., and Emani-Naeini, A. (1994). *Feedback Control of Dynamic Systems*. Addison-Wesley, New York.
208. Astrom, K.J. and Hagglund, T. (1995). *PID Controllers: Theory, Design, and Tuning*, 2nd edn. ISA, Research Triangle Park, NC.
209. Bhattacharyya, S.P., Chapellat, H., and Keel, L.H. (1995). *Robust Control: The Parametric Approach*. Prentice Hall, Englewood Cliffs, NJ.
210. Zhou, K., Doyle, J., and Glover, K. (1996). *Robust and Optimal Control*. Prentice Hall, Englewood Cliffs, NJ.
211. Dutton, K., Thompson, S., and Barraclough, B. (1997). *The Art of Control Engineering*. Prentice-Hall, Englewood Cliffs, NJ.
212. Datta, A., Ho, M.T., and Bhattacharyya, S.P. (2000). *Structure and Synthesis of PID Controllers*. Springer-Verlag, Berlin, Germany.
213. Dorato, P. (2000). *Analytic Feedback System Design*. Brooks-Cole, Pacific Groove, CA.
214. Kailath, T., Sayed, A.H., and Hassibi, B. (2000). *Linear Estimation*. Information and System Sciences Series. Prentice Hall, Englewood Cliffs, NJ.
215. Lurie, B.J. and Enright, P.J. (2000). *Classical Feedback Control with MATLAB*. Marcel Dekker, New York.
216. Rosenwasser, E. and Yusupov, R. (2000). *Sensitivity of Automatic Control Systems*. CRC Press, Boca Raton, FL.
217. Pintelon, R. and Shoukens, J. (2001). *System Identification: A Frequency Domain Approach*. IEEE Press, Piscataway, NJ.
218. D'Azzo, J.J., Houpis, C.H., and Sheldon, S.N. (2003). *Linear Control System Analysis and Design with Matlab*, 5th edn. Marcel Dekker, New York.
219. Skogestad, S. and Postlethwaite, I. (2005). *Multivariable Feedback Control. Analysis and Design*, 2nd edn. John Wiley & Sons Ltd, New York.
220. Verhaegen, M. and Verdult, V. (2007). *Filtering and System Identification: An Introduction*. Cambridge University Press, Cambridge, U.K.
221. Ogata, K. (2010). *Modern Control Engineering*. Prentice-Hall, Englewood Cliffs, NJ.
222. Dorf, R.C. and Bishop, R.H. (2011). *Modern Control Systems*. Prentice Hall, Englewood Cliffs, NJ.

General MIMO Systems

223. McMillan, B. (1952). Introduction to formal realizability theory—I, II. *Bell Syst. Tech. J.*, 31(2, 3), 217–279, 541–600.
224. Bristol, E.H. (1966). On a new measure of interaction for multi-variable process control. *IEEE Trans. Autom. Control*, AC-11(1), 133–134.
225. Hsu, C.-H. and Chen, C.-T. (1968). A proof of the stability of multivariable feedback systems. *Proc. IEEE*, 56(11), 2061–2062.
226. Rosenbrock, H.H. (1969). Design of multivariable control systems using the inverse Nyquist array. *Proc. IEE*, 116(11), 1929–1936.
227. Luyben, W.L. (1970). Distillation decoupling. *AIChE J.*, 16, 198–203.
228. Mayne, D.Q. (1973). The design of linear multivariable systems, *Automatica*, 9(2), 201–207.
229. Rosenbrock, H.H. (1973). The zeros of a system. *Int. J. Control*, 18(2), 297–299.
230. Barman, J.F. and Katzenelson, J. (1974). A generalized Nyquist-type stability criterion for multivariable feedback systems. *Int. J. Control*, 20(4), 593–622.
231. Davison, E. and Wang, S.H. (1974). Properties and calculation of transmission zeros of linear multivariable systems. *Automatica*, 10(6), 643–658.
232. Desoer, C.A. and Schulman, J.D. (1974). Zeros and poles of matrix transfer functions and their dynamical interpretation. *IEEE Trans. Circuits Syst.*, CAS-21(1), 3–8.
233. Rosenbrock, H.H. (1974). Corrections to 'the zeros of a system.' *Int. J. Control*, 20(3), 525–527.
234. MacFarlane, A.G.J. and Karcanias, N. (1976). Poles and zeros of linear multivariable systems: A survey of the algebraic, geometric and complex-variable theory. *Int. J. Control*, 24(1), 33–74.
235. Shaked, U., Horowitz, I., and Golde, S. (1976). Synthesis of multivariable, basically non-interacting systems with significant plant uncertainty. *Automatica*, 12(1), 61–71.
236. MacFarlane, A.G.J. and Postlethwaite, I. (1977). The generalized Nyquist stability criterion and multivariable root loci. *Int. J. Control*, 25(1), 81–127.
237. Postlethwaite, I. (1977). A generalized inverse Nyquist stability criterion. *Int. J. Control*, 26(3), 325–340.
238. Witcher, M.F. and McAvoy, T.J. (1977). Interacting control systems: Steady state and dynamic measurement of interaction. *ISA Trans.*, 16(3), 35–41.
239. Doyle, J.C. (1978). Robustness of multi-loop linear feedback systems. In *Proceedings of the IEEE Conference on Decision and Control, 17th Symposium on Adaptive Processes*, Fort Lauderdale, FL, pp. 12–18.
240. MacFarlane, A.G.J. and Karcanias, N. (1978). Relationships between state space and frequency response concepts. In *Proceedings of the 7th IFAC Congress*, Lisbon, Portugal, pp. 1771–1779.
241. MacFarlane, A.G.J. and Scott-Jones, D.F.A. (1979). Vector gain. *Int. J. Control*, 29(1), 65–91.
242. Mayne, D.Q. (1979). Sequential design of linear multivariable systems. *Proc. IEE*, 126(6), 568–572.
243. Desoer, C. and Wang, Y.-T. (1980). On the generalized Nyquist stability criterion. *IEEE Trans. Autom. Control*, 25(2), 187–196.
244. Wall, J.E., Doyle, J.C., and Harvey, C.A. (1980). Tradeoffs in the design of multivariable feedback systems. In *Proceedings of the 18th Allerton Conference*, Monticello, IL, pp. 715–725.
245. Weischedel, K. and McAvoy, T.J. (1980). Feasibility of decoupling in conventionally controlled distillation columns. *Ind. Eng. Chem. Fundam.*, 19, 379–384.
246. MacFarlane, A.G.J. and Hung, Y.S. (1981). Use of parameter groups in the analysis and design of multivariable feedback systems. In *Proceedings of the 20th IEEE Conference on Decision and Control, Symposium on Adaptive Processes*, San Diego, CA, pp. 1492–1494.
247. Mees, A.I. (1981). Achieving diagonal dominance. *Syst. Control Lett.*, 1(3), 155–158.
248. Postlethwaite, I., Edmunds, J., and MacFarlane, A. (1981). Principal gains and principal phases in the analysis of linear multivariable feedback systems. *IEEE Trans. Autom. Control*, 26(1), 32–46.
249. Doyle, J. (1982). Analysis of feedback systems with structured uncertainties. *IEE Proc. Control Theory Appl.*, Part D, 129(6), 242–250.

250. Doyle, J.C., Wall, J.E., and Stein, G. (1982). Performance and robustness analysis for structured uncertainty. In *Proceedings of the 21st IEEE Conference on Decision and Control*, Orlando, FL, pp. 629–636.
251. Arkun, Y., Manousiouthakis, B., and Palazoglu, A. (1984). Robustness analysis of process control systems. A case study of decoupling control in distillation. *Ind. Eng. Chem. Process Des. Dev.*, 23, 93–101.
252. Bryant, G.F. (1985). Direct methods in multivariable control. I. Gauss elimination revisited. In *Proceedings of the International Conference—Control 85* (Conference Publication No. 252), Cambridge, U.K. http://www.eng.cam.ac.uk
253. Grosdidier, P., Morari, M., and Holt, B.R. (1985). Closed-loop properties from steady-state gain information. *Ind. Eng. Chem. Fundam.*, 24(2), 221–235.
254. Marino-Galarraga, M., Marlin, T.E., and McAvoy, T.J. (1985). Using the relative disturbance gain to analyse process operability. In *Proceedings of the 1985 American Control Conference* (Catalogue No. 85CH2119-6), Boston, MA, pp. 1078–1083.
255. Stanley, G., Marino-Galarraga, M., and McAvoy, T.J. (1985). Shortcut operability analysis. I. The relative disturbance gain. *Ind. Eng. Chem. Process Des. Dev.*, 24(4), 1181–1188.
256. Grosdidier, P. and Morari, M. (1986). Interaction measures for systems under decentralized control. *Automatica*, 22(3), 309–319.
257. Manousiouthakis, V., Savage, R., and Arkun, Y. (1986). Synthesis of decentralized process control structures using the concept of block relative gain. *AIChE J.*, 32(6), 991–1003.
258. Mijares, G., Cole, J.D., Naugle, N.W., Preisig, H.A., and Holland, C.D. (1986). New criterion for the pairing of control and manipulated variables. *AIChE J.*, 32(9), 1439–1449.
259. Slaby, J. and Rinard, I.H. (1986). Complete interpretation of the dynamic relative gain array. In *American Institute of Chemical Engineers, Annual Meeting*, Miami, FL. http://www.aiche.org
260. Grosdidier, P. and Morari, M. (1987). A computer aided methodology for the design of decentralized controllers. *Comput. Chem. Eng.*, 11(4), 423–433.
261. Nett, C.N. and Spang, H.A. (1987). Control structure design: A missing link in the evolution of modem control theories. In *Proceedings of the American Control Conference*, Minneapolis, MN. http://ieeexplore.ieee.org
262. Skogestad, S. and Morari, M. (1987). Implications of large RGA elements on control performance. *Ind. Eng. Chem. Res.*, 26(11), 2323–2330.
263. Skogestad, S. and Morari, M. (1987). Effect of disturbance directions on closed-loop performance. *Ind. Eng. Chem. Res.*, 26(10), 2029–2035.
264. Chiu, M.S. and Arkun, Y. (1990). Decentralized control structure selection based on integrity considerations. *Ind. Eng. Chem. Res.*, 29(3), 369–373.
265. Yu, C.-C. and Fan, M.K.H. (1990). Decentralized integral controllability and D-stability. *Chem. Eng. Sci.*, 45(11), 3299–3309.
266. Chang, J.-W. and Yu, C.-C. (1992). Relative disturbance gain array. *AIChE J.*, 38(4), 521–534.
267. Hovd, M. and Skogestad, S. (1992). Simple frequency-dependent tools for control system analysis, structure selection and design. *Automatica*, 28(5), 989–996.
268. Campo, P.J. and Morari, M. (1994). Achievable closed-loop properties of systems under decentralized control: Conditions involving the steady-state gain. *IEEE Trans. Autom. Control*, 39(5), 932–943.
269. Franchek, M.A., Herman, P.A., and Nwokah, O.D.I. (1995). Robust multivariable control of distillation columns using a non-diagonal controller matrix. In *1995 ASME International Mechanical Engineering Congress and Exposition*, Part 1 (of 2), November 12–17, 1995, San Francisco, CA. http://www.asmeconferences.org
270. Van de Wal, M. and de Jager, B. (1995). Control structure design: A survey. In *Proceedings of the American Control Conference*, Seattle, WA, pp. 225–229.
271. Skogestad, S. and Havre, K. (1996). The use of RGA and condition number as robustness measures. *Comput. Chem. Eng.*, 20, S1005–S1010.
272. Franchek, M.A., Herman, P., and Nwokah, O.D.I. (1997). Robust non-diagonal controller design for uncertain multivariable regulating systems. *Trans. ASME. J. Dyn. Syst. Meas. Control*, 119(1), 80–85.
273. Wade, H.L. (1997). Inverted decoupling: A neglected technique. *ISA Trans.*, 36, 3–10.

Papers Related to Fragility

274. Keel, L.H. and Bhattacharyya, S.P. (1997). Robust, fragile, or optimal? *IEEE Trans. Autom. Control*, 42, 1098–1105.
275. Dorato, P. (1998). Non-fragile controller design, an overview. In *Proceedings of the American Control Conference*, Philadelphia, PA, pp. 2829–2931.
276. Famularo, P.D.D., Abdallah, C.T., Jadbabaie, A., and Haddad, W. (1998). Robust non-fragile LQ controllers: The static state feedback case. In *Proceedings of the American Control Conference*, Philadelphia PA, pp. 1109–1113.
277. Haddad, W. M. and Corrado, J.R. (1998). Robust resilient dynamic controllers for systems with parametric uncertainty and controller gain variations. In *Proceedings of the American Control Conference*, Philadelphia, PA, pp. 2837–2841.
278. Jadbabaie, A., Chaouki, T., Famularo, D., and Dorato, P. (1998). Robust, non-fragile and optimal controller design via linear matrix inequalities. In *Proceedings of the American Control Conference*, Philadelphia, PA, pp. 2842–2846.
279. Keel, L.H. and Bhattacharyya, S.P. (1998). Authors' reply. *IEEE Trans. Autom. Control*, 43, 1268.
280. Keel, L.H. and Bhattacharyya, S.P. (1998). Stability margins and digital implementation of controllers. In *Proceedings of the American Control Conference*, Philadelphia, PA, pp. 2852–2856.
281. Mäkilä, P.M. (1998). Comments on "Robust, fragile, or optimal?" *IEEE Trans. Autom. Control*, 43, 1265–1267.
282. Corrado, J.R. and Haddad, W.M. (1999). Static output feedback controllers for systems with parametric uncertainty and controller gain variation. In *Proceedings of the 1999 American Control Conference*, San Diego, CA, pp. 915–919.
283. Yang, G.H., Wang, J.L., and Lin, C. (1999). H∞ control for linear systems with controller uncertainty. In *Proceedings of the 1999 American Control Conference*, San Diego, CA, pp. 3377–3381.
284. Paattilammi, J. and Mäkilä, P.M. (2000). Fragility and robustness: A case study on paper machine headbox control. *IEEE Control Syst. Mag.*, February, 13–22.

Papers Related to Hybrid and Switching Control Systems

285. Narendra, K. and Goldwyn, R. (1964). A geometrical criterion for the stability of certain nonlinear nonautonomous systems. *IEEE Trans. Circuits Syst.*, 11, 406–408.
286. Willems, J. (1973). The circle criterion and quadratic Lyapunov functions for stability analysis. *IEEE Trans. Autom. Control*, 18, 184.
287. Molchanov, A.P. and Pyatnitskii, E.S. (1989). Criteria of asymptotic stability of differential and difference inclusions encountered in control theory. *Syst. Control Lett.*, 13, 59–64.
288. Feuer, A., Goodwin, G.V., and Salgado, M. (1997). Potential benefits of hybrid control for linear time invariant plants. In *Proceedings of the American Control Conference*, Albuquerque, NM, pp. 2790–2794.
289. Seron, M.M., Braslavsky, J.H., and Goodwin, G.C. (1997). *Fundamental Limitations in Filtering and Control*. Springer, London, U.K.
290. Dayawansa, W.P. and Martin, C.F. (1999). A converse Lyapunov theorem for a class of dynamical systems which undergo switching. *IEEE Trans. Autom. Control*, 44, 751–760.
291. Liberzon, D. and Morse, A.S. (1999). Basic problems in stability and design of switched systems. *IEEE Control Syst. Mag.*, 19, 59–70.
292. Decarlo, R.A., Branicky, M.S., Pettersson, S., and Lennartson, B. (2000). Perspectives and results on the stability and stabilizability of hybrid systems. *Proc. IEEE*, 88, 1069–1082.
293. Mcclamroch, N.H. and Kolmanovsky, I. (2000). Performance benefits of hybrid control design for linear and nonlinear systems. *Proc. IEEE*, 88, 1083–1096.
294. Shorten, R.N., Mason, O., O'Cairbre, F., and Curran, P. (2004). A unifying framework for the SISO circle criterion and other quadratic stability criteria. *Int. J. Control*, 77, 1–8.
295. Shorten, R., Wirth, F., Mason, O., Wulff, K., and King, C. (2007). Stability criteria for switched and hybrid systems. *SIAM Rev.*, 49(4), 545–592.

Miscellaneous: Control

296. Tustin, A. (1947). A method of analyzing the behavior of linear systems in terms of time series. *JIEE*, 94, Pt IIA.
297. Powell, J.D., Parsons, E., and Tashka, G. (1976). A comparison of flight control design methods. In *Proceedings of the Guidance and Control Conference*, San Diego, CA. http://www.aiaa.org
298. Kannai, Y. (1982). Causality and stability of linear systems described by partial differential operators. *SIAM J. Control Optim.*, 10(5), 669–674.
299. Anderson, B.D.O. and Parks, P.C. (1985). Lumped approximation of distributed systems and controllability questions. *IEE Proc.*, 132(3), 89–94.
300. Collins Jr., E.G., King, J.A., and Bernstein, D.S. (1991). Robust control design for a benchmark problem using the maximum entropy approach. In *Proceedings of the American Control Conference*, Boston, MA, 1935–1936.
301. Wie, B. and Bernstein, D. (1992). Benchmark problems for robust control design. *J. Guidance Control Dyn.*, 15, 1057–1059.

Part II: Wind Turbine Control

Books Related to Wind Energy

302. Putnam, P.C. (1948). *Power from the Wind*. Van Nostrand Reinhold, New York.
303. Freris, L.L. (1990). *Wind Energy Conversion Systems*. Prentice Hall, Englewood Cliffs, NJ.
304. Burton, T., Sharpe, D., Jenkins, N., and Bossanyi, E. (2001). *Wind Energy Handbook*. Wiley, London, U.K.
305. International Electro-technical Commission (IEC) 61400-21. (2001). Wind Turbine Generator Systems, Part 21: Measurement and Assessment of Power Quality Characteristics of Grid Connected Wind Turbines, Ed. 1.
306. Manwell, J.F., McGowan, J.G., and Rogers, A.L. (2002). *Wind Energy Explained. Theory, Design and Application*. Wiley, New York.
307. Lubosny, Z. (2003). *Wind Turbine Operation in Electric Power Systems*. Springer, New York.
308. Ackermann, T. (2005). *Wind Power in Power Systems*. Wiley, London, U.K.
309. European Wind Energy Association (EWEA). (2005). Large Scale Integration of Wind Energy in the European Power Supply: Analysis, Issues and Recommendations.
310. Bianchi, F.D., De Battista, H., and Mantz, R.J. (2006). *Wind Turbine Control Systems: Principles, Modelling and Gain Scheduling Design*. Springer, Berlin, Germany.
311. Hau, E. (2006). *Wind Turbines. Fundamentals, Technologies, Application, Economics*, 2nd edn. Springer, Berlin, Germany.
312. Heier, S. (2006). *Grid Integration of Wind Energy Conversion Systems*, 2nd edn. Wiley, London, U.K.
313. American Wind Energy Association (AWEA). (2008). Wind energy basics. http://www.awea.org/newsroom/pdf/Wind_Energy_Basics.pdf
314. Munteanu, I., Bratcu, A.I., Cutululis, N., and Ceanga, E. (2008). *Optimal Control of Wind Energy Systems: Towards a Global Approach*. Springer, Berlin, Germany.
315. U.S. Department of Energy. (2008). Energy Efficiency and Renewable Energy. 20% Wind Energy by 2030. Increasing Wind Energy's Contribution to U.S. Electricity Supply, DOE/GO-102008-2567, July 2008.
316. European Wind Energy Association (EWEA). (2009). *Wind Energy—The Facts*. Earthscan, Oxford, U.K.
317. Hansen, M.O.L. (2009). *Aerodynamics of Wind Turbines*, 2nd edn. Earthscan, Oxford, U.K.
318. National Renewable Energy Laboratory (NREL). (2010). Large-Scale Offshore Wind Power in the United States. NREL/TP-500-49229, September 2010, NREL, Golden, CO.

Books Related to Engineering

319. Farlow, S.J. (1937). *Partial Differential Equations for Scientists and Engineers*. Dover, Mineola, NY.
320. Lanczos, C. (1986). *The Variational Principles of Mechanics*. Dover, Mineola, NY.
321. Areal, A. (1987). *An Introduction to Composite Materials*. Cambridge University Press, Cambridge, U.K.
322. Chapman, A.J. (1987). *Fundamentals of Heat Transfer*. MacMillan, London, U.K.
323. Kundur, P. (1994). *Power System Stability and Control*. EPRI, Mc-Graw-Hill, London, U.K.
324. Mohan, N., Undeland, T.M., and Robbins, W.P. (1995). *Power Electronics. Converters, Applications and Design*. Wiley, New York.
325. Bose, B.K. (1997). *Power Electronics and Variable Frequency Drives*. IEEE Press, New York.
326. Hall, C. (1998). *Handbook of Composites*, 2nd edn. Springer-Verlag, Berlin.
327. Bergen, A.R. and Vittal, V. (2000). *Power Systems Analysis*, 2nd edn. Prentice Hall, Englewood Cliffs, NJ.
328. Leonhard, W. (2001). *Control of Electrical Drives*. Springer, New York.
329. Riso and DNV. (2002). *Guidelines for Design of Wind Turbines*, 2nd edn. DNV/Riso, Copenhagen, Denmark.
330. Trinks, W., Mawhinney, M.H., Shannon, R.A., Reed, R.J., and Garvey, J.R. (2003). *Industrial furnaces*, 6th edn. Wiley, New York.
331. Fowles, G.R. and Cassiday, G.L. (2004). *Analytical Mechanics*, 7th edn. Bruce Cole, Chicago, IL.
332. Gere, J.M. and Timoshenko, S.P. (2006). *Mechanics of Materials*. CBS Publishers, New Delhi, India.
333. Rossmann, J.S. and Dym, C.L. (2009). *Introduction to Mechanics: A Continuum Approach*. Taylor & Francis, Boca Raton, FL.
334. Adams, M.L. (2010). *Rotating Machinery Vibration*, 2nd edn. Taylor & Francis, Boca Raton, FL.
335. Boylestad, R.L. (2010). *Introductory Circuit Analysis*, 12th edn. Prentice Hall, Englewood Cliffs, NJ.
336. Matlab/Simulink (1984–2012). The Mathworks Inc. http://www.mathworks.com

Associations, Agencies, Labs, and Industry Related to Wind Energy

337. International Electro-Technical Commission. http://www.iec.ch/
338. American Wind Energy Association (AWEA). http://www.awea.org/
339. European Wind Energy Association (EWEA). http://www.ewea.org/
340. MEASNET. http://www.measnet.org/
341. U.S. Department of Energy (DOE). http://www.energy.gov/
342. National Renewable Energy Lab (NREL; USA). http://www.nrel.gov/
343. CENER National Laboratory (Spain). http://www.cener.com/
344. Sandia National Laboratory (USA). http://windpower.sandia.gov/
345. Riso National Laboratory (Denmark). http://www.risoe.dk/
346. DEWI National Laboratory (Germany). http://www.dewi.de/
347. Germanischer Lloyd Certification GmbH, http://www.gl-group.com/
348. Garrad Hassan. http://www.gl-garradhassan.com/
349. M.Torres Group. Spain, http://www.mtorres.es/
350. Airborne Wind Energy Consortium. http://www.aweconsortium.org/

Papers Related to Wind Turbines: Modeling, Design, and Control

351. Glauert, H. (1926). The Analysis of Experimental Results in the Windmill Brake and Vortex Ring States of an Airscrew. Report 1026. *Aeronautical Research Committee Reports and Memoranda*, Her Majesty's Stationery Office, London, U.K.
352. Bongers, P.M.M. and Van Engelen, T.G. (1987). A theoretical model and simulation of a wind turbine. *Wind Eng.*, 11(6), 344–350.
353. De la Salle, S.A., Reardon, D., Leithead, W.E., and Grimble, M.J. (1990). Review of wind turbine control. *Int. J. Control*, 52(6), 1295–1310.

354. Sheinman, Y. and Rosen, A. (1991). A dynamic model for performance calculations of grid-connected horizontal axis wind turbines. Part I: Description model. *Wind Eng.*, 15(4), 211–228.
355. Sheinman, Y. and Rosen, A. (1991). A dynamic model for performance calculations of grid-connected horizontal axis wind turbines. Part II: Validation. *Wind Eng.*, 15(4), 229–239.
356. Leithead, W.E. (1992). Effective wind speed models for simple wind turbine simulation. In *Proceedings of the British Wind Energy Conference*, Nottingham, U.K., pp. 321–326.
357. Carlson, O., Hylander, J., and Thorborg, K. (1996). Survey of variable speed operation of wind turbines. In *Proceedings of the EWEC'96*, Bedford, Great Britain, pp. 406–409.
358. Grauers, A. (1996). Directly driven wind turbine generators. In *Proceedings of the International Conference on Electrical Machines, ICEM 96*, Vigo, Spain, pp. 417–422.
359. Leith, D.J. and Leithead, W.E. (1996). Appropriate realization of gain-scheduled controllers with application to wind turbine regulation. *Int. J. Control*, 65(2), 223–248.
360. Leithead, W.E. and Rogers, M.C.M. (1996). Drive-train characteristics of constant speed HAWT's: Part I— Representation by simple dynamic models. *Wind Eng.*, 20(3), 149–174.
361. Leithead, W.E. and Rogers, M.C.M. (1996). Drive-train characteristics of constant speed HAWT's: Part II—Simple characterization of dynamics. *Wind Eng.*, 20(3), 175–201.
362. Mercer, A.S. and Bossanyi, E.A. (1996). Stall regulation of variable speed HAWTS. In *Proceedings of the European Wind Energy Conference*, Göteborg, Sweden, pp. 828–828.
363. Kendall, L., Balas, M., Lee, Y., and Fingersh, L. (1997). Application of proportional–integral and disturbance accommodating control to variable speed variable pitch horizontal axis wind turbines. *Wind Eng.*, 21, 21–38.
364. Leith, D.J. and Leithead, W.E. (1997). Implementation of wind turbine controllers. *Int. J. Control*, 66, 349–380.
365. Vilsboell, N., Pinegin, A.L., Fischer, T., and Bugge, J. (1997). Analysis of advantages of the double supply machine with variable rotation speed application in wind energy converters. *DEWI Mag.*, 11, 50–65.
366. Garcia-Sanz, M. (1998–2008). Control systems for TWT-1.65/70-Class Ia, TWT-1.65/77-Class IIa, TWT-1.65/82-Class IIIa, TWT-2.5/90-Class Ia, TWT-2.5/100-Class IIa, and TWT-2.5/103-Class IIIa wind turbines. M.Torres internal reports.
367. Quarton, D.C. (1998). The evolution of wind turbine design analysis: A twenty year progress review. *Wind Energy*, 1, 5–24.
368. Hudson, G. and Underwood, C.P. (1999). A simple building modeling procedure for Matlab/Simulink. In *Proceedings of the 6th International Conference on Building Performance Simulation* (IBPSA '99), Kyoto-Japan, pp. 777–783.
369. Bossanyi, E.A. (2000). The design of closed loop controllers for wind turbines. *Wind Energy*, 3, 149–163.
370. Leithead, W.E. and Connors, B. (2000). Control of variable speed wind turbines: Dynamic models. *Int. J. Control*, 73(13), 1173–1188.
371. Slootweg, J.G., Polinder, H., and Kling, W.L. (2001). Dynamic modelling of a wind turbine with doubly fed induction generator. In *Proceedings IEEE Power Engineering Society, Summer Meeting*, Vancouver, Canada, pp. 1–6.
372. Leishman, J.G. (2002). Challenges in modeling the unsteady aerodynamics of wind turbines. In *Proceedings of the 21st ASME Wind Energy Symposium and the 40th AIAA Aerospace Sciences Meeting*, Reno, NV. http://www.asmeconferences.org and http://www.aiaa.org
373. Mutschler, P. and Hoffmann, R. (2002). Comparison of wind turbines regarding their energy generation. In *Proceedings of the Power Electronics Specialists Conference*, Cairns, Queensland, Australia, pp. 6–11.
374. Andrew, C.M., Henderson, R., Smith, B., Sorensen, H.C., Barthelmie, J., and Boesmans, B. (2003). Offshore wind energy in Europe—A review of the state-of-the-art. *Wind Energy*, 6, 35–52.
375. Barreras, M. and Garcia-Sanz, M. (2003). Model identification of a multivariable industrial furnace. In *Proceedings of the 13th IFAC Symposium on System Identification*, Rotterdam, the Netherlands. http://www.sysid2003.nl

376. Bossanyi, E.A. (2003). Wind turbine control for load reduction. *Wind Energy*, 6, 229–244.
377. Hooft, E.L., Schaak, P., and Engelen, T.G. (2003). Wind turbine control algorithms. DOWEC-F1W1-EH-03-094/0. ECN-C-03-111, Energy Research Centre of the Netherlands (*ECN Win denergie*).
378. Stol, K. and Balas, M. (2003). Periodic disturbance accommodating control for blade load mitigation in wind turbines. *J. Solar Energy Eng.*, 125(4), 379–385.
379. Garcia-Sanz, M. and Torres E. (2004). Control y experimentación del aerogenerador sincrono multipolar de velocidad variable TWT1650, *RIAI*, 1(3), 53–62 (in Spanish).
380. Hansen, A.D., Iov, F., Blaabjerg, F., and Hansen, L.H. (2004). Review of contemporary wind turbine concepts and their market penetration. *Wind Eng.*, 28, 247–263.
381. Johnson, K.E., Fingersh, L.J., Balas, M.J., and Pao, L.Y. (2004). Methods for increasing Region 2 power capture on a variable-speed wind turbine. *J. Sol. Energy Eng.*, 126(4), 1092–1101.
382. Spruce, C.J. (2004). Power control of active stall wind turbines. In *Proceedings of the European Wind Energy Conference*, London, U.K. http://www.ewea.org
383. Van der Hooft, E.L. and Van Engelen, T.G. (2004). Estimated wind speed feed forward control for wind turbine operation optimization. In *Proceedings of the European Wind Energy Conference*, London, U.K. http://www.ewea.org
384. Wright, A.D. and Balas, M.J. (2004). Design of controls to attenuate loads in the controls advanced research turbine. In *Proceedings of the ASME Wind Energy Symposium*, Reno, NV, pp. 76–86.
385. Madsen, H.A., Larsen, G.C., and Thomsen, K. (2005). Wake flow characteristics in low ambient turbulence conditions. In *Proceedings of Copenhagen Offshore Wind 2005*, Copenhagen, Denmark. http://www.ewea.org
386. Garcia-Sanz, M. and Barreras, M. (2006). Non-diagonal MIMO robust control for a 4×4 electrical industrial furnace. In *Proceedings of the IFAC Robust Control Design Conference*, Toulouse, France. http://www.laas.fr/rocond06
387. Johnson, K.E., Pao, L.Y., Balas, M.J., and Fingersh, L.J. (2006). Control of variable-speed wind turbines: Standard and adaptive techniques for maximizing energy capture. *IEEE Control Syst. Mag.*, 26(3), 70–81.
388. Kallesoe, B.S. (2006). A low-order model for analyzing effects of blade fatigue load control. *Wind Energy*, 9, 421–436.
389. Kuik, G., Ummels, B.C., and Hendricks, R. (2006). Perspectives of wind energy. In *Proceedings of the Advances in New and Sustainable Energy Conversion and Storage Technologies*, Dubrovnik, Croatia, pp. 61–79.
390. Van Engelen, T. (2007). Control design based on aero-hydro-servo-elastic linear models from turbu (ECN). In *Proceedings of the European Wind Energy Conference*, Milan, Italy. http://www.ewea.org
391. Larsen, G.C., Madsen, H.A., Larsen, T.J., and Troldborg, N. (2008). Wake modeling and simulation. Risø-R-1653(EN). Risø National Laboratory for Sustainable Energy, Wind Energy Division, Roskilde, Denmark.
392. Garcia-Sanz, M. (2009). Wind turbines: New challenges and advanced control solutions. *Int. J. Robust Nonlinear Control*, 19(1), 1–116.

Papers Related to Electrical Modeling of WTs

393. Pena, R., Clare, J.C., and Asher, G.M. (1996). Doubly fed induction generator using back-to-back PWM converters and its application to variable speed wind-energy generation. *Proc. Inst. Elect. Eng.*, 143(3), 231–241.
394. Pena, R., Clare, J.C., and Asher, G.M. (1996). A doubly fed induction generator using back-to-back PWM converters supplying an isolated load from a variable speed wind turbine. *Proc. Inst. Elect. Eng.*, 143(5), 380–387.
395. Muller, S., Deicke, M., and De Doncker, R.W. (2002). Doubly fed induction generator systems for wind turbine. *IEEE Ind. Appl. Mag.*, 8(3), 26–33.
396. Ekanayake, J.B., Holdsworth, L., Wu, X.G., and Jenkins, N. (2003). Dynamic modelling of doubly fed induction generator wind turbines. *IEEE Trans. Power Syst.*, 1(2), 803–809.

Papers Related to WTs Grid Integration and Energy Storage

397. Chen, Z. and Spooner, E. (2001). Grid power quality with variable speed wind turbines. *IEEE Trans. Energy Conv.*, 6(2), 148–154.
398. Cartwright, P. and Xu, L. (2004). The integration of large scale wind power generation into transmission network using power electronics. In *Proceedings CIGRE General Session*, Paris, France. http://www.cigre.org
399. Cartwright, P., Xu, L., and Sasse, S. (2004). Grid integration of large offshore wind farms using hybrid HVDC transmission. In *Proceedings of Nordic Wind Power Conference* (NWPC'04), Gothenburg, Sweden, pp. 1–2.
400. Millais, C. and Teske, S. (2004). *Wind Force 12: A Blueprint to Achieve 12% of the World's Electricity from Wind Power by 2020*. European Wind Energy Association, Brussels, Belgium.
401. Abbey, C. and Joos, G. (2005). Short-term energy storage for wind energy applications. In *Proceedings of the 40th IAS Annual Meeting, 2005 Industry Applications Conference, IEEE*, Hong Kong, China, Vol. 3, pp. 2035–2042.
402. IEEE. (2005). Working with wind: Integrating wind into the power system. *IEEE Power Energy Mag.*, Special Issue, 3(6), 2005.
403. Piwko, R., Miller, N., Sanchez-Gasca, J., Xiaoming, Y., Renchang, D., and Lyons, J. (2005). Integrating large wind farms into weak power grids with long transmission lines. In *Proceedings of the 2005 IEEE/PES Transmission and Distribution Conference and Exhibition: Asia and Pacific*, Dalian, China, pp. 1–7.
404. Cimuca, G.O., Saudemont, C., Robyns, B., and Radulescu, M.M. (2006). Control and performance evaluation of a flywheel energy-storage system associated to a variable-speed wind generator. *IEEE Trans. Ind. Electron.*, 53(4), 1074–1085.
405. Muljadi, E. and Parsons, B. (2006). Comparing single and multiple turbine representations in a wind farm simulation. In *Proceedings of the European Wind Energy Conference*, Athens, Greece, 13 pp.
406. Abbey, C. and Joos, G. (2007). Supercapacitor energy storage for wind energy applications. *IEEE Trans. Ind. Appl.*, 43(3), 769–776.
407. Senjyu, T., Kikunaga, Y., Yona, A., Sekine, H., Saber, A.Y., and Funabashi, T. (2008). Coordinate control of wind turbine and battery in wind power generator system. In *Proceedings of the 2008 IEEE Power and Energy Society General Meeting—Conversion and Delivery of Electrical Energy in the 21st Century, IEEE*, Piscataway, NJ, pp. 1–7.
408. Teleke, S., Baran, M.E., Huang, A.Q., Bhattacharya, S., and Anderson, L. (2009). Control strategies for battery energy storage for wind farm dispatching. *IEEE Trans. Energy Conv.*, 24(3), 725–732.
409. Greigarn, T. and Garcia-Sanz, M. (2011). Control of flywheel energy storage systems for wind farm power fluctuation mitigation. In *Proceedings of the IEEE EnergyTech Conference*, Cleveland, OH. http://ieeexplore.ieee.org

Papers Related to Offshore Wind Energy: Modeling and Control

410. Cordle, A. (2002–2006). State-of-the-art in design tools for floating offshore wind turbines. Project UpWind. *European Commission under the 6th (EC) RTD Framework Programme*, Integrated Wind Turbine Design.
411. Kirby, N.M., Xu, L., Luckett, M., and Siepman, W. (2002). HVDC transmission for large offshore wind farms. *IEE Power Eng. J.*, 16(3), 135–141.
412. Byrne, W. and Houlsby, G.T. (2003). Foundations for offshore wind turbines. *Phil. Trans. R. Soc. Lond.*, 361, 2909–2930.
413. Henderson, A.R. and Patel, M.H. (2003). On the modeling of a floating offshore wind turbine. *Wind Energy*, 6, 53–86.
414. Butterfield, S., Musial, W., Jonkman, J., Sclavounos, P., and Wayman, L. (2005). Engineering challenges for floating offshore wind turbines. In *Proceedings of the 2005 Copenhagen Offshore Wind Conference*, Copenhagen, Denmark. http://www.ewea.org

415. Bozhko, S.V., Blasco-Gimenez, R., Li, R., Clare, J.C., and Asher, G.M. (2006). Control of offshore DFIG-based wind farm grid with line-commutated HVDC connection. In *IEEE, EPE-PEMC*, Portoro, Slovenia, pp. 71–78.
416. Jonkman, J.M. and Sclavounos, P.D. (2006). Development of fully coupled aeroelastic and hydrodynamic models for offshore wind turbines. In *Proceedings of the 2006 ASME Wind Energy Symposium*, Reno, NV, 24 pp.
417. Musial, W., Butterfield, S., and Ram, B. (2006). Energy from offshore wind. In *Proceedings of the 2006 Offshore Technology Conference*, Houston, TX. http://www.infield.com/otc_houston_2006.htm
418. Nielsen, F.G., Hanson, T.D., and Skaare, B. (2006). Integrated dynamic analysis of floating offshore wind turbines. *OMAE 2006*, Hamburg, Germany. 25th International Conference on Offshore mechanics and arctic engineering, http://www.omae2006.com
419. Wayman, E.N., Sclavounos, P.D., Butterfield, S., Jonkman, J., and Musial, W. (2006). Coupled dynamic modeling of floating wind turbine systems. In *Proceedings of the 2006 Offshore Technology Conference*, Houston, TX. http://www.infield.com/otc_houston_2006.htm
420. Xiang, D., Ran, L., Bumby, J.R., Tavner, P.J., and Yang, S. (2006). Coordinated control of an HVDC link and doubly fed induction generator in a large offshore wind farm. *IEEE Trans Power Delivery*, 21(1), 463–471.
421. Bozhko, S.V., Blasco-Gimenez, R., Li, R., Clare, J.C., and Asher, G.M. (2007). Control of offshore DFIG-based wind farm grid with line-commutated HVDC connection. *IEEE Trans Energy Conv.*, 22(1), 71–78.
422. Jonkman, J.M. and Buhl Jr., M.L. (2007). Development and verification of a fully coupled simulator for offshore wind turbines. In *Proceedings of the 45th AIAA Aerospace Sciences Meeting and Exhibit, Wind Energy Symposium*, Reno, NV. http://www.aiaa.org
423. Jonkman, J.M. and Buhl Jr., M.L. (2007). Loads analysis of a floating offshore wind turbine using fully coupled simulation. In *Proceedings of the WindPower 2007 Conference and Exhibition*, Los Angeles, CA. http://www.awea.org
424. Larsen, T.J. and Hanson, T.D. (2007). A method to avoid negative damped low frequent tower vibrations for a floating, pitch controlled wind turbine. *J. Phys.*, Conference Series 75, 012073, pp. 1–11.
425. Li, G., Yin, M., Zhou, M., and Zhao, C. (2007). Decoupling control for multi-terminal VSC-HVDC based wind farm interconnection. In *IEEE Power Engineering Society General Meeting*, Tampa, FL, pp. 1–6.
426. Li, R., Bozhko, S., and Asher, G.M. (2008). Frequency control design for offshore wind farm grid with LCC-HVDC link connection. *IEEE Trans. Power Electron.*, 23(3), 1085–1092.
427. Namik, H., Stol, K., and Jonkman, J. (2008). State-space control of tower motion for deepwater floating offshore wind turbines. In *Proceedings of the 46th AIAA Aerospace Sciences Meeting and Exhibition*, Reno, NV, 18 pp.
428. Jonkman, J.M. (2009). Dynamics of offshore floating wind turbines: Model development and verification. *Wind Energy*, 12, 459–492.
429. Namik, H. and Stol, K. (2009). Disturbance accommodating control of floating offshore wind turbines. In *Proceedings of the 47th AIAA Aerospace Sciences Meeting*, Orlando, FL, 15 pp.
430. Namik, H. and Stol, K. (2010). Individual blade pitch control of floating offshore wind turbines. *Wind Energy*, 13(1), 74–85.
431. Roddier, D., Cermelli, C., Aubault, A., and Weinstein, A. (2010). WindFloat: A floating foundation for offshore wind turbines. *J. Renewable Sustainable Energy, Am. Inst. Phys.*, 2, 033104/1–033104/34.
432. Faley, K. and Garcia-Sanz, M. (2011). Controller design to reduce mechanical fatigue in offshore wind turbines affected by ice and tide. In *ASME International Mechanical Engineering Congress and Exposition, IMECE*, Denver, CO. http://www.asmeconferences.org

Papers Related to Smart Wind Turbine Blades

433. Theodorsen, T. (1935). General theory of aerodynamic instability and the mechanism of flutter. Technical Report NACA Report 497, National Advisory Committee for Aeronautics (NACA).

434. Hammond, C.E. (1983). Wind tunnel results showing rotor vibratory loads reduction using higher harmonic pitch. *J. Am. Helicopter Soc.*, 28, 10–15.
435. Barret, R. (1990). Intelligent rotor blade actuation through directionally attached piezoelectric crystals. In *Proceedings of the 46th Annual Forum and Technology Display*, Washington, DC. American Helicopter Society, http://www.vtol.org
436. Leishman, J.G. (1994). Unsteady lift of a flapped airfoil by indicial concepts. *J. Aircr.*, 31, 288–297.
437. Verhaegen, M. (1994). Identification of the deterministic part of MIMO state space models given in innovations form from input–output data. *Automatica*, 30, 61–74.
438. Miller, L.S. (1995). Experimental investigation of aerodynamic devices for wind turbine rotational speed control. Technical Report NREL/TP-441-6913, NREL, Golden, CO.
439. Migliore, P.G., Quandt, G.A., and Miller, L.S. (1995). Wind turbine trailing edge aerodynamic brakes. Technical Report NREL/TP-441-7805, NREL, Golden, CO.
440. Straub, F.K. (1996). A feasibility study of using smart materials for rotor control. *Smart Mater. Struct.*, 5(1), 1–10.
441. Stuart, J., Wright, A., and Butterffeld, C. (1996). Considerations for an integrated wind turbine controls capability at the national wind technology center: An aileron control case study for power regulation and load mitigation. Technical Report NREL/TP-440-21335, NREL, Golden, CO.
442. Stuart, J., Wright, A., and Butterffeld, C. (1997). Wind turbine control systems: Dynamic model development using system identification and the fast structural dynamics code. In *Proceedings of the 35th AIAA/ASME*, Reno, NV. http://www.asmeconferences.org and http://www.aiaa.org
443. Miller, L.S., Quandt, G.A., and Huang, S. (1998). Atmospheric tests of trailing edge aerodynamic devices. Technical Report NREL/SR-500-22350, NREL, Golden, CO.
444. Lau, E. and Krener, A.J. (1999). LPV control of two dimensional wing flutter. In *Proceedings of the 38th Conference on Decision and Control*, Phoenix, AR, pp. 3005–3010.
445. Chopra, I. (2000). Status of application of smart structures technology to rotorcraft systems. *J. Am. Helicopter Soc.*, 45, 228–252.
446. Chopra, I. and Gessow, A. (2000). Recent progress on the development of a smart rotor system. In *Proceedings of the 26th European Rotorcraft Forum*, The Hague, the Netherlands. http://www.erf2011.org/
447. Breitbach, E.J., Anhalt, C., and Monner, H.P. (2001). Overview of adaptronics in aeronautical applications. *Air Space Europe*, 3, 148–151.
448. Stanewsky, E. (2001). Adaptive wing and flow control technology. *Prog. Aerosp. Sci.*, 37, 583–667.
449. Yen, D.T., Van Dam, C.P., Smith, R.L., and Collins, S.D. (2001). Active load control for wind turbine blades using MEM translational tabs. In *Proceedings of the 39th AIAA/ASME*, Salt Lake City, UT. http://www.asmeconferences.org and http://www.aiaa.org
450. Chopra, I. (2002). Review of state of art of smart structures and integrated systems. *AIAA J.*, 40, 2145–2187.
451. Enenkl, B., Klopper, V., and Preibler, D. (2002). Full scale rotor with piezoelectric actuated blade flaps. In *Proceedings of the 28th European Rotorcraft Forum*, Bristol, U.K. http://www.erf2011.org/
452. Leishman, J.G. (2002). Challenges in modelling the unsteady aerodynamics of wind turbines. *Wind Energy*, 5, 85–132.
453. Nakafuji, D.T.Y., Van Dam, C.P., Michel, J., and Morrison, P. (2002). Load control for wind turbines—A non-traditional microtab approach. In *Proceedings of the 40th AIAA/ASME*, Reno, NV. http://www.asmeconferences.org and http://www.aiaa.org
454. Bossanyi, E.A. (2003). Individual blade pitch control for load reduction. *Wind Energy*, 6, 119–128.
455. Marrant, B.A.H. and Van Holten, T. (2003). Smart dynamic control of large offshore wind turbines. *OWEMES-2003*, Naples, Italy. Conference on offshore wind and other marine renewable energy in Mediterranean and European seas, http://www.owemes.org
456. Molenaar, D.P. (2003). Cost effective design and operation of variable speed wind turbines. PhD thesis, Technical University of Delft, Delft, the Netherlands.
457. Van der Hooft, E.L. and Van Engelen, T.G. (2003). Feedforward control of estimated wind speed. Technical Report ECN-C-03-137, ECN, Petten, the Netherlands.

458. Veers, P.S., Ashwill, T.D., Sutherland, H.J., Laird, D.L., Lobitz, D.W., Griffin, D.A., Mandell, J.F., et al. (2003). Trends in the design, manufacture and evaluation of wind turbine blades. *Wind Energy*, 6, 245–259.
459. Collis, S.S., Joslin, R.D., Seifert, A., and Theolis, V. (2004). Issues in active flow control: Theory, control, simulation and experiment. *Prog. Aerosp. Sci.*, 40, 237–289.
460. Baker, J.P., Standish, K.J., and Van Dam, C.P. (2005). Two-dimensional wind tunnel and computational investigation of a microtab modified s809 airfoil. In *Proceedings of the 43rd AIAA/ASME*, Cincinnati, OH. http://www.asmeconferences.org and http://www.aiaa.org
461. Basualdo, S. (2005). Load alleviation on wind turbine blades using variable airfoil geometry. *Wind Eng.*, 29, 169–182.
462. Bossanyi, E.A. (2005). Further load reductions with individual pitch control. *Wind Energy*, 8, 481–485.
463. Buhl, T., Gaunaa, M., and Bak, C. (2005). Load reduction potential using airfoils with variable trailing edge geometry. In *Proceedings of the 43rd AIAA/ASME*, Cincinnati, OH. http://www.asmeconferences.org and http://www.aiaa.org
464. Buhl, T., Gaunaa, M., and Bak, C. (2005). Potential of load reduction using airfoils with variable trailing edge geometry. *J. Solar Energy Eng.*, 127, 503–516.
465. Joncas, S., Bergsma, O., and Beukers, A. (2005). Power regulation and optimization of offshore wind turbines through trailing edge flap control. In *Proceedings of the 43rd AIAA/ASME*, Cincinnati, OH. http://www.asmeconferences.org and http://www.aiaa.org
466. Larsen, T.J., Madsen, H.A., and Thomsen, K. (2005). Active load reduction using individual pitch, based on local blade flow measurements. *Wind Energy*, 8, 67–80.
467. Standish, K.J. and Van Dam, C.P. (2005). Computational analysis of a microtab-based aerodynamic load control system for rotor blades. *J. Am. Helicopter Soc.*, 50, 249–258.
468. Torben, H.A.M., Larsen, J., and Thomsen, K. (2005). Active load reduction using individual pitch, based on local blade flow measurements. *Wind Energy*, 8, 67–80.
469. Troldborg, N. (2005). Computational study of the Riso-B1-18 airfoil with a hinged flap providing variable trailing edge geometry. *Wind Energy*, 29, 89–113.
470. Van Engelen, T.G. and Van der Hooft, E.L. (2005). Individual pitch control inventory. Technical Report ECN-C-03-138, ECN, Petten, the Netherlands.
471. Andersen, P.B., Gaunaa, M., Bak, C., and Buhl, T. (2006). Load alleviation on wind turbine blades using variable airfoil geometry. In *Proceedings of the 2006 European Wind Energy Conference and Exhibition*, Athens, Greece, 8 pp.
472. Engelen, T.G. (2006). Design model and load reduction assessment for multi-rotational mode individual pitch control. In *Proceedings of the European Wind Energy Conference, EWEC 2006*, Athens, Greece. http://www.ewea.org
473. Gaunaa, M. (2006). Unsteady 2D potential-flow forces on a thin variable geometry airfoil undergoing arbitrary motion. Technical Report Risø-R-1478(EN), Risø, Roskilde, Denmark.
474. International Energy Agency (IEA). (2009). *Proceedings, IEA Topical Expert Meeting* on the application of smart structures for large wind turbine rotor blades, http://www.iea.org
475. Marrant, B.A.H. and Van Holten, T.H. (2006). Comparison of smart rotor blade concepts for large offshore wind turbines. In *Proceedings of the Offshore Wind Energy and Other Renewable Energies in Mediterranean and European Seas*, Rome, Italy. http://www.owemes.org
476. McCoy, T. and Griffin, D.A. (2006). Active control of rotor aerodynamics and geometry: Status, methods and preliminary results. In *Proceedings of the 44th AIAA/ASME*, http://www.aiaa.org
477. Roth, D., Enenkl, B., and Dieterich, O. (2006). Active rotor control by flaps for vibration reduction—Full scale demonstrator and first flight test results. In *Proceedings of the 32nd European Rotorcraft Forum*, Maastricht, the Netherlands.
478. Schroeder, K., Ecke, W., Apitz, J., Lembke, E., and Lenschow, G. (2006). A fibre bragg grating sensor system monitors operational load in a wind turbine rotor blade. *Meas. Sci. Technol.*, 17(0), 11167–1172.
479. Schulz, M.J. and Sundaresan, M.J. (2006). Smart sensor system for structural condition monitoring of wind turbines. Report NREL/SR-500-40089, NREL, Golden, CO.

480. Van Engelen, T. (2006). Design model and load reduction assessment for multi-rotational mode individual pitch control (higher harmonics control). In *Proceedings of the European Wind Energy Conference*, Athens, Greece. http://www.ewea.org
481. Van Engelen, T., Markou, H., Buhl, T., and Marrant, B. (2006). Morphological study of aeroelastic control concepts for wind turbines. Technical Report, ECN, Petten, the Netherlands.
482. Van Kuik, G.A.M. (2006). Introductory note. In *Proceedings of the IEA Topical Expert Meeting on the Application of Smart Structures for Large Wind Turbine Rotor Blades*, Delft, the Netherlands. http://www.iea.org
483. Berg, D.E., Zayas, J.R., Lobitz, D.W., vanDam, C.P., Chow, R., and Baker, J.P. (2007). Active aerodynamic load control for wind turbine blades. In *Proceedings of the 5th Joint ASME/JSME Fluids Engineering Conference, FEDSM2007-37604*, San Diego, CA, 012080, p. 20.
484. Bak, C., Gaunaa, M., Andersen, P.B., Buhl, T., Hansen, P., Clemmensen, K., and Moeller, R. (2007). Wind tunnel test on wind turbine airfoil with adaptive trailing edge geometry. In *Proceedings of the 45th AIAA Aerospace Sciences Meeting and Exhibit*, Reno, NV. http://www.asmeconferences.org and http://www.aiaa.org
485. Barlas, T.K. (2007). Smart rotor blades and rotor control, state of the art. Knowledge base report for upwind project. Technical report, *UpWind* WP 1B3, European Wind Energy Association (EWEA).
486. Barlas, T.K. and Van Kuik, G.A.M. (2007). State of the art and prospectives of smart rotor control for wind turbines. *J. Phys.*, Conference Series 75, 012080, pp. 20.
487. Bersee, H. (2007). Smart rotor blades and rotor control. *EWEC-2007*, Upwind Workshop Track.
488. Chow, R. and Van Dam, C.P. (2007). Computational investigations of deploying load control microtabs on a wind turbine airfoil. In *Proceedings of the 45th AIAA/ASME*, Reno, NV. http://www.asmeconferences.org and http://www.aiaa.org
489. Hand, M.M. and Balas, M.J. (2007). Blade load mitigation control design for a wind turbine operating in the path of vortices. *Wind Energy*, 10, 339–355.
490. Hand, M.M., Wright, A.D., Fingersh, L.J., and Harris, M. (2007). Advanced wind turbine controllers attenuate loads when upwind velocity measurements are inputs. In *Proceedings of the 44th AIAA/ASME*, Reno, NV. http://www.asmeconferences.org and http://www.aiaa.org
491. Hulskamp, A.W., Beukers, A., Bersee, H.E.N., Van Wingerden, J.W., and Barlas, T. (2007). Design of a wind tunnel scale model of an adaptive wind turbine blade for active aerodynamic load control experiments. In *Proceedings of the 16th International Conference on Composite Materials*, Tokyo, Japan. http://www.jscm.gr.jp/conference/iccm-16
492. Oerlemans, S. (2007). Wind tunnel aero-acoustic tests of six airfoils for use on small wind turbines. Technical Report NREL/SR-500-35339, NREL, Golden, CO.
493. Bossanyi, E. and Wright, A. (2008). Field testing of individual pitch control on the NREL cart-2 wind turbine. Technical Report under the *6th Framework Integrated Project "Upwind."* European Wind Energy Association (EWEA).
494. Thomsen, S.C., Niemann, H., and Poulsen, N.K. (2008). Individual pitch control of wind turbines using local inflow measurements. In *Proceedings of the 17th World Congress of the International Federation of Automatic Control*, Seoul, Korea. http://ifac.org
495. Van Wingerden, J.W., Hulskamp, A.W., Barlas, T., Marrant, B., Van Kuik, G.A.M., Molenaar, D.P., and Verhaegen, M. (2008). On the proof of concept of a 'smart' wind turbine rotor blade for load alleviation. *Wind Energy*, 11, 265–280.
496. Wilson, D.G., Berg, D.E., Lobitz, D.W., and Zayas, J.R. (2008). Optimized active aerodynamic blade control for load alleviation on large wind turbines. In *Proceedings of the AWEA Windpower 2008 Conference & Exhibition*, Houston, TX. http://www.awea.org
497. Selvam, K., Kanev, S., Van Wingerden, J.W., Van Engelen, T., and Verhaegen, M. (2009). Feedback-feedforward individual pitch control for wind turbine load reduction. *Int. J. Robust Nonlinear Control*, 19, 72–91.
498. Wilson, D.G., Berg, D.E., Resor, B.R., Barone, M.F., and Berg, J.C. (2009). Combined individual pitch control and active aerodynamic load controller investigation for the 5mw upwind turbine. In *Proceedings of the AWEA Windpower 2009 Conference & Exhibition*, Chicago, IL. http://www.awea.org

Papers Related to Airborne Wind Turbines

499. Loyd, M.L. (1980). Crosswind kite power. *J. Energy*, 4(3), 106–111.
500. Canale, M., Fagiano, L., Milanese, M., and Ippolito, M. (2006). Control of tethered airfoils for a new class of wind energy generator. In *Proceedings of the 45th IEEE Conference on Decision and Control*, San Diego, CA, pp. 4020–4026.
501. Houska, B. and Diehl, M. (2006). Optimal control of towing kites. In *Proceedings of the 45th IEEE Conference on Decision and Control*, San Diego, CA, pp. 2693–2697.
502. Canale, M., Fagiano, L., and Milanese, M. (2007). Power kites for wind energy generation. *IEEE Control Syst. Mag.*, 27(6), 25–38.
503. Houska, B. and Diehl, M. (2007). Optimal control for power generating kites. In *Proceedings of the 9th European Control Conference*, Kos, Greece, pp. 3560–3567.
504. Ilzhöfer, A., Houska, B., and Diehl, M. (2007). Nonlinear MPC of kites under varying wind conditions for a new class of large-scale wind power generators. *Int. J. Robust Nonlinear Control*, 17, 1590–1599.
505. Lansdorp, B., Ruiterkamp, R., and Ockels, W. (2007). Towards flight testing of remotely controlled surf-kites for wind energy generation. In *Proceedings of the AIAA Atmospheric Flight Mechanics Conference and Exhibit*, Hilton Head, CA. http://www.aiaa.org
506. Roberts, B.W., Shepard, D.H., Caldeira, K., Cannon, M.E., Eccles, D.G., Grenier, A.J., and Freidin, J.F. (2007). Harnessing high–altitude wind power. *IEEE Trans. Energy Conv.*, 22(1), 136–144.
507. Williams, P., Lansdorp, B., and Ockels, W. (2008). Optimal crosswind towing and power generation with tethered kites. *J. Guidance Control Dyn.*, 31, 81–93.
508. Garcia-Sanz, M. and White, N. (2010). Airborne wind turbine. US Patent, Provisional Application No. 61/387,432.
509. White, N., Tierno, N., and Garcia-Sanz, M. (2011). A novel approach to airborne wind energy-design and modeling. In *Proceedings of the IEEE EnergyTech Conference*, Cleveland, OH. http://ieeexplore.ieee.org
510. Garcia-Sanz, M., White, N., and Tierno, N., A novel approach to airborne wind energy: Overview of the EAGLE system project. In *Proceedings of the 63rd National Aerospace and Electronics Conference*, NAECON, July 2011, Dayton, OH. http://www.naecon.org
511. Tierno, N., White, N., and Garcia-Sanz, M. (2011). Longitudinal flight control for a novel airborne wind energy system: Nonlinear and robust MIMO control design techniques. In *ASME International Mechanical Engineering Congress & Exposition, IMECE*, Denver, CO. http://www.asmeconferences.org
512. Joby Energy Inc., Santa Cruz, CA, www.jobyenergy.com
513. Makani Power Inc., Alameda, CA, www.makanipower.com
514. Magenn Power Inc., Moffett Field, CA, www.magenn.com
515. SkyWindpower Corporation, Oroville, CA, www.skywindpower.com
516. Ampyx Power, The Hague, the Netherlands, www.ampyxpower.com
517. KITEnrg, Torino, Italy, www.kitenergy.net

Papers Related to Wind Energy Economics

518. EWEA. (2004). *Wind Power Economics. Wind Energy Costs, Investment Factors.* European Wind Energy Association, Brussels, Belgium.
519. Fingersh, L., Hand, M., and Laxson, A. (2006). Wind turbine design cost and scaling model. Technical Report, NREL/TP-500-40566, NREL, Golden, CO.
520. Wiser, R. and Bolinger, M. (2007). Annual Report on U.S. Wind Power Installation, Cost, and Performance Trends: 2006. U.S. Department of Energy, Washington, DC.
521. Wiser, R. and Bolinger, M. (2008). Annual Report on U.S. Wind Power Installation, Cost, and Performance Trends: 2007. U.S. Department of Energy, Washington, DC.

Index

C

D

E

N

Q

U

Z

For Product Safety Concerns and Information please contact our EU representative GPSR@taylorandfrancis.com Taylor & Francis Verlag GmbH, Kaufingerstraße 24, 80331 München, Germany

T - #0069 - 160425 - C8 - 254/178/34 [36] - CB - 9781439821794 - Gloss Lamination